T0338539

Applied Quantum Mechanics

Featuring new coverage of quantum engineering and quantum information processing, the third edition of this bestselling textbook continues to provide a uniquely practical introduction to the fundamentals of quantum mechanics.

Key features:

- Straightforward explanations of quantum effects, suitable for readers from all backgrounds, equipping students with a robust understanding of underlying theoretical principles.
- Real-world engineering problems showcasing the application of theory to practice, providing a relevant and accessible introduction to cutting-edge quantum applications.
- Over 60 worked examples using MATLAB®, enabling deeper understanding through computational exploration and visualization.
- A new chapter on quantum engineering, introducing state-of-the-art concepts in quantum information processing and quantum device design.

Revised and updated throughout, and supported online by downloadable MATLAB code, exam questions, and solutions to over 150 homework problems for instructors, this is the ideal textbook for engineers and scientists studying a first course in quantum mechanics.

A. F. J. Levi is Professor of Electrical and Computer Engineering and Physics and Astronomy, and Chair of the Ming Hsieh Department of Electrical and Computer Engineering – Electrophysics, at the University of Southern California (USC). He joined USC after a decade at AT&T Bell Laboratories, Murray Hill, New Jersey, and has over 25 years' experience in teaching applied quantum mechanics to engineers and scientists. He is the inventor of hot electron spectroscopy, discovered ballistic electron transport in transistors, created the first microdisk laser, and carried out groundbreaking work in optimal design of small electronic and photonic systems.

"*Applied Quantum Mechanics*, third edition, by A. F. J. Levi, is an essential update in the rapidly evolving field of quantum mechanics. With a clear and concentrated focus on practical equation manipulation, using computational tools, and an intuitive understanding of applications, this edition skillfully integrates the latest advancements in electronics and quantum engineering. An indispensable resource for students and researchers, it strikes the perfect balance between conciseness and comprehensiveness."

Daryoosh Vashaee, *North Carolina State University*

"The new third edition of this textbook provides an essential introduction to quantum mechanics and its many technological and scientific applications that I would recommend to all undergraduate students. In particular, the focus on technological applications and devices and the many worked example problems make the material more concrete and illustrate the power of quantum mechanics."

Aaron M. Lindenberg, *Stanford University*

"*Applied Quantum Mechanics* by Levi stands out because of its unique blend of subjects, encompassing both fundamental and applied aspects of quantum mechanics. The new edition incorporates the subject of quantum circuits and information, which is absolutely timely and highly needed. I have used the previous editions over the years and I look forward to using the new edition. The textbook is especially well suited for upper-class undergraduates and/or first-year graduate students pursuing degrees in engineering or natural sciences."

Ridwan Sakidja, *Missouri State University*

"This third edition of *Applied Quantum Mechanics* by A. F. J. Levi will be a great asset for students in physics, engineering, and materials science who seek a solid introduction to the practical aspect of quantum mechanics that will allow them to investigate the electrical and optical properties of nanoscale devices. The addition of the section on quantum information processing will also give the students a birds-eye view of the rapidly growing field of quantum computing. Quantum engineering is an important breadth of knowledge that any engineer and scientist interested in the working of devices at the nanoscale must acquire, and this book offers a solid introduction to it."

Marc Cahay, *University of Cincinnati*

Applied Quantum Mechanics

Third Edition

A. F. J. Levi
University of Southern California

CAMBRIDGE
UNIVERSITY PRESS

Shaftesbury Road, Cambridge CB2 8EA, United Kingdom

One Liberty Plaza, 20th Floor, New York, NY 10006, USA

477 Williamstown Road, Port Melbourne, VIC 3207, Australia

314–321, 3rd Floor, Plot 3, Splendor Forum, Jasola District Centre, New Delhi - 110025, India

103 Penang Road, #05-06/07, Visioncrest Commercial, Singapore 238467

Cambridge University Press is part of Cambridge University Press & Assessment, a department of the University of Cambridge.

We share the University's mission to contribute to society through the pursuit of education, learning and research at the highest international levels of excellence.

www.cambridge.org
Information on this title: www.cambridge.org/highereducation/isbn/9781009308076

DOI: 10.1017/9781009308083

First published 2003
Second edition published 2006
Third edition published 2024

A catalogue record for this publication is available from the British Library

Library of Congress Cataloging-in-Publication Data
Names: Levi, A. F. J. (Anthony Frederic John), 1959- author.
Title: Applied quantum mechanics / A.F.J. Levi.
Description: Third edition. | Cambridge ; New York : Cambridge University
    Press, 2022. | Includes index.
Identifiers: LCCN 2023008118 (print) | LCCN 2023008119 (ebook) |
    ISBN 9781009308076 (hardback) | ISBN 9781009308083 (epub)
Subjects: LCSH: Quantum theory. | Quantum theory–Industrial applications.
Classification: LCC QC174.12 .L44 2022 (print) | LCC QC174.12 (ebook) |
    DDC 530.12–dc23/eng20230711
LC record available at https://lccn.loc.gov/2023008118
LC ebook record available at https://lccn.loc.gov/2023008119

ISBN 978-1-009-30807-6 Hardback

Additional resources for this publication at www.cambridge.org/Levi3e

Dass ich erkenne, was die Welt
Im Innersten zusammenhält

Goethe
(*Faust*, I.382–3)

Contents

Contents

Contents

Contents

Preface to the Third Edition

A great deal of progress has been made in applied quantum mechanics since the first and second editions of this book were published. While there is a continued focus on three main themes – practicing manipulation of equations and analytic problem solving in quantum mechanics, utilizing the availability of modern computer power to numerically solve problems, and developing an intuition for applications of quantum mechanics – the need for an accessible introductory book about applied quantum mechanics is even greater now than it was in 2003. In the US there is renewed emphasis on research in electronics, particularly transistors and photonics, quantum information processing, and quantum engineering. The third edition of *Applied Quantum Mechanics* sets out to address these interests. To accommodate a new chapter called "Toward Quantum Engineering," the previous first chapter – introducing background material on classical mechanics and electromagnetism – has been moved to an appendix. In addition, the text in the book has been made more concise. This has created room for some additional material whose aim is to maintain reader interest by broadening the range of applications and concepts.

The book content remains designed for a one-semester course. For those on a quarter system or those wishing to focus on core elements of the book, Chapter 9 on time-independent perturbation theory and Chapter 10 on the hydrogen atom can be skipped.

Changes in the third edition include the addition of problems to each chapter. Chapter 11 is new and addresses device optimization, control, and provides an introduction to quantum information processing.

Cambridge University Press has a website with supporting material for both students and instructors who use the book. This includes MATLAB code used to create figures and solutions to exercises. The website is: www.cambridge.org/Levi3e.

Preface to the Second Edition

Following the remarkable success of the first edition and not wanting to give up on a good thing, the second edition of this book continues to focus on three main themes: practicing manipulation of equations and analytic problem solving in quantum mechanics, utilizing the availability of modern compute power to numerically solve problems, and developing an intuition for applications of quantum mechanics. Of course there are many books which address the first of the three themes. However, the aim here is to go beyond that which is readily available and provide the reader with a richer experience of the possibilities of the Schrödinger equation and quantum phenomena.

Changes in the second edition include the addition of problems to each chapter. These also appear on the Cambridge University Press website. To make space for these problems and other additions, previously printed listing of MATLAB code has been removed from the text. Chapter 1 now has a section on harmonic oscillation of a diatomic molecule. Chapter 2 has a new section on quantum communication. In Chapter 3 the discussion of numerical solutions to the Schrödinger now includes periodic boundary conditions. The tight binding model of band structure has been added to Chapter 4 and the numerical evaluation of density of states from dispersion relation has been added to Chapter 5. The discussion of occupation number representation for electrons has been extended in Chapter 7. Chapter 11 is a new chapter in which quantization of angular momentum and the hydrogenic atom are introduced.

Cambridge University Press has a website with supporting material for both students and teachers who use the book. This includes MATLAB code used to create figures and solutions to exercises. The website is: www.cambridge.org/9780521860963

Many thanks to Omid Nohadani for help with formatting the current version of the book.

Preface to the First Edition

The theory of quantum mechanics forms the basis for our present understanding of physical phenomena on an atomic and sometimes macroscopic scale. Today, quantum mechanics can be applied to most fields of science. Within engineering, important subjects of practical significance include semiconductor transistors, lasers, quantum optics, and molecular devices. As technology advances, an increasing number of new electronic and opto-electronic devices will operate in ways which can only be understood using quantum mechanics. Over the next thirty years, fundamentally quantum devices such as single-electron memory cells and photonic signal processing systems may well become commonplace. Applications will emerge in any discipline that has a need to understand, control, and modify entities on an atomic scale. As nano- and atomic-scale structures become easier to manufacture, increasing numbers of individuals will need to understand quantum mechanics in order to be able to exploit these new fabrication capabilities. Hence, one intent of this book is to provide the reader with a level of understanding and insight that will enable him or her to make contributions to such future applications, whatever they may be.

The book is intended for use in a one-semester introductory course in applied quantum mechanics for engineers, material scientists, and others interested in understanding the critical role of quantum mechanics in determining the behavior of practical devices. To help maintain interest in this subject, I felt it was important to encourage the reader to solve problems and to explore the possibilities of the Schrödinger equation. To ease the way, solutions to example exercises are provided in the text, and the enclosed CD-ROM contains computer programs written in the MATLAB language that illustrate these solutions. The computer programs may be usefully exploited to explore the effects of changing parameters such as temperature, particle mass, and potential within a given problem. In addition, they may be used as a starting point in the development of designs for quantum mechanical devices.

The structure and content of this book are influenced by experience teaching the subject. Surprisingly, existing texts do not seem to address the interests or build on the computing skills of today's students. This book is designed to better match such student needs.

Some material in the book is of a review nature, and some material is merely an introduction to subjects that will undoubtedly be explored in depth by those interested in pursuing more advanced topics. The majority of the text, however, is an essentially self-contained study of quantum mechanics for electronic and opto-electronic applications.

There are many important connections between quantum mechanics and classical mechanics and electromagnetism. For this and other reasons, Chapter 1 is devoted to a review of classical concepts. This establishes a point of view with which the predictions of quantum mechanics can be compared. In a classroom situation it is also a convenient way in which to establish a uniform minimum knowledge base. In Chapter 2 the Schrödinger wave equation is introduced and used to motivate qualitative descriptions of atoms, semiconductor crystals, and a heterostructure diode. Chapter 3 develops the more systematic use of the one-dimensional Schrödinger equation to describe a particle in simple potentials. It is in this chapter that the quantum mechanical phenomenon of tunneling is introduced. Chapter 4 is devoted to developing and using the propagation matrix method to calculate electron scattering from a one-dimensional potential of arbitrary shape. Applications include resonant electron tunneling and the Kronig–Penney model of a periodic crystal potential. The generality of the method is emphasized by applying it to light scattering from a dielectric discontinuity. Chapter 5 introduces some related mathematics, the generalized uncertainty relation, and the concept of density of states. Following this, the quantization of conductance is introduced. The harmonic oscillator is discussed in Chapter 6 using the creation and annihilation operators. Chapter 7 deals with fermion and boson distribution functions. This chapter shows how to numerically calculate the chemical potential for a multi-electron system. Chapter 8 introduces and then applies time-dependent perturbation theory to ionized impurity scattering in a semiconductor and spontaneous light-emission from an atom. The semiconductor laser diode is described in Chapter 9. Finally, Chapter 10 discusses the (still useful) time-independent perturbation theory.

Throughout this book, I have tried to make applications to systems of practical importance the main focus and motivation for the reader. Applications have been chosen because of their dominant roles in today's technologies. Understanding is, after all, only useful if it can be applied

Note on MATLAB Programs

If you have not already installed the MATLAB®[1] language on your computer, you will need to purchase a copy and do so. MATLAB is available from MathWorks (www.mathworks.com/).

After verifying that MATLAB has been correctly installed, download the directory AppliedQMmatlab from www.cambridge.org/Levi3e and copy to a convenient location in your computer user directory.

Launch MATLAB using the icon on the desktop or from the start menu. The MATLAB command window will appear on your computer screen. From the MATLAB command window, use the path browser to set the path to the location of the AppliedQMmatlab directory. Type the name of the file you wish to execute in the MATLAB command window (do not include the ".m" extension). Press the enter key on the keyboard to run the program.

You will find that some programs prompt for input from the keyboard. Most programs display results graphically with intermediate results displayed in the MATLAB command window.

To edit values in a program, or to edit the program itself, double-click on the file name to open the file editor.

You should note that the computer programs in the AppliedQMmatlab directory are not optimized. They are written in a very simple way to minimize any possible confusion or sources of error. The intent is that these programs be used as an aid to the study of applied quantum mechanics. When required, integration is performed explicitly, and in the simplest way possible. However, for exercises involving matrix diagonalization, use is made of special MATLAB functions.

Some programs make use of the functions chempot.m, fermi.m, mu.m, runge4.m, and solve_schM.m.

---

[1] MATLAB is a registered trademark of MathWorks, Inc.

1 Toward Quantum Mechanics

1.1 Introduction

Quantum mechanics is a basis for understanding physical phenomena on an atomic scale. An electron point particle of rest mass m_0, charge magnitude e, and quantized spin magnitude $\hbar/2$, can behave as a wave. A light wave of radial frequency ω can behave as if containing one or more quantized particles called photons, each with spin magnitude $\hbar$. This wave–particle duality of electrons and photons, the linear superposition of particle states, the existence of identical indistinguishable particles, quantum entanglement of particles, and measurement whose outcome is noncausal are all unique to quantum mechanics. Particle energy $E = \hbar\omega$ is quantized on a scale set by Planck's constant $\hbar = 1.0545 \times 10^{-34}$ J s.[1]

The basic physical building blocks of nature may be categorized into particles of matter and carriers of force between matter. All known elementary constituents of matter and transmitters of force are quantized. For example, energy, momentum, and angular momentum can take on discrete quantized values. The electron is an elementary particle of matter and the photon is an elementary transmitter of force. Neutrons, protons, and atoms are composite particles made up of elementary particles of matter and transmitters of force. These composite particles are also quantized. Classical mechanics cannot explain quantization and so quantum mechanics is required to understand the microscopic properties of atoms – which, for example, make up solids such as crystalline semiconductors.

Quantum mechanics has applications in engineering, including semiconductor transistors, lasers, and quantum optics. The purpose and intent of this book is to provide the readers with a level of understanding and insight that will enable them to appreciate and to make contributions to the development of future applications of quantum phenomena.

The theory of quantum mechanics has been established by experiment. The most important early experiments involved light. Long before it was realized that light waves are quantized into particles called photons, key experiments on the wave properties of light were performed. For example, it was established that the color of visible light is associated with different wavelengths of light (red: 760–622 nm, orange: 622–597 nm, yellow: 597–577 nm, green: 577–492 nm, blue: 472–455 nm, violet: 455–390 nm).

The connection between optical and electrical phenomena was established by Maxwell in 1864. This extended the concept of light to include the complete electromagnetic spectrum. A great deal of effort was, and continues to be, spent gathering information on the behavior of electromagnetic radiation. Table 1.1 shows the frequencies and wavelengths corresponding to different parts of the electromagnetic spectrum.

---

[1] Sometimes $\hbar$ is called Planck's *reduced* constant to distinguish it from $h = 2\pi\hbar$.

Table 1.1 Spectrum of electromagnetic radiation

Name	Wavelength (m)	Frequency (Hz)
Radio	$> 10^{-1}$	$< 3 \times 10^9$
Microwave	$10^{-1} - 10^{-4}$	$3 \times 10^9 - 3 \times 10^{12}$
Infrared	$10^{-4} - 7 \times 10^{-7}$	$3 \times 10^{12} - 4.3 \times 10^{14}$
Visible	$7 \times 10^{-7} - 4 \times 10^{-7}$	$4.3 \times 10^{14} - 7.5 \times 10^{14}$
Ultraviolet	$4 \times 10^{-7} - 10^{-9}$	$7.5 \times 10^{14} - 3 \times 10^{17}$
X-rays	$10^{-9} - 10^{-11}$	$3 \times 10^{17} - 3 \times 10^{19}$
Gamma rays	$< 10^{-11}$	$> 3 \times 10^{19}$

1.1.1 Diffraction and Interference of Light

Among the important properties of light waves are diffraction, linear superposition, and interference. The empirical observation of these phenomena was neatly summarized by the work of Young in 1803. Today, the famous Young's slits experiment can be performed using a single-wavelength, visible laser light source, a screen with two slits cut in it, and a viewing screen (see Fig. 1.1). Light passing through the slits interferes with *itself*, and an intensity interference pattern is observed on the viewing screen. The interference pattern is due to the superposition of waves, for which each slit is an effective coherent source. Hence, the Young's slits interference experiment can be understood using the principle of linear superposition. The wave source at each diffracting slit (Huygens' principle) interferes to create an interference pattern in which intensity maxima correspond to electric fields adding coherently in phase and intensity minima correspond to electric fields adding coherently out of phase.

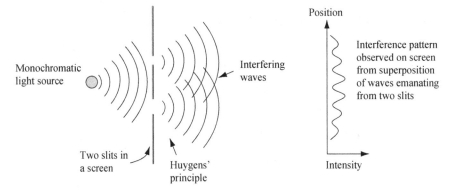

Fig. 1.1 Illustration of Young's slits experiment. Light from a monochromatic source passing through the slits interferes with itself, and an intensity interference pattern is observed on the viewing screen. The interference pattern is due to the superposition of waves, for which each slit is an effective source. Intensity maxima correspond to electric fields adding coherently in phase and intensity minima correspond to electric fields adding coherently out of phase.

The linear superposition of two waves at exactly the same frequency can give rise to interference because of a relative phase delay between the waves. For convenience, consider two plane waves labeled $j = 1$ and $j = 2$, propagating with wave vector $\mathbf{k}_{\mathrm{ph}}$, with wavelength $\lambda = 2\pi/k_{\mathrm{ph}}$ where $k_{\mathrm{ph}} = |\mathbf{k}_{\mathrm{ph}}|$, amplitude $|\mathbf{E}_j|$, phase ϕ_j, and frequency ω. Generality is not lost by only considering plane waves because in a linear system any wave can be made from a linear superposition of plane waves. The two plane waves can be represented as

$$\mathbf{E}_1 = \tilde{\mathbf{e}_1}|\mathbf{E}_1|e^{i(\mathbf{k}_{\mathrm{ph}}\cdot\mathbf{r}-\omega t)}e^{i\phi_1} \tag{1.1}$$

and

$$\mathbf{E}_2 = \tilde{\mathbf{e}_2}|\mathbf{E}_2|e^{i(\mathbf{k}_{\mathrm{ph}}\cdot\mathbf{r}-\omega t)}e^{i\phi_2}, \tag{1.2}$$

respectively, where $\mathbf{r}$ is position and $\tilde{\mathbf{e}_j}$ is the unit vector in the direction of the electric field $\mathbf{E}_j$. Equations (1.1) and (1.2) are examples of plane waves with the electric field linearly polarized in the $\tilde{\mathbf{e}_j}$ direction.

The intensity due to the linear superposition of $\mathbf{E}_1$ and $\mathbf{E}_2$ with $\tilde{\mathbf{e}_1} = \tilde{\mathbf{e}_2}$ is just

$$\boxed{|\mathbf{E}|^2 = |\mathbf{E}_1 + \mathbf{E}_2|^2 = |\mathbf{E}_1|^2 + |\mathbf{E}_2|^2 + 2|\mathbf{E}_1||\mathbf{E}_2|\cos(\phi),} \tag{1.3}$$

where $\phi = \phi_2 - \phi_1$ is the relative phase between the waves. The expression for $|\mathbf{E}|^2$ is called the *interference equation*. The linear superposition of the two waves gives a sinusoidal interference pattern in the intensity as a function of phase delay, ϕ.

Plotting the interference equation in Fig. 1.2 for the case in which $|\mathbf{E}_1| = |\mathbf{E}_2| = |\mathbf{E}_0|$ in Eq. (1.3), intensity maxima $|\mathbf{E}|^2_{\mathrm{max}} = 4|\mathbf{E}_0|^2$ that are four times the intensity of the individual wave occur when $\phi = 2n\pi$, where n is an integer. Intensity minima $|\mathbf{E}|^2_{\mathrm{min}} = 0$ occur when $\phi = (2n - 1)\pi$. In the more general case, when $|\mathbf{E}_1| \neq |\mathbf{E}_2|$, the interference pattern is still periodic in ϕ, but the intensity maximum has a value $(|\mathbf{E}_1| + |\mathbf{E}_2|)^2$ and the intensity minimum has a value $(|\mathbf{E}_1| - |\mathbf{E}_2|)^2$.

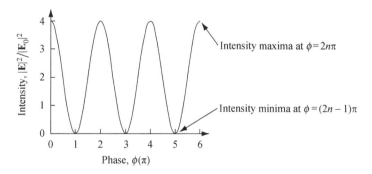

Fig. 1.2 The linear superposition of two waves at exactly the same frequency can give rise to interference if there is a relative phase delay between the waves. The figure illustrates the sinusoidal interference pattern in intensity as a function of phase delay, ϕ, between two equal amplitude waves. Intensity maxima occur when $\phi = 2n\pi$, where n is an integer.

1.1.2 Black-Body Radiation and Evidence for Quantization of Light

Experimental evidence for the quantization of light into photons initially came from measurement of the emission spectrum of thermal light (called black-body radiation).

The equipartition theorem of classical statistical thermodynamics assigns an average energy of $k_{\mathrm{B}}T/2$ to each degree of freedom of a system in thermal equilibrium. For electromagnetic radiation, the degrees of freedom are *normal modes* of frequency $\omega = c k_{\mathrm{ph}}$. It follows that the electromagnetic field radiative energy density emitted from a black body in free space at absolute temperature T is

$$U_{\mathrm{S}}(\omega) = \frac{k_{\mathrm{B}}T}{\pi^2 c^3} \omega^2. \tag{1.4}$$

Radiative energy density is the energy per unit volume per unit angular frequency, and it is measured in $\mathrm{J\,s\,m^{-3}}$. Equation (1.4) is the Rayleigh–Jeans formula that predicts a physically impossible infinite radiative energy density as $\omega \to \infty$. This divergence in radiative energy density with decreasing wavelength is called the classical ultraviolet catastrophe. The impossibility of infinite energy as $\omega \to \infty$ was indeed a disaster (or catastrophe) for classical physics. The only way out was to invent a new type of physics: quantum physics.

The black-body radiation spectrum was explained by Planck in 1900. He implicitly (and later explicitly) assumed emission and absorption of discrete energy quanta of electromagnetic radiation, so that $E = \hbar\omega$, where ω is the angular frequency of the electromagnetic wave and $\hbar$ is a constant. This gives a radiative energy density:

$$U_{\mathrm{S}}(\omega) = \frac{\hbar\omega^3}{\pi^2 c^3} \frac{1}{e^{\hbar\omega/k_{\mathrm{B}}T} - 1}. \tag{1.5}$$

Equation (1.5) solves the ultraviolet catastrophe and agrees with the classical Rayleigh–Jeans result in the limit of long wavelength electromagnetic radiation ($\omega \to 0$). In both approaches, thermal *equilibrium* between the radiation and the material bodies (made of atoms) is assumed. Hence, the radiation has a radiative energy density distribution that is characteristic of *thermal light*.

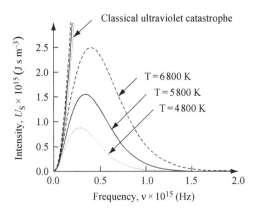

Fig. 1.3 Radiative energy density of black-body radiation emitted from the unit surface area into a fixed direction from a black body as a function of frequency ($\nu = \omega/2\pi$) for three different absolute temperatures, $T = 4\,800\,\mathrm{K}$, $T = 5\,800\,\mathrm{K}$, and $T = 6\,800\,\mathrm{K}$. The predictions of the classical Rayleigh–Jeans formula are also plotted to show the ultraviolet catastrophe.

Figure 1.3 shows the energy density of radiation emitted by unit surface area into a fixed direction from a black body as a function of frequency for three different temperatures. The Sun has a surface temperature of near $T = 5\,800\,\text{K}$, which has a peak in radiative energy density at visible frequency.

Additional experimental evidence for the quantization of electromagnetic energy in such a way that $E = \hbar\omega$ comes from the photoelectric effect.

1.1.3 Photoelectric Effect and the Photon Particle

When light of angular frequency ω is incident on a metal, electrons can be emitted from the metal surface if $\hbar\omega > e\phi_W$, where ϕ_W is the work function of the metal. The minimum energy for an electron to escape the metal into vacuum is $+e\phi_W$. In addition, such photoelectric-effect experiments show that the *number* of electrons leaving the surface depends upon the intensity of the incident electromagnetic field.

As illustrated in Fig. 1.4(a), a particle of light is incident on a metal, and the collision can cause an electron particle to be ejected from the surface. The maximum excess kinetic energy of the electron leaving the surface is observed in experiments to be $T_{\max} = \hbar\omega - e\phi_W$, where $\hbar$ is the slope of the curve in Fig. 1.4(b). The maximum kinetic energy of any ejected electron depends only upon the angular frequency, ω, of the light particle with which it collided, and this energy is *independent* of light intensity. Each light particle has energy $\hbar\omega$, and light intensity is given by the particle flux. This is different from the classical case, which predicts that energy E is proportional to light intensity. If the electromagnetic field of amplitude A oscillating at frequency ω_0 is treated as a classical harmonic oscillator then $E \propto A^2$, the intensity of the electromagnetic field.

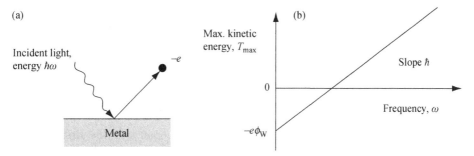

Fig. 1.4 (a) Light of energy $\hbar\omega$ can cause electrons to be emitted from the surface of a metal. (b) The maximum kinetic energy of emitted electrons is proportional to the frequency of light, ω, and *independent* of light intensity. The proportionality constant is $\hbar$.

In 1905, Einstein explained the photoelectric effect[2] by postulating that light behaves as a particle and that (in agreement with Planck's work) it is quantized in energy, so that

$$\boxed{E = \hbar\omega,} \tag{1.6}$$

---

[2] In 1921, Einstein received the Nobel Prize in physics "for his services to Theoretical Physics, and especially for his discovery of the law of the photoelectric effect."

where $\hbar = 1.054\,592 \times 10^{-34}$ J s is *Planck's constant*. It is important to notice that the key result, $E = \hbar\omega$, comes directly from experiment. Also notice that for $\hbar\omega$ to have the dimensions of energy, $\hbar$ must have dimensions of J s. A quantity of this type is called an action. The units J s can also be expressed as kg m^2 s^{-1}.

The quantum of light is the *photon*. It has zero mass and is an example of an elemental quantity in quantum mechanics.

From classical electrodynamics it is known that electromagnetic *plane waves carry momentum* of magnitude $p = U/c$, where U is the electromagnetic energy density. This means that if a photon has energy E then the magnitude of its momentum is $p = E/c$. Because experiment shows that light is quantized in energy so that $E = \hbar\omega$, it seems natural that momentum should also be quantized. Following this line of thought, $p = \hbar\omega/c$ or, since $\omega = c2\pi/\lambda_{ph} = ck_{ph}$, a photon in a plane wave normal mode with wave vector $\mathbf{k}_{ph}$ has photon momentum

$$\boxed{\mathbf{p} = \hbar\mathbf{k}_{ph}.} \tag{1.7}$$

This result could have been guessed from dimensional analysis. Planck's constant $\hbar$ is measured in units of kg m^2 s^{-1}, which, dividing by a length, gives units of momentum. Since, for a plane wave of wavelength λ_{ph}, there is only one natural inverse length scale $k_{ph} = 2\pi/\lambda_{ph}$, it is reasonable to suggest that momentum is just $\mathbf{p} = \hbar\mathbf{k}_{ph}$.

Using the relationship $E = \hbar\omega$, it follows that if the energy of a photon measured in eV is known, then the photon wavelength λ_{ph} in free space measured in nm is given by the expression

$$\lambda_{ph}(\text{nm}) = \frac{1\,240}{E(\text{eV})}. \tag{1.8}$$

The energy of $\lambda_{ph} = 1\,000$ nm wavelength light is $E = 1.24$ eV. Compared with room temperature thermal energy $k_B T = 25$ meV, the quantized energy of $\lambda_{ph} = 1\,000$ nm wavelength light is large and the magnitude of photon momentum is $p = \hbar k_{ph} = 6.63 \times 10^{-28}$ kg m s^{-1}.

1.1.4 An Experiment to Prove the Photon Exists

Many years after the initial suggestion that light is quantized the first laboratory experiments were performed that proved the existence of the photon. Famously, Kimble, Dagenais, and Mandel published a paper in 1977 showing that light is made up of discrete photons, each of which can create a single click in a detector.[3] In the 1980s Grangier, Roger, and Aspect were able to refine these experiments[4] and also show the interference of a single photon, thereby demonstrating in complementary experiments the particle and wave nature of the photon.

It is possible to perform experiments similar to those of Kimble, Dagenais, and Mandel using laser diode-based single-photon sources, fiber-optic components, and single-photon detectors as illustrated in Fig. 1.5. An optical source emits single photons that are, on average, spaced apart in time by $\langle \tau_{ph} \rangle$. The photon flux is incident on an ideal lossless symmetric 50:50 beam splitter with linear response. Single-photon detectors

---

[3] H. J. Kimble, M. Dagenais, and L. Mandel, *Phys. Rev. Lett.* **39**, 691 (1977).
[4] P. Grangier, G. Roger, and A. Aspect, *Europhys. Lett.* **1**, 173 (1986).

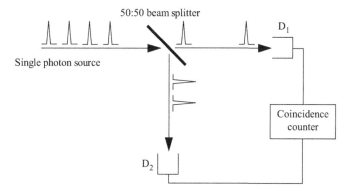

Fig. 1.5 An optical source emits single photons that are incident on an ideal lossless symmetric 50:50 beam splitter. Single-photon detectors D_1 and D_2 are placed at the two output ports of the beam splitter. Because the photon is an indivisible elementary particle, it must either be detected by D_1 or D_2, but not both.

D_1 and D_2 are placed at the two output ports of the beam splitter. In the simplest configuration the optical path length between the beam splitter output port and the associated detector is the same and each detector response time is very much smaller than $\langle \tau_{ph} \rangle$. The photon path taken through the beam splitter can only be inferred *after* it is detected by D_1 or D_2. The inferred photon path selection is purely *random* and hence noncausal. The photon is either detected by D_1 or D_2, but not both, and it is fundamentally not known beforehand at which output port the photon will be detected. When properly implemented, the experiment reveals that there are no coincidence counts between the two detectors thereby proving that the photon is an indivisible quantized particle.

The absence of single-photon coincidence counts in the experiment is something that cannot be explained by a classical wave model of light. Maxwell's electromagnetic waves appear at both detectors at the same time and so give rise to coincidence detection – something that is not observed experimentally. A classical description using Maxwell's equations is accurate when there are a large number of incoherent photons associated with a particular electromagnetic field. If there are very few photons, special conditions involving identical indistinguishable photons, or a coherent superposition of photons, then a quantum description is appropriate.

1.1.5   Random Number Generation and Stochastic Computing

A single photon incident on a lossless symmetric 50:50 beam splitter with linear response has an exactly 50% chance of being detected at one of the two output ports. This pure random behavior is *guaranteed* by quantum mechanics and can be used as a mechanism to generate random numbers for applications that include computation.

In *stochastic computing*, numbers are represented as probabilities p of a binary 1 or 0 signal in a clocked bit-stream of length n_{bits}. As $n_{bits} \to \infty$ the average value of the signal is p distributed in the interval $[0, 1]$. A circuit that can perform multiplication followed by addition on stochastic data is illustrated in Fig. 1.6. Since basic linear algebra operations involve multiplication of matrix $\mathbf{A}$ with vector $\mathbf{s}$, stochastic computing might be tasked

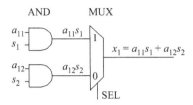

Fig. 1.6 Illustration of the AND and MUX functions to perform matrix element multiplication followed by addition in stochastic computing. Numbers are represented as probabilities of a binary 1 or 0 signal in a clocked bit-stream of length n_{bits}. Select (SEL) has a random value of binary 1 or 0 each clock cycle and the output is scaled to fall in the interval [0,1].

with computing both $\mathbf{x} = \mathbf{As}$ and $\mathbf{s} = \mathbf{A}^{-1}\mathbf{x}$. For the simplest 2×2 matrix, $\mathbf{x} = \mathbf{As}$ may be written as

$$
\begin{bmatrix} x_1 \\ x_2 \end{bmatrix} = \begin{bmatrix} a_{11}\ a_{12} \\ a_{21}\ a_{22} \end{bmatrix} \begin{bmatrix} s_1 \\ s_2 \end{bmatrix},
$$

so that evaluation of $x_1 = a_{11} \times s_1 + a_{12} \times s_2$ requires multiplication and addition (also known as multiply and accumulate, or MAC). In the limit $n_{bits} \to \infty$, the output of the AND function on stochastic input streams a_{11} and s_1 is the product of probabilities $p(a_{11})$ and $p(s_1)$. The sum of the two AND outputs is found by multiplexing using a select (SEL) that has random value of binary 1 or 0 each clock cycle. In this way, the average value of the MAC output is scaled to fall in the interval [0,1].

This use of random bit streams to represent numbers has the advantage that the multiplication and addition circuits are very simple to implement. There is also some inherent robustness to random errors in the bit stream. However, accuracy of the calculation is sensitive to unintended correlations between random number generators, the finite number of bits used to represent a number, and the condition of the matrix. While the use of a single photon source can, in principle, be used to physically guarantee random number generation, the number of bits, n_{bits}, and matrix condition number are also important considerations when solving $\mathbf{s} = \mathbf{A}^{-1}\mathbf{x}$ for which the determinant of matrix $\mathbf{A}$ must be calculated (see Problem 1.16).

1.1.6 Photons in Classical RF Wireless Communication

Before photons were introduced, properties of many electromagnetic phenomena were successfully described using Maxwell's equations. It seems appropriate to ask under what circumstances this works. If there are a large number of incoherent photons associated with a particular electromagnetic field, for example in an RF wireless communication link, the classical description might give accurate results. If there are few photons or there are special conditions involving a coherent superposition of photons, then a quantum description will likely be needed.

If single photons are used for communication then the quantized photon energy and momentum indicated in Eqs. (1.6) and (1.7) should be described using quantum mechanics. Section 1.1.7 discusses the application of single photon transmitter and receivers to secure quantum communication.

1.1.7 Secure Quantum Communication

As discussed in Section 1.1.1, electromagnetic waves can be polarized and linearly superimposed. It is also known that an electromagnetic wave of angular frequency ω consists of elementary particles called photons. Each photon has energy $\hbar\omega$. Remarkably, the combination of these facts can be used to create a secure communication channel.

Each photon in a particular optical mode that arrives at a perfect beam splitter will be detected at only one output port. If, on average, the beam splitter directs half the photons in one direction and half in another, then on average half the energy in a flux of photons is likewise redirected. There is a probability of one half that a given photon will be measured to emerge in one of the beam splitter's outputs. It is an essential feature of quantum mechanics that the detection of an independent photon at one of the two ports of the beam splitter is a random process.

Now, as illustrated in Fig. 1.7(a), a horizontally polarized optical wave passes through a birefringent medium (such as a calcite crystal) before being detected by the horizontal polarization detector. The detector can measure as little as one photon within a measurement time period τ. If the same flux of photons is of such low intensity that it contains one photon per detector measurement period then this measurement process

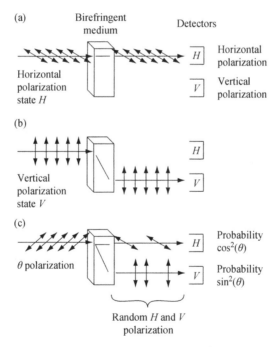

Fig. 1.7 (a) A horizontally polarized electromagnetic wave (H) and (b) a vertically polarized electromagnetic wave (V) can be unambiguously distinguished using a birefringent medium and detectors. If the electromagnetic wave is polarized at angle θ with respect to the horizontal direction, then each photon has an inferred post-detection path through the birefringent medium that is randomly selected so that, on average, horizontal polarization is detected with probability $\cos^2(\theta)$ and the vertical polarization is detected with probability $\sin^2(\theta)$. The total probability of detecting the electromagnetic field intensity of each photon is $\sin^2(\theta) + \cos^2(\theta) = 1$.

confirms the arrival of an individual horizontally polarized photon (H). Likewise, as indicated in Fig. 1.7(b), a vertically polarized photon (V) is detected using the birefringent medium and a second photodetector. However, if, as illustrated in Fig. 1.7(c), the optical wave is polarized at angle θ with respect to the horizontal direction then, post-detection, it may be inferred that each photon randomly chooses its path (H or V) through the birefringent medium and, on average, is detected by the horizontal photodetector with probability $\cos^2(\theta)$ and the vertical photon detector with probability $\sin^2(\theta)$.

The polarization state of each horizontally (H) or vertically (V) polarized photon can be used to reliably carry one bit of information. This is because the polarizations H and V are orthogonal. Hence, the combination of the birefringent medium and detectors can unambiguously detect the polarization state of each arriving photon. Communication via such a channel is insecure because it is always possible to build a repeater that reliably detects each arriving photon in orthogonal state H or V and then retransmits the same signal.

The situation changes dramatically if single photon communication is via nonorthogonal states. For example, a linear superposition state consisting of equal weights of H and V can be used to create a diagonally polarized state $D_+ = (H + V)/\sqrt{2}$ in which the $+$ sign indicates $\theta = +45°$ in Fig. 1.7(c). The state orthogonal to D_+ is $D_- = (H - V)/\sqrt{2}$, where the $-$ sign indicates $\theta = -45°$ in Fig. 1.7(c). When a single (independent) photon in state D_+ (or D_-) interacts with the birefringent medium, the inferred post-detection path taken is *randomly* selected between H and V. If single photons are used to communicate information in the nonorthogonal basis of V and D_+ in a quantum channel and use is made of a classical (many photons per bit of information) channel, it is possible to construct secure communication between two parties.[5] To see how this works, consider Alice and Bob as depicted in Fig. 1.8.

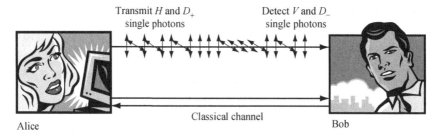

Transmit H and D_+ single photons

Detect V and D_- single photons

Classical channel

Alice

Bob

Fig. 1.8 Alice can transmit information to Bob via a quantum communication channel that uses single photons and nonorthogonal polarization states. Alice and Bob can also communicate via a classical channel.

Alice can transmit information to Bob via a quantum communication channel that uses single photons and nonorthogonal polarization states. Alice and Bob can also communicate via a classical channel. The method or protocol that Alice uses to communicate with Bob starts with Alice generating a secret random binary number sequence. Using the B92 protocol,[6] Alice agrees with Bob through the public channel that she will transmit using

[5] S. J. Wiesner, *SIGACT News* **15**, 78 (1983); C. H. Bennett and G. Brassard, *Proc. IEEE Int. Conf. on Computers, Systems and Signal Processing* (Bangalore, India, December 10–12, 1984) pp. 175–179; C. H. Bennett and S. J. Wiesner, *Phys. Rev. Lett.* **69**, 2881 (1992).
[6] C. H. Bennett, *Phys. Rev. Lett.* **68**, 3121 (1992).

the nonorthogonal basis, D_+ and H for binary bit 1 and bit 0, respectively. Bob agrees with Alice to test each arriving single photon by randomly setting his detector to V polarization for binary 1 or D_- polarization for binary 0. Notice, Bob detects photons in a nonorthogonal basis. As illustrated in Fig. 1.9, by detecting single photons, on average Bob identifies $1/2 \times 1/2 = 1/4$ of Alice's random bit sequence in an otherwise lossless channel. Each photon that Bob detects corresponds to a bit that Alice sent. Bob then tells Alice through the public channel the position of each bit he detected but not its value. As illustrated in Fig. 1.10, because Alice now knows which bits of her random binary number sequence Bob has successfully received, she discards all the other bits, keeping only the reconciled (or *sifted*) bits as the key for a provably secure one-time pad cipher. She then uses the key and the one-time pad cipher to transmit data via the classical channel to Bob.

Alice's bit value	$H=0$	$H=0$	$D_+=1$	$D_+=1$
	◀▶	◀▶	◀▶ + ↕	◀▶ + ↕
Bob tests with	$V=1$	$D_-=0$	$V=1$	$D_-=0$
	↕	◀▶ − ↕	↕	◀▶ − ↕
Observation probability	$P=0$	$P=\frac{1}{2}$	$P=\frac{1}{2}$	$P=0$

Fig. 1.9 Transmission and detection using a nonorthogonal basis ensures security. On average, Bob identifies $1/2 \times 1/2 = 1/4$ of Alice's random bit sequence in an otherwise lossless channel. This inefficiency is the overhead that is paid for security using quantum key distribution (QKD).

Alice generates a random number sequence and transmits binary bit 0 and bit 1 as H and D_+, respectively

Bit	1	1	1	0	1	0	0	0	1	0	1	1	1	1	0	1	1
Photon state	D_+	D_+	D_+	H	D_+	H	H	H	D_+	H	D_+	D_+	D_+	D_+	H	D_+	D_+

Bob tests with randomly chosen D_- or V for bit 0 and bit 1 respectively

Detector state	V	V	D_-	D_-	D_-	D_-	V	D_-	D_-	V	V	D_-	D_-	D_-	D_-	V	D_-
Probability Bob detects $\frac{1}{2}$	$\frac{1}{2}$	$\frac{1}{2}$	-	$\frac{1}{2}$	-	$\frac{1}{2}$	-	$\frac{1}{2}$	-	-	$\frac{1}{2}$	-	-	-	$\frac{1}{2}$	$\frac{1}{2}$	-
Bob detects	1	1	-	0	-	0	-	0	-	-	1	-	-	-	0	1	-
Alice keeps bits	1	1	-	0	-	0	-	0	-	-	1	-	-	-	0	1	-

Fig. 1.10 Alice generates a secret random binary number sequence and agrees with Bob through the public channel that she will transmit using a nonorthogonal basis, D_+ and H for binary bit 1 and bit 0, respectively. Bob agrees with Alice to test each arriving single photon by randomly setting his detector to V polarization for binary 1 or D_- polarization for binary 0. Each photon that Bob detects corresponds to a bit that Alice sent (for illustrative purposes only; in the figure Bob's bit detection probability of $1/2$ has been set to unity). Bob tells Alice through the public channel the position of each bit he detected but not its value. Because Alice now knows which bits of her random binary number sequence Bob has successfully received, she discards all the other bits, keeping only the reconciled (or *sifted*) bits as the key for a provably secure one-time pad cipher. She then uses the key and the one-time pad cipher to transmit data via the classical channel to Bob.

Because the key is transmitted using a quantum channel, the method is called quantum key distribution (QKD).

Suppose there is an eavesdropper, conveniently called Eve, trying to listen in on the communication between Alice and Bob. Because each bit of information transmitted in the classical channel contains many photons, Eve can easily listen in to all classical messages by tapping a small percentage of the photons in each bit of information. However, in the quantum channel, Eve cannot tap a fraction of each photon in the key transmission because a photon is an indivisible (elementary) particle. Using a simple photodiode, Eve cannot measure the photon without destroying it and because information is being transmitted using nonorthogonal states, she cannot create another photon with exactly the same properties as the one she destroyed.[7] The photon that Eve destroys will not be detected by Bob, increasing the bit error ratio detected by Bob in the quantum channel. A bit error ratio can easily be established between Alice and Bob by exchanging information on the values of each bit in a sequence after they have been received. In a lossless channel, *even if Eve uses the same photon detection settings as Bob*, she only has a probability of 1/2 of detecting photons in a given state and so, statistically, will *never* have the same key as Bob.

It follows that Alice and Bob cannot be tricked into using a key that is not secret. Even if Eve eavesdrops, Alice and Bob can still agree on a key only known to them. Once the QKD has been successfully completed, use of a one time cipher (also know as Vernam encryption[8]) can be used to pass secret messages.

For a fixed number of binary data bits and the same number of binary key bits, encryption is achieved by applying an exclusive OR (XOR) operation to the data and key. Decryption is achieved by applying the XOR operation to the encrypted data and the same key. In this way the message in the data is only available to those who have access to the encryption key.

Quantum key distribution has been implemented for fiber optic and free-space optical links.[9] However, to date, poor link performance severely limits practical applications. Challenges that must be overcome before widespread adoption can take place include increasing bit transmission rates and increasing the communication distance achieved.

The reason why photon particles often appear as waves and show the quantum effects of particle linear superposition and particle interference is because they do not scatter strongly among themselves. This fact makes it rather simple to create photons that have a well-defined wavelength and, when propagating in free space, a long coherence time and length. So what about other particles such as the electron? That is addressed in Section 1.1.8.

1.1.8 A Connection between Quantization of Photons and Other Particles

If photons are particles with energy $E = \hbar\omega$ and quantized momentum $\mathbf{p} = \hbar\mathbf{k}_{\text{ph}}$ in a plane wave normal mode of wavelength λ_{ph} then there may be other particles that are

---

[7] The fact that nonorthogonal states can never be copied precisely is called the "no cloning theorem" and is a basic feature of quantum mechanics. More about the no cloning theorem is presented in Section 4.6.

[8] Named after Gilbert Vernam's practical approach to implementation and patented in 1919. See also C. Shannon, *Bell System Tech. J.* **28**, 656 (1949).

[9] For example, see N. Gisin et al., *Rev. Mod. Phys.* **74**, 145 (2002) and J. C. Bienfang et al., *Optics Express* **12**, 2011 (2004).

also characterized by energy, plane wave normal mode wavelength, and momentum. The essential link between quantization of photons and quantization of other particles such as electrons is momentum. In general, *interaction between particles involves exchange of momentum*. Photons and electrons have momentum and they can interact with each other. Such interaction must involve exchange of momentum. If photon momentum is quantized, it is natural to assume that electron momentum is quantized. The uncomfortable alternative is to have two types of momentum, quantized momentum for photons and classical momentum for electrons!

In a photoelectric effect experiment, a photon with quantized momentum $\hbar\mathbf{k}_{\mathrm{ph}}$ and energy $E = \hbar\omega$ collides with an electron in a metal. The photon energy is *absorbed*, and the electron is ejected from the metal. The collision process may be described using a diagram similar to Fig. 1.11.

Experimental verification that electron momentum is quantized in a way similar to that of a photon would require showing that an electron is characterized by a wavelength. For example, measurement of an electron on the right time or length scale might exhibit wave-related interference effects similar to those seen for photons. Because electron kinetic energy is related to momentum and quantized momentum is related to wavelength, the wavelength λ_{e} of an electron with mass m_0 and energy E in free space may be found. The answer is $\lambda_{\mathrm{e}} = 2\pi\hbar/\sqrt{2m_0 E}$, which for an electron with $E = 1\,\mathrm{eV}$ gives a quite small wavelength $\lambda_{\mathrm{e}} = 1.226\,\mathrm{nm}$. In addition, unlike photons, electrons can interact quite strongly via the coulomb potential. An interference experiment with electrons involves preparing a beam of electrons with well-defined energy and then scattering the electrons with a coulomb potential analogous to the slits used by Young in his experiments with light.

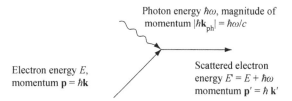

Photon energy $\hbar\omega$, magnitude of momentum $|\hbar\mathbf{k}_{\mathrm{ph}}| = \hbar\omega/c$

Electron energy E, momentum $\mathbf{p} = \hbar\mathbf{k}$

Scattered electron energy $E' = E + \hbar\omega$ momentum $\mathbf{p}' = \hbar\,\mathbf{k}'$

Fig. 1.11 The momentum and energy exchange between a photon and an electron may be described in a scattering diagram in which time flows from left to right. The electron has initial wave vector $\mathbf{k}$ and scattered wave vector $\mathbf{k}'$. The quantized momentum carried by the photon is $\hbar\mathbf{k}_{\mathrm{ph}}$.

1.1.9 Diffraction and Interference of Electrons

In 1927, Davisson and Germer[10] reported that an almost monoenergetic beam of electrons of kinetic energy $E = p^2/2m_0$ incident on a crystal of nickel gives rise to Bragg scattering peaks for electrons emerging from the metal. As illustrated in Fig. 1.12, the periodic array of atoms that forms a nickel crystal with FCC lattice constant $L = 0.352\,\mathrm{nm}$ creates a periodic coulomb potential from which the electrons scatter in a manner similar to light scattering from Young's slits. The observation of intensity maxima for electrons emerging from the crystal showed that electrons behave as waves. The electron waves exhibited the key features of diffraction, linear superposition, and interference. The experiment of

[10]C. Davisson and L. H. Germer, *Phys. Rev.* **30**, 705 (1927).

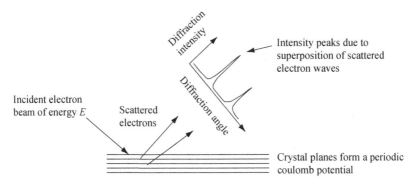

Fig. 1.12 A monoenergetic beam of electrons scattered from a metal crystal showing intensity maxima. The periodic array of atoms that forms the metal crystal creates a periodic coulomb potential from which electrons scatter. The observation of intensity maxima for electrons emerging from the crystal is evidence that electrons behave as waves.

Davisson and Germer supported the idea put forward by de Broglie[11] of electron *waves* in atoms. An electron of momentum $\mathbf{p} = \hbar\mathbf{k}$ (where $|\mathbf{k}| = 2\pi/\lambda_e$) has wavelength

$$\lambda_e = 2\pi\hbar\frac{1}{p} = \frac{2\pi\hbar}{\sqrt{2m_0 E}}. \tag{1.9}$$

It is important to note that if electrons of kinetic energy $E = p^2/2m_0$ behave as waves in such a way that

$$\psi(\mathbf{r}, t) \sim e^{-i(Et/\hbar - \mathbf{k}\cdot\mathbf{r})}, \tag{1.10}$$

where $\mathbf{p} = \hbar\mathbf{k}$, then a corresponding *wave equation* must exist.

1.1.10 When Is a Particle a Wave?

Photons, electrons, and other atomic-scale entities such as neutrons and protons can appear to behave either as particles or waves. There seems to be a *complementarity* in the way their behavior may be treated. They are both particle and wave. Neutrons, protons, and electrons can seem like particles, with a mass and position. However, if probed on the appropriate length or time scale, they might exhibit the key characteristics of waves, such as superposition and interference. Even though localized detection suggests the presence of a particle, the particle dynamics can evolve with wave character. While electrons and photons exhibit both particle and wave behavior, they are not called *wavicles*; they are called particles.

An isolated system of particles with little scattering can often exhibit wave-like behavior. This is because the states of the system are long-lived. It is easier to measure the consequences of wave-like behavior such as superposition and interference when states are long-lived. The reason for this is simply that coherent wave effects that last a relatively long time can be less spectrally broadened and are hence easier to interpret. In addition, long-lived states give the experimenter more time to perform some types of measurement.

[11]L. de Broglie, *Ann. de Physique* **3**, 22 (1925).

Obviously, when considering wave–particle duality, an important scale is set by the particle wavelength and its corresponding energy.

Photon energy is quantized as $E = \hbar\omega$, photon mass is zero (or *very* close to zero), photon momentum is $\mathbf{p} = \hbar\mathbf{k}_{ph}$, and photon wavelength is $\lambda_{ph} = 2\pi/k_{ph}$. The dispersion relationship for a photon moving long distances at the speed of light in free space is $E = \hbar ck_{ph}$ or, more simply, $\omega = ck_{ph}$.

Likewise, electron momentum is quantized as $\mathbf{p} = \hbar\mathbf{k}$, where $\mathbf{k}$ is electron wave vector, electron rest mass is $m_0 = 9.109\,565 \times 10^{-31}$ kg, and electron energy is $E = p^2/2m_0$. The dispersion relationship for an isolated electron moving at a nonrelativistic speed in free space is $E = \hbar^2 k^2/2m_0$. If the energy E of the electron measured in eV is known, then the electron wavelength λ_e in free space measured in units of nm is given by the expression

$$\lambda_e(\text{nm}) = \frac{1.226}{\sqrt{E(\text{eV})}}. \tag{1.11}$$

This means that an electron with a kinetic energy of $E = 100\,\text{eV}$ would have a wavelength of $\lambda_e = 0.1226$ nm. To measure this wavelength, electrons can be scattered from a structure with a coulomb potential that has a similar characteristic length scale (for example a nickel crystal with FCC lattice constant $L = 0.352$ nm).

Similarly, other finite-mass particles, for example the neutron, have a wavelength that is inversely related to the square root of the particle's kinetic energy. For the neutron, the free-space wavelength is

$$\lambda_n(\text{nm}) = \frac{0.0286}{\sqrt{E(\text{eV})}}. \tag{1.12}$$

Clearly, a neutron of kinetic energy $E = 100\,\text{eV}$ has a very small wavelength $\lambda_n = 0.002\,86$ nm, which is quite difficult to observe in an experiment.

1.1.11 Feynman Paths

A particle with wave properties moving from point a to b in free space might take a path s_n and arrive at b with an associated amplitude magnitude A_n and phase ϕ_n, where n labels the path. If the particle leaves point a at time $t = 0$ then the total amplitude at point b at time t is the *sum* of amplitudes $\mathbf{A}_{tot} = \mathbf{A}_1 + \mathbf{A}_2 + \cdots$ at b for all allowed paths and momenta. For a specific path in which amplitude magnitude does not change then $\mathbf{A}_n(t) = A_n(t = 0)e^{i\phi_n}$. If a particle has energy $E = \hbar\omega$ and momentum $\hbar k$, and making the simplification that at any instant $\mathbf{k}$ is parallel to the path taken, s, then the total amplitude at b at time t is $\mathbf{A}_{tot}(t) = A_1 e^{iks_1 - i\omega t} + A_2 e^{iks_2 - i\omega t} + \cdots$. In quantum mechanics the probability of detecting the particle at position b is $P_b(t) = \mathbf{A}_{tot}^* \mathbf{A}_{tot} = |\mathbf{A}_{tot}|^2$, which is the magnitude of the total probability amplitude $\mathbf{A}_{tot}$ squared.

As shown in Fig. 1.13, the shortest path between a and b for a particle of energy $E = \hbar\omega$ moving in free space is $s(n_0)$. A path such as $s(n_1)$ might have a small difference in length compared to $s(n_0)$ and so almost the same phase at b as the particle that took path $s(n_0)$. A different path such as $s(n_2)$ is much longer compared to $s(n_0)$ and hence can have a very different phase at b.

The paths between a and b are $s(n)$, where n labels each path in the ordered sequence of increasing path length. For the case being considered, the particle is free to propagate at energy $\hbar\omega$ using any of the infinite number of allowed paths between a and b. In free space,

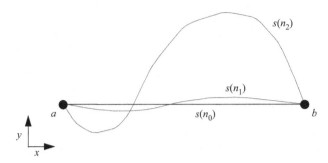

Fig. 1.13 Illustration of paths s a particle can take in free space between position a and b. The shortest path is a straight line, $s(n_0)$. Other paths such as $s(n_1)$ can make small deviations from $s(n_0)$ and so have almost the same phase at b as $s(n_0)$. Paths such as $s(n_2)$ are much longer compared to $s(n_0)$ and hence can have a very different phase at b.

each path has the same amplitude magnitude A_0 and so the total probability amplitude at position b is an integral over all paths. The resulting path integral[12] is

$$\mathbf{A}_{\text{tot}} = A_0 e^{-i\omega t} \int e^{iks(n)} dn. \tag{1.13}$$

Because $s(n_0)$ has the minimum length corresponding to a straight-line path, then

$$\frac{ds(n_0)}{dn} = 0. \tag{1.14}$$

Expanding $s(n)$ in a Taylor series for small variations in path length about this minimum results in

$$s(n) = s(n_0) + \frac{ds(n_0)}{dn}(n - n_0) + O\left((n - n_0)^2\right). \tag{1.15}$$

Hence, to first order, all paths near $s(n_0)$ have the same length and phase and the phases directly add in the integral to make a large contribution to $\mathbf{A}_{\text{tot}}$. Other paths can result in phases that vary considerably and that, when added, tend to cancel each other out. It is in this way, and in contrast to the classical case, that quantum mechanics allows a particle moving between point a and b in free space to explore *all* paths.

While this path integral description shows how the classical straight-line path of a particle in free space moving between point a and b *emerges* from the sum of all quantum paths, in general, the method is not an efficient way to solve practical problems and so a different approach is needed.

The development of a theory powerful and efficient enough to describe the properties of electrons, atoms, and other quantized particles was first developed by Heisenberg.[13] Even though this work, which was published in 1925, is insightful and interesting, a more

---

[12]R. P. Feynman and A. R. Hibbs, *Quantum Mechanics and Path Integrals*, New York, McGraw-Hill, 1965 (ISBN 978-0-07-020650-2).
[13]W. Heisenberg, *Zeitschrift für Physik* **33**, 879 (1925).

intuitive approach was developed in 1926 by Schrödinger, who found a wave equation associated with particles that behave as waves (i.e., a wave equation for *wavicles*).[14]

1.2 The Schrödinger Wave Equation

A wave equation is sought to describe the dynamics of particles with wavy character. Based on previous experience with classical mechanics and classical electrodynamics, it is assumed that time, t, is a continuous, smooth parameter and that position, $\mathbf{r}$, is a continuous, smooth variable. Any theory able to describe the dynamics of wavy particles must predict quantities such as particle position $\mathbf{r}$ and momentum $\mathbf{p}$ as a function of time. In the appropriate limit, the theory should incorporate the results of classical physics.

Since waviness is associated with a particle, it is sensible to introduce a wave function ψ that carries the appropriate information. Young's slits experiments suggest that such a wave function, which depends upon position and time, can be formed from a linear superposition of plane waves. Under these conditions, it seems reasonable to consider the special case of plane waves without loss of generality so that

$$\psi(\mathbf{r}, t) = A e^{i(\mathbf{k} \cdot \mathbf{r} - \omega t)}. \tag{1.16}$$

The wave function $\psi(\mathbf{r}, t)$ cannot be used to represent the particle directly because it is complex and this is at variance with everyday experience that measured quantities are real. The easiest way to guarantee a real value is to measure its intensity, $\psi(\mathbf{r}, t)^* \psi(\mathbf{r}, t) = |\psi(\mathbf{r}, t)|^2$. As with the Young's slits experiment, the intensity gives a measure of the entity's presence in different regions of space. For a particle that is confined to a finite-sized spatial domain (a bound state), the probability of finding the particle in volume $\mathrm{d}^3 r$ at position $\mathbf{r}$ in space at time t is proportional to $|\psi(\mathbf{r}, t)|^2$. Recognizing that for such a bound state the particle is definitely in some part of space, the intensity can be *normalized* so that integration of $|\psi(\mathbf{r}, t)|^2$ over the domain is unity. Hence, $|\psi(\mathbf{r}, t)|^2$ is a *probability density* for finding the particle in volume $\mathrm{d}^3 r$ at position $\mathbf{r}$ in space at time t. To find the most likely position of a wavy particle in space, the probability distribution is weighted with position $\mathbf{r}$ to obtain the average position $\langle \mathbf{r} \rangle$ so that

$$\langle \mathbf{r} \rangle = \int_{-\infty}^{\infty} \psi^*(\mathbf{r}, t) \mathbf{r} \psi(\mathbf{r}, t) \mathrm{d}^3 r = \int_{-\infty}^{\infty} \mathbf{r} |\psi(\mathbf{r}, t)|^2 \mathrm{d}^3 r. \tag{1.17}$$

In quantum mechanics, the average value of position is $\langle \mathbf{r} \rangle$ and is called the *expectation value* of the position *operator*, $\mathbf{r}$.

To understand and track particle dynamics, other quantities such as the particle momentum must be known. Associated with momentum is the operator $\mathbf{p} = \hbar \mathbf{k}$. To find the average value of momentum $\langle \mathbf{p} \rangle$, an integral over all space is performed so that

$$\langle \mathbf{p} \rangle = \int_{-\infty}^{\infty} \psi^*(\mathbf{k}, t) \hbar \mathbf{k} \, \psi(\mathbf{k}, t) \mathrm{d}^3 k, \tag{1.18}$$

---

[14]E. Schrödinger, *Annalen der Physik* **79**, 361 (1926).

in which $\psi(\mathbf{k}, t)$ is the Fourier transform of $\psi(\mathbf{r}, t)$. Restricting motion to the x direction, the expectation value of the observable p_x associated with the momentum operator $\hat{p}_x = \hbar k_x$ is

$$\langle p_x \rangle = \int_{-\infty}^{\infty} \psi^*(k_x) \hbar k_x \, \psi(k_x) \, \mathrm{d}k_x. \tag{1.19}$$

Often, the momentum operator $\hat{p}_x$ is written with a hat, $\wedge$, to indicate that it is a quantum operator, but the expectation value does not have a hat as it is just a real number. The time dependence $\mathrm{e}^{-\mathrm{i}\omega t}$ of the wave function is ignored when evaluating $\psi^*\psi$ since the time-dependent terms cancel. Taking the Fourier transform of $\psi(x)$ to obtain $\psi(k_x)$ gives

$$\langle p_x \rangle = \frac{1}{2\pi} \int_{-\infty}^{\infty} \mathrm{d}k_x \left(\int_{-\infty}^{\infty} \psi^*(x') \mathrm{e}^{\mathrm{i}k_x x'} \, \mathrm{d}x' \right) \hbar k_x \left(\int_{-\infty}^{\infty} \psi(x) \mathrm{e}^{-\mathrm{i}k_x x} \, \mathrm{d}x \right). \tag{1.20}$$

Integrating the far right-hand term in the brackets by parts using $\int U V' \, \mathrm{d}x = UV - \int U'V \, \mathrm{d}x$ with $U = \psi(x)$ and $V' = \mathrm{e}^{-\mathrm{i}k_x x}$ gives

$$\int_{-\infty}^{\infty} \psi(x) \mathrm{e}^{-\mathrm{i}k_x x} \mathrm{d}x = \left[\frac{1}{-\mathrm{i}k_x} \mathrm{e}^{-\mathrm{i}k_x x} \psi(x) \right]_{-\infty}^{\infty} + \int_{-\infty}^{\infty} \frac{1}{\mathrm{i}k_x} \frac{\mathrm{d}\psi(x)}{\mathrm{d}x} \mathrm{e}^{-\mathrm{i}k_x x} \mathrm{d}x. \tag{1.21}$$

The function in the square brackets is assumed (or regularized) to be zero in the limit $x \to \pm\infty$, so the expression for $\langle p_x \rangle$ given by Eq. (1.20) becomes

$$\langle p_x \rangle = \frac{\hbar}{2\mathrm{i}\pi} \int_{-\infty}^{\infty} \mathrm{d}k_x \int_{-\infty}^{\infty} \psi^*(x') \mathrm{e}^{\mathrm{i}k_x x'} \mathrm{d}x' \int_{-\infty}^{\infty} \mathrm{e}^{-\mathrm{i}k_x x} \left(\frac{\mathrm{d}}{\mathrm{d}x} \psi(x) \right) \mathrm{d}x, \tag{1.22}$$

which may be rewritten as

$$\langle p_x \rangle = -\mathrm{i}\hbar \int_{-\infty}^{\infty} \mathrm{d}x' \int_{-\infty}^{\infty} \psi^*(x') \frac{1}{2\pi} \mathrm{d}x \int_{-\infty}^{\infty} \mathrm{e}^{-\mathrm{i}k_x(x-x')} \frac{\mathrm{d}}{\mathrm{d}x} \psi(x) \, \mathrm{d}k_x. \tag{1.23}$$

Recognizing that $\frac{1}{2\pi} \int_{-\infty}^{\infty} \mathrm{e}^{-\mathrm{i}k_x(x-x')} \, \mathrm{d}k_x = \delta(x - x')$ results in

$$\langle p_x \rangle = -\mathrm{i}\hbar \int_{-\infty}^{\infty} \mathrm{d}x' \int_{-\infty}^{\infty} \psi^*(x') \delta(x - x') \left(\frac{\mathrm{d}}{\mathrm{d}x} \psi(x) \right) \mathrm{d}x, \tag{1.24}$$

so that

$$\langle p_x \rangle = -\mathrm{i}\hbar \int_{-\infty}^{\infty} \psi^*(x) \left(\frac{\mathrm{d}}{\mathrm{d}x} \psi(x) \right) \mathrm{d}x. \tag{1.25}$$

The important conclusion is that if $\hat{p}_x = \hbar k_x$ in k-space (momentum space) then the momentum operator in real space is a *spatial* derivative,

$$\hat{p}_x = -i\hbar \frac{d}{dx}. \tag{1.26}$$

The momentum operator and the position operator are said to form a *conjugate pair* linked by a Fourier transform. In this sense there is a full symmetry between the position and momentum operator.

While position and momentum are important examples of operators in quantum mechanics, so is potential. If potential is a scalar function of position only, the quantum mechanical operator for potential is just $\hat{V}(x)$. However, in the more general case, potential is also a function of time, $\hat{V}(x, t)$.

Summarizing, in quantum mechanics, every particle can be described by using a wave function $\psi(\mathbf{r}, t)$, where for bound states $|\psi(\mathbf{r}, t)|^2$ is the probability of finding the particle in the volume d^3r at position $\mathbf{r}$ at time t. The wave function and its spatial derivative are continuous, finite, and single valued.

Quantum operators are often associated with classical variables. Table 1.2 lists the classical variables along with the corresponding quantum operators mentioned thus far.

Because time is a *parameter* used to measure system evolution and hence not an operator, it does not appear in Table 1.2. Time is not a dynamical variable, and so it does not have an expectation value. This fact exposes a weakness in the theory. In Table 1.2 the energy operator for the wave function $\psi(x, t)$ is listed as $i\hbar(\partial/\partial t)$ (see Exercise 1.6). This cannot strictly be true since time is not an operator. This inconsistency is a hint that there exists a more complete theory for which time is not just a parameter.

The average or *expectation* value of an observable associated with the operator $\hat{A}$, such as those listed in Table 1.2, is

$$\langle A \rangle = \int_{-\infty}^{\infty} \psi^* \hat{A} \psi \, d^3r. \tag{1.27}$$

Table 1.2 Classical variables and quantum operators for $\psi(\mathbf{r}, t)$

Description	Classical theory	Quantum theory
Position	$\mathbf{r}$	$\mathbf{r}$
Potential	$V(\mathbf{r}, t)$	$V(\mathbf{r}, t)$
Momentum	p_x	$-i\hbar \dfrac{d}{dx}$
Energy	E	$i\hbar \dfrac{\partial}{\partial t}$

In classical mechanics, the total energy function or Hamiltonian of a particle mass m moving in potential V is

$$H = T + V = \frac{p^2}{2m} + V. \tag{1.28}$$

Substituting the corresponding expressions for quantum mechanical momentum and potential into the equation gives a wave equation,

$$\hat{H}\psi(x,t) = \frac{-\hbar^2}{2m}\frac{\mathrm{d}^2}{\mathrm{d}x^2}\psi(x,t) + V(x,t)\psi(x,t). \tag{1.29}$$

In three dimensions, this equation is written

$$\hat{H}\psi(\mathbf{r},t) = \frac{-\hbar^2}{2m}\nabla^2\psi(\mathbf{r},t) + V(\mathbf{r},t)\psi(\mathbf{r},t), \tag{1.30}$$

where

$$\nabla^2\psi(\mathbf{r},t) = \frac{\mathrm{d}^2\psi}{\mathrm{d}x^2} + \frac{\mathrm{d}^2\psi}{\mathrm{d}y^2} + \frac{\mathrm{d}^2\psi}{\mathrm{d}z^2}. \tag{1.31}$$

Replacing the Hamiltonian with the energy operator,

$$\hat{H}\psi(\mathbf{r},t) = i\hbar\frac{\partial}{\partial t}\psi(\mathbf{r},t), \tag{1.32}$$

where

$$\hat{H} = \left(\frac{-\hbar^2}{2m}\nabla^2 + V(\mathbf{r},t)\right) \tag{1.33}$$

is the *Hamiltonian operator* for a particle of mass m. The equation

$$\boxed{\left(\frac{-\hbar^2}{2m}\nabla^2 + V(\mathbf{r},t)\right)\psi(\mathbf{r},t) = i\hbar\frac{\partial}{\partial t}\psi(\mathbf{r},t)} \tag{1.34}$$

is called the *Schrödinger equation*. This equation is first-order in the time derivative indicating that the wave function $\psi(\mathbf{r},t)$ evolves *deterministically* from a single initial condition.

If before measurement the system is in a state consisting of a superposition of eigenstates then after measurement the system can only be found in one eigenstate. Quantum mechanics forces the result of measurement to be associated with a single eigenstate of the system. Hence, measurement is *noncausal* since, for a general superposition state ψ, it is not possible to know the result of the measurement beforehand. The probability of finding the system in a particular eigenstate after measurement is established experimentally by recording the average outcome of a measurement on each of many identically prepared systems.

Consider the special case of a closed system in which energy is conserved and potential energy is time independent in such a way that $V = V(\mathbf{r})$. Separation of the wave function into orthogonal spatial components allows the solution of one-dimensional problems. In quantum mechanics, the Hamiltonian is separable if the potential is separable. This is the case if, for example, $V(\mathbf{r}) = V(x)V(y)V(z)$.

If the one-dimensional system has a potential in which $V = V(x)$, then the time and spatial parts of the wave function may be separated out using the method of separation of variables. Assuming the wave function can be expressed as a product, $\psi(x,t) = \psi(x)\phi(t)$, then substitution into the one-dimensional Schrödinger equation gives

$$\frac{-\hbar^2}{2m}\frac{\mathrm{d}^2}{\mathrm{d}x^2}\psi(x)\phi(t) + V(x)\psi(x)\phi(t) = i\hbar\frac{\partial}{\partial t}\psi(x)\phi(t). \tag{1.35}$$

Dividing both sides by $\psi(x)\phi(t)$, the left-hand side is a function of x only and the right-hand side is a function of t only. This is true if both sides are equal to a constant E. It follows that

$$\boxed{E\phi(t) = i\hbar\frac{\partial}{\partial t}\phi(t)} \tag{1.36}$$

is the *time-dependent Schrödinger equation* and

$$\boxed{\left(\frac{-\hbar^2}{2m}\frac{\mathrm{d}^2}{\mathrm{d}x^2} + V(x)\right)\psi(x) = E\psi(x)} \tag{1.37}$$

is the *time-independent Schrödinger equation* in one dimension. The constant E is just the *energy eigenvalue* of the particle described by the wave function.

It is important to remember that these two equations apply when the potential may be considered independent of time. Of course, the potential is not truly time-independent in the sense that it must have been created at some time in the past. However, it is assumed that the transients associated with this creation have negligible effect on the calculated results.

The solution to the time-dependent Schrödinger equation is of simple harmonic form

$$\phi(t) = \mathrm{e}^{-i\omega t}, \tag{1.38}$$

where $E = \hbar\omega$. Notice that Schrödinger's equation *requires* the exponential form $\mathrm{e}^{-i\omega t}$. Alternatives such as $\mathrm{e}^{+i\omega t}$ or trigonometric functions such as $\sin(\omega t)$ are not allowed. Unlike oscillatory solutions in classical mechanics, Schrödinger's equation gives no choice for the form of the time dependence appearing in the wave function.

The solution to the time-independent Schrödinger equation is $\psi(x)$, which is, of course, independent of time. A particle in such a time-independent state will remain in that state until acted upon by some external entity that forces it out of that state.

The time-independent Schrödinger equation is a second-order differential eigenequation whose nth eigenvalue and eigenfunction solution is found by applying boundary and initial conditions.

If the particle is confined by a potential to a local region of space, the probability of finding a particle somewhere in space is unity and, in this case, it makes sense to require

21

wave functions that are solutions to the time-independent Schrödinger equation be square integrable and normalized such that

$$\int_{-\infty}^{\infty} \psi_n^*(\mathbf{r})\psi_n(\mathbf{r})\, \mathrm{d}^3 r = 1. \tag{1.39}$$

Here, a notation in which the integer n labels the wave function associated with a given energy eigenvalue is adopted. Wave functions with different quantum numbers, say n and m, have the mathematical property of orthogonality, so that for $n \neq m$,

$$\int_{-\infty}^{\infty} \psi_n^*(\mathbf{r})\psi_m(\mathbf{r})\, \mathrm{d}^3 r = 0. \tag{1.40}$$

Eqs. (1.39) and (1.40) may be summarized by stating that the wave functions are *orthonormal* or

$$\int_{-\infty}^{\infty} \psi_n^*(\mathbf{r})\psi_m(\mathbf{r})\, \mathrm{d}^3 r = \delta_{nm}, \tag{1.41}$$

where δ_{nm} is the Kronecker-delta function, which has a value of unity when $n = m$ and otherwise is zero.

1.2.1 The Gaussian Electron Wave Packet and Dispersion

Perhaps the simplest application of the Schrödinger equation is its use in describing an electron mass m_0 moving unimpeded through space. In this case, the time-independent Schrödinger equation is

$$\hat{H}\psi_n(\mathbf{r}) = \frac{-\hbar^2}{2m_0}\nabla^2\psi_n(\mathbf{r}) + V(\mathbf{r})\psi_n(\mathbf{r}) = E_n\psi_n(\mathbf{r}), \tag{1.42}$$

where for free space the potential $V(\mathbf{r}) = 0$. Here, E_n are *energy eigenvalues* and ψ_n are *eigenstates* so that

$$\psi_n(\mathbf{r}, t) = \psi_n(\mathbf{r})e^{-i\omega t}. \tag{1.43}$$

The wave functions and corresponding energy eigenvalues are labeled by the quantum number n. For an electron in free space, $V(\mathbf{r}) = 0$, and so

$$E_n = \frac{\hat{p}_n^2}{2m_0} = \frac{\hbar^2 k_n^2}{2m_0} = \hbar\omega_n \tag{1.44}$$

and

$$\psi_n(\mathbf{r}, t) = \left(Ae^{i\mathbf{k}\cdot\mathbf{r}} + Be^{-i\mathbf{k}\cdot\mathbf{r}}\right)e^{-i\omega_n t}, \tag{1.45}$$

where the first term in the parentheses is a plane wave of amplitude A traveling left to right and the second term is a plane wave of amplitude B traveling from right to left.

The term $e^{-i\omega_n t}$ gives the time dependence of the wave function. Selecting a boundary condition with $B = 0$, and considering the case of motion in the x direction only, the wave function becomes

$$\psi_n(x,t) = Ae^{ik_x x}e^{-i\omega_n t}. \tag{1.46}$$

To find the momentum associated with this wave function, the momentum operator is applied to the wave function:

$$\hat{p}_x\psi_n(x,t) = -i\hbar\frac{\mathrm{d}}{\mathrm{d}x}\psi_n(x,t) = \hbar k_x Ae^{ik_x x}e^{-i\omega_n t} = \hbar k_x\psi_n(x,t). \tag{1.47}$$

Since, in quantum mechanics, the real eigenvalue of an operator can, at least in principle, be measured, the x-directed momentum of the particle is just $p_x = \hbar k_x$.

From these results, it may be concluded that an electron with mass m_0 and energy E in free space has momentum with magnitude of $p = \hbar k = \sqrt{2m_0 E}$ and a nonlinear dispersion relation $E_k = \hbar\omega = \hbar^2 k^2/2m_0$, or

$$\boxed{\omega(k) = \frac{\hbar k^2}{2m_0}.} \tag{1.48}$$

This dispersion relation is illustrated in Fig. 1.14(a). Phase velocity is defined as $v_\mathrm{p} = \omega/k$ and group velocity is $v_\mathrm{g} = \partial\omega/\partial k$.

The Schrödinger equation does not allow an electron in free space to have just any energy and wavelength; rather, the electron may only have values that satisfy the dispersion relation.

An electron can be constrained to occupy a finite region of space by forming a *wave packet* from a continuum of plane–wave eigenstates. This consists of the superposition of many eigenstates that destructively interfere everywhere except in some localized region

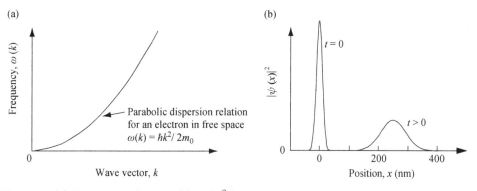

Fig. 1.14 (a) Dispersion relation $\omega(k) = \hbar k^2/2m_0$ for an electron moving in the x direction in free space. The magnitude of phase velocity is $v_\mathrm{p} = \omega/k$ and that of group velocity $v_\mathrm{g} = \partial\omega/\partial k$. (b) Snapshots of $|\psi(x)|^2$ for a Gaussian wave packet showing the effect of dispersion at different times.

of space. A Gaussian pulse may be formed by modulating a plane wave of momentum $\hbar k_0$ in the x direction in such a way that at time $t = 0$,

$$\psi(x, t = 0) = A e^{ik_0 x} e^{-x^2/4\sigma_x^2}, \tag{1.49}$$

where $A = 1/\left(2\pi\sigma_x^2\right)^{1/4}$, the mean position at time $t = 0$ is $x_0 = 0$, and a measure of the spatial spread in the wave function at time $t = 0$ is given by σ_x. The probability density at time $t = 0$ is just a normalized Gaussian function:

$$\psi^*(x, t = 0)\psi(x, t = 0) = |\psi(x, t = 0)|^2 = A^2 e^{-x^2/2\sigma_x^2}, \tag{1.50}$$

with full width at half maximum of $\text{FWHM} = 2\sqrt{2\ln(2)}\,\sigma_x$. The Gaussian pulse contains a continuum of momentum components centered about the original plane–wave momentum $\hbar k_0$. The Fourier transform of wave function $\psi(x, t = 0)$ gives the k-space representation:

$$\psi(k, t = 0) = \frac{1}{A\sqrt{\pi}} e^{-(k-k_0)^2\sigma_x^2}. \tag{1.51}$$

The corresponding probability density in k-space (momentum space) is given by

$$|\psi(k, t = 0)|^2 = \frac{1}{A^2\pi} e^{-(k-k_0)^2 2\sigma_x^2} = \frac{1}{A^2\pi} e^{-(k-k_0)^2/2\sigma_k^2}, \tag{1.52}$$

where k_0 is the average value of k, and a measure of the spread in the distribution of k at time $t = 0$ is given by the standard deviation $\sigma_k = 1/2\sigma_x$. Because the product $\sigma_k\sigma_x = 1/2$ is a constant, localizing the Gaussian pulse in real space will increase the spread of the corresponding distribution in k-space. Conversely, localizing the Gaussian pulse in k-space increases the spatial extent of the pulse in real space. Recognizing that momentum $p = \hbar k$, then, for a Gaussian pulse at time $t = 0$ with standard deviation Δp in momentum and Δx in position,

$$\Delta p\Delta x = \frac{\hbar}{2}, \tag{1.53}$$

which is an example of the *uncertainty principle. Conjugate pairs of operators cannot be measured to arbitrary accuracy.* In this case, it is not possible to simultaneously know the exact position of a particle and its momentum.

Since each plane wave has a time dependence of the form $e^{-i\omega(k)t}$, for time $t > 0$,

$$\psi(k, t) = \frac{1}{A\sqrt{\pi}} e^{-(k-k_0)^2\sigma_x^2} e^{-i\omega(k)t}, \tag{1.54}$$

where $\omega(k)$ is the dispersion relation. A Taylor expansion about k_0 for an arbitrary dispersion relation is, to second order, given by

$$\omega(k) = \omega(k_0) + \frac{d\omega(k_0)}{dk}(k - k_0) + \frac{1}{2!}\frac{d^2\omega(k_0)}{dk^2}(k - k_0)^2 + \cdots \tag{1.55}$$

To find the effect of dispersion on the Gaussian pulse as a function of time in real space, the Fourier transform of $\psi(k, t)$ is taken to obtain $\psi(x, t)$ and, specializing to the case of a freely propagating electron of mass m_0 with parabolic dispersion $\omega(k) = \hbar k^2/2m_0$ and initial mean position x_0, this becomes

$$\psi(x,t) = \left(\frac{\sigma_x^2}{2\pi}\right)^{\frac{1}{4}} \frac{e^{i(k_0(x-x_0)-\omega(k_0)t)}}{\sqrt{\sigma_x^2 + \frac{i\hbar t}{2m_0}}} e^{-\left(x-x_0-\frac{\hbar k_0 t}{m_0}\right)^2 / \left(4\sigma_x^2 + \frac{i2\hbar t}{m_0}\right)}. \tag{1.56}$$

The term $e^{i(k_0(x-x_0)-\omega(k_0)t)}$ is a plane wave oscillating at frequency $\omega(k_0) = \hbar k_0^2/2m_0$ with momentum $\hbar k_0$ and a *phase velocity* $v_{\mathrm{p}} = \hbar k_0/2m_0$. The term $-(x - x_0 - \hbar k_0 t/m_0)^2$ shows that the peak of the wave packet moves a distance $\hbar k_0 t/m_0$ in time t, indicating a *group velocity* for the wave packet of $v_{\mathrm{g}} = \hbar k_0/m_0$. If a classical particle is described by such a wave packet then the velocity of the particle is given by the group velocity of the wave packet.

The corresponding Gaussian wave packet probability density,

$$|\psi(x,t)|^2 = \frac{1}{\sqrt{2\pi\Delta x^2(t)}} e^{-\left(x-x_0-\frac{\hbar k_0 t}{m_0}\right)^2 / (2\Delta x^2(t))}, \tag{1.57}$$

has a standard deviation that increases with time as

$$\Delta x(t) = \sqrt{\sigma_x^2 + \frac{\hbar^2 t^2}{4m_0^2\sigma_x^2}}. \tag{1.58}$$

The wave packet delocalizes as a function of time because of the parabolic dispersion $\omega = \hbar k^2/2m_0$. Figure 1.14(b) illustrates the effect this has by showing snapshots of $|\psi(x)|^2$ at different times.

Figure 1.15(a) is a plot of the real part of the wave function, $\mathrm{Re}(\psi(x))$, calculated at time $t = 0$ and $t = 0.5\,\mathrm{ps}$ showing that the leading edge of a time-evolved Gaussian wave packet contains higher spatial frequencies than the trailing edge. The relatively larger phase velocity in the leading edge causes the pulse to broaden with increasing time. Figure 1.15(b) plots the time-independent wave vector spectrum.

The characteristic time $\Delta\tau_x$ for the width of a wave packet subject to parabolic dispersion to double may be found from Eq. (1.58) and used to illustrate why classical particles always appear particle-like.

Consider a particle of mass one gram ($m = 10^{-3}\,\mathrm{kg}$), the position of which is known to an accuracy of one micron so that $\Delta x = 10^{-6}\,\mathrm{m}$. Modeling the particle as a Gaussian wave packet gives a characteristic time $\Delta\tau_x$ which is many times the age of the universe![15] The constant $\hbar$ sets an absolute scale that ensures that the macroscopic classical particle remains a classical particle for all time.

This result may now be contrasted with an atomic-scale entity. Consider an electron of mass m_0 in a circular orbit of radius $a_{\mathrm{B}} = 0.529 \times 10^{-10}\,\mathrm{m}$ around a proton. Adopting the semiclassical picture of a hydrogen atom discussed in Section 1.2.4, the electron orbits in the coulomb potential of the proton with tangential speed $2.2 \times 10^6\,\mathrm{m\,s^{-1}}$. For purposes of illustration, the electron is described by a Gaussian wave packet and its position is known to an accuracy of $\Delta x = 10^{-11}\,\mathrm{m}$. In this case, the characteristic time $\Delta\tau_x$ is significantly shorter than the time to complete one orbit ($\tau_{\mathrm{orbit}} \sim 1.5 \times 10^{-16}\,\mathrm{s}$). Long before the electron wave packet can complete one orbit it has lost all of its particle character and has

---

[15]The age of the universe is thought to be in the range 10–20 billion years.

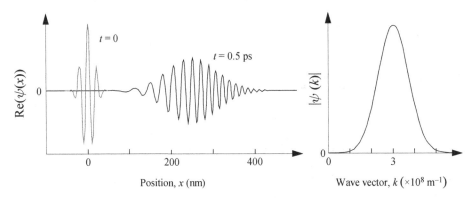

Fig. 1.15 (a) Calculated $\mathrm{Re}(\psi(x))$ at time $t = 0$ and $t = 0.5$ ps showing that the leading edge of a time-evolved Gaussian wave function contains higher spatial frequencies than the trailing edge. The relatively larger phase velocity in the leading edge causes the pulse to broaden with increasing time. (b) The wave vector spectrum $|\psi(k)|$ is independent of time. Parameters: particle mass $0.07 \times m_0$, $\sigma_x = 10$ nm, $E = 50$ meV.

completely delocalized over the circular trajectory. The concept of a semiclassical wave packet describing an isolated particle-like electron orbiting the proton of a hydrogen atom just does not make any sense.

The effect of dispersion in the wave components of a Gaussian pulse propagating in a medium with parabolic dispersion is to increase the spatial extent of the pulse as a function of time. This has profound influence on the ability to assign wave or particle character to an object such as an electron of mass m_0. For a freely propagating electron pulse, the momentum of the plane–wave components does not change (the k-space spectrum is time independent), so the product $\Delta p \Delta x$ must be increasing with time from its minimum value at time $t = 0$ due to $\Delta x(t)$ given by Eq. (1.58). The uncertainty relation for momentum and position may be written more generally as

$$\boxed{\Delta p \Delta x \geq \frac{\hbar}{2}.} \tag{1.59}$$

While this relationship limits the precision with which particle position and momentum can be known, it is pulse shape, dispersion relation, and potential that control the temporal evolution of a pulse. So, for example, if the dispersion is linear and lossless in a constant potential then a Gaussian pulse will not disperse with time.

1.2.2 Measure of Wave Packet Dispersion

Wave packets do not always have a convenient Gaussian shape, so evaluation of pulse width is not a good measure of dispersion. If pulse shape at time $t = 0$ is $|\psi(x, t = 0)|^2$ then, assuming time is a continuously evolving real parameter, dispersion can be quantified as a scalar quantity that is the spatial overlap of the initial pulse with the dispersed pulse at time t appropriately time-shifted to $t = 0$.

1.2.2.1 The Nondispersive Airy Electron Wave Packet

The solution to the Schrödinger equation for an electron in a uniform electric field is an Airy wave function. It is possible to construct a hypothetical electron wave packet based on the Airy function that does not disperse in the constant potential of free space.[16] However, such a wave packet turns out to have infinite energy, is not square integrable, and hence not physical. Any modification of the pure Airy wave function to make it physically meaningful results in a dispersive wave packet.

1.2.3 The Hydrogen Atom

In 1911, Rutherford showed experimentally that electrons appear to orbit the nucleus of atoms. With this in mind, consider a single electron moving in the coulomb potential of the single proton in a hydrogen atom.

As illustrated in Fig. 1.16(a), the proton is at the origin and the electron is at position **r**. The electron has charge $-e$ and mass $m_0 = 9.109381 \times 10^{-31}$ kg. The proton charge is equal and opposite to that of the electron, and the proton mass is $m_p = 1.672621 \times 10^{-27}$ kg. Because the ratio of proton mass to electron mass is more than 1800, the electron can be considered as orbiting an infinite-mass proton.

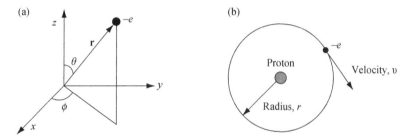

Fig. 1.16 (a) A hydrogen atom consists of an electron and a proton. A spherical coordinate system describes a single electron moving in the coulomb potential $V(\mathbf{r}) = V(r, \theta, \phi) = -e^2/4\pi\epsilon_0 r$ of the single proton. (b) A classical circular orbit of an electron mass m_0 and charge $-e$ moving with velocity v in the coulomb potential of a proton mass m_p and charge $+e$. This classical view predicts that hydrogen is unstable.

If the electron is in a curved orbit similar to that sketched in Fig. 1.16(b) it is accelerating and hence must, according to classical electrodynamics, radiate electromagnetic waves. As the electron energy radiates away, the radius of the orbit decreases until the electron collapses into the proton nucleus. This should all happen within a time of about 0.1 ns. Experiment shows that the classical theory does not work! Hydrogen is observed to be stable. In addition, the spectrum of hydrogen is observed to consist of discrete spectral lines – which, again, is a feature not predicted by classical models.

Having established the wavy nature of the electron when travelling in free space and when scattered, as shown in the Davisson and Germer experiment, it seems reasonable to insist that an electron in a circular orbit must also exhibit a wavy character. This is interesting, because the geometry forces the wave to fold back upon itself. Since a

[16]M. V. Berry and M. L. Balazs, *Am. J. Phys.* **47**, 264 (1979) and J. Lekner, *Eur. J. Phys.* **30**, L43–L46 (2009).

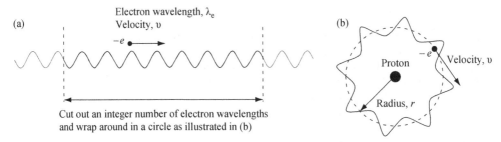

Fig. 1.17 (a) Illustration of electron wave with wavelength λ_e. (b) Illustration of an electron wave wrapped around in a circular orbit about a proton. The single-valueness of the electron wave function suggests that only an integer number of electron wavelengths can fit into a circular orbit of radius r.

wave function describing the electron might be expected to be single-valued, only integer wavelengths can be fitted into a circular orbit of a given circumference. Figure 1.17 illustrates the idea with an integer number, n, wavelengths of an electron wave function wrapped around a circular orbit of radius $r = n\lambda_e/2\pi$, where n is a nonzero positive integer.

It is possible to put together an ad hoc explanation of the properties of the hydrogen atom using just a few postulates. In 1913, Bohr showed that the spectral properties of hydrogen may be described quite accurately if the following rules are adopted:

1. Electrons exist in stable circular orbits around the proton.

2. Electrons can transition between orbits by emission or absorption of a photon of energy $\hbar\omega$.

3. The angular momentum of the electron in a given orbit is quantized according to $p_\theta = n\hbar$, where n is a nonzero positive integer.

Postulate 3 admits the wavy nature of the electron. Stable, long-lived orbits only exist when an integer number of wavelengths can fit into the classical orbit.

Classically, the atom can radiate electromagnetic waves when there is a net oscillating dipole moment. This involves net oscillatory current flow due to the movement of charge. However, when an integer number of wavelengths fit into the classical orbit this may be thought of as two counter-propagating waves whose associated currents exactly cancel. In this case, there is no net oscillatory current, and the electron state is long lived.

The postulates of Bohr allow many parameters to be calculated, such as the average radius of a stable electron orbit and the energy difference between stable orbits.

1.2.3.1 Calculation of the Average Radius of an Electron Orbit in Hydrogen

Equating the electrostatic force and the centripetal force:

$$\frac{-e^2}{4\pi\varepsilon_0 r^2} = -\frac{m_0 v^2}{r}, \tag{1.60}$$

where the left-hand term is the electrostatic force and the right-hand term is the centripetal force. Here, an infinite proton mass and electron mass m_0 is assumed instead of using the reduced mass m_r such that $1/m_r = 1/m_p + 1/m_0$. Because angular momentum is quantized, $p_\theta = n\hbar = m_0 v r_n$, in which n is a nonzero positive integer. The equation for angular momentum may be rewritten as $m_0^2 v^2 = n^2 \hbar^2 / r_n^2$ or $m_0 v^2 / r_n = (1/r_n m_0)(n^2 \hbar^2 / r_n^2)$. Substituting into the expression for centripetal force gives

$$\frac{e^2}{4\pi\varepsilon_0 r_n^2} = \frac{1}{r_n m_0} \frac{n^2 \hbar^2}{r_n^2} \tag{1.61}$$

and hence the quantized value of the nth electron orbit radius in hydrogen is

$$r_n = \frac{4\pi\varepsilon_0 n^2 \hbar^2}{m_0 e^2}. \tag{1.62}$$

The spatial scale is set by the radius for $n = 1$, giving the *Bohr radius*

$$r_1 = a_B = \frac{4\pi\varepsilon_0 \hbar^2}{m_0 e^2} = 0.529\,177 \times 10^{-10}\,\text{m}. \tag{1.63}$$

Notice that if the reduced mass were used then $r_1 = 0.529\,889 \times 10^{-10}\,\text{m}$.

1.2.3.2 Calculation of Energy Difference between Electron Orbits in Hydrogen

The energy difference between orbits is important, since it allows prediction of the optical spectra of excited hydrogen atoms.

If the electron is assumed to exist in a stable circular orbit around the proton, as shown schematically in Fig. 1.16(b), the momentum of the electron is $m_0 v_n = n\hbar/r_n$, and

$$v_n = \frac{n\hbar}{m_0 r_n} = \frac{m_0 e^2}{4\pi\varepsilon_0 n^2 \hbar^2} \frac{n\hbar}{m_0} = \frac{e^2}{4\pi\varepsilon_0 n\hbar} \tag{1.64}$$

for the magnitude of electron velocity in the nth orbit. The value of v_n for $n = 1$ is $v_1 = 2.2 \times 10^6\,\text{m s}^{-1}$. The kinetic energy of the electron is

$$T = \frac{1}{2} m_0 v_n^2 = \frac{1}{2} m_0 \frac{e^4}{(4\pi\varepsilon_0)^2 n^2 \hbar^2}. \tag{1.65}$$

The potential energy is just the force times the distance between charges, so

$$V = \frac{-e^2}{4\pi\varepsilon_0 r_n} = -m_0 \frac{e^4}{(4\pi\varepsilon_0)^2 n^2 \hbar^2}. \tag{1.66}$$

The total energy for the nth orbit is

$$E_n = T + V = -\frac{1}{2} m_0 \frac{e^4}{(4\pi\varepsilon_0)^2 n^2 \hbar^2}. \tag{1.67}$$

The fact that $T = -V/2$ is a specific example of the more general "virial theorem," which states that $\langle T \rangle = \gamma \langle V \rangle / 2$ for a system in a stationary state and with a potential proportional to r^γ.

29

The energy difference between orbits with quantum number n_1 and n_2 is

$$E_{n_2} - E_{n_1} = \frac{m_0}{2} \frac{e^4}{(4\pi\varepsilon_0)^2\hbar^2} \left(\frac{1}{n_1^2} - \frac{1}{n_2^2}\right).$$ (1.68)

The pre-factor in this equation is a natural energy scale for the system. It corresponds to $n = 1$ in Eq. (1.67) and is the lowest energy state, E_1:

$$E_1 = \mathrm{Ry} = \frac{-m_0}{2} \frac{e^4}{(4\pi\varepsilon_0)^2\hbar^2} = -13.6058\,\mathrm{eV},$$ (1.69)

where Ry is called the Rydberg constant.

When the electron orbiting the proton of a hydrogen atom is excited to a high energy state, it can lose energy by emitting a photon of energy $\hbar\omega$. Because the energy levels of the hydrogen atom are quantized, photon emission spectra consist of a discrete number of spectral lines. In Fig. 1.18, the emission lines correspond to transitions from high energy levels to lower energy levels.

It is also possible for the reverse process to occur. In this case, photons with the correct energy can be absorbed, causing an electron to be excited from a low energy level to a higher energy level. This absorption process could be represented in Fig. 1.18 by changing the direction of the arrows on the vertical lines.

Different groups of energy transitions result in emission of photons of energy, $\hbar\omega$. The characteristic emission line spectra have been given the names (Lyman, Balmer, Paschen) of those who first observed them.

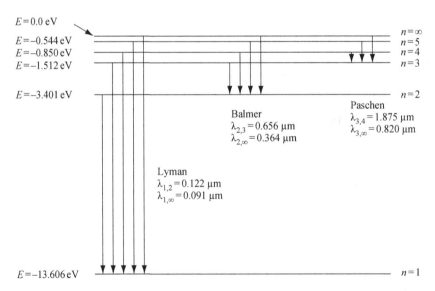

Fig. 1.18 Photon emission spectra of excited hydrogen consist of a discrete number of spectral lines corresponding to transitions from high energy levels to lower energy levels. Different groups of characteristic emission line spectra have been given the names of those who first observed them.

The Bohr model of the hydrogen atom is an ad hoc hybrid between classical and quantum ideas. A more quantum approach is to use the Schrödinger equation to describe an electron moving in the spherically symmetric coulomb potential of the proton charge.

In spherical coordinates, the time-independent solutions to the Schrödinger equation are of the form

$$\psi_{nlm}(r, \theta, \phi) = R_{nl}(r)\Theta_l^m(\theta)\Phi_m(\phi), \tag{1.70}$$

where the r-, θ-, and ϕ-dependent parts to the wave equation have been separated out. The resulting three equations have wave functions that are quantized with integer quantum numbers n, l, and m and are separately normalized. For example, it can be shown that the function $\Phi_m(\phi)$ must satisfy

$$\frac{\mathrm{d}^2}{\mathrm{d}\phi^2}\Phi_m(\phi) + m^2\Phi_m(\phi) = 0, \tag{1.71}$$

where m is the quantum number for the wave function,

$$\Phi_m(\phi) = A\mathrm{e}^{\mathrm{i}m\phi}. \tag{1.72}$$

The normalization constant A can be found from

$$\int_0^{2\pi} \Phi_m^*(\phi)\Phi_m(\phi)\mathrm{d}\phi = A^2 \int_0^{2\pi} \mathrm{e}^{-\mathrm{i}m\phi}\mathrm{e}^{\mathrm{i}m\phi}\mathrm{d}\phi = A^2 \int_0^{2\pi} \mathrm{d}\phi = 2\pi A^2 = 1. \tag{1.73}$$

Hence, $A = 1/\sqrt{2\pi}$ and

$$\Phi_m(\phi) = \frac{1}{\sqrt{2\pi}}\mathrm{e}^{\mathrm{i}m\phi}. \tag{1.74}$$

It is reasonable to expect that $\Phi_m(\phi)$ should be single valued, thereby forcing the function to repeat itself every 2π. This happens if m is an integer. The other quantum numbers that specify the state of the electron are also integers with a special relationship to one another:

$$n = 1, 2, 3, \ldots,$$
$$l = 0, 1, 2, \ldots, (n - 1),$$
$$m = \pm l, \ldots, \pm 2, \pm 1, 0.$$

The principal quantum number n specifies the energy of the Bohr orbit. The quantum numbers l and m relate to the quantization of orbital angular momentum. The orbital angular momentum quantum number is l, and the azimuthal quantum number is m. Because, in this description, the energy level for given n is independent of quantum numbers l and m, there is an n^2 degeneracy in states of energy E_n, since

$$\sum_{l=0}^{l=n-1} (2l + 1) = n^2. \tag{1.75}$$

In addition to n, l, m, the electron has a spin quantum number $\sigma_\mathrm{s} = \pm 1/2$. Electron spin is an intrinsic property of the electron that arises due to the influence of special relativity on the behavior of the electron. In 1928, Dirac showed that electron spin emerges as a natural consequence of rotational symmetry in a relativistic treatment of the Schrödinger equation.

1.2.4 Periodic Table of Elements

The hydrogen atom discussed in Section 1.2.3 is only one type of atom. There are over one hundred other atoms, each with their own unique characteristics. The incredible richness of nature can, in part, be thought of as due to the fact that there are many ways to form different combinations of atoms in the gas, liquid, and solid states. Chemistry sets out to introduce a methodology that predicts the behavior of different combinations of atoms. It is convenient to use a periodic table in which atoms are grouped according to the similarities in their chemical behavior. For example, H, Li, Na, and other atoms form the column IA elements of the periodic table because they all have a single electron available for chemical reaction with other atoms. The rules of quantum mechanics can help explain why atoms behave the way they do and why chemists can group atoms according to the number of electrons available for chemical reaction.

1.2.4.1 The Pauli Exclusion Principle and the Properties of Atoms

It is an experimental fact that no two electrons in a noninteracting system can have the same quantum numbers n, l, m, σ_s. Each electron state is assigned a unique set of values of n, l, m, σ_s because such identical half-odd integer spin particles cannot occupy the same state. This is the *Pauli exclusion principle*, which determines many properties of atoms in the periodic table, including the formation of electron shells. For a given n in an atom there are only a finite number of values of l, m, and σ_s that an electron may have. If there is an electron assigned to each of these values, then a complete shell is formed. Completed shells occur in the chemically inert noble elements of the periodic table, which are He, Ne, Ar, Kr, Xe, and Rn. All other atoms apart from H use one of these atoms as a core and add additional electron states to incomplete subshell states. Electrons in these incomplete subshells are available for chemical reaction with other atoms and therefore dominate the chemical activity of the atom. This simple version of the shell model, summarized in Table 1.3, works quite well for low atomic number atoms.

The lowest energy state of an atom is also called the *ground state*. The quantum numbers n, l, and m are used but $l = 0$ is replaced with s, $l = 1$ with p, $l = 2$ with d, and $l = 3$ with f. This naming convention comes from early work that labeled spectroscopic lines as sharp, principal, diffuse, and fundamental.

For example, the electron configuration for $Si(z = 14)$ is $1s^2 2s^2 2p^6 3s^2 3p^2$. In this notation, $2p^6$ means that $n = 2, l = 1$, and there are six electrons in the 2p shell. In this description used by chemists, there are four electrons in the outer $n = 3$ shell and the $n = 1$ and $n = 2$ shells are completely full. Full shells are chemically inert, and in this case $1s^2 2s^2 2p^6$ is the Ne ground state, so the ground state configuration for Si may be written as $[Ne]3s^2 3p^2$. Silicon consists of an inert (chemically inactive) neon core and four chemically active electrons in the $n = 3$ shell. In a crystal formed from Si atoms, it is the chemically active electrons in the $n = 3$ shell that participate in the formation of chemical

Table 1.3 Electron shell states

n	l	m	$2\sigma_s$	Allowable states in subshell	Allowable states in complete shell
1	0	0	±1	2	2
2	0	0	±1	2	8
		−1	±1		
	1	0	±1	6	
		1	±1		
3	0	0	±1	2	18
		−1	±1		
	1	0	±1	6	
		1	±1		
		−2	±1		
		−1	±1		
	2	0	±1	10	
		1	±1		
		2	±1		

bonds that hold the crystal together. These electrons are called *valence electrons*, and they occupy valence electron states of the crystal. Table 1.4 illustrates this classification method.

Technologically important examples of ground state atomic configurations include:

Al$[Ne]3s^2 3p^1$ group IIIB

Si$[Ne]3s^2 3p^2$ group IVB

P$[Ne]3s^2 3p^3$ group VB

Ga$[Ar]3d^{10} 4s^2 p^1$ group IIIB

Ge$[Ar]3d^{10} 4s^2 p^2$ group IVB

As$[Ar]3d^{10} 4s^2 p^3$ group VB

In$[Kr]4d^{10} 5s^2 p^1$ group IIIB

Table 1.5 is the periodic table of elements giving the ground state configuration for each element along with the atomic mass.

Table 1.4 Electron ground state for first 18 elements of the periodic table

Atomic number	Element	$n = 1$ $l = 0$ 1s	$n = 2$ $l = 0$ 2s	$n = 2$ $l = 1$ 2p	$n = 3$ $l = 0$ 3s	$n = 3$ $l = 1$ 3p	Shorthand notation
1	H	1					$1s^1$
2	He	2					$1s^2$
3	Li	[He] core	1				[He] $2s^1$
4	Be	2 electrons	2				[He] $2s^2$
5	B		2	1			[He] $2s^2 2p^1$
6	C		2	2			[He] $2s^2 2p^2$
7	N		2	3			[He] $2s^2 2p^3$
8	O		2	4			[He] $2s^2 2p^4$
9	F		2	5			[He] $2s^2 2p^5$
10	Ne		2	6			[He] $2s^2 2p^6$
11	Na	[Ne] core			1		[Ne] $3s^1$
12	Mg	10 electrons			2		[Ne] $3s^2$
13	Al				2	1	[Ne] $3s^2 3p^1$
14	Si				2	2	[Ne] $3s^2 3p^2$
15	P				2	3	[Ne] $3s^2 3p^3$
16	S				2	4	[Ne] $3s^2 3p^4$
17	Cl				2	5	[Ne] $3s^2 3p^5$
18	Ar				2	6	[Ne] $3s^2 3p^6$

1.2.5 Electronic Properties of Bulk Semiconductors and Heterostructures

The energy states of an electron in a hydrogen atom are quantized and may only take on discrete values. The same is true for electrons in all atoms. There are discrete energy levels that an electron in a single atom may have, and all other energy values are not allowed.

In a single-crystal solid, electrons from the many atoms that make up the crystal can interact with one another. Under these circumstances, the discrete energy levels of single-atom electrons disappear, and instead there are finite and continuous ranges or bands of energy states with contributions from many individual atom electronic states.

The lowest energy state of a semiconductor crystal exists when electrons occupy all available valence-band states. In a pure bulk semiconductor crystal there is a range of energy that does not have states that can propagate and this is called the energy band gap. A typical band-gap energy is $E_g = 1\,\text{eV}$. The band-gap energy separates valence-band states from conduction-band states by an energy of at least E_g. In this

Table 1.5 The periodic table of elements

IA	IIA	IIIB	IVB	VB	VIB	VIIB	VIII	VIII	VIII	IB	IIB	IIIA	IVA	VA	VIA	VIIA	Noble
Hydrogen H $1s^1$ 1.0079																	Helium He $1s^2$ 4.0026
Lithium Li $1s^2 2s^1$ 6.941	Beryllium Be $1s^2 2s^2$ 9.0122											Boron B $1s^2 2s^2 2p$ 10.81	Carbon C $1s^2 2s^2 2p^2$ 12.01	Nitrogen N $1s^2 2s^2 2p^3$ 14.007	Oxygen O $1s^2 2s^2 2p$ 15.999	Fluorine F $1s^2 2s^2 2p^5$ 18.998	Neon Ne $1s^2 2s^2 2p^6$ 20.18
Sodium Na $[Ne]3s^1$ 22.9898	Magnesium Mg $[Ne]3s^2$ 24.305											Aluminum Al $[Ne]3s^2 3p^1$ 26.982	Silicon Si $[Ne]3s^2 3p^2$ 28.086	Phosphorous P $[Ne]3s^2 3p^3$ 30.974	Sulfur S $[Ne]3s^2 3p^4$ 32.064	Chlorine Cl $[Ne]3s^2 3p^5$ 35.453	Argon Ar $[Ne]3s^2 3p^6$ 39.948
Potassium K $[Ar]4s^1$ 39.09	Calcium Ca $[Ar]4s^2$ 40.08	Scandium Sc $[Ar]3d^1 4s^2$ 44.956	Titanium Ti $[Ar]3d^2 4s^2$ 47.90	Vanadium V $[Ar]3d^3 4s^2$ 50.942	Chromium Cr $[Ar]3d^5 4s^1$ 52.00	Manganese Mn $[Ar]3d^5 4s^2$ 54.938	Iron Fe $[Ar]3d^6 4s^2$ 55.85	Cobalt Co $[Ar]3d^7 4s^2$ 58.93	Nickel Ni $[Ar]3d^8 4s^2$ 58.71	Copper Cu $[Ar]3d^{10} 4s^1$ 63.55	Zinc Zn $[Ar]3d^{10} 4s$ 65.38	Gallium Ga $[Ar]3d^{10} 4s^2 4p$ 69.72	Germanium Ge $[Ar]3d^{10} 4s^2 4p$ 72.59	Arsenic As $[Ar]3d^{10} 4s^2 4p^3$ 74.922	Selenium Se $[Ar]3d^{10} 4s^2 4p$ 78.96	Bromine Br $[Ar]3d^{10} 4s^2 4p$ 79.91	Krypton Kr $[Ar]3d^{10} 4s^2 4p^6$ 83.80
Rubidium Rb $[Kr]5s^1$ 85.47	Strontium Sr $[Kr]5s^2$ 87.62	Yttrium Y $[Kr]4d^1 5s^2$ 88.91	Zirconium Zr $[Kr]4d^2 5s^2$ 91.22	Niobium Nb $[Kr]4d^4 5s$ 92.91	Molybdenum Mo $[Kr]4d^5 5s^1$ 95.94	Technetium Tc $[Kr]4d^5 5s$ 98.91	Ruthenium Ru $[Kr]4d^7 5s$ 101.07	Rhodium Rh $[Kr]4d^8 5s$ 102.90	Palladium Pd $[Kr]4d^{10}$ 106.40	Silver Ag $[Kr]4d^{10} 5s$ 107.87	Cadmium Cd $[Kr]4d^{10} 5s^2$ 112.40	Indium In $[Kr]4d^{10} 5s^2 5p^1$ 114.82	Tin Sn $[Kr]4d^{10} 5s^2 5p^2$ 118.69	Antimony Sb $[Kr]4d^{10} 5s^2 5p^3$ 121.75	Tellurium Te $[Kr]4d^{10} 5s^2 5p^4$ 127.60	Iodine I $[Kr]4d^{10} 5s^2 5p^5$ 126.90	Xenon Xe $[Kr]4d^{10} 5s^2 5p^6$ 131.30
Cesium Cs $[Xe]6s^1$ 132.91	Barium Ba $[Xe]6s^2$ 137.34	Lanthanum La $[Xe]5d^1 6s^2$ 138.91	Hafnium Hf $[Xe]4f^{14} 5d^2 6s^2$ 178.49	Tantalum Ta $[Xe]4f^{14} 5d^3 6s^2$ 180.95	Tungsten W $[Xe]4f^{14} 5d^4 6s^2$ 183.85	Rhenium Re $[Xe]4f^{14} 5d^5 6s^2$ 186.2	Osmium Os $[Xe]4f^{14} 5d^6 6s^2$ 190.20	Iridium Ir $[Xe]4f^{14} 5d^7 6s^2$ 192.22	Platinum Pt $[Xe]4f^{14} 5d^9 6s^1$ 195.09	Gold Au $[Xe]4f^{14} 5d^{10} 6s^1$ 196.97	Mercury Hg $[Xe]4f^{14} 5d^{10} 6s^2$ 200.59	Thallium Tl $[Xe]4f^{14} 5d^{10} 6s^2 6p$ 204.37	Lead Pb $[Xe]4f^{14} 5d^{10} 6s^2 6p$ 207.19	Bismuth Bi $[Xe]4f^{14} 5d^{10} 6s^2 6p^3$ 208.98	Polonium Po $[Xe]4f^{14} 5d^{10} 6s^2 6p^4$ 210	Astatine At $[Xe]4f^{14} 5d^{10} 6s^2 6p$ 210	Radon Rn $[Xe]4f^{14} 5d^{10} 6s^2 6p^6$ 222
Francium Fr $[Rn]7s^1$ 223	Radium Ra $[Rn]7s^2$ 226	Actinium Ac $[Rn]6d^1 7s^2$ 227															

Rare earths

Lanthanides

Cerium Ce $[Xe]4f^2 5d^0 6s^2$ 140.12	Praseodymium Pr $[Xe]4f^3 5d^0 6s^2$ 140.91	Neodymium Nd $[Xe]4f^4 5d^0 6s^2$ 144.24	Promethium Pm $[Xe]4f^5 5d^0 6s$ 145	Samarium Sm $[Xe]4f^6 5d^0 6s$ 150.35	Europium Eu $[Xe]4f^7 5d^0 6s$ 151.96	Gadolinium Gd $[Xe]4f^7 5d^1 6s^2$ 157.25	Terbium Tb $[Xe]4f^9 5d^0 6s^2$ 158.92	Dysprosium Dy $[Xe]4f^{10} 5d^0 6s^2$ 162.50	Holmium Ho $[Xe]4f^{11} 5d^0 6s^2$ 164.93	Erbium Er $[Xe]4f^{12} 5d^0 6s^2$ 167.26	Thulium Tm $[Xe]4f^{13} 5d^0 6s^2$ 168.93	Ytterbium Yb $[Xe]4f^{14} 5d^0 6s^2$ 173.04	Lutetium Lu $[Xe]4f^{14} 5d^1 6s^2$ 174.97

Actinides

Thorium Th $[Rn]6d^2 7s^2$ 232.04	Protactinium Pa $[Rn]5f^2 6d^1 7s^2$ 231	Uranium U $[Rn]5f^3 6d^1 7s^2$ 238.03	Neptunium Np $[Rn]5f^4 6d^1 7s^2$ 237.05	Plutonium Pu $[Rn]5f^6 6d^0 7s^2$ 244	Americium Am $[Rn]5f^7 6d^0 7s^2$ 243	Curium Cm $[Rn]5f^7 6d^1 7s^2$ 247	Berkelium Bk $[Rn]5f^8 6d^1 7s^2$ 247	Californium Cf $[Rn]5f^{10} 6d^0 7s^2$ 251	Einsteinium Es	Fermium Fm	Mendelevium Md	Nobelium No	Lawrencium Lw

VIIB   VIII   →

way, the semiconductor retains a key feature of an atom – there are *allowed* states (that can propagate in the bulk crystal) and *disallowed* states (that cannot propagate). The existence of an electron energy band gap in a pure bulk semiconductor crystal in which no propagating electron states exist may only be explained by quantum mechanics.

Single crystals of the group IV element Si or the group III–V binary compound GaAs are examples of technologically important semiconductors. A pure crystalline semiconductor is an electrical insulator at low temperature. In the lowest energy state, or *ground state*, of a pure semiconductor, all electron states are occupied in the *valence band* and there are no electrons in the *conduction band*. The periodic array of atoms that forms the semiconductor crystal creates a periodic coulomb potential. If an electron is free to move in the material, its motion is influenced by the presence of the periodic potential. Typically, electrons with energy near the conduction-band minimum or energy near the valence-band maximum have an electron dispersion relation that may be characterized by a parabola, $\omega(k) \propto k^2$. The kinetic energy of the electron in the bulk crystal may therefore be written as $E_k = \hbar\omega = \hbar^2 k^2 / 2m_e^*$, where m_e^* is the *effective electron mass*. The value of m_e^* can be greater or less than the mass of a "bare" electron moving in free space. Because different semiconductors have different periodic coulomb potentials, different semiconductors have different values of effective electron mass. For example, the conduction-band effective electron mass in GaAs is approximately $m_e^* = 0.07 \times m_0$, and in InAs it is near $m_e^* = 0.02 \times m_0$, where m_0 is the bare electron mass.

As illustrated in Fig. 1.19, at finite temperatures the electrically insulating ground state of the semiconductor changes and electrical conduction can take place due to thermal excitation of a few electrons from the valence band into the conduction band. Electrons excited into the conduction band and absences of electrons (or *holes*) in the valence band are free to move in response to an electric field; they can thereby contribute to electrical conduction. The thermal excitation process involves lattice vibrations that collide via the coulomb interaction with electrons. In addition, electrons may scatter among themselves. The various collision processes allow electrons to equilibrate to a temperature T, which is the same as the temperature of the lattice.

The statistical energy distribution of many electrons in thermal equilibrium at absolute temperature T is typically described by the Fermi–Dirac distribution function,

$$f_k(E_k) = \frac{1}{e^{(E_k - \mu)/k_B T} + 1}. \tag{1.76}$$

In this expression, the chemical potential μ is defined by classical thermodynamics, and $f_k(E_k)$ is the probability of occupancy of a given electron state of energy E_k, and so it has a value between 0 and 1. The Fermi–Dirac distribution is driven by the Pauli

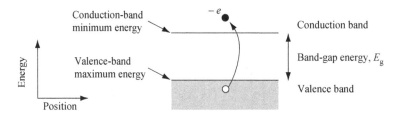

Fig. 1.19 Energy–position diagram of a semiconductor showing the valence band, the conduction band, and the band-gap energy E_g. Also shown is the excitation of an electron from the valence band into the conduction band.

exclusion principle, which states that *identical indistinguishable noninteracting half-odd-integer spin particles cannot occupy the same state.*

The distribution function for electrons in the limit of low temperature ($T \rightarrow 0\,\mathrm{K}$) becomes a step function, with electrons occupying all available states up to energy $\mu_{T=0}$ and no states with energy greater than $\mu_{T=0}$. This low-temperature limit of the chemical potential is so important that it has a special name: it is called the Fermi energy, E_F. For electrons with effective mass m_e^*, the Fermi energy $E_\mathrm{F} = \hbar^2 k_\mathrm{F}^2/2m_\mathrm{e}^*$, where k_F is called the Fermi wave vector.

At finite temperatures, and in the limit of electron energies that are large compared with the chemical potential, the distribution function takes on the Boltzmann form $f_\mathrm{k}(E \rightarrow \infty) = \mathrm{e}^{-E/k_\mathrm{B}T}$. This describes the high-energy tail of an electron distribution function at finite temperatures.

As shown schematically in Fig. 1.19, a valence-band electron with enough energy to surmount the semiconductor band-gap energy can enter the conduction band. The promotion of an electron from the valence band to the conduction band is an example of an *excitation* of the system. In the presence of such excitations, the lattice vibrations can collide with electrons in such a way that a distribution of electrons exists in the conduction band and a distribution of *holes* (absences of electrons) exists in the valence band. The energy distribution of occupied states is usually described by the Fermi–Dirac distribution function given by Eq. (1.76), in which energy is measured from the conduction-band minimum for electrons and from the valence-band maximum for holes. The concept of electron and hole distribution functions to describe an excited semiconductor is illustrated in Fig. 1.20.

A typical value of the band-gap energy is $E_\mathrm{g} = 1\,\mathrm{eV}$. For example, at room temperature pure crystalline silicon has $E_\mathrm{g} = 1.12\,\mathrm{eV}$. A measure of conduction-band electron occupation probability at room-temperature energy $k_\mathrm{B}T = 25\,\mathrm{meV}$ can be obtained using the Boltzmann factor. For $E_\mathrm{g} = 1\,\mathrm{eV}$, this gives $prob \sim \mathrm{e}^{-E_\mathrm{g}/k_\mathrm{B}T} = \mathrm{e}^{-40} = 4 \times 10^{-18}$, and the resulting electrical conductivity of such *intrinsic* material is small. Semiconductor technology makes use of *extrinsic* methods to increase electrical conductivity to a carefully controlled and predetermined value.

Electrons that are free to move in the material may be introduced into the conduction or valence band by adding impurity atoms. This extrinsic process, called *substitutional*

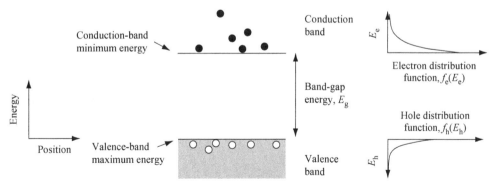

Fig. 1.20 Energy–position diagram of a semiconductor showing the valence band, the conduction band, and the band-gap energy E_g. The thermal excitation of electrons from the valence band into the conduction band creates an electron distribution function $f_\mathrm{e}(E_\mathrm{e})$ in the conduction band and a hole distribution function $f_\mathrm{h}(E_\mathrm{h})$ in the valence band. Electron energy E_e is measured from the conduction-band minimum and hole energy E_h is measured from the valence-band maximum.

doping, replaces atoms of the pure semiconductor with dopant atoms, the effect of which is to add mobile charge carriers. In the case of n-type doping, electrons are added to the conduction band. To maintain overall charge neutrality of the semiconductor, for every negatively charged electron added to the conduction band there is a positively charged impurity atom located at a substitutional doping lattice site in the crystal. As the density of mobile charge carriers is increased, the electrical conductivity of the semiconductor can increase dramatically.

The exact amount of impurity concentration or doping level in a thin layer of semiconductor can be precisely controlled using crystal growth techniques such as molecular beam epitaxy (MBE) or metalorganic chemical vapor deposition (MOCVD). In addition to controlling the impurity concentration spatial profile, the same epitaxial semiconductor crystal growth techniques may be used to grow thin layers of different semiconductor materials on top of one another to form a *heterostructure*. Single-crystal heterostructures can be achieved if the different semiconductors have the same crystal symmetry and lattice constant. In fact, even nonlattice matched semiconductor crystal layers may be grown, as long as the strained layer is thin and the lattice mismatch is not too great. The interface between the two different materials is called a *heterointerface*. Semiconductor expitaxial layer thickness is controllable to within a monolayer of atoms, and quite complex structures can be grown with many heterointerfaces.

A basic question concerns the relative position of the band gaps or "band-structure line up." At a heterointerface consisting of material with two different band-gap energies, part of the band-gap energy difference appears as a potential step in the conduction band. Figure 1.21(a) shows a conduction-band potential energy step ΔE_c and a valence-band potential energy step ΔE_v created at a heterointerface between GaAs and $Al_{0.3}Ga_{0.7}As$. By carefully designing a multilayer heterostructure semiconductor it is possible to create a specific potential as a function of position in the conduction band. Typically, atomically abrupt changes are possible at heterointerfaces and more gradual changes in potential may be achieved by using either changes in doping concentration

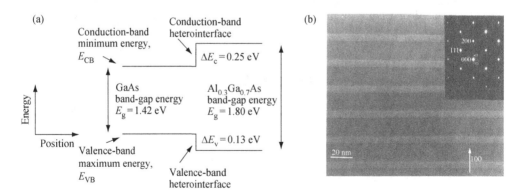

Fig. 1.21 (a) Conduction-band potential energy step ΔE_c and the valence-band potential energy step ΔE_v at a heterointerface between GaAs and $Al_{0.3}Ga_{0.7}As$. The band-gap energy of the two semiconductors is indicated. The conduction-band edge energy is E_{CB} and the valence-band edge energy is E_{VB}. (b) High-resolution transmission electron microscope cross-section image of epitatially grown single-crystal (100)-oriented Si and Ge layers. The inset electron diffraction pattern confirms the high perfection of the superlattice periodicity. Image courtesy of R. Hull and J. Bean, University of Virginia.

or by forming semiconductor alloys, the band gaps of which change as a function of position in the crystal growth direction. Figure 1.21(b) shows an electron microscope cross-section image of an epitatially grown single-crystal semiconductor consisting of a sequence of Ge and Si layers. The periodic layered heterostructure forms a super-lattice that can be measured by electron diffraction (Fig. 1.21(b), inset). The fact that a monoenergetic electron beam creates a diffraction pattern is, in itself, both a manifestation of the wavy nature of electrons and a measure of the quality of the superlattice.

1.2.5.1 The Single-Crystal Heterostructure Diode

Epitaxial crystal growth techniques can be used to create a heterostructure diode. To illustrate the current–voltage characteristics of such a diode, the special case of a unipolar n-type device formed using the heterointerface between GaAs and $Al_{0.3}Ga_{0.7}As$ is considered. The GaAs is doped heavily n-type, and the wider band gap $Al_{0.3}Ga_{0.7}As$ is more lightly doped n-type. The n-type $Al_{0.3}Ga_{0.7}As$ is depleted due to the presence of a conduction band potential energy step ΔE_c at the heterointerface. As shown in Fig. 1.22, the depletion region exposes the positive charge of the substitutional dopant atoms. The potential energy profile $eV(x)$ and the value of the depletion thickness w in the $Al_{0.3}Ga_{0.7}As$ region with uniform impurity concentration n may be found by solving Poisson's equation, $\nabla \cdot \mathbf{E} = \rho/\varepsilon_0\varepsilon_{r0} = -\nabla^2 V$. Considering an electric field $\mathbf{E}$ in one dimension and noting that the charge density is $\rho = en$, this gives

$$\frac{dE_x}{dx} = \frac{en}{\varepsilon_0\varepsilon_{r0}} = -\frac{d^2}{dx^2}V(x), \tag{1.77}$$

where $V(x)$ is the potential profile of the conduction band minimum, ϵ_{r0} is the low-frequency relative permittivity, and E_x is the electric field in the x direction. The potential profile is found by integrating Eq. (1.77), and the depletion region thickness for a built-in potential energy barrier $eV_0 = \Delta E_c - eV_n$ is approximately

$$w = \left(\frac{2\varepsilon_0\varepsilon_r}{en}V_0\right)^{1/2}. \tag{1.78}$$

Under an external bias voltage of V_{bias} applied to the GaAs, Eq. (1.78) becomes

$$w = \left(\frac{2\varepsilon_0\varepsilon_r}{en}(V_0 - V_{bias})\right)^{1/2}. \tag{1.79}$$

In Fig. 1.22, an effective potential energy barrier of approximately eV_0 always exists for electrons trying to move left-to-right from the heavily doped GaAs into the less doped $Al_{0.3}Ga_{0.7}As$ region. This limits left-to-right current flow to a constant value, I_0. However, electrons of energy E_e moving right-to-left from $Al_{0.3}Ga_{0.7}As$ to GaAs see an effective barrier $e(V_0 - V_{bias})$, which depends upon the external bias voltage V_{bias} applied to the GaAs. In a simple thermionic emission model, this current flow can have an exponential dependence upon voltage bias because, as illustrated in Fig. 1.22, the high-energy tail of the Femi–Dirac distribution function has an exponential Boltzmann form, e^{-E_e/k_BT}. Only those electrons in the undepleted $Al_{0.3}Ga_{0.7}As$ region with enough energy to surmount

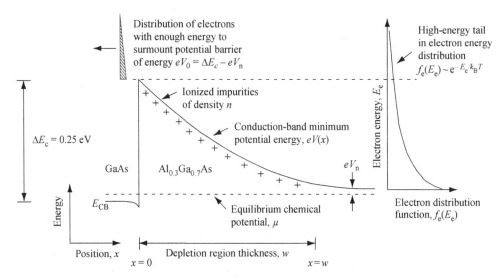

Fig. 1.22 Sketch of the conduction-band edge potential profile of a unipolar n-type GaAs/Al$_{0.3}$Ga$_{0.7}$As heterostructure diode. The GaAs is heavily doped, and the Al$_{0.3}$Ga$_{0.7}$As is lightly doped. The conduction-band offset at the heterointerface is $\Delta E_c = 0.25\,\text{eV}$. The depletion region thickness is w. The chemical potential μ used to describe the electron distribution under zero-bias equilibrium conditions is shown.

the potential barrier can contribute to the right-to-left current. It follows that the voltage dependence of right-to-left current is $I_0 e^{eV_{\text{bias}}/k_{\text{B}}T}$.

The net current flow across the diode is just the sum of right-to-left and left-to-right current, which gives

$$I = I_0 \left(e^{eV_{\text{bias}}/n_{\text{id}}k_{\text{B}}T} - 1 \right). \tag{1.80}$$

In this expression a phenomenological factor n_{id}, called the ideality factor, is included, which takes into account the fact that there may be deviations from the predictions of the simple thermionic emission model. A nonideal diode has $n_{\text{id}} > 1$. Figure 1.23 shows the current–voltage characteristics of a diode for the ideal case when $n_{\text{id}} = 1$.

It is clear from Fig. 1.23 that a diode has a highly nonlinear current–voltage characteristic. The exponential increase in current with positive or forward voltage bias, and the essentially constant current with negative or reverse voltage bias, are very efficient mechanisms for controlling current flow. It is this characteristic *exponential sensitivity* of an output (current) to an input control signal (voltage) that makes the diode such an important device and a basic building block for constructing other, more complex, devices.

The concept of exponential sensitivity can often be used as a guide when trying to decide whether a new device or device concept is likely to find practical applications.

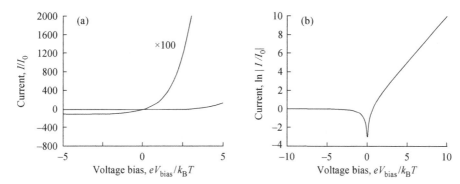

Fig. 1.23 (a) Current–voltage characteristics of an ideal diode plotted on a linear scale. (b) Current–voltage characteristics of an ideal diode. Natural logarithm of normalized current magnitude is plotted on the vertical axis. For $n_{\mathrm{id}} = 1$ and $T = 300\,\mathrm{K}$, the change in forward bias voltage to increase the current tenfold is $\ln(10)k_{\mathrm{B}}T/e = 60\,\mathrm{mV}$.

1.3 Example Exercises

Exercise 1.1
(a) Given that Planck's radiative energy density spectrum for thermal light is

$$U_{\mathrm{S}}(\omega) = \frac{\hbar\omega^3}{\pi^2 c^3}\frac{1}{\mathrm{e}^{\hbar\omega/k_{\mathrm{B}}T} - 1},$$

show that in the low-frequency (long-wavelength) limit this reduces to the Rayleigh–Jeans spectrum,

$$U_{\mathrm{S}}(\omega) = \frac{k_{\mathrm{B}}T}{\pi^2 c^3}\omega^2,$$

in which the Planck constant does not appear. This is an example of the *correspondence principle*, in which as $\hbar \to 0$ the classical result is obtained. Show that the high-frequency (short-wavelength) limit of Planck's radiative energy density spectrum reduces to the Wien spectrum,

$$U_{\mathrm{S}}(\omega) = \frac{\hbar\omega^3}{\pi^2 c^3}\mathrm{e}^{-\hbar\omega/k_{\mathrm{B}}T}.$$

(b) Find the energy of *peak* radiative energy density, $\hbar\omega_{\mathrm{peak}}$, in Planck's expression for $U_{\mathrm{S}}(\omega)$, and compare the *average* value $\hbar\omega_{\mathrm{average}}$ with $k_{\mathrm{B}}T$.
(c) Show that the total radiative energy density for thermal light is

$$U_{\mathrm{total}} = \frac{\pi^2 k_{\mathrm{B}}^4 T^4}{15c^3\hbar^3}.$$

(d) The Sun has a surface temperature of 5800 K and an average radius 6.96×10^8 m. Assuming that the mean Sun–Earth distance is 1.50×10^{11} m, what is the maximum radiative power per unit area incident on the upper Earth atmosphere facing the Sun?

41

Exercise 1.2

Show that the de Broglie wavelength of an electron of kinetic energy $E(\text{eV})$ is $\lambda_\text{e} = 1.23/\sqrt{E(\text{eV})}$ nm.

(a) Calculate the wavelength and momentum associated with an electron of kinetic energy $1\,\text{eV}$.

(b) Calculate the wavelength and momentum associated with a photon that has the same energy.

(c) Show that constructive interference for a monochromatic plane wave of wavelength λ scattering from two planes separated by distance d occurs when $n\lambda = 2d\cos(\theta)$, where θ is the incident angle of the wave measured from the plane normal and n is an integer. What energy electrons and what energy photons should be used to observe Bragg scattering peaks when performing electron or photon scattering measurements on a nickel crystal with lattice constant $L = 0.352\,\text{nm}$?

Exercise 1.3

An expression for the energy levels of a hydrogen atom is $E_n = -13.6(1/n^2)\,\text{eV}$, where n is an integer, such as $n = 1, 2, 3\ldots$

(a) Using this expression, draw an energy level diagram for the hydrogen atom.

(b) Derive the expression for the energy (in units of eV) and wavelength (in units of nm) of emitted light from transitions between energy levels.

(c) Calculate the three longest wavelengths (in units of nm) for transitions terminating at $n = 2$.

Exercise 1.4

If a photon of energy 2 eV is reflected from a metal mirror, how much momentum is exchanged? Why can the reflection not be modeled as a collision of the photon with a single electron in the metal?

Exercise 1.5

Consider a helium atom (He) with one electron missing. Estimate the energy difference between the ground state and the first excited state. Express the answer in units of eV. The binding energy of the electron in the hydrogen atom is $13.6\,\text{eV}$.

Exercise 1.6

(a) Show that in k-space position is a differential operator, $i\hbar\,\text{d}/\text{d}p_x$, by evaluating the expectation value

$$\langle x \rangle = \int_{-\infty}^{\infty} \psi^*(x)\hat{x}\psi(x)\,\text{d}x$$

in terms of $\phi(k_x)$, which is the Fourier transform of $\psi(x)$.

(b) The wave function for a particle in real space is $\psi(x,t)$. Usually, it is assumed that position x and time t are continuous and smoothly varying. Given that particle energy is quantized so that $E = \hbar\omega$, show that the energy operator for the wave function $\psi(x,t)$ is $i\hbar\,\partial/\partial t$.

Exercise 1.7

Show the degeneracy of states ψ_{nlm} in a hydrogen atom is n^2 by proving

$$\sum_{l=0}^{l=n-1} (2l+1) = n^2.$$

Solutions

Solutions 1.1

(a) Expanding the exponential in the expression for Planck's radiative energy density spectrum for thermal light gives $e^{\hbar\omega/k_B T} \sim 1 + \hbar\omega/k_B T + \cdots$ in the low-frequency limit, so that

$$U_S(\omega) = \frac{\hbar\omega^3}{\pi^2 c^3}\frac{1}{e^{\hbar\omega/k_B T}-1} \sim \frac{\hbar\omega^3}{\pi^2 c^3}\frac{1}{1+\hbar\omega/k_B T + \cdots -1} = \frac{k_B T}{\pi^2 c^3}\omega^2,$$

which is the Rayleigh–Jeans spectrum. This is an example of the *correspondence principle* in which as $\hbar \to 0$ the classical result is obtained.

In the high-frequency limit $\omega \to \infty$, so that the exponential term in $U_S(\omega)$ becomes $1/(e^{\hbar\omega/k_B T}-1) \to e^{-\hbar\omega/k_B T}$, it follows that the high-frequency (short-wavelength) limit of Planck's radiative energy density spectrum reduces to the Wien spectrum,

$$U_S(\omega) = \frac{\hbar\omega^3}{\pi^2 c^3}e^{-\hbar\omega/k_B T}.$$

(b) The *peak* radiative energy density, $\hbar\omega_{\text{peak}}$, in Planck's expression for $U_S(\omega)$ occurs when

$$\frac{d}{d\omega}U_S(\omega) = \frac{d}{d\omega}\left(\frac{\hbar\omega^3}{\pi^2 c^3}\frac{1}{e^{\hbar\omega/k_B T}-1}\right) = 0,$$

$$\frac{3\hbar\omega^2}{\pi^2 c^3}\frac{1}{e^{\hbar\omega/k_B T}-1} - \frac{\hbar\omega^3}{\pi^3 c^3}\left(\frac{\hbar}{k_B T}\right)\frac{e^{\hbar\omega/k_B T}}{(e^{\hbar\omega/k_B T}-1)^2} = 0,$$

$$\frac{3\hbar\omega^2}{\pi^2 c^3} - \frac{\hbar\omega^3}{\pi^2 c^3}\left(\frac{\hbar}{k_B T}\right)\frac{e^{\hbar\omega/k_B T}}{e^{\hbar\omega/k_B T}-1} = 0.$$

Dividing by $\hbar\omega^2/\pi^2 c^3$, letting $x = \hbar\omega/k_B T$, and multiplying by $(e^x - 1)$ gives $3(e^x - 1) - xe^x = 0$ or $3(1 - e^{-x}) - x = 0$, which may be solved numerically to give $x \sim 2.82$, so that $\hbar\omega_{\text{peak}} \sim 2.82 \times k_B T$. For the Sun with $T = 5\,800\,\text{K}$, this gives a peak in $U_S(\omega)$ at energy $\hbar\omega_{\text{peak}} = 1.41\,\text{eV}$, or, equivalently, frequency $\nu = 341\,\text{THz}$ ($\omega = 2.14 \times 10^{15}\,\text{rad s}^{-1}$).

To find an average value of z distributed as $f(z)$ the expectation value $\langle f(z)\rangle = \int zf(z)dz/f(z)dz$ is sought. In this case the integral is weighted by energy $\hbar\omega$, so

$$\hbar\omega_{\text{average}} = \frac{\int\limits_0^\infty \hbar\omega U_S(\omega)d\omega}{\int\limits_0^\infty U_S(\omega)d\omega} = \frac{\int\limits_0^\infty \frac{\hbar^2\omega^4}{\pi^2c^3}\frac{1}{e^{\hbar\omega/k_BT}-1}d\omega}{\int\limits_0^\infty \frac{\hbar\omega^3}{\pi^2c^3}\frac{1}{e^{\hbar\omega/k_BT}-1}d\omega}.$$

Introducing $x = \hbar\omega/k_BT$ so that $dx = (\hbar/k_BT)d\omega$, $\omega = x(k_BT/\hbar)$, and $d\omega = dx(k_BT/\hbar)$ gives

$$\hbar\omega_{\text{average}} = \frac{\frac{\hbar^2}{\pi^2c^3}\frac{k_BT}{\hbar}\left(\frac{k_BT}{\hbar}\right)^4\int\limits_0^\infty \frac{x^4}{e^x-1}dx}{\frac{\hbar}{\pi^2c^3}\frac{k_BT}{\hbar}\left(\frac{k_BT}{\hbar}\right)^3\int\limits_0^\infty \frac{x^3}{e^x-1}dx} = k_BT\frac{\int\limits_0^\infty \frac{x^4}{e^x-1}dx}{\int\limits_0^\infty \frac{x^3}{e^x-1}dx}.$$

To solve the integrals, use is made of[17]

$$\int\limits_0^\infty \frac{x^{p-1}}{e^{rx}-q}dx = \frac{1}{qr^p}\Gamma(p)\sum_{k=1}^\infty \frac{q^k}{k^p},$$

where $\Gamma(p)$ is the gamma function.[18]

For the numerator, $\Gamma(p = 5) = 24$, and the integral is approximately 25. For the denominator, $\Gamma(p = 4) = 6$, and the integral is approximately 6.5. Hence, $\hbar\omega_{\text{average}} \sim k_BT(25/6.5) = 3.85k_BT$. Notice that the average value is greater than the peak value since the function $U_S(\omega)$ is not symmetric. For the Sun with $T = 5800\,\text{K}$ this gives $\hbar\omega_{\text{average}} = 1.92\,\text{eV}$, which may be compared with $\hbar\omega_{\text{peak}} = 1.41\,\text{eV}$.

(c) To find the total radiative energy density for thermal light, $U_{\text{total}}(\omega)$, the expression for $U_S(\omega)$ is integrated over all frequencies. This gives

$$U_{\text{total}} = \int\limits_{\omega=0}^{\omega=\infty} U_S(\omega)d\omega = \int\limits_{\omega=0}^{\omega=\infty} \frac{\hbar\omega^3}{\pi^2c^3}\frac{1}{e^{\hbar\omega/k_BT}-1}d\omega = \frac{k_B^4T^4}{\pi^2c^3\hbar^3}\int\limits_{x=0}^{x=\infty} \frac{x^3}{e^3-1}dx,$$

where $x = \hbar\omega/k_BT$. The integral on the right-hand side is standard,

$$\int\limits_{x=0}^{x=\infty} \frac{x^3}{e^x-1}dx = \frac{\pi^4}{15},$$

giving the final result,

$$U_{\text{total}} = \frac{\pi^2k_B^4T^4}{15c^3\hbar^3}.$$

The fact that the total energy density is proportional to T^4 is known as the Stefan–Boltzmann radiation law.

[17] I. S. Gradshteyn and I. M. Ryzhik, *Table of Integrals, Series, and Products*, San Diego, Academic Press, 1980, p. 326 (ISBN 0 12 294760 6).

[18] M. Abramowitz and I. A. Stegun, *Handbook of Mathematical Functions*, New York, Dover Publications, 1974, pp. 267–273 (ISBN 0 486 61272 4).

(d) The Sun has a surface temperature of 5800 K, so the total isotropic radiative energy density of the Sun is

$$U_{total} = \frac{\pi^2 k_B^4 T^4}{15 c^3 \hbar^3} = \frac{\pi^2 \times (1.3807 \times 10^{-23})^4 (5800)^4}{15 \times (3 \times 10^8)^3 (1.054 \times 10^{-34})^3} = 0.857 \, \mathrm{J\,m^{-3}}.$$

The average radius of the Sun is $r_{Sun} = 6.96 \times 10^8$ m, and the mean Sun–Earth distance is $R_{Sun-Earth} = 1.50 \times 10^{11}$ m. The maximum radiative power per unit area incident on the upper Earth atmosphere facing the Sun is given by

$$\frac{U_{total}}{4} \times c \times \left(\frac{r_{Sun}}{R_{Earth-Sun}}\right)^2 = \frac{0.857}{4} \times (3 \times 10^8) \times \left(\frac{6.96 \times 10^8}{1.50 \times 10^{11}}\right)^2 = 1.4 \, \mathrm{kW\,m^{-2}},$$

where division by 4 is to take into account that fraction of the energy density flux moving from the Sun in the direction of the Earth.

Solutions 1.2

The fact that the de Broglie wavelength of an electron of kinetic energy $E(\mathrm{eV})$ is given by

$$\lambda_e = \frac{1.23}{\sqrt{E(\mathrm{eV})}} \, \mathrm{nm}$$

follows directly from electron energy dispersion relation $E = \hbar^2 k^2 / 2m$, where $k = 2\pi/\lambda_e$.

(a) The electron wavelength is $\lambda_e(1\,\mathrm{eV}) = 1.23$ nm and the momentum of the electron is

$$p_e(1\,\mathrm{eV}) = \hbar k = \frac{2\pi\hbar}{\lambda_e} = 5.40 \times 10^{-25} \, \mathrm{kg\,m\,s^{-1}}.$$

(b) The wavelength of a photon is $\lambda_{ph} = c/\nu = 2\pi c/\omega$. Hence, for a photon of energy $E = \hbar\omega = 1\,\mathrm{eV}$,

$$\frac{2\pi\hbar c}{\hbar\omega} = \frac{6.626 \times 10^{-34} \times 3.000 \times 10^8}{1.602 \times 10^{-19}} = \lambda_{ph}(1\,\mathrm{eV}) = 1\,241 \, \mathrm{nm}.$$

The momentum of the photon is $p_{ph} = \hbar k_{ph}$, so that

$$p_{ph}(1\,\mathrm{eV}) = \frac{2\pi\hbar}{\lambda_{ph}} = 5.34 \times 10^{-28} \, \mathrm{kg\,m\,s^{-1}},$$

which is about 1 000 times less than that of an electron of energy 1 eV.

(c) A monochromatic plane wave of wavelength λ scatters from two planes separated by distance d. As shown in Fig. 1.E24, constructive interference occurs when the extra path length $n\lambda = 2d\cos(\theta)$, where θ is the incident angle of the wave measured from the plane-normal and n is an integer.

If $n = 1$ and $\theta = 0$, then $\lambda = 2d$. For a nickel crystal with lattice constant $L = 0.352$ nm, the kinetic energy of a monochromatic beam of electrons should be greater than

$$E_{electron} = \left(\frac{1.23}{2 \times L}\right)^2 = 3.05 \, \mathrm{eV},$$

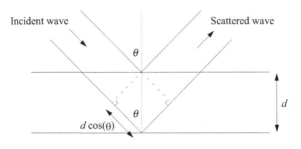

Fig. 1.E24

and for photons the energy should be greater than

$$E_{\text{photon}} = \frac{1\,241}{2 \times L} = 1.76 \times 10^3 \, \text{eV}$$

to observe Bragg scattering peaks from the crystal.

Solutions 1.3

(a) Using the expression for the energy levels of a hydrogen atom $E_n = -13.6/n^2 \, \text{eV}$, where n is an integer such as $n = 1, 2, 3, \ldots$, the energy level diagram for the hydrogen atom can be drawn. The energy levels for $n = 1, 2, 3, 4, 5$, and ∞ are shown in Fig. 1.18. The lowest energy (ground state) corresponds to $n = 1$.

(b) Electrons can make transitions between the energy levels of (a) by emitting light. The energy transitions are given by

$$\Delta E = E_{n_2} - E_{n_1} = -13.6 \left(\frac{1}{n_2^2} - \frac{1}{n_1^2} \right) \text{eV}.$$

Since $\lambda = 2\pi c/\omega$, the wavelength (in units of nm) of emitted light from transitions between energy levels is given by

$$\lambda = \frac{2\pi c\hbar}{\Delta E} = -\frac{2\pi c\hbar}{13.6} \frac{1}{\dfrac{1}{n_2^2} - \dfrac{1}{n_1^2}} = \frac{91.163 \, \text{nm}}{\dfrac{1}{n_2^2} - \dfrac{1}{n_1^2}}.$$

(c) Adopting the notation λ_{nm} for a transition from the quantum state labeled by quantum number n to the quantum state labeled by m, from part (b),

$\lambda_{32} = 656.375 \, \text{nm}$,

$\lambda_{42} = 486.204 \, \text{nm}$,

$\lambda_{52} = 434.110 \, \text{nm}$.

Solutions 1.4

A photon of energy $E = 2 \, \text{eV}$ is incident from free-space at angle θ normal to a flat metal mirror. The magnitude of photon momentum is $p_{\text{ph}} = \hbar k_{\text{ph}} = \hbar \omega/c$ where frequency ω is given by the quanta of photon energy $E = \hbar \omega$. The momentum change upon reflection is $(2\hbar\omega/c)\cos(\theta)$, which, for $\theta = 0$, gives a value $2E/c = 2 \times 2 \times 1.6 \times 10^{-19}/3 \times 10^8 = 2.1 \times 10^{-27}$ kg m s^{-1}. Reflection cannot be modeled as a photon collision with a single electron in the metal because it is not possible to conserve both energy and momentum *and* have the photon maintain its wavelength. Collision with a single electron would take energy from the photon and cause a change in photon wavelength. Since, by definition, light reflected from a mirror has the same wavelength as the incident light, photon energy cannot change.

Solutions 1.5

Estimating the energy difference between the ground state and the first excited state of a helium atom (He) with one electron missing proceeds by noting it is essentially the same system as the hydrogen atom except that the nucleus has a charge $2e$.

The helium ion potential for an electron in the nth orbit is

$$V = \frac{Ze^2}{4\pi\varepsilon_0 r_n} = \frac{-m}{n^2\hbar^2}\left(\frac{Ze^2}{4\pi\varepsilon_0}\right)^2,$$

where $Z = 2$ and $\mathrm{Ry} = -m(e^2/4\hbar\pi\varepsilon_0)^2 = -13.6\,\mathrm{eV}$, so that $E_n = \frac{Z^2}{n^2}\mathrm{Ry}$. Hence, for $n_1 = 1$ and $n_2 = 2$ the energy difference is

$$E_{n_2} - E_{n_1} = 4\mathrm{Ry}\left(\frac{1}{n_2^2} - \frac{1}{n_1^2}\right) = 4\mathrm{Ry}\left(\frac{1}{4} - 1\right) = -3\mathrm{Ry} = 40.8\,\mathrm{eV}.$$

This is four times the value of $E_2 - E_1$ for the hydrogen atom. The reason is that the energy levels are proportional to Z^2 for hydrogen-like ions.

Solutions 1.6

(a) The expectation value of the position operator in k-space is found by evaluating

$$\langle x \rangle = \int_{-\infty}^{\infty} \psi^*(x)\hat{x}\psi(x)\,\mathrm{d}x$$

in terms of $\phi(k)$, which is the Fourier transform of $\psi(x)$. Hence,

$$\langle x \rangle = \frac{1}{2\pi}\int_{-\infty}^{\infty}\mathrm{d}x\left(\int_{-\infty}^{\infty}\phi^*(k')e^{-ik'x}\,\mathrm{d}k'\right)x\left(\int_{-\infty}^{\infty}\phi(k)e^{ikx}\,\mathrm{d}k\right),$$

which may be rewritten as

$$\langle x \rangle = \frac{1}{2\pi}\int_{-\infty}^{\infty}\mathrm{d}x\int_{-\infty}^{\infty}\phi^*(k')e^{-ik'x}\,\mathrm{d}k'\,x\int_{-\infty}^{\infty}\phi(x)\frac{\mathrm{d}}{\mathrm{d}k}\left(\frac{e^{ikx}}{ix}\right)\mathrm{d}k.$$

1 Toward Quantum Mechanics

Integrating by parts gives

$$\langle x \rangle = \frac{1}{2\pi} dx \int_{-\infty}^{\infty} dk' \phi^*(k') e^{-ik'x} x \left(\left[\phi(x) \frac{e^{ikx}}{ix} \right]_{-\infty}^{\infty} - \int_{-\infty}^{\infty} dk \left(\frac{d}{dk} \phi(k) \right) \frac{e^{ikx}}{ix} \right).$$

The limits $\pm\infty$ are required to be zero so that

$$\langle x \rangle = \frac{-1}{2\pi i} \int_{-\infty}^{\infty} dx \int_{-\infty}^{\infty} dk' \phi^*(k') e^{-ik'x} \int_{-\infty}^{\infty} dk \left(\frac{d}{dk} \phi(k) \right) e^{ikx}$$

$$= \frac{i}{2\pi} \int_{-\infty}^{\infty} dx e^{i(k-k')x} \int_{-\infty}^{\infty} \int_{-\infty}^{\infty} dk \, dk' \phi^*(k') \frac{d}{dk} \phi(k).$$

Making use of the fact that a delta function can be expressed as

$$\frac{1}{2\pi} \int_{-\infty}^{\infty} dx e^{i(k-k')x} = \delta(k - k') \text{ gives}$$

$$\langle x \rangle = i \int_{\infty}^{\infty} \int_{-\infty}^{\infty} dk \, dk' \phi^*(k') \frac{d}{dk} \phi(k) \delta(k - k') = i\hbar \int_{-\infty}^{\infty} dk \phi^*(k) \frac{d}{dp} \phi(k).$$

Hence, in k-space position is a differential operator, $i\hbar \frac{d}{dp}$.

(b) The expectation value of the energy operator is found in a similar way only now $\phi(t)$ is the Fourier transform of $\psi(\omega)$. Hence,

$$\langle E \rangle = \int_{-\infty}^{\infty} d\omega \, \psi^*(\omega) \hbar\omega \, \psi(\omega) = \frac{1}{2\pi} \int_{-\infty}^{\infty} d\omega \left(\int_{-\infty}^{\infty} dt' \phi^*(t') e^{-i\omega t'} \right) \hbar\omega \left(\int_{-\infty}^{\infty} dt \phi(t) e^{i\omega t} \right)$$

$$= \frac{1}{2\pi} \int_{-\infty}^{\infty} d\omega \int_{-\infty}^{\infty} dt' \phi^*(t') e^{-i\omega t'} \hbar\omega \int_{-\infty}^{\infty} dt \phi(t) \frac{\partial}{\partial t} \left(\frac{e^{i\omega t'}}{i\omega} \right)$$

$$= \frac{1}{2\pi} \int_{-\infty}^{\infty} d\omega \int_{-\infty}^{\infty} dt' \phi^*(t') e^{-i\omega t'} \hbar\omega \left(\left[\phi(t) \frac{e^{i\omega t}}{i\omega} \right]_{-\infty}^{\infty} - \int_{-\infty}^{\infty} dt \left(\frac{\partial}{\partial t} \phi(t) \right) \frac{e^{i\omega t}}{i\omega} \right),$$

so that

$$\langle E \rangle = \frac{-\hbar}{2\pi i} \int_{-\infty}^{\infty} d\omega \int_{-\infty}^{\infty} dt' \phi^*(t') e^{-i\omega t'} \int_{-\infty}^{\infty} dt \left(\frac{\partial}{\partial t} \phi(t) \right) e^{i\omega t}$$

$$= \frac{i\hbar}{2\pi} \int\limits_{-\infty}^{\infty} d\omega \, e^{i\omega(t-t')} \int\limits_{-\infty}^{\infty} \int\limits_{-\infty}^{\infty} dt \, dt' \phi^*(t') \frac{\partial}{\partial t} \phi(t)$$

$$= i\hbar \int\limits_{-\infty}^{\infty} \int\limits_{-\infty}^{\infty} dt \, dt' \phi^*(t') \frac{\partial}{\partial t} \phi(t) \delta(t - t') = i\hbar \int\limits_{-\infty}^{\infty} dt \, \phi^*(t) \frac{\partial}{\partial t} \phi(t),$$

and it may be concluded that energy is a differential operator, $i\hbar \frac{\partial}{\partial t}$.

Solutions 1.7

Electron energy in a hydrogen atom state ψ_{nlm} depends upon n but not on l or m. For a given n, the allowed values of l are $l = 0, 1, 2, \ldots, (n-1)$ and the allowed values of m are $\pm l, \ldots, \pm 2, \pm 1, 0$. Hence, the degeneracy of a state with principle quantum number n is a sum over n terms:

$$\sum_{l=0}^{l=n-1} (2l + 1) = 1 + 3 + 5 + \cdots + (2n - 5) + (2n - 3) + (2n - 1).$$

Reordering the right-hand side as

$$\sum_{l=0}^{l=n-1} (2l + 1) = (2n - 1) + (2n - 3) + (2n - 5) + \cdots + 5 + 3 + 1$$

suggests adding the two equations. Doing this gives

$$2 \sum_{l=0}^{l=n-1} (2l + 1) = 2n + 2n + 2n + \cdots.$$

Since there are n terms on the right-hand side, this may be written as

$$2 \sum_{l=0}^{l=n-1} (2l + 1) = n2n \,, \text{ and so the degeneracy of the state } \psi_{nlm} \text{ is } \sum_{l=0}^{l=n-1} (2l + 1) = n^2.$$

1.4 Problems

Problem 1.1

(a) The Sun has a surface temperature of $5\,800$ K and an average radius of 6.96×10^8 m. Assuming the mean Sun–Mars distance is 2.28×10^{11} m, what is the total radiative power per unit area incident on the upper Mars atmosphere facing the Sun?

(b) If the surface temperature of the Sun was $6\,800$ K, by how much would the total radiative power per unit area incident on Mars increase?

Problem 1.2

(a) A positron has charge $+e$ and the same mass m_0 as a bare electron. The energy of a particle with rest mass m_0 moving at velocity v with momentum $p = \gamma_{\text{Lorentz}} m_0 v$ is

$E = \gamma_{\text{Lorentz}} m_0 c^2$, where c is the speed of light and γ_{Lorentz} is the Lorentz factor such that $\gamma_{\text{Lorentz}}^2 = c^2/(c^2 - v^2)$. Why can a single high energy photon not decay into an electron and a positron?

(b) *Two* colliding real photons γ_1 and γ_2 can create particles that have mass.[19] Describe the conditions under which these photons decay into a positron and an electron.

Problem 1.3
Consider a lithium atom (Li) with two electrons missing.

(a) Draw an energy level diagram for the Li^{++} ion.

(b) Derive the expression for the energy (in eV) and wavelength (in nm) of emitted light from transitions between energy levels.

(c) Calculate the three longest wavelengths (in nm) for transitions terminating at $n = 2$.

(d) If the lithium ion were embedded in a dielectric with relative permittivity $\varepsilon_{\text{r}} = 10$, what would be the expression for the energy (in eV) and wavelength (in nm) of emitted light from transitions between energy levels?

Problem 1.4
A particle mass m confined by a real potential is described by wave function $\psi(x, t)$.

(a) Write down the expression for the average value of the particle position $\langle x \rangle$ and then make use of the Schrödinger equation to show that the average value of momentum is

$$\langle p \rangle = m \frac{\mathrm{d}}{\mathrm{d}t} \langle x \rangle = \frac{\mathrm{i}\hbar}{2} \int_{-\infty}^{\infty} x \frac{\mathrm{d}}{\mathrm{d}x} \left(\psi^* \frac{\mathrm{d}\psi}{\mathrm{d}x} - \frac{\mathrm{d}\psi^*}{\mathrm{d}x} \psi \right) \mathrm{d}x.$$

(b) Evaluate the integral in part (a) and show that

$$\langle p \rangle = \int_{-\infty}^{\infty} \psi^* \left(-\mathrm{i}\hbar \frac{\mathrm{d}}{\mathrm{d}x} \right) \psi \, \mathrm{d}x,$$

so that the momentum operator may be identified as $\hat{p} = -\mathrm{i}\hbar \frac{\mathrm{d}}{\mathrm{d}x}$.

Problem 1.5
Create a simple model of a heterostructure diode that predicts current increases exponentially with increasing forward voltage bias. Explain the assumptions made to develop the model. Under what conditions will this predicted behavior fail? By how much is voltage bias across an ideal diode increased to change current by a factor of 10 at room temperature $(T = 300 \, \text{K})$? Does this represent a *fundamental* limit to power dissipation in electronic switching devices operating at room temperature?

Problem 1.6
Write down the Hamiltonian operator for:

[19]D. L. Burke et al. *Phys. Rev. Lett.* **79**, 1626 (1997).

(a) a one-dimensional simple harmonic oscillator,
(b) a helium atom,
(c) a hydrogen molecule, and
(d) a molecule with n_n nuclei and n_e electrons.

Problem 1.7

Calculate the classical velocity of the electron in the nth orbit of a Li^{++} ion. If this electron is described as a wave packet and its position is known to an accuracy of $\Delta x = 1\,pm$, calculate the characteristic time $\Delta \tau_{\Delta x}$ for the width of the wave packet to double. Compare $\Delta \tau_{\Delta x}$ with the time to complete one classical orbit. Should the electron be described as a particle or a wave?

Problem 1.8

What is the Bohr radius of an electron with effective mass $m_e^* = 0.021 \times m_0$ in InAs that has low-frequency relative permittivity $\varepsilon_{r0} = 14.55$? What is the effective Rydberg energy for an electron describing a hydrogenic orbit in the medium?

Problem 1.9

Because electromagnetic radiation possesses momentum, it can exert a force. If completely absorbed by matter, the absorbed electromagnetic energy flux density divided by the speed of light is a pressure called radiation pressure.

(a) If the maximum radiative power per unit area incident on the upper Earth atmosphere facing the Sun is $1.4\,kW\,m^{-2}$, what is the corresponding radiation pressure?

(b) Estimate the photon flux needed to create the pressure in (a).

(c) Compare the result in (a) with the pressure due to one atmosphere.

(d) What is the normal fluctuation in pressure per unit time?

Problem 1.10

(a) As described in Section 1.1.7, Alice can transmit information to Bob via a quantum communication channel that uses single photons and nonorthogonal polarization states. Explain Bob's choice of test basis in Fig. 1.10.

(b) In the absence of a single photon source, optical quantum key distribution (QKD) uses light from an attenuated laser. In a particular system the probability of single photon emission per laser pulse is 0.09. There is a $-3\,dB$ coupling loss from the laser to glass fiber. The link operates with a clock rate of 1.25 GHz (bit time $\tau = 800\,ps$), optical loss in the 50 km long glass fiber is $-10\,dB$, and time jitter in the photodetector requires that only every second time interval be used for photon detection. What is the maximum sustained data rate for guaranteed secure QKD in the system?

(c) No light can pass between two linear polarizers if their respective polarizations are oriented at $90°$ to each other. If a third linear polarizer oriented at a $45°$ angle is placed between the two linear polarizers, what is the maximum fraction of incident light intensity that can pass through the system?

Problem 1.11

A particle mass m confined by a real potential is described by bound-state wave function $\psi(x,t)$ and Schrödinger's equation. Show that

$$\frac{d}{dt} \int_{-\infty}^{\infty} |\psi(x,t)|^2 dx = 0,$$

so that if the wave function $\psi(x,t)$ is normalized it remains so for all time.

Problem 1.12

A sphere has fixed uniform charge density $\rho = e \times 1.5 \times 10^{28} \, \text{m}^{-3}$.

(a) Calculate the force on an electron initially placed at the surface of the sphere.

(b) Assuming the electron in (a) is free to move in the volume of the sphere, what is the potential seen by the electron and what is its subsequent motion?

(c) What photon energy and wavelength can be absorbed by the system?

(d) At what radius R_c does an electron of mass m_0 have a predicted peak magnitude of electron velocity that exceeds the speed of light in vacuum?

(e) If the total system, consisting of the sphere of positive charge density and the electron, is charge neutral what is the value of R_c? How does electron dispersion limit the validity of the model?

Problem 1.13

The wave function at time $t = 0$ for an electron localized as a Gaussian wave packet in one dimension centered at $x = x_0$ with spatial extent σ_x and momentum $\hbar k_0$ is

$$\psi(x, t = 0) = A e^{ik_0 x} e^{-(x-x_0)^2/(4\sigma_x^2)}.$$

(a) Find the normalization constant A and show that $\langle x \rangle = x_0$ and $\sqrt{\langle x^2 \rangle - \langle x \rangle^2} = \sigma_x$.

(b) Take the Fourier transform of $\psi(x, t = 0)$ to determine $\psi(k, t = 0)$. Making the substitution $2\sigma_k = 1/\sigma_x$, compare the result to that of part (a). Explain the significance of $\sigma_k \sigma_x = 1/2$. Hint: use Cauchy's integral theorem, which states that $\oint f(z)dz = 0$ where z is a complex variable and $f(z)$ has no poles within the closed integration loop.

(c) Find $\psi(x, t = 0)$ and $\psi(k, t = 0)$ in the limit $\sigma_x \to 0$ and describe how the electron wave packet evolves in time. If the electron's location is measured with absolute certainty at time $t = 0$, where is the electron located at any subsequent time? Explain the result.

(d) If a localized single *photon* in free space can be described by the same initial Gaussian wave packet $\psi(x, t = 0)$, how will it evolve in time? Explain the difference in the predicted behavior of the photon compared to the electron.

Problem 1.14

Optical QKD protocol may be viewed as a sensor to detect eavesdropping on an optic link. Alice and Bob use the B92 protocol and measure error rate to detect the presence of Eve, an eavesdropper, in an otherwise lossless optical QKD link.

(a) What is the average error rate generated by eavesdropping?

(b) What can Alice and Bob infer about the methods used by Eve?

Problem 1.15
The $\pm\hbar/2$ spin of an electron charge e and mass m_0 emerges as a consequence of rotational symmetry and a relativistic treatment of the Schrödinger equation.

(a) *Suppose* spin to be a classical concept due to rotation of a spherical electron of classical radius $r_e = e^2/\left(4\pi\epsilon_0 m_0 c^2\right)$ about an axis passing through its center. Assuming a uniform classical electron density in the sphere, find the electron's moment of inertia.

(b) If the angular momentum of the classical electron is $\hbar/2$ what is the speed of the surface of the electron at its equator? Compare this to the speed of light, c, and comment.

(c) If all the electron density is in a thin torus of vanishingly small cross-section making a ring at the equator, calculate the radius at which the ring speed is 1% of c. Calculate the current in the ring and the magnetic field through the ring.

(d) Calculate the magnitude and direction of the classical force due to the coulomb interaction experienced by the electron charge density of the ring in (c).

(e) What happens to the force calculated in (d) if, as in quantum mechanics, the electron is a point particle?

Problem 1.16
Random generation of binary 1 or 0 can be physically guaranteed by quantum mechanics and so there is interest in using this as a resource for stochastic computing. Numbers can be represented as probabilities of a binary 1 or 0 signal in a clocked bit-stream of length n_{bits} such that as $n_{bits} \to \infty$ the average value of the signal is a number distributed in the interval $[0,1]$. For a 2×2 mixing matrix $\mathbf{A}$ and average signal vector $\mathbf{s}$, the vector $\mathbf{x} = \mathbf{As}$ may be written as

$$\begin{bmatrix} x_1 \\ x_2 \end{bmatrix} = \begin{bmatrix} a_{11} \ a_{12} \\ a_{21} \ a_{22} \end{bmatrix} \begin{bmatrix} s_1 \\ s_2 \end{bmatrix},$$

so that evaluation of $\mathbf{x}$ requires both multiplication and addition. A circuit that can perform this multiply-accumulate function on stochastic data with output scaled to fall in the interval $[0,1]$ is shown in Fig. 1.6.

(a) When evaluating the determinant $|\mathbf{A}|$ how does root-mean-square (RMS) error scale as a function of n_{bits} and the two-norm condition number of the matrix?

(b) Multiplication of two stochastic bit-streams can be achieved using the AND operation and addition can be achieved using the MUX function with a random select. What circuit elements can perform (i) subtraction and (ii) division of two stochastic bit-streams?

Problem 1.17
(a) Show that a Gaussian wave function describing an electron with effective mass m_e^* moving in the x-direction at group velocity $v_g = d\omega(k_0)/dk$, central momentum $\hbar k_0 = m_e^* v_g$, and standard deviation σ_x about x_0 at time $t = 0$ is

$$\psi(x,t) = \left(\frac{\sigma_x^2}{2\pi}\right)^{\frac{1}{4}} \frac{1}{\sqrt{\sigma_x^2 + \frac{i\hbar t}{2m_e^*}}} e^{\left(i\frac{m_e^* v_g}{\hbar}\left(x - x_0 - \frac{v_g t}{2}\right) - \frac{(x - x_0 - v_g t)^2}{(4\sigma_x^2 + i2\hbar t/m_e^*)}\right)}.$$

53

For time $0 \le t \le 0.6$ ps, find $D(t)$, the overlap area of $|\psi(x,t)|^2$ with $|\psi(x,0)|^2$, when the mean position of $|\psi(x,t)|^2$ is shifted to $x = 0$. Obtain the quantitative measure of pulse dispersion $D(t)$ using parameters $E = 50$ meV, $m_e^* = 0.07 \times m_0$, and $\sigma_x = 10$ nm.

(b) Show that the pulse dispersion measure is bounded such that $0 \le D \le 1$.

Problem 1.18

The shortest path length of a particle with wave properties moving in free space in the x-direction from point a to point b is $s(n_0)$. The particle has energy $E = \hbar\omega$ and momentum $\hbar k$. Assuming that at any instant $\mathbf{k}$ is parallel to the path taken then, since in free space $A_n = A_0$, the total amplitude at b at time t is the sum over all possible paths

$$\mathbf{A}_{\text{tot}} = \sum_n A_n e^{iks(n)-i\omega t} = A_0 e^{-i\omega t} \sum_{n=-n_{\max}}^{n=n_{\max}} e^{iks(n)} = A_0 e^{-i\omega t} e^{iks(n_0)} \Phi(n).$$

For the equilateral triangle paths in the x-y plane with constant step increase in midpoint height Δy shown in Fig. 1.P25, plot $\text{Im}(\Phi(n))$ as a function of $\text{Re}(\Phi(n))$ using parameters $s(n_0) = 1$, $k = \pi$, $\Delta y = 0.005$, and $n_{\max} = 2\,000$. Explain the results obtained.

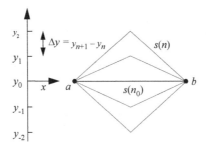

Fig. 1.P25

2 Using the Schrödinger Wave Equation

2.1 Introduction

According to the Schrödinger equation, a particle with wave character and mass m in the presence of a potential may be described as a state that is a function of space and time. Space and time are assumed to be smooth and continuous. The potential can localize the particle to one region of space forming a *bound state*. Alternatively, a particle that is not bound to a local region of space is in an *unbound state* (also called a scattering state).

The time-independent Schrödinger equation for a particle of mass m in a potential $V(\mathbf{r})$, which is a function of space only, was introduced in Section 1.2. The second-order differential equation is

$$\hat{H}\psi_n(\mathbf{r}) = \left(-\frac{\hbar^2}{2m}\nabla^2 + V(\mathbf{r})\right)\psi_n(\mathbf{r}) = E_n\psi_n(\mathbf{r}), \tag{2.1}$$

where $\hat{H}$ is the Hamiltonian operator, $E_n = \hbar\omega_n$ are energy eigenvalues, and $\psi_n(\mathbf{r})$ are time-independent stationary states, such that

$$\psi_n(\mathbf{r}, t) = \psi_n(\mathbf{r})e^{-i\omega_n t}. \tag{2.2}$$

The *eigenfunctions* $\psi_n(\mathbf{r})$ are found for a given potential by using the Schrödinger wave equation and applying *boundary conditions*. The wave equation contains a vast number of possible solutions. Often, results for a particular set of circumstances can most easily be found by specifying the potential $V(\mathbf{r})$ and applying boundary conditions.[1]

When the potential is independent of time, solutions to Schrödinger's equation are called *stationary states* because the probability density $|\psi(\mathbf{r}, t)|^2$ is independent of time. A simple bound-state standing wave function $\psi(x)$ with energy $\hbar\omega$ and time dependence $e^{-i\omega t}$ is stationary in the sense that it does not propagate a flux through space. In contrast to a standing wave, a traveling wave (scattering state) of the form $e^{i(kx-\omega t)}$ carries a constant flux. Such functions can also be solutions to the time-independent Schrödinger equation. In this case, the state is only stationary in the weaker sense that there is no time-varying flux component.

Proper solutions to the Schrödinger wave equation require that $\psi(\mathbf{r})$ and $\nabla\psi(\mathbf{r})$ are *continuous* everywhere. To provide some practice with wave functions, the effect of not complying with this requirement is described in the next section.

---

[1] This is the conventional way of solving the *forward physical system problem*. Alternatively, eigenfunctions or eigenvalues may be known and the task is to find the potential and boundary conditions. This is called an *inverse problem* because it is known that a solution exists. The inverse problem is a subset of the more general optimization problem of finding the potential that most closely results in desired eigenfunctions or eigenvalues. In this case, it is possible that the desired objective function may only be approached to within some *distance*, meaning a solution that exactly meets the objective is not accessible. Numerical techniques such as machine learning are a distinct and separate data-driven approach to finding characteristic patterns in measured data that can then be used to predict behavior.

2.1.1 The Effect of Discontinuity in the Wave Function and its Slope

The wave function and the spatial derivative of the wave function should be continuous. The argument used relies on the notion that energy is a constant of the motion and hence a well-behaved quantity. For ease of discussion, consider a one-dimensional wave function that is real and with the potential energy set to zero.

The *discontinuity in a wave function* at position x_0 illustrated in Fig. 2.1 results in a nonphysical infinite contribution to the expectation value of particle kinetic energy, T. Approximating $\psi(x)$ as a piecewise function and evaluating the contribution to the expectation value of kinetic energy in the limit $x \to x_0$ gives

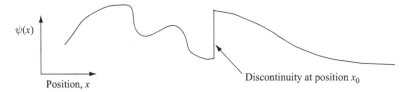

$\psi(x)$

Position, x

Discontinuity at position x_0

Fig. 2.1 Illustration of a discontinuity in wave function $\psi(x)$ at position $x = x_0$. The discontinuity is characterized by an infinite slope in the wave function.

$$\Delta\langle T\rangle = \lim_{\eta\to\infty} -\frac{\hbar^2}{2m} \int_{x_0-\frac{1}{\eta}}^{x_0+\frac{1}{\eta}} \psi^* \frac{d^2\psi}{dx^2} dx, \tag{2.3}$$

where η is a dummy variable. Since $\psi(x)$ is assumed to be real-valued, the complex conjugate is dropped. Integration by parts gives

$$\Delta\langle T\rangle = \lim_{\eta\to\infty} \int_{x_0-\frac{1}{\eta}}^{x_0+\frac{1}{\eta}} \psi\left(\frac{-\hbar^2}{2m}\frac{d^2\psi}{dx^2}\right) dx = \frac{-\hbar^2}{2m} \lim_{\eta\to\infty} \left(\psi\frac{d\psi}{dx}\Big|_{x_0-\frac{1}{\eta}}^{x_0+\frac{1}{\eta}} - \int_{x_0-\frac{1}{\eta}}^{x_0+\frac{1}{\eta}} \left(\frac{d\psi}{dx}\right)^2 dx\right). \tag{2.4}$$

For $\psi(x)$ *discontinuous*, the term on the far right of Eq. (2.4) is infinite. This may be shown by using the median value theorem for functions $f(x)$ and $g(x)$:

$$\int_a^b f(x)g(x)dx = f(\xi) \int_a^b g(x)dx, \tag{2.5}$$

where $a < \xi < b$, and in this case with $f(x) = g(x) = \frac{d\psi}{dx}$, so that

$$\int_{x_0-\frac{1}{\eta}}^{x_0+\frac{1}{\eta}} \left(\frac{d\psi}{dx}\right)^2 dx = \frac{d\psi}{dx}\Big|_\xi \int_{x_0-\frac{1}{\eta}}^{x_0+\frac{1}{\eta}} \frac{d\psi}{dx} dx = \frac{d\psi}{dx}\Big|_\xi \left[\psi\left(x_0+\frac{1}{\eta}\right) - \psi\left(x_0-\frac{1}{\eta}\right)\right]_{x_0-\frac{1}{\eta}}^{x_0+\frac{1}{\eta}} \tag{2.6}$$

for $x_0 - 1/\eta < \xi < x_0 + 1/\eta$. The term in the rectangular brackets is a nonzero constant and within the limit $\eta \to \infty$, $\xi \to x_0$, so that

$$\int_{x_0-\frac{1}{\eta}}^{x_0+\frac{1}{\eta}} \left(\frac{\mathrm{d}\psi}{\mathrm{d}x}\right)^2 \mathrm{d}x = \frac{\mathrm{d}\psi}{\mathrm{d}x}\bigg|_{x \to x_0} \left[\psi\left(x_0 + \frac{1}{\eta}\right) - \psi\left(x_0 - \frac{1}{\eta}\right)\right]_{x_0-\frac{1}{\eta}}^{x_0+\frac{1}{\eta}} = \infty, \qquad (2.7)$$

because $\mathrm{d}\psi/\mathrm{d}x|_{x \to x_0} = \psi'|_{x \to x_0} = \infty$.

Hence, the contribution to the expectation value of kinetic energy due to a discontinuity in the wave function is infinite, $\Delta\langle T\rangle \to \infty$, because of the infinite slope of the wave function. To avoid such an *unphysical* result requires that the wave function be continuous.

A discontinuity in the spatial derivative of a real continuous wave function makes a finite contribution to the expectation value of kinetic energy. Such a discontinuity in the spatial derivative corresponds to a kink in the real wave function that adds $\Delta\langle T\rangle = -(\hbar^2/2m)\psi(\Delta\psi')$ to the expectation value of the kinetic energy, where ψ is the wave function at the kink and $\Delta\psi'$ is the difference between the slopes of the wave function on the two sides of the kink. Figure 2.2 illustrates a wave function, $\psi(x)$, with a kink at position x_0.

The contribution of the kink at position x_0 to the expectation value of particle kinetic energy is

$$\Delta\langle T\rangle = \lim_{\eta \to \infty} \int_{x_0-\frac{1}{\eta}}^{x_0+\frac{1}{\eta}} \psi\left(\frac{-\hbar^2}{2m}\frac{\mathrm{d}^2\psi}{\mathrm{d}x^2}\right)\mathrm{d}x = \frac{-\hbar^2}{2m}\lim_{\eta \to \infty}\left(\psi\frac{\mathrm{d}\psi}{\mathrm{d}x}\bigg|_{x_0-\frac{1}{\eta}}^{x_0+\frac{1}{\eta}} - \int_{x_0-\frac{1}{\eta}}^{x_0+\frac{1}{\eta}}\left(\frac{\mathrm{d}\psi}{\mathrm{d}x}\right)^2\mathrm{d}x\right).$$

$$(2.8)$$

Since $\psi(x)$ is continuous, $\mathrm{d}\psi/\mathrm{d}x$ has finite values everywhere. In this case, the integral on the far right-hand side of Eq. (2.8) vanishes as $\eta \to \infty$, leaving

$$\Delta\langle T\rangle = \frac{-\hbar^2}{2m}\lim_{\eta \to \infty}\psi\frac{\mathrm{d}\psi}{\mathrm{d}x}\bigg|_{x_0-\frac{1}{\eta}}^{x_0+\frac{1}{\eta}}$$

$$= \frac{-\hbar^2}{2m}\lim_{\eta \to \infty}\left(\psi\left(x_0 + \frac{1}{\eta}\right)\psi'\left(x_0 + \frac{1}{\eta}\right) - \psi\left(x_0 - \frac{1}{\eta}\right)\psi'\left(x_0 - \frac{1}{\eta}\right)\right). \qquad (2.9)$$

Position, x

Fig. 2.2 Illustration of a kink in the wave function $\psi(x)$ at position $x = x_0$. The kink is characterized by a discontinuity in the slope of the wave function.

57

Because $\psi(x)$ is continuous, it follows that as $\eta \to \infty$,

$$\psi\left(x_0 + \frac{1}{\eta}\right) = \psi\left(x_0 - \frac{1}{\eta}\right) = \psi(x_0). \tag{2.10}$$

If

$$\Delta\psi = \psi'\left(x_0 + \frac{1}{\eta}\right) - \psi'\left(x_0 - \frac{1}{\eta}\right) \tag{2.11}$$

then

$$\Delta\langle T \rangle = \frac{-\hbar^2}{2m}\psi(x_0)\Delta\psi. \tag{2.12}$$

So it may be concluded that a kink in the wave function at position x_0 makes a contribution to the expectation value of the kinetic energy that is proportional to the value of the wave function multiplied by the difference in slope of the wave function at x_0. Allowing solutions to the wave function that contain *arbitrary* kinks would add arbitrary values of kinetic energy. This, and other complications, can be avoided by requiring that the spatial derivative of the wave function describing a particle moving in a well-behaved potential be continuous. This is a significant constraint on the number and type of wave functions that are allowed.

Solutions to Schrödinger's time-independent equation (Eq. (2.1)) must be consistent with the potential $V(\mathbf{r})$. If a potential is badly behaved and possesses a delta function singularity, there is a kink in the wave function and the spatial derivative of the wave function is discontinuous.

If ψ is continuous and smooth (ψ and ψ' are continuous), the expectation value of kinetic energy, $\langle T \rangle$, is proportional to the integral over all space of the value of the wave function multiplied by the wave function curvature. Hence, the energy of a wave function increases if it has more "wiggles" or, equivalently, more regions of high slope.

2.2 Bound-State Wave Function Normalization and Completeness

The value of $|\psi_n(\mathbf{r})|^2 \mathrm{d}^3 r$ for bound states is interpreted as the probability of finding a particle in state $\psi_n(\mathbf{r})$ in the volume element $\mathrm{d}^3 r$ at position $\mathbf{r}$ in space. For this reason it is also convenient to require that eigenstates $\psi_n(\mathbf{r})$ be *normalized* such that integration of $|\psi_n(\mathbf{r})|^2$ over all space yields unity. Eigenstates are also *orthogonal* and hence *orthonormal* such that

$$\boxed{\int_{-\infty}^{\infty} \psi_n^*(\mathbf{r})\psi_m(\mathbf{r})\mathrm{d}^3 r = \delta_{nm},} \tag{2.13}$$

where the Kronecker delta function $\delta_{nm} = 0$ unless $n = m$, in which case $\delta_{nn} = 1$. In addition, the orthogonal wave functions ψ_n are *complete*, so that any arbitrary wave function $\psi(\mathbf{r})$ can be expressed as a sum of orthonormal wave functions weighted by coefficients a_n in such a way that

$$\boxed{\psi(\mathbf{r}) = \sum_n a_n \psi_n(\mathbf{r}).}$$

(2.14)

2.3 Inversion Symmetry in the Potential

There are a number of simplifying concepts that may be used when solving problems involving the Schrödinger equation. If potential energy $V(\mathbf{r})$ has *inversion symmetry* in space such that

$$V(\mathbf{r}) = V(-\mathbf{r}),$$

(2.15)

then eigenfunctions with different eigenvalues have either *odd* or *even parity*, so that $\psi_n(\mathbf{r}) = \pm\psi_n(-\mathbf{r})$, where

$$\psi_n(\mathbf{r}) = +\psi_n(-\mathbf{r})$$

(2.16)

has even parity, and

$$\psi_n(\mathbf{r}) = -\psi_n(-\mathbf{r})$$

(2.17)

has odd parity.

Suppose there are a pair of eigenfunctions that have the same eigenvalues *and* do not obey this parity requirement; that is, they are *linearly independent*. Let $\psi(\mathbf{r})$ be a function that does not obey the parity requirement. In this situation, $\psi(\mathbf{r})$ and $\psi(-\mathbf{r})$ are linearly independent, so

$$\psi_+(\mathbf{r}) = \psi(\mathbf{r}) + \psi(-\mathbf{r})$$

(2.18)

can be constructed with even parity and

$$\psi_-(\mathbf{r}) = \psi(\mathbf{r}) - \psi(-\mathbf{r})$$

(2.19)

with odd parity; these *do* obey the required parity.

Summarizing, inversion symmetry in the potential will result in wave functions with definite parity.

2.3.1 One-Dimensional Rectangular Potential Well with Infinite Barrier Energy

A simple potential with inversion symmetry is a one-dimensional, rectangular potential well with infinite barrier energy of the type illustrated in Fig. 2.3(a). A natural coordinate system for the one-dimensional potential well that admits inversion symmetry in the potential such that $V(x) = V(-x)$ has $x = 0$ in the middle of the potential well, $V(-L/2 < x < L/2) = 0$, and $V(x \leq -L/2, x \geq L/2) = \infty$.

The *boundary condition* $\psi = 0$ for $x \leq -L/2$ and $x \geq L/2$ causes the wave function to vanish at the edges of the potential well because the potential barrier has infinite energy.

In the potential well,

$$\frac{-\hbar^2}{2m}\frac{d^2}{dx^2}\psi(x) = E\psi(x), \tag{2.20}$$

and $\psi(x) = Ae^{ikx} + Be^{-ikx}$ is of sinusoidal form. At the $x = -L/2$ boundary,

$$\psi(x = -L/2) = Ae^{-ikL/2} + Be^{ikL/2} = 0, \tag{2.21}$$

and at the $x = L/2$ boundary,

$$\psi(x = L/2) = Ae^{ikL/2} + Be^{-ikL/2} = 0. \tag{2.22}$$

Since there is only one solution for each wave function, the determinant of coefficients for the two equations (Eqs. (2.21) and (2.22)) is zero:

$$\begin{vmatrix} e^{-ikL/2} & e^{ikL/2} \\ e^{ikL/2} & e^{-ikL/2} \end{vmatrix} = e^{-ikL} - e^{ikL} = 0. \tag{2.23}$$

It follows that the boundary conditions require solutions in which $\sin(k_n L) = 0$, where $k_n L = n\pi$ and n is a positive nonzero integer. The corresponding wave functions $\psi(x)$ are sinusoidal with

$$\psi_n(x, t) = \sqrt{\frac{2}{L}} \sin\left(k_n\left(x + \frac{L}{2}\right)\right), \tag{2.24}$$

where n is a positive nonzero integer, $\sqrt{2/L}$ is a normalization constant, the wave vector

$$k_n = \frac{2\pi}{\lambda_n} = n\frac{2\pi}{2L} = \frac{n\pi}{L}, \text{ for } n = 1, 2, 3, \ldots, \tag{2.25}$$

has wavelength $\lambda_n = 2L/n$, and the energy eigenvalues are $E_n = \frac{\hbar^2 k_n^2}{2m}$. Hence,

$$\boxed{E_n = \frac{\hbar^2}{2m}\frac{n^2\pi^2}{L^2}.} \tag{2.26}$$

The *lowest energy state* or *ground state* of the system has energy eigenvalue

$$E_1 = \frac{\hbar^2\pi^2}{2mL^2}. \tag{2.27}$$

This is the *zero-point energy* for the bound state. There is no classical analog for zero-point energy that arises due to the wavy nature of a particle. The lowest energy bound state in a confining potential has minimum curvature and hence a minimum energy that is a measure of how confined the particle is in real space. In this sense, the wavy nature of particles in quantum mechanics imposes a relationship between energy and space. A particle confined to any region of space cannot sustain an average energy below the ground state or zero-point energy.

Wave functions in this simple symmetric one-dimensional potential well are of either *even* or *odd parity*. Figure 2.3(a) is a sketch of a one-dimensional, rectangular potential

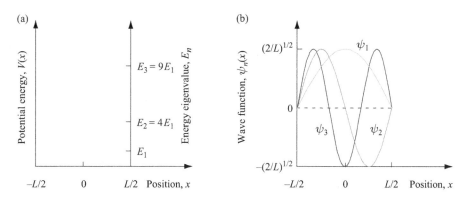

Fig. 2.3 (a) Sketch of a one-dimensional, rectangular potential well with infinite barrier energy showing the energy eigenvalues E_1, E_2, and E_3. The thickness of the well is L and the potential energy is such that $V(-L/2 < x < L/2) = 0$ and $V(x \leq -L/2, x \geq L/2) = \infty$. (b) Sketch of the eigenfunctions ψ_1, ψ_2, and ψ_3 for the potential shown in (a). The kinks in the wave functions at $x = \pm L/2$ do not contribute to particle energy because $\psi(x = \pm L/2) = 0$ in the expression $\Delta T = -\hbar^2 \psi(x) \Delta \psi / 2m$ (Eq. (2.12)).

well with infinite barrier energy showing energy eigenvalues E_1, E_2, and E_3. Figure 2.3(b) sketches the first three eigenfunctions, ψ_1, ψ_2, and ψ_3, for the potential shown in Fig. 2.3(a). The kink in the wave functions at $x = \pm L/2$ does *not* contribute to particle energy because $\psi(x = \pm L/2) = 0$ in the expression $\Delta T = -\hbar^2 \psi(x) \Delta \psi / 2m$ (Eq. (2.12)).

The time dependence of the eigenfunctions $\psi(x, t) = \psi(x)\phi(t)$ is found by solving the time-dependent Schrödinger equation

$$E\phi(t) = i\hbar \frac{\partial}{\partial t} \phi(t). \tag{2.28}$$

The solutions are of the form $\phi_n(t) = e^{-iE_n t/\hbar} = e^{-i\omega_n t}$, where $E_n = \hbar\omega_n$. Hence, the solution for a wave function with quantum number n is

$$\psi_n(x, t) = \psi_n(x)e^{-i\omega_n t}, \tag{2.29}$$

which can be written as

$$\psi_n(x, t) = \sqrt{\frac{2}{L}} \sin\left(k_n\left(x + \frac{L}{2}\right)\right) e^{-i\omega_n t}, \tag{2.30}$$

where $k_n = n\pi/L$ is the wave vector and $\omega_n = \hbar k_n^2/2m$ is the angular frequency of the nth eigenstate. In Eq. (2.30), the wave function $\psi_n(x, t)$ consists of a spatial part and an oscillatory time-dependent part. The *frequency of oscillation* of the time-dependent part is given by the energy eigenvalue $E_n = \hbar\omega_n$ and is found using the *spatial* solution to the time-independent Schrödinger equation. This link between spatial and temporal solutions can be traced back to the separation of variables in Schrödinger's equation (Eqs. (1.36) and (1.37) in Chapter 1). Such separation in variables is possible when the potential is time independent.

2.4 Numerical Solution of the Schrödinger Equation

The time-independent Schrödinger equation (Eq. (2.1)) describing a particle of mass m constrained to motion in a time-independent, one-dimensional potential such that $V(0 < x < L) \neq \infty$ and $V(x \leq 0, x \geq L) = \infty$ is

$$\hat{H}\psi(x) = \left(-\frac{\hbar^2}{2m} \frac{d^2}{dx^2} + V(x) \right) \psi(x) = E\psi(x).$$ (2.31)

Approximate, *bound-state* solutions to this equation may be found numerically if the continuous functions are discretized. A first approach to do this samples the wave function and potential at a set of $N+1$ equally spaced points in such a way that position $x_j = j \times h_0$, where the index $j = 0, 1, 2, \ldots, N$, and $h_0 = x_{j+1} - x_j$ is the interval between adjacent sampling points. Such sampling of a particle wave function and potential is illustrated in Fig. 2.4.

At each sampling point, the wave function has value $\psi_j = \psi(x_j)$ and the potential is $V_j = V(x_j)$. The first derivative of the discretized wave function $\psi(x_j)$ in the two-point finite-difference approximation is

$$\frac{d}{dx}\psi(x_j) = \frac{\psi(x_j) - \psi(x_{j-1})}{h_0}.$$ (2.32)

A three-point finite difference approximation of the second derivative of the discretized wave function is

$$\frac{d^2}{dx^2}\psi(x_j) = \frac{\psi(x_{j-1}) - 2\psi(x_j) + \psi(x_{j+1})}{h_0^2}.$$ (2.33)

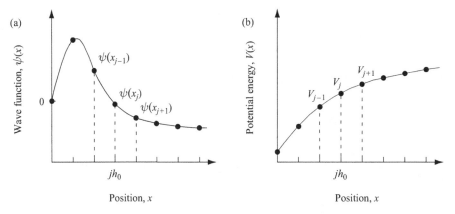

Fig. 2.4 (a) Example of sampling the wave function $\psi(x)$ at positions x_j in such a way that $x_j = jh_0$, where $j = 0, 1, 2, \ldots, N$ and the fixed interval between sampling points is $h_0 = x_{j+1} - x_j$. (b) Sampling the potential $V(x)$ at equally spaced intervals.

A discretized *approximation* of the Schrödinger equation is obtained by substituting Eq. (2.33) into Eq. (2.31) to give the matrix equation

$$(\mathbf{H} - E\mathbf{1})\psi(x) = \sum_{j=0}^{N}(-u_j\psi(x_{j-1}) + (d_j - E)\psi(x_j) - u_{j+1}\psi(x_{j+1})) = 0, \tag{2.34}$$

where the Hamiltonian $\mathbf{H}$ is a symmetric tridiagonal matrix and $\mathbf{1}$ is the identity (or unit) matrix. The diagonal matrix elements are

$$d_j = \frac{\hbar^2}{mh_0^2} + V_j \tag{2.35}$$

and the adjacent off-diagonal matrix elements are

$$u_j = \frac{\hbar^2}{2mh_0^2}. \tag{2.36}$$

As discussed in Section 2.3.1, to find the eigenenergies and eigenstates of a particle of mass m in a one-dimensional rectangular potential well with infinite barrier energy, the boundary conditions require $\psi_0(x_0) = \psi_N(x_N) = 0$, so that Eq. (2.34) may be written as

$$(\mathbf{H} - E\mathbf{1})\psi \tag{2.37}$$

$$= \begin{bmatrix} (d_1 - E) & -u_2 & 0 & 0 & \cdot & & \cdot \\ -u_2 & (d_2 - E) & -u_3 & 0 & \cdot & & \cdot \\ 0 & -u_3 & (d_3 - E) & -u_4 & \cdot & & \cdot \\ \cdot & \cdot & \cdot & \cdot & \cdot & & \cdot \\ \cdot & \cdot & \cdot & \cdot & -u_{N-1} & (d_{N-1} - E) \end{bmatrix} \begin{bmatrix} \psi_1 \\ \psi_2 \\ \psi_3 \\ \cdot \\ \psi_{N-1} \end{bmatrix} = 0.$$

The solutions to this equation may be found using conventional numerical methods. Programing languages such as MATLAB® also have routines that efficiently diagonalize the tridiagonal matrix and solve for the eigenvalues and eigenfunctions.

In the preceding example, the situation in which the particle is confined to one-dimensional motion in a region of length $L = Nh_0$ was considered. Outside this region, the potential is infinite and the wave function is zero. Particle motion is localized to one region of space, so only bound states can exist as solutions to the Schrödinger equation. Because the particle is not transmitted beyond the domain defined between boundary positions $x = x_0 = 0$ and $x = x_N = L$, this is a quantum nontransmitting boundary problem.

An extension is to consider the *periodic boundary condition* in which $\psi(x) = \psi(x + L)$ so that $\psi_0(x_0) = \psi_N(x_N)$ and $d\psi_0/dx|_{x=x_0} = d\psi_N/dx|_{x=x_N}$. In this situation, Eq. (2.37) is modified to

$$(\mathbf{H} - E\mathbf{1})\psi \tag{2.38}$$

$$= \begin{bmatrix} (d_1 - E) & -u_2 & 0 & \cdot & & \cdot & -u_1 \\ -u_2 & (d_2 - E) & -u_3 & \cdot & & \cdot & \cdot \\ 0 & -u_3 & (d_3 - E) & \cdot & & & \cdot \\ \cdot & \cdot & \cdot & \cdot & & & \cdot \\ \cdot & \cdot & \cdot & -u_{N-1} & (d_{N-1} - E) & -u_N & \cdot \\ -u_1 & \cdot & \cdot & \cdot & -u_N & (d_N - E) \end{bmatrix} \begin{bmatrix} \psi_1 \\ \psi_2 \\ \psi_3 \\ \cdot \\ \psi_{N-1} \\ \psi_N \end{bmatrix} = 0.$$

If there are transmitting boundaries at positions $x = x_0 = 0$ and $x = x_N = L$, then it is possible that $\psi_0(x_0) \neq \psi_N(x_N)$ and $d\psi_0/dx|_{x=x_0} \neq d\psi_N/dx|_{x=x_N}$. In this situation, unbound particle states as well as sources and sinks of particle flux may exist. An approach to dealing with this case is called the quantum transmitting boundary method.[2]

2.5 Current Flow

From Maxwell's equations or by elementary consideration of current conservation, a change in charge density ρ is related to the divergence of current density $\mathbf{J}$ through

$$\frac{\partial}{\partial t}\rho(\mathbf{r}, t) = -\nabla \cdot \mathbf{J}(\mathbf{r}, t). \tag{2.39}$$

This is the classical expression for *current continuity*. The time dependence of charge density (the *temporal* dependence of a scalar field) is related to the net current into or out of a region of space (the *spatial* dependence of a vector field). For a particle of charge e, density $\rho = e|\psi|^2 = e\psi^*\psi$, so that

$$\frac{\partial \rho}{\partial t} = e\frac{\partial}{\partial t}(\psi^*\psi) = e\left(\psi^*\frac{\partial \psi}{\partial t} + \psi\frac{\partial \psi^*}{\partial t}\right). \tag{2.40}$$

The time-dependent Schrödinger equation for a particle of mass m is

$$i\hbar\frac{\partial}{\partial t}\psi(\mathbf{r}, t) = \left(-\frac{\hbar^2}{2m}\nabla^2 + V(\mathbf{r})\right)\psi(\mathbf{r}, t) = \hat{H}\psi(\mathbf{r}, t). \tag{2.41}$$

Multiplying both sides by $\psi^*(\mathbf{r}, t)$ gives

$$i\hbar\psi^*(\mathbf{r}, t)\frac{\partial}{\partial t}\psi(\mathbf{r}, t) = -\frac{\hbar^2}{2m}\psi^*(\mathbf{r}, t)\nabla^2\psi(\mathbf{r}, t) + \psi^*(\mathbf{r}, t)V(\mathbf{r})\psi(\mathbf{r}, t), \tag{2.42}$$

which, when multiplied by $e/i\hbar$, is the first term in the expression for $\partial\rho/\partial t$. The second term is found by taking the complex conjugate of Eq. (2.41) and multiplying both sides by $\psi(\mathbf{r}, t)$. In effect, $\psi(\mathbf{r}, t)$ and $\psi^*(\mathbf{r}, t)$ are interchanged and the sign of i in Eq. (2.42) is changed. Because $V(\mathbf{r})$ is assumed real, $\psi^*(\mathbf{r}, t)V(\mathbf{r})\psi(\mathbf{r}, t) = \psi(\mathbf{r}, t)V(\mathbf{r})\psi^*(\mathbf{r}, t)$, the rate of change of charge density is

$$\frac{\partial \rho}{\partial t} = -\psi^*\frac{e\hbar}{i2m}\nabla^2\psi + \psi\frac{e\hbar}{i2m}\nabla^2\psi^* + \frac{e\psi^*\psi}{i\hbar}(V(\mathbf{r}) - V(\mathbf{r}))$$

$$= \frac{ie\hbar}{2m}\left(\psi^*\nabla^2\psi - \psi\nabla^2\psi^*\right) = \frac{ie\hbar}{2m}\nabla \cdot (\psi^*\nabla\psi - \psi\nabla\psi^*) = -\nabla \cdot \mathbf{J}. \tag{2.43}$$

Hence, the *current density* for a particle described by state ψ is

$$\boxed{\mathbf{J} = -\frac{ie\hbar}{2m}(\psi^*\nabla\psi - \psi\nabla\psi^*) = \frac{e\hbar}{m}\,\mathrm{Im}(\psi^*\nabla\psi),} \tag{2.44}$$

---

[2] C. S. Lent and D. J. Kirkner, *J. Appl. Phys.* **67**, 6353 (1990); C. L. Fernando and W. R. Frensley, *J. Appl. Phys.* **76**, 2881 (1994); Z. Shano, W. Porod, C. S. Lent, and D. J. Kirkner, *J. Appl. Phys.* **78**, 2177 (1995).

or, in one dimension,

$$J_x = -\frac{ie\hbar}{2m}\left(\psi^*(x)\frac{\mathrm{d}}{\mathrm{d}x}\psi(x) - \psi(x)\frac{\mathrm{d}}{\mathrm{d}x}\psi^*(x)\right) = \frac{e\hbar}{m}\,\mathrm{Im}\left(\psi^*(x)\frac{\mathrm{d}}{\mathrm{d}x}\psi(x)\right). \tag{2.45}$$

The above derivation of the current density requires that the potential be *real*. If the potential is complex, it can have the effect of absorbing or creating nonquantized continuum density. This can be a useful feature that helps when solving some types of problems in the many particle *semiclassical* thermodynamic limit.

A point worth highlighting is the obvious symmetry in the expression for the current operator. This symmetry has an important influence on the types of wave functions that can carry current.

2.5.1 Current in a Rectangular Potential Well with Infinite Barrier Energy

The one-dimensional rectangular potential centered at $x = 0$ with infinite barrier energy described in Section 2.3.1 has a $n = 1$ ground state energy

$$E_1 = \hbar\omega_1 = \frac{\hbar^2\pi^2}{2mL^2} \tag{2.46}$$

and the ground state wave function is a standing wave,

$$\psi_1(x,t) = \left(\frac{2}{L}\right)^{1/2}\sin\left(\frac{\pi}{L}\left(x + \frac{L}{2}\right)\right)e^{-i\omega_1 t}. \tag{2.47}$$

The current density in this case is

$$J_x = -\frac{ie\hbar}{2m}\left(\psi_1^*(x)\frac{\mathrm{d}}{\mathrm{d}x}\psi_1(x) - \psi_1(x)\frac{\mathrm{d}}{\mathrm{d}x}\psi_1^*(x)\right) = 0, \tag{2.48}$$

from which it may be concluded that current is not carried by a single standing wave. This should come as no surprise, since a standing wave can be thought of as two counter-propagating traveling waves whose individual contributions to current flow exactly cancel each other out.

A linear superposition state consisting of a simple combination of the ground state wave function, ψ_1, and the first excited state wave function, ψ_2, is

$$\psi(x,t) = \frac{1}{\sqrt{2}}(\psi_1(x,t) + \psi_2(x,t)), \tag{2.49}$$

which has current density

$$J_x = -\frac{2e\pi\hbar}{mL^2}\left(\cos\left(\frac{\pi x}{L}\right)\cos\left(\frac{2\pi x}{L}\right) + \frac{1}{2}\sin\left(\frac{\pi x}{L}\right)\sin\left(\frac{2\pi x}{L}\right)\right)\sin((\omega_2 - \omega_1)t). \tag{2.50}$$

The ability of a linear superposition of stationary bound states to carry current is due to the presence of different frequency components. The current has an oscillatory time dependence, which in this case is given by the difference frequency, $\omega_2 - \omega_1$, between the time dependence of the eigenfunctions ψ_1 and ψ_2. In general, for a finite number of eigenfunctions contributing to the superposition state, there is a time period after which the time-dependent current repeats itself. This is called the full *revival time*.

2.5.2 Current Flow due to a Traveling Wave

A particle with charge e and mass m that is in an unbound (scattering) state described by a wave function that is a plane wave traveling from left to right of the form $\psi(x) = e^{i(kx-\omega t)}$ has current

$$
\begin{aligned}
J_x &= -\frac{ie\hbar}{2m}\left(\psi^*(x)\frac{d}{dx}\psi(x) - \psi(x)\frac{d}{dx}\psi^*(x)\right) \\
&= -\frac{ie\hbar}{2m}\left(e^{-ikx}ike^{ikx} + e^{ikx}ike^{-ikx}\right) = -\frac{ie\hbar}{2m}2ik = \frac{e\hbar k}{m}.
\end{aligned}
\tag{2.51}
$$

Since momentum in the x-direction is $p_x = mv_x = \hbar k$, the current associated with the traveling wave may be written in the familiar form $J_x = ev_x$, where e is the particle charge and v_x is its velocity.

2.6 Degeneracy as a Consequence of Symmetry

The number of states with the same energy eigenvalue is called the *degeneracy* of the state and is most often a direct consequence of symmetry in the potential. In situations in which this is *not* the case, the degeneracy is said to be accidental.

2.6.1 Bound States in Three Dimensions and Degeneracy of Eigenvalues

To illustrate how degeneracy arises from symmetry in a potential, the case of a box-shaped potential of side L with zero interior potential energy and infinite exterior potential energy is considered. For a potential box in three dimensions with infinite barrier energy,

$$
V(x, y, z) = 0 \quad \text{for} \quad -\frac{L}{2} < x, y, z < \frac{L}{2},
\tag{2.52}
$$

$$
V(x, y, z) = \infty \quad \text{for} \quad -\frac{L}{2} \geq x, y, z \geq \frac{L}{2}.
\tag{2.53}
$$

The Hamiltonian is of separable form,

$$
\hat{H} = \hat{H}_x + \hat{H}_y + \hat{H}_z,
\tag{2.54}
$$

where $\hat{H}_x = \dfrac{-\hbar^2}{2m}\dfrac{d^2}{dx^2} + V(x)$, etc. Consequently, the wave functions $\psi(x, y, z)$ that are solutions to the time-independent Schrödinger equation,

$$
\begin{aligned}
\hat{H}\psi(x, y, z) &= \left(\frac{-\hbar^2}{2m}\left(\frac{d^2}{dx^2} + \frac{d^2}{dy^2} + \frac{d^2}{dz^2}\right) + V(x) + V(y) + V(z)\right)\psi(x, y, z) \\
&= E\psi(x, y, z),
\end{aligned}
\tag{2.55}
$$

can be factored as

$$\psi(x, y, z) = \psi_x(x)\psi_y(y)\psi_z(z). \tag{2.56}$$

The wave function is of this product form because there are no terms in the Hamiltonian that involve a dependency on combinations of x, y, or z. The energy eigenvalues for the three-dimensional potential well are

$$E_{n_x,n_y,n_z} = \frac{\hbar^2 k^2}{2m} = \frac{\hbar^2}{2m}\left(k_x^2 + k_y^2 + k_z^2\right), \tag{2.57}$$

where $k^2 = k_x^2 + k_y^2 + k_z^2$. For the box-shaped potential, it follows that

$$k_x = \frac{n_x \pi}{L}, k_y = \frac{n_y \pi}{L}, k_z = \frac{n_z \pi}{L}, \tag{2.58}$$

where n_x, n_y, and n_z are nonzero positive integers. The corresponding energy eigenvalues are given by

$$E_{n_x,n_y,n_z} = \frac{\hbar^2 \pi^2}{2mL^2}\left(n_x^2 + n_y^2 + n_z^2\right), \tag{2.59}$$

where the lowest energy value is

$$E_{1,1,1} = \frac{3\hbar^2 \pi^2}{2mL^2}. \tag{2.60}$$

The energy eigenvalues are labeled by three nonzero, positive-integer quantum numbers that correspond to orthogonal eigenfunctions $\psi_{n_x}(x)$, $\psi_{n_y}(y)$, and $\psi_{n_z}(z)$. Because n_x, n_y, and n_z label independent contributions to the total energy eigenvalue, many of these energy eigenvalues are *degenerate* in energy. Such degeneracy exists because combinations of *different* quantum numbers result in the same value of energy. For example, $E_{n_x=1,n_y=2,n_z=3}$ can be rearranged in a number of ways:

$$(1, 2, 3), (1, 3, 2), (2, 1, 3), (2, 3, 1), (3, 1, 2), (3, 2, 1).$$

All six states have the same energy eigenvalue $14E_{1,1,1}/3$, and so the state is said to be six-fold degenerate.

The lowest energy state is the ground state, which has quantum numbers $n_x = n_y = n_z = 1$ and energy $E_{1,1,1}$. Only one combination of quantum numbers has this energy, and thus the state is nondegenerate.

A conclusion is that degeneracy of these energy eigenvalues is a consequence of symmetry in the potential. If the potential is changed in such a way that the symmetry is broken, a new set of energy eigenvalues is created that will have a different degeneracy. Usually, reducing or breaking the level of symmetry will reduce the degeneracy.

2.7 Symmetric Finite-Barrier Potential

Bound states can also exist in a potential well with finite barrier energy such as the one-dimensional, rectangular potential well $V(-L < x < L) = 0$ and $V(x \le -L, x \ge L) = V_0$

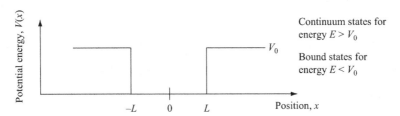

Fig. 2.5 Sketch of a simple symmetric one-dimensional, rectangular potential well of thickness $2L$. The potential energy is such that $V(x) = 0$ for $-L < x < L$ and $V(x) = V_0$ elsewhere. The value of V_0 is a finite, positive constant.

shown in Fig. 2.5. It is clear from the figure that any bound state in the system must have energy eigenvalue $E < V_0$. A particle with energy $E > V_0$ will belong to the class of *continuum* unbound states. The *binding energy* of a particular bound state is the minimum energy required to promote the particle from that bound state to an unbound state. A particle with energy eigenvalue $E = V_0$ is in a *critically bound state*.

By symmetry, $\psi(x) = \pm\psi(-x)$, and for $|x| < L$ the wave function either has even parity

$$\psi(x) = A\cos(kx) \tag{2.61}$$

or odd parity

$$\psi(x) = B\sin(kx). \tag{2.62}$$

The magnitude of the wave vectors in Eqs. (2.61) and (2.62) is

$$k = \sqrt{\frac{2mE}{\hbar^2}}. \tag{2.63}$$

For $|x| > L$, the wave function is

$$\psi(x) = Ce^{-\kappa|x|}, \tag{2.64}$$

and

$$\kappa = \sqrt{\frac{2m(V_0 - E)}{\hbar^2}} \tag{2.65}$$

for $E < V_0$. The solution $\psi = Ce^{\kappa|x|}$ is ill-behaved since $\psi(|x| \to \infty) \to \infty$ and so is not allowed.

The probabilistic interpretation associated with the bound-state wave function requires $\int \psi^*(x)\psi(x)\mathrm{d}x$ to be finite. Technically, ψ must be square-integrable.

From Eq. (2.63) $k^2 = 2mE/\hbar^2$ and from Eq. (2.65) $\kappa^2 = 2m(V_0 - E)/\hbar^2$. Introducing $L^2K_0^2$,

$$L^2K_0^2 = L^2k^2 + L^2\kappa^2 = \frac{(2mE + 2mV_0 - 2mE)}{\hbar^2}L^2 = \frac{2mV_0L^2}{\hbar^2}. \tag{2.66}$$

The condition relating k and κ is a circle in the $(L\kappa, Lk)$ plane of radius

$$LK_0 = \sqrt{\frac{2mV_0L^2}{\hbar^2}} \tag{2.67}$$

and must be satisfied when obtaining solutions for $\psi(x)$.

Since $V(x)$ and $\mathrm{d}^2\psi(x)/\mathrm{d}x^2$ are finite everywhere, $\psi(x)$ and $\mathrm{d}\psi(x)/\mathrm{d}x$ must be continuous everywhere, including at $x = \pm L$. Hence, the boundary conditions,

$$\psi(x)|_{x=L+\delta} = \psi(x)|_{x=L-\delta} \tag{2.68}$$

and

$$\frac{\mathrm{d}}{\mathrm{d}x}\psi(x)\bigg|_{x=L+\delta} = \frac{\mathrm{d}}{\mathrm{d}x}\psi(x)\bigg|_{x=L-\delta}, \tag{2.69}$$

lead directly to solutions of $\psi(x)$ at $x = L$.

Even parity solutions must satisfy

$$A\cos(kL) = Ce^{-\kappa L} \tag{2.70}$$

from Eq. (2.68), and

$$-kA\sin(kL) = -\kappa Ce^{-\kappa L} \tag{2.71}$$

from Eq. (2.69), so that *even parity* solutions require

$$kL\tan(kL) = \kappa L \tag{2.72}$$

and *odd parity* solutions require

$$kL\cot(kL) = -\kappa L. \tag{2.73}$$

Solutions that simultaneously satisfy the equation for a circle of radius $LK_0 = \sqrt{2mV_0L^2}/\hbar$ in the $(\kappa L, kL)$ plane *and* Eq. (2.72) or Eq. (2.73) can be found by graphical means.

2.7.1 Calculation of Bound States in a Symmetric Finite-Barrier Potential

Solutions for bound-states confined to a potential well similar to that sketched in Fig. 2.5 occur when the equation for a circle of radius LK_0 and Eq. (2.72) or Eq. (2.73) is simultaneously satisfied in the $(\kappa L, kL)$ plane.

When $LK_0 = 1$, then $V_0L^2 = \hbar^2/2m$, and, as shown in Fig. 2.6(a), there is one solution with even parity.

When $LK_0 = 5$ there are four bound-state solutions, two with even parity and two with odd parity. In this case, $V_0L^2 = 25\hbar^2/2m$. It is clear from Fig. 2.6(b) that there will always be at least four solutions if $2mV_0L^2/\hbar^2 > (3\pi/2)^2$.

To illustrate the wave functions for the bound states of an electron in a finite potential well, they are sketched in Fig. 2.7 for the case in which $2m_e^*V_0L^2/\hbar^2 = 23.0$. Here the effective electron mass is taken to be $m_e^* = 0.07 \times m_0$, which is the value for an electron

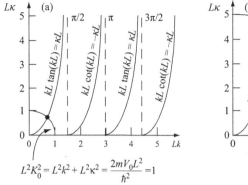
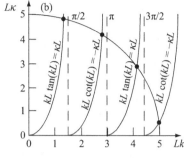

$$L^2 K_0^2 = L^2 k^2 + L^2 \kappa^2 = \frac{2mV_0 L^2}{\hbar^2} = 1$$

Fig. 2.6 Illustration of graphical method to find solution for simple symmetric, one-dimensional, rectangular potential well with finite barrier energy such that $V(x) = 0$ for $-L < x < L$ and $V(x) = V_0$ elsewhere. In (a) the value of V_0 results in $2mV_0 L^2/\hbar^2 = 1$, and there is one solution. In (b) the value of V_0 results in $2mV_0 L^2/\hbar^2 = 25$, and there are four bound-state solutions.

in the conduction band of the semiconductor GaAs. In the figure, the boundaries of the potential at position $x = \pm L$ are indicated to help identify the penetration of the wave function into the barrier region.

Conclusions may now be drawn concerning the nature of solutions obtained for a particle in a simple rectangular potential well with finite barrier energy. First, the fact that wave functions have alternating even and odd parity is a direct consequence of the symmetry of the potential. Second, there is at least one lowest energy state solution that is *always* of *even parity*. This is a consequence of the potential's symmetry. For the half-space potential in which $V(x \leq 0) = \infty$, $V(0 < x < L) = 0$, and $V(x \geq L) = V_0$, there exists no bound state in the potential well if $L^2 K_0^2 \leq (\pi/2)^2$. Third, for a finite potential well, the wave function always extends into the barrier region. In quantum mechanics, a wave function penetrating a potential barrier is said to *tunnel* into the barrier. For the case of the simple potential considered, tunneling *always* has the effect of reducing the energy of the eigenvalue E_n compared with the infinite potential well case. This reduction in energy may be understood by recognizing that tunneling decreases the spatial curvature of the wave function and hence the energy associated with it.

In the limit $V_0 \to \infty$, notice that $L^2 K_0^2 = L^2 k^2 + L^2 \kappa^2 = 2mV_0 L^2/\hbar^2 \to \infty$. The intersection of the circle of radius $LK_0 \to \infty$ with curves defined by $kL \tan(kL) = \kappa L$ and $kL \cot(kL) = -\kappa L$ now occurs at $Lk = n\pi/2$, where $n = 1, 2, 3, \ldots$. The energy eigenvalues are

$$E_n = \frac{\hbar^2}{2m} n^2 k^2 = \frac{\hbar^2}{2m} \frac{n^2 \pi^2}{4L^2}, \tag{2.74}$$

which is the same result previously derived in Section 2.3.1 for the energy levels of a particle in a rectangular potential well with infinite barrier energy. The only difference between Eq. (2.74) and Eq. (2.26) is the factor 4 in the denominator, which arises from the fact that the rectangular potential well of infinite barrier energy had a total well thickness of L and the finite barrier energy potential being considered has total well thickness of $2 \times L$.

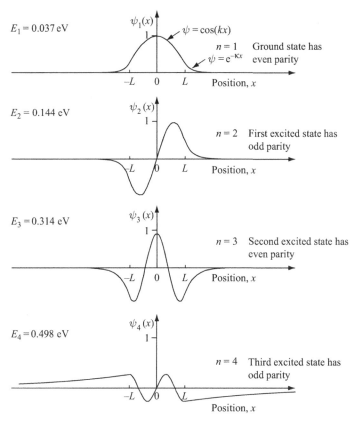

Fig. 2.7 Sketch of the eigenfunctions ψ_1, ψ_2, ψ_3, and ψ_4 for a simple symmetric one-dimensional rectangular potential well of total thickness $2 \times L$ with finite barrier energy such that $V(x) = 0$ for $-L < x < L$ and $V(x) = V_0$ elsewhere. Effective electron mass is $m_e^* = 0.07 \times m_0$, barrier energy is $V_0 = 0.5\,\mathrm{eV}$, and total well thickness is $2 \times L = 10\,\mathrm{nm}$. The four bound energy eigenvalues measured from the bottom of the potential well are $E_1 = 0.037\,\mathrm{eV}$, $E_2 = 0.144\,\mathrm{eV}$, $E_3 = 0.314\,\mathrm{eV}$, and $E_4 = 0.498\,\mathrm{eV}$. Because E_4 is close in value to V_0, the wave function ψ_4 is not well confined by the potential and so extends well into the barrier region.

2.8 Transmission and Reflection of Unbound States

A classical particle of mass m is transmitted if its energy is greater than a potential step and is reflected if its energy is less than the potential step. In quantum mechanics the situation is different. The wavy nature of an unbound state ensures that it feels the presence of changes in potential even if particle energy is greater than the potential step. Physically, this manifests itself as a scattering event with particle transmission and reflection probabilities. A simple situation to consider is transmission and reflection of unbound states (scattering states) from a potential step in one dimension.

A particle of energy $E > V_2$ and mass m is incident from the left on a step change in potential of energy $V_0 = V_2 - V_1$ at position x_0. Figure 2.8 is a sketch of the potential energy. Solutions of the time-independent Schrödinger equation $(\frac{-\hbar^2}{2m}\frac{d^2}{dx^2} + V(x))\psi(x) = E\psi(x)$ are of the form

$$\psi_1(x) = Ae^{ik_1x} + Be^{-ik_1x} \qquad \text{for } x < x_0, \tag{2.75}$$

where k_1 is the wave vector in region 1 and

$$\psi_2(x) = Ce^{ik_2x} + De^{-ik_2x} \qquad \text{for } x \geq x_0, \tag{2.76}$$

where k_2 is the wave vector in region 2. These equations describe left- and right-traveling waves in the two regions. Setting $V_1 = 0$, the energy of the particle is

$$E = \frac{\hbar^2 k_1^2}{2m}, \tag{2.77}$$

so that the wave vector in region 1 is

$$k_1 = \frac{\sqrt{2mE}}{\hbar}. \tag{2.78}$$

When $E < V_2$ for the potential step shown in Fig. 2.8, k_2 is imaginary. This results in $\psi_2(x \to \infty) = Ce^{-k_2x}|_{x\to\infty} = 0$ and $|A|^2 = |B|^2 = 1$, which has the physical meaning that the particle is completely reflected at the potential step (see Exercise 2.1).

In general, a particle of energy E moving in a one-dimensional potential with value V_j in the jth region has a wave vector

$$k_j = \frac{\sqrt{2m_j(E - V_j)}}{\hbar}. \tag{2.79}$$

For $E > V_j$, the wave vector k_j is real, and for $E < V_j$ the wave vector k_j is imaginary.

For electrons moving across a semiconductor heterostructure interface, there exists the possibility that the effective electron mass can change. Since, in some models, it is possible for the electron to have a different effective mass in each region, two situations should be considered, one where $m_1 = m_2$ and one where $m_1 \neq m_2$.

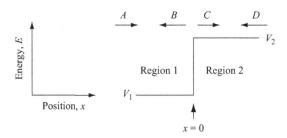

Fig. 2.8 Sketch of a one-dimensional potential energy step. In region 1 the potential energy is V_1, and in region 2 the potential energy is V_2 with $V_2 > V_1$. The coefficients A and C are for waves traveling left to right in regions 1 and 2, respectively. The coefficients B and D are for waves traveling right to left in regions 1 and 2, respectively. The transition between region 1 and region 2 occurs at position $x = 0$

2.8.1 Scattering from a Potential Step when $m_1 = m_2$

In this case, it is assumed that m_j has one value in all regions.

At the boundary between regions 1 and 2 at $x = 0$, the wave functions are linked by the constraint that the wave function ψ and the derivative $d\psi/dx$ are continuous, so that

$$\psi_1(0) = \psi_2(0), \tag{2.80}$$

$$\frac{d}{dx}\psi_1(0) = \frac{d}{dx}\psi_2(0). \tag{2.81}$$

These conditions and Eqs. (2.75) and (2.76) give, for $x = 0$,

$$A + B = C + D, \tag{2.82}$$

$$A - B = \frac{k_2}{k_1}C - \frac{k_2}{k_1}D. \tag{2.83}$$

Suppose the particle is incident from the left. This initial condition requires $|A|^2 = 1$. Also, $|D|^2 = 0$, since there is no left-traveling wave in region 2. Substituting $A = 1$ and $D = 0$ into Eqs. (2.82) and (2.83) results in

$$1 + B = C, \tag{2.84}$$

$$1 - B = \frac{k_2}{k_1}C. \tag{2.85}$$

Adding Eqs. (2.84) and (2.85) gives

$$2 = (1 + k_2/k_1)C, \tag{2.86}$$

$$C = \frac{2}{(1 + k_2/k_1)}, \tag{2.87}$$

and from Eq. (2.84),

$$B = C - 1 = \frac{2}{(1 + k_2/k_1)} - 1 = \frac{(2 - 1 - k_2/k_1)}{(1 + k_2/k_1)} = \frac{(1 - k_2/k_1)}{(1 + k_2/k_1)}. \tag{2.88}$$

Identifying electron velocity in the jth region as $v_j = \hbar k_j/m$, the transmission probability $|C|^2$ and reflection probability $|B|^2$ may be written as

$$|C|^2 = \frac{4}{(1 + k_2/k_1)^2} = \frac{4}{(1 + v_2/v_1)^2}, \tag{2.89}$$

$$|B|^2 = \frac{(1 - k_2/k_1)^2}{(1 + k_2/k_1)^2} = \left(\frac{v_1 - v_2}{v_1 + v_2}\right)^2. \tag{2.90}$$

Because $V_2 > V_1$, it follows that $k_2 < k_1$, so that $k_2/k_1 < 1$. For $E > V_2$ it can easily be seen that $|C|^2$ is always greater than 1 (hence larger than $|A|^2$), which means the probability amplitude for finding the particle anywhere in region 2 is greater than the probability of finding it anywhere in region 1. To understand this, consider the velocity of the particle in each region $v_j = \hbar k_j/m$, where $k_j = \sqrt{2m(E - V_j)}/\hbar$ so $v_1 > v_2$, because $V_2 > V_1$. Classically, when a particle moves faster it spends less time at any given position and the probability of finding it in each infinitesimal part of space dx is smaller. However,

73

care should be taken about this conclusion and not to confuse it with the concept of transmission and reflection that is meaningful only when considering probability current density. In fact, when dealing with traveling waves the probability current is a more useful concept than probability itself – as opposed to bound states, in which the opposite is true.

Before considering probability current density, it is helpful to find solutions for transmission and reflection probability in the situation in which $m_1 \neq m_2$.

2.8.2 Scattering from a Potential Step when $m_1 \neq m_2$

In this case, m_j varies from region to region. At the boundary between regions 1 and 2 at $x = 0$ continuity in the wave function ψ and the derivative $(1/m_j)\mathrm{d}\psi/\mathrm{d}x$ is required, so that

$$\psi_1(0) = \psi_2(0), \tag{2.91}$$

$$\frac{1}{m_1}\frac{\mathrm{d}}{\mathrm{d}x}\psi_1(0) = \frac{1}{m_2}\frac{\mathrm{d}}{\mathrm{d}x}\psi_2(0). \tag{2.92}$$

Inadequacies in the model forces a choice in boundary condition that ensures conservation of current $j_x \propto ep_x/m$ rather than continuity in the derivative of the wave function. (More accurate models satisfy both of these conditions.)

These conditions and Eqs. (2.75) and (2.76) give, for $x = 0$,

$$A + B = C + D, \tag{2.93}$$

$$A - B = \frac{m_1 k_2}{m_2 k_1}C - \frac{m_1 k_2}{m_2 k_1}D. \tag{2.94}$$

Since the particle is incident from the left, then $A = 1$ and $D = 0$, and

$$1 + B = C, \tag{2.95}$$

$$1 - B = \frac{m_1 k_2}{m_2 k_1}C. \tag{2.96}$$

Solving for the transmission probability $|C|^2$ and the reflection probability $|B|^2$ gives

$$|C|^2 = \frac{4}{\left(1 + \dfrac{m_1 k_2}{m_2 k_1}\right)^2}, \tag{2.97}$$

$$|B|^2 = \frac{\left(1 - \dfrac{m_1 k_2}{m_2 k_1}\right)^2}{\left(1 + \dfrac{m_1 k_2}{m_2 k_1}\right)^2}. \tag{2.98}$$

Compared with Eqs. (2.89) and (2.90), the ratio m_1/m_2 appearing in Eqs. (2.97) and (2.98) gives an extra degree of freedom in determining transmission and reflection probability.

2.8.3 Probability Current Density for Scattering at a Potential Step

Probability current density for transmission and reflection is different from transmission and reflection probability. Incident current $\mathbf{J}_I$, reflected current $\mathbf{J}_R$, and transmitted current $\mathbf{J}_T$, are shown schematically in Fig. 2.9.

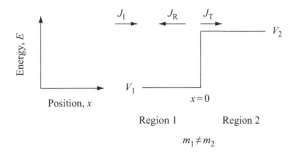

Fig. 2.9 Sketch of a one-dimensional, rectangular potential energy step. In region 1 the potential energy is V_1 and the particle mass is m_1. In region 2 the potential energy is V_2 and the particle mass is m_2. The transition between regions 1 and 2 occurs at position $x = 0$. Incident probability current density $\mathbf{J}_I$, reflected probability current density $\mathbf{J}_R$, and transmitted probability current density $\mathbf{J}_T$ are indicated.

For a particle incident from the left on a piecewise constant potential step, $|A|^2 = 1, |D|^2 = 0$, and $|C|^2$ and $|B|^2$ are given by Eqs. (2.97) and (2.98), respectively.

Current density is calculated using $\mathbf{J} = -i\frac{e\hbar}{2m}(\psi^*\nabla\psi - \psi\nabla\psi^*)$. The incident current is

$$\mathbf{J}_I = \frac{e\hbar k_1}{m_1}|A|^2. \tag{2.99}$$

The reflected current is

$$\mathbf{J}_R = -\frac{e\hbar k_1}{m_1}|B|^2 \tag{2.100}$$

and the transmitted current is

$$\mathbf{J}_T = \frac{e\hbar k_2}{m_2}|C|^2. \tag{2.101}$$

The reflection coefficient for the particle flux is

$$Refl = -\frac{\mathbf{J}_R}{\mathbf{J}_I} = -\left|\frac{B}{A}\right|^2 = -\left(\frac{1 - m_1 k_2/m_2 k_1}{1 + m_1 k_2/m_2 k_1}\right)^2, \tag{2.102}$$

where the minus sign indicates current flowing in the negative x direction. This is the same as the reflection probability given by Eq. (2.98), because the ratio of velocity terms that contribute to particle flux is unity. The transmission coefficient for the particle flux is

$$Trans = \frac{\mathbf{J}_T}{\mathbf{J}_I} = \frac{m_1 k_2}{m_2 k_1}\left|\frac{C}{A}\right|^2 = \frac{4k_1 k_2/m_1 m_2}{\left(\frac{k_1}{m_1} + \frac{k_2}{m_2}\right)^2} = 1 - Refl, \tag{2.103}$$

where $Trans + Refl = 1$ is required due to current conservation.

2.8.4 Impedance Matching Unbound States for Unity Transmission across a Potential Step

The flux transmission probability for a particle of energy $E > V_2$ approaching the potential step shown in Fig. 2.9 is given by Eq. (2.103). Since momentum $p = \hbar k = mv$, the velocity $v_j = \hbar k_j / m_j$ is the physically significant quantity in the expression for the transmission coefficient

$$Trans = \frac{m_1 k_2}{m_2 k_1} \left| \frac{C}{A} \right|^2 = \frac{m_1 k_2}{m_2 k_1} \frac{4}{(1 + v_2/v_1)^2} = \frac{v_2}{v_1} \frac{4}{(1 + v_2/v_1)^2}. \tag{2.104}$$

If $Trans = 1$, then this can be rewritten as

$$1 = \frac{v_2}{v_1} \frac{4}{(1 + v_2/v_1)^2}, \tag{2.105}$$

which shows that unity transmission occurs when the velocity of the particle in the two regions is matched[3] in such a way that $v_2/v_1 = 1$. In microwave transmission line theory, this is called an impedance matching condition. Since impedance matching occurs when $v_2/v_1 = 1$ then

$$\frac{v_2}{v_1} = \frac{m_1 k_2}{m_2 k_1} = \frac{m_1}{m_2} \left(\frac{2m_2 \hbar^2 (E - V_2)}{2m_1 \hbar^2 (E - V_1)} \right)^{1/2} = \left(\frac{m_1 (E - V_2)}{m_2 (E - V_1)} \right)^{1/2} = 1, \tag{2.106}$$

which may be rewritten as

$$\boxed{1 = \frac{m_1}{m_2} \frac{E - V_2}{E - V_1}} \tag{2.107}$$

or

$$\boxed{\frac{m_2}{m_1} = \frac{E - V_2}{E - V_1}.} \tag{2.108}$$

For an electron incident on the potential step with energy E, the value of E for which $Trans = 1$ depends upon the ratio of effective electron mass in the two regions and the difference in potential energy between the steps.

In the (nonrelativistic) limit when energy E becomes large,

$$v_2/v_1|_{E \to \infty} = \sqrt{\frac{m_1}{m_2}} \tag{2.109}$$

and

$$Trans|_{E \to \infty} = \lim_{E \to \infty} \frac{4(v_2/v_1)}{(1 + v_2/v_1)^2}. \tag{2.110}$$

If $m_1/m_2 \neq 1$ the model predicts $Trans < 1$ as $E \to \infty$.

---

[3] T. H. Chiu and A. F. J. Levi, *Appl. Phys. Lett.* **55**, 1891 (1989).

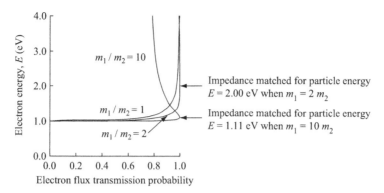

Fig. 2.10 Electron flux transmission probability as a function of energy for unbound electron-state motion across the potential energy step illustrated in Fig. 2.9, with $V_2 - V_1 = 1\,\mathrm{eV}$. In region 1 the electron has mass m_1, and in region 2 the electron has mass m_2.

The result of calculating electron flux transmission probability as a function of energy for electron motion across the 1 eV potential step is shown in Fig. 2.10 for the cases when $m_1 = m_2$, $m_1 = 2 \times m_2$, and $m_1 = 10 \times m_2$.

2.8.5 The Reflectionless sech^2 Potential

Impedance matching described in the previous section results in zero reflection and unity transmission at only one value of particle energy. However, nontrivial one-dimensional potentials exist in which a particle of mass m is impedance matched (and so has zero reflection and unity transmission) for *all* positive incident particle energies. The one-dimensional potential is

$$V(x) = -\frac{\hbar^2 k_0^2}{2m}\nu(\nu + 1)\,\mathrm{sech}^2\left(k_0 x\right), \tag{2.111}$$

where the parameter $1/k_0$ sets a length scale, ν is an integer, and $\mathrm{sech}^2 = 1/\cosh^2$. As illustrated in Fig. 2.11(a), this potential well is symmetric such that $V(x) = V(-x)$.

Figure 2.11(b) shows a contour plot of particle transmission as a function of ν and positive particle energy, E. Unity particle transmission occurs for all positive E only when ν is an integer. For integer ν and $\nu \neq 0, -1$, there is a critically bound state of infinite extent that has zero energy eigenvalue. Discrete bound-state energy eigenvalues (curves with $E < 0$ in Fig. 2.11(b)) exist in the potential well when particle energy is negative.

When $\nu = 0$ or $\nu = -1$ in Eq. (2.111), the potential is $V(x) = 0$ and there is, of course, perfect impedance matching resulting in reflectionless states for all positive incident particle energies.[4]

---

[4] In fact, *supersymmetry* connects the transmission properties of the sech^2 potentials that have different integer ν. For more on the properties of the sech^2 potential, see J. Lekner, *Am. J. Phys.* **75**, 1151 (2007), and for more on supersymmetry scattering states, see A. Abouzaid and A. F. J. Levi, *Europhy. Lett. EPL* **135**, 40002 (2021).

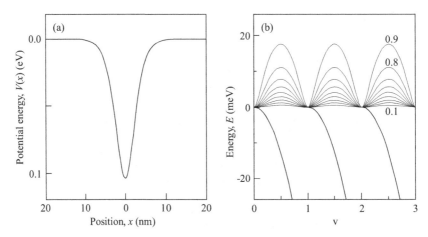

Fig. 2.11 (a) The $V(x) = -\frac{\hbar^2 k_0^2}{2m}\nu(\nu + 1)\,\mathrm{sech}^2(k_0 x)$ potential for parameters $\nu = 1$, $k_0 = 3.09 \times 10^8\,\mathrm{m}^{-1}$, and particle mass $m = 0.07 \times m_0$. (b) Contour plot of particle transmission in 0.1 steps as a function of ν and *positive* particle energy, E. Unity particle transmission occurs for all positive E only when ν is an integer. For integer ν and $\nu \neq 0, -1$, unity particle transmission occurs when there is a critically bound state of infinite extent that has zero energy eigenvalue. Discrete *bound-state* energy eigenvalues (curves with $E < 0$) exist in the potential well when particle energy is *negative*.

2.9 Impedance Matching Bound States across a Potential Step

Impedance matching a particle either side of a potential step changes the nature of the wave function and hence the probability density.

To illustrate impedance matching for *bound states*,[5] Fig. 2.12 is a sketch of probability density $|\psi|^2$ for three types of wave function in a step potential with $V = \infty$ outside regions 1 and 2. Wave function ψ_1 corresponds to a particle with energy $E < V_2$, and so it is near zero in region 2 apart from an exponentially decaying contribution in the potential barrier near $x = 0$. Wave function ψ_2 corresponds to a particle with energy $E > V_2$, which has finite transmission probability. In this case, the probability density in region 2 is greater than in region 1. Wave function ψ_3 corresponds to a particle of energy $E > V_2$ that is impedance matched across the potential step. In this situation, the amplitude of ψ_3 is the *same* on either side of the potential step.

2.10 Particle Tunneling

If electrons or other particles of mass m and kinetic energy E impinge on a rectangular potential barrier of energy V_0 and thickness L in such a way that $E < V_0$, they penetrate into the barrier. If the potential barrier is thin, there can be a significant chance that the particle is transmitted through the barrier. Such tunneling of a finite mass particle is *fundamentally quantum mechanical*.

[5] J. F. Müller, A. F. J. Levi, and S. Schmitt-Rink, *Phys. Rev. B* **38**, 9843 (1988).

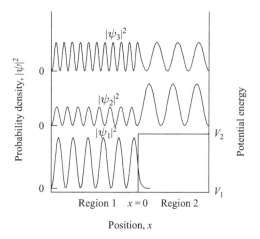

Fig. 2.12 Sketch of probability density $|\psi|^2$ for three types of *bound-state* wave function in a step potential with infinite potential energy, $V = \infty$, *outside* regions 1 and 2. Wave function ψ_1 corresponds to a particle with energy $E < V_2$, and so it is essentially zero in region 2 far from $x = 0$. Wave function ψ_2 corresponds to a particle with energy $E > V_2$. In this case, the probability density in region 2 is greater than in region 1. Wave function ψ_3 corresponds to a particle of energy $E > V_2$ that is impedance matched across the potential step. In this situation, the amplitude of ψ_3 is the same either side of the potential step.

Analogies are often made between tunneling in quantum mechanics and effects that occur in electromagnetism. However, evanescent field coupling in classical electromagnetism is, in fact, the large photon number limit (the thermodynamic limit) of quantum mechanical photon particle tunneling.

For electrons, the tunnel current density depends exponentially upon the barrier thickness, L, and barrier energy, V_0, and so it tends not to be important when $V_0 L$ is large.

For a particle of energy $E < V_0$, the solution for the wave function in regions 1 and 2 contains right- and left-propagating terms with the coefficients indicated in Fig. 2.13. In region 1, the wave function is of the form $\psi_1(x) = Ae^{ikx} + Be^{-ikx}$, and in region 2 the wave function is of the form $\psi_2(x) = Ce^{-\kappa x} + De^{\kappa x}$, where $k = \sqrt{2mE}/\hbar$ and $\kappa = \sqrt{2m(V_0 - E)}/\hbar$. The particle wave function has left- and right-propagating exponentially decaying solutions inside the barrier. This is required if tunnel current is

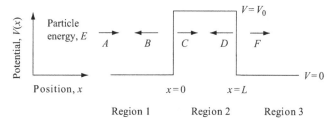

Fig. 2.13 Illustration of a particle of energy E and mass m incident on a one-dimensional, rectangular potential barrier of energy $V = V_0$ and thickness L. The incident wave has amplitude $A = 1$, and the transmitted wave has amplitude F. Wave reflection at positions $x = 0$ and $x = L$ can give resonances in transmission as a function of particle energy when $E > V_0$.

79

to flow through the barrier. Because a particle is incident from the left, in region 3 only a right-propagating wave need be considered. To find the general solution requires that the wave function and its derivative be continuous at $x = 0$ and $x = L$. This condition gives four equations:

$$A + B = C + D,$$ (2.112)

$$(A - B) = \frac{i\kappa}{k}(C - D),$$ (2.113)

$$Ce^{-\kappa L} + De^{\kappa L} = Fe^{ikL},$$ (2.114)

$$-Ce^{-\kappa L} + De^{\kappa L} = \frac{ik}{\kappa}Fe^{ikL}.$$ (2.115)

The tunneling transmission probability, $|F/A|^2$, is found by eliminating the other coefficients. Adding Eqs. (2.112) and (2.113) gives

$$2A = \left(1 + \frac{i\kappa}{k}\right)C + \left(1 - \frac{i\kappa}{k}\right)D.$$ (2.116)

Subtracting Eq. (2.115) from Eq. (2.114) gives

$$2Ce^{-\kappa L} = \left(1 - \frac{ik}{\kappa}\right)Fe^{ikL}.$$ (2.117)

Adding Eqs. (2.114) and (2.115) gives

$$2De^{\kappa L} = \left(1 + \frac{ik}{\kappa}\right)Fe^{ikL}.$$ (2.118)

Using Eqs. (2.117) and (2.118) to eliminate C and D from Eq. (2.116) leads to

$$2A = \left(1 + \frac{i\kappa}{k}\right)\left(1 - \frac{ik}{\kappa}\right)\frac{F}{2}e^{ikL+\kappa L} + \left(1 - \frac{i\kappa}{k}\right)\left(1 + \frac{ik}{\kappa}\right)\frac{F}{2}e^{ikL-\kappa L},$$ (2.119)

$$\frac{F}{A} = \frac{4k\kappa e^{-ikL}}{(k + i\kappa)(\kappa - ik)e^{\kappa L} + (k - i\kappa)(\kappa + ik)e^{-\kappa L}}$$

$$= \frac{4ik\kappa}{\left(-(\kappa - ik)^2 e^{\kappa L} + (\kappa + ik)^2 e^{-\kappa L}\right)}e^{-ikL}.$$ (2.120)

Dividing the top and bottom of Eq. (2.120) by κ^2 gives

$$\frac{F}{A} = \frac{4ik/\kappa}{\left(-(1 - ik/\kappa)^2 e^{\kappa L} + (1 + ik/\kappa)^2 e^{-\kappa L}\right)}e^{-ikL}.$$ (2.121)

The magnitude squared of Eq. (2.121) is

$$\frac{F^* F}{A^* A} = \frac{16(k/\kappa)^2}{(1 + ik/\kappa)^2((1 + ik/\kappa)^*)^2(e^{-2\kappa L} + e^{2\kappa L}) - ((1 + ik/\kappa)^*)^4 - (1 + ik/\kappa)^4}$$

$$\tag{2.122}$$

$$\left|\frac{F}{A}\right|^2 = \frac{16(k/\kappa)^2}{(1 - ik/\kappa)^2(1 + ik/\kappa)^2((e^{-\kappa L} - e^{\kappa L})^2 + 2) - (1 - ik/\kappa)^4 - (1 + ik/\kappa)^4}$$

$$= \frac{16(k/\kappa)^2}{(1 - ik/\kappa)^2(1 + ik/\kappa)^2(4\sinh^2(\kappa L) + 2) - (1 - ik/\kappa)^4 - (1 + ik/\kappa)^4}. \tag{2.123}$$

Dealing with the last two terms in the parentheses in the denominator, it is easy to check that

$$(1 - ik/\kappa)^4 + (1 + ik/\kappa)^4 = -16(k/\kappa)^2 + 2(1 + ik/\kappa)^2(1 - ik/\kappa)^2. \tag{2.124}$$

Substituting Eq. (2.124) into Eq. (2.123) gives

$$\left|\frac{F}{A}\right|^2 = \frac{(k/\kappa)^2}{\frac{1}{4}((1 + ik/\kappa)^2(1 - ik/\kappa)^2)\sinh^2(\kappa L) + (k/\kappa)^2}$$

$$= \frac{(k/\kappa)^2}{\frac{1}{4}(1 + (k/\kappa)^2)^2\sinh^2(\kappa L) + (k/\kappa)^2}. \tag{2.125}$$

So the tunneling transmission probability is

$$\boxed{\left|\frac{F}{A}\right|^2 = \frac{1}{1 + ((k^2 + \kappa^2)/2k\kappa)^2\sinh^2(\kappa L)}.} \tag{2.126}$$

An immediate goal is to plot the probability density distribution for an electron of energy E traveling from left to right and incident on a rectangular potential barrier of energy V_0 such that $E < V_0$. To obtain this probability distribution, C, D, and B need to be found. Rewriting Eqs. (2.117) and (2.118) gives

$$C = \left(1 - \frac{ik}{\kappa}\right)\frac{F}{2}e^{ikL}e^{\kappa L}, \tag{2.127}$$

$$D = \left(1 + \frac{ik}{\kappa}\right)\frac{F}{2}e^{ikL}e^{-\kappa L}. \tag{2.128}$$

Using Eqs. (2.112), (2.113), (2.127), and (2.128) an expression for B is

$$B = -i\frac{1}{2}Fe^{ikL} \times \frac{k^2 + \kappa^2}{k\kappa}\sinh(\kappa L). \tag{2.129}$$

The calculation of scattering-state wave functions is now complete. Unfortunately, because these scattering states are traveling waves, they cannot be normalized. To understand why, consider the expression for the absolute value squared of the wave function in regions 1, 2, and 3, which is given by the following three equations:

$$|\psi_1|^2 = \left(|A|^2 + |B|^2 + 2\mathrm{Re}(AB^*\mathrm{e}^{2ikx})\right),$$ (2.130)

$$|\psi_2|^2 = \left(|C|^2\mathrm{e}^{-2\kappa x} + |D|^2\mathrm{e}^{2\kappa x} + 2\mathrm{Re}(CD^*)\right),$$ (2.131)

$$|\psi_3|^2 = |F|^2.$$ (2.132)

Normalization of the total wave function ψ requires evaluation of the integral, as follows:

$$\int_{-\infty}^{\infty} |\psi|^2 \mathrm{d}x = \int_{-\infty}^{0} \left(|A|^2 + |B|^2 + 2\mathrm{Re}(AB^*\mathrm{e}^{2ikx})\right) \mathrm{d}x$$

$$+ \int_{0}^{L} \left(|C|^2\mathrm{e}^{-2\kappa x} + |D|^2\mathrm{e}^{2\kappa x} + 2\mathrm{Re}(CD^*)\right) \mathrm{d}x + \int_{L}^{\infty} |F|^2 \mathrm{d}x.$$ (2.133)

This integral is infinite, so a value for A that satisfies the normalization condition $\int |\psi|^2 \mathrm{d}x = 1$ cannot be found. However, *relative* probability $|\psi|^2/|A|^2$ can be found by keeping A as an unknown constant.

In Fig. 2.14, $A = 1$ for convenience and in (a) the real part of ψ and in (b) the relative probability $|\psi|^2$ is plotted as a function of position. The relative probability in region 3 is constant with position because the electron is described as a traveling wave moving from left to right.

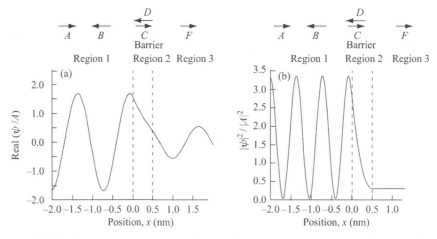

Fig. 2.14 (a) Real part of wave function ψ as a function of position for an electron of energy $E = 0.9\,\mathrm{eV}$ traveling from left to right and incident on a rectangular potential barrier of energy $V_0 = 1\,\mathrm{eV}$ and thickness $L = 0.5\,\mathrm{nm}$. (b) Relative probability for electron in (a).

The description of the electron in region 1 is a little more complicated. Rewriting Eq. (2.133),

$$|\psi_1|^2 = \left(|A|^2 + |B|^2 + 2\mathrm{Re}(AB^*e^{2ikx})\right) = 1 + |B|^2 + 2|B|\cos(2kx - \theta), \qquad (2.134)$$

where $A = 1$ and $B = |B|e^{i\theta}$, with θ being a phase determined by the boundary conditions. It is apparent from Eq. (2.134) that $|\psi_1|^2$ will oscillate with position and take on a value between $(1+|B|)^2$ and $(1-|B|)^2$, or 3.375 and 0.027 in this particular example. The reason for the oscillation in electron relative probability in region 1 is that the incoming right-traveling component of the wave function ψ_1 with amplitude A interferes with the left-traveling component of the wave function with amplitude B. The left-traveling component that causes electron *self-interference* is created by reflection from the potential barrier.

Region 2 has left- and right-propagating, exponentially decaying solutions of amplitude C and D, respectively. These two components *must* exist for current to flow through the potential barrier (see Exercise 2.1(c)).

2.10.1 Electron Tunneling Limit to Reduction in Size of CMOS Transistors

Tunneling of electrons is not an abstract theoretical concept of no practical significance. It can dominate performance of many components, and it is of importance for many electronic devices. For example, tunneling of electrons between source and drain and through a gate oxide barrier can contribute directly to leakage current in metal-oxide-semiconductor (MOS) field-effect transistors (FETs). Because of this, tunneling is a *fundamental limit* to reduction in size (*scaling*) of complimentary metal oxide semiconductor (CMOS) transistors. As gate length and minimum feature size of such a transistor is reduced (see Fig. 2.15), the gate oxide thickness, t_{ox}, should be reduced to maintain device performance. This is because in simple models transconductance is proportional to gate capacitance. For very small minimum feature sizes, the transistor can no longer function because t_{ox} is so small that electrons efficiently tunnel from the metal gate into the source-drain channel.

Silicon dioxide, SiO_2, which is used as an oxide for CMOS transistors, also has a minimum thickness due to the fact that the constituents of the oxide are discrete particles (atoms). Transistors with a minimum feature size of 50 nm have a gate oxide thickness of $t_{ox} = 1.3$ nm, which is so thin that on average it has only five Si-atom-containing layers. Transistors, with a minimum feature size of 25 nm, have a gate oxide that is only four Si-atom layers thick, corresponding to $t_{ox} = 0.7$ nm. Because there are two interfaces – one for the metal gate and one for the channel – they are not completely oxidized and the actual oxide may be thought of as only two Si-atom layers thick. This leaves a physically thin tunnel barrier for electrons. The absence of a single monolayer of oxide atoms over a significant area under the transistor gate can lead to undesirable leakage current due to electron tunneling between the metal gate and the source drain channel.

The fact that SiO_2 is made up of discrete atoms is a *structural limit*. The thickness of the oxide cannot be varied by, say, one quarter of a monolayer because the minimum layer thickness is, by definition, one atomic layer. The trivial conclusion is that, in addition to tunneling, there are *structural* limits to scaling conventional devices such as MOS transistors.[6]

---

[6] M. Schulz, *Nature* **399**, 729 (1999).

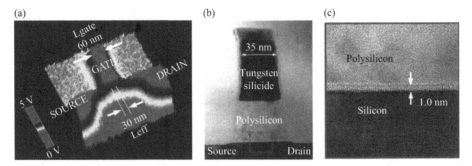

Fig. 2.15 (a) Scanning capacitance micrograph of the two-dimensional doping profile in a 60 nm gate length n-type MOSFET. The effective channel length is measured to be only $L_{\text{eff}} = 30$ nm. (b) Transmission electron microscope cross-section through a 35 nm gate length MOSFET. The channel length is only about 100 silicon lattice sites long. An enlargement of the channel region delineated is shown in (c). The gate oxide thickness estimated from the image is only about 1.0 nm. Images courtesy of G. Timp, University of Illinois.

To circumvent these difficulties, there is interest in developing a gate insulator for CMOS that would allow a thicker dielectric layer to be used in transistor fabrication. For constant gate capacitance, a way to achieve this is to increase the value of the dielectric's relative permittivity, ε_{r}. To see this, consider a parallel plate capacitor of area A_{area}. Capacitance is

$$C = \frac{\varepsilon_0 \varepsilon_{\text{r}} A_{\text{area}}}{t_{\text{ox}}}. \tag{2.135}$$

So, for constant C and area A_{area}, increasing thickness t_{ox} requires an increase in ε_{r}. Unfortunately, using known material properties, there are limits to how far this approach may be exploited. Nevertheless, the use of *high-k* (corresponding to a high value of ε_{r}) dielectrics is now an established technology.[7]

2.11 Example Exercises

Exercise 2.1
(a) Classically (i.e., from Maxwell's equations), change in charge density, ρ, is related to divergence of current density, **J**. Use this fact and the time-dependent Schrödinger wave equation for particles of mass m and charge e to derive the current density,

$$\mathbf{J} = -\frac{ie\hbar}{2m}(\psi^*\nabla\psi - \psi\nabla\psi^*) = \frac{e\hbar}{m}\,\text{Im}\,(\psi^*\nabla\psi).$$

(b) If a wave function can be expressed as $\psi(x,t) = Ae^{i(kx-\omega t)} + Be^{i(-kx-\omega t)}$, show that particle flux is proportional to $|A|^2 - |B|^2$.

---

[7] D. Buchanan, *IBM J. Res. Develop.* **43**, 245 (1999).

(c) For current to flow through a tunnel barrier, the wave function must contain *both* left- and right-propagating exponentially decaying solutions with a phase difference. If $\psi(x,t) = Ae^{\kappa x - i\omega t} + Be^{-\kappa x - i\omega t}$, show that particle flux is proportional to $\text{Im}(AB^*)$. Hence, show that if $\psi(x,t) = Be^{-\kappa x - i\omega t}$, then the particle flux is zero.

(d) Show that a particle of energy E and mass m moving from left to right in a one-dimensional potential in such a way that $V(x) = 0$ for $x < 0$ and $V(x) = V_0$ for $x \geq 0$ has unity reflection probability if $E < V_0$.

Exercise 2.2

(a) Find the values of the first three energy eigenvalues of a conduction band electron in GaAs with effective electron mass $m_e^* = 0.07 \times m_0$, assuming a rectangular, infinite, one-dimensional potential well of thickness (i) $L = 10\,\text{nm}$ and (ii) $L = 20\,\text{nm}$. Find an expression for the difference in energy levels, and compare it with room temperature thermal energy $k_B T$.

(b) For a quantum box in GaAs with no occupied electron states, estimate the size below which the conduction band quantum energy level spacing becomes larger than the classical coulomb blockade energy. Assume the confining potential may be approximated by a potential barrier of infinite energy for $|x| \geq L/2$, $|y| \geq L/2$, $|z| \geq L/2$, and zero energy elsewhere.

Exercise 2.3

Electrons in the conduction band of different semiconductors have different effective electron mass. In addition, the effective electron mass, m_j, is almost never the same as the free electron mass, m_0. These facts are usually accommodated in simple models of semiconductor heterojunction barrier transmission and reflection by requiring $\psi_1|_{0-\delta} = \psi_2|_{0+\delta}$ and $(1/m_1)\nabla\psi_1|_{0-\delta} = (1/m_2)\nabla\psi_2|_{0+\delta}$ at the heterojunction interface at position $x = 0$. Why are these boundary conditions used? Calculate the transmission and reflection coefficient for an electron of energy E, moving from left to right, impinging normally to the plane of a semiconductor heterojunction potential barrier of energy V_0. The effective electron mass on the left-hand side is m_1, and the effective electron mass on the right-hand side is m_2. Express the condition for no reflection in terms of electron velocities on either side of the interface.

Exercise 2.4

An asymmetric one-dimensional potential well of thickness L has an infinite potential energy barrier on the left-hand side and a finite constant potential of energy V_0 on the right-hand side. Find the minimum value of L for which an electron has at least one bound state when $V_0 = 1\,\text{eV}$.

Exercise 2.5

(a) Consider a one-dimensional potential well approximated by a delta function in space so that $V(x) = -b\delta(x = 0)$. Show that there is one bound state for a particle of mass m, and find its energy and eigenstate.

(b) Show that any one-dimensional delta function potential with $V(x) = \pm b\delta(x = 0)$ always introduces a kink in the wave function describing a particle of mass m.

Exercise 2.6

Using the method outlined in Section 2.4, write a computer program to solve the Schrödinger wave equation for the first four eigenvalues and eigenstates of an electron with effective mass $m_e^* = 0.07 \times m_0$ confined to a rectangular potential well of thickness $L = 10\,\text{nm}$ bounded by infinite barrier potential energy.

Solutions

Solutions 2.1

(a) The change in charge density ρ is related to the divergence of current density $\mathbf{J}$ through

$$\frac{\partial}{\partial t}\rho(\mathbf{r}, t) = -\nabla \cdot \mathbf{J}(\mathbf{r}, t).$$

In quantum mechanics, it is assumed that the charge density of a particle with charge e and wave function ψ can be written $\rho = e|\psi|^2$, so that

$$\frac{\partial \rho}{\partial t} = e\frac{\partial}{\partial t}(\psi^*\psi) = e\left(\psi^*\frac{\partial \psi}{\partial t} + \psi\frac{\partial \psi^*}{\partial t}\right).$$

We make use of the fact that the time-dependent Schrödinger equation for a particle of mass m moving in a real potential is

$$i\hbar\frac{\partial}{\partial t}\psi(\mathbf{r}, t) = \left(-\frac{\hbar^2}{2m}\nabla^2 + V(\mathbf{r})\right)\psi(\mathbf{r}, t) = \hat{H}\psi(\mathbf{r}, t).$$

Multiplying both sides by $\psi^*(\mathbf{r}, t)$ gives

$$i\hbar\psi^*(\mathbf{r}, t)\frac{\partial}{\partial t}\psi(\mathbf{r}, t) = -\frac{\hbar^2}{2m}\psi^*(\mathbf{r}, t)\nabla^2\psi(\mathbf{r}, t) + \psi^*(\mathbf{r}, t)V(\mathbf{r})\psi(\mathbf{r}, t),$$

which, when multiplied by $e/i\hbar$, is the first term on the right-hand side in our expression for $\partial\rho/\partial t$. Taking the complex conjugate of the time-dependent Schrödinger equation and multiplying both sides by $\psi(\mathbf{r}, t)$ gives

$$\frac{\partial \rho}{\partial t} = -\psi^*\frac{e\hbar}{i2m}\nabla^2\psi + \psi\frac{e\hbar}{i2m}\nabla^2\psi^* + \frac{e\psi^*\psi}{i\hbar}(V(\mathbf{r}) - V(\mathbf{r})) = \frac{ie\hbar}{2m}\left(\psi^*\nabla^2\psi - \psi\nabla^2\psi^*\right)$$

$$= \frac{ie\hbar}{2m}\nabla \cdot (\psi^*\nabla\psi - \psi\nabla\psi^*) = -\nabla \cdot \mathbf{J}.$$

Because $V(\mathbf{r})$ is taken to be real, $\psi^*(\mathbf{r}, t)V(\mathbf{r})\psi(\mathbf{r}, t) = \psi(\mathbf{r}, t)V(\mathbf{r})\psi^*(\mathbf{r}, t)$. It follows that an expression for the current density for a particle charge e described by state ψ is

$$\mathbf{J} = -\frac{ie\hbar}{2m}(\psi^*\nabla\psi - \psi\nabla\psi^*).$$

(b) Consider a particle of mass m and charge e that is described by a wave function $\psi(x, t) = Ae^{i(kx - \omega t)} + Be^{i(-kx - \omega t)}$, where k is real. The aim is to show that particle flux is proportional to $|A|^2 - |B|^2$.

Starting with the current density in one dimension and substituting in the expression for the wave function gives

$$J_x = -\frac{ie\hbar}{2m}\left(\psi^*\frac{\mathrm{d}}{\mathrm{d}x}\psi - \psi\frac{\mathrm{d}\psi^*}{\mathrm{d}x}\right),$$

$$\psi^*\frac{\mathrm{d}\psi}{\mathrm{d}x} = (A^*e^{-ikx} + B^*e^{ikx})(ikAe^{ikx} - ikBe^{-ikx})$$

$$= ik|A|^2 - ikA^*Be^{-2ikx} + ikB^*Ae^{2ikx} - ik|B|^2,$$

and

$$\psi\frac{\mathrm{d}\psi^*}{\mathrm{d}x} = (Ae^{ikx} + Be^{-ikx})(-ikA^*e^{-ikx} + ikB^*e^{ikx})$$

$$= -ik|A|^2 - ikBA^*e^{-2ikx} + ikAB^*e^{2ikx} + ik|B|^2,$$

so that, finally,

$$J_x = -\frac{ie\hbar}{2m}\left(ik2\left(|A|^2 - |B|^2\right)\right) = \frac{e\hbar k}{m}\left(|A|^2 - |B|^2\right).$$

This is not an unexpected result, since the wave function $\psi(x,t)$ is made up of two traveling waves. The first traveling wave Ae^{ikx} might consist of an electron probability density $\rho = |A|^2$ moving from left to right at velocity $v_+ = \hbar k/m$ carrying current density $j_+ = ev_+\rho = e\hbar k|A|^2/m$. The second traveling wave Be^{-ikx} consists of an electron probability density $\rho = |B|^2$ moving from right to left at velocity $v_- = -\hbar k/m$ carrying current density $j_- = ev_-\rho = -e\hbar k|B|^2/m$. The net current density is just the sum of the currents $J_x = j_+ + j_- = e\hbar k\left(|A|^2 - |B|^2\right)/m$.

(c) A particle of mass m and charge e is described by the wave function $\psi(x,t) = Ae^{\kappa x - i\omega t} + Be^{-\kappa x - i\omega t}$, where κ is real. As in part (b), the expression for current density in one dimension gives

$$J_x = -\frac{ie\hbar}{2m}\left(\psi^*\frac{\mathrm{d}}{\mathrm{d}x}\psi - \psi\frac{\mathrm{d}\psi^*}{\mathrm{d}x}\right),$$

$$\psi^*\frac{\mathrm{d}\psi}{\mathrm{d}x} = (A^*e^{\kappa x} + B^*e^{-\kappa x})(A\kappa e^{\kappa x} - B\kappa e^{-\kappa x})$$

$$= \kappa|A|^2e^{2\kappa x} + \kappa B^*A - \kappa A^*B - \kappa|B|^2e^{-2\kappa x},$$

and

$$\psi\frac{\mathrm{d}\psi^*}{\mathrm{d}x} = (Ae^{\kappa x} + Be^{-\kappa x})(A^*\kappa e^{\kappa x} - B^*\kappa e^{-\kappa x})$$

$$= \kappa|A|^2e^{2\kappa x} - \kappa AB^* + \kappa BA^* - \kappa|B|^2e^{-2\kappa x}.$$

so that

$$J_x = \frac{ie\hbar}{2m}2\kappa(A^*B - B^*A) = \frac{2e\hbar\kappa}{m}(\mathrm{Im}(AB^*)).$$

This last equality may be shown by explicitly writing the real and imaginary parts of the terms in the brackets:

$$(A^*B - B^*A) = (A_{Re} - iA_{Im})(B_{Re} + iB_{Im}) - (B_{Re} - iB_{Im})(A_{Re} + iA_{Im}).$$

Multiplying out the terms on the right-hand side gives

$$(A^*B - B^*A) = A_{Re}B_{Re} + iA_{Re}B_{Im} - iA_{Im}B_{Re} + A_{Im}B_{Im}$$
$$- A_{Re}B_{Re} - iB_{Re}A_{Im} + iB_{Im}A_{Re} - A_{Im}B_{Im}.$$

This simplifies to

$$(A^*B - B^*A) = 2iA_{Re}B_{Im} - 2iA_{Im}B_{Re} = 2i(A_{Re}B_{Im} - A_{Im}B_{Re}).$$

Taking the imaginary part of the terms on the right-hand side allows addition of the real terms $A_{Re}B_{Re}$ and $A_{Im}B_{Im}$, so that the expression may be factored,

$$(A^*B - B^*A) = -2i(Im(A_{Re}B_{Re} - iA_{Re}B_{Im} + iA_{Im}B_{Re} + A_{Im}B_{Im}))$$
$$= -2i(Im((A_{Re} + iA_{Im})(B_{Re} - iB_{Im}))) = -2i(Im(AB^*)),$$
$$J_x = \frac{ie\hbar}{2m}2\kappa(A^*B - B^*A) = \frac{2e\hbar\kappa}{m}(Im(AB^*)).$$

The significance of this result is that current can only flow through a one-dimensional tunnel barrier in the presence of *both* exponentially decaying terms in the wave function $\psi(x) = Ae^{\kappa x} + Be^{-\kappa x}$. In addition, the complex coefficients A and B *must differ in phase*. The term $Ae^{\kappa x}$ can only carry current if the term $Be^{-\kappa x}$ exists. This second term indicates that the tunnel barrier is not opaque and that transmission of current via tunneling is possible.

Suppose a particle of mass m and charge e is described by wave function $\psi(x,t) = Be^{-\kappa x - i\omega t}$. Substitution into the expression for the one-dimensional current density gives

$$J_x = -\frac{ie\hbar}{2m}\left(\psi^*\frac{d}{dx}\psi - \psi\frac{d\psi^*}{dx}\right) = -\frac{ie\hbar}{2m}(B^*e^{-\kappa x}(-B\kappa e^{-\kappa x}) - Be^{-\kappa x}(-B^*\kappa e^{-\kappa x}))$$
$$= -\frac{ie\hbar}{2m}(-B^*B\kappa e^{-2\kappa x} + BB^*\kappa e^{-2\kappa x}) = 0.$$

(d) To show that a particle of energy E and mass m moving in a one-dimensional potential in such a way that $V(x) = 0$ for $x < 0$ in region 1 and $V(x) = V_0$ for $x \geq 0$ in region 2 has unity reflection probability if $E < V_0$, the solutions of the time-independent Schrödinger equation $\left(\frac{-\hbar^2}{2m}\frac{d^2}{dx^2} + V(x)\right)\psi(x) = E\psi(x)$ are written down. The solutions that describe left- and right-traveling waves in the two regions are

$$\psi_1(x) = Ae^{ik_1 x} + Be^{-ik_1 x}, \qquad \text{for } x < 0,$$

and

$$\psi_2(x) = Ce^{ik_2 x} + De^{-ik_2 x}, \qquad \text{for } x \geq 0.$$

When $E < V_0$, the value of k_2 is imaginary, and so $k_2 \rightarrow ik_2$. The value of D is zero to avoid a wave function with infinite value as $x \rightarrow \infty$. The value of C is finite, but the contribution to the wave function $\psi_2(x \to \infty) = Ce^{-k_2 x}|_{x \to \infty} = 0$. Hence, it may be concluded that the transmission probability in the limit $x \to \infty$ must be zero, since $|\psi_2(x \to \infty)|^2 = 0$. From this it follows that $|A|^2 = |B|^2 = 1$, which has the physical meaning that the particle is completely reflected at the potential step when $E < V_0$.

Alternatively, the reflection probability could be calculated from the square of the amplitude coefficient for a particle of energy E and mass m incident on a one-dimension potential step

$$B = \left(\frac{1 - k_2/k_1}{1 + k_2/k_1}\right),$$

When particle energy $E < V_0$, the value of k_2 is imaginary, so that $k_2 \to i k_2$, which gives

$$|B|^2 = B^* B = \left(\frac{1 + i k_2/k_1}{1 - i k_2/k_1}\right)\left(\frac{1 - i k_2/k_1}{1 + i k_2/k_1}\right) = 1.$$

From this it may be concluded that the particle has a unity reflection probability.

Solutions 2.2

(a) Differences in energy eigenvalues for an electron in the conduction band of GaAs that has an effective electron mass $m_e^* = 0.07 \times m_0$ and is confined by a rectangular infinite one-dimensional potential well of thickness (i) $L = 10$ nm or (ii) $L = 20$ nm are to be found.

The expression for energy eigenvalues of an electron confined in a one-dimensional, infinite, rectangular potential of thickness L is:

$$E_n = \frac{\hbar^2}{2m_e^*}\frac{n^2 \pi^2}{L^2}.$$

For $m_e^* = 0.07 \times m_0$, this gives

$$E_n = \frac{n^2}{L^2} \times 8.607 \times 10^{-37} \text{ J} = \frac{n^2}{L^2} \times 5.37 \times 10^{-18} \text{ eV},$$

so that the first three energy eigenvalues for $L = 10$ nm are $E_1 = 54$ meV, $E_2 = 215$ meV, and $E_3 = 484$ meV, and the differences in energy between adjacent energy eigenvalues are $\Delta E_{12} = 161$ meV and $\Delta E_{23} = 269$ meV.

Increasing the value of the potential well thickness by a factor of two to $L = 20$ nm reduces the energy eigenvalues by a factor of four. The energy eigenvalues scale as $1/L^2$, and so they are quite sensitive to the thickness of the potential well. For practical systems, ΔE_{nm} should be large, typically $\Delta E_{nm} \gg k_B T$. However, although E_n scales as n^2/L^2, the *difference* in energy levels *scales linearly* with increasing eigenvalue, n. For adjacent energy levels, the difference in energy increases linearly with increasing eigenvalue as

$$\Delta E_{n+1,n} = \frac{\hbar^2}{2m_e^*}\frac{\pi^2}{L^2}\left((n+1)^2 - n^2\right) = \frac{\hbar^2}{2m_e^*}\frac{\pi^2}{L^2}(2n+1).$$

At room temperature $k_B T = 25$ meV. Thus for the case $n = 1$ the condition is

$$L^2 \ll \frac{\hbar^2}{2m_e^*}\frac{\pi^2}{k_B T}(2n+1) = \frac{3\hbar^2}{2m_e^*}\frac{\pi^2}{k_B T} = \frac{3}{0.025} \times 5.37 \times 10^{-18} = 6.44 \times 10^{-16} \text{ m}^2.$$

So, for the situation in which $\Delta E_{nm} \gg k_B T$ then $L \ll 25$ nm.

(b) Using the results of (a), the size of a quantum box in GaAs with no occupied electron states at which there is a crossover from classical coulomb blockade energy to quantum energy eigenvalues dominating electron energy levels can be calculated.

For a one-dimensional, rectangular potential well of thickness L and infinite barrier energy the eigenenergy levels are given by

$$E_{n_x} = \frac{\hbar^2}{2m_e^*} \frac{n_x^2 \pi^2}{L^2}$$

and the differences in energy levels are

$$\Delta E_{n_x+1, n_x} = \frac{\hbar^2}{2m_e^*} \frac{\pi^2}{L^2} (2n_x + 1).$$

The eigenenergy levels of a quantum box are

$$E_n = \frac{\hbar^2}{2m_e^*} \frac{(n_x^2 + n_y^2 + n_z^2)\pi^2}{L^2},$$

so that the difference in energy is expected to scale as $1/L^2$. Because the classical coulomb blockade energy $\Delta E = e^2/2C$ scales as $1/L$ with decreasing size, there must be a crossover to quantum energy eigenvalues dominating electron dynamics.

For a quantum box in GaAs with no occupied electron states, a simple estimate for the crossover size is determined by $E_{n_x=1, n_y=1, n_z=1} = e^2/2C$. Approximating the capacitance as $C = 4\pi\varepsilon_0\varepsilon_r(L/2)$ gives

$$L = \frac{3\pi^2\hbar^2}{2m_e^*} \frac{4\pi\varepsilon_0\varepsilon_r}{e^2}.$$

For GaAs, using $m_e^* = 0.07 \times m_0$ and $\varepsilon_r = 13.1$, a characteristic thickness value for the crossover of $L = 147$ nm and a ground state energy $E_1 = 6.75$ meV is obtained. Clearly, for quantum dots in GaAs characterized by a size $L \ll 150$ nm, quantization in energy states will dominate. Care should also be taken to include the physics behind the coulomb blockade by calculating a *quantum capacitance* using electron wave functions and a quantum coulomb blockade by self-consistently solving for the wave functions and Maxwell's equations.

Solutions 2.3

To accommodate the fact that electrons in the conduction band of different semiconductors labeled 1 and 2 have differing effective electron mass, simple models of semiconductor heterojunction barrier transmission and reflection require that the electron wave function at the boundary between regions 1 and 2 at position $x = 0$ satisfy

$$\psi_1(0) = \psi_2(0)$$

and

$$\frac{1}{m_1} \frac{d}{dx} \psi_1(0) = \frac{1}{m_2} \frac{d}{dx} \psi_2(0).$$

This leads to

$$A + B = C + D.$$

and

$$A - B = \frac{m_1 k_2}{m_2 k_1} C - \frac{m_1 k_2}{m_2 k_1} D.$$

If the particle is incident from the left, then $A = 1$ and $D = 0$, giving

$$1 + B = C,$$

$$1 - B = \frac{m_1 k_2}{m_2 k_1} C.$$

Recognizing velocity $v_1 = \hbar k_1/m_1$ and $v_2 = \hbar k_2/m_2$ and solving for the transmission probability $|C|^2$ and the reflection probability $|B|^2$ gives

$$|C|^2 = \frac{4}{\left(1 + \dfrac{m_1 k_2}{m_2 k_1}\right)^2} = \frac{4}{\left(1 + \dfrac{v_2}{v_1}\right)^2},$$

$$|B|^2 = \frac{\left(1 - \dfrac{m_1 k_2}{m_2 k_1}\right)^2}{\left(1 + \dfrac{m_1 k_2}{m_2 k_1}\right)^2} = \frac{\left(1 - \dfrac{v_2}{v_1}\right)^2}{\left(1 + \dfrac{v_2}{v_1}\right)^2}.$$

From this it is clear that there is no reflection when the particle velocity is the same in each of the two regions. This is an example of an impedance-matching condition that, in this case, will occur when particle energy $E = V_0/(1 - m_2/m_1)$.

It is worth making a few comments about the approach to solving this exercise. The boundary conditions were selected to guarantee current conservation and no discontinuity in the electron wave function. However, with this choice it is not possible to avoid the possibility of kinks in the wave function. Such an approach was first discussed by Bastard.[8] Unfortunately, the Schrödinger equation cannot be solved correctly using different effective electron masses for electrons on either side of an interface. A more complex, correct way to proceed is to use the atomic potentials of atoms forming the heterointerface and solve using the bare electron mass and the usual boundary conditions,

$$\psi_1(0) = \psi_2(0)$$

and

$$\frac{d\psi_1(0)}{dx} = \frac{d\psi_2(0)}{dx}.$$

A simplified version of this situation might use the Kronig–Penney potential, which will be discussed in Chapter 3.

There are a number of additional reasons why the model used should only be considered approximate. For example, the electron can be subject to other types of scattering. Such

[8] G. Bastard, *Phys. Rev. B* **24**, 5693 (1981).

scattering may take the form of diffuse electron scattering due to interface roughness or a large phonon emission probability. The presence of nonradiative electron recombination due to impurities and traps, or, typically in direct band gap semiconductors, radiative processes may be important. These effects may result in a non-unity sum of transmission and reflection coefficients $(R + T \neq 1)$.

Situations often arise when an effective electron mass cannot be used to describe electron motion. The electron dispersion relation may be nonparabolic over the energy range of interest. In addition, when the character (or s, p, d, etc., symmetry) of an electron wave function is very different on either side of a potential step, the simple approach used above is inappropriate and can lead to incorrect results.

Solutions 2.4

The upper part of the Fig. 2.E16 sketches four bound-state wave functions for a particle of mass m in a symmetric potential for which $V_0 L^2 = 25\hbar^2/2m$, so that $LK_0 = L\sqrt{2mV_0}/\hbar = 5$. As expected, the wave functions have alternating even and odd parity, with even parity for the lowest energy state, ψ_1.

The lower part the Fig. 2.E16 shows the asymmetric potential that arises when the potential for $x < 0$ is infinite. Because the wave function must be zero for $x < 0$, the only

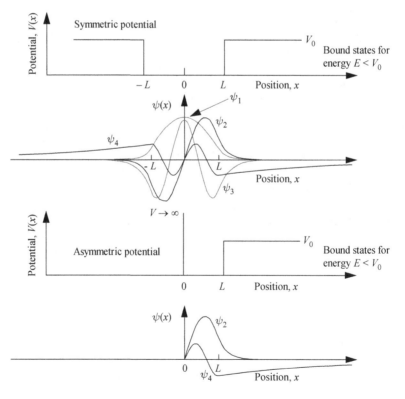

Fig. 2.E16

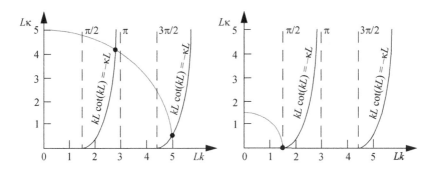

Fig. 2.E17

allowed solutions for bound states are of the form ψ_2 and ψ_4 previously obtained for the symmetric case and $x > 0$. These correspond to the odd-parity wave functions that are found using the graphical method sketched in Fig. 2.E17 when $V_0 L^2 = 25\hbar^2/2m$, so that $LK_0 = L\sqrt{2mV_0}/\hbar = 5$.

The lowest energy solution exists where the radius of the arc $LK_0 = L\sqrt{2mV_0}/\hbar$ intersects the function $kL\cot(kL) = -\kappa L$. Vanishing binding energy will occur when the particle is in a potential so that $LK_0 \rightarrow \pi/2$ and $\kappa \ll k$ (Fig. 2.E17). Since $LK_0 = L\sqrt{2mV_0}/\hbar$, the limit of vanishing binding energy occurs at $L\sqrt{2mV_0}/\hbar = \pi/2$, so that

$$L \leq \pi\hbar/2\sqrt{2mV_0}.$$

To find the minimum value of L, the values of V_0 and m must be known. For the case in which the step potential energy is $V_0 = 1\,\text{eV}$, the minimum value of L for an electron of mass $m = m_0$ to ensure that at least one bound state exists is $L_{\min} = \pi\hbar/2(2m_0V_0)^{1/2} \sim 0.3\,\text{nm}$.

Solutions 2.5
(a) The eigenfunction and eigenvalue of a particle in a delta-function potential well, $V(x) = -b\delta(x = 0)$, where b is a measure of the strength of the potential, is sought. Due to the symmetry of the potential the wave function is expected to be an even function. The delta function will introduce a kink in the wave function. Schrödinger's equation for a particle subject to a delta-function potential is

$$\left(\frac{-\hbar^2}{2m}\frac{d^2}{dx^2} - b\delta(x = 0) \right)\psi(x) = E\psi(x).$$

Integrating between $x = 0-$ and $x = 0+$,

$$\int\limits_{x=0-}^{x=0+} \left(\frac{d^2}{dx^2}\psi(x) \right) dx = \int\limits_{x=0-}^{x=0+} \left(\frac{-2m}{\hbar^2}(E + b\delta(x = 0))\psi(x) \right) dx,$$

results in

$$\psi'(0+) - \psi'(0-) = \frac{-2mb}{\hbar^2}\psi(0),$$

where $d\psi/dx = \psi'$. For an even function the spatial derivative is odd, so that $\psi'(0+) = -\psi'(0-)$, and

$$\psi'(0+) - \psi'(0-) = 2\psi'(0+) = \frac{-2mb}{\hbar^2}\psi(0).$$

This is the boundary condition.

The Schrödinger equation is $(-\hbar^2/2m)d^2\psi(x)/dx^2 = E\psi(x)$ for $x > 0$ and $x < 0$. For a bound state, $\psi(x) \to 0$ as $x \to \infty$ and $\psi(x) \to 0$ as $x \to -\infty$ is required so the only possibility is

$$\psi(x) = Ae^{-\kappa|x|},$$

where $\kappa = \sqrt{2m|E|/\hbar^2}$. Substituting this wave function into Schrödinger's equation gives $-\hbar^2 k^2/2m = E$, but from previous work

$$2\psi'(0+) = -2A\kappa e^{-\kappa x}|_{x=0} = \frac{-2mb}{\hbar^2}\psi(0) = \frac{-2mb}{\hbar^2}Ae^{-\kappa x}|_{x=0},$$

so that $\kappa = mb/\hbar^2$. Hence,

$$E = \frac{-mb^2}{2\hbar^2}$$

is the energy of the bound state.

To find the complete expression for the wave function, A needs to be evaluated. Normalization of the wave function requires

$$\int_{x=-\infty}^{x=\infty} |\psi(x)|^2 dx = 1,$$

and this gives $A = \sqrt{\kappa}$ so that $\psi(x) = \sqrt{mb/\hbar^2}e^{-\frac{mb}{\hbar^2}|x|}$.

The spatial decay (the e^{-1} distance) for this wave function is just $\Delta x = 1/\kappa = \hbar^2/mb$. Physically, it is reasonable to use the delta function potential approximation when the spatial extent of the potential is much smaller than the spatial decay of the wave function. This might occur, for example, in the description of certain single-atom defects in a crystal.

(b) It follows from (a) that any delta function in the potential of the form $V(x) = \pm b\delta(x = 0)$ introduces a kink in the wave function. To show this, Schrödinger's equation is integrated for a particle subject to a delta function potential,

$$\left(\frac{-\hbar^2}{2m}\frac{d^2}{dx^2} \pm b\delta(x = 0)\right)\psi(x) = E\psi(x),$$

between $x = 0-$ and $x = 0+$ in such a way that

$$\int\limits_{x=0-}^{x=0+} \left(\frac{d^2}{dx^2} \psi(x) \right) dx = \int\limits_{x=0-}^{x=0+} \left(\frac{-2m}{\hbar^2} (E \mp b\delta(x))\psi(x) \right) dx,$$

$$\frac{d}{dx}\psi(0+) - \frac{d}{dx}\psi(0-) = \frac{\pm 2mb}{\hbar^2}\psi(0).$$

So, for finite $\psi(0)$ and b there is always a difference in the slope of the wave functions at $\psi(0+)$ and $\psi(0-)$, and hence a kink in the wave function about $x = 0$. A delta function potential introduces a kink in the wave function.

Solutions 2.6

The method outlined in Section 2.4 may be used to numerically solve the discretized one-dimensional Schrödinger wave equation for an electron of effective mass $m_e^* = 0.07 \times m_0$ in a rectangular potential well of thickness $L = 10\,\text{nm}$ bounded by infinite barrier energy.

The solution is a MATLAB computer program which consists of two parts. The first part deals with input parameters such as the length L, effective electron mass, the number of discretization points, N, and the plotting routine. It is important to choose a high enough value of N so that the wave function does not vary too much between adjacent discretization points and so that the three-point finite-difference approximation used in Eq. (2.33) is reasonably accurate.

The second part of the computer program solves the discretized Hamiltonian matrix (Eq. (2.37)) and is a function, solve_schM, called from the main routine, Chapt2Exercise6. The diagonal matrix element given by Eq. (2.35) has potential values $V_j = 0$. The adjacent off-diagonal matrix elements are given by Eq. (2.36).

In this particular exercise, the first four energy eigenvalues are $E_0 = 0.0537\,\text{eV}$, $E_1 = 0.2149\,\text{eV}$, $E_2 = 0.4834\,\text{eV}$, and $E_3 = 0.8592\,\text{eV}$.

2.12 Problems

Problem 2.1

(a) Using the method outlined in Exercise 2.6 as a starting point, calculate numerically the first five energy eigenvalues and eigenfunctions for an electron with effective mass $m_e^* = 0.07 \times m_0$ confined to the asymmetric potential well sketched in the Fig. 2.P18 and bounded by barriers of infinite energy at $x \leq 0\,\text{nm}$ and $x \geq 50\,\text{nm}$. The value of the step change in potential energy in the figure is $V_{step} = 0.2\,\text{eV}$.

The solution should include plots of the eigenfunctions $\psi_n(x)$ and $|\psi_n(x)|^2$ along with a listing of the computer program used to calculate the eigenfunctions and eigenvalues.

(b) Explain the change in shape of each wave function with increasing eigenenergy.

(c) If an eigenenergy coincides with the value of V_{step}, what is the shape of the eigenfunction?

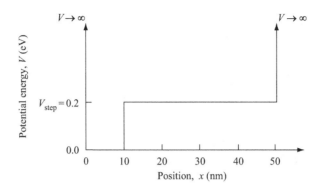

Fig. 2.P18

Problem 2.2

(a) Use a Taylor expansion to show that the second derivative of a wave function $\psi(x)$ sampled at positions $x_j = jh_0$, where j is an integer and h_0 is a small fixed increment in position x, may be approximated as

$$\frac{d^2}{dx^2}\psi(x_j) = \frac{\psi(x_{j-1}) - 2\psi(x_j) + \psi(x_{j+1})}{h_0^2}.$$

(b) By keeping additional terms to order h_0^4 in the expansion, show that a more accurate approximation of the second derivative is

$$\frac{d^2}{dx^2}\psi(x_j) = \frac{-\psi(x_{j-2}) + 16\psi(x_{j-1}) - 30\psi(x_j) + 16\psi(x_{j+1}) - \psi(x_{j+2})}{12h_0^2}.$$

Problem 2.3

(a) Using the method outlined in Exercise 2.6 as a starting point, calculate numerically the first four eigenfunctions and corresponding energy eigenvalues for an electron with effective mass $m_e^* = 0.07 \times m_0$ confined to a potential $V(x) = 0$ of thickness $L = 10\,\text{nm}$ with *periodic* boundary conditions. Periodic boundary conditions require that the wave function at position $x = 0$ is connected (wrapped around) to position $x = L$. The eigenfunction and its first derivative are continuous and smooth at this connection.

The solution should include plots of the eigenfunctions and a listing of the computer program used to calculate the eigenfunctions and eigenvalues.

(b) Explain the change in shape of each eigenfunction with increasing eigenenergy.

Problem 2.4

(a) Using the method outlined in Exercise 2.6 as a starting point, calculate numerically the first three energy eigenvalues and eigenfunctions for an electron (not including spin) with effective mass $m_e^* = 0.07 \times m_0$ confined to a triangular potential well of thickness $L = 10\,\text{nm}$ bounded by barriers of infinite energy at $x \leq 0$ and $x \geq L$. The triangular potential well as a function of position x is given by $V(x) = V_0 \times (x - L/2)/L$ where $V_0 = 1\,\text{eV}$.

(b) Plot the first three energy eigenvalues as a function of $0 \leq V_0 \leq 1\,\text{eV}$.

Problem 2.5

Calculate the transmission and reflection flux coefficient for an electron of energy E, moving from left to right, impinging normal to the plane of a semiconductor heterojunction potential barrier of energy V_0, where the effective electron mass on the left-hand side is m_1 and the effective electron mass on the right-hand side is m_2.

If the potential barrier energy is $V_0 = 1.5\,\text{eV}$ and the ratio of effective electron mass on either side of the heterointerface is $m_1/m_2 = 3$, at what particle energy is the transmission flux coefficient unity? What is the transmission flux coefficient in the limit $E \to \infty$?

Problem 2.6

A particle mass m is confined to motion in one dimension. The potential energy is $V(x) = 0$ for $0 < x < L$ and $V(x) = \infty$ elsewhere. Find the eigenenergies and normalized eigenfunctions $\psi(x, t)$ for the system.

Note the relationship $2\sin(x)\sin(y) = \cos(x - y) - \cos(x + y)$.

Problem 2.7

Using the method outlined in Exercise 2.6 as a starting point, calculate numerically the first six energy eigenvalues, eigenfunctions, and expectation values of position for an electron (not including spin) with effective mass $m_e^* = 0.07 \times m_0$ confined to the double potential well sketched in Fig 2.P19 and bounded by barriers of infinite energy for $x \le 0\,\text{nm}$ and $x \ge 100\,\text{nm}$.

The solution should include plots of the eigenfunctions and magnitude of eigenfunctions squared. Explain the shape of the ground state and first excited state wave functions. Explain the shape of each eigenfunction and the expectation value of position associated with each eigenfunction.

Problem 2.8

(a) A particle mass m_1 moves in a one-dimensional double barrier potential of energy $V_0 = 0.2$ eV sketched in Fig. 2.P19. The ground state and first few excited states of the particle are $E_1 = 0.063$ eV, $E_2 = 0.098$ eV, $E_3 = V_0 = 0.200$ eV, $E_4 = 0.206$ eV, $E_5 = 0.217$ eV, and $E_6 = 0.234$ eV. Sketch and explain the shapes of the corresponding

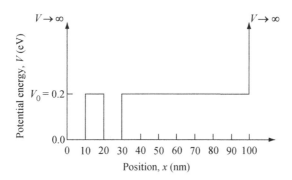

Fig. 2.P19

97

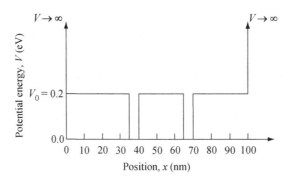

Fig. 2.P20

eigenfunctions. Estimate and explain the behavior of the expectation value of position for each state.

(b) A particle mass m_2 moves in a one-dimensional double barrier potential of energy $V_0 = 0.2$ eV sketched in Fig. 2.P20 and is bounded by barriers of infinite energy for $x \leq 0$ nm and $x \geq 100$ nm. The ground state, first excited state, and second excited state eigenenergies of the particle are $E_1 = 0.06955$ eV, $E_2 = 0.06956$ eV, and $E_3 = V_0 = 0.200$ eV, respectively. Sketch and explain the shapes of the corresponding eigenfunctions. What is the expectation value of position for each state?

(c) Explain the differences between the results in part (a) and (b).

Problem 2.9
(a) Find the eigenfunctions $\psi_n(x, t)$ and energy eigenvalues E_n for a particle mass m moving in one dimension and confined by the potential $V(x) = 0$ for $0 < x < L$ and $V(x) = \infty$ elsewhere.

(b) Repeat part (a) but now with constant *complex* potential $V(x) = V_1 + iV_2$, where $V_1 = 0$ V and $V_2 \neq 0$ V for $0 < x < L$ and $V(x) = \infty$ elsewhere. Explain the result.

(c) At time $t = 0$ ps the particle in (b) is in the ground state. What is the probability of finding the particle in the same state at time $t = 1$ ps if $V_2 = -0.001$ V?

(d) What objections are there to the use of complex potentials in quantum mechanics?

Problem 2.10
Six atoms are arranged symmetrically on the circumference of a ring of radius 0.15 nm.

(a) Using periodic boundary conditions, determine the wave vectors and eigenenergies for free electrons confined to the ring. If each atom contributes a single free electron to the ring, calculate the sum of the ground-state energies of these electrons.

(b) Repeat part (a) but with the atoms arranged in a linear chain, assuming an infinite potential outside of the chain.

(c) Assuming molecules in (a) and (b) have the same work function, obtain an estimate for the free-electron contribution to the energy required to break the ring of atoms into a linear chain.

(d) A four-stroke internal combustion engine consuming benzene fuel at $0.1 \, \text{cm}^3$ per cycle runs at $2\,400$ rpm. How much power is generated if combustion converts benzene rings to chains as in (c)? How might the power generated by the engine be increased?

Note that C_6H_6 has density $879\,\mathrm{kg\,m^{-3}}$ and molecular weight $78.11\,\mathrm{g\,mol^{-1}}$.

Problem 2.11

An electron has wave function at time $t = 0$ that is

$$\Psi(x, t = 0) = \frac{1}{\sqrt{2}}(\psi_1(x) + \psi_2(x)),$$

where $\psi_1(x)$ is the ground state and $\psi_2(x)$ is the first excited state of the particle in a one-dimensional potential well of thickness $L = 10\,\mathrm{nm}$ and infinite barrier energy.

(a) What is the average energy of the particle at time $t = 0$?

(b) Find the state $\Psi(x, t)$ and average particle energy for time $t > 0$. Compare the result with the value obtained in (a). Find the current, $J(x, t)$.

Problem 2.12

Adiabatic quantum computing assumes it is possible to evolve a system from an initial to final configuration (potential) while remaining in the ground state. The adiabatic theorem guarantees this is possible if the system evolves slowly enough. The shortest evolution time between initial configuration, A, and final configuration, B, is normally achieved when the difference in energy, Δ, between the ground and first excited state is maximized over the complete path. Time-efficient adiabatic quantum computing is enabled by finding the optimal path.

To show that paths exist with different minimum energy gap, $\Delta_{\min}$, consider a system that consists of a particle mass $m = 0.07 \times m_0$ confined to motion in one dimension in a potential whose initial configuration is $V(x) = 0$ for $0 < x < L = 25\,\mathrm{nm}$ and is infinite elsewhere and the final configuration is a potential consisting of two Gaussian peaks such that $V(x) = V_0 + V_1 e^{-((x-x_1)/\sigma_1)^2} + V_2 e^{-((x-x_2)/\sigma_2)^2}$ where $0 \le V_1 \le 0.35\,\mathrm{eV}$ and $0 \le V_2 \le 0.35\,\mathrm{eV}$.

(a) Plot the energy gap Δ as a function of V_1 and V_2 for the case when $\sigma_1 = \sigma_2 = 2.5\,\mathrm{nm}$, $x_1 = 0.3 \times L$, and $x_2 = 0.7 \times L$. Explain the features of the Δ landscape. Show that an optimal path exists from initial configuration $A(V_1 = V_2 = 0)$ to any final configuration $B(V_1 \ge 0, V_2 \ge 0)$.

(b) Repeat (a) only now for the case when $\sigma_1 = 4.5\,\mathrm{nm}$ and $\sigma_2 = 1.5\,\mathrm{nm}$. Explain the change in the features of the Δ landscape.

Problem 2.13

The Schrödinger equation may be numerically integrated in real space and real time using the finite-difference time-domain (FDTD) method. To illustrate this, consider the motion of an electron, mass m_0, moving in the x-direction such that

$$\frac{\partial}{\partial t}\psi(x, t) = \left(\frac{i\hbar}{2m_0}\frac{d^2}{dx^2} - \frac{i}{\hbar}V(x)\right)\psi(x, t).$$

(a) Rewrite the wave function in terms of real and imaginary components so that

$$\psi(x, t) = \psi_{\mathrm{Re}}(x, t) + i\psi_{\mathrm{Im}}(x, t),$$

99

and show that two coupled equations are obtained:

$$\frac{\partial}{\partial t}\psi_{\text{Re}}(x,t) = \left(\frac{-\hbar}{2m_0}\frac{\mathrm{d}^2}{\mathrm{d}x^2} + \frac{1}{\hbar}V(x)\right)\psi_{\text{Im}}(x,t)$$

and

$$\frac{\partial}{\partial t}\psi_{\text{Im}}(x,t) = \left(\frac{\hbar}{2m_0}\frac{\mathrm{d}^2}{\mathrm{d}x^2} - \frac{1}{\hbar}V(x)\right)\psi_{\text{Re}}(x,t).$$

(b) Assuming time steps increment by t_0 and space steps increment by h_0, find the simplest finite-difference expression for the time and spatial derivatives appearing in (a).

(c) The jth position in space is $x_j = j \times h_0$ where j is an integer. The real part of the wave function increments in time as $t_n = n \times t_0$ where n is an integer and the imaginary part of the wave function increments in time in half-integer steps. Show that this gives two equations,

$$\psi_{\text{Re}}(x_j, t_{n+1}) = \psi_{\text{Re}}(x_j, t_n)$$
$$- \frac{\hbar t_0}{2m_0 h_0^2}\left(\psi_{\text{Im}}(x_{j+1}, t_{n+1/2}) - 2\psi_{\text{Im}}(x_j, t_{n+1/2}) + \psi_{\text{Im}}(x_{j-1}, t_{n+1/2})\right)$$
$$+ \frac{t_0}{\hbar}V(x_j)\psi_{\text{Im}}(x_j, t_{n+1/2})$$

and

$$\psi_{\text{Im}}(x_j, t_{n+3/2}) = \psi_{\text{Im}}(x_j, t_{n+1/2})$$
$$+ \frac{\hbar t_0}{2m_0 h_0^2}\left(\psi_{\text{Re}}(x_{j+1}, t_{n+1}) - 2\psi_{\text{Re}}(x_j, t_{n+1}) + \psi_{\text{Re}}(x_{j-1}, t_{n+1})\right)$$
$$- \frac{t_0}{\hbar}V(x_j)\psi_{\text{Re}}(x_j, t_{n+1}).$$

(d) An electron is confined to a region of space $0 \leq x \leq 100\,\text{nm}$ in which $V(x) = 0$. Consider an initial wave function that is a sinusoid modulated by a Gaussian such that it has the form

$$\psi_{\text{Re}}(x) = e^{-((x-x_0)/\sigma_x)^2}\cos\left(2\pi((x - x_0)/\lambda_0)\right)$$

and

$$\psi_{\text{Im}}(x) = e^{-((x-x_0)/\sigma_x)^2}\sin\left(2\pi((x - x_0)/\lambda_0)\right),$$

with $x_0 = 10\,\text{nm}$, $\lambda_0 = 5\,\text{nm}$, and $\sigma_x = 5\,\text{nm}$. Integrate the Schrödinger equation to find the subsequent electron motion using the FDTD method with $t_0 = 0.01\,\text{fs}$ and $h_0 = 0.1\,\text{nm}$. Plot $\psi_{\text{Re}}(x)$, $\psi_{\text{Im}}(x)$, and $|\psi(x)|^2$ at time $t = 400\,\text{fs}$ and explain the results. What happens if the electron wavelength is changed to $\lambda_0 = 10\,\text{nm}$? Explain what happens if the sine and cosine are swapped in the definition of the initial wave function.

Problem 2.14

Symplectic finite-difference time-domain (SFDTD) integration can be more accurate than the FDTD(2,2) method described in Problem 2.13. For SFDTD(3,4) the explicit equation for the real part of the wave function in one dimension is

$$\psi_{Re}(x_j, t_{n+l,m})$$

$$= \psi_{Re}(x_j, t_{n+(l-1),m}) + \frac{t_0}{\hbar} V(x_j) \psi_{Im}(x_j, t_{n+1,m})$$

$$- \frac{\hbar t_0}{2m_0 h_0^2} \left\{ \frac{4}{3} c_l (\psi_{Im}(x_{j+1}, t_{n+l,m}) - 2\psi_{Im}(x_j, t_{n+l,m}) + \psi_{Im}(x_{j-1}, t_{n+l,m})) \right\}$$

$$+ \frac{\hbar t_0}{2m_0 h_0^2} \left\{ \frac{1}{12} c_l (\psi_{Im}(x_{j+2}, t_{n+l,m}) - 2\psi_{Im}(x_j, t_{n+l,m}) + \psi_{Im}(x_{j-2}, t_{n+l,m})) \right\},$$

where j, n, l, and m are integers, $(n + l, m)$ denotes the lth time stage after n time steps, and m is the total number of 'time stages.' Here, j and n have the same meaning as in Problem 2.13. Choose $m = 3$ so there are three time stages and $1 \leq l \leq 3$. The coupled set of equations is integrated in three ($m = 3$) stages per time step. The set of numbers c_l weight the symplectic numerical integration of the real part of the wave function. The corresponding set of weights for the imaginary component of the wave function is d_l and $d_l = c_{m-l+1}$. SFDTD(3,4) yields $c_1 = 0.26833010$, $c_2 = -0.18799162$, and $c_3 = 0.91966152$.

(a) Write down the corresponding equation for the imaginary component of the wave function.

(b) Modify the FDTD code in Problem 2.13 to incorporate the higher-order integrator. Introduce the higher-order accuracy in time by embedding a time stage loop within the time integration loop. The code structure should resemble the following:

```
for t=1:tsteps
for timestage=1:3
for x=1:xsteps
[code for real component]
end
for x=1:xsteps
[code for imaginary component]
end
end
end
```

Run the code and verify that for short times the same results are achieved as with the conventional FDTD(2,2) scheme. Take care to properly modify the various spatial loop limits for the higher-order method, and run the code to propagate the Gaussian wave packet up to time $t = 400\,\text{fs}$.

(c) Integrate the probability distribution in space to calculate the total probability for time t_n,

$$P(t_n) = |\psi(t_n)|^2 = \sum_{j=1}^{xsteps} \left(\psi_{Re}^2(x_j, t_n) + \psi_{Im}^2(x_j, t_n) \right).$$

Using $P(t_1)$ as the normalization for the probability distribution over the lattice, calculate the relative probability error, E_P, and error accumulation, $S(E_P)$, as functions of time,

$$E_P(t_n) = 10 \log \left| \frac{P(t_1) - P(t_n)}{P(t_1)} \right|,$$

$$S(t_n) = \log \left(\sum_{j=1}^{n} \left| \frac{P(t_1) - P(t_n)}{P(t_1)} \right| \right).$$

Do this for FDTD(2,2) and SFDTD(3,4) up to time $t = 2\,000$ fs and demonstrate the dramatic difference in conservation of total probability. Repeat for larger space and time increments and demonstrate that the SFDTD(3,4) method remains more accurate than the FDTD(2,2) method for coarser space-time grids.

Problem 2.15
The finite difference approximation used to obtain the second spatial derivative in the Schrödinger equation in Problem 2.13 does not use all the information available in the uniformly-sampled real-space wave function. An alternative approach, called the *split-operator method*, does this by exploiting the properties of Fourier transforms. To see how this works, consider a particle mass m_0 constrained to motion in the x-direction and moving in a real potential $V(x)$ such that

$$\frac{\partial}{\partial t} \psi(x, t) = \left(\frac{i\hbar}{2m_0} \frac{d^2}{dx^2} - \frac{i}{\hbar} V(x) \right) \psi(x, t).$$

The wave function and potential are discretized uniformly so that the jth position in space is $x_i = j \times h_0$ where j is an integer and h_0 is a constant spatial increment. The wave function increments in time as $t_n = n \times t_0$ where n is an integer and t_0 is a constant time step.

(a) Show that the time derivative in the Schrödinger equation may be approximated to second order as

$$\frac{\partial}{\partial t} \psi(x_j, t_n) = \frac{\psi(x_j, t_{n+1}) - \psi(x_j, t_{n-1})}{2t_0}.$$

(b) Show that the second spatial derivative of the real-space wave function in the Schrödinger equation can be obtained by taking the inverse Fourier transform of the wave function in k-space multiplied by $-k^2$.

(c) Describe the functional elements of a numerical algorithm that increments the solution of the Schrödinger equation from wave function $\psi(t_n)$ to $\psi(t_{n+1})$. Include a description of how the solution is initialized. Comment on what is learned.

Problem 2.16
An electron of energy E, mass m_0, and initially moving freely in the x-direction, is incident on a potential step such that $V(x \geq 0) = V_0 = 2\,\text{eV}$ and $V(x < 0) = 0\,\text{eV}$. Calculate the real and imaginary parts of the electron wave function $\psi(-2\,\text{nm} < x < 2\,\text{nm})$, the probability density $|\psi(-2\,\text{nm} < x < 2\,\text{nm})|^2$, the transmission probability flux, and the reflection probability flux for the cases when
  (a) $E = 1\,\text{eV}$,
  (b) $E = 2\,\text{eV}$,

(c) $E = 3\,\text{eV}$,

(d) $E = 300\,\text{eV}$.

Explain the results obtained.

(e) How might a beam of electrons, each of energy $E = 300\,\text{eV}$, be used to detect the presence of the scattering potential $V_0 = 2\,\text{eV}$? Explain the approach to this problem.

Problem 2.17

An electron, constrained to motion in the x-direction by a potential $V(x) = 0$ for $-L/2 < x < L/2$ and $V(x) = \infty$ elsewhere, is in a superposition state of the ground and third excited state such that

$$\psi(x,t) = \frac{1}{\sqrt{2}}(\psi_1(x,t) + \psi_4(x,t)).$$

Setting $L = 20\,\text{nm}$, find expressions for:

(a) Probability density, $|\psi(x,t)|^2$, and the numerical value of the full revival time (the repeat period).

(b) Average particle position, $\langle x(t) \rangle$, and its largest numerical value.

(c) Current flux, $J_x(x,t)$, and the maximum numerical value of current flux.

Note: $2\sin(\theta)2\cos(\phi) = \sin(\theta + \phi) + \sin(\theta - \phi)$.

Problem 2.18

The lowest energy bound-state wave function of a particle mass m in a one-dimensional potential well,

$$V(x) = -\frac{\hbar^2 k_0^2}{2m}\,\text{sech}^2\,(k_0 x),$$

is

$$\psi_0(x) = A\,\text{sech}(k_0 x),$$

where $1/k_0$ is a characteristic length.

(a) Find the normalization constant A.

(b) Substitute $\psi_0(x)$ into the time-independent Schrödinger equation and find the ground-state energy eigenvalue.

(c) If the particle is in an unbound state with wave function

$$\psi(x) = A\left(\frac{ik - k_0\tanh(k_0 x)}{ik + k_0}\right)e^{ikx}$$

and positive energy eigenvalue

$$E = \frac{\hbar^2 k^2}{2m}$$

find the transmission and reflection coefficients of the particle as a function of E.

Problem 2.19

An electron of mass m_0 and energy E moving left-to-right in the x-direction is incident on a rectangular potential barrier of energy $V_0 = 0.5\,\text{eV}$ and thickness $L = 1.5\,\text{nm}$.

(a) Plot electron energy $0 < E \leq 1.5\,\text{eV}$ as a function of transmission probability.

(b) Plot the real and imaginary parts of the wave function $\psi(x)$ for the resonant transmission condition when the electron energy is near $E = 0.667\,\text{eV}$.

(c) Plot electron energy $0 < E \leq 1.5\,\text{eV}$ as a function of wave function phase at a position to the right of the potential barrier. Repeat the calculation setting the energy of the potential barrier to zero. Plot electron energy as a function of the difference in phase in the presence and absence of the potential barrier.

3 Electron Propagation

3.1 Introduction

The propagation method[1] can be used to describe a particle with wave character moving in an arbitrary one-dimensional potential, $V(x)$. This is done by approximating the potential as a series of potential steps. For a particle of energy E incident from the left, transmission and reflection at the first step is calculated along with phase accumulated propagating to the step and expressed as a 2×2 matrix. The wave function coefficients for a particle traversing a one-dimensional potential consisting of a number of such regions may be calculated by multiplying together the appropriate 2×2 matrices. This is a numerically efficient way to calculate transmission, reflection, and wave function of an electron moving in a one-dimensional potential, $V(x)$.

3.2 The Propagation Matrix Method

Suppose an electron of energy E and mass m_0 is incident from the left on a one-dimensional, continuous, smooth potential energy profile, $V(x)$. In this case the solutions to the Schrödinger equation may be found by dividing the potential into a number of small-length piecewise-constant potential energy steps.[2] A particle of energy E impinging on a single potential step at position x_{j+1} is illustrated in Fig. 3.1.

Figure 3.2 shows the detail of the potential step at position index $j + 1$. The electron has wave vector

$$k_j = \frac{\sqrt{2m_0(E - V_j)}}{\hbar} \tag{3.1}$$

in region j, and the wave functions, which are solutions to the Schrödinger equation in regions j and $j + 1$, are

$$\psi_j = A_j e^{ik_j x} + B_j e^{-ik_j x}, \tag{3.2}$$

$$\psi_{j+1} = C_{j+1} e^{ik_{j+1}x} + D_{j+1} e^{-ik_{j+1}x}. \tag{3.3}$$

The convention is to use A and C as coefficients for the wave function traveling left-to-right in regions j and $j + 1$, respectively; B and D are the corresponding right-to-left traveling-wave coefficients.

---

[1] E. O. Kane, *Tunneling Phenomena in Solids*, edited by E. Burstein and S. Lundqvist, New York, Plenum, 1969, p. 1.
[2] Accuracy is controlled by adjusting the length of the piecewise-constant potential energy steps.

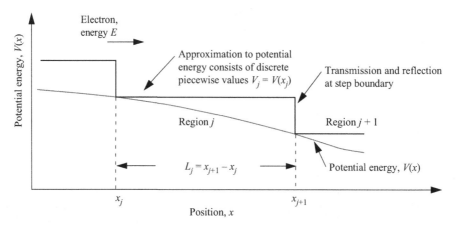

Fig. 3.1 Diagram illustrating approximation of a smoothly varying one-dimensional potential energy profile $V(x)$ with a series of potential energy steps. In this approach, the potential between position x_j and x_{j+1} in region j is approximated by a value V_j. Associated with the potential step at x_j and free propagation distance $L_j = x_{j+1} - x_j$ is a 2×2 matrix that carries all of the amplitude and phase information about the particle.

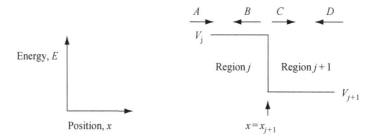

Fig. 3.2 Sketch of a one-dimensional potential step. In region j the potential energy is V_j and in region $j + 1$ the potential energy is V_{j+1}. The coefficients A and C correspond to waves traveling left to right in regions j and $j + 1$, respectively. The coefficients B and D correspond to waves traveling right to left in regions j and $j + 1$, respectively. The transition between region 1 and region 2 occurs at position $x = x_{j+1}$.

The two wave functions given by Eqs. (3.2) and (3.3) should be continuous in ψ and $\mathrm{d}\psi/\mathrm{d}x$ (see Section 2.1.1) so that at the boundary between regions j and $j + 1$,

$$\psi_j|_{x=x_{j+1}} = \psi_{j+1}|_{x=x_{j+1}} \tag{3.4}$$

and

$$\frac{\mathrm{d}\psi_j}{\mathrm{d}x}\bigg|_{x=x_{j+1}} = \frac{\mathrm{d}\psi_{j+1}}{\mathrm{d}x}\bigg|_{x=x_{j+1}}. \tag{3.5}$$

Substituting Eqs. (3.2) and (3.3) into Eqs. (3.4) and (3.5) gives two equations:

$$A_j e^{ik_j x} + B_j e^{-ik_j x} = C_{j+1} e^{ik_{j+1} x} + D_{j+1} e^{-ik_{j+1} x}, \tag{3.6}$$

$$A_j e^{ik_j x} - B_j e^{-ik_j x} = \frac{k_{j+1}}{k_j} C_{j+1} e^{ik_{j+1} x} - \frac{k_{j+1}}{k_j} D_{j+1} e^{-ik_{j+1} x}. \tag{3.7}$$

For typical semiconductor heterostructures with different effective electron mass, current continuity may be achieved by replacing all factors (k_{j+1}/k_j) in Eq. (3.7) with $(m_j k_{j+1}/m_{j+1} k_j)$.

Equations (3.6) and (3.7) for a potential step at position $x_{j+1} = 0$ may be written as a *matrix equation*:

$$\begin{bmatrix} 1 & 1 \\ 1 & -1 \end{bmatrix} \begin{bmatrix} A_j \\ B_j \end{bmatrix} = \begin{bmatrix} 1 & 1 \\ \frac{k_{j+1}}{k_j} & -\frac{k_{j+1}}{k_j} \end{bmatrix} \begin{bmatrix} C_{j+1} \\ D_{j+1} \end{bmatrix}. \tag{3.8}$$

Since the inverse of $\begin{bmatrix} 1 & 1 \\ 1 & -1 \end{bmatrix}$ in Eq. (3.8) is $\frac{1}{2}\begin{bmatrix} 1 & 1 \\ 1 & -1 \end{bmatrix}$,

$$\begin{bmatrix} A_j \\ B_j \end{bmatrix} = \frac{1}{2}\begin{bmatrix} 1 & 1 \\ 1 & -1 \end{bmatrix} \begin{bmatrix} 1 & 1 \\ \frac{k_{j+1}}{k_j} & -\frac{k_{j+1}}{k_j} \end{bmatrix} \begin{bmatrix} C_{j+1} \\ D_{j+1} \end{bmatrix} = \mathbf{p}_{j\text{step}} \begin{bmatrix} C_{j+1} \\ D_{j+1} \end{bmatrix}, \tag{3.9}$$

where $\mathbf{p}_{j\text{step}}$ is the 2×2 matrix describing wave propagation at a potential step, so

$$\mathbf{p}_{j\text{step}} = \frac{1}{2} \begin{bmatrix} 1 + \frac{k_{j+1}}{k_j} & 1 - \frac{k_{j+1}}{k_j} \\ 1 - \frac{k_{j+1}}{k_j} & 1 + \frac{k_{j+1}}{k_j} \end{bmatrix}. \tag{3.10}$$

Propagation between potential steps separated by distance L_j carries phase information only, so that $A_j e^{ik_j L_j} = C_{j+1}$ and $B_j e^{-ik_j L_j} = D_{j+1}$. This may be expressed in matrix form as

$$\begin{bmatrix} A_j \\ B_j \end{bmatrix} = \mathbf{p}_{j\text{free}} \begin{bmatrix} C_{j+1} \\ D_{j+1} \end{bmatrix}, \tag{3.11}$$

where

$$\mathbf{p}_{j\text{free}} = \begin{bmatrix} e^{-ik_j L_j} & 0 \\ 0 & e^{ik_j L_j} \end{bmatrix}. \tag{3.12}$$

To find the combined effect of $\mathbf{p}_{j\text{free}}$ and $\mathbf{p}_{j\text{step}}$, the two matrices are multiplied together so that propagation across the complete jth element consisting of a free propagation region and a step is

$$\mathbf{p}_j = \mathbf{p}_{j\text{free}} \mathbf{p}_{j\text{step}} = \begin{bmatrix} p_{11} & p_{12} \\ p_{21} & p_{22} \end{bmatrix}. \tag{3.13}$$

Multiplying out the matrices $\mathbf{p}_{j\text{free}} \mathbf{p}_{j\text{step}}$ given by Eqs. (3.12) and (3.10), respectively, gives the propagation matrix for the jth region:

$$
\mathbf{p}_j = \frac{1}{2} \left[\begin{matrix} \left(1 + \dfrac{k_{j+1}}{k_j}\right) \mathrm{e}^{-\mathrm{i}k_j L_j} & \left(1 - \dfrac{k_{j+1}}{k_j}\right) \mathrm{e}^{-\mathrm{i}k_j L_j} \\[2mm] \left(1 - \dfrac{k_{j+1}}{k_j}\right) \mathrm{e}^{\mathrm{i}k_j L_j} & \left(1 + \dfrac{k_{j+1}}{k_j}\right) \mathrm{e}^{\mathrm{i}k_j L_j} \end{matrix} \right],
\tag{3.14}
$$

where

$$
p_{11} = p_{22}^*
\tag{3.15}
$$

and

$$
p_{21} = p_{12}^*
\tag{3.16}
$$

due to time-reversal symmetry. See Exercise 3.6.

For the general case of N_{step} potential steps, the propagation matrix for each region within the *domain* is multiplied out to obtain the *total* propagation matrix,

$$
\mathbf{P} = \mathbf{p}_1 \mathbf{p}_2 \cdots \mathbf{p}_j \cdots \mathbf{p}_{N_{\text{step}}} = \prod_{j=1}^{j=N_{\text{step}}} \mathbf{p}_j.
\tag{3.17}
$$

If the particle is introduced from the left of the domain then $A = 1$, and if there is no reflection at the far right of the domain $D = 0$. Hence,

$$
\begin{bmatrix} A \\ B \end{bmatrix} = \left(\prod_{j=1}^{j=N_{\text{step}}} \mathbf{p}_j \right) \begin{bmatrix} C \\ D \end{bmatrix} = \mathbf{P} \begin{bmatrix} C \\ D \end{bmatrix}
\tag{3.18}
$$

may be rewritten as

$$
\begin{bmatrix} 1 \\ B \end{bmatrix} = \begin{bmatrix} P_{11} & P_{12} \\ P_{21} & P_{22} \end{bmatrix} \begin{bmatrix} C \\ 0 \end{bmatrix}.
\tag{3.19}
$$

In this case, because $1 = P_{11}C$, it follows that

$$
|C|^2 = \left| \frac{1}{P_{11}} \right|^2.
\tag{3.20}
$$

Because the propagation method solves the Schrödinger equation, it is possible to construct the particle wave function. After the transmission and reflection coefficients, $C(N_{\text{step}})$ and $D(N_{\text{step}})$, have been obtained, the values of A_j and B_j may be found by back substitution and the wave function calculated.

Numerical solutions using the propagation matrix method and its application to electron transport across semiconductor heterostructures have been studied in some detail.[3]

---

[3] See, for example, M. Steslicka and R. Kucharczyk, *Vacuum* **45**, 211 (1994); B. Jonsson and S. T. Eng, *IEEE J. Quant. Electron.* **26**, 2025 (1990); M. O. Vassell, J. Lee, and H. F. Lockwood, *J. Appl. Phys.* **54**, 5206 (1983); G. Bastard, *Phys. Rev. B* **24**, 5693 (1981).

3.3 Current Conservation and the Propagation Matrix

Classically, the temporal change in charge density ρ is related to the spatial divergence of current density $\mathbf{J}$. The relationship between charge density and current density is determined by conservation of charge and current. In Chapter 2, the current density for an electron with effective mass m_e^* and charge e was given by

$$\mathbf{J} = -\frac{ie\hbar}{2m_e^*}(\psi^*\nabla\psi - \psi\nabla\psi^*) = \frac{e\hbar}{m_e^*}\,\mathrm{Im}\,(\psi^*\nabla\psi). \tag{3.21}$$

Figure 3.3 shows electron propagation coefficients across a one-dimensional potential step. The values of $|C/A|^2$ and $|B/A|^2$ are ratios, so that absolute normalization of wave functions is not important. If the spatial part of the wave function $\psi(x)$ is a traveling wave, then $\psi^*(x)$ is a time-reversed solution. The wave vector k_j describes a plane wave and is real because the present discussion is restricted to simple traveling waves and because the particle energy $E > V_2$. For k_j real, the wave function and its complex conjugate in region 1 and in region 2 may be expressed as

$$\psi_1 = \frac{1}{\sqrt{k_1}}\left(Ae^{ik_1x} + Be^{-ik_1x}\right), \qquad \psi_1^* = \frac{1}{\sqrt{k_1}}\left(A^*e^{-ik_1x} + B^*e^{ik_1x}\right), \tag{3.22}$$

$$\psi_2 = \frac{1}{\sqrt{k_2}}\left(Ce^{ik_2x} + De^{-ik_2x}\right), \qquad \psi_2^* = \frac{1}{\sqrt{k_2}}\left(C^*e^{-ik_2x} + D^*e^{ik_2x}\right). \tag{3.23}$$

The factor $1/\sqrt{k_j}$ normalizes for *unit flux*. Hence, coefficient $A \to A/\sqrt{k_1}$, etc.

Applying the current density (Eq. (3.21)) to wave functions given by Eqs. (3.22) and (3.23) when the particle energy $E > V_2$ gives

$$J_x = \frac{e\hbar}{m}\left(|A|^2 - |B|^2\right) \text{ for } x < 0 \tag{3.24}$$

and

$$J_x = \frac{e\hbar}{m}\left(|C|^2 - |D|^2\right) \text{ for } x > 0. \tag{3.25}$$

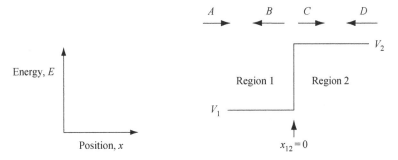

Fig. 3.3 Sketch of a one-dimensional potential step. In region 1 the potential energy is V_1, and in region 2 the potential energy is V_2. The transition between region 1 and region 2 occurs at position $x = x_{12}$.

It follows that current conservation requires

$$A^*A - B^*B = C^*C - D^*D, \tag{3.26}$$

so the net current flow left to right in region 1 must be equal to the net flow in region 2.

This current conservation requires

$$|\mathbf{P}| = 1 \tag{3.27}$$

when k_j is real and

$$|\mathbf{P}| = \pm\mathrm{i} \tag{3.28}$$

for imaginary k_j, where the $\pm$ depends on whether the incoming wave is from the left or right. See Exercise 3.6.

3.4 The Rectangular Potential Barrier

Figure 3.4 is a sketch of a rectangular potential barrier. A particle with wave character incident on the barrier from the left with amplitude A sees a potential step-up in energy of V_0 at $x = 0$, a barrier propagation region of length L, and a potential step-down at $x = L$. A particle of energy E, mass m, and charge e has wave number k_1 outside the barrier and k_2 in the barrier region $0 < x < L$.

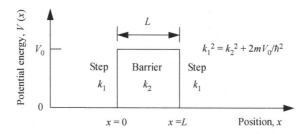

Fig. 3.4 Sketch of the potential of a one-dimensional rectangular barrier of energy V_0. The thickness of the barrier is L. A particle mass m incident from the left of energy E has wave vector k_1. In the barrier region, the wave vector is k_2. The wave vectors k_1 and k_2 are related through $k_1^2 = k_2^2 + 2mV_0/\hbar^2$.

3.4.1 Transmission Probability for a Rectangular Potential Barrier

A particle impinging on a step change in potential between two regions in which the wave vector changes from k_1 to k_2 due to the potential step-up is shown in Fig. 3.3. The corresponding wave function changes from ψ_1 to ψ_2. Solutions of the Schrödinger equation for a step change in potential are

$$\psi_1 = \frac{A}{\sqrt{k_1}}\mathrm{e}^{\mathrm{i}k_1x} + \frac{B}{\sqrt{k_1}}\mathrm{e}^{-\mathrm{i}k_1x}, \tag{3.29}$$

$$\psi_2 = \frac{C}{\sqrt{k_2}}e^{ik_2x} + \frac{D}{\sqrt{k_2}}e^{-ik_2x}. \tag{3.30}$$

The wave functions ψ_1 and ψ_2 are related to each other by the constraint that the wave function and its derivative must be continuous at the step change in potential, so that

$$\psi_1|_{\text{step}} = \psi_2|_{\text{step}} \tag{3.31}$$

and

$$\left.\frac{\mathrm{d}\psi_1}{\mathrm{d}x}\right|_{\text{step}} = \left.\frac{\mathrm{d}\psi_2}{\mathrm{d}x}\right|_{\text{step}}. \tag{3.32}$$

This gives

$$\frac{A}{\sqrt{k_1}} + \frac{B}{\sqrt{k_1}} = \frac{C}{\sqrt{k_2}} + \frac{D}{\sqrt{k_2}}, \tag{3.33}$$

$$\frac{A}{\sqrt{k_1}} - \frac{B}{\sqrt{k_1}} = \frac{k_2}{k_1}\frac{C}{\sqrt{k_2}} - \frac{k_2}{k_1}\frac{D}{\sqrt{k_2}}. \tag{3.34}$$

Rewritten in matrix form, these equations become

$$\frac{1}{\sqrt{k_1}}\begin{bmatrix} 1 & 1 \\ 1 & -1 \end{bmatrix}\begin{bmatrix} A \\ B \end{bmatrix} = \frac{1}{\sqrt{k_2}}\begin{bmatrix} 1 & 1 \\ \frac{k_2}{k_1} & -\frac{k_2}{k_1} \end{bmatrix}\begin{bmatrix} C \\ D \end{bmatrix}. \tag{3.35}$$

The inverse of the left-hand matrix is

$$\frac{k_1}{2\sqrt{k_1}}\begin{bmatrix} 1 & 1 \\ 1 & -1 \end{bmatrix}. \tag{3.36}$$

Hence, Eq. (3.35) may be rewritten as

$$\begin{bmatrix} A \\ B \end{bmatrix} = \frac{k_1}{2}\frac{1}{\sqrt{k_1}}\begin{bmatrix} 1 & 1 \\ 1 & -1 \end{bmatrix}\begin{bmatrix} 1 & 1 \\ \frac{k_2}{k_1} & -\frac{k_2}{k_1} \end{bmatrix}\frac{1}{\sqrt{k_2}}\begin{bmatrix} C \\ D \end{bmatrix} = \frac{1}{2\sqrt{k_1 k_2}}\begin{bmatrix} 1 & 1 \\ 1 & -1 \end{bmatrix}\begin{bmatrix} k_1 & k_1 \\ k_2 & -k_2 \end{bmatrix}\begin{bmatrix} C \\ D \end{bmatrix}. \tag{3.37}$$

Multiplying out the two square matrices gives the 2×2 matrix describing propagation at the step-up in potential,

$$\begin{bmatrix} A \\ B \end{bmatrix} = \frac{1}{2\sqrt{k_1 k_2}}\begin{bmatrix} k_1 + k_2 & k_1 - k_2 \\ k_1 - k_2 & k_1 + k_2 \end{bmatrix}\begin{bmatrix} C \\ D \end{bmatrix}. \tag{3.38}$$

Since the rectangular potential barrier illustrated in Fig. 3.4 consists of a step up and a step down, the 2×2 matrix for the step down is found by simply interchanging k_1 and k_2. The total propagation matrix for the rectangular potential barrier of thickness L consists of the step-up 2×2 matrix multiplied by the propagation matrix from the barrier thickness L multiplied by the step-down matrix. Hence, the propagation matrix for the domain becomes,

$$
\mathbf{P} = \frac{1}{2\sqrt{k_1 k_2}} \begin{bmatrix} k_1 + k_2 & k_1 - k_2 \\ k_1 - k_2 & k_1 + k_2 \end{bmatrix} \begin{bmatrix} e^{-ik_2 L} & 0 \\ 0 & e^{ik_2 L} \end{bmatrix} \frac{1}{2\sqrt{k_1 k_2}} \begin{bmatrix} k_2 + k_1 & k_2 - k_1 \\ k_2 - k_1 & k_2 + k_1 \end{bmatrix}
$$

$$
= \frac{1}{4 k_1 k_2} \begin{bmatrix} (k_1 + k_2) e^{-ik_2 L} & (k_1 - k_2) e^{ik_2 L} \\ (k_1 - k_2) e^{-ik_2 L} & (k_1 + k_2) e^{ik_2 L} \end{bmatrix} \begin{bmatrix} k_2 + k_1 & k_2 - k_1 \\ k_2 - k_1 & k_2 + k_1 \end{bmatrix}. \tag{3.39}
$$

To find the matrix elements of $\mathbf{P}$, the matrices in Eq. (3.39) are multiplied out. For example, P_{12} becomes

$$
P_{12} = -\frac{k_2^2 - k_1^2}{4 k_1 k_2} \left(e^{ik_2 L} - e^{-ik_2 L} \right). \tag{3.40}
$$

Obtaining P_{11} from Eq. (3.39),

$$
P_{11} = \frac{(k_2 + k_1)(k_1 + k_2) e^{-ik_2 L} + (k_1 - k_2)(k_2 - k_1) e^{ik_2 L}}{4 k_1 k_2}
$$

$$
= -\frac{1}{2} \frac{(k_2^2 + k_1^2) \left(e^{ik_2 L} - e^{-ik_2 L} \right)}{2 k_1 k_2} + \frac{1}{2} \left(e^{-ik_2 L} + e^{ik_2 L} \right), \tag{3.41}
$$

which can be used to calculate the transmission probability of an electron incident on the rectangular potential barrier of energy V_0 and thickness L shown in Fig. 3.4.

3.4.1.1 Transmission when $E \geq V_0$

Specializing to the case in which the energy of the incident particle is greater than the potential barrier energy, $E \geq V_0$, and so k_2 is real, Eq. (3.41) becomes

$$
P_{11} = -i \frac{k_2^2 + k_1^2}{2 k_1 k_2} \sin(k_2 L) + \cos(k_2 L). \tag{3.42}
$$

Hence the transmission probability $Trans = 1/(P_{11} P_{11}^*)$ is

$$
Trans = \frac{1}{|P_{11}|^2} = \left(\left(\frac{k_2^2 + k_1^2}{2 k_1 k_2} \right)^2 \sin^2(k_2 L) + \cos^2(k_2 L) \right)^{-1}. \tag{3.43}
$$

This equation can be rearranged by noting $\cos^2(\theta) = 1 - \sin^2(\theta)$ so that

$$
Trans(E \geq V_0) = \left(1 + \left(\frac{k_1^2 - k_2^2}{2 k_1 k_2} \right)^2 \sin^2(k_2 L) \right)^{-1}. \tag{3.44}
$$

For particle energy $E \geq V_0$, this equation predicts unity transmission when $\sin^2(k_2 L) = 0$. In this case, $Trans_{\max} = 1$ and $k_2 L = n\pi$ for $n = 1, 2, 3, \ldots$. Such resonances in transmission correspond to the situation in which the scattered waves originating from $x = 0$ and $x = L$ interfere and *exactly cancel* any reflection from the potential barrier.

3.4.1.2 Transmission when $E < V_0$

For $E < V_0$, the wave number k_2 becomes imaginary such that $k_2 = i\kappa = \sqrt{2m(E - V_0)}/\hbar$ and

$$Trans(E < V_0) = \left(1 + \left(\frac{k_1^2 + \kappa^2}{2k_1\kappa}\right)^2 \sinh^2(\kappa L)\right)^{-1}, \tag{3.45}$$

where use has been made of the fact that $\sin(k_j x) = (e^{ik_j x} - e^{-ik_j x})/2i$, $\sinh(k_j x) = (e^{k_j x} - e^{-k_j x})/2$, $\sin(ik_j x) = i \sinh(k_j x)$, and $\sin^2(ik_j x) = -\sinh^2(k_j x)$.

In this case, a particle with energy less than V_0 can only be transmitted through the barrier via quantum mechanical tunneling. The results of Eq. (3.45) are identical to those obtained previously in Chapter 2 and summarized by Eq. (2.126).

3.4.2 Transmission as a Function of Energy

For incoming particle energy greater than or equal to the barrier energy $E \geq V_0$, the transmission function is given by Eq. (3.44). Using the relations $k_1^2 = 2mE/\hbar^2$ and $k_2^2 = 2m(E - V_0)/\hbar^2$ gives

$$Trans(E \geq V_0) = \left(1 + \frac{1}{4}\left(\frac{V_0^2}{E(E - V_0)}\right) \sin^2(k_2 L)\right)^{-1}. \tag{3.46}$$

For the case when the incoming particle has energy less than the barrier energy $E < V_0$, the value of κ is related to particle energy E by

$$\kappa^2 = \frac{2m(V_0 - E)}{\hbar^2} \tag{3.47}$$

and the transmission function given by Eq. (3.45) is

$$Trans(E < V_0) = \left(1 + \frac{1}{4}\left(\frac{V_0^2}{E(V_0 - E)}\right) \sinh^2(\kappa L)\right)^{-1}. \tag{3.48}$$

3.4.3 Transmission Resonances

It can be shown (Exercise 3.1(a)) that Eq. (3.44) may be written as

$$Trans(E \geq V_0) = \left(1 + \frac{1}{4}\frac{b^2}{(b + k_2^2 L^2)}\frac{\sin^2(k_2 L)}{k_2^2 L^2}\right)^{-1}, \tag{3.49}$$

where $b = 2mV_0 L^2/\hbar^2$.

The minimum of the *Trans* function occurs when $\sin^2(k_2 L) = 1$ or, equivalently, when $k_2 L = (2n-1)\pi/2$ for $n = 1, 2, 3, \ldots$, and may be written (Exercise 3.1(b)) as

$$Trans_{min} = 1 - \frac{b^2}{\left(2k_2^2 L^2 + b\right)^2} = 1 - \frac{V_0^2}{(2E - V_0)^2}. \tag{3.50}$$

Maximum transmission, $Trans = 1$, occurs when $\sqrt{2m_0(E - V_0)} \times L/\hbar = n\pi$, where $n = 1, 2, 3 \ldots$ (Exercise 3.1(c)). The value of $Trans(E \to V_0) = \left(1 + \frac{mV_0 L^2}{2\hbar^2}\right)^{-1}$ (Exercise 3.1(d)).

Figure 3.5(a) shows the potential energy of a one-dimensional rectangular barrier with energy V_0 and thickness L. Figure 3.5(b) shows the transmission spectrum of an electron incident on the potential barrier. For electron energy less than V_0, there is an exponential decrease in transmission probability with decreasing E because transmission is dominated by tunneling. For electron energy greater than V_0, there are resonances with unity transmission.

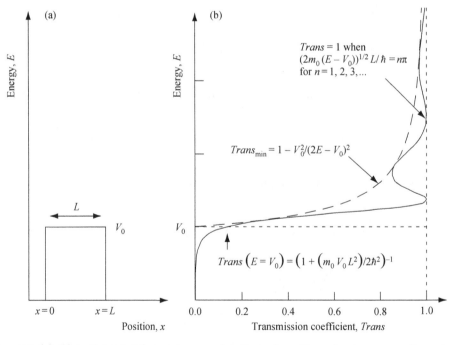

Fig. 3.5 (a) Plot of potential energy as a function of position showing a one-dimensional rectangular barrier of energy V_0 and thickness L. (b) Solid curve is plot of transmission probability, *Trans*, for an electron incident from the left on the potential barrier shown in (a). Dashed curve is $Trans_{min}$ and dotted line is a guide to the eye for unity transmission and energy V_0. In this example, electron mass is m_0, and the rectangular potential barrier has energy $V_0 = 1.0\,\mathrm{eV}$ and thickness $L = 1\,\mathrm{nm}$.

3.4.4 Electron Wave Packet Tunneling

The difference in time that a particle spends interacting with and without a local potential is called the collision lifetime.[4] For an electron incident on a tunnel barrier, this measure may not be useful.[5] However, electron wave packet dynamics at a tunnel barrier *is* worth studying.

A wave function at time $t = 0$ that consists of a plane wave with momentum $\hbar k_0$ in the x-direction modulated by a Gaussian spatial function is

$$\psi(x, t = 0) = \left(\frac{1}{2\pi\sigma_x^2} \right)^{\frac{1}{4}} e^{ik_0 x} e^{\frac{-x^2}{4\sigma_x^2}}. \tag{3.51}$$

The mean position at time $t = 0$ is $x_0 = 0$ and a measure of the spatial spread of the wave packet is the standard deviation σ_x.

The Fourier transform of the real-space wave function is the k-space representation that is also a Gaussian wave packet with standard deviation $\sigma_k = 1/2\sigma_x$, which evolves as a function of time t such that

$$\psi(k, t) = \left(\frac{2\sigma_x^2}{\pi} \right)^{\frac{1}{4}} e^{-(k - k_0)^2 \sigma_x^2} e^{-i\omega(k)t}, \tag{3.52}$$

where $\omega(k) = \hbar k^2 / 2m_0$ is the electron dispersion relation in the constant zero potential of free space.

An electron described by a Gaussian wave function initially propagating in a constant potential that is incident on a rectangular potential barrier has, in general, a probability of being transmitted or reflected. If the electron with expectation energy E is moving from left to right in the x-direction and is incident on the rectangular potential barrier of thickness L and energy $V_0 > E$, it can tunnel and subsequently propagate as a pulse and be detected at some distance to the right of the barrier.

Potential energy tunnel barriers can be created with atomic layer precision in semiconductor heterostructures such as GaAs/AlGaAs. Figure 3.6 shows that the peak of a Gaussian electron wave packet in the conduction band of GaAs that has tunneled and is measured at some fixed position, x_R, far to the right of an AlGaAs potential barrier, arrives at time t_p *before* the peak of the electron wave packet propagating in the absence of the barrier. In this case, for an incident single-electron wave packet with expectation energy $E < V_0$, tunneling acts as a *high-pass filter* of plane-wave pulse components with momentum $\hbar k$. Higher energy components at the leading edge of the incident pulse have a greater probability of tunneling. The transmitted pulse is no longer a Gaussian wave packet; rather it contains k-states with amplitudes that peak at a value of $k' > k_0$ and the *group velocity of the transmitted pulse is greater than that of the incident pulse*. The effect of k-space filtering is illustrated in Fig. 3.7(a). As shown in Fig. 3.7(b), for an incident pulse with fixed k_0

---

[4] F. T. Smith, *Phys. Rev.* **118**, 349 (1960).

[5] The existence of a characteristic time that describes a tunneling event is controversial. See, for example, L. A. MacColl, *Phys. Rev.* **40**, 621 (1932); E. P. Wigner, *Phys. Rev.* **98**, 145 (1955); T. E. Hartman, *J. Appl. Phys.* **33**, 3427 (1962); M. Bttiker and R. Landauer, *Phys. Rev. Lett.* **49**, 1739 (1982); E. H. Hauge and J. A. Stvneng, *Rev. Mod. Phys.* **61**, 917 (1989); R. Landauer and T. Martin, *Rev. Mod. Phys.* **66**, 217 (1994); H. G. Winful, *Phys. Rev. Lett.* **91**, 260401 (2003); T. Rivlin, E. Pollak, and R. S. Dumont, *Phys. Rev. A* **103**, 012225 (2021).

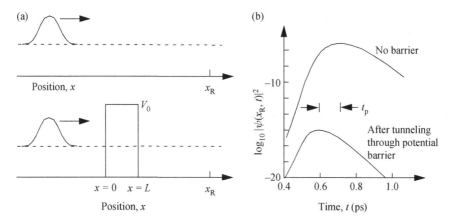

Fig. 3.6 (a) Schematic showing electron wave packet with expectation energy E propagating in a region of constant potential (upper sketch) and incident on a rectangular potential barrier of energy $V_0 = 0.35\,\mathrm{eV} > E$ (lower sketch). Measurement position, x_R, is far to the right of the potential barrier. (b) Probability of measuring the electron at a fixed position, x_R, as a function of time in the presence and absence of a potential barrier. The incident electron is a Gaussian wave packet moving from left to right that is initiated at position $x_0 = -400\,\mathrm{nm}$. In the presence of a rectangular potential barrier with potential energy V_0 and thickness L, the electron with expectation energy $E < V_0$ has a peak detection probability that occurs at time $t_p = 0.120\,\mathrm{ps}$ *before* the peak for an electron propagating in the absence of the potential barrier. Parameters are $E = 0.15\,\mathrm{eV}$, $\sigma_k = 8 \times 10^7\,\mathrm{m}^{-1}$, $V_0 = 0.35\,\mathrm{eV}$, $L = 20\,\mathrm{nm}$, $x_0 = -400\,\mathrm{nm}$, $x_R = 210\,\mathrm{nm}$, and $m_e^* = 0.07 \times m_0$ give $t_p = 0.120\,\mathrm{ps}$.

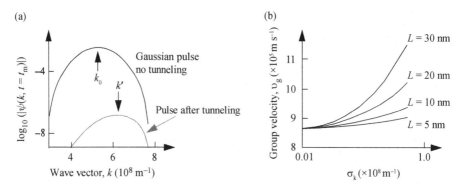

Fig. 3.7 (a) An incident electron wave packet with expectation energy E consists of a Gaussian distribution of k-space amplitudes with a standard deviation $\sigma_k = 8 \times 10^7\,\mathrm{m}^{-1}$ centered on wave vector $k_0 = 5.3 \times 10^8\,\mathrm{m}^{-1}$. Because higher energy components of the wave packet have a higher probability of tunneling, the tunnel barrier of thickness $L = 20\,\mathrm{nm}$ acts as a high-pass filter for wave vector amplitudes. The transmitted pulse measured at time $t = t_m$ is no longer a Gaussian wave packet, contains time-independent k-states with amplitudes that peak at a value $k' = 6.2 \times 10^8\,\mathrm{m}^{-1} > k_0$, and the group velocity of the transmitted pulse is *greater* than that of the incident pulse. (b) Group velocity of the transmitted pulse depends on the value of σ_k and barrier thickness L. Parameters are $E = 0.15\,\mathrm{eV}$, $V_0 = 0.35\,\mathrm{eV}$, and $m_e^* = 0.07 \times m_0$.

the group velocity v_g of the transmitted pulse *increases* with increasing σ_k and increasing L. This demonstrates that changing the standard deviation of an incident particle of energy E or changing the barrier thickness controls the group velocity of the transmitted pulse.

In semiconductors, the high-pass filter action of tunneling through a single barrier can be related to complex band structure (see Section 3.6.5). The imaginary wave vectors of magnitude κ in a semiconductor band gap determine the probability amplitude attenuation of the pulse components that tunnel. Electron tunneling is exponentially sensitive to details of the imaginary band structure since, for an almost opaque rectangular potential barrier of thickness L, tunneling probability amplitude scales as $\mathrm{e}^{-\kappa L}$.

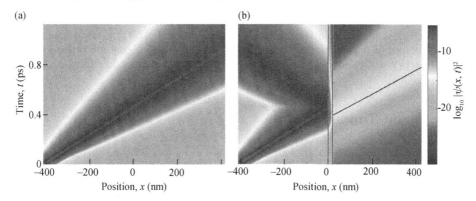

Fig. 3.8 (a) Space-time plot of a Gaussian pulse describing electron probability $|\psi(x,t)|^2$ in a constant potential. Peak pulse position is shown as a line whose slope is inversely proportional to group velocity, v_{g}. (b) Gaussian pulse incident on a rectangular potential barrier of energy V_0. Vertical lines indicate the position of the potential barrier. The effects of self-interference due to reflection from the potential barrier are visible. After transmission through the barrier the electron wave packet has a group velocity that is *greater* than that of the pulse propagating in the absence of the barrier. The peak in the attenuated transmitted pulse describing an electron that has tunneled through the potential barrier arrives *before* the peak of the pulse propagating in a constant potential. Parameters are $E = 0.15\,\mathrm{eV}$, $\sigma_{\mathrm{k}} = 8 \times 10^7\,\mathrm{m}^{-1}$, $V_0 = 0.35\,\mathrm{eV}$, $L = 20\,\mathrm{nm}$, $x_0 = -400\,\mathrm{nm}$, and $m_{\mathrm{e}}^* = 0.07 \times m_0$.

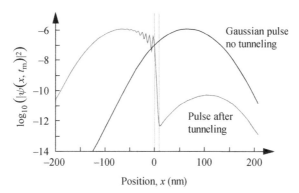

Fig. 3.9 Calculated electron probability $\log_{10}|\psi(x,t_{\mathrm{m}})|^2$ at time t_{m} in a constant potential and in the presence of a rectangular barrier of energy V_0. Vertical lines indicate position of the potential barrier. The effects of self-interference due to reflection from the potential barrier are visible. After transmission through the barrier, the electron wave packet has a group velocity that is greater than that of the pulse propagating in the absence of the barrier. The peak in the attenuated transmitted pulse describing an electron that has tunneled through the potential barrier arrives *before* the peak of the pulse propagating in a constant potential. Parameters are $E = 0.15\,\mathrm{eV}$, $\sigma_{\mathrm{k}} = 8 \times 10^7\,\mathrm{m}^{-1}$, $V_0 = 0.35\,\mathrm{eV}$, $L = 10\,\mathrm{nm}$, and $m_{\mathrm{e}}^* = 0.07 \times m_0$.

The reflected portion of the single-electron pulse interferes with itself and strongly influences the behavior of the peak in $|\psi(x,t)|^2$ near the barrier during the collision. These effects can be illustrated using the time-space plots shown in Fig. 3.8.

Figure 3.8(a) shows $|\psi(x,t)|^2$ for an electron wave packet propagating in a constant potential and Fig. 3.8(b) is the case when there is a potential barrier. The solid lines with finite slope indicate the peak of the incident and transmitted pulses. The lines have different slopes because the group velocity of the incident and transmitted pulse is not the same. The offset at the tunnel barrier is due the k-space filtering effect.

Figure 3.9 is a plot of spatial probability distribution $|\psi(x,t_\mathrm{m})|^2$ at a specific time t_m. Note the use of a $\log_{10}$ scale to show self-interference from particle reflection at the potential barrier and exponential suppression of particle transmission tunnel probability.

3.5 Resonant Tunneling

To obtain unity transmission probability for particle energy $E < V_0$ a different type of potential must be considered. A simple example is the double rectangular potential barrier sketched in Fig. 3.10(a). As with the finite potential well of Section 2.7, there is *always* a symmetric lowest-energy resonant wave function associated with this symmetric potential well. The lowest-energy resonance has energy E_0 and is indicated by a broken line in Fig. 3.10(a). The wave function at energy E_0 is symmetric because the potential is symmetric. Due to the finite thickness and energy of the potential barrier on either side of the potential well, there is a finite probability that a particle can tunnel out of the potential well.

Figure 3.10(b) shows the transmission probability of an electron of mass m_0 incident from the left on the potential given in Fig. 3.10(a). With increasing electron energy, there is an overall smooth increase in background transmission because the effective potential barrier seen by the electron decreases. At an incident energy $E = E_0 = 0.358\,\mathrm{eV}$ there

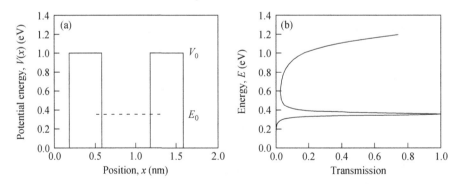

Fig. 3.10 (a) Symmetric double rectangular potential barrier with barrier thickness $L_\mathrm{b} = 0.4\,\mathrm{nm}$, well thickness $L_\mathrm{w} = 0.6\,\mathrm{nm}$, and barrier energy $V_0 = 1\,\mathrm{eV}$. There is a resonance at energy E_0 for an electron mass m_0 in such a potential. This resonance energy is indicated by the broken line. (b) Transmission probability of an electron mass m_0 incident from the left on the potential given in (a). There is an overall increase in background transmission with increasing energy, and a unity transmission resonance occurs at energy $E = E_0 = 357.9\,\mathrm{meV}$ with $\gamma = 30.7\,\mathrm{meV}$ and $\tau = 21\,\mathrm{fs}$.

is a unity peak in transmission probability that stands out above the smoothly varying electron transmission background.

The resonance has a finite width in energy because if a particle is placed in the potential well it can tunnel out. The lifetime of the resonance at energy E_0 may be calculated approximately as

$$\tau = \hbar/\gamma, \qquad (3.53)$$

where γ is the full width at half maximum in the transmission peak when fitted to a Lorentzian line shape. The value of τ is the characteristic response time of the system assuming that the resonance may be described by a Lorentzian line shape. This is always an approximation for a symmetric potential and electron mass m because, for such a case, the transmission line shape is *always* asymmetric. The asymmetry exists because of the overall increase in background transmission with increasing energy.

Figure 3.11(a) shows calculated real and imaginary parts of the electron wave function at the resonance energy E_0. The exact cancellation of back-scattering is illustrated in Fig. 3.11(b), which shows that the magnitude of the wave function squared, $|\psi(x)|^2$, is unity either side of the potential barrier. On resonance the value of $|\psi(x)|^2$ peaks to its maximum value in the center of the structure. Electron probability is enhanced and stored by the resonator for a characteristic time τ.

Crystal growth techniques such as molecular beam epitaxy may be used to make a double-barrier potential similar to that shown in Fig. 3.10(a). The potential might be formed in the conduction band of a semiconductor heterostructure. For example, epitaxially grown AlAs may be used to form the two tunnel barriers, and GaAs may be used to form the potential well of a single crystal n-type unipolar diode. The resonant tunneling heterostructure diode current–voltage characteristics of such a device are expected to be proportional to electron transmission probability. While energy is not identical to applied voltage and current is not identical to the transmission coefficient shown in Fig. 3.10(b), the trends are similar. In particular, a region of negative differential resistance exists that

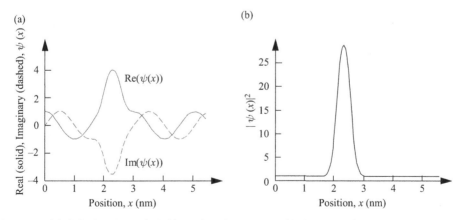

Fig. 3.11 (a) Calculated real (solid curve) and imaginary (dashed curve) electron wave function, $\psi(x)$, at resonant energy $E_0 = 0.358\,\mathrm{eV}$ for the potential shown in Fig. 3.10(a). (b) $|\psi(x)|^2$ for the electron in (a). Parameters are $m = m_0$, $V_0 = 1\,\mathrm{eV}$, $L_b = 0.4\,\mathrm{nm}$, and $L_w = 0.6\,\mathrm{nm}$.

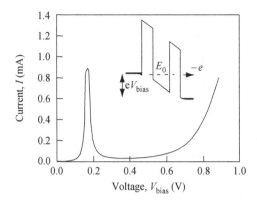

Fig. 3.12 Illustration of resonant tunnel diode current–voltage characteristics. There is a peak in current when the diode is voltage biased in such a way that electrons can resonantly tunnel through the potential well. A region of negative differential resistance also exists for some values of V_{bias}.

is of interest as an element in the design of electronic circuits. Figure 3.12 illustrates the current–voltage characteristics of a resonant tunneling heterojunction diode. The inset is the conduction band minimum profile under voltage bias V_{bias}.

3.5.1 Heterostructure Bipolar Transistor with Resonant Tunnel Barrier

It is possible to gain more insight into electron transport in nanoscale structures by combining what is known about resonant tunnel devices with a heterostructure bipolar transistor (HBT). The inset in Fig. 3.13 shows the conduction and valence band profile of an n-p-n AlAs/GaAs HBT with an AlAs tunnel emitter and an AlAs tunnel barrier at the collector–subcollector interface. The separation between the two tunnel barriers is $x_B + x_C = 40\,\text{nm}$, where x_B is the thickness of the transistor base and x_C is the thickness of the collector. The p-type GaAs base region of the transistor consists of a thin two-dimensional sheet of substitutional Be atoms of average concentration $6 \times 10^{13}\,\text{cm}^{-2}$. The value of x_B is less than 10 nm and is determined by the spatial extent of the wave function in the x direction associated with the charge carriers. The emitter and subcollector are heavily doped n-type so that they have a Fermi energy E_{Fe} at low temperature.

The n-p-n HBT shown in Fig. 3.13 may be viewed as a three-terminal resonant tunnel device with emitter current, I_E, and collector current, I_C. The third terminal is the p-type base, which allows independent control of the energy of conduction band electrons that are tunnel-injected from the emitter into the base collector region, where resonant electron interference effects can take place. Because this is a transistor, it is possible to fix the base current, I_B, and measure common-emitter current gain $\beta = I_C/I_B$ as a function of collector–emitter voltage bias V_{CE}.

Figure 3.13 shows the results of measuring β at a temperature $T = 4.2\,\text{K}$. As might be expected for a resonant tunnel structure, there are strong resonances in β associated with quantization of energy levels in the 40 nm thick base collector region. This confirms that quantum mechanical reflection and resonance phenomena play an important role in determining device performance.

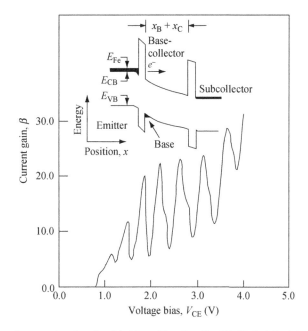

Fig. 3.13 Measured current gain, β with bias, V_{CE} for the HBT sketched in the inset. Lattice temperature is $T = 4.2\,\mathrm{K}$, emitter area is $7.8 \times 10^{-5}\,\mathrm{cm}^2$, and base current is $I_B = 0.1\,\mathrm{mA}$. The AlAs tunnel emitter is 8 nm thick, the p-type base is delta doped with a Be sheet concentration of $6 \times 10^{13}\,\mathrm{cm}^{-2}$, the AlAs collector barrier is 5 nm thick, and the emitter–collector barrier separation is $x_B + x_C = 40\,\mathrm{nm}$. The conduction-band minimum, E_{CB}, the valence-band maximum, E_{VB}, and the emitter Fermi energy, E_{Fe}, are indicated.

The presence of elastic and inelastic electron scattering in the collector and p-type base has the effect of breaking the coherence of the wave function associated with resonant transport of an electron through the structure. The resulting decoherence reduces resonant interference and introduces an incoherent background contribution to current flow between emitter and collector.

Inelastic electron scattering presents a *fundamental* limit to transistor performance. However, such scattering rates cannot be evaluated independently of either V_{CE} or V_{CB} because the resonances depend upon the potential profile and hence the voltage bias across the device.

The existence of resonant effects illustrates that electron transport in the base and the collector may no longer be treated separately. For this nanoscale transistor, the electron cannot be considered as a classical point-particle; rather it is a *nonlocal*, truly quantum-mechanical object because its wave function extends across the base and collector.

3.5.2 Resonant Tunneling between Two Quantum Wells

If the potential barrier separating two identical one-dimensional potential wells is of sufficient strength to isolate the ground state eigenfunctions, then the ground state eigenenergies are degenerate. This situation is depicted schematically in Fig. 3.14(a) for the case of rectangular potential wells each of thickness L and barrier energy V_0. Each

121

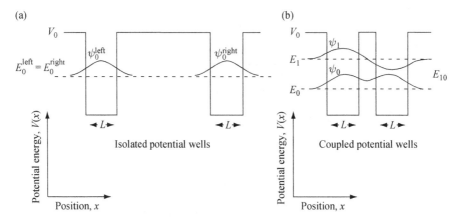

(a)

(b)

Isolated potential wells

Coupled potential wells

Fig. 3.14 (a) Two identical isolated potential wells with thickness L and barrier potential energy V_0 have identical (degenerate) ground state eigenenergies (indicated by the dotted line). The wave functions ψ_0^{left} and ψ_0^{right} associated with the left and right potential wells are symmetric and centered about their respective potential wells. (b) The lowest energy wave functions for two identical potential wells coupled by a thin tunnel barrier are ψ_0 and ψ_1. Wave function ψ_0 is symmetric and has an eigenenergy E_0. Wave function ψ_1 is antisymmetric and has an eigenenergy E_1, which is greater in value than E_0 and $E_{10} = E_1 - E_0$.

ground state wave function ψ_0 is symmetric about its potential well and has an associated eigenenergy, $E_0^{\text{left}} = E_0^{\text{right}}$. The wave functions ψ_0^{left} and ψ_0^{right} are identical apart from a spatial translation.

Figure 3.14(b) is a sketch of what happens when the rectangular potential wells are no longer isolated by a strong potential barrier. Particle states originally associated with one well can interact via quantum-mechanical tunneling with states in the other well. Now, the lowest energy wave functions for two identical potential wells coupled by a thin tunnel barrier are ψ_0 and ψ_1. The function ψ_0 is symmetric and has an eigenenergy E_0. The function ψ_1 is antisymmetric and has an eigenenergy E_1, which is greater in value than E_0. The reason $E_0 < E_1$ is that the symmetric wave function ψ_0 has less integrated change in slope (curvature) multiplied by the value of the wave function than the wave function ψ_1 (see Section 2.1.1).

A particle in eigenstate ψ_0 or ψ_1 can be found in either quantum well because the wave functions are spread out or *delocalized* across both potential wells. The coupling between the original states via tunneling breaks degeneracy and causes splitting in energy levels $E_{10} = E_1 - E_0$.

For the three barriers and two potential wells sketched in Fig. 3.15(a), as shown in Fig. 3.15(b), there are two resonances separated in energy by E_{10}, one associated with an antisymmetric wave function and one with a symmetric wave function. Each wave function may be viewed as being formed from a linear combination of two lowest-energy symmetric wave function contributions for a single potential well, one from the left and one from the right potential well. These individual contributions are initially considered as degenerate bound states subsequently coupled via tunneling through the central potential barrier. The separation in energy between the two resonances E_{10} depends upon the coupling strength through the central potential barrier. Again, the lowest energy resonance in this double potential well is associated with a symmetric wave function, and the upper-energy

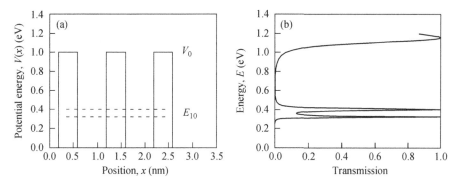

Fig. 3.15 (a) Symmetric triple rectangular potential barriers with barrier thickness $L_b = 0.4$ nm, well thickness $L_b = 0.6$ nm, and barrier energy $V_0 = 1$ eV. For an electron of mass m_0 in such a potential, there are two resonant states separated in energy by energy E_{10}. The energy of these resonances is indicated by the dashed line. (b) Transmission probability of an electron mass m_0 incident from the left on the potential given in (a). There is an overall increase in background transmission with increasing energy and unity transmission resonances occur at energy $E = 321.5$ meV ($\gamma = 12.2$ meV) and $E = 401.5$ meV ($\gamma = 20.8$ meV). The splitting in energy is $E_{10} = 80$ meV.

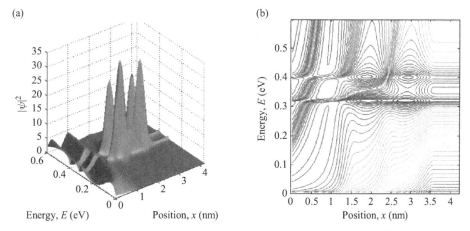

Fig. 3.16 (a) Calculated magnitude of electron wave function squared, $|\psi(x)|^2$, as a function of position and energy for the symmetric triple rectangular potential barrier shown in Fig. 3.15. The electron has mass m_0, barrier thickness $L_b = 0.4$ nm, well thickness $L_w = 0.6$ nm, and barrier energy $V_0 = 1$ eV. (b) contour plot of (a) using negative logarithmic scale.

resonance is associated with an antisymmetric wave function. The lowest energy resonance has a smaller full width at half maximum γ because the tunnel barrier energy seen by a particle is greater than that of the higher energy resonance.

Figure 3.16(a) is calculated magnitude of electron wave function squared, $|\psi(x)|^2$, as a function of position and energy for the symmetric triple-rectangular potential barrier shown in Fig. 3.15. With the exception of the resonances, the three-dimensional plot shows the incident wave creating interference on reflection from the potential. The parabolic dispersion for electrons causes the wavelength of the standing waves to depend on energy. Figure 3.16(b) is a contour plot of Fig. 3.16(a) using a negative logarithmic scale.

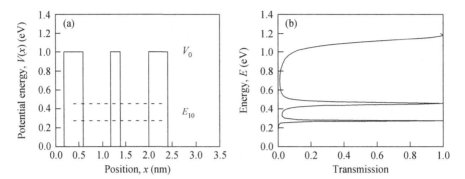

Fig. 3.17 (a) Symmetric triple rectangular potential barriers with outer barrier thickness 0.4 nm, central barrier thickness 0.2 nm, well thickness 0.6 nm, and barrier energy $V_0 = 1$ eV. For an electron of mass m_0 in such a potential, there are two resonances separated in energy by energy E_{10}. The energy of these resonances is indicated by the broken line. (b) Transmission probability of an electron of mass m_0 incident from the left on the potential given in (a). There is an overall increase in background transmission with increasing energy, and unity transmission resonances occur at energy $E = 271.9$ meV ($\gamma = 8.9$ meV) and $E = 455.9$ meV ($\gamma = 27.5$ meV). The splitting in energy is $E_{10} = 184$ meV.

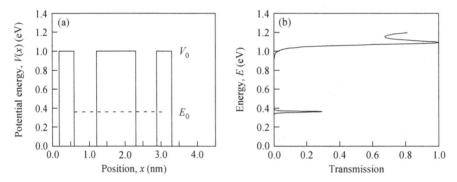

Fig. 3.18 (a) Symmetric triple-rectangular potential barriers with outer barrier thickness 0.4 nm, central barrier thickness 1.1 nm, well thickness 0.6 nm, and barrier energy $V_0 = 1$ eV. For an electron of mass m_0 in such a potential, there is a resonance at energy E_0, indicated by the dashed line. (b) Transmission probability of an electron mass m_0 incident from the left on the potential given in (a). There is an overall increase in background transmission with increasing energy and non-unity transmission resonance at energy $E = E_0 = 360.2$ meV ($\gamma = 10.6$ meV). The value of transmission probability at energy E_0 is 0.308.

Decreasing the thickness of the central barrier breaks the symmetry of the potential, coupling between the left and right potential well increases and E_{10} increases as shown in Fig. 3.17. Ultimately, as the thickness of the middle barrier vanishes, the result for a single potential well of twice the original individual well thickness is obtained.

Figure 3.18(b) is a plot of transmission probability of an electron mass m_0 incident from the left on the symmetric triple-rectangular potential barriers given in Fig. 3.18(a). There is an overall increase in background transmission with increasing energy and *non-unity* transmission resonance at energy $E = E_0 = 360.2$ meV.

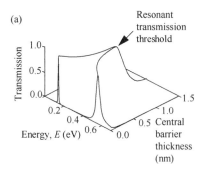
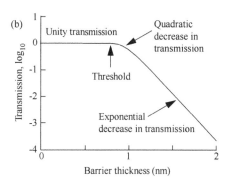

Fig. 3.19 (a) Three-dimensional plot of transmission coefficient for an electron of mass m_0 as a function of incident particle energy, E, and central barrier thickness for symmetric triple-rectangular potential barriers. The outer barrier thickness is 0.4 nm, well thickness is 0.6 nm, and barrier energy is $V_0 = 1\,\mathrm{eV}$. The resonant transmission *threshold* occurs when the central barrier thickness is 0.8 nm. (b) Logarithm of maximum transmission as a function of central barrier thickness on a linear scale showing the threshold in barrier thickness beyond which unity electron transmission no longer occurs.

Figure 3.19(a) is a three-dimensional plot of the transmission on a linear scale for an electron of mass m_0 as a function of both incident particle energy, E, and central barrier thickness for symmetric triple-rectangular potential barriers. In Fig. 3.19(b) maximum particle transmission is plotted on a logarithm scale. In the logarithmic plot, unity transmission corresponds to zero and low transmission corresponds to a large negative number.

With increasing central barrier thickness, the coupling between the two wells decreases and the value of the energy splitting also decreases. Eventually, beyond a threshold value of central barrier thickness the energy splitting is so small that it disappears because it becomes smaller than the line width of an individual resonance ($E_{10} \leq \gamma$). For central barrier thickness greater than this threshold value, the remaining resonant transmission peak has a value less than unity. Physically, less than unity transmission occurs because a particle trapped in one of the two potential wells escapes by tunneling through the thin barrier much more readily than it is able to tunnel through the thick barrier. With increasing central barrier thicknesses above the resonant transmission threshold, conduction of electrons by a transmission resonance is initially suppressed quadratically and then exponentially.

The symmetry of the potential may also be broken by changing barrier energy. Figure 3.20(a) shows nonsymmetric triple-rectangular potential barriers. As indicated in Fig. 3.20(b), there is still a splitting in the resonance, but unity transmission is *not* obtained.

Figure 3.21 is a detail of the transmission peak showing that neither resonant peak of Fig. 3.20(b) has unity transmission.

Resonant tunnel diodes manufactured in single-crystal GaAs/AlGaAs with band edge potential profiles similar to Fig. 3.6(a) have been investigated in some detail.[6] Practical implementation of these devices using semiconductor structures requires consideration of asymmetry in the potential when voltage bias is applied. The asymmetry in band-edge potential profile under bias reduces the peak resonance transmission. In addition, for a

---

[6] For a survey, see, N. G. Einspruch and W. R. Frensley (eds), *Heterostructures and Quantum Devices*, San Diego, Academic Press, 1994 (ISBN 0 12 234124 4).

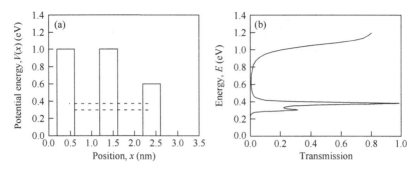

Fig. 3.20 (a) Nonsymmetric triple-rectangular potential barriers with barrier thickness 0.4 nm, well thickness 0.6 nm, left and central barrier energy 1 eV, and right barrier energy 0.6 eV. (b) Transmission probability of an electron of mass m_0 incident from the left on the potential given in (a). There is an overall increase in background transmission with increasing energy and nonunity transmission resonance at two energies indicated by the broken lines.

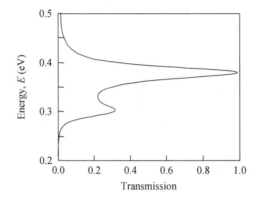

Fig. 3.21 Detail of transmission probability of an electron of mass m_0 incident from the left on the potential given in Fig. 3.16(a). There are non-unity transmission resonance peaks at two energies.

device that passes electrical current, space charging effects can be important. If electron density in the potential well is great enough, it will distort the band-edge potential profile. For this reason, transmission probability should, in principle, be found by simultaneously solving Schrödinger's equation for the electron wave functions and Poisson's equation for charge density. An additional design consideration is the fact that resonant tunnel diodes must have low capacitance and low series resistance if they are to operate at high frequencies. This is particularly difficult to achieve in semiconductor heterostructures where tunnel barrier thickness is typically on a nm scale.

3.6 Energy Bands in a Periodic Potential

3.6.1 Bloch's Theorem

One-dimensional motion of an electron moving in free space is described by a wave function $\psi_{k_1}(x) = Ae^{i(k_1 x - \omega t)}$ and, as expected, the probability of finding the particle at any position in space is uniform. No particular location in space is more or less significant than any other location. The underlying symmetry may be expressed by stating that probability is translationally invariant over all space.

In the presence of a periodic potential $V(x)$ such that $V(x) = V(x + nL)$, where L is the minimum spatial period of the potential and n is an integer, electron probability is modulated spatially by the same periodicity as the potential. The isotropic electron probability symmetry of free space is broken and replaced by a new probability symmetry which is translationally invariant over a space spanned by a unit cell size of L. Electron probability is the same in each unit cell and described by the function $|U(x)|^2$. Under these conditions an electron wave function is identical to $U_k(x)$ to within a phase factor e^{ikx}. To find the phase factor, it is necessary to solve for the eigenstates of the one-electron time-independent Schrödinger equation,

$$\hat{H}\psi(x) = \left(\frac{-\hbar^2}{2m_0} \frac{\mathrm{d}^2}{\mathrm{d}x^2} + V(x) \right) \psi(x) = E\psi(x), \tag{3.54}$$

where $V(x) = V(x + nL)$ for integer n. The electron wave function in band j can be a *Bloch function* of the form

$$\boxed{\psi_{j,k}(x) = U_{j,k}(x)e^{ikx}} \tag{3.55}$$

where $U_{j,k}(x+nL) = U_{j,k}(x)$ has the same periodicity as the potential and n is an integer. The term e^{ikx} carries the phase information between unit cells via what is called the *Bloch wave vector k*.

The electron wave function changes from position x to $x + L$ such that

$$\psi_{j,k}(x + L) = U_{j,k}(x + L)e^{ik(x+L)}, \tag{3.56}$$

$$\psi_{j,k}(x + L) = U_{j,k}(x)e^{ikx}e^{ikL}, \tag{3.57}$$

$$\boxed{\psi_{j,k}(x + L) = \psi_{j,k}(x)e^{ikL}.} \tag{3.58}$$

Equation (3.58) is an expression of Bloch's theorem, which states that a potential with period L has wave functions in the jth band that can be separated into a part with the same period as the potential and a phase-wave term e^{ikL}.

Equation (3.55) shows that, in general, electron probability $|\psi_{j,k}(x)|^2 = |U_{j,k}(x)|^2$ depends upon k and j but only contains the cell-periodic part of the wave function.

The consequences of Bloch's theorem are quite dramatic when it comes to describing electron motion. Such motion in a periodic potential must involve propagation of the phase through the factor kx in Eq. (3.55). Associated with wave vector k is a *crystal momentum* $\hbar k$. This momentum is not the same as the momentum of an actual electron because the $U_{j,k}(x)$ term in Eq. (3.55) means that the electron has a wide range of momenta. The Bloch wave function $\psi_{j,k}(x) = U_{j,k}(x)e^{ikx}$ is not an eigenfunction of the momentum operator $\hat{p} = -i\hbar\, \mathrm{d}/\mathrm{d}x$. Crystal momentum $\hbar k$ is an effective momentum of the electron and is an extremely useful way of describing electron motion in a periodic potential.

3.6.2 Periodic Array of Rectangular Potential Energy Barriers

Electron dispersion in a periodic potential can be analyzed by considering a lattice of rectangular potential barriers of the type sketched in Fig. 3.22. This potential can be used in a Kronig–Penney model.

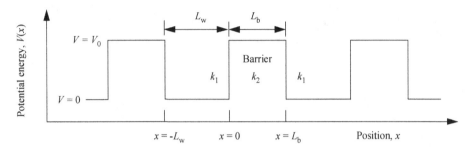

Fig. 3.22 Periodic array of rectangular potential barriers with energy V_0 and thickness L_b. The potential wells have thickness L_w and the cell repeats in distance $L = L_b + L_w$.

An electron of mass m_0 and energy $E > 0$ is constrained to motion in one dimension and experiences a periodic piecewise-constant potential with lattice constant L and a unit cell consisting of a rectangular barrier of thickness L_b, potential energy $V = V_0$, a potential well of thickness L_w, and potential energy $V = 0$. If a potential well is labeled region 1 and the adjacent potential barrier is labeled region 2, the wave function for an electron of energy E in the unit cell is

$$\psi_1(-L_w \le x < 0) = a\,e^{ik_1 x} + b\,e^{-ik_1 x} \tag{3.59}$$

and

$$\psi_2(0 \le x < L_b) = c\,e^{ik_2 x} + d\,e^{-ik_2 x}, \tag{3.60}$$

where the value of k_1 in the potential well is $k_1 = \sqrt{2m_0 E}/\hbar$ and the value of k_2 in the potential barrier region is $k_2 = \sqrt{2m_0(E - V_0)}/\hbar$. The amplitudes a, b, c, and d may be found by applying boundary conditions that enforce continuity, smoothness, and the phase factor e^{ikL} required by Blochs theorem, where k is the Bloch wave vector. This gives

$$\psi_1(x = 0) = \psi_2(x = 0), \tag{3.61}$$

$$\frac{\mathrm{d}}{\mathrm{d}x}\psi_1(x = 0) = \frac{\mathrm{d}}{\mathrm{d}x}\psi_2(x = 0), \tag{3.62}$$

$$e^{ikL}\psi_1(x = -L_w) = \psi_2(x = L_b), \tag{3.63}$$

$$e^{ikL}\frac{\mathrm{d}}{\mathrm{d}x}\psi_1(x = -L_w) = \frac{\mathrm{d}}{\mathrm{d}x}\psi_2(x = L_b). \tag{3.64}$$

Applying these conditions to the wave function in each region results in four linear homogeneous equations,

$$a + b = c + d, \tag{3.65}$$

$$(a - b)k_1 = (c - d)k_2, \tag{3.66}$$

$$e^{ikL}\left(a e^{-ik_1 L_w} + b e^{ik_1 L_w}\right) = c e^{ik_2 L_b} + d e^{-ik_2 L_b}, \tag{3.67}$$

$$e^{ikL}\left(a e^{-ik_1 L_w} - b e^{ik_1 L_w}\right)k_1 = (c e^{ik_2 L_b} - d e^{-ik_2 L_b})k_2, \tag{3.68}$$

which may be written as a matrix equation:

$$\begin{bmatrix} 1 & 1 & -1 & -1 \\ k_1 & -k_1 & -k_2 & k_2 \\ -e^{ikL-ik_1 L_w} & -e^{ikL+ik_1 L_w} & e^{ik_2 L_b} & e^{-ik_2 L_b} \\ -k_1 e^{ikL-ik_1 L_w} & k_1 e^{ikL+ik_1 L_w} & k_2 e^{ik_2 L_b} & -k_2 e^{-ik_2 L_b} \end{bmatrix} \begin{bmatrix} a \\ b \\ c \\ d \end{bmatrix} = \mathbf{A} \begin{bmatrix} a \\ b \\ c \\ d \end{bmatrix} = \mathbf{0}. \tag{3.69}$$

If the inverse $\mathbf{A}^{-1}$ exists then $a + b + c + d = 0$. Interesting solutions require that the inverse $\mathbf{A}^{-1}$ does *not* exist and this happens if the characteristic determinant is zero, so that

$$\begin{vmatrix} 1 & 1 & -1 & -1 \\ k_1 & -k_1 & -k_2 & k_2 \\ -e^{ikL-ik_1 L_w} & -e^{ikL+ik_1 L_w} & e^{ik_2 L_b} & e^{-ik_2 L_b} \\ -k_1 e^{ikL-ik_1 L_w} & k_1 e^{ikL+ik_1 L_w} & k_2 e^{ik_2 L_b} & -k_2 e^{-ik_2 L_b} \end{vmatrix} = |\mathbf{A}| = 0. \tag{3.70}$$

For an electron of energy E, real or imaginary solutions can exist for Bloch wave vector k. Real values of k correspond to electrons that have *delocalized propagating wave states* in the crystal. Imaginary values of k are associated with *localized electron states*. *Complex band structure* describes electron dispersion that can have real Bloch wave vector k or imaginary wave vector values.

3.6.3 Real Band Structure

The solution for real Bloch wave vector k is

$$\boxed{\cos(kL) = \cos(k_2 L_b)\cos(k_1 L_w) - \frac{(k_2^2 + k_1^2)}{2k_1 k_2}\sin(k_2 L_b)\sin(k_1 L_w) = \theta_B.} \tag{3.71}$$

The real band structure for propagating electron states of energy E and normalized wave vector kL/π requires $|\theta_B| < 1$ so that

$$\boxed{kL = \frac{\arccos(\theta_B)}{\pi}.} \tag{3.72}$$

3.6.4 Imaginary Band Structure

The imaginary band structure describing localized electron states has $|\theta_B| \geq 1$ and

$$\boxed{\kappa L = \frac{\operatorname{arccosh}(\theta_B)}{\pi},} \tag{3.73}$$

where $\kappa = |ik|$.

A consequence of *analytic continuity* is that the delocalized (real k) and localized (imaginary k) states in the dispersion relation connect smoothly. The complex band structure is usually *not* symmetric in the band gap energy.

3.6.5 Complex Band Structure

Since both real and imaginary wave vectors can be present in the crystal, the band structure is *complex*. The complex dispersion relation shown in Fig. 3.23 plots electron energy, E_k, as a function of the Bloch wave vector, k.

The allowed energy bands have real Bloch wave vector and may be thought of as arising from resonant transmission through all cell-periodic potentials that are coupled through tunneling when electron energy is less than barrier energy. The large number of such resonances overlap in energy to create a continuous energy band of allowed propagating states in a bulk crystal.

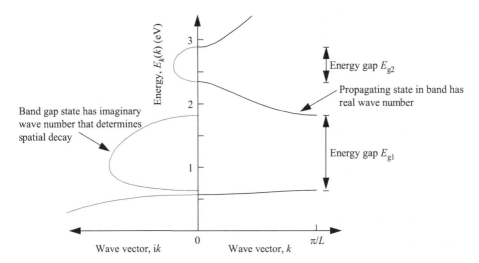

Fig. 3.23 Electron dispersion relation in the reduced-zone representation. Real wave vectors, k, associated with propagating states in a bulk crystal have values from $k = 0$ to $k = \pi/L$. In regions where there is an energy gap, E_g, wave vectors associated with non-propagating states in the bulk crystal take on complex values. The states in the energy gap E_{g1} have Im $k \neq 0$ and Re $k = \pi/L$. The states in energy gap E_{g2} have Im $k \neq 0$ and Re $k = 0$. The complex band structure is, in general, not symmetric in the band gap energy. This dispersion relation is an example of complex band structure with both real and imaginary wave vectors that are solution to Bloch's equation. Parameters are: electron mass $m = m_0$, potential barrier energy $V_0 = 2\,\mathrm{eV}$, lattice constant $L = 1\,\mathrm{nm}$, barrier thickness $L_b = 0.5\,\mathrm{nm}$, and well thickness $L_w = 0.5\,\mathrm{nm}$.

Energy gap regions form where there are no resonances and an imaginary Bloch wave vector in the complex band structure describes a non-propagating localized state in the crystal. The wave vector is in the complex plane. Such localized states can exist at a defect in a bulk crystal, at a crystal surface, or at a crystal heterointerface. An analysis of complex band structure is an *essential* first step towards understanding and developing accurate models of many nanoscale electronic devices, including those involving electron tunneling.

At the boundary between the allowed and forbidden gap region of bulk-crystal complex band structure, the electron wave functions are standing waves and have definite symmetry.

3.6.6 The Tight Binding Approximation

There are other ways to calculate the electron dispersion relation in periodic potentials. One approach is called the tight binding approximation.

The method is illustrated by initially considering an electron that exists locally in the lowest energy s-state $\phi(\mathbf{r})$, or *s-atomic orbital*, of an isolated atom. The electronic structure of a periodic array of such atoms is then developed by allowing a small overlap of electronic wave functions between adjacent atoms.

For simplicity, consider a one-dimensional crystal with a primitive cell that contains one atom and with lattice sites at positions $x_n = nL$ where n is an integer and L is the nearest-neighbor atom spacing. The model assumes there are integer N atoms and the lattice has periodic boundary conditions. If the potential of the nth atom at position $x_n = nL$ is $V(x - x_n)$, then the total potential for the crystal is the sum over single-atom potentials. The time-independent Schrödinger equation for an electron at position x is

$$\hat{H}\psi_k(x) = \left(\frac{-\hbar^2}{2m_0} \frac{\mathrm{d}^2}{\mathrm{d}x^2} + \sum_n V(x - x_n) \right) \psi_k(x) = E\psi_k(x), \tag{3.74}$$

where $\psi_k(x)$ is a wave function that must satisfy the Bloch condition $\psi_k(x + L) = \psi_k(x)e^{ikL}$ given by Eq. (3.58) and k is the Bloch wave vector.

Previously, *delocalized* Bloch functions $\psi_k(x) = U_k(x)e^{ikx}$ in which $U_k(x) = U_k(x+L)$ were used. However, *localized* states (Wannier functions) $\phi(x)$ can also satisfy Bloch's theorem. The Wannier functions are localized around the lattice site x_n and are orthogonal for different lattice points so that

$$\int \phi^*(x - x_m)\phi(x - x_n)\mathrm{d}x = \delta_{mn}. \tag{3.75}$$

The Wannier functions $\phi(x)$ are not restricted in the same way as the $U_k(x)$ of Bloch functions for which $U_k(x) = U_k(x + L)$.

The Wannier functions are related to the Bloch functions via the direct sum over N lattice sites

$$\psi_k(x) = \frac{1}{\sqrt{N}} \sum_{n=1}^{N} e^{ikx_n} \phi(x - x_n), \tag{3.76}$$

and this expression for $\psi_k(x)$ satisfies Bloch's theorem (see Problem 3.9).

To first order, the expectation value of electron energy is (ignoring the normalization term $1/\sqrt{N}$),

$$E_k = \int \psi_k^*(x)\hat{H}\psi_k(x)\mathrm{d}x = \sum_{x_m}\sum_{x_n} e^{ik(x_n - x_m)}\int \phi_k^*(x - x_m)\hat{H}\phi_k(x - x_n)\mathrm{d}x. \qquad (3.77)$$

If there is little overlap between atomic electron wave functions that are separated by two or more nearest-neighbor atom spacings, then only two integrals need to be kept. The first is the shifted (renormalized) atom energy level,

$$-E_0 = \int \phi_k^*(x)\hat{H}\phi_k(x)\mathrm{d}x, \qquad (3.78)$$

and the second is the contribution from overlaps of nearest neighbors,

$$-t_{\mathrm{hop},1} = \int \phi_k^*(x - x_m)\hat{H}\phi_k(x - x_n)\mathrm{d}x, \qquad (3.79)$$

in which $x_m = x_n \pm L$. In Eq. (3.79), $t_{\mathrm{hop},1}$ is the nearest-neighbor overlap (or hopping) integral with the sign convention that t_{hop} is negative for s-atomic orbitals and positive for p-atomic orbitals. Hence, the electron dispersion relation for a linear chain of s-atomic orbitals with equal atom spacing L in the limit $N \to \infty$ is

$$E_k = -E_0 - t_{\mathrm{hop},1}(e^{ikL} + e^{-ikL}) = -E_0 - 2t_{\mathrm{hop},1}\cos(kL). \qquad (3.80)$$

This cosine tight binding band for nearest-neighbor interactions in one dimension has an energy band width of $E_b = 4t_{\mathrm{hop},1}$. For convenience $E_0 = 0$, giving

$$E_k = -2t_{\mathrm{hop},1}\cos(kL). \qquad (3.81)$$

If next nearest-neighbor interactions are included then there is an additional overlap integral $t_{\mathrm{hop},2}$ involving sites $x_m = x_n \pm 2L$ that modifies the dispersion relation to give

$$E_k = -2t_{\mathrm{hop},1}\cos(kL) - 2t_{\mathrm{hop},2}\cos(2kL). \qquad (3.82)$$

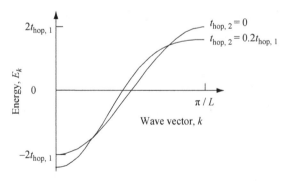

Fig. 3.24 Electron dispersion relation E_k for a one-dimensional crystal with $N \to \infty$, nearest neighbor atom spacing L, and s-atomic orbitals at lattice sites. The dispersion relation E_k is calculated in the tight binding approximation for nearest neighbor interaction ($t_{\mathrm{hop},2} = 0$) giving a cosine band of energy width $E_b = 4t_{\mathrm{hop},1}$. When next nearest neighbor interactions are included (in this case with $t_{\mathrm{hop},2} = 0.2 \times t_{\mathrm{hop},1}$) the dispersion relation E_k is modified.

The effect on E_k of including $t_{hop,2} = 0.2 \times t_{hop,1}$ is illustrated in Fig. 3.24. If the dispersion relation for nearest-neighbor interactions given by Eq. (3.81) is generalized to a cubic lattice with $L = L_x = L_y = L_z$ and Bloch wave vector components k_x, k_y, and k_z, then

$$E_k = -2t_{hop,1}(\cos(k_x L) + \cos(k_y L) + \cos(k_z L)). \tag{3.83}$$

In this case, the three-dimensional cosine tight binding band has an energy band width of $E_b = 12t_{hop,1}$.

3.6.7 Crystal Momentum and Effective Electron Mass

The presence of a periodic potential changes the way an electron moves in a bulk crystal. There are energy bands described by the dispersion relation $\omega = \omega(k)$ in which an electron wave packet can propagate freely at group velocity

$$v_g = \frac{\partial}{\partial k}\omega(k). \tag{3.84}$$

There are also energy band gaps where the electron cannot propagate freely. Clearly, when compared to free space, the presence of a periodic potential dramatically changes electron dynamics.

The influence a periodic potential $V(x)$ has on the response of an electron to an external force requires the introduction of *crystal momentum*. To show this, consider the effect on electron motion of an external electric field $\mathbf{E}$ applied in the x direction. Classical mechanics suggests that an electron subject to an external force $e\mathbf{E}$ causes a change in its momentum. This electron motion in quantum mechanics is described by the total Hamiltonian,

$$\hat{H} = \frac{\hat{p}_x^2}{2m_0} + \hat{V}(x) - e|\mathbf{E}|\hat{x}. \tag{3.85}$$

The electron wave function $\psi(x,t)$ evolves in time according to the time-dependent Schrödinger equation

$$\hat{H}\psi(x,t) = i\hbar\frac{\partial}{\partial t}\psi(x,t). \tag{3.86}$$

For an electron in state $\psi_0(x, t = 0)$ at time $t = 0$, Eq. (3.86) may be rewritten as

$$\psi(x,t) = e^{-i\hat{H}t/\hbar}\psi_0(x, t = 0), \tag{3.87}$$

which for the Hamiltonian considered is

$$\psi(x,t) = e^{-i\left(\frac{\hat{p}_x^2}{2m_0} + \hat{V}(x) - e|\mathbf{E}|\hat{x}\right)t/\hbar}\psi_0(x, t = 0). \tag{3.88}$$

Replacing x with $(x + L)$ gives

$$\psi((x+L),t) = e^{-i\left(\frac{\hat{p}_x^2}{2m_0} + \hat{V}(x+L) - e|\mathbf{E}|(\hat{x}+L)\right)t/\hbar}\psi_0((x+L), t = 0). \tag{3.89}$$

133

Using the fact that $V(x) = V(x + L)$ for a periodic potential, and using Bloch's theorem, $\psi_{j,k}(x + L) = \psi_{j,k}(x)e^{ikL}$, for an electron in a Bloch state, this gives

$$\psi((x + L), t) = e^{ie|\mathbf{E}|Lt/\hbar}e^{ik(t=0)L}e^{-i\left(\frac{\hat{p}_x^2}{2m_0} + \hat{V}(x) - e|\mathbf{E}|\hat{x}\right)t/\hbar}\psi_0(x, t = 0), \qquad (3.90)$$

and hence, comparing with Eq. (3.88),

$$\psi((x + L), t) = e^{ik(t)L}\psi(x, t), \qquad (3.91)$$

where

$$k(t) = \frac{e|\mathbf{E}|t}{\hbar} + k(t = 0). \qquad (3.92)$$

The time derivative of Eq. (3.92) can be written as

$$\frac{\mathrm{d}}{\mathrm{d}t}(\hbar k) = e|\mathbf{E}|. \qquad (3.93)$$

A conclusion is that the effect of an external force is to change a quantity $\hbar k$. This is called the *crystal momentum*. Electrons accelerate according to the rate of change of crystal momentum in the periodic potential.

In analogy with classical mechanics, the acceleration of a particle in a given allowed band of propagating states due to an external force can be described by

$$\frac{\mathrm{d}}{\mathrm{d}t}\left(\frac{\mathrm{d}x}{\mathrm{d}t}\right) = \frac{\mathrm{d}}{\mathrm{d}t}\frac{\partial\omega(k)}{\partial k} = \frac{\partial^2\omega(k)}{\partial k^2}\frac{\mathrm{d}k}{\mathrm{d}t} = \frac{1}{\hbar}\frac{\partial^2\omega(k)}{\partial k^2}\frac{\mathrm{d}}{\mathrm{d}t}(\hbar k) = \frac{1}{\hbar}\frac{\partial^2\omega(k)}{\partial k^2}e|\mathbf{E}|, \qquad (3.94)$$

where use has been made of Eqs. (3.84) and (3.93).

The result given by Eq. (3.94) can be expressed in the usual Newtonian form of force equals mass times acceleration if an electron effective mass $m_e^*(k)$ is introduced for the particle so that

$$\boxed{m_e^*(k) = \frac{\hbar}{\frac{\partial^2}{\partial k^2}\omega(k)}.} \qquad (3.95)$$

Equation (3.95) indicates that effective mass is inversely proportional to the curvature of the dispersion relation for a given allowed band of propagating states.

Usually, when discussing effective electron mass, a quantity $m_{\mathrm{eff}}(k)$ is used, which is the effective electron mass normalized with respect to the bare electron mass m_0, so that

$$m_e^*(k) = m_{\mathrm{eff}}(k) \times m_0. \qquad (3.96)$$

For an electron in the conduction band of GaAs, the value $m_{\mathrm{eff}} = 0.07$ is often used, so the effective electron mass is $m_e^* = 0.07 \times m_0$.

To illustrate how a dispersion relation influences particle group velocity and effective mass, consider a particular allowed band that is described by a cosine function (see Eq. (3.81)) of wave vector k in such a way that

$$E(k) = \frac{E_b}{2}(1 - \cos(kL)) = E_b \sin^2\left(\frac{kL}{2}\right), \qquad (3.97)$$

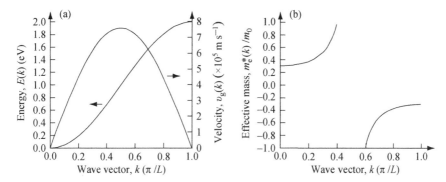

Fig. 3.25 (a) Electron dispersion relation $E(k) = E_b(1 - \cos(kL))/2\hbar$ and group velocity $v_g(k)$ as a function of wave vector k. Band width $E_b = 2\,\mathrm{eV}$, and lattice constant $L = 0.5\,\mathrm{nm}$. (b) Effective electron mass $m_e^*(k)$ as a function of wave vector k. The free electron mass is m_0.

where E_b is the energy band width. Using this prototype dispersion relation, the group velocity is

$$v_g(k) = \frac{\partial}{\partial k} \frac{E(k)}{\hbar} = \frac{E_b L}{2\hbar} \sin(kL) \tag{3.98}$$

and an effective mass is

$$m_e^*(k) = \frac{\hbar}{\frac{\partial^2}{\partial k^2} w(k)} = \frac{2\hbar^2}{E_b L^2 \cos(kL)}. \tag{3.99}$$

Equations (3.97), (3.98), and (3.99) are plotted in Fig. 3.25, assuming an electron band width energy $E_b = 2\,\mathrm{eV}$ and a periodic potential lattice constant $L = 0.5\,\mathrm{nm}$. A quite remarkable prediction evident in Fig. 3.25(b) is the negative effective electron mass $m_e^*(k)$ for certain values of wave vector k. The physical meaning of this can be seen in Fig. 3.25(a) – electron velocity $v_g(k)$ can decrease with increasing wave vector k.

3.6.7.1 The Band Structure of GaAs

So far, a relatively simple model of a periodic potential has been considered. While the electron dispersion relations of actual crystalline solids are more complex, they do retain the important features, such as band gaps. Figure 3.26 shows part of the calculated electron dispersion relation of GaAs along two of the principal symmetry directions, $\Gamma - X$ (the [100] direction, indicated by Δ) and $\Gamma - L$ (the [111] direction, indicated by Λ). The compound GaAs has the zinc blende crystal structure with a low-temperature lattice constant of $L = 0.565\,\mathrm{nm}$. There is interest in GaAs and other III–V compound semiconductors[7] in part because they can be used to make laser diodes and high-speed transistors.

---

[7] For additional information on the electronic band structure of semiconductors, see M. L. Cohen and J. R. Chelikowsky, *Electronic Structure and Optical Properties of Semiconductors*, Springer Series in Solid-State Sciences **75**, ed. M. Cardona, New York, Springer-Verlag, 1989 (ISBN 0 387 51391 4). For complex band structure, see Y-C Chang, *Phys. Rev. B* **25**, 605 (1982).

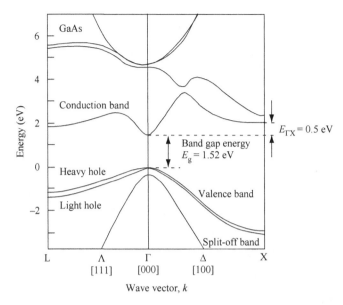

Fig. 3.26 Calculated low-temperature band structure of GaAs from the Γ symmetry point toward the L and X crystal symmetry points. The conduction band, heavy-hole band, light-hole band, and split-off band are indicated. The conduction band effective electron mass near the Γ symmetry point is approximately $m_e^* = 0.07 \times m_0$. The heavy-hole mass is $m_{hh} = 0.5 \times m_0$, and the light-hole mass is $m_{lh} = 0.08 \times m_0$. The band gap energy $E_g = 1.52\,\text{eV}$ is indicated, as is the energy $E_{\Gamma X}$ separating the minimum of the conduction band at Γ and the subsidiary minimum near the X symmetry point.

As may be seen in the dispersion relation of Fig. 3.26, there is a conduction band and a valence band separated in energy by a band gap. At temperatures close to $0\,\text{K}$, the band gap energy is $E_g^{0\,\text{K}} = 1.52\,\text{eV}$. At room temperature, this value reduces to $E_g^{300\text{K}} = 1.42\,\text{eV}$.

3.7 The Nonequilibrium Electron Transistor

What is known about quantum mechanical transmission and reflection of electrons at potential-energy steps can be applied to the design of a new type of semiconductor transistor. The techniques described in Section 1.2.5 can be used to create a conduction band minimum potential profile for a semiconductor transistor that operates by injecting electrons from a lightly n-type doped, wide band-gap emitter in such a way that they have high velocity while traversing a thin, heavily n-type doped base region. Electrons that successfully traverse the base of thickness x_B without scattering emerge as current flowing in the lightly n-type doped collector. Figure 3.27(a) is a schematic diagram of this double heterojunction unipolar transistor. Indicated in the figure is the emitter current I_E, base current I_B, collector current I_C, and voltages V_{BE} and V_{CE} for the transistor biased in the common emitter configuration.

This nonequilibrium electron transistor (NET) makes use of relatively high-velocity electron transmission through the base with little or no scattering. Extreme nonequilibrium

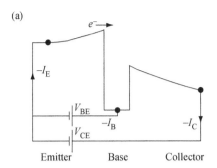
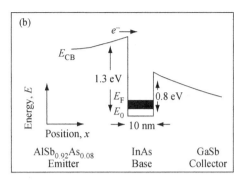

Fig. 3.27 (a) Schematic diagram of a double heterojunction unipolar transistor. Emitter current I_E, base current I_B, collector current I_C, and voltages V_{BE} and V_{CE} for the transistor biased in the common emitter configuration are indicated. The large dots indicate electrical contacts between the transistor and the leads that connect to batteries. (b) Schematic diagram of the conduction-band edge potential profile, E_{CB}, of an AlSb$_{0.92}$As$_{0.08}$/InAs/GaSb double heterojunction unipolar NET under bias. The confinement energy E_0 and the Fermi energy E_F of the occupied two-dimensional electron states in the InAs base are indicated. Electrons represented by e^- are injected from the forward-biased AlSb$_{0.92}$As$_{0.08}$ emitter into the InAs base region with a large excess kinetic energy. The effective electron mass near E_{CB} is $m^*_{InAs} = 0.021 \times m_0$ in the base and $m^*_{GaSb} = 0.048 \times m_0$ in the collector.

electron transport or ballistic electron transport occurs in the base. If the distance a nonequilibrium electron travels before scattering is on average λ_B, then the probability of electron transmission through the base region without scattering is e^{-x_B/λ_B}.

To illustrate some other design considerations, Fig. 3.27(b) shows a schematic diagram of the conduction-band edge potential (E_{CB}) of an AlSb$_{0.92}$As$_{0.08}$As/InAs/GaSb double heterojunction unipolar NET under bias. The transistor base consists of a 10 nm thick epilayer of InAs with a high carrier concentration of two-dimensionally confined electrons, $n \sim 2 \times 10^{12}$ cm^{-2}. The collector arm is a 350 nm thick layer of Te doped ($n \sim 1 \times 10^{16}$ cm^{-3}) GaSb, and the entire structure is grown by molecular beam epitaxy on a (001)-oriented n-type GaSb substrate. To support high-current density necessary for high-speed operation and to avoid space-charging effects in the emitter and collector arms of a NET, it is necessary to dope them to an n-type impurity concentration density $n > j/ev_{av}$, where v_{av} is an appropriate average electron velocity and j is the current density.

The potential energy step at the emitter base heterointerface is $\phi_{EB} = 1.3$ eV, and the potential energy step at the base-collector interface is $\phi_{BC} = 0.8$ eV. The device can operate at room temperature because both the emitter barrier energy ϕ_{EB} and collector barrier energy ϕ_{BC} are much greater than the ambient thermal energy $k_BT \sim 0.025$ eV, so that reverse (leakage) currents are small.

The lowest-energy bound-electron state in the finite rectangular potential well of the transistor base is indicated as E_0. At low temperatures, electrons in the potential well occupy quantum states up to a Fermi energy $E_F = \hbar^2 k_F^2/2m^*_e$, where k_F is the Fermi wave vector.

Because emitter electrons injected into the base have high kinetic energy $E_k \sim \phi_{EB}$, they also have a high velocity. A typical nonequilibrium electron velocity is near 10^8 cm s^{-1}, so the majority of injected electrons with a large component of momentum in the x direction traverse the base of thickness $x_B = 10$ nm in the very short time of near

137

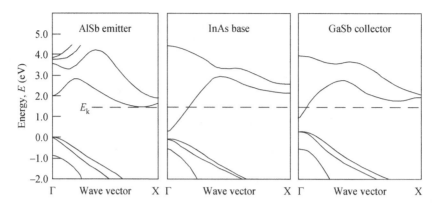

Fig. 3.28 Dispersion curves for electron motion through the three materials forming the AlSb/InAs/GaSb double heterostructure unipolar NET. The dashed line indicates the approximate value of energy for an electron moving through the device. Electron states used in the transmission of an electron energy E through (100)-oriented semiconductor layers are near the points where the dashed line intersects the solid curves. The value of effective electron mass near the conduction band minimum (E_{CB}) in the $\Gamma-X$ direction in the AlSb emitter is close to the free-electron mass, m_0. At an energy E_k, which is above E_{CB} in the base and collector, the effective electron masses are greater than their respective values near E_{CB}.

10 fs (assuming, of course, that an electron may be described as a point particle with a well-defined *local* velocity).

It is important to inject electrons in a narrow range of energies close to E_k. Optimum device performance occurs when quantum reflections from ϕ_{BC} are minimized, and this may only be achieved for a limited range of injection energies. Reflections from the abrupt change in potential at ϕ_{BC} are minimized when the nonequilibrium electron group velocity (the slope $\partial\omega/\partial k$ at energy E_k in Fig. 3.28) is the same on each side of the base-collector heterointerface. As discussed in Section 2.8, this impedance-matching condition is $m^*_{InAs}/m^*_{GaSb} = E/(E - \phi_{BC})$, where m^*_{InAs} and m^*_{GaSb} are appropriate effective electron masses in the base and collector, respectively. Therefore, by choosing E_k, ϕ_{BC}, base and collector materials quantum reflections from ϕ_{BC} may be eliminated for a small range ($\sim0.5\,\mathrm{eV}$) of E_k. At a real heterojunction interface, impedance matching also requires that the symmetry of the electron wave function in the base and collector be similar.[8]

Although it is known that degradation in device performance due to quantum mechanical reflection can be designed out of the device, there are other considerations that should be addressed. For example, an injected electron, with energy E_k and wave vector $\mathbf{k}$, moving in the x direction is able to scatter via the coulomb interaction into high-energy states in the base, changing energy by an amount $\hbar\omega$ and changing wave vector by $\mathbf{q}$. Inelastic processes of this type include the electron scattering from lattice vibrations of the semiconductor crystal. Another possible inelastic process involves the high-energy injected electron scattering from low-energy (thermal) electrons in the transistor base. Because these scattering processes can change the direction of travel of the electron, they can also dramatically reduce the number of electrons that traverse the base and flow as

---

[8] M. D. Stiles and D. R. Hamann, *Phys. Rev. B* **38**, 2021 (1988). If the symmetry of the electron wave function is not matched on either side of the interface, the electron can suffer *complete* quantum mechanical reflection *even* if the electron velocity is precisely matched!

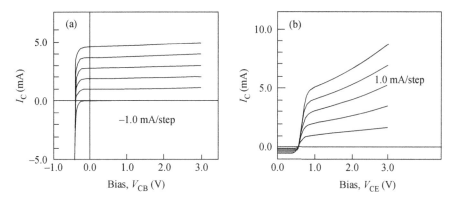

Fig. 3.29 (a) Room-temperature ($T = 300$ K) common-base current gain characteristics of the device shown schematically in Fig. 3.27(b). Curves were taken in steps beginning with an injected emitter current of zero. The emitter area is 7.8×10^{-5} cm^2, I_C is collector current, and V_{CB} is collector-base voltage bias. (b) Room-temperature common emitter current gain characteristics. Curves were taken in steps of $I_B = 0.1$ mA, beginning with an injected base current of zero, and V_{CE} is collector-emitter voltage bias. Current gain $\beta \sim 10$.

current in the collector. Such dynamical constraints imposed by inelastic scattering are quite difficult to deal with and thus make NET design a significant task. Nevertheless, some NETs do have quite good device characteristics.

Figure 3.29 shows measured common base current gain $\alpha = I_C/I_E$ and common emitter current gain $\beta = I_C/I_B$ for a device maintained at a temperature $T = 300$ K.[9] As may be seen in Fig. 3.29(b), the room-temperature value of β increases from $\beta = 10$ at $V_{CE} = 1.0$ V to $\beta = 17$ at $V_{CE} = 3.0$ V. At the liquid nitrogen temperature of $T = 77$ K, the current gain $\beta = 12$ at $V_{CE} = 1.0$ V and $\beta = 40$ at $V_{CE} = 3.0$ V. The reduction in base collector potential barrier energy ϕ_{BC}, with increasing collector voltage bias V_{CE}, improves collector efficiency for incoming nonequilibrium electrons and, in agreement with simple calculations, contributes to the observed slope in the common emitter saturation characteristics.

It is possible to devise experiments that further explore the role of quantum mechanical reflection from the abrupt change in potential energy ϕ_{BC} seen by a nonequilibrium electron approaching the base collector heterostructure. Figure 3.30(a) shows the band diagram for a NET device with a 15 nm thick AlSb tunnel emitter that may be used to inject an essentially monoenergetic beam of electrons perpendicular to the heterointerface. After traversing the 10 nm thick base, electrons impinge on the GaSb collector. If the injection energy E_k is less than the base-collector barrier energy, $\phi_{BC} \sim 0.8$ eV, no electrons are collected, and all the injected current I_E flows in the base. For $E_k > \phi_{BC}$, some electrons traverse the base and subsequently contribute to the collector current I_C.

An important scattering mechanism determining collector efficiency is quantum mechanical reflection from the step change in potential energy ϕ_{BC}. This reflection is determined in part by electron velocity mismatch across the abrupt InAs/GaSb heterointerface. For electron velocities v_{InAs} in InAs and v_{GaSb} in GaSb, quantum mechanical reflection is

[9] A. F. J. Levi and T. H. Chiu, *Appl. Phys. Lett.* **51**, 984 (1987).

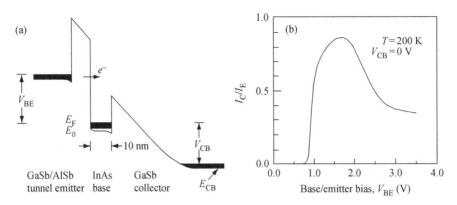

Fig. 3.30 (a) Conduction-band diagram of an $AlSb_{0.92}As_{0.08}$/InAs/GaSb tunnel emitter hetero-junction unipolar NET under base emitter bias V_{BE} and collector base bias V_{CB}. The conduction-band minimum E_{CB} is indicated, as are the confinement energy E_0 and the Fermi energy E_F of the occupied two-dimensional electron states in the 10 nm thick InAs base. The energy of a beam of electrons traversing the base and impinging on the base-collector heterostructure is controlled by varying V_{BE}. (b) Measured ratio of the collector and emitter currents, I_C/I_E, as functions of base emitter bias, V_{BE}. The ratio I_C/I_E is related to the electron transmission probability across the base-collector heterostructure. Measurement performed with sample at temperature $T = 200$ K.

$Refl = ((v_{InAs} - v_{GaSb})/(v_{InAs} + v_{GaSb}))^2$. In this simplified example, it is assumed that there is no contribution to $Refl$ from a mismatch in the character (symmetry) of the electron wave function across the interface. Since $E_k \propto V_{BE}$ to within ± 0.1 eV, electron collection efficiency may be explored as a function of injection energy by plotting the ratio I_C/I_E with base/emitter voltage bias V_{BE}. Typical results of such a measurement at a temperature $T = 200$ K are shown in Fig. 3.30(b).[10] As may be seen, for injection energy E_k less than ϕ_{BC}, no electrons are collected. For $E_k > \phi_{BC}$, the ratio I_C/I_E increases rapidly with decreasing velocity mismatch at either side of the heterointerface. Maximum base transport efficiency occurs for $E_k \sim 1.5$ eV. With further increase in E_k, collection efficiency decreases, and finally for $E_k \sim 2.5$ eV less than 50% of electrons are collected. At energies greater than 2.5 eV, electrons are injected into electronic states above the lowest conduction band in which both velocity and wave function character mismatch creating high quantum mechanical reflection. Hence, for $E_k \geq 2.5$ eV, electron collection efficiency decreases dramatically.

The transistor conduction band profile in Fig. 3.30(a) shows a tunnel emitter. If there is interest in designing such a transistor for high-speed operation then it seems natural to develop a measure of the electron tunneling time through the emitter. This, as described in Section 3.4.4, is a nontrivial task with a rather involved answer. It is not possible, for example, to relate the tunneling rate that comes from tunneling probability to a tunneling time.

To summarize, using a limited amount of quantum mechanics there is much that can be understood about the design of a new type of transistor. The device operates by injecting nonequilibrium electrons that are transported through a region of semiconductor only 10 nm thick. The existence of extreme nonequilibrium electron transport is a *necessary*

[10]T. H. Chiu and A. F. Levi, *Appl. Phys. Lett.* **55**, 1891 (1989).

condition for successful device operation. Transistor performance is improved by carefully matching the velocity across the base collector semiconductor heterojunction.

It is quite surprising that understanding key elements of this transistor design is successful without the need to develop a more sophisticated approach. In fact, the understanding is quite superficial. There are a number of complex and subtle issues concerning the description of nonequilibrium electron motion in the presence of strong elastic and inelastic scattering that are the subject of continued study.[11]

3.8 Other Engineering Applications

The propagation matrix method can be used in a number of other engineering applications. By way of example, the matrix method is applied to the propagation of an electromagnetic wave normal to an inhomogeneous dielectric medium consisting of dielectric layers.

The propagation matrix is defined for a domain as $\begin{bmatrix} A \\ B \end{bmatrix} = \mathbf{P} \begin{bmatrix} C \\ D \end{bmatrix}$, where $\mathbf{P} = \mathbf{P}_1 \mathbf{P}_2 \cdots \mathbf{P}_j \cdots \mathbf{P}_{N_{\text{step}}}$ and

$$\mathbf{P}_j = \frac{1}{2} \begin{bmatrix} \left(1 + \frac{k_{j+1}}{k_j}\right) e^{-ik_j L_j} & \left(1 - \frac{k_{j+1}}{k_j}\right) e^{-ik_j L_j} \\ \left(1 - \frac{k_{j+1}}{k_j}\right) e^{ik_j L_j} & \left(1 + \frac{k_{j+1}}{k_j}\right) e^{ik_j L_j} \end{bmatrix}. \tag{3.100}$$

For the optical problem, the refractive index $n_r(x)$ controls the wavelength in the medium. As an example, consider an electromagnetic wave of wavelength $\lambda_{\text{vac}} = 1\,\mu\text{m}$ propagating in free space and incident on a lossless dielectric with relative permittivity $\varepsilon_r > 1$ and relative permeability $\mu_r = 1$. The refractive index for vacuum is $n_{\text{vac}} = 1$, and for the dielectric it is $n_r = \sqrt{\varepsilon_r} > 1$. Example values for a dielectric refractive index are 1.45 for SiO_2 and 3.55 for Si. To find the transmission and reflection of the electromagnetic wave, the propagation matrix of Eq. (3.100) may be used directly, only now $k_j \to 2\pi/\lambda_j$, where $\lambda_j = \lambda_{\text{vac}}/n_{r,j}$ and λ_{vac} is the wavelength in vacuum. Hence,

$$\mathbf{P}_j = \frac{1}{2} \begin{bmatrix} \left(1 + \frac{\lambda_j}{\lambda_{j+1}}\right) e^{-i(2\pi L_j)/\lambda_j} & \left(1 - \frac{\lambda_j}{\lambda_{j+1}}\right) e^{-i(2\pi L_j)/\lambda_j} \\ \left(1 - \frac{\lambda_j}{\lambda_{j+1}}\right) e^{i(2\pi L_j)/\lambda_j} & \left(1 + \frac{\lambda_j}{\lambda_{j+1}}\right) e^{i(2\pi L_j)/\lambda_j} \end{bmatrix}. \tag{3.101}$$

This may be rewritten as

$$\mathbf{P}_j = \frac{1}{2} \begin{bmatrix} \left(1 + \frac{n_{r,j+1}}{n_{r,j}}\right) e^{-i(2\pi n_{r,j} L_j)/\lambda_{\text{vac}}} & \left(1 - \frac{n_{r,j+1}}{n_{r,j}}\right) e^{-i(2\pi n_{r,j} L_j)/\lambda_{\text{vac}}} \\ \left(1 - \frac{n_{r,j+1}}{n_{r,j}}\right) e^{i(2\pi n_{r,j} L_j)/\lambda_{\text{vac}}} & \left(1 + \frac{n_{r,j+1}}{n_{r,j}}\right) e^{i(2\pi n_{r,j} L_j)/\lambda_{\text{vac}}} \end{bmatrix}. \tag{3.102}$$

If the electromagnetic wave is incident on the domain from the left, $A = 1$, and, assuming no reflection at the far right of the domain, $D = 0$. Thus,

$$\begin{bmatrix} A \\ B \end{bmatrix} = \mathbf{P} \begin{bmatrix} C \\ D \end{bmatrix} \text{ becomes}$$

---

[11]For example, see A. F. J. Levi, *Essential Electron Transport for Device Physics*, Melville, New York, AIP Publishing, 2020 (ISBN 978 0 7354 2158 5); A. F. J. Levi, *Electron. Lett.* **24**, 1273 (1988).

$$\begin{bmatrix} 1 \\ B \end{bmatrix} = \begin{bmatrix} P_{11} & P_{12} \\ P_{21} & P_{22} \end{bmatrix} \begin{bmatrix} C \\ 0 \end{bmatrix}.$$

The transmission coefficient is obtained from $1 = P_{11}C$ and the reflection coefficient from $B = P_{21}C$. Transmission intensity is $|C|^2 = |1/P_{11}|^2$ and reflected intensity is $|B^2| = |P_{21}/P_{11}|^2$.

From this, it follows that reflected electromagnetic power from a step change in refractive index for a plane wave incident from vacuum is

$$r = |B|^2 = \left| \frac{k_{\text{vac}} - k_1}{k_{\text{vac}} + k_1} \right|^2 = \left| \frac{1 - k_1/k_{\text{vac}}}{1 + k_1/k_{\text{vac}}} \right|^2 = \left(\frac{1 - n_{\text{r}}}{1 + n_{\text{r}}} \right)^2. \tag{3.103}$$

This is just the well-known result from standard classical optics. The reflection in optical power may be thought of as due to a velocity mismatch. To show this, note that the velocity of light, v_j, in a homogeneous isotropic medium of refractive index $n_{\text{r},j}$ is $v_j = c/n_{\text{r},j}$. Hence, reflection due to a step change in refractive index from $n_{\text{r},j}$ to $n_{\text{r},j+1}$ is

$$r = \left| \frac{k_j - k_{j+1}}{k_j + k_{j+1}} \right|^2 = \left| \frac{n_{\text{r},j} - n_{\text{r},j+1}}{n_{\text{r},j} + n_{\text{r},j+1}} \right|^2 = \left(\frac{1/v_j - 1/v_{j+1}}{1/v_j + 1/v_{j+1}} \right)^2 = \left(\frac{v_{j+1} - v_j}{v_{j+1} + v_j} \right)^2. \tag{3.104}$$

This is the same result obtained previously in Section 2.8 for the transmission of an electron energy E over a potential energy step V_0 where $E > V_0$.

The propagation matrix method may be used to calculate not only transmission and reflection from a single dielectric step but also transmission and reflection from a dielectric with much more complex spatial variation in refractive index, such as occurs in multilayer dielectric optical coatings. Exercise 3.3 makes use of the propagation matrix method to design multilayer dielectric mirrors for a laser.

3.9 The WKB Approximation

The methods used in this chapter are not the only approach to solving transmission and reflection of particles due to spatial changes in potential. By way of example, this section describes what is known as the WKB approximation. In their papers published in 1926, Wentzel, Kramers, and Brillouin introduced into quantum mechanics what is in essence a semiclassical method.[12] In fact, their basic approach to solving differential equations had already been introduced years earlier, in 1837, by Liouville.[13]

The underlying idea may be explained by noting that a particle of mass m and energy $E = \hbar\omega$ moving in free space has a wave function $\psi(\mathbf{r}, t) = A e^{i(\mathbf{k}\cdot\mathbf{r} - \omega t)}$ for which the wave vector $\mathbf{k}$ does not vary. If the particle now encounters a potential and the potential is slowly varying, then it should be possible to obtain a *local* value of $\mathbf{k}$ from the local kinetic energy. For a potential that is varying slowly enough, this is usually a good approximation, and for $E > V$ then $|\mathbf{k}(\mathbf{r})| = \sqrt{2m(E - V(\mathbf{r}))}/\hbar$. Of course, the phase of the wave function must also change from $\mathbf{k}\cdot\mathbf{r}$ to an integral $\int \mathbf{k}(\mathbf{r})\cdot d\mathbf{r}$ to take into account an assumed local wave vector that is a function of space.

[12] G. Wentzel, *Z. Physik.* **38**, 518 (1926); H. A. Kramers, *Z. Physik.* **39**, 828 (1926); L. Brillouin, *Compt. Rend.* **183**, 24 (1926).
[13] J. Liouville, *J. de Math.* **2**, 16, 418 (1837).

Consider a particle moving in one dimension. A slowly and smoothly varying potential $V(x)$ is assumed so that the approximation $k = k(x)$ may be made and the exponent of the wave function may be integrated. In this semiclassical approximation, the spatial part of the wave function is of the form

$$\psi(x) = \frac{a}{\sqrt{k(x)}} e^{i \int^x k(x') dx'} \tag{3.105}$$

and

$$k(x) = \sqrt{\frac{2m(E - V(x))}{\hbar^2}}, \qquad \text{for } E > V, \tag{3.106}$$

$$k(x) = i\sqrt{\frac{2m(V(x) - E)}{\hbar^2}}, \qquad \text{for } E < V. \tag{3.107}$$

In general, there are left- and right-propagating waves so that

$$\psi_{\text{WKB}}(x) = \frac{1}{\sqrt{k(x)}} \left(a\, e^{i \int^x k(x') dx'} + b\, e^{-i \int^x k(x') dx'} \right). \tag{3.108}$$

The accuracy of the WKB method relies on the fractional change in wave vector $k(x)$ being very much less than unity over the distance λ. Obviously, at the classical turning points, where $E = V(x)$ and $k(x) \to 0$, the particle wavelength is infinite. The way around this problem typically is achieved by using an appropriate connection formula.

3.9.1 Tunneling through a High-Energy Barrier of Finite Thickness

Consider the potential for a rectangular barrier of energy V_0 and thickness L depicted in Fig. 3.31. The transmission coefficient for a particle of energy $E < V_0$ and mass m is (Eq. (3.48))

$$Trans(E < V_0) = \frac{1}{1 + \frac{1}{4} \frac{V_0^2}{E(V_0 - E)} \sinh^2(\kappa L)}, \tag{3.109}$$

where $k_2 = i\kappa$ for real κ and $\kappa^2 = (2m(V_0 - E))/\hbar$ for a rectangular barrier.

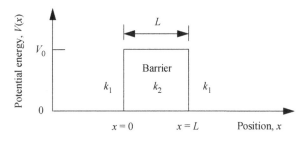

Fig. 3.31 Sketch of the potential of a one-dimensional rectangular potential barrier of energy V_0. The thickness of the barrier is L. A particle of mass m, incident from the left, of energy E, has wave vector k_1. In the barrier region the wave vector is k_2. If $E < V_0$ then $k_2 = i\kappa$ for real κ.

3 Electron Propagation

If the barrier is of high energy ($E \ll V_0$) and finite thickness L, so that it is almost opaque to particle transmission, then

$$\kappa L = \sqrt{\frac{2m(V_0 - E)}{\hbar^2}} L \gg 1. \tag{3.110}$$

Approximating the sinh term in Eq. (3.109) by

$$\sinh^2(\kappa L) = \left(\frac{e^{\kappa L} - e^{-\kappa L}}{2} \right)^2 \sim \frac{1}{4} e^{2\kappa L}, \tag{3.111}$$

the transmission probability when $E \ll V_0$ becomes

$$Trans \cong \frac{1}{1 + \frac{1}{16} \frac{V_0^2}{E(V_0 - E)} e^{2\kappa L}} \cong 16 \frac{E(V_0 - E)}{V_0^2} e^{-2\kappa L}. \tag{3.112}$$

This approximation leads to an expression for transmission probability that has no backward-traveling component, thereby violating conditions for current flow established in Exercise 2.1(c).

The result given by Eq. (3.112) was obtained for the nearly opaque, rectangular potential barrier. Generalizing to a nearly opaque potential barrier of arbitrary shape, such as that shown schematically in Fig. 3.32, the WKB approximation for transmission becomes

$$\boxed{Trans \sim e^{-2 \int_{x_1}^{x_2} \sqrt{\frac{2m(V(x) - E)}{\hbar^2}} \, dx}} \tag{3.113}$$

The physical meaning of this is that V_0 has been replaced by a weighted "average" barrier energy V_{av}, so that

$$\kappa L = \int_{x_1}^{x_2} \sqrt{\frac{2m(V(x) - E)}{\hbar^2}} \, dx = \sqrt{\frac{2m(V_{\mathrm{av}} - E)L}{\hbar^2}}. \tag{3.114}$$

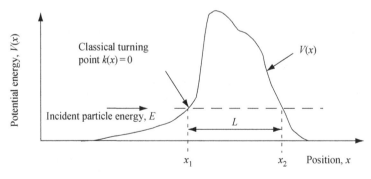

Fig. 3.32 Plot of potential energy as a function of position showing a one-dimensional barrier of energy $V(x)$. An electron of energy E, mass m, incident on the potential barrier has classical turning points at positions x_1 and x_2. The distance an electron tunnels through such a barrier is $L = x_2 - x_1$.

3.10 Example Exercises

Exercise 3.1
(a) Show that Eq. (3.44), describing transmission of a particle mass m and energy $E \geq V_0$ moving normal to a rectangular barrier of energy V_0 and thickness L, may be written as

$$Trans(E \geq V_0) = \left(1 + \frac{1}{4}\frac{b^2}{(b + k_2^2 L^2)}\frac{\sin^2(k_2 L)}{k_2^2 L^2}\right)^{-1},$$

where parameter $b = 2mV_0 L^2/\hbar^2$.
   (b) Show that the function $Trans$ in (a) has a minimum when

$$Trans_{\min} = 1 - \frac{b^2}{(2k_2^2 L^2 + b)^2} = 1 - \frac{V_0^2}{(2E - V_0)^2}.$$

   (c) Show that the function $Trans$ in (a) is unity when $\sqrt{2m(E - V_0)}L/\hbar = n\pi$.
   (d) Show that $Trans(E \rightarrow V_0) = \left(1 + \frac{mV_0 L^2}{2\hbar^2}\right)^{-1}$.

Exercise 3.2
(a) Find transmission as a function of energy for a particle mass m_0 through 12 identical one-dimensional potential barriers each of energy 10 eV, thickness 0.1 nm, sequentially placed every 0.5 nm (so that the potential well between each barrier has thickness 0.4 nm). What are the allowed (band) and disallowed (band gap) ranges of energy for particle transmission through the structure? How does the velocity of the transmitted particle vary as a function of energy?
   (b) How do these bands compare with the situation in which there are only three barriers, each with 10 eV barrier energy, 0.1 nm barrier thickness, and 0.4 nm well thickness?

Exercise 3.3
(a) Use the propagation matrix to design a high-reflectivity Bragg mirror for linearly polarized electromagnetic radiation with center wavelength $\lambda_0 = 980$ nm incident normal to the surface of an AlAs/GaAs periodic dielectric layer stack consisting of 25 identical layer pairs. See Fig. 3.E33. Each individual dielectric layer has a thickness $\lambda/4n_r$, with n_r being the refractive index of the dielectric. Use $n_{AlAs} = 3.0$ for the refractive index of AlAs and $n_{GaAs} = 3.5$ for GaAs. Calculate and plot optical reflectivity in the wavelength range 900 nm $< \lambda < 1\,100$ nm.
   (b) Extend the design of the Bragg reflector to a two-mirror structure similar to that used in the design of a vertical-cavity surface-emitting laser (VCSEL). See Fig. 3.E33. This may be achieved by increasing the number of pairs to 50 and making the thickness of the central AlAs layer one wavelength. Recalculate and plot the reflectivity over the same wavelength range as in (a). Using high wavelength resolution, find the bandwidth of this optical pass band filter near $\lambda = 980$ nm.
   The results should include a printout of the computer program used and a computer-generated plot of transmission as a function of incident wavelength.

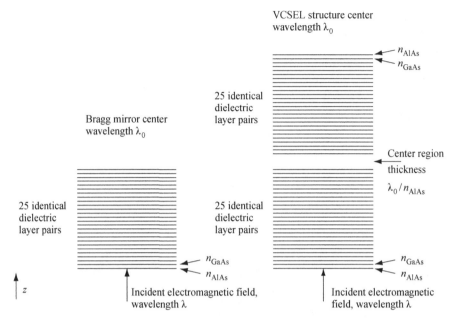

Fig. 3.E33

Exercise 3.4
Show that the Bloch wave function $\psi_k(x) = U_k(x)e^{ikx}$ is not an eigenfunction of the momentum operator $\hat{p} = -i\hbar\,d/dx$.

Exercise 3.5
Using the method outlined in Section 2.4, write a computer program to solve the Schrödinger wave equation for the first two eigenvalues and eigenstates of an electron of mass m_0 confined to a double-rectangular potential well sketched in Fig 3.E34. Each well is of thickness 0.6 nm and they are separated by 0.4 nm. The barrier potential energy is 1 eV. In the region $x \leq 0$ nm and $x \geq 5$ nm the potential is infinite.

Exercise 3.6
(a) Show that time-reversal symmetry requires $p_{11} = p_{22}^*$ and $p_{21} = p_{12}^*$.

(b) Show that current conservation at the jth step of a piecewise-constant potential in one dimension requires $|\mathbf{P}| = 1$ when k_j is real and $|\mathbf{P}| = \pm i$ for imaginary k_j, where the $\pm$ depends on whether the incoming wave is from the left or right.

Exercise 3.7
(a) Find p_{11}, p_{21}, and the propagation matrix $\mathbf{P}$ for an electron with motion in one dimension that is incident on a delta function potential barrier.

(b) Analyze and plot transmission of an electron incident from the left on the delta function potential barrier of (a).

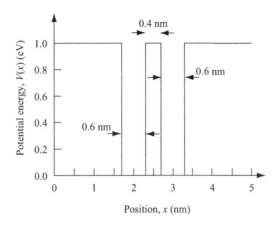

Fig. 3.E34

Exercise 3.8

Derive and analyze the complex band structure of a one-dimensional periodic array of delta function potential energy barriers with a nearest neighbor separation of L.

Solutions

Solutions 3.1

(a) Starting with the expression

$$Trans(E \geq V_0) = \left(1 + \left(\frac{k_1^2 - k_2^2}{2k_1k_2}\right)^2 \sin^2(k_2 L)\right)^{-1},$$

multiplying the terms in k out, gives

$$Trans = \left(1 + \frac{1}{4}\frac{k_1^4 + k_2^4 - 2k_1^2 k_2^2}{k_1^2 k_2^2} \sin^2(k_2 L)\right)^{-1}$$

and substituting for $k_1^2 = k_2^2 + 2mV_0/\hbar^2$ gives

$$Trans = \left(1 + \frac{1}{4}\frac{\left(k_2^2 + \frac{2mV_0}{\hbar^2}\right)^2 + k_2^4 - 2\left(k_2^2 + \frac{2mV_0}{\hbar^2}\right)k_2^2}{k_2^2 \left(k_2^2 + \frac{2mV_0}{\hbar^2}\right)} \sin^2(k_2 L)\right)^{-1}.$$

Expanding the terms in the numerator gives

$$2k_2^4 + \left(\frac{2mV_0}{\hbar^2}\right)^2 + 2k_2^2 \left(\frac{2mV_0}{\hbar^2}\right) - 2k_2^4 - 2k_2^2 \left(\frac{2mV_0}{\hbar^2}\right),$$

147

so that

$$Trans = \left(1 + \frac{1}{4} \frac{\left(\frac{2mV_0}{\hbar^2}\right)^2}{\left(\frac{2mV_0}{\hbar^2} + k_2^2\right)k_2^2} \sin^2(k_2 L)\right)^{-1}.$$

Multiplying the both the top and bottom of second term by L^4, so that the terms in k_2^4 are dimensionless, gives

$$Trans = \left(1 + \frac{1}{4} \frac{\left(\frac{2mV_0 L^2}{\hbar^2}\right)^2}{\left(\frac{2mV_0 L^2}{\hbar^2} + k_2^2 L^2\right)k_2^2 L^2} \sin^2(k_2 L)\right)^{-1}.$$

Substituting in the parameter $b = 2mV_0 L^2/\hbar^2$ gives the desired result:

$$Trans(E \geq V_0) = \left(1 + \frac{1}{4} \frac{b^2}{(b + k_2^2 L^2)} \frac{\sin^2(k_2 L)}{k_2^2 L^2}\right)^{-1}.$$

(b) The minimum of the *Trans* function occurs when $\sin^2(k_2 L) = 1$. When this happens, $k_2 L = (2n-1)\pi/2$ for $n = 1, 2, 3, \ldots$. Substituting this value into the expression obtained in (a) gives

$$Trans_{min} = \left(1 + \frac{1}{4} \frac{b^2}{(b + k_2^2 L^2)} \frac{1}{k_2^2 L^2}\right)^{-1} = \left(1 + \frac{1}{4} \frac{b^2}{(b + k_2^2 L^2)} \frac{1}{k_2^2 L^2}\right)^{-1}$$

$$= \left(\frac{4(b + k_2^2 L^2) k_2^2 L + b^2}{4(b + k_2^2 L^2) k_2^2 L^2}\right)^{-1} = \frac{4bk_2^2 L^2 + 4k_2^4 L^4}{b^2 + 4(bk_2^2 L^2 + k_2^4 L^4)}$$

$$= \frac{4bk_2^2 L^2 + 4k_2^4 L^4 + b^2 - b^2}{(2k_2^2 L^2 + b)^2} = \frac{(2k_2^2 L^2 + b)^2 - b^2}{(2k_2^2 L^2 + b^2)} = 1 - \frac{b^2}{(2k_2^2 L^2 + b)^2}.$$

Rewriting this in terms of energy by substituting for $b = 2mV_0 L^2/\hbar^2$ and $k_2^2 = 2m\left((E - V_0)/\hbar^2\right)$ gives

$$Trans_{min} = 1 - \frac{\left(\frac{2mV_0 L^2}{\hbar^2}\right)^2}{\left(2\frac{2m(E - V_0)}{\hbar^2} L^2 + \frac{2mV_0 L^2}{\hbar^2}\right)^2} = 1 - \frac{V_0^2}{(2(E - V_0) + V_0)^2} = 1 - \frac{V_0^2}{(2E - V_0)^2}.$$

(c) It is clear from (a) that *Trans* is a maximum when $\sin^2(k_2 L) = 0$. When this happens, *Trans* $= 1$ and $k_2 L = n\pi$ for $n = 1, 2, 3, \ldots$. This corresponds to resonances in transmission that occur when particle-waves back-scattered from the step change in barrier potential at positions $x = 0$ and $x = L$ interfere and exactly cancel each other, resulting in zero reflection from the potential barrier.

148

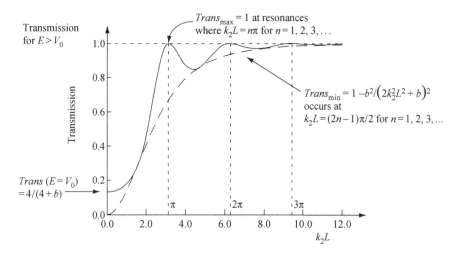

Fig. 3.E35

(d) When $E \rightarrow V_0$ or $k_2 L \rightarrow 0$ then $\sin^2(k_2 L) = k_2^2 L^2 + \cdots$ may be substituted into the expression for $Trans$:

$$Trans = \left(1 + \frac{1}{4}\frac{b^2}{(b + k_2^2 L^2)}\frac{\sin^2(k_2 L)}{k_2^2 L^2}\right)^{-1},$$

$$Trans(E \rightarrow V_0) = \left(1 + \frac{1}{4}\frac{b^2}{(b + k_2^2 L^2)}\frac{k_2^2 L^2}{k_2^2 L^2}\right)^{-1}\Bigg|_{k_2 L \rightarrow 0}$$

$$= \left(1 + \frac{b}{4}\right)^{-1} = \frac{4}{4 + b} = \left(1 + \frac{m V_0 L^2}{2\hbar^2}\right)^{-1}.$$

In Fig. 3.E35, $Trans$, $Trans_{max}$, and $Trans_{min}$ are plotted as a function of $k_2 L$ for an electron of mass m_0 incident on a rectangular potential barrier of energy $V_0 = 1.0\,\mathrm{eV}$ and thickness $L = 1\,\mathrm{nm}$. The incident electron has energy $E \geq V_0$.

Solutions 3.2

(a) Fig. 3.E36 shows the one-dimensional potential as a function of position, x, and calculated particle transmission as a function of particle energy, E. Particle transmission is plotted on a negative natural logarithm scale. This means that unity transmission is zero and low transmission corresponds to a large number.

As is clear from Fig. 3.E36, there are bands in energy of near unity transmission and bands of energy where transmission is suppressed. The former correspond to conduction bands and the latter to band gaps in a semiconductor crystal. The approximate allowed (band, E_b) and disallowed (band gap, E_g) ranges of energy for particle transmission

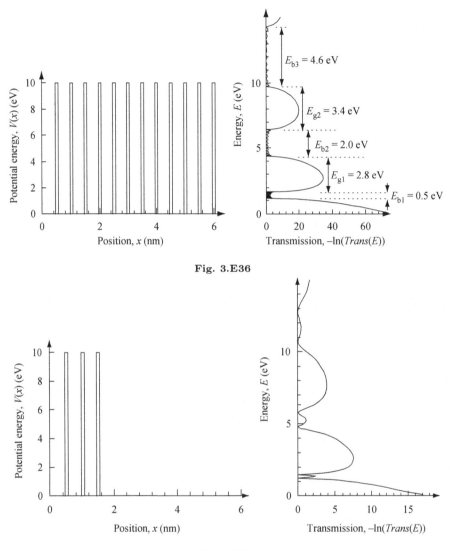

Fig. 3.E36

Fig. 3.E37

through the structure are indicated in Fig. 3.E36. Disallowed non-propagating states with zero velocity exist in the band gap.

For 11 potential wells, there are 11 transmission resonances. When the finite width resonances overlap, they form a continuum transmission band.

(b) The results of (a) are now compared with the situation in which there are only three barriers, each with 10 eV barrier energy, 0.1 nm barrier thickness, and 0.4 nm well thickness. As can be seen in Fig. 3.E37, even with three potential barriers (and two potential wells), the basic structure of allowed and disallowed energy ranges has formed. For two potential wells, there are two transmission resonances. The resonances at higher energy are broader because the scattering strength of the potential barrier experienced by an electron at high energy is smaller.

Solutions 3.3

The propagation matrix can be used to design a high-reflectivity Bragg mirror for electromagnetic radiation with center wavelength $\lambda_0 = 980\,$nm incident normally to the surface of an AlAs/GaAs periodic dielectric layer stack consisting of 25 identical layer pairs. Each individual dielectric layer has thickness $\lambda/4n_r$, where n_r is the refractive index of the dielectric. The refractive index is taken as $n_{\text{AlAs}} = 3.0$ for AlAs and $n_{\text{GaAs}} = 3.5$ for GaAs.

In part (a) of the exercise, the task is to calculate and plot optical reflectivity in the wavelength range $900\,$nm $< \lambda < 1\,100\,$nm. In Fig. 3.E38(a) results of performing the calculation are shown. The mirror has a reflectivity of close to unity over a wavelength bandwidth $\Delta\lambda = 120\,$nm centered at $\lambda = 980\,$nm.

For the VCSEL design in part (b) there is a 25 layer-pair mirror above and a 25 layer-pair mirror below a central AlAs layer that is one wavelength long. The two mirrors form a high-Q resonant optical cavity. Again, the reflectivity is close to unity over a similar wavelength bandwidth, with the addition of an optical band pass with near unity transmission and a FWHM of 13 pm centered at wavelength $\lambda = 980\,$nm.

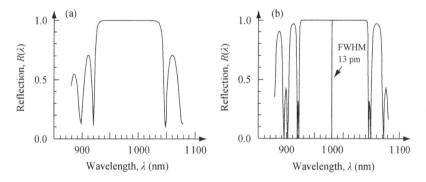

Fig. 3.E38

Solutions 3.4

To show that the Bloch wave function $\psi_k(x) = U_k(x)e^{ikx}$ is not an eigenfunction of the momentum operator $\hat{p} = -i\hbar\,d/dx$, the wave function is operated on as follows:

$$\hat{p}\psi_k(x) = -i\hbar\frac{d}{dx}U_k(x)e^{ikx} = \hbar k\,\psi_k(x) - i\hbar\,e^{ikx}\frac{d}{dx}U_k(x).$$

Solutions 3.5

The task is to use the method outlined in Section 2.4 to numerically solve the one-dimensional Schrödinger wave equation for an electron of mass m_0 in a symmetric double-well structure with finite barrier potential energy.

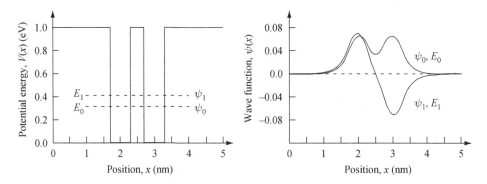

Fig. 3.E39

The main computer program called deals with input parameters such as the length L, the electron mass, the number of discretization points, N_{steps}, and the plotting routine. Because a nontransmitting boundary condition is used, it is important to choose L large enough so that the wave function is approximately zero at the boundaries $x_0 = 0$ and $x_{N_{\text{steps}}} = L$. Also, a large enough value of N_{steps} should be chosen so that the wave function does not vary significantly between adjacent discretization points. This ensures that the three-point finite-difference approximation used is accurate.

The call to solve_schM.m solves the discretized Hamiltonian matrix.

In this exercise, the first two energy eigenvalues are $E_0 = 0.3226\,\text{eV}$ and $E_1 = 0.4053\,\text{eV}$. The separation in energy is $E_{10} = 0.0827\,\text{eV}$. The eigenfunctions generated by the program and plotted in Fig. 3.E39 are not normalized.

Solutions 3.6

(a) Conventionally, a particle with wave character is initially traveling from left to right and incident on a potential step at position $x = x_{j+1}$. As illustrated in Fig. 3.2, the incident wave has a coefficient A, and the forward scattered wave has coefficient C. The wave reflected at the potential step has a coefficient B, and a wave incident traveling from right to left has coefficient D. The corresponding wave functions on either side of the potential step are

$$\psi_j(x, t) = \left(A_j e^{ik_j x} + B_j e^{-ik_j x}\right) e^{-i\omega t}$$

and

$$\psi_{j+1}(x, t) = \left(C_{j+1} e^{ik_{j+1} x} + D_{j+1} e^{-ik_{j+1} x}\right) e^{-i\omega t}.$$

If the potential step is real and does not change with time, then there is no way a solution based on a left-to-right propagating incident wave can exist without also allowing an associated time-reversed solution. This may be shown by writing the time-dependent Schrödinger equation

$$\hat{H}\psi(x, t) = i\hbar \frac{\partial}{\partial t}\psi(x, t).$$

Reversing time requires changing the sign of the parameter $t \to -t$. Performing the sign change and taking the complex conjugate gives

$$\hat{H}\psi^*(x, -t) = i\hbar \frac{\partial}{\partial t}\psi^*(x, -t).$$

This equation is the same as the previous one except for the replacement of $\psi(x, t)$ with $\psi^*(x, -t)$. This means that if $\psi(x, t)$ is a solution then the time-reversed solution $\psi^*(x, -t)$ also applies.

The propagation matrix links coefficients A and B to C and D via a matrix

$$\begin{bmatrix} A \\ B \end{bmatrix} = \begin{bmatrix} p_{11} & p_{12} \\ p_{21} & p_{22} \end{bmatrix} \begin{bmatrix} C \\ D \end{bmatrix},$$

where rows and columns have terms that contain left-to-right traveling waves of the form e^{ikx} and right-to-left traveling waves of the form e^{-ikx}. Since time reversal of the wave function $\psi_j(x, t) = \left(A_j e^{ik_j x} + B_j e^{-ik_j x} \right) e^{-i\omega t}$ gives $\psi_j^*(x, -t) = \left(A_j^* e^{-ik_j x} + B_j^* e^{ik_j x} \right) e^{-i\omega t}$ and does not change the energy of the system, the time-reversed form of the previous equation is

$$\begin{bmatrix} B^* \\ A^* \end{bmatrix} = \begin{bmatrix} p_{11} & p_{12} \\ p_{21} & p_{22} \end{bmatrix} \begin{bmatrix} D^* \\ C^* \end{bmatrix}.$$

The trick now is to take the complex conjugate of both sides and interchange rows so that a direct comparison can be made. This gives a time-reversed solution,

$$\begin{bmatrix} A \\ B \end{bmatrix} = \begin{bmatrix} p_{22}^* & p_{21}^* \\ p_{12}^* & p_{11}^* \end{bmatrix} \begin{bmatrix} C \\ D \end{bmatrix},$$

confirming

$$p_{11} = p_{22}^*$$

and

$$p_{21} = p_{12}^*.$$

This relationship may be thought of as a direct consequence of the time-reversal symmetry built into the Schrödinger equation when the potential is real and does not change with time.

There are situations when time-reversal symmetry is broken. An example is when there is dissipation in the system. Dissipation might be modeled semiclassically by introducing an imaginary part to the potential in the Schrödinger equation.

(b) The transmission and reflection coefficients across the potential step in Fig. 3.3 are related to each other by current conservation $A^*A - B^*B = C^*C - D^*D$ and the 2×2 propagation matrix,

$$\begin{bmatrix} A \\ B \end{bmatrix} = \begin{bmatrix} p_{11} & p_{12} \\ p_{21} & p_{22} \end{bmatrix} \begin{bmatrix} C \\ D \end{bmatrix},$$

so that

$$A = p_{11}C + p_{12}D, \qquad A^* = p_{11}^*C^* + p_{12}D^*,$$
$$B = p_{21}C + p_{22}D, \qquad B^* = p_{21}^*C^* + p_{22}^*D^*.$$

Hence, for no incoming wave from the right $D = 0$,

$$A^*A - B^*B = (p_{11}^*p_{11} - p_{21}^*p_{21})C^*C.$$

However, since for $D = 0$ current conservation requires

$$A^*A - B^*B = C^*C,$$

the only nontrivial way to simultaneously satisfy these two equations is if

$$p_{11}^*p_{11} - p_{21}^*p_{21} = 1.$$

Making use of the fact that $p_{11} = p_{22}^*$ and $p_{21} = p_{12}^*$, then

$$p_{11}p_{22} - p_{12}p_{21} = 1,$$

or, in more compact form, for any domain the determinant of matrix $\mathbf{P}$ is

$$|\mathbf{P}| = 1.$$

This is an expression of *current conservation* when k_j is *real*. For imaginary k_j, the determinant of matrix $\mathbf{P}$ is $|\mathbf{P}| = \pm i$, where the $\pm$ depends upon whether the incoming wave is from the left or the right. ✎

Solutions 3.7
A delta function potential energy barrier may be considered as the limit of a simultaneously infinite rectangular barrier energy $(eV_0 \to \infty)$ and zero barrier thickness $(L \to 0)$. This idea is illustrated in Fig. 3.E40, in which (a) is a sketch of a one-dimensional rectangular barrier of energy V_0. The thickness of the barrier is L. A particle of mass m incident from the left of energy E has wave vector k_1. In the barrier region, the wave vector is k_2. The wave vectors k_1 and k_2 are related via $k_1^2 = k_2^2 + 2mV_0/\hbar^2$. In Fig. 3.E40(b), potential energy as a function of position is shown, illustrating a one-dimensional rectangular potential barrier in the delta function limit. An incident particle with wave vector k_1 and energy $E = \hbar^2 k_1^2/2m$ has wave vector ik_2 in the barrier.

Since $V_0 \to \infty, L \to 0$, the product $V_0 L \to constant$. Because $V_0 \to \infty$, this means that $k_2 \to ik_2$ for all incident particle energies $E = \hbar^2 k_1^2/2m$.

Now $V_0 = E + \hbar^2 k_2^2/2m$, and since $V_0 \gg E$, it follows that $V_0 = \hbar^2 k_2^2/2m$, or

$$k_2^2 L = \frac{2mV_0 L}{\hbar^2}.$$

Substituting $k_2 \to ik_2$ into Eq. (3.41) results in

$$P_{11} = -\frac{1}{2} \frac{(k_1^2 - k_2^2)\left(e^{-k_2 L} - e^{k_2 L}\right)}{i2k_1 k_2} + \frac{1}{2}\left(e^{k_2 L} + e^{-k_2 L}\right),$$

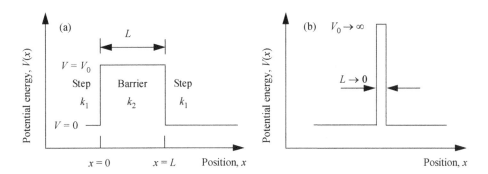

Fig. 3.E40

which can be simplified in the limit $k_2 L \to 0$, so that $e^{k_2 L} = 1 + k_2 L + \cdots$ and $e^{-k_2 L} = 1 - k_2 L + \cdots$. This expansion allows P_{11} to be written in the limit $k_2 L \to 0$ as

$$P_{11} = \frac{1}{2} \frac{\left(k_1^2 - k_2^2\right)\left(2k_2 L\right)}{i2k_1 k_2} + 1 = 1 - \frac{i\left(k_1^2 - k_2^2\right) k_2 L}{2k_1 k_2}.$$

Since $k_2 L \to 0$,

$$P_{11} = 1 - i\left(\frac{k_1 L}{2} - \frac{k_2^2 L}{2k_1}\right) = 1 - i\left(0 - \frac{2m_e^* V_0 L}{2\hbar^2 k_1}\right) = 1 + i\frac{mV_0 L}{\hbar^2 k_1},$$

so that in the limit $k_2 L \to 0$, the expression for P_{11} is

$$\boxed{P_{11} = 1 + i\frac{k_0}{k_1},}$$

where

$$\boxed{k_0 = mV_0 L/\hbar^2}$$

is a constant.

The transmission probability $|C|^2$ for a particle of mass m incident from the left on a delta function potential barrier is of a simple form that does *not* depend exponentially on the energy, E, of the incident particle,

$$|C|^2 = \left|\frac{1}{P_{11}}\right|^2 = \left|\frac{1}{1 + i(k_0/k_1)}\right|^2 = \frac{E}{E + \left(\hbar^2 k_0^2/2m\right)},$$

where the electron wave vector $k_1 = \sqrt{2mE}/\hbar$. The equation for $|C|^2$ is plotted as a function of energy, E, in the Fig. 3.E41. In this particular case, the parameter k_0 is chosen so that $E_0 = \hbar^2 k_0^2/2m = 1\,\text{eV}$. As may be seen in the figure, under these circumstances the particle has a transmission probability of 0.5 when the particle energy is $E = E_0 = 1\,\text{eV}$.

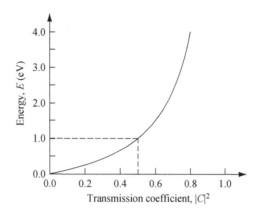

Fig. 3.E41

The other matrix element to be found is P_{12}. Previously, from Eq. (3.40),

$$P_{12} = -\frac{(k_2^2 - k_1^2)}{4k_1 k_2}\left(e^{ik_2 L} - e^{-ik_2 L}\right),$$

so that in the delta function limit ($k_2 \to ik_2$ and $k_2 L \to 0$),

$$\boxed{P_{12} = i\frac{k_0}{k_1}.}$$

Because $P_{11} = P_{22}^*$ and $P_{12} = P_{21}^*$, the propagation matrix $\mathbf{P}$ for a potential energy barrier in the delta function limit is known.

Solutions 3.8

Figure 3.E42 shows a schematic of the system that will be considered.[14] It is a one-dimensional periodic array of delta function potential energy barriers with nearest neighbor separation of L so that $V(x) = V(x + L)$. This separation defines the unit cell of the periodic potential. Also shown in the figure are wave amplitudes A, B, C, and D for an electron scattering from a unit cell.

It follows from Bloch's theorem that the coefficients A, B, C, and D are related to each other by a phase factor kL in such a way that

$$C = Ae^{ikL}$$

and

$$D = Be^{ikL},$$

where k is the *Bloch wave vector*. These equations may be expressed in matrix form as

$$\begin{bmatrix} A \\ B \end{bmatrix} = e^{-ikL}\begin{bmatrix} C \\ D \end{bmatrix} = \mathbf{P}\begin{bmatrix} C \\ D \end{bmatrix}$$

[14]R. de L. Kronig and W. J. Penney, *Proc. Royal. Soc. A* **130**, 499 (1930).

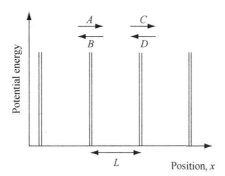

Fig. 3.E42

so that

$$\begin{bmatrix} p_{11} - e^{-ikL} & p_{12} \\ p_{21} & p_{22} - e^{-ikL} \end{bmatrix} \begin{bmatrix} C \\ D \end{bmatrix} = 0.$$

An interesting solution to these linear homogeneous equations exists if the determinant of the 2×2 matrix is zero, that is

$$\left(p_{11} - e^{-ikL} \right)(p_{22} - e^{-ikL}) - p_{12}p_{21} = \quad p_{11}p_{22} + e^{-2ikL} - p_{22}e^{-ikL} - p_{11}e^{-ikL} - p_{12}p_{21} = 0.$$

Since $p_{11} = p_{22}^*$ and $p_{12} = p_{21}^*$, this can be rewritten

$$e^{-2ikL} - p_{11}e^{-ikL} - p_{11}^*e^{-ikL} = -p_{11}p_{22} + p_{12}p_{21} = -|\mathbf{p}|.$$

From current continuity, the determinant $|\mathbf{p}| = 1$ for real k. Taking the real part of both sides for terms in p:

$$e^{-2ikL} - 2\mathrm{Re}(p_{11})e^{-ikL} = -1.$$

Note that $e^{ix} = (\cos(x) - i\sin(x))$ so that

$$\cos(2kL) - i\sin(2kL) - 2\mathrm{Re}(p_{11})(\cos(kL) - i\sin(kL)) = -1.$$

Taking the imaginary part of this expression gives

$$\sin(2kL) - 2\mathrm{Re}(p_{11})\sin(kL) = 0,$$

and since $2\sin(x)\cos(x) = \sin(2x)$, this can be written

$$2\sin(kL)(\cos(kL) - \mathrm{Re}(p_{11})) = 0.$$

Hence,

$$\boxed{\cos(kL) = \mathrm{Re}(p_{11}),}$$

and so the Bloch wave vector k can only be real if the magnitude of $\mathrm{Re}(P_{11})$ is less than 1, i.e.,

$$\boxed{|\mathrm{Re}(p_{11})| < 1.}$$

In general, for a bulk crystal this gives rise to bands of allowed propagating states with real values of the Bloch wave vector k.

In the delta function limit the matrix coefficient for a single barrier is $p_{11} = (1+\mathrm{i}(k_0/k_1))$ (Exercise 3.7), where $k_0 = mV_0 L/\hbar^2$ and k_1 is the wave vector outside the delta function barrier.

In this model of a periodic potential, delta function barriers are separated by free-propagation regions of length L. The total propagation matrix for a cell of length L becomes a product of the delta function potential barrier matrix (Exercise 3.7) and the free-propagation matrix (Eq. (3.12)),

$$\mathbf{p} = \mathbf{p}_{\delta-\text{barrier}}\mathbf{p}_{\text{free}} = \begin{bmatrix} 1+\mathrm{i}\frac{k_0}{k_1} & \mathrm{i}\frac{k_0}{k_1} \\ -\mathrm{i}\frac{k_0}{k_1} & 1-\mathrm{i}\frac{k_0}{k_1} \end{bmatrix} \begin{bmatrix} \mathrm{e}^{-\mathrm{i}k_1 L} & 0 \\ 0 & \mathrm{e}^{\mathrm{i}k_1 L} \end{bmatrix} = \begin{bmatrix} p_{11} & p_{12} \\ p_{21} & p_{22} \end{bmatrix},$$

so the new p_{11} for the cell is

$$p_{11} = \left(1+\mathrm{i}\frac{k_0}{k_1}\right)\mathrm{e}^{-\mathrm{i}k_1 L}.$$

Since $\mathrm{e}^{-\mathrm{i}x} = \cos(x) - \mathrm{i}\sin(x)$, the matrix element can be written as

$$p_{11} = \cos(k_1 L) - \mathrm{i}\sin(k_1 L) + \mathrm{i}\frac{k_0}{k_1}\cos(k_1 L) + \frac{k_0}{k_1}\sin(k_1 L).$$

Taking the real part and making use of $\mathrm{Re}(p_{11}) = \cos(kL)$ gives

$$\mathrm{Re}(p_{11}) = \cos(k_1 L) + \frac{k_0 L}{k_1 L}\sin(k_1 L) = \theta_{\mathrm{B}} = \cos(kL).$$

Because $|\mathrm{Re}(p_{11})| < 1$ (or, more compactly, $|\theta_{\mathrm{B}}| < 1$) for allowed real values of the Bloch wave vector k in which $\theta_{\mathrm{B}} = \cos(kL)$, it may be concluded that

$$\boxed{-1 < \cos(k_1 L) + k_0 L\frac{\sin(k_1 L)}{k_1 L} < 1.}$$

It is now possible to plot the function $\theta_{\mathrm{B}} = \cos(k_1 L) + k_0 L\frac{\sin(k_1 L)}{k_1 L}$ and establish regions of allowed propagating states with real values of Bloch wave vector k. This has been done in Fig. 3.E43 for the case when $k_0 L = 8$.

The allowed bands of propagating states with real values of the Bloch wave vector k are indicated as solid curves in shaded regions. Also apparent in the figure is the fact that the boundary between the upper edge of bands with propagating states and the lower edge of bands with non-propagating states occurs at values $k_1 L = n\pi$, where n is a nonzero positive integer, $n = 1, 2, 3, \ldots$.

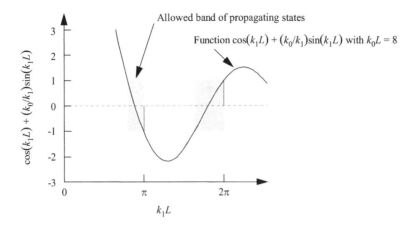

Fig. 3.E43

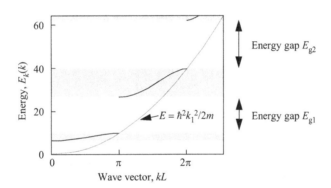

Fig. 3.E44

In summary, propagating electron states of energy E and the normalized real Bloch wave vector kL/π require the solution to

$$\boxed{\cos(kL) = \theta_B}$$

with $|\theta_B| < 1$ so that

$$\boxed{kL = \frac{\arccos(\theta_B)}{\pi}.}$$

Figure 3.E44 shows the dispersion relation $E_k(k)$ as a function of the normalized real Bloch wave vector kL for the case when $k_0 L = 8$. In the limit of high electron energy, the plane-wave term e^{ikL} in Bloch's theorem becomes the same as for free electrons, $E = \hbar^2 k_1^2/2m_0$. As $k_1 \to \infty$, $k_1 \to k$ so that $E \to \hbar^2 k^2/2m_0$.

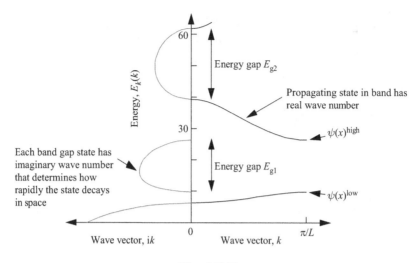

Fig. 3.E45

Imaginary Bloch wave vectors can be solutions to $|\theta_B| \geq 1$ so that

$$\kappa L = \frac{\mathrm{arccosh}(\theta_B)}{\pi},$$

where $\kappa = |ik|$. The corresponding electron states are non-propagating and localized. Such states contribute to the imaginary band structure and occur in the energy gap regions.

The band structure is *complex* since both real and imaginary values of the Bloch wave vector can be present in a crystal.

Because the potential is periodic in L, the complex dispersion relation may be redrawn in compact form as a function of the Bloch wave vector in the reduced zone. This is shown in Fig. 3.E45. The wave functions at the Brillouin zone separating the first band gap are $\psi^{\mathrm{low}}_{k=\pi/L}(x)$ and $\psi^{\mathrm{high}}_{k=\pi/L}(x)$. In regions where there is an energy gap, E_g, wave vectors are associated with non-propagating states in a bulk crystal and take imaginary values. The dispersion relation is an example of complex band structure with both real and imaginary wave vectors that are solutions to Bloch's equation. For the band structure shown, $k_0 L = 8$, and the vertical scale is in units of $(k_0 L)^2$.

The allowed energy bands have real Bloch wave vector that may be thought of as arising from resonant transmission through all the potential wells (each of thickness L bounded by a delta function potential energy barrier), which are coupled through tunneling. There are a large number of such resonances that overlap in energy to create a continuous energy band of allowed propagating states in a bulk crystal.

Energy gap regions form where there are no resonances, and an imaginary Bloch wave vector in the complex band structure describes a non-propagating localized state in the crystal. The wave vector is in the complex plane. Such localized states can exist at a defect in a bulk crystal, at a crystal surface, or at a crystal heterointerface. An analysis of complex band structure is an essential first step towards understanding and developing

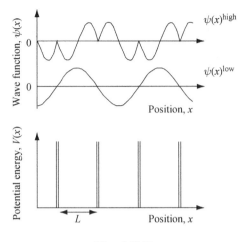

Fig. 3.E46

accurate models of many nanoscale electronic devices, including those involving electron tunneling.

Electron wave functions are standing waves and have definite symmetry at the boundary between the allowed and forbidden gap region of bulk-crystal complex band structure. The energy gap region labeled E_{g1} in Fig. 3.E45 occurs at the Brillouin zone boundary $k = \pi/L$ and is bound by a low energy band-edge eigenstate $\psi^{low}_{k=\pi/L}(x)$ and a high energy band-edge eigenstate $\psi^{high}_{k=\pi/L}(x)$. The electron wave function is a Bloch function of the form $\psi_k(x) = U_k(x)e^{ikx}$. In both cases, the cell periodic part of the wave function $U_k(x)$ is modulated by a Bloch phase term e^{ikx}. At the Brillouin zone boundary the Bloch wave vector has the value $k = \pi/L$. For the delta function potential and in the lowest energy allowed band the cell periodic part of the wave function, $U_k(x)$, is essentially a half-period sine wave in each potential well. Modulation at the Brillouin zone boundary by the Bloch phase term changes the phase of $U_k(x)$ in adjacent potential wells giving the wave function $\psi^{low}_{k=\pi/L}(x)$ shown in Fig. 3.E46. Similarly, in the next allowed energy band $U_k(x)$ is essentially a single period sine wave in each potential well. Modulation at the Brillouin zone boundary by the Bloch wave vector changes the phase of $U_k(x)$ in adjacent potential wells giving the wave function $\psi^{high}_{k=\pi/L}(x)$ shown in Fig. 3.E46. The states in the energy gap E_{g1} have Im $k \neq 0$ and Re $k = \pi/L$. The states in energy gap E_{g2} have Im $k \neq 0$ and Re $k = 0$.

3.11 Problems

Problem 3.1
Write a computer program in MATLAB that uses the propagation matrix method to find the transmission resonances of a particle of mass $m = 0.07 \times m_0$ incident on a parabolic potential energy barrier with $V(x) = x^2/L^2$ eV for $|x| \leq L = 5$ nm and $V(x) = 0$ eV for $|x| > L$. What happens to the resonant transmission energy levels if the particle mass is $m = 0.14 \times m_0$?

The solution should include plots of transmission as a function of energy, a list of the resonant energy level values and associated spectral line widths, and a listing of the computer program used.

Problem 3.2

(a) Using the method outlined in Section 2.4, write a computer program to solve the Schrödinger wave equation for the first three eigenvalues and eigenstates of an electron of mass m_0 confined to a triple-rectangular potential well sketched in Fig. 3.P47. Each well is of thickness 0.125 nm and the barrier is of thickness 0.125 nm. The barrier potential energy is 50 eV. In the region $x \leq 0$ nm and $x \geq 1$ nm the potential is infinite. Compare the energy eigenvalues with those obtained using the propagation method for the same potential except in the region $x \leq 0$ nm and $x \geq 1$ nm where the potential is zero.

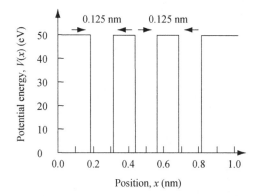

Fig. 3.P47

(b) If (a) models an electron in a linear molecule, which state is likely to bond the molecule together?

The results should include: (i) a printout of the computer program used; (ii) a computer-generated plot of the potential; (iii) a list of the energy eigenvalues; (iv) a computer-generated plot of the eigenfunctions.

Problem 3.3

Write a program in MATLAB that uses the propagation matrix method to calculate transmission of an electron with effective mass $m_e^* = 0.07 \times m_0$ as a function of the potential energy drop -2 eV $< \Delta V < 0$ eV caused by the application of a uniform electric field across a single potential barrier structure as shown in Fig. 3.P48. Calculate the specific case of initial particle energy $E = 0.025$ eV with the particle incident on the structure from the left-hand side. Explain the results obtained.

Problem 3.4

(a) Write a computer program to solve the Schrödinger wave equation for the first 17 eigenvalues of an electron with effective mass $m_e^* = 0.07 \times m_0$ confined to a periodic potential *with periodic boundary conditions*. Each of the eight quantum wells is of thickness

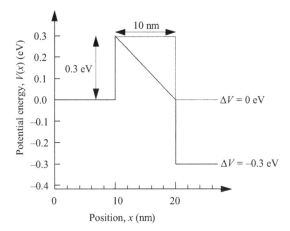

Fig. 3.P48

6.25 nm. Each quantum well is separated by a potential barrier of thickness 3.75 nm. The barrier potential energy is 0.9 eV. How many energy band gaps are present in the first 17 eigenvalues and what are their values? Plot the highest energy eigenfunction of the first band and the lowest energy eigenfunction of the second band.

(b) How does dispersion in (a) change if the quantum well thickness is changed to 8.75 nm and the potential barrier thickness to 1.25 nm?

Problem 3.5
Use the results of Problem 3.4 with periodic boundary conditions to approximate a periodic one-dimensional delta function potential with period 10 nm by considering eight potential barriers with energy 20 eV and thickness 0.25 nm. Plot the lowest and highest energy eigenfunction of the first band. Explain the difference in the wave functions obtained.

Problem 3.6
(a) Explain why an *abrupt* step change in potential by an amount V_0 at position $x = 0$ *always* causes quantum mechanical reflection for a particle mass m and energy $E > V_0$ and so can *never* give the classical result of no reflection when $E > V_0$.

(b) Quantum mechanical reflection in (a) can be reduced if, as illustrated in Fig. 3.P49, the step change in potential is replaced by a smoothly varying potential to ensure an adiabatic transition near position $x = 0$. A suitable potential (which becomes a step potential in the limit $\alpha \to \infty$) is

$$V(x) = \frac{V_0}{e^{-\alpha x} + 1}.$$

Using this potential profile, find an expression for the reflection coefficient when $E > V_0$ and show that the reflection coefficient is zero in the limit $\hbar \to 0$ for finite positive α.

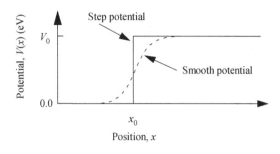

Fig. 3.P49

Problem 3.7

What value (in eV) of the overlap integral t should be used in the nearest neighbor tight binding model to reproduce the effective electron mass $m_e^* = 0.07 \times m_0$ near the conduction band minimum of s-orbital atoms in a 1D lattice with lattice constant $L = 0.565\,\text{nm}$?

Problem 3.8

Explain why the energy bandwidth of allowed electron states in a crystal should decrease as the lattice spacing between atoms increases.

Problem 3.9

A crystal with identical atoms at lattice sites $x_n = nL$, where n is an integer and L is the nearest neighbor atom spacing, has wave function $\psi_{k_x}(x)$ that can be expressed as a direct lattice sum of Wannier functions $\phi(x)$ localized around each lattice site x_n. Show that

$$\psi_{k_x}(x) = \sum_n e^{ik_x x_n} \phi(x - x_n)$$

satisfies the Bloch condition

$$\psi_{k_x}(x + L) = \psi_{k_x}(x) e^{ik_x L}.$$

Problem 3.10

The propagation matrix method divides a one-dimensional potential $V(x)$ into N potential steps and then solves the Schrödinger equation for a particle of energy E in a piecewise fashion. The wave function across the jth step is

$$\psi_j(x) = A_j e^{ik_j x} + B_j e^{-ik_j x},$$

and because A_j and B_j can be calculated it is possible to obtain $\psi_j(x)$.

(a) Use the propagation matrix method to reproduce the transmission characteristics shown in Fig. 3.11(b) for an electron incident from the left on the potential shown in Fig. 3.11(a). Verify the FWHM of the lowest energy resonant transmission peaks.

(b) Calculate and plot the modulus of the wave function squared, $|\psi_j(x)|^2$, for the cases when the electron has energy $E = 321.6$ meV and energy $E = 401.5$ meV. Comment on the results.

164

(c) Plot $|\psi_j(x)|^2$ as a function of position and energy. Explain the features in the 3D plot or contour plot.

Problem 3.11
Starting from the same potential as Problem 3.10, plot the transmission, reflection, and probability density at the first two resonances for the cases when the potential in the *wells* is of complex form $V = V_1 + iV_2$ such that
   (a) $V_1 = 0\,\mathrm{eV}$ and $V_2 = 0.01$ eV.
   (b) $V_1 = 0\,\mathrm{eV}$ and $V_2 = -0.01$ eV.
How do the FWHMs of the resonances change compared to those in Fig. 3.11(b)? Explain the results. It can be helpful to plot the real and imaginary parts of the wave function on resonance.
   (c) What objections are there to use of complex potentials in quantum mechanics?

Problem 3.12
(a) Write a computer program in MATLAB that reproduces Fig. 3.19(a) of the text using the propagation matrix method. Use an electron effective mass $m_e^* = 0.07 \times m_0$, 1.5 nm outer barrier thickness, 1 eV barrier energy, and 2 nm well thickness. Consider electron energy in the range of 0 eV to 1 eV and vary the central barrier thickness from 0 nm to 6 nm. Include a three-dimensional plot of the transmission coefficient versus the injection energy and central barrier thickness. Describe why energy splitting occurs. Plot the logarithm of maximum transmission as a function of central barrier thickness on a linear scale and determine the approximate value of barrier thickness beyond which unity resonant electron transmission no longer occurs.
   (b) Modify the code from (a) to include calculation of the wave function given a specific central barrier thickness and injection energy. Plot the electron probability density at the transmission maxima for central barrier thicknesses of 0 nm and 6 nm. Because the propagation matrix is used to calculate the wave functions with a right-propagating wave initial condition, only one of the degenerate wave functions is calculated for the localized electron. Sketch by hand the other degenerate probability density.
   (c) Plot the probability density at the transmission maximum when the central barrier thickness is 3.5 nm. Explain the result and compare with the expected probability density when the central barrier thickness is less than 3 nm.

Problem 3.13
(a) Using the method outlined in Exercise 2.6 as a starting point, write a computer program that will numerically evaluate the time evolution of both $|\psi(x,t)|^2$ and the current density of a superposition of states for an electron confined within a constant potential well with infinite potential barriers. Use this program to generate a movie of $|\psi(x,t)|^2$ and the current density of a superposition of the first and second excited states of an infinite potential well that is 11 nm thick containing a 1 nm thick, 0.6 eV energy barrier centered within the well. Use an effective electron mass of $m_e^* = 0.07 \times m_0$. Submit snapshots of $|\psi(x,t)|^2$ and current density at times $t = 0$, $t = 3$ fs, and $t = 7.5$ fs along with a printout of the computer program. Describe how the superposition state evolves in time.
   (b) Repeat part (a) using a potential well that is 30 nm thick containing a 20 nm thick barrier of potential energy 0.6 eV centered within the well. Submit snapshots of $|\psi(x,t)|^2$

and current density at times $t = 0$, $t = 2.5$ fs, and $t = 6.5$ fs. Describe how the superposition state evolves in time.

(c) Compare the results from parts (a) and (b) and explain their differences.

Problem 3.14

Calculated transmission as a function of energy for an electron incident on an array of eight AlGaAs barriers is shown as a solid curve in Fig. 3.P50. The conduction band electron has mass $m_e^* = 0.07 \times m_0$, and each AlGaAs barrier is 1 nm thick and has barrier energy 0.3 eV relative to the GaAs wells, each of which are 2.5 nm thick.

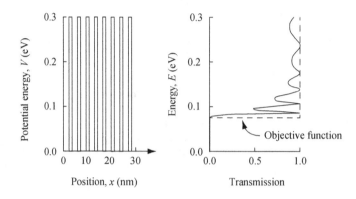

Fig. 3.P50

(a) Keeping the total thickness of each barrier-well pair constant and by independently varying the relative thickness of each barrier-well pair, write a MATLAB program to find the optimal potential profile that minimizes transmission for electron energies less than 75 meV and maximizes electron transmission for energies in the range 75 meV to 300 meV. This step-function objective is shown as the dashed line in Fig. 3.P50.

(b) What physically limits finding a solution that *exactly* matches the objective function?

(c) The optimization in (a) is formulated using a one-dimensional physical model (the electron is confined to motion in the x-direction). If the physical model allows surface roughness in the potential and electron scattering in the y- and z-directions, how does this change the optimal result in (a)?

Problem 3.15

Six atoms, each with a coulomb potential, form a ring with equal nearest-neighbor spacing between atoms. Treating this as a one-dimensional problem with spatial coordinate along the circumference of the ring, sketch and explain the solution for the first seven lowest-energy electron wave functions in the system.

Problem 3.16

(a) Plot the first four real and first four complex bands $E_k(k)$ in the reduced zone for a Kronig–Penney model with $k_0 L = 8$ (see Solution 3.8).

(b) Explain the change in energy band width, energy band gap, and effective electron mass at the real and complex band extrema as a function of increasing energy, E_k.

(c) How is the spatial extent of mid-gap states expected to change with increasing energy, E_k?

Problem 3.17

Using the results of Problem 3.10, write a MATLAB program that demonstrates the propagation of a single-electron Gaussian wave packet superposition state $\psi(x, t)$ constrained to motion in the x-direction and initially moving left-to-right in (a) a constant potential $V(x) = 0\,\text{eV}$ and (b) with peak position $x = 0$ at $t = 0$ and incident on a rectangular potential barrier of energy $V(25\,\text{nm} \leq x \leq 30\,\text{nm}) = 0.6\,\text{eV}$. The electron has effective electron mass $m_e^* = 0.07 \times m_0$, the standard deviation of the Gaussian wave packet is $\sigma_k = 3 \times 10^8\,\text{m}^{-1}$, and its central energy is 0.5 eV.

Problem 3.18

The Schrödinger equation describing the behavior of a single particle mass m incident on a potential barrier $V(x)$ is similar to the Helmholtz equation describing propagation of an electromagnetic field incident on a lossless dielectric with refractive index $n_r(x)$.

(a) Consider a particle with positive energy E and mass m moving in the x-direction that is incident on a rectangular potential barrier $V(x) = V_0$ for $0 < x < L$ and $V(x) = 0$ elsewhere. Show that the transmission probability $|C|^2 \to 0$ as $\omega \to 0$.

(b) Consider linearly polarized electromagnetic radiation of frequency ω and wave vector k normally incident on a lossless dielectric slab of thickness L and refractive index n_r. Show that the transmission probability $|C|^2 \to 1$ as $\omega \to 0$ and explain the difference in this result compared to (a).

(c) If, in the low-frequency limit, the lossless dielectric slab in (b) has dispersion $\omega = ak^\alpha$ where a and α are real constants, find the values of α for which $|C|^2 \to 0$ as $\omega \to 0$. Use MATLAB to make a three-dimensional plot of $|C(k, \alpha)|^2$ using parameter values in the range $0 < k < 0.005$ and $-1 < \alpha < 2.5$.

(d) If group velocity $v_g = \partial\omega/\partial k$ in the dielectric in part (c) is limited to be less than or equal to the speed of light in vacuum, what further constraints are placed on allowed values of α? Explain the limitations of the physical model used.

Problem 3.19

The wave function of an electron energy E_k constrained to motion in a one-dimensional periodic potential is a Bloch function $\psi_k(x) = U_k(x)e^{ikx}$.

(a) Substitute $\psi_k(x)$ into the one-electron time-independent Schrödinger equation and show that the cell-periodic term $U_k(x)$ satisfies

$$\left(\frac{\hbar^2}{2m_0} \left(-i\frac{\mathrm{d}}{\mathrm{d}x} + k \right)^2 + V(x) \right) = E_k U_k(x).$$

(b) Find $U_k(x)$ and E_k when the potential is a constant such that $V(x) = V_0$.

Problem 3.20

If the crystal structure of a simple one-dimensional metal with a potential periodic in L is deformed by a strain field such that the potential becomes periodic in $2L$ as illustrated

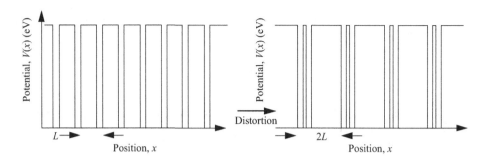

Fig. 3.P51

in Fig. 3.P51, what changes to the band structure and properties of the metal can be predicted?

Problem 3.21

The propagation matrix method can be used to find energy eigenvalues for bound states. To show this, consider a symmetric potential well of thickness L_w and energy V_0 illustrated in Fig. 3.P52. The wave function describing a particle mass m and energy E has wave vector $k_1 = \sqrt{2m(E - V_0)}/\hbar$ outside and $k_2 = \sqrt{2mE}/\hbar$ inside the potential well.

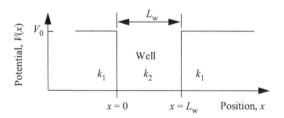

Fig. 3.P52

(a) The wave function at position $x = \pm\infty$ is connected via the propagation matrix such that

$$\begin{bmatrix} A \\ B \end{bmatrix}_{x=-\infty} = \mathbf{P} \begin{bmatrix} C \\ D \end{bmatrix}_{x=\infty} = \begin{bmatrix} P_{11} & P_{12} \\ P_{21} & P_{22} \end{bmatrix} \begin{bmatrix} C \\ D \end{bmatrix}_{x=\infty}.$$

Explain why bound-state solutions require $A = D = 0$.

(b) For the case when $L_w = 10\,\mathrm{nm}$, $m = 0.07 \times m_0$, and $V_0 = 0.5\,\mathrm{eV}$, plot P_{11} as a function of particle energy in the range $0 < E < V_0$ and find the bound-state energy eigenvalues.

(c) Explain the behavior of P_{11} when either $E < 0$ or $E > 0$.

(d) Explain how to extend the propagation matrix method calculation of bound state energy eigenvalues and eigenfunctions to an arbitrary potential well.

Problem 3.22

An electron of energy E and mass m_e^* is incident at angle θ_1 from the normal of a planar rectangular potential barrier of energy V_0 and thickness L.

(a) For $m_e^* = 0.07 \times m_0$, $V_0 = 0.15\,\text{eV}$, and $L = 15\,\text{nm}$, calculate and plot electron transmission as a function of $0 \leq \theta_1 < 60°$ and $0 < E \leq 0.35\,\text{eV}$. Explain the results.

(b) Identify and plot the critical angle $\theta_{\text{crit}}(V_0 < E \leq 0.35\,\text{eV})$. Why is transmission not zero for $\theta > \theta_{\text{crit}}$?

Problem 3.23

The dispersion relation of an electron moving in the conduction band of a one-dimensional periodic potential with lattice spacing $L = 0.5\,\text{nm}$ is described by a nearest neighbor tight-binding model.

(a) Find the value of the hopping integral t_{hop} that results in an effective electron mass of $m_e^* = 0.07 \times m_0$ for an electron with energy $E = 50\,\text{meV}$ above the conduction band minimum.

(b) The dispersion relation in (a) is $\omega(k)$ with first derivative $d\omega/dk = \omega'(k)$ and second derivative $d^2\omega/dk^2 = \omega''(k)$. A Gaussian electron wave function propagates in the band according to

$$\psi(x,t) = \left(\frac{\sigma_x^2}{2\pi}\right)^{\frac{1}{4}} \frac{e^{i(k_0(x-x_0)-\omega(k_0)t)}}{\sqrt{\sigma_x^2 + \frac{i\omega''(k_0)t}{2}}}\, e^{-(x-x_0-\omega'(k_0)t)^2/(4\sigma_x^2 + i2\omega''(k_0)t)},$$

where $\hbar\omega(k_0)$ is the central energy, x_0 is the initial mean position, and a measure of the initial spatial spread of the wave packet is the standard deviation σ_x. Plot $|\psi(x)|^2$ for the electron in (a) at time $t = 0$ and $t = 0.5\,\text{ps}$ with $\sigma_x = 10\,\text{nm}$. How does the wave packet evolve in time if the central energy of the electron in the band is $E = 2 \times t_{\text{hop}}$?

4 Eigenstates and Operators

4.1 Introduction

Quantum mechanics is a very successful description of atomic scale systems. The mathematical formalism relies on the algebra of noncommuting linear Hermitian operators. Postulates provide a logical framework with which to make contact with the results of experimental measurements.

4.1.1 The Postulates of Quantum Mechanics

There are four assumptions or postulates for quantum mechanics.

4.1.1.1 Postulate 1

Associated with every physical observable is a corresponding operator $\hat{A}$ from which results of measurement of the observable may be deduced.

Each operator is linear and satisfies an eigenvalue equation of the form $\hat{A}\psi_n = a_n\psi_n$, in which the eigenvalues a_n are real numbers and the eigenfunctions ψ_n form a complete orthogonal set in state-function space. The eigenvalues, which may take on discrete values or exist for a continuous range of values, are guaranteed to be real (and hence measurable) if the corresponding operator is Hermitian.[1] In general, the eigenfunctions themselves are complex and hence not directly measurable.

4.1.1.2 Postulate 2

The only possible result of a measurement on a single system of a physical observable associated with the operator $\hat{A}$ is an eigenvalue of the operator $\hat{A}$.

In this way, the result of measurement is related to the eigenvalue of the mathematical operator $\hat{A}$. The act of measurement on the system gives an eigenvalue a_n, which is a real number. The eigenfunction associated with this eigenvalue is stationary. As a consequence, after the measurement has been performed, the state of the system remains in the measured eigenstate unless acted upon by a force.

---

[1] Requiring Hermitian Hamiltonians in quantum mechanics is sufficient to ensure real eigenvalues but may not be necessary (R. Haydock and M. J. Kelly, *J. Phys. C: Solid State Physics* **8**, L290 (1975)) and possibly replaceable by a parity-time symmetry requirement (C. M. Bender and S. Boettcher, *Phys. Rev. Lett.* **80**, 5243 (1998)) or a pseudo-Hermitian Hamiltonian (A. Mostafazadeh, *J. Math. Phys.* **43**, 205 (2002)).

4.1.1.3 Postulate 3

For every system there always exists a state-function Ψ that contains all of the information that is knowable about the system.

The state-function Ψ contains all of the information on all observables in the system. It may be used to find the relative probability of obtaining eigenvalue a_n associated with operator $\hat{A}$ for a particular system at a given time.

4.1.1.4 Postulate 4

The time evolution of Ψ is determined by $i\hbar \frac{\partial \Psi}{\partial t} = \hat{H}\Psi$, where $\hat{H}$ is the Hamiltonian operator for the system.

Time evolution of the state function is determined by Schrödinger's equation (Eq. (1.34)).

The postulates of quantum mechanics are the underlying assumptions on which the theory is built. They may only be justified to the extent that results of physical experiments do not contradict them. The postulates, which are a connection between mathematics and the physical aspects of the model, contain the strangeness of quantum mechanics.

The probabilistic interpretation of measurement and the associated collapse of the state function to an eigenfunction are *physical* aspects of the model. Measurement is fundamentally noncausal since, in general, it is impossible to know which eigenstate the system is in before measurement but it is known after measurement. Between measurements the time-evolution of the complete system described by state vector Ψ(t) evolves deterministically according to Schrödinger's equation. These and other aspects of the physical model contribute to the nonintuitive nature of quantum mechanics.

4.2 One-Particle Wave Function Space

The probabilistic interpretation of the one-particle bound-state wave function means that $|\psi(\mathbf{r}, t)|^2 d^3 r$ represents the probability of finding the particle at time t in volume $d^3 r$ about the point $\mathbf{r}$ in space. Physical experience suggests it is reasonable to assume that the total probability of finding the particle somewhere in space is certain so that the integral of the magnitude of the wave function squared over all space is unity.

The wave functions $\psi(\mathbf{r}, t)$ must be defined, continuous, and differentiable. Also, the wave functions exist in a wave-function space that is linear. The integrands for which this equation converges are square integrable functions. This is a set called $\mathcal{L}^2$ and it has the structure of Hilbert space, $\mathcal{H}$.

There are analogies between an ordinary N-dimensional vector space consisting of N orthonormal unit vectors and the eigenfunction space in quantum mechanics. They are, for example, both linear spaces. However, an important difference becomes apparent when considering scalar products.

If the ordinary vector

$$\mathbf{A} = \sum_{j}^{N} a_j \mathbf{a}_j, \tag{4.1}$$

where a_j is the jth coefficient and $\mathbf{a}_j$ is the jth orthonormal unit vector, and similarly the vector

$$\mathbf{B} = \sum_j^N b_j \mathbf{b}_j, \tag{4.2}$$

then the scalar product of the two vectors $\mathbf{A}$ and $\mathbf{B}$ is just

$$\mathbf{A} \cdot \mathbf{B} = \sum_j^N a_j b_j. \tag{4.3}$$

In quantum mechanics there are wave functions such as $\psi_A(\mathbf{r})$ and $\psi_B(\mathbf{r})$. In this case, the scalar product is an integral,

$$\int \psi_A^*(\mathbf{r}) \psi_B(\mathbf{r}) \, \mathrm{d}^3 r.$$

The integral is needed because wave function space is continuously infinite dimensional. This is one of the characteristics of Hilbert space that distinguishes it from an ordinary N-dimensional vector space. In direct analogy with two vectors $\mathbf{A}$ and $\mathbf{B}$ in Euclidean space for which $|\mathbf{A} \cdot \mathbf{B}|^2 \leq |\mathbf{A}|^2 |\mathbf{B}|^2$, any two states $\psi_A(\mathbf{r})$ and $\psi_B(\mathbf{r})$ in Hilbert space satisfy $|\int \psi_A^*(\mathbf{r}) \psi_B(\mathbf{r}) \mathrm{d}^3 r|^2 \leq \int \psi_A^*(\mathbf{r}) \psi_A(\mathbf{r}) \mathrm{d}^3 r \int \psi_B^*(\mathbf{r}) \psi_B(\mathbf{r}) \mathrm{d}^3 r$, which is called the Schwarz inequality.

4.3 Properties of Linear Operators

A linear operator $\hat{A}$ associates with every function $\psi(\mathbf{r})$ another function $\phi(\mathbf{r})$ in a linear way. This associativity may be expressed mathematically as

$$\phi(\mathbf{r}) = \hat{A}\psi(\mathbf{r}). \tag{4.4}$$

A linear operator $\hat{A}$ commutes with constants and is distributive. Hence, letting the wave function $\psi(\mathbf{r})$ be a linear combination $\psi(\mathbf{r}) = \lambda_1 \psi_1(\mathbf{r}) + \lambda_2 \psi_2(\mathbf{r})$, where λ_1 and λ_2 are numbers that weight the contribution of $\psi_1(\mathbf{r})$ and $\psi_2(\mathbf{r})$, then

$$\hat{A}(\lambda_1 \psi_1(\mathbf{r}) + \lambda_2 \psi_2(\mathbf{r})) = \lambda_1 \hat{A}\psi_1(\mathbf{r}) + \lambda_2 \hat{A}\psi_2(\mathbf{r}). \tag{4.5}$$

The momentum operator for a particle moving in one dimension is $\hat{A} = \hat{p}_x = -i\hbar \, \mathrm{d}/\mathrm{d}x$. If the wave function $\psi(x) = \lambda_1 \psi_1(x) + \lambda_2 \psi_2(x)$, then, applying the momentum operator, the expression for $\phi(x)$ in Eq. (4.4) becomes

$$\phi(x) = -i\hbar \frac{\mathrm{d}}{\mathrm{d}x}(\lambda_1 \psi_1(x) + \lambda_2 \psi_2(x)) = -\lambda_1 i\hbar \frac{\mathrm{d}}{\mathrm{d}x}\psi_1(x) - \lambda_2 i\hbar \frac{\mathrm{d}}{\mathrm{d}x}\psi_2(x). \tag{4.6}$$

4.3.1 Product of Operators

The product of operators is associative so that, for example,

$$\hat{A}\hat{B}\hat{C} = \hat{A}\left(\hat{B}\hat{C}\right) = \left(\hat{A}\hat{B}\right)\hat{C}. \tag{4.7}$$

The product of operators $\hat{A}$ and $\hat{B}$ acting upon the function $\psi(\mathbf{r})$ is

$$\left(\hat{A}\hat{B}\right)\psi(\mathbf{r}) = \hat{A}\left(\hat{B}\psi(\mathbf{r})\right). \tag{4.8}$$

Equation (4.8) indicates that operator $\hat{B}$ acts first upon $\psi(\mathbf{r})$ to give $\phi(\mathbf{r}) = \hat{B}\psi(\mathbf{r})$. Operator $\hat{A}$ then acts upon the new function $\phi(\mathbf{r})$. The order in which operators act upon a function is critical because, in general, the operators $\hat{A}$ and $\hat{B}$ do not commute, that is, $\hat{A}\hat{B} \neq \hat{B}\hat{A}$.

This important property may be illustrated by considering the one-dimensional momentum operator in real space $\hat{A} = \hat{p}_x = -i\hbar\,\mathrm{d}/\mathrm{d}x$ and the one-dimensional position operator $\hat{B} = \hat{x}$. Let the wave function $\psi = \psi(x)$. Then

$$\hat{A}\hat{B}\psi(x) = -i\hbar\frac{\mathrm{d}}{\mathrm{d}x}(x\psi(x)) = -i\hbar\psi(x) - i\hbar x\frac{\mathrm{d}}{\mathrm{d}x}\psi(x) \tag{4.9}$$

and

$$\hat{B}\hat{A}\psi(x) = -i\hbar x\frac{\mathrm{d}}{\mathrm{d}x}\psi(x). \tag{4.10}$$

Comparing Eqs. (4.9) and (4.10), $\hat{A}\hat{B} \neq \hat{B}\hat{A}$ and so the order in which operators are applied is important. The sensitivity of pairs of operators to the order in which a product is applied is quantified using commutator algebra, as discussed in Section 4.3.5.

4.3.2 Properties of Hermitian Operators

The results of physical measurements are real numbers. This means that a physical model of reality is restricted to prediction of real numbers. Hermitian operators play a special role in quantum mechanics because these operators guarantee real eigenvalues. Hence, a physical system described using a Hermitian operator will provide information on measurable quantities.

The adjoint $\hat{A}^{\dagger}$ of an operator $\hat{A}$ is defined by

$$\boxed{\int \psi^*(\mathbf{r})\hat{A}^{\dagger}\phi(\mathbf{r})\,\mathrm{d}^3r = \left(\int \phi^*(\mathbf{r})\hat{A}\psi(\mathbf{r})\,\mathrm{d}^3r\right)^*.} \tag{4.11}$$

It follows from the definition of Hermitian operators that an operator $\hat{A}$ is *Hermitian* when it is its own adjoint – that is $\hat{A}^{\dagger} = \hat{A}$. On the other hand, an operator $\hat{A}$ is *anti-Hermitian* if $\hat{A}^{\dagger} = -\hat{A}$. The adjoint of a complex number a is its complex conjugate – that is $a^{\dagger} = a^*$.

If $\hat{A}$ is a Hermitian operator then $\hat{A}^{\dagger} = \hat{A}$ and, in matrix notation,

$$\boxed{A_{nm}^* = A_{mn},} \tag{4.12}$$

where the matrix elements $A_{nm}^* = \int \left(\phi_n^*(\mathbf{r}) \hat{A} \psi_m(\mathbf{r}) \right)^* d^3r$ and $A_{mn} = \int \psi_m^*(\mathbf{r}) \hat{A} \phi_n(\mathbf{r}) d^3r$.

For two operators $\hat{A}$ and $\hat{B}$ and complex number a the following relations hold: $\left(a\hat{A} \right)^\dagger = a^* \hat{A}^\dagger$, $\left(\hat{A}^\dagger \right)^\dagger = \hat{A}$, and $\left(\hat{A}\hat{B} \right)^\dagger = \hat{B}^\dagger \hat{A}^\dagger$.

4.3.2.1 Real Eigenvalues and Orthogonal Eigenfunctions

To show that the eigenvalues of a Hermitian operator are real and that the associated eigenfunctions are orthogonal, consider the Hermitian operator $\hat{A}$ such that

$$\hat{A}\phi_n = a_n \phi_n, \tag{4.13}$$

where ϕ_n is an eigenfunction of $\hat{A}$ and a_n is the corresponding eigenvalue. Multiplying both sides of Eq. (4.13) by ϕ_m^* and integrating over all space gives

$$\int \phi_m^* \hat{A} \phi_n \, d^3r = a_n \int \phi_m^* \phi_n \, d^3r. \tag{4.14}$$

Similarly, interchanging the subscripts m and n gives

$$\int \phi_n^* \hat{A} \phi_m \, d^3r = \int \left(\hat{A}\phi_n \right)^* \phi_m \, d^3r = a_m \int \phi_n^* \phi_m \, d^3r. \tag{4.15}$$

Taking the complex conjugate,

$$\int \phi_m^* \hat{A} \phi_n \, d^3r = a_m^* \int \phi_m^* \phi_n \, d^3r. \tag{4.16}$$

Subtracting Eq. (4.14) from Eq. (4.16) gives

$$0 = (a_n - a_m^*) \int \phi_m^* \phi_n \, d^3r. \tag{4.17}$$

For the case when $n = m$,

$$0 = (a_n - a_n^*) \int \phi_n^* \phi_n \, d^3r. \tag{4.18}$$

Since $|\phi_n|^2$ is finite, $a_n = a_n^*$ for Eq. (4.18) to be valid and so the conclusion is that *eigenvalues of Hermitian operators are real numbers*. This is of practical importance in quantum mechanics because it guarantees that the eigenvalue of a Hermitian operator results in a real measurable quantity.

For the case in which $n \neq m$, the integral is zero provided $a_n \neq a_m$. Hence, the nondegenerate eigenfunctions of Hermitian operators are *orthogonal* to each other, and

$$0 = \int \phi_m^* \phi_n \, d^3r \tag{4.19}$$

for $n \neq m$. Since $e^{\pm i\pi} = -1$, the phase of nondegenerate orthogonal eigenfunctions is arbitrary to within $\pm\pi$.

4.3.3 Normalization of Eigenfunctions

Because eigenvalue equations involve linear operators, eigenfunctions may be specified to within an arbitrary constant. It is convention that the constant is chosen for bound states in such a way that the integral of the magnitude of an eigenfunction squared over all space is unity. Because such eigenfunctions are orthogonal and normalized they are called *orthonormal*. The orthonormal properties of Hermitian operator eigenfunctions can be expressed as

$$\int \phi_n^* \phi_m \, \mathrm{d}^3 r = \delta_{nm},$$
(4.20)

where the Kronecker delta $\delta_{nm} = 0$ if $n \neq m$ and $\delta_{nm} = 1$ if $n = m$.

4.3.4 Completeness of Eigenfunctions

The completeness property of eigenfunctions $\phi_n(\mathbf{r})$ means they can be used as *basis functions* with which to expand an *arbitrary function* $\psi(\mathbf{r})$ so that

$$\psi(\mathbf{r}) = \sum_n a_n \phi_n(\mathbf{r}).$$
(4.21)

In the case of continuum states, the sum becomes an integral.[2] The expansion coefficient a_m is obtained by multiplying both sides of the equation by $\phi_m^*(\mathbf{r})$ and integrating:

$$\int \phi_m^*(\mathbf{r})\psi(\mathbf{r}) \, \mathrm{d}^3 r = \sum_n a_n \int \phi_m^*(\mathbf{r})\phi_n(\mathbf{r}) \, \mathrm{d}^3 r.$$
(4.22)

Using the fact that $\int \phi_m^*(\mathbf{r})\phi_n(\mathbf{r}) \, \mathrm{d}^3 r = \delta_{mn}$,

$$\int \phi_m^*(\mathbf{r})\psi(\mathbf{r}) \, \mathrm{d}^3 r = a_m,$$
(4.23)

so that a_m is the projection of $\psi(\mathbf{r})$ on $\phi_m(\mathbf{r})$, $\psi(\mathbf{r})$ is an arbitrary function, and $\sum_n a_n \phi_n(\mathbf{r})$ is the expansion of that function in terms of the orthonormal basis eigenfunctions $\phi_n(\mathbf{r})$.

4.3.5 Commutator Algebra

The commutator for the pair of operators $\hat{A}$ and $\hat{B}$ is defined as

$$\left[\hat{A}, \hat{B}\right] = \hat{A}\hat{B} - \hat{B}\hat{A}.$$
(4.24)

---

[2] Notice the similarity to the Fourier transform. If $\phi_n(x) \sim \mathrm{e}^{ikx}$ then an arbitrary function $\psi(x)$ is just $\psi(x) = \frac{1}{\sqrt{2\pi}} \int_{k=-\infty}^{k=\infty} a(k)\mathrm{e}^{ikx} \mathrm{d}k$, where the coefficients $a(k)$ are found using the Fourier transform $a(k) = \frac{1}{\sqrt{2\pi}} \int_{x=-\infty}^{x=\infty} \mathrm{e}^{-ikx} \psi(x)\mathrm{d}x$.

Mathematically, quantum mechanics may be thought of as the description of physical systems with noncommuting operators. As with matrix algebra, in general $\hat{A}\hat{B} \neq \hat{B}\hat{A}$.

By way of an example, the one-dimensional momentum operator in real space is $\hat{A} = \hat{p}_x = -i\hbar \, d/dx$ and the one-dimensional position operator is $\hat{B} = \hat{x}$. These are the same operators used in Eqs. (4.9) and (4.10), which, when substituted into Eq. (4.24), give the commutator

$$[\hat{p}_x, \hat{x}] = -i\hbar. \tag{4.25}$$

The fact that the right-hand side of this equation is nonzero means that the pair of linear operators used in this particular example are noncommuting. Of course, if the order in which the operators appear in Eq. (4.25) is interchanged, then the sign of the commutator is reversed. In the example this gives $[\hat{x}, \hat{p}_x] = i\hbar$. In general, this anti-symmetry can be expressed as

$$\left[\hat{A}, \hat{B}\right] = -\left[\hat{B}, \hat{A}\right]. \tag{4.26}$$

There are other useful relationships that be derived from Eq. (4.24). For example, linearity requires

$$\left[\hat{A}, \hat{B} + \hat{C} + \hat{D} + \cdots\right] = \left[\hat{A}, \hat{B}\right] + \left[\hat{A}, \hat{C}\right] + \left[\hat{A}, \hat{D}\right] + \cdots. \tag{4.27}$$

The distributive nature of linear operators requires

$$\left[\hat{A}, \hat{B}\hat{C}\right] = \hat{A}\hat{B}\hat{C} - \hat{B}\hat{C}\hat{A} = \left(\hat{A}\hat{B} - \hat{B}\hat{A}\right)\hat{C} + \hat{B}\left(\hat{A}\hat{C} - \hat{C}\hat{A}\right) = \hat{B}\left[\hat{A}, \hat{C}\right] + \left[\hat{A}, \hat{B}\right]\hat{C}, \tag{4.28}$$

so that if $\hat{B} = \hat{C}$ then

$$\left[\hat{A}, \hat{B}^2\right] = \hat{B}\left[\hat{A}, \hat{B}\right] + \left[\hat{A}, \hat{B}\right]\hat{B}. \tag{4.29}$$

The Jacobi identity,

$$\left[\hat{A}, \left[\hat{B}, \hat{C}\right]\right] + \left[\hat{B}, \left[\hat{C}, \hat{A}\right]\right] + \left[\hat{C}, \left[\hat{A}, \hat{B}\right]\right] = 0, \tag{4.30}$$

follows since

$$\left[\hat{A}, \left[\hat{B}, \hat{C}\right]\right] = \left[\hat{A}, \hat{B}\hat{C}\right] - \left[\hat{A}, \hat{C}\hat{B}\right] = \hat{B}\left[\hat{A}, \hat{C}\right] + \left[\hat{A}, \hat{B}\right]\hat{C} - \hat{C}\left[\hat{B}, \hat{A}\right] - \left[\hat{A}, \hat{C}\right]\hat{B}$$
$$= \left[\hat{B}, \left[\hat{A}, \hat{C}\right]\right] + \left[\left[\hat{A}, \hat{B}\right], \hat{C}\right] = -\left[\hat{B}, \left[\hat{C}, \hat{A}\right]\right] - \left[\hat{C}, \left[\hat{A}, \hat{B}\right]\right]. \tag{4.31}$$

If operators $\hat{A}$ and $\hat{B}$ are Hermitian then $\hat{A}^\dagger = \hat{A}$, $\hat{B}^\dagger = \hat{B}$, and

$$\left[\hat{A}, \hat{B}\right]^\dagger = \left(\hat{A}\hat{B} - \hat{B}\hat{A}\right)^\dagger = \hat{B}^\dagger\hat{A}^\dagger - \hat{A}^\dagger\hat{B}^\dagger = -\left(\hat{A}\hat{B} - \hat{B}\hat{A}\right) = -\left[\hat{A}, \hat{B}\right], \tag{4.32}$$

so that the commutator of two Hermitian operators is anti-Hermitian.

4.4 Dirac Notation

So far the discussion has focused on single particle quantum systems using continuous wave functions $\psi(\mathbf{r}, t)$. This is a real space representation. Taking the Fourier transform to obtain $\psi(\mathbf{k}, t)$ gives a momentum space representation. Of course, the physical state of the system should be independent of the coordinate representation. In the basis-independent notation introduced by Dirac, state vectors, ψ, are called ket vectors and depicted by the symbol $|\psi\rangle$. They are elements of a linear Hilbert space. The complex conjugate of ψ is ψ^* and is represented by the bra symbol $\langle\psi|$. If the quantum state $|\psi\rangle$ is known then it is always possible to find the wave function in real space, momentum space, or other coordinate representation. For example, the spatial wave function of a particle at time t is given by $\langle\mathbf{r}, t|\psi\rangle = \psi(\mathbf{r}, t)$. The scalar inner product $(\phi, \psi) = \int \phi^*(\mathbf{r}, t)\psi(\mathbf{r}, t)\, \mathrm{d}^3 r$ is represented by the bra-ket symbol $\langle\phi(\mathbf{r}, t)|\psi(\mathbf{r}, t)\rangle$ and so

$$\int \phi^*(\mathbf{r}, t)\psi(\mathbf{r}, t)\, \mathrm{d}^3 r \equiv \langle\phi(\mathbf{r}, t)|\psi(\mathbf{r}, t)\rangle = \langle\psi(\mathbf{r}, t)|\phi(\mathbf{r}, t)\rangle^*. \tag{4.33}$$

The act of operating with $\hat{A}$ on the state $|\psi\rangle$ is represented as

$$\hat{A}|\psi\rangle. \tag{4.34}$$

In Dirac notation the time-independent Schrödinger equation is written as

$$\boxed{\hat{H}|n\rangle = E_n|n\rangle,} \tag{4.35}$$

where the set $\{|n\rangle\}$ is the Hilbert-space basis. The time-dependence of the state vector is

$$|n, t\rangle = |n\rangle\mathrm{e}^{-\mathrm{i}E_n t/\hbar}, \tag{4.36}$$

where $|n, t\rangle$ is in the $\{|n, t\rangle\}$ basis. Likewise, the Schrödinger equation, which describes the time evolution of the quantum state $|\psi\rangle$, is written as

$$\boxed{\hat{H}|\psi\rangle = \mathrm{i}\hbar\frac{\partial}{\partial t}|\psi\rangle.} \tag{4.37}$$

For every ket $|\chi\rangle = \hat{A}|\psi\rangle$ there is an associated bra such that $\langle\chi| = \langle\psi|\hat{A}^\dagger$. Using this with the property of a scalar product $\langle\chi|\phi\rangle = \langle\phi|\chi\rangle^*$ then $\langle\psi|\hat{A}^\dagger|\phi\rangle = \langle\chi|\phi\rangle = \langle\phi|\chi\rangle^* = \langle\phi|\hat{A}|\psi\rangle^*$ so that the adjoint $\hat{A}^\dagger$ of an operator $\hat{A}$ can be defined as

$$\langle\psi|\hat{A}^\dagger|\phi\rangle = \langle\phi|\hat{A}|\psi\rangle^*. \tag{4.38}$$

Alternatively, noting that $|\hat{A}\psi\rangle = \hat{A}|\psi\rangle = |\chi\rangle$ and $\langle\hat{A}\psi| = \langle\psi|\hat{A}^\dagger = \langle\chi|$ and $(\hat{A}^\dagger)^\dagger = \hat{A}$ so that $\langle\hat{A}^\dagger\phi| = \langle\phi|(\hat{A}^\dagger)^\dagger = \langle\phi|\hat{A}$ and $\langle\hat{A}^\dagger\phi|\psi\rangle = \langle\phi|\hat{A}\psi\rangle$, which can also be used to define the adjoint $\hat{A}^\dagger$ of an operator $\hat{A}$.

The operator $\hat{A}$ is *Hermitian* when it is its own adjoint $\hat{A}^\dagger$, that is, $\hat{A}^\dagger = \hat{A}$ or $\langle\psi|\hat{A}|\phi\rangle = \langle\phi|\hat{A}|\psi\rangle^*$ or $\langle\psi|\hat{A}\phi\rangle = \langle\hat{A}\psi|\phi\rangle$.

The orthonormal condition is expressed as

$$\int \phi_n^*(\mathbf{r})\phi_m(\mathbf{r})\, d^3r = \langle \phi_n|\phi_m\rangle = \langle n|m\rangle = \delta_{nm}. \tag{4.39}$$

The projection of $\psi(\mathbf{r})$ on the basis eigenfunction $\phi_m(\mathbf{r})$ is expressed as

$$a_m = \langle \phi_m|\psi\rangle, \tag{4.40}$$

and the expansion of an arbitrary state-function $|\psi\rangle$ is

$$|\psi\rangle = \sum_n b_n |\phi_n\rangle. \tag{4.41}$$

If $|\phi_i\rangle$ forms an orthonormal set then

$$\sum_i |\phi_i\rangle\langle\phi_i| = \hat{1} \tag{4.42}$$

is the identity (unit) operator since

$$\sum_i |\phi_i\rangle\langle\phi_i|\phi_j\rangle = \sum_i |\phi_i\rangle\delta_{ij} = |\phi_j\rangle. \tag{4.43}$$

The Schwarz inequality for states $|\psi\rangle$ and $|\phi\rangle$ is

$$|\langle\phi|\psi\rangle|^2 \leq \langle\phi|\phi\rangle\langle\psi|\psi\rangle. \tag{4.44}$$

4.4.1 Linear Algebra

The mathematics used to describe quantum phenomena is the linear algebra of non-commuting operators (matrices). The state of a system is described by a complex vector $|\psi\rangle$ in Hilbert space, $\mathcal{H}$.

The state ket-vector $|\psi\rangle$ in a given basis can be written as a column vector of N complex probability amplitude elements, and its adjoint is a bra-vector $\langle\psi|$ written as a row vector so that

$$|\psi\rangle = \begin{bmatrix} a_1 \\ a_2 \\ \vdots \\ a_N \end{bmatrix} \quad \text{and} \quad \langle\psi| = [a_1^*, a_2^*, \ldots, a_N^*]. \tag{4.45}$$

Normalization of the state vector requires

$$\langle\psi|\psi\rangle = \sum_{j=1}^{N} |a_j|^2 = 1. \tag{4.46}$$

4.4.1.1 Inner Product

The inner product $\langle\phi|\psi\rangle = \langle\psi|\phi\rangle^*$ between two states $|\psi\rangle$ and $|\phi\rangle$ is a bra-ket and is independent of the basis used. If

$$|\psi\rangle = \begin{bmatrix} a_1 \\ a_2 \\ \vdots \\ a_N \end{bmatrix} \quad \text{and} \quad \langle\phi| = [b_1^*, b_2^*, \ldots, b_N^*] \quad \text{then}$$

$$\langle\phi|\psi\rangle = [b_1^*, b_2^*, \ldots, b_N^*] \begin{bmatrix} a_1 \\ a_2 \\ \vdots \\ a_N \end{bmatrix} = \sum_{j=1}^{N} b_j^* a_j. \tag{4.47}$$

Orthogonal states require $\langle\phi|\psi\rangle = 0$.

4.4.1.2 Outer Product

The outer product $|\psi\rangle\langle\phi|$ is a linear operator (a matrix) so that

$$|\psi\rangle\langle\phi| = \begin{bmatrix} a_1 \\ a_2 \\ \vdots \\ a_N \end{bmatrix} [b_1^*, b_2^*, \ldots, b_N^*] = \begin{bmatrix} a_1 b_1^* & \cdots & a_1 b_N^* \\ \vdots & & \vdots \\ a_N b_1^* & \cdots & a_N b_N^* \end{bmatrix}. \tag{4.48}$$

A linear operator $\hat{A}$ transforms one state (or function) into another state so that $|\phi\rangle = \hat{A}|\psi\rangle$ and the operator is a matrix.

4.4.1.3 Expectation Value

The expectation value associated with an operator $\hat{A}$ acting on the state $|\psi\rangle$ is

$$\langle A\rangle = \langle\psi|\hat{A}|\psi\rangle = \sum_{ij} a_i^* a_{ij} a_j. \tag{4.49}$$

If the operator $\hat{A}$ is Hermitian, then $\langle A\rangle$ is a real number that is the average result of repeated measurements on identically prepared systems in state $|\psi\rangle$. This measure of average value (the first moment of a distribution function) is most useful *if* the distribution is symmetric and strongly peaked.

4.4.1.4 Matrix Element

The matrix element between states $|\psi_i\rangle$ and $|\psi_j\rangle$ in a given orthonormal basis and associated with operator $\hat{A}$ is

$$\langle\psi_i|\hat{A}|\psi_j\rangle = a_{ij} \tag{4.50}$$

and

$$\hat{A} = \sum_{ij} a_{ij} |\psi_i\rangle \langle \psi_j|. \tag{4.51}$$

4.4.1.5 Hermitian Operators

The Hermitian conjugate of operator $\hat{A}$ is the complex conjugate of the transpose (a reflection of matrix elements about the diagonal) of $\hat{A}$ so that if, in a given basis,

$$\hat{A} = \begin{bmatrix} a_{11} & a_{12} & & \\ a_{21} & a_{22} & & \\ & & \ddots & \\ & & & a_{NN} \end{bmatrix} = [a_{ij}] \tag{4.52}$$

then

$$\hat{A}^\dagger = [a_{ji}^*]. \tag{4.53}$$

It follows that $(|\psi\rangle)^\dagger = \langle\psi|$ and $(|\psi\rangle\langle\phi|)^\dagger = |\phi\rangle\langle\psi|$.
Hermitian operators are those that satisfy $[a_{ij}] = [a_{ji}^*]$ or

$$\boxed{\hat{A} = \hat{A}^\dagger.} \tag{4.54}$$

They are normal since $\hat{A}^\dagger \hat{A} = \hat{A}^2 = \hat{A}\hat{A}^\dagger$ and the eigenvalues are always real. Normal operators are diagonalizable.

4.4.1.6 Identity Operator

The identity (or unit) operator, $\hat{1}$, leaves the function it operates on unchanged:

$$\hat{1} = \begin{bmatrix} 1 & 0 & & \\ 0 & 1 & & \\ & & \ddots & \\ & & & 1 \end{bmatrix} = \sum_i |\psi_i\rangle\langle\psi_i|. \tag{4.55}$$

4.4.1.7 Trace

The trace of a matrix operator is the sum of the leading diagonal elements,

$$\text{Tr}(\hat{A}) = \sum_j \langle \psi_j | \hat{A} | \psi_j \rangle = \sum_j A_{jj}, \tag{4.56}$$

and is independent of the basis used. A traceless operator has $\text{Tr}(\hat{A}) = 0$. The trace of a product of operators is cyclic such that $\text{Tr}(\hat{A}\hat{B}\hat{C}) = \text{Tr}(\hat{C}\hat{A}\hat{B}) = \text{Tr}(\hat{B}\hat{C}\hat{A})$.

4.4.1.8 Unitary Operators

Unitary operators are such that

$$\hat{U}^\dagger \hat{U} = \hat{U}\hat{U}^\dagger = \hat{1}$$ (4.57)

with eigenvalue equation $\hat{U}|\psi_j\rangle = u_j|\psi_j\rangle$. The eigenvalues of unitary operators have unit norm so that they may be written as $u_j = e^{i\theta_j}$ with $0 \le \theta_j < 2\pi$ real.

Inverse operators are such that $\hat{A}^{-1}\hat{A} = \hat{1}$ and so for a unitary operator,

$$\hat{U}^{-1} = \hat{U}^\dagger.$$ (4.58)

It follows that $\left(\hat{U}|\psi_i\rangle\right)^\dagger \left(\hat{U}|\psi_j\rangle\right) = \langle\psi_i|\hat{U}^\dagger\hat{U}|\psi_j\rangle = \langle\psi_i|\psi_j\rangle = \delta_{ij}$.

Length and inner product are conserved under unitary transformation. To show the latter, consider $|f_2\rangle = \hat{U}|f_1\rangle$ and $|g_2\rangle = \hat{U}|g_1\rangle$. It follows that $\langle g_2|f_2\rangle = \langle g_1|\hat{U}^\dagger\hat{U}|f_1\rangle = \langle g_1|\hat{U}^{-1}\hat{U}|f_1\rangle = \langle g_1|\hat{1}|f_1\rangle = \langle g_1|f_1\rangle$, and hence the inner product of functions $|f\rangle$ and $|g\rangle$ is invariant under unitary transformation.

Unitary operators change basis so that if $|j\rangle$ is an orthonormal basis then so is $\hat{U}|j\rangle$.

Since matrix elements satisfy $\langle g_2|\hat{B}|f_2\rangle = (|g_2\rangle)^\dagger \hat{B}|f_2\rangle = (\hat{U}|g_1\rangle)^\dagger \hat{B}(\hat{U}|f_1\rangle) = \langle g_1|\hat{U}^\dagger\hat{B}\hat{U}|f_1\rangle = \langle g_1|\hat{A}|f_1\rangle$, it follows that a unitary matrix can be used to change the representation of an operator from $\hat{A}$ to $\hat{B} = \hat{U}\hat{A}\hat{U}^\dagger$. This may also be written $\hat{A} = \hat{U}^\dagger\hat{B}\hat{U}$ or $\hat{U}\hat{A}\hat{U}^\dagger = \left(\hat{U}\hat{U}^\dagger\right)\hat{B}\left(\hat{U}\hat{U}^\dagger\right) = \hat{B}$.

4.4.1.9 Tensor Product

A tensor product combines Hilbert spaces to create a higher dimensional Hilbert space. If $|\psi\rangle$ has amplitudes $(a_1, \dots, a_{N_1})$ and $|\phi\rangle$ has amplitudes $(b_1, \dots, b_{N_2})$, the *product state* $|\psi\rangle \otimes |\phi\rangle$ in $\mathcal{H}_1 \otimes \mathcal{H}_2$ is a column vector of length $N_1 \times N_2$,

$$|\psi\rangle \otimes |\phi\rangle = \begin{bmatrix} a_1|\phi\rangle \\ a_2|\phi\rangle \\ \vdots \\ a_{N_1}|\phi\rangle \end{bmatrix} = \begin{bmatrix} a_1 b_1 \\ a_1 b_2 \\ \vdots \\ a_1 b_{N_2} \\ a_2 b_1 \\ \vdots \\ a_{N_1} b_{N_2} \end{bmatrix}.$$ (4.59)

For example,

$$\begin{bmatrix} a_1 \\ a_2 \end{bmatrix} \otimes \begin{bmatrix} b_1 \\ b_2 \end{bmatrix} = \begin{bmatrix} a_1 b_1 \\ a_1 b_2 \\ a_2 b_1 \\ a_2 b_2 \end{bmatrix}.$$

If operator matrix $\hat{A} = [a_{ij}]$ and operator matrix $\hat{B} = [b_{ij}]$, then the tensor product is

$$
\hat{A} \otimes \hat{B} =
\begin{bmatrix}
a_{11}\hat{B} & \cdots & a_{1N_1}\hat{B} \\
\vdots & & \vdots \\
a_{N_11}\hat{B} & \cdots & a_{N_1N_1}\hat{B}
\end{bmatrix}.
\tag{4.60}
$$

For example,

$$
\begin{bmatrix}
a_{11} & a_{12} \\
a_{21} & a_{22}
\end{bmatrix}
\otimes
\begin{bmatrix}
b_{11} & b_{12} \\
b_{21} & b_{22}
\end{bmatrix}
=
\begin{bmatrix}
a_{11}\hat{B} & a_{11}\hat{B} \\
a_{21}\hat{B} & a_{22}\hat{B}
\end{bmatrix}
$$

$$
=
\begin{bmatrix}
a_{11}b_{11} & a_{11}b_{12} & a_{12}b_{11} & a_{12}b_{12} \\
a_{11}b_{21} & a_{11}b_{22} & a_{12}b_{21} & a_{12}b_{22} \\
a_{21}b_{11} & a_{21}b_{12} & a_{22}b_{11} & a_{22}b_{12} \\
a_{21}b_{21} & a_{21}b_{22} & a_{22}b_{21} & a_{22}b_{22}
\end{bmatrix}.
$$

4.5 Measurement of Real Numbers

In quantum mechanics, each type of physical observable is usually associated with a Hermitian operator. As described in Section 4.3.2, the use of Hermitian operators is *sufficient* to ensure that the associated eigenvalue is a real quantity. In this way, the result of a measurement is a real number that corresponds to one of the set of continuous or discrete eigenvalues for the system

$$
\hat{A}|n\rangle = a_n|n\rangle,
\tag{4.61}
$$

where $\hat{A}$ is a Hermitian operator, $|n\rangle$ is an eigenfunction in the basis $\{|n\rangle\}$, and a_n is its eigenvalue.

If there are two different physical observables with eigenvalues a_n and b_n, respectively, then there are two different associated operators $\hat{A}$ and $\hat{B}$. For a given system, measurement of $\hat{A}$ followed by measurement of $\hat{B}$ is denoted by $\hat{B}\hat{A}$, and the result may be different for $\hat{A}\hat{B}$. If the measurements interfere with each other, then the commutator

$$
\left[\hat{A}, \hat{B}\right] = \hat{A}\hat{B} - \hat{B}\hat{A} \neq 0.
\tag{4.62}
$$

Measurement of position and momentum are examples of measurements that interfere with each other. The commutation relation for the position operator $\hat{x}$ and the momentum operator $\hat{p}_x = -i\hbar\, d/dx$ for a particle moving in one dimension is

$$
[\hat{x}, \hat{p}_x] = i\hbar.
\tag{4.63}
$$

The momentum and position operators do not commute. A measurement on one observable influences the value of the other. The coupling between the two observables through the commutation relation has the physical consequence that the observable quantities cannot be measured simultaneously with arbitrary accuracy. This fact constrains approaches to control quantum systems.

Measurements that are independent of each other have corresponding operators that commute. If two operators commute, then they possess common eigenfunctions. In this case, $\hat{B}\hat{A} = \hat{A}\hat{B}$, and $\hat{A}\hat{B}\phi_B = \hat{B}\hat{A}\phi_B = \hat{A}b\phi_B = b\hat{A}\phi_B$. The function $\hat{A}\phi_B$ is thus an eigenfunction of $\hat{B}$ with eigenvalue b. If there is only one eigenfunction of $\hat{B}$ associated with eigenvalue b, then $\hat{A}\phi_B = c\phi_B$, where c is a constant, so that ϕ_B is an eigenfunction of $\hat{A}$.

4.5.1 Time Dependence of Expectation Value

The expectation value of the observable A associated with operator $\hat{A}$ for a system in state $|\psi\rangle$ is

$$\langle A \rangle = \langle \psi | \hat{A} | \psi \rangle. \tag{4.64}$$

Making note of the fact that the Schrödinger equation (Eq. (4.37)) is

$$\frac{-i}{\hbar} \hat{H} |\psi\rangle = \left| \frac{\partial \psi}{\partial t} \right\rangle, \tag{4.65}$$

the time derivative of Eq. (4.64) is found using the product rule for differentiation and substituting in Eq. (4.65):

$$
\begin{aligned}
\frac{d}{dt}\langle A \rangle &= \left\langle \frac{\partial \psi}{\partial t} \left| \hat{A} \right| \psi \right\rangle + \left\langle \psi \left| \frac{\partial}{\partial t}\hat{A} \right| \psi \right\rangle + \left\langle \psi \left| \hat{A} \right| \frac{\partial \psi}{\partial t} \right\rangle \\
&= \frac{i}{\hbar} \left\langle \hat{H}\psi \left| \hat{A} \right| \psi \right\rangle - \frac{i}{\hbar} \left\langle \psi \left| \hat{A}\hat{H} \right| \psi \right\rangle + \left\langle \psi \left| \frac{\partial}{\partial t}\hat{A} \right| \psi \right\rangle \\
&= \frac{i}{\hbar} \left\langle \psi \left| \hat{H}\hat{A} \right| \psi \right\rangle - \frac{i}{\hbar} \left\langle \psi \left| \hat{A}\hat{H} \right| \psi \right\rangle + \left\langle \psi \left| \frac{\partial}{\partial t}\hat{A} \right| \psi \right\rangle \\
&= \frac{i}{\hbar} \left\langle \psi \left| \hat{H}\hat{A} - \hat{A}\hat{H} \right| \psi \right\rangle + \left\langle \psi \left| \frac{\partial}{\partial t}\hat{A} \right| \psi \right\rangle,
\end{aligned}
\tag{4.66}
$$

where the Hermitian property of $\hat{H}$ is used such that $\langle \hat{H}\psi | \phi \rangle = \langle \psi | \hat{H}\phi \rangle$. Hence,

$$\boxed{\frac{d}{dt}\langle A \rangle = \frac{i}{\hbar} \left\langle \left[\hat{H}, \hat{A} \right] \right\rangle + \left\langle \frac{\partial}{\partial t}\hat{A} \right\rangle.} \tag{4.67}$$

If the operator $\hat{A}$ has no explicit time dependence, then $\langle \frac{\partial}{\partial t}\hat{A} \rangle = 0$ and

$$\frac{d}{dt}\langle A \rangle = \frac{i}{\hbar} \left\langle \left[\hat{H}, \hat{A} \right] \right\rangle. \tag{4.68}$$

4.5.1.1 Time Dependence of Position Operator of Particle Moving in Free Space

To check this result, consider a particle of mass m moving in free space in such a way that the Hamiltonian describing motion in the x direction is

$$\hat{H} = -\frac{\hbar^2}{2m}\frac{d^2}{dx^2}. \tag{4.69}$$

To evaluate the time dependence of the expectation value of the observable x associated with the position operator $\hat{x}$,

$$\frac{\mathrm{d}}{\mathrm{d}t}\langle x\rangle = \frac{\mathrm{i}}{\hbar}\left\langle\left[\hat{H},\hat{x}\right]\right\rangle. \tag{4.70}$$

The commutator operating on the wave function $\psi(x,t)$ that describes the particle gives

$$
\begin{aligned}
\frac{\mathrm{i}}{\hbar}[\hat{H},\hat{x}]\psi &= -\frac{\hbar^2}{2m}\frac{\mathrm{i}}{\hbar}\left(\frac{\mathrm{d}}{\mathrm{d}x}\left(\frac{\mathrm{d}}{\mathrm{d}x}\hat{x}\psi\right) - \hat{x}\frac{\mathrm{d}}{\mathrm{d}x}\left(\frac{\mathrm{d}}{\mathrm{d}x}\psi\right)\right)\\
&= -\frac{\mathrm{i}\hbar}{2m}\left(\frac{\mathrm{d}}{\mathrm{d}x}\psi + \frac{\mathrm{d}}{\mathrm{d}x}\left(\hat{x}\frac{\mathrm{d}}{\mathrm{d}x}\psi\right) - \hat{x}\frac{\mathrm{d}}{\mathrm{d}x}\left(\frac{\mathrm{d}}{\mathrm{d}x}\psi\right)\right)\\
&= -\frac{\mathrm{i}\hbar}{2m}\left(\frac{\mathrm{d}}{\mathrm{d}x}\psi + \frac{\mathrm{d}}{\mathrm{d}x}\psi + \hat{x}\frac{\mathrm{d}}{\mathrm{d}x}\left(\frac{\mathrm{d}}{\mathrm{d}x}\psi\right) - \hat{x}\frac{\mathrm{d}}{\mathrm{d}x}\left(\frac{\mathrm{d}}{\mathrm{d}x}\psi\right)\right)\\
&= -\frac{\mathrm{i}\hbar}{m}\frac{\mathrm{d}}{\mathrm{d}x}\psi. \tag{4.71}
\end{aligned}
$$

The Hamiltonian does not commute with the position operator. Using the fact that the wave function of a free particle moving in the x direction is of the form $\psi = e^{\mathrm{i}(k_x x - \omega t)}$, it follows that

$$\frac{\mathrm{d}}{\mathrm{d}t}\langle x\rangle = \frac{\hbar k_x}{m}. \tag{4.72}$$

As expected, this is just the x component of momentum divided by the mass or, equivalently, the speed of the particle in the x direction.

4.5.2 Uncertainty of Expectation Value

Let $\hat{A}$ be an operator corresponding to an observable A when the system is in state $\psi(\mathbf{r},t)$. The mean (expectation) value of measurements on identically prepared systems is

$$\langle A\rangle = \int \psi^*(\mathbf{r},t)\hat{A}\psi(\mathbf{r},t)\,\mathrm{d}^3 r. \tag{4.73}$$

The spread in values of the observable A can be defined in terms of the mean of squares of the deviations (the variance):

$$
\begin{aligned}
(\Delta A)^2 &= \left\langle\left(\hat{A}-\langle A\rangle\right)^2\right\rangle = \left\langle\hat{A}^2 + \langle A\rangle^2 - 2\hat{A}\langle A\rangle\right\rangle\\
&= \left\langle A^2\right\rangle + \langle A\rangle^2 - 2\langle A\rangle\langle A\rangle, \tag{4.74}
\end{aligned}
$$

where $\langle A\rangle$ is the mean of the measured value A and $(\Delta A)^2$ is the square of the deviations. It follows that the variance is

$$\boxed{\Delta A^2 = \left\langle A^2\right\rangle - \langle A\rangle^2} \tag{4.75}$$

and the square root of the variance (the standard deviation) is

$$\Delta A = \sqrt{\langle A^2 \rangle - \langle A \rangle^2}. \tag{4.76}$$

Expressed in integral form,

$$\Delta A^2 = \int \psi^*(\mathbf{r}, t) \hat{A}^2 \psi(\mathbf{r}, t) \, d^3 r - \left(\int \psi^*(\mathbf{r}, t) \hat{A} \psi(\mathbf{r}, t) \, d^3 r \right)^2. \tag{4.77}$$

The physical meaning of this is that $\langle A \rangle$ is the average value of many observations on identically prepared systems, and ΔA is a measure of the *standard deviation*, or spread, in the values of the measurement. This is a good measure *if* the distribution is strongly peaked and symmetric around the average value $\langle A \rangle$.[3] Of course, there are other ways to measure a spread in the values of a measurement. However, the above approach based on standard deviation is the most commonly used.

As an example, the solution to Exercise 4.4 shows how to calculate the expectation value of position and momentum for a particle mass m confined by a rectangular potential with infinite barrier energy.

4.5.3 The Generalized Uncertainty Relation

There is an important concept in quantum mechanics that links the uncertainty in results of measurement between a given pair of associated noncommuting operators. The spread in results of one set of measurements associated with one operator, $\hat{A}$, is related to the spread in measured values of the associated noncommuting operator, $\hat{B}$.

Consider an operator $\hat{A}$:

$$\left\langle \hat{A} \hat{A}^\dagger \right\rangle \geq 0 \tag{4.78}$$

because

$$\left\langle \hat{A} \hat{A}^\dagger \right\rangle = \langle \psi | \hat{A}^\dagger \hat{A} | \psi \rangle = \left\langle \hat{A} \psi \middle| \hat{A} \psi \right\rangle \geq 0 \tag{4.79}$$

from the definition of a Hermitian conjugate. Or, in terms of integrals,

$$\left\langle \hat{A} \hat{A}^\dagger \right\rangle = \int \psi^* \left(\hat{A}^\dagger \hat{A} \psi \right) d^3 r = \int \left(\hat{A} \psi \right)^* \left(\hat{A} \psi \right) d^3 r = \int \left(\hat{A} \psi \right)^2 d^3 r \geq 0. \tag{4.80}$$

A linear combination $\hat{A} + i\hat{B}$ can be created, so that

$$\left\langle \hat{A} + i\hat{B} \right\rangle = \left\langle \hat{A} \right\rangle + i \left\langle \hat{B} \right\rangle. \tag{4.81}$$

---

[3] The standard deviation is defined as $\sigma_x = \sqrt{\langle x^2 \rangle - \langle x \rangle}$ where $\langle x \rangle$ is the mean and $\langle x^2 \rangle$ is the second raw moment. The second central moment about the mean is the variance, σ_x^2. The standard deviation is a commonly used measure of statistical dispersion. The normal probability distribution $f(x) = \frac{1}{\sigma_x \sqrt{2\pi}} \exp\left(\frac{1}{2} \left(\frac{x - \langle x \rangle}{\sigma_x} \right)^2 \right)$ is symmetric about $\langle x \rangle$ and is defined in terms of σ_x and $\langle x \rangle$.

If $\hat{A}$ and $\hat{B}$ are Hermitian, then $\left(\hat{A}+i\hat{B}\right)^{\dagger}=\hat{A}-i\hat{B}$. If an operator $\left(\hat{A}+i\lambda\hat{B}\right)$ is created, where λ is real and $\hat{A}$ and $\hat{B}$ are Hermitian operators, then

$$\left\langle \left(\hat{A}+i\lambda\hat{B}\right)\left(\hat{A}+i\lambda\hat{B}\right)^{\dagger}\right\rangle = \left\langle \left(\hat{A}+i\lambda\hat{B}\right)\left(\hat{A}^{\dagger}-i\lambda\hat{B}^{\dagger}\right)\right\rangle \geq 0, \tag{4.82}$$

$$\left\langle A^{2}\right\rangle + \lambda^{2}\left\langle B^{2}\right\rangle - i\lambda\left\langle \hat{A}\hat{B}-\hat{B}\hat{A}\right\rangle \geq 0. \tag{4.83}$$

The minimum value of λ is found by taking the derivative with respect to λ such that

$$0 = \frac{d}{d\lambda}\left(\left\langle A^{2}\right\rangle + \lambda^{2}\left\langle B^{2}\right\rangle - i\lambda\left\langle \hat{A}\hat{B}-\hat{B}\hat{A}\right\rangle\right)$$
$$= 2\lambda_{\min}\left\langle B^{2}\right\rangle - i\left\langle \hat{A}\hat{B}-\hat{B}\hat{A}\right\rangle = 2\lambda_{\min}\left\langle B^{2}\right\rangle - i\left\langle \left[\hat{A},\hat{B}\right]\right\rangle, \tag{4.84}$$

$$\lambda_{\min} = \frac{i}{2}\frac{\left\langle \left[\hat{A},\hat{B}\right]\right\rangle}{\left\langle B^{2}\right\rangle}. \tag{4.85}$$

Substituting the minimum value $\lambda_{\min}$ into Eq. (4.83) gives

$$\left\langle A^{2}\right\rangle - \frac{\left\langle \left[\hat{A},\hat{B}\right]\right\rangle^{2}\left\langle B^{2}\right\rangle}{4\left\langle B^{2}\right\rangle^{2}} + \frac{\left\langle \left[\hat{A},\hat{B}\right]\right\rangle^{2}}{2\left\langle B^{2}\right\rangle} \geq 0, \tag{4.86}$$

so that

$$\boxed{\langle A^{2}\rangle\langle B^{2}\rangle \geq -\frac{\langle[\hat{A},\hat{B}]\rangle^{2}}{4}.} \tag{4.87}$$

The product of the expectation value of the square of a Hermitian operator with the square of another has a minimum value that is proportional to the square of the expectation value of the commutator of the two operators. To show that this applies to the standard deviation, a new set of operators is created in such a way that

$$\hat{A} \rightarrow \hat{A} - \langle A\rangle \equiv \delta\hat{A}, \tag{4.88}$$
$$\hat{B} \rightarrow \hat{B} - \langle B\rangle \equiv \delta\hat{B}, \tag{4.89}$$

so that

$$\left\langle (\delta A)^{2}\right\rangle = \left\langle \left(\hat{A}-\langle A\rangle\right)^{2}\right\rangle = \left\langle A^{2}\right\rangle - \left\langle 2\hat{A}\langle A\rangle\right\rangle + \langle A\rangle^{2} = \left\langle A^{2}\right\rangle - \langle A\rangle^{2} = \Delta A^{2}, \tag{4.90}$$

where ΔA is the standard deviation. $\left[\delta\hat{A},\delta\hat{B}\right]$ can be related to operators $\hat{A}$ and $\hat{B}$:

$$\left[\delta\hat{A},\delta\hat{B}\right] = \hat{A}\hat{B} - \hat{A}\langle B\rangle - \langle A\rangle\hat{B} + \langle A\rangle\langle B\rangle - \hat{B}\hat{A} + \langle B\rangle\hat{A} + \hat{B}\langle A\rangle - \langle B\rangle\langle A\rangle$$
$$= \hat{A}\hat{B} - \hat{B}\hat{A} = \left[\hat{A},\hat{B}\right]. \tag{4.91}$$

Substituting $\delta\hat{A}$ and $\delta\hat{B}$ into the expression $\langle A^2 \rangle \langle B^2 \rangle \geq -\left\langle \left[\hat{A},\hat{B} \right] \right\rangle^2 / 4$,

$$\left\langle \delta A^2 \right\rangle \left\langle \delta B^2 \right\rangle \geq -\left\langle \left[\hat{A},\hat{B} \right] \right\rangle^2 / 4. \tag{4.92}$$

Using $\langle \delta A^2 \rangle = \Delta A^2$ from Eq. (4.90) and the relation given by Eq. (4.91) allows Eq. (4.92) to be rewritten as

$$\Delta A^2 \Delta B^2 \geq -\left\langle \left[\hat{A},\hat{B} \right] \right\rangle^2 / 4, \tag{4.93}$$

or

$$\boxed{\Delta A \Delta B \geq \left| \frac{1}{2} \left\langle \left[\hat{A},\hat{B} \right] \right\rangle \right|,} \tag{4.94}$$

which is the generalized uncertainty relation. This relationship between a conjugate pair of noncommuting linear operators may be considered a consequence of the mathematics that is built into the description of quantum phenomena. It arises from the linear algebra of noncommuting Hermitian operators.

As a specific example of the uncertainty relation, consider a particle moving in one dimension. The operator $\hat{A} = \hat{p}_x = -i\hbar \, (d/dx)$, is the x component of the momentum operator, and $\hat{B} = \hat{x}$ is the x-position operator. Then, from the commutation relation,

$$\langle [\hat{p}_x, \hat{x}] \rangle \equiv [\hat{p}_x, \hat{x}] = -i\hbar \tag{4.95}$$

and the uncertainty relation for position and momentum operators can be found from Eq. (4.94):

$$\Delta p_x \Delta x \geq \frac{\hbar}{2} = \left| \frac{1}{2} \langle [\hat{p}_x, \hat{x}] \rangle \right|. \tag{4.96}$$

Suppose the particle is an electron confined to some region of space. If a measurement is performed to determine electron position once and then the measurement is repeated in a large number of identically prepared systems containing an electron, a Gaussian distribution of position results with spread of, say, $\Delta x = 1$ nm might be obtained. The uncertainty relation given by Eq. (4.96) means that in this case it is not possible to know the momentum of the electron to an accuracy better than $\Delta p = \hbar/2\Delta x = 5.27 \times 10^{-26}$ kg m s^{-1}. The spread in momentum has a corresponding spread in velocity, which is $\Delta v = \Delta p/m_0 = 5.7 \times 10^4$ m s^{-1}. This value of velocity would cause a classical particle to traverse a distance of 1 nm in just 17 fs.

4.6 The No Cloning Theorem

The description of secure quantum communication in Section 1.1.7 made use of the fact that nonorthogonal states can never be precisely copied. This is called the no cloning theorem and is a basic feature that arises due to the linear algebra that describes quantum mechanical systems.[4] The existence of the no cloning theorem removes the general use of copying states to control information loss in a system.

To prove the no cloning theorem, suppose it is possible make a copy of a pure state $|\psi_1\rangle$ by applying a unitary to an initial state $|\psi_0\rangle$ so that $\hat{U}(|\psi_1\rangle \otimes |\psi_0\rangle) = |\psi_1\rangle \otimes |\psi_1\rangle$ or

$$|\psi_1\rangle \rightarrow |\psi_1\rangle \otimes |\psi_1\rangle, \tag{4.97}$$

where $|\psi_1\rangle \otimes |\psi_1\rangle$ is a product state. For another orthogonal state $|\psi_2\rangle$,

$$|\psi_2\rangle \rightarrow |\psi_2\rangle \otimes |\psi_2\rangle. \tag{4.98}$$

In each case, the information contained in the wave function describing the particle is used to create an additional independent, identical particle. The resulting two-particle wave function is a product of the independent particle wave functions.

If an attempt is made to copy a state

$$|\psi\rangle = a_1|\psi_1\rangle + a_2|\psi_2\rangle \tag{4.99}$$

that is nonorthogonal and a linear combination of the states $|\psi_1\rangle$ and $|\psi_2\rangle$ then, using Eqs. (4.97) and (4.98),

$$|\psi\rangle \rightarrow a_1^2|\psi_1\rangle \otimes |\psi_1\rangle + a_2^2|\psi_2\rangle \otimes |\psi_2\rangle. \tag{4.100}$$

However, a clone of the linear superposition state requires

$$|\psi\rangle \rightarrow (a_1|\psi_1\rangle + a_2|\psi_2\rangle) \otimes (a_1|\psi_1\rangle + a_2|\psi_2\rangle)$$
$$= a_1^2|\psi_1\rangle \otimes |\psi_1\rangle + a_2^2|\psi_2\rangle \otimes |\psi_2\rangle + a_1 a_2(|\psi_1\rangle \otimes |\psi_2\rangle + |\psi_2\rangle \otimes |\psi_1\rangle). \tag{4.101}$$

It follows that only pure orthogonal states can be copied and not nonorthogonal linear superposition states.

The no cloning theorem highlights the fact that quantum information is different from classical information. For example, it is not possible to make precise backup copies of quantum information contained in nonorthogonal states. The same idea showed up when considering secure quantum communication in Section 1.1.7. Because an eavesdropper cannot make a precise copy of the nonorthogonal quantum state carrying the information, there is always some signature of the eavesdropper's presence impressed on the signal that can subsequently be detected.

4.7 Density of States

The density of states concept is important because of its role in controlling and changing the occupation probability of specific states in a system.

4.7.1 Density of Electron States

Suppose an electron is known to occupy a particular eigenstate up to a certain moment in time. The probability of occupation may be changed at some later time by changing the

---

[4] W. K. Wootters and W. H. Zurek, *Nature* **299**, 802 (1982); D. Dieks, *Phys. Lett. A* **92**, 271 (1982).

potential seen by the electron. This change in potential causes the electron to occupy other eigenstates, which, in general, have different energy eigenvalues. It might be expected that the number of distinct states available in a given energy range has an influence on the probability of the electron changing its state. This is one reason why the *density of states* – the number of electron states per unit energy interval – is a useful quantity. Density of states can be used to control electron scattering and current. As an application, it can be shown how a one-dimensional density of states leads to quantization of electron conduction. Of course, the density of electron states is important in many other applications. In a semiconductor laser diode, it plays a key role in determining device behavior.

So far, the concept of energy eigenvalues and eigenfunctions as solutions to the time-independent Schrödinger equation has been introduced. In general, the number of eigenstates varies per unit energy interval and this can complicate the calculation of density of states. Sometimes it is more convenient to calculate the density of electron states using solutions to Schrödinger's equation in k-space.

Consider a particle mass m that is free to move in space. The particle is described by a wave function $\psi(\mathbf{r}, t)$ of the form $e^{i\mathbf{k}\cdot\mathbf{r}}e^{-i\omega t}$ and a nonlinear dispersion relation $\omega = \hbar k^2/2m$. The energy of the particle is $E = \hbar^2 k^2/2m$. The volume of space in which the particle moves is large and defined by a cube of side L along the cartesian coordinates x, y, and z so that volume $V_{\mathrm{vol}} = L_x L_y L_z = L^3$ with $L_x = L_y = L_z$. Applying *periodic boundary conditions* to the particle wave function, ensuring that $\psi(x), \psi(y)$, and $\psi(z)$ are continuous and smooth, discretizes the wave vector components such that $k_x = 2n_x\pi/L_x$, $k_y = 2n_y\pi/L_y$, and $k_z = 2n_z\pi/L_z$, where n_x, n_y, and n_z are integers. See Fig. 4.1.

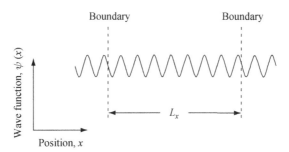

Fig. 4.1 Illustration showing periodic boundary conditions applied to the wave function $\psi(x)$ such that $\psi(x) = \psi(x + L_x)$ and $\dfrac{\mathrm{d}}{\mathrm{d}x}\psi(x) = \dfrac{\mathrm{d}}{\mathrm{d}x}(\psi(x + L_x))$.

Because each wave vector component is linear in the integer n, the quantum states are *equally spaced* in *k-space*. For this reason, it is often easier to count particle states in k-space than counting states in energy space. See Fig. 4.2(a). As illustrated in Fig. 4.2(b), each k-state takes up a volume $(2\pi)^3/L^3$ or $(2\pi)^3$ when normalizing to unit length.

The density of states in k-space is the *number of states* between k and $k + dk$ per *unit volume*. For a large number of equally spaced states in k-space in three dimensions, this gives

$$D_3(k)\mathrm{d}k = \frac{1}{V_{\mathrm{vol}}}\frac{L^3}{(2\pi)^3}4\pi k^2\mathrm{d}k = \frac{4\pi k^2}{(2\pi)^3}\mathrm{d}k, \tag{4.102}$$

189

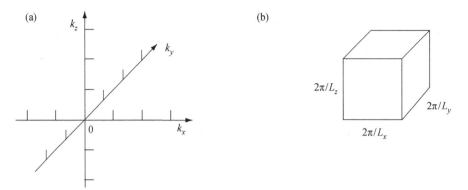

Fig. 4.2 (a) Counting particle states in k-space is often easier than counting states in energy space. In this example, the particle of mass m can move in the positive direction with velocity $\hbar k/m$ or in the negative direction with velocity $-\hbar k/m$. (b) The volume of k-space occupied by one k-state is $(2\pi)^3/L^3$ where $L_x L_y L_z = L^3$.

where $D_3(k)$ is the density of states of a free particle of mass m and the subscript 3 indicates three dimensions. In Eq. (4.102), the volume of a shell of radius k and thickness dk is divided by the volume occupied by each k-state to obtain the number of states in the shell, and divided by the volume V_{vol} to obtain the density of states.

If the particle being considered has spin, this can be included as a multiplicative term in Eq. (4.102). For example, an electron of spin quantum number $\sigma_s = \pm 1/2$ (or eigenvalue $\pm \hbar/2$) multiplies the density of states by a factor 2 because there are two possible spin states the electron could be in.

The energy of a particle of mass m is $E = \hbar^2 k^2/2m$ and so $dE = (\hbar^2 k/m)dk$. Hence,

$$D_3(E)dE = D_3(k)\frac{dk}{dE}dE = \frac{4\pi k^2}{(2\pi)^3}\frac{m}{\hbar^2 k}dE = \frac{1}{2\pi^2}\frac{km}{\hbar^2}dE, \tag{4.103}$$

so that the density of states *per spin* in three dimensions is

$$\boxed{D_3(E) = \frac{1}{4\pi^2}\left(\frac{2m}{\hbar^2}\right)^{3/2}\sqrt{E}.} \tag{4.104}$$

It is straightforward to show that in two dimensions the density of states per spin is

$$\boxed{D_2(E) = \frac{m}{2\pi\hbar^2}} \tag{4.105}$$

and that for one dimension the density of states per spin is

$$\boxed{D_1(E) = \frac{2}{4\pi}\left(\frac{2m}{\hbar^2}\right)^{1/2}\frac{1}{\sqrt{E}}.} \tag{4.106}$$

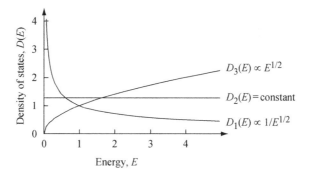

Fig. 4.3 Density of states of a free particle of mass m as a function of energy plotted for one, two, and three dimensions.

The results may be summarized by plotting the density of states of a free particle of mass m for different dimensions. This is done in Fig. 4.3. The divergent behavior of $D_1(E \to 0)$ arises from the $1/\sqrt{E}$ term in Eq. (4.106) and is called a Van Hove singularity.

For a particle of mass m, the densities of states in three and one dimensions are equal when the particle energy has a value $E = \pi\hbar^2/m$. At this value of particle energy, the density of states is $m/\hbar^2\sqrt{2}\pi^{3/2}$. The density of states in three and two dimensions are equal with value $m/2\pi\hbar^2$ when the particle energy is $E = \pi^2\hbar^2/2m$.

Typically, a *quantum well* potential formed from a semiconductor heterostructure such as epitaxially grown thin layers of GaAs and AlGaAs has a two-dimensional density of electron states for low-energy electrons. Quantum wells formed from heterostructures are of practical importance in many semiconductor devices. For example, such quantum wells are often used as the active region of a semiconductor laser diode. The reason for this is that the small volume of the quantum well reduces the current needed to achieve lasing and, over some range of emission wavelengths, differential optical gain can be larger than bulk values.

The atomic precision with which quantum wells can be fabricated is well illustrated in Fig. 4.4(a). The figure shows a transmission electron micrograph of an InGaAs quantum well that is just three monolayers thick sandwiched between InP barrier layers. The spots in the image represent tunnels between pairs of atoms. The minimum separation between tunnels in InP is 0.34 nm.

A laterally patterned quantum well can be made to form a *quantum wire* that has a one-dimensional density of electronic states. Structures of this type can be designed to exhibit quantized electrical conductance. Quantized conductance is not predicted classically and is an example of an effect that may play an important role in future very small (scaled) electronic structures such as transistors.

The density of states in "zero dimensions" arises due to states confined by a potential in all three dimensions similar to an isolated atom. In semiconductor devices, the structures that give rise to such a potential are called *quantum dots*. An example of InP quantum dots imaged by an atomic force microscope (AFM) is shown in Fig. 4.4(b). In this case most of the quantum dots have a diameter in the range 15–20 nm.

Quantum wells, wires, and dots formed in semiconductor structures do not have potentials with infinite barrier energies. However, there are still bound states that exist in potential minima formed by potentials with finite barrier energy.

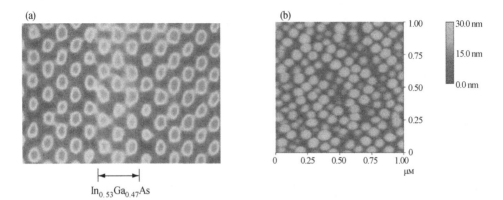

(a)

(b)

$In_{0.53}Ga_{0.47}As$

Fig. 4.4 (a) Transmission electron micrograph showing an InGaAs *quantum well* in cross-section that is three monolayers thick and is sandwiched between InP barrier layers. The spots in the image represent tunnels between pairs of atoms. The minimum separation between tunnels in InP is 0.34 nm. Image courtesy of M. Gibson. (b) Area view of InP self-assembled *quantum dots* grown using low-pressure MOCVD on an InAlP matrix layer lattice-matched to a GaAs substrate. As measured from AFM images, areal density of quantum dots is 1.5×10^{10} cm^{-2} and dominant size is in the range of 15–20 nm for a 15-monolayer "planar-growth-equivalent" deposition time at a growth temperature of 650 °C. Dominant sizes are controllable by changing the deposition time. Image courtesy of R. Dupuis.

4.7.2 Quantum Conductance

Electron motion through a region with a one-dimensional density of states can result in quantized conductance. Figure 4.5 shows schematically a one-dimensional conduction region created by a confining potential that is attached to electrodes placed on the left and right. The transverse electron wave number for an electron moving in the x direction is quantized by the confining potential. Each quantized level defines a channel for electron transmission. In the following, it is assumed that each channel is independent and hence uncorrelated.

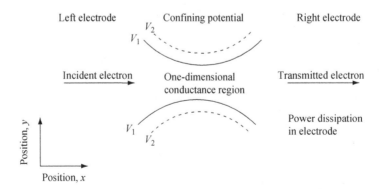

Fig. 4.5 Diagram showing top view of left and right electrodes connected via a one-dimensional conductance region defined by a confining potential. Contours of constant potential energy V_1 and V_2 where $V_1 < V_2$ are shown. Transverse electron wave number is quantized by the confining potential.

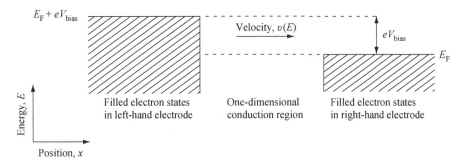

Fig. 4.6 At low temperatures, electrons in the left and right electrodes occupy states up to the Fermi energy, E_F. A potential energy eV_{bias} applied between the left and right electrodes allows occupied electron states in the left-hand electrode to traverse the one-dimensional region and enter unoccupied electron states in the right electrode.

At low temperature, electrons in the electrodes occupy states up to the Fermi energy, E_F. If a voltage, V_{bias}, is applied between the electrodes, current flows. As shown in Fig. 4.6, potential energy eV_{bias} applied between the left and right electrodes allows occupied electron states in the left electrode to traverse the one-dimensional region and enter unoccupied electron states in the right electrode. The current that flows is proportional to the applied voltage, electron velocity, v, the one-dimensional transmission coefficient, T, and the one-dimensional density of electron states, D_1.

The evaluation of v, T, and D_1 depends on electron energy E. The velocity of an electron characterized by positive wave vector k and mass m moving from left to right in the one-dimensional region is $v(E) = \sqrt{2mE}/m$, or, more generally, $v = \partial\omega/\partial k$. The density of electron states with positive k in the one-dimensional region is $D_1(E) = m/(\pi\hbar\sqrt{2mE})$, or, more generally, $D_1(E) = 1/(\pi\hbar v(E))$. A factor of 2 is included in $D_1(E)$ because there are two possible spins an electron can have per positive k value.

The current at voltage bias V_{bias} is given by the integral

$$I = e \int_{E_F}^{E_F+eV_{bias}} v(E)T(E)D_1(E)\,\mathrm{d}E = e \int_{E_F}^{E_F+eV_{bias}} \frac{1}{\pi\hbar}T(E)\,\mathrm{d}E. \qquad (4.107)$$

The key point is that terms in the expressions for v and D_1 cancel so that $vD_1 = 1/\pi\hbar$. Simplifying further by only considering small voltage bias V_{bias}, the transmission coefficient $T(E) \rightarrow T(E_F)$, so that this may be taken out of the integral, and current becomes $I = e^2 T(E_F)V_{bias}/\pi\hbar$. In this situation, the conductance $G_{cond} = I/V_{bias}$ is

$$G_n = \frac{e^2}{\pi\hbar}T(E_F). \qquad (4.108)$$

This is sometimes called the Landauer formula.[5] Conductance G_n has a subscript n because conductance is quantized. To see this, all one need do is consider the situation in which the transmission coefficient has its maximum value $T(E_F) = 1$. In

[5] R. Landauer, *IBM J. Res. Dev.* **1**, 223 (1957); R. Landauer, *Philo. Mag.* **21**, 863 (1970); M. Buttiker, Y. Imry, R. Landauer, and S. Pinhas, *Phys. Rev. B* **31**, 6207 (1985).

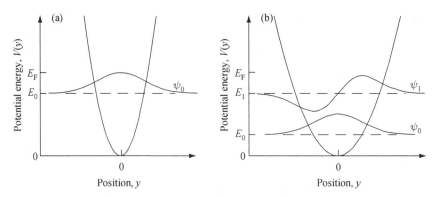

Fig. 4.7 At low temperatures, electrons occupy states up to the Fermi energy, E_F. For the strong one-dimensional confining potential shown in (a), electrons occupy the lowest-energy transverse state ψ_0, which has energy $E_0 < E_F$. This limits maximum conductance per electron per spin to 25.8 kΩ^{-1}. Conductance may be increased by increasing the number of parallel paths an electron can access in traversing the confining potential. As shown in (b), one way to achieve this is to maintain the Fermi energy while increasing the width of the one-dimensional potential channel, thereby fitting more transverse electron waveguide modes through.

this case, the maximum conductance is $e^2/2\pi\hbar = 1/R_K = 1/(25.8 \text{ k}\Omega)$ *per electron spin.*[6]

The only way to increase maximum conduction is to increase the number of parallel paths an electron can take from left to right through the region between the electrodes. Electrical conduction will then increase in a step-wise fashion to a value proportional to the number of parallel electron paths available between the electrodes. As illustrated in Fig. 4.7, one way to increase the number of parallel paths is to increase the width of the one-dimensional potential channel, thereby fitting more transverse *electron waveguide modes* through. The description of electron waveguide modes in the confining one-dimensional potential arises because the wave nature of the electron suggests an analogy with classical electromagnetic waveguides.

Conduction is not limited by electron scattering or dissipation; rather it is limited by the quantum mechanical wavy nature of the electron. In this simple model system, no electron scattering takes place in the one-dimensional conduction region. The one-dimensional potential acts as a lossless electron waveguide. Electron scattering and power dissipation take place in the electrodes.

For electronic devices in which electrons are constrained to move through regions comparable to the electron wavelength, it is necessary to consider quantum conductance. The value of quantum conductance per electron per spin of only $(25.8 \text{ k}\Omega)^{-1}$ limits the current drive performance of small devices and hence the speed at which these devices can operate.

From a practical point of view, it might be more productive to consider devices that do not operate in the linear, near-equilibrium regime. In this case, a different description of electrical conductivity should be adopted.

---

[6] The value of R_K is known as the von Klitzing constant (see Appendix A).

4.7.3 Calculating Density of States from a Dispersion Relation

It is also possible to calculate the density of states $N(E)$ in a crystal by considering the dispersion relation E_k. In general, the dispersion relation is not that of a free particle of mass m; rather it takes on a form that is due to the crystal potential. Each state described by the Bloch wave vector k contributes to the density of states. A way to numerically evaluate $N(E_k)$ is to sample k-space using equally spaced discrete values of k. Each value of k has an associated energy in the dispersion relation E_k. If each state is broadened in energy by γ, then the electron density of states is

$$N(E) = \sum_k \frac{\gamma/\pi}{(E - E_k)^2 + (\gamma/2)^2}. \tag{4.109}$$

4.7.4 Density of Tight-Binding States

An *isolated* linear chain with a primitive cell containing one atom has lattice sites $x_j = jL$ where $j = 1, 2, \ldots, N_\text{atom}$, the nearest-neighbor atom spacing is L, and N_atom is the number of atoms in the chain. In the tight-binding approximation a nearest-neighbor s-orbital overlap integral, $t_{\text{hop},1}$, is a matrix element coupling adjacent atomic sites. The Hamiltonian for the chain of atoms is a tridiagonal constant $N_\text{atom} \times N_\text{atom}$ matrix with upper and lower diagonal elements $t_{\text{hop},1}$. The corresponding Schrödinger equation has energy eigenvalues E_k and a density of states $N(E)$ given by Eq. (4.109). The energy eigenvalues of an isolated one-dimensional tight-binding chain with nearest-neighbor interaction is found by diagonalizing a tridiagonal constant $N_\text{atom} \times N_\text{atom}$ matrix of Toeplitz form,

$$\begin{bmatrix} a & b & 0 & & & \\ c & a & b & & & \\ 0 & c & a & & & \\ & & & \ddots & & \\ & & & c & a & b \\ & & & & c & a \end{bmatrix},$$

which, for the physical system of interest, has *on-site* term $a = 0$, and off-diagonal terms $b = c = t_{\text{hop},1} = -1$. The analytic expression for the eigenvalues of this matrix is

$$E_j(N_\text{atom}) = a + 2\sqrt{bc} \cos\left(\frac{\pi j}{N_\text{atom} + 1}\right) \tag{4.110}$$

with $j = 1, \ldots, N_\text{atom}$. The first few energy eigenvalues for chains containing N_atom atoms are:

$E(N_\text{atom} = 1) = 0$,

$E(N_\text{atom} = 2) = \pm t_{\text{hop},1}$,

$E(N_\text{atom} = 3) = (0, \pm\sqrt{2})t_{\text{hop},1}$,

$E(N_\text{atom} = 4) = 2(\pm 0.809, \pm 0.309)t_{\text{hop},1}$.

In the limit of a chain with lattice constant L containing a large number of identical atoms, there are many energy eigenvalues bounded by minimum and maximum value $E_{\text{max,min}}(N_{\text{atom}} \to \infty) = \pm 2t_{\text{hop,1}}$.

Application of *periodic* boundary conditions changes the isolated open-ended chain to a loop so atom position $x_{j=-1}$ occupies position $x_{j=N_{\text{atom}}}$ and atom position $x_{j=1+N_{\text{atom}}}$ occupies position $x_{j=1}$. The tridiagonal-constant matrix describing nearest-neighbor tight-binding interaction is modified by inclusion of corner elements $(1, N_{\text{atom}})$ and $(N_{\text{atom}}, 1)$. This sparse matrix has the form

$$\begin{bmatrix} 0 & t_{\text{hop,1}} & 0 & & & & t_{\text{hop,1}} \\ t_{\text{hop,1}} & 0 & t_{\text{hop,1}} & & & & \\ 0 & t_{\text{hop,1}} & 0 & & & & \\ & & & \cdots & \cdots & & \\ & & & t_{\text{hop,1}} & 0 & t_{\text{hop,1}} \\ t_{\text{hop,1}} & & & & 0 & t_{\text{hop,1}} & 0 \end{bmatrix}.$$

In general, if all atoms interact then all non-lead diagonal matrix elements of the Hamiltonian have nonzero entries.

Figure 4.8 shows the results of calculating $N(E)$ for a one-dimensional lattice of atoms with s-orbital electrons in the tight-binding approximation, nearest-neighbor interaction strength $t_{\text{hop,1}}$, $N_{\text{atom}} \to \infty$, and *periodic* boundary conditions. The dispersion relation $E_k = -2t_{\text{hop,1}} \cos(kL)$ gives a symmetric density of states. As with $D_1(E)$ in Eq. (4.106), there is a Van Hove singularity at the band-edge energies of $\pm 2|t_{\text{hop,1}}|$. If dispersion may be approximated as a parabolic function of wave number, then $N(E)$ approaches the analytic expression $D(E)$ obtained using an electron effective mass.

Inclusion of next nearest-neighbor interactions changes the electron dispersion relation of the one-dimensional lattice to $E_k = -2t_{\text{hop,1}} \cos(kL) - 2t_{\text{hop,2}} \cos(2kL)$ and the

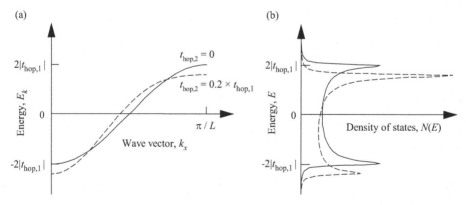

Fig. 4.8 (a) Dispersion relation E_k in the tight-binding approximation of a one-dimensional crystal with lattice spacing L, periodic boundary conditions, s-atomic orbitals, and $N_{\text{atom}} \to \infty$. With nearest-neighbor interaction only, there is a cosine band of energy width $E_b = 4t_{\text{hop,1}}$ (solid curve) and when next nearest-neighbor interaction is included (dash curve), the dispersion is modified. (b) Density of states $N(E_k)$ calculated using Eq. (5.94) with Lorentzian broadening $\gamma = |t_{\text{hop,1}}|/10$. For nearest-neighbor interaction (solid curve, $t_{\text{hop,1}}$) the density of states is symmetric. When next nearest-neighbor interactions are included (dashed curve, $t_{\text{hop,2}} = 0.2 \times t_{\text{hop,1}}$) the density of states is asymmetric.

(a) 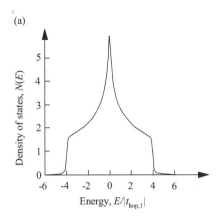 (b)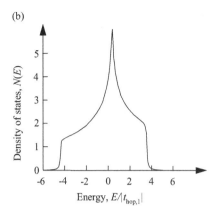

Fig. 4.9 (a) Calculated tight-binding density of states for a homogeneous square lattice with nearest-neighbor interactions only and $t_{\mathrm{hop},1} = -1$. (b) Same as (a) but including a next nearest-neighbor term with $t_{\mathrm{hop},2} = 0.1 \times t_{\mathrm{hop},1}$. $N(E)$ is in arbitrary units, energy is in units of $t_{\mathrm{hop},1}$, and $\gamma = 0.1 \times |t_{\mathrm{hop},1}|$.

corresponding density of states is asymmetric. The broken curves in Fig. 4.8 show this for the case $t_{\mathrm{hop},2} = 0.2 \times t_{\mathrm{hop},1}$. In general, the values of the nth neighbor interaction strength $t_{\mathrm{hop},n}$ not only control electron dispersion but also the density of states.

A homogeneous *square lattice* with lattice constant L, *periodic* boundary conditions, and only nearest-neighbor interactions can have dispersion relation $E_k = 2t_{\mathrm{hop},1}(\cos(k_x L) + \cos(k_y L))$, where $t_{\mathrm{hop},1}$ is the contribution of nearest-neighbor electron hopping between adjacent atom lattice sites. As illustrated in Fig. 4.9(a), the corresponding density of states is symmetric in energy. Inclusion of a next-nearest-neighbor hopping term, $t_{\mathrm{hop},2}$, contributing to electron dispersion results in a density of states shown in Fig. 4.9(b) that is nonsymmetric in energy because of the additional Fourier components in the expression for E_k.

A *cubic lattice* with lattice constant L, *periodic* boundary conditions, and only nearest-neighbor interactions can have a dispersion relation $E_k = 2t_{\mathrm{hop},1}(\cos(k_x L) + \cos(k_y L) + \cos(k_z L))$, where $t_{\mathrm{hop},1}$ is the contribution of nearest-neighbor electron hopping between adjacent atom lattice sites. The corresponding density of states shown in Fig. 4.10(a) is symmetric in energy. Again, as shown in Fig. 4.10(b), inclusion of a next-nearest-neighbor hopping term, $t_{\mathrm{hop},2}$, contributing to electron dispersion results in a nonsymmetric density of states.

In general, the band-edge energy for a simple homogeneous cubic lattice of dimension d, with nearest-neighbor atom interactions only, occurs at $\pm 2d$.

4.7.4.1 Surface States

Surface and heterointerface states can be spatially localized and isolated from states that exist in the body of a crystal. The potential experienced by an electron at a surface can be different enough to create isolated localized states. Some properties of surface states may be illustrated by considering a number of s-orbital atomic states in an isolated one-

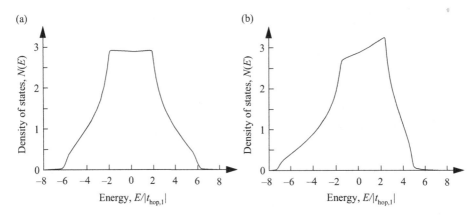

Fig. 4.10 (a) Calculated tight-binding density of states for a homogeneous cubic lattice with nearest-neighbor interactions only and $t_{\text{hop},1} = -1$. (b) Same as (a) but including a next nearest-neighbor term with $t_{\text{hop},2} = 0.1 \times t_{\text{hop},1}$. $N(E)$ is in arbitrary units, energy is in units of $t_{\text{hop},1}$, and $\gamma = 0.1 \times |t_{\text{hop},1}|$.

dimensional chain with atom sites at positions x_j with $j = 1, 2, \ldots, N_{\text{atom}}$. The nearest-neighbor electron hopping integral may be set to $t_{\text{hop},1} = -1$ and, in this example, on-site real potential ν_j is zero except at the end points of the lattice located at x_1 and $x_{N_{\text{atom}}}$, respectively, where it can take on a value in the range $0 \le \nu < 3.5$.

Figure 4.11(a) plots the energy eigenvalues as a function of $\nu_1 = \nu_{N_{\text{atom}}} = \nu$ with $(0 \le \nu < 3.5)$ for the case when $N_{\text{atom}} = 9$. As ν increases, a degenerate pair of surface states with odd and even symmetry emerges. For large values of ν these states can have

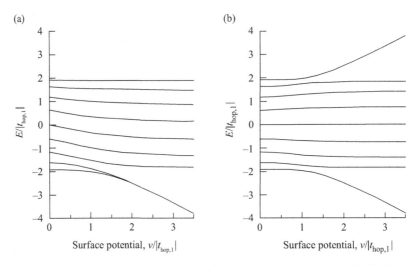

Fig. 4.11 (a) Energy eigenvalues as a function of $\nu_1 = \nu_{N_{\text{atom}}} = \nu \times t_{\text{hop},1}$ with $(0 \le \nu < 3.5)$ for a one-dimensional chain of $N_{\text{atom}} = 9$ atoms with nearest-neighbor hopping integral $t_{\text{hop},1} = -1$. (b) Same as (a) but with $\nu = \nu_1 = -\nu_{N_{\text{atom}}}$.

long lifetimes as they become isolated in energy eigenvalue and isolated spatially from other states by localizing at the surface. If sufficiently isolated from scattering with other states, they can form an independent subspace.

Figure 4.11(b) plots the energy eigenvalues as a function of $\nu = \nu_1 = -\nu_{N_{\text{atom}}}$ with $(0 \leq \nu < 3.5)$ for $N_{\text{atom}} = 9$. For large values of ν there is a localized high-energy surface state and a corresponding low-energy surface state.

4.7.5 Density of Photon States

Photons, like electrons, may often be characterized by a wavelength, λ, or k-state in which $k = 2\pi/\lambda$. The three-dimensional density of states in k-space, $D_3^{\text{opt}}(k)$, follows directly and is given by Eq. (4.102). This density of states may be expressed in terms of angular frequency ω if the relationship between ω and k is known. Since the dispersion of polarized light propagating in three-dimensional free space is $\omega = ck$ (where c is the speed of light), it follows that

$$D_3^{\text{opt}}(\omega)d\omega = \frac{4\pi}{(2\pi)^3} \frac{\omega^2}{c^2} \frac{1}{c} d\omega = \frac{\omega^2}{2\pi^2 c^3} d\omega. \tag{4.111}$$

In general, since a photon has a spin quantum number of $\sigma_s = \pm 1$ corresponding to angular momentum eigenvalue $\pm\hbar$ (there are two orthogonal polarizations of light in free space), this density of states should be multiplied by a factor of 2. Hence, the density of photon states (or field modes) in three-dimensional free space in the frequency range ω to $\omega + d\omega$ is

$$\boxed{D_3^{\text{opt}}(\omega) = \frac{\omega^2}{\pi^2 c^3}.} \tag{4.112}$$

In an isotropic lossless dielectric medium characterized by a refractive index n_r, the dispersion relation is modified to $\omega = ck/n_r$, and the expression for $D_3^{\text{opt}}(\omega)$ becomes

$$D_3^{\text{opt}}(\omega) = \frac{\omega^2 n_r^3}{\pi^2 c^3}. \tag{4.113}$$

In direct analogy with the electron density of states discussed in Section 4.7.1, there is interest in understanding the behavior of photons in situations in which large changes in the density of states and highly nonlinear dispersion relations exist in lossless dielectric and active semiconductor nanostructures. It is straightforward to show that the photon density of states is modified by using dielectric structures that vary with a half-wavelength period in space. Such structures are called photonic crystals[7] and belong to a larger class of meta-materials.[8] For typical infrared laser light at a wavelength near 1500 nm and an effective refractive index near 1.5, this implies periods of approximately 500 nm and features with sizes less than this. Such nanoscale dielectrics are easily fabricated in two dimensions using existing semiconductor fabrication techniques. Figure 4.12 shows a scanning electronmicroscope image of a two-dimensional photonic crystal with a triangular

---

[7] The concept of dispersion in photonic crystals was first introduced by K. Ohtaka, *Phys. Rev. B* **19**, 5057 (1979). For an introduction, see J. D. Joannopoulos, R. D. Meade, and J. N. Winn, *Photonic Crystals*, Princeton, Princeton University Press, 1995 (ISBN 0 691 03744 2).

[8] Meta-materials are purely artificial structures consisting of sub-wavelength structures whose aggregate response is sometimes remarkable and not usually found in nature.

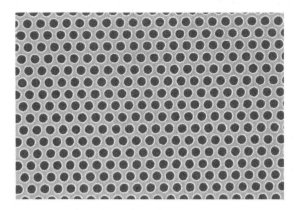

Fig. 4.12 Scanning electron microscope image of two-dimensional photonic crystal in plan view. The triangular lattice has 500 nm period with 350 nm holes etched in a 400 nm thick silicon layer bonded to a 2 μm thick silica layer.

lattice created by etching 350 nm diameter holes into 400 nm thick single-crystal silicon. The thin silicon layer is bonded to a 2 μm thick layer of silica grown on a silicon substrate. Light of wavelength near 1500 nm that is waveguided in the plane sees large periodic changes in refractive index from silicon, with $n_r = 3.47$, to air, with $n_r = 1$. The periodicity in refractive index can sometimes result in dramatic changes in photon density of states, including ranges of photon energy, called photonic band gaps, where no photons can propagate.

While the description of light propagation in periodic dielectrics is simplified by the existence of spatial symmetry, analysis of the photon density of states in nonperiodic nanoscale dielectrics is a significantly more challenging task.

4.8 Example Exercises

Exercise 4.1
Show that if wave functions $\psi_1(\mathbf{r})$ and $\psi_2(\mathbf{r})$ belong to the space of linear square-integrable wave functions, the linear combination $\psi(\mathbf{r}) = \lambda_1 \psi_1(\mathbf{r}) + \lambda_2 \psi_2(\mathbf{r})$, where λ_1 and λ_2 are complex numbers, is also square integrable.

Exercise 4.2
Show that the density of states for a free electron of mass m in one and two dimensions is

$$D_1(E) = \frac{2}{4\pi} \left(\frac{2m}{\hbar^2} \right)^{1/2} \frac{1}{\sqrt{E}}$$

per spin, and

$$D_2(E) = \frac{m}{2\pi\hbar^2}$$

per spin, respectively.

Exercise 4.3

An electron in a one-dimensional, rectangular potential well of thickness L with infinite barrier energy is in the simple superposition state consisting of the ground and first excited state so that

$$\psi(x,t) = \frac{1}{\sqrt{2}}(\psi_1(x,t) + \psi_2(x,t)).$$

Find expressions for:
(a) The probability density, $|\psi(x,t)|^2$.
(b) The average particle position, $\langle x(t) \rangle$.
(c) The momentum probability density, $|\psi(p_x,t)|^2$.
(d) The average momentum, $\langle p_x(t) \rangle$.
(e) The current flux, $J_x(x,t)$.

Exercise 4.4

Find $\langle x \rangle$, $\langle x^2 \rangle$, and Δx^2 for a particle of mass m confined by the potential $V(x) = 0$ for $0 < x < L$ and $V(x) = \infty$ elsewhere. Show that as the state number $n \to \infty$ the average values approach those obtained from classical mechanics. Calculate the average particle momentum $\langle p_x \rangle$, $\langle p_x^2 \rangle$, and Δp_x^2 as a function of state n. How does the product $\Delta x \Delta p$ depend upon n?

Exercise 4.5

Find $\psi(k)$ for a particle with state function $\psi(x) = 1/\sqrt{2L}$ for $|x| < L$ and $\psi(x) = 0$ for $|x| > L$. Show that the uncertainty (standard deviation) in its momentum Δp_x is infinite, and plot $|\psi(k)|^2$ and $|\psi(x)|^2$. Calculate the uncertainty in position, Δx.

Exercise 4.6

(a) Show that the momentum operator $\hat{p}_x = -i\hbar \frac{d}{dx}$ commutes with $\hat{p}_x^2$.
(b) Show that a smoothly varying potential $\hat{V}(x)$ does not commute with $\hat{p}_x$.
(c) The motion of a particle mass m is described by the Hamiltonian $\hat{H} = (\hat{p}_x^2/2m) + \hat{V}(x)$. Show that Newton's definition of force acting on the particle can be expressed in quantum mechanics by Ehrenfest's theorem,

$$\frac{d}{dt}\langle p_x \rangle = -\left\langle \frac{d}{dx}V(x) \right\rangle.$$

Under what circumstances does

$$\frac{d}{dt}\langle p_x \rangle = -\frac{d}{d\langle x \rangle}V(\langle x \rangle)?$$

Exercise 4.7

Prove that the expectation value of the (Hermitian) momentum operator in Cartesian coordinates is real. Show that $-i\hbar(\partial/\partial r)$ is not a Hermitian operator in radial coordinates. Show that the radial momentum operator $\hat{p}_r = -i\hbar(1/r)(\partial/\partial r)r$ is Hermitian.

Exercise 4.8
Consider a particle of mass m in a finite, one-dimensional, rectangular potential well for which $V(x) = 0$ for $-L < x < L$ and $V(x) = V_0$ elsewhere. The value of V_0 is a finite positive constant. Calculate the average kinetic energy of the particle ground state, and show that the contribution from the region outside the quantum well is negative.

Exercise 4.9
Show that a consequence of the hermiticity of the Hamiltonian is that probability $\langle \psi(t)|\psi(t)\rangle$ is conserved.

Exercise 4.10
Discuss the similarities and differences between classical electrodynamics and quantum mechanics.

Solutions

Solutions 4.1
To illustrate the nature of linear square-integrable functions, consider wave functions $\psi_1(\mathbf{r})$ and $\psi_2(\mathbf{r})$ that belong to this space. In a linear space, linear combinations may be formed,

$$\psi(\mathbf{r}) = \lambda_1 \psi_1(\mathbf{r}) + \lambda_2 \psi_2(\mathbf{r}),$$

where λ_1 and λ_2 are complex numbers. To show that $\psi(\mathbf{r})$ is also square-integrable, expand:

$$|\psi(\mathbf{r})|^2 = |\lambda_1|^2|\psi_1(\mathbf{r})|^2 + |\lambda_2|^2|\psi_2(\mathbf{r})|^2 + \lambda_1^*\lambda_2\psi_1^*(\mathbf{r})\psi_2(\mathbf{r}) + \lambda_1\lambda_2^*\psi_1(\mathbf{r})\psi_2^*(\mathbf{r}).$$

The last two terms have the same modulus with an upper limit:

$$|\lambda_1||\lambda_2|\left(|\psi_1(\mathbf{r})|^2 + |\psi_2(\mathbf{r})|^2\right).$$

Therefore, $|\psi(\mathbf{r})|^2$ is smaller than a function the integral of which converges, since $\psi_1(\mathbf{r})$ and $\psi_2(\mathbf{r})$ are square-integrable.

Solutions 4.2
Starting with the expression for the one-dimensional density of k-states per spin, allowing for positive and negative k,

$$D_1(k)\,dk = \frac{2dk}{2\pi},$$

$$D_1(E)\,dE = D_1(k)\frac{dk}{dE}dE = \frac{m}{\hbar^2 k}\frac{2}{(2\pi)}dE = \frac{2m}{\hbar^2(2\pi)}\frac{\hbar\,dE}{\sqrt{2mE}} = \frac{2}{4\pi}\left(\frac{2m}{\hbar^2}\right)^{1/2}\frac{1}{\sqrt{2mE}}dE.$$

The density of states per spin in two dimensions is just

$$D_2(E)dE = D_2(k)\frac{dk}{dE}dE = \frac{2\pi k}{(2\pi)^2}\frac{m}{\hbar^2 k}dE = \frac{m}{2\pi\hbar^2}dE,$$

where wave number $k = \sqrt{2mE/\hbar^2}$, energy $E = \hbar^2 k^2 / 2m$, and the energy increment $dE = \hbar^2 k \, dk / m$.

Solutions 4.3

An electron is a one-dimensional, rectangular potential well of thickness L with infinite barrier energy. The electron is in the simple superposition state consisting of the ground and first excited state so that

$$\psi(x,t) = \frac{1}{\sqrt{2}} (\psi_1(x,t) + \psi_2(x,t)).$$

(a) To obtain the probability density $|\psi(x,t)|^2$, expressions for the wave functions $\psi_1(x,t)$ and $\psi_2(x,t)$ must be found. The first two lowest-energy wave functions for a particle of mass m confined to the potential well of thickness L centered at $x = 0$ are

$$\psi_1(x,t) = \sqrt{\frac{2}{L}} \cos\left(\frac{\pi x}{L}\right) e^{-i\omega_1 t}$$

and

$$\psi_2(x,t) = \sqrt{\frac{2}{L}} \sin\left(\frac{2\pi x}{L}\right) e^{-i\omega_2 t},$$

where $E_n = \hbar \omega_n = \frac{\hbar^2}{2m} \frac{n^2 \pi^2}{L^2}$ and n is a positive nonzero integer. The expression for the probability density,

$$|\psi|^2 = \frac{1}{2} \left(\psi_1 \psi_1^* + \psi_1 \psi_2^* + \psi_2 \psi_1^* + \psi_2 \psi_2^* \right),$$

$$|\psi(x,t)|^2 = \frac{1}{2} \left(|\psi_1(x)|^2 + |\psi_2(x)|^2 + 2\mathrm{Re} \left(\psi_1(x) \psi_2^*(x) \right) \cos \left((\omega_1 - \omega_2) t \right) \right),$$

shows an oscillatory solution in which the average position of the particle moves from one side of the well to the other. The sinusoidal oscillation frequency is $(\omega_2 - \omega_1) = 3\hbar \pi^2 / 2mL^2$.

(b) The average position of the particle is

$$\langle x(t) \rangle = \int_{-L/2}^{L/2} \psi^* x \psi \, dx = \frac{1}{2} \int_{-L/2}^{L/2} \left(|\psi_1|^2 + |\psi_2|^2 + 2\mathrm{Re} \left(\psi_1 \psi_2^* \right) \cos \left((\omega_1 - \omega_2) t \right) \right) x \, dx$$

$$= \frac{1}{2} \int_{-L/2}^{L/2} \left(2\mathrm{Re} \left(\psi_1 \psi_2^* \right) \cos \left((\omega_1 - \omega_2) t \right) \right) x \, dx$$

$$= (\cos((\omega_1 - \omega_2)t)) \left(\frac{2}{L} \right) \int_{-L/2}^{L/2} x \cos\left(\frac{\pi x}{L}\right) \sin\left(\frac{2\pi x}{L}\right) dx$$

$$= (\cos((\omega_1 - \omega_2)t)) \left(\frac{1}{L} \right) \left(2 - \frac{2}{9} \right) \frac{L^2}{\pi^2} = \left(16L/9\pi^2 \right) \cos((\omega_2 - \omega_1)t).$$

(c) The momentum probability density $|\psi(p_x, t)|^2$ is found using the Fourier transform,

$$\psi(p_x, t) = \frac{1}{\sqrt{2\pi\hbar}} \int \psi(x, t) e^{-ip_x x} dx = \frac{1}{\sqrt{4\pi\hbar}} \int_{-L/2}^{L/2} \left(\psi_1(x, t) e^{-ip_x x} + \psi_2(x, t) e^{-ip_x x} \right) dx$$

$$= \sqrt{\frac{2}{4\pi\hbar L}} \left(e^{-i\omega_1 t} \int_{-L/2}^{L/2} \cos\left(\frac{\pi x}{L}\right) e^{-ip_x x} dx + e^{-i\omega_4 t} \int_{-L/2}^{L/2} \sin\left(\frac{2\pi x}{L}\right) e^{-ip_x d} dx \right),$$

$$|\psi(p_x, t)|^2 = \frac{1}{2\pi\hbar L} \left(\left| \int_{-L/2}^{L/2} \cos\left(\frac{\pi x}{L}\right) e^{-ip_x x} dx \right|^2 + \left| \int_{-L/2}^{L/2} \sin\left(\frac{2\pi x}{L}\right) e^{-ip_x x} dx \right|^2 \right.$$

$$\left. + 2\mathrm{Re}\left(e^{i(\omega_4 - \omega_1)t} \int_{-L/2}^{L/2} \cos\left(\frac{\pi x}{L}\right) e^{-ip_x x} \sin\left(\frac{2\pi x}{L}\right) e^{ip_x x} dx \right) \right).$$

Use is made of the standard indefinite integrals[9]

$$\int e^{ax} \cos(bx) dx = \frac{e^{ax}(a\cos(bx) + b\sin(bx))}{a^2 + b^2},$$

$$\int e^{ax} \sin(bx) dx = \frac{e^{ax}(a\sin(bx) - b\cos(bx))}{a^2 + b^2},$$

$$|\psi(p_x, t)|^2 = \frac{1}{2\pi\hbar L} \left(\left| \frac{L\pi \left(e^{ip_x L/2} + e^{-ip_x L/2} \right)}{\pi^2 - p_x^2 L^2} \right|^2 + \left| \frac{2L\pi \left(e^{ip_x L/2} - e^{-ip_x L/2} \right)}{4\pi^2 - p_x^2 L^2} \right|^2 \right.$$

$$\left. + 2\mathrm{Re}\left(e^{i(\omega_4 - \omega_1)t} \frac{L\pi \left(e^{ip_x L/2} + e^{-ip_x L/2} \right)}{\pi^2 - p_x^2 L^2} \frac{2L\pi \left(e^{-ip_x L/2} - e^{ip_x L/2} \right)}{4\pi^2 - p_x^2 L^2} \right) \right),$$

$$|\psi(p_x, t)|^2 = A\cos^2\left(\frac{p_x L}{2}\right) + B\sin^2\left(\frac{p_x L}{2}\right) + C\sin(p_x L)\sin((\omega_4 - \omega_1)t),$$

$$A = \frac{2L\pi}{\hbar(\pi^2 - L^2 p_x^2)^2}, \quad B = \frac{2L\pi}{\hbar((4\pi)^2 - L^2 p_x^2)^2}, \quad C = \frac{-4L\pi}{\hbar(\pi^2 - L^2 p_x^2)((2\pi)^2 - L^2 p_x^2)}.$$

(d) The average momentum of the particle is $\langle p_x(t) \rangle = -(8\hbar/(3L))\sin((\omega_2 - \omega_1)t)$.

(e) The current flux is

$$J_x(x, t) = -\frac{2e\pi\hbar}{mL^2} \left(\cos\left(\frac{\pi x}{L}\right) \cos\left(\frac{2\pi x}{L}\right) + \frac{1}{2}\sin\left(\frac{\pi x}{L}\right) \sin\left(\frac{2\pi x}{L}\right) \right) \sin((\omega_2 - \omega_1)t).$$

[9] I. S. Gradshteyn and I. M. Ryzhik, *Table of Integrals, Series, and Products*, San Diego, California, Academic Press, 1980, p. 196 (ISBN 0 12 294760 6).

Solutions 4.4

Figure 4.E13(a) is a sketch of the one-dimensional potential to be considered. The rectangular potential well is surrounded by infinite barrier energy and the first few energy eigenvalues are E_1, E_2, and E_3. To simplify the expression for the eigenfunctions, the position $x = 0$ is chosen to be the left-hand boundary of the potential, so that

$$V(x) = 0, \qquad 0 < x < L,$$

and

$$V(x) = \infty \qquad \text{elsewhere.}$$

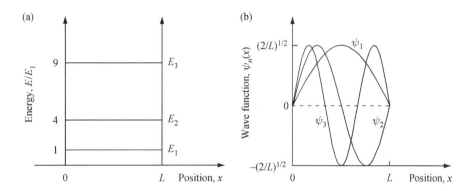

Fig. 4.E13

The expectation value of the particle position and the uncertainty in the position when the particle is in the nth energy state is found by considering the time-independent Schrödinger equation:

$$\left(-\frac{\hbar^2}{2m}\frac{d^2}{dx^2} + V(x)\right)\psi_n(x) = E_n\psi_n(x).$$

The boundary conditions are $\psi_n(x) = 0$ at $x = 0$ and $x = L$. Solutions to the wave function are

$$\psi_n = A_n\sin(k_nx),$$

where $k_n = n\pi/L$ and n is a nonzero positive integer $n = 1, 2, 3, \ldots$. Figure 4.E13(b) sketches the eigenfunctions ψ_1, ψ_2, and ψ_3 for the potential shown in (a).

The normalization constant A_n is found from the normalization condition

$$\int_{x=0}^{x=L} \psi_n^*(x)\psi_n(x)\,dx = A_n^2\int_{x=0}^{x=L} \sin^2(k_nx)\,dx = 1,$$

$$\frac{1}{A_n^2} = \int_{x=0}^{x=L}\left(\frac{1}{2} - \frac{1}{2}\cos(2k_nx)\right)dx = \left[\frac{x}{2} + \frac{1}{4k_n}\sin(2k_nx)\right]_0^L = \frac{L}{2} + 0,$$

205

where the relation $2\sin(x)\sin(y) = \cos(x - y) - \cos(x + y)$ is used. Hence, $A_n = \sqrt{2/L}$, and the wave function may be written as

$$\psi_n(x) = \sqrt{\frac{2}{L}} \sin\left(\frac{n\pi x}{L}\right).$$

To find the expectation value of x,

$$\langle x_n \rangle = \int \psi_n^*(x) x \psi_n(x)\, dx = A_n^2 \int x\, \sin^2(k_n x)\, dx = A_n^2 \int x \left(\frac{1}{2} - \frac{1}{2}\cos(2k_n x)\right) dx.$$

Written in the form shown on the right-hand side, the integral may be found by inspection of odd and even functions to give

$$\langle x_n \rangle = A_n^2 \left[\frac{x^2}{4}\right]_0^L.$$

Alternatively, the equation for $\langle x_n \rangle$ may be solved by integrating by parts using $\int UV'\, dx = UV - \int U'V\, dx$.

In this case, $U = x$ and $V' = \left(\frac{1}{2} - \frac{1}{2}\cos(2k_n x)\right)$, so that $V = \frac{x}{2} + \frac{1}{4k_n}\sin(2k_n x)$ and

$$\langle x_n \rangle = A_n^2 \left(\left[\frac{x^2}{2} + \frac{x}{4k_n}\sin(2k_n x)\right]_0^L - \int \left(\frac{x}{2} + \frac{1}{4k_n}\sin(2k_n x)\right) dx\right)$$

$$= A_n^2 \left[\frac{x^2}{2} + \frac{x}{4k_n}\sin(2k_n x) - \frac{x^2}{4} + \frac{1}{8k_n^2}\cos(2k_n x)\right]_0^L.$$

But $k_n = n\pi/L$ and $n = 1, 2, 3, \ldots$, so that

$$\langle x_n \rangle = A_n^2 \left(\frac{L^2}{2} + 0 - \frac{L^2}{4} + 0\right) = A_n^2 \frac{L^2}{4}.$$

And, since $A_n^2 = 2/L$,

$$\langle x_n \rangle = \frac{L}{2}.$$

In classical mechanics, the particle in the potential well moves at constant velocity, v, and traverses the well in time $\tau = L/v$. The average position is given by

$$\langle x \rangle_{\text{classical}} = \int_{t=0}^{t=\tau} \frac{vt\, dt}{\tau} = \frac{1}{2}v\frac{\tau^2}{\tau} = \frac{1}{2}v\tau = \frac{1}{2}v\frac{L}{v} = \frac{L}{2}.$$

Hence $\langle x \rangle = L/2$, which is quite satisfying since it is the same as the quantum result.

The classical expectation value of the observable x^2 associated with the quantum mechanical operator $\hat{x}^2$ is

$$\left\langle x_n^2 \right\rangle = A_n^2 \int x^2 \sin^2(k_n x)\, dx = A_n^2 \int \left(\frac{x^2}{2} - \frac{x^2}{2} \cos(2k_n x) \right) dx$$

$$= A_n^2 \left(\left[\frac{x^3}{6} - \frac{x^2}{2} \frac{1}{2k_n} \sin(2k_n x) \right]_0^L + \int \frac{x}{2k_n} \sin(2k_n x)\, dx \right)$$

$$= A_n^2 \left(\left[\frac{x^3}{6} - \frac{x^2}{2} \frac{1}{2k_n} \sin(2k_n x) + \frac{x}{2k_n} \left(-\frac{1}{2k_n} \right) \cos(2k_n x) \right]_0^L \right.$$
$$\left. + \int \frac{1}{4k_n^2} \cos(2k_n x)\, dx \right)$$

$$= A_n^2 \left[\frac{x^3}{6} - \frac{x^2}{4k_n} \sin(2k_n x) - \frac{x}{4k_n^2} \cos(2k_n x) + \frac{1}{8k_n^3} \sin(2k_n x) \right]_0^L ,$$

where the second and fourth terms contribute zero, since $k_n = \frac{n\pi}{L}$ and $n = 1, 2, 3, \ldots$. Hence,

$$\left\langle x_n^2 \right\rangle = A_n^2 \left(\frac{L^3}{6} - \frac{L}{4k_n^2} \right) = A_n^2 \left(\frac{L^3}{6} - \frac{L}{4} \frac{L^2}{n^2\pi^2} \right) = A_n^2 \left(\frac{L^3}{6} - \frac{L^3}{4n^2\pi^2} \right),$$

but $A_n^2 = 2/L$, so

$$\boxed{\left\langle x_n^2 \right\rangle = \frac{L^2}{3} - \frac{L^2}{2n^2\pi^2}.}$$

A measure of the uncertainty in the position of the particle in the nth state is given by the standard deviation $\Delta x_n = \left(\left\langle x_n^2 \right\rangle - \left\langle x_n \right\rangle^2 \right)^{1/2}$, which is calculated using

$$\Delta x_n^2 = \left\langle x_n^2 \right\rangle - \left\langle x_n \right\rangle^2 = \frac{L^2}{3} - \frac{L^2}{2n^2\pi^2} - \frac{L^2}{4} = \frac{L^2}{12} \left(4 - \frac{6}{n^2\pi^2} - 3 \right),$$

$$\boxed{\Delta x_n^2 = \frac{L^2}{12} \left(1 - \frac{6}{n^2\pi^2} \right).}$$

These results may be compared with those for a classical particle. The classical particle moves at constant velocity v and traverses the well in time $\tau = L/v$. Hence, classically,

$$\langle x \rangle_{\text{classical}} = \int_{t=0}^{t=\tau} \frac{vt}{\tau}\, dt = \frac{1}{2} v\tau = \frac{L}{2} \quad \text{and} \quad \left\langle x^2 \right\rangle_{\text{classical}} = \int_{t=0}^{t=\tau} \frac{(vt)^2}{\tau}\, dt = \frac{1}{3} v^2\tau^2 = \frac{L^2}{3}.$$

Thus, as $n \to \infty$ the quantum results approach the classical solution.

The average values $\langle p_x \rangle$ and $\langle p_x^2 \rangle$ are

$$\langle p_x \rangle = 0 \quad \text{and} \quad \left\langle p_x^2 \right\rangle = 2mE_n,$$

and since

$$E_n = \frac{\hbar^2 k_n^2}{2m} = \frac{\hbar^2 n^2 \pi^2}{2mL^2},$$

then

$$\left\langle p_x^2 \right\rangle = \frac{\hbar^2 n^2 \pi^2}{L^2}.$$

Hence,

$$\Delta x^2 = \frac{L^2}{12} \left(1 - \frac{6}{n^2 \pi^2} \right),$$

$$\Delta p_x^2 = \frac{\hbar^2 \pi^2 n^2}{L^2},$$

$$\Delta x \Delta p_x = \frac{\hbar}{\sqrt{12}} \left(n^2 \pi^2 - 6 \right)^{1/2}.$$

Solutions 4.5
Taking the Fourier transform of $\psi(x) = 1/\sqrt{2L}$ for $|x| < L$ and $\psi(x) = 0$ for $|x| > L$:

$$\psi(k) = \frac{1}{\sqrt{2\pi}} \int_{-L}^{L} \frac{1}{\sqrt{2L}} e^{-ikx} dx = \left[\frac{1}{\sqrt{2\pi L}} \frac{-1}{ik} e^{-ikx} \right]_{-L}^{L} = \frac{1}{\sqrt{\pi L}} \frac{1}{k} \sin(kL),$$

$$|\psi(k)|^2 = \frac{\sin^2(kL)}{\pi L k^2}.$$

The wave functions $\psi(x)$ and $\psi(k)$ are plotted in Fig. 4.E14.
To show that the uncertainty (standard deviation) in particle momentum Δp is infinite, it is necessary to calculate $\langle p_x \rangle$ and $\langle p_x^2 \rangle$:

$$\langle p_x \rangle = \int_{-\infty}^{\infty} \psi^*(k) \hbar k \psi(k) dk = \frac{\hbar}{L\pi} \int_{-\infty}^{\infty} \frac{1}{k} \sin^2(kL) dk = 0$$

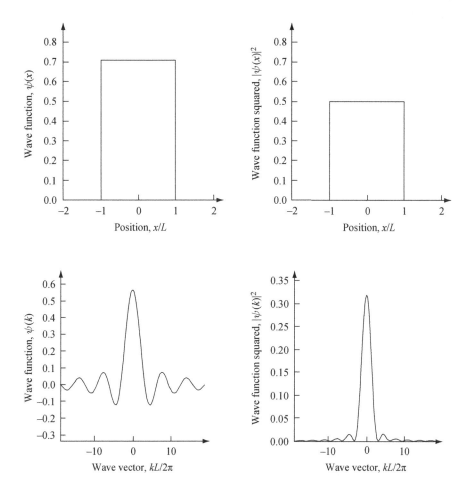

Fig. 4.E14

by symmetry, and

$$\left\langle p_x^2 \right\rangle = \int\limits_{-\infty}^{\infty} \psi^*(k)\hbar^2 k^2\, \psi(k)\mathrm{d}k = \frac{\hbar^2}{L\pi} \int\limits_{-\infty}^{\infty} \sin^2(kL)\mathrm{d}k = \frac{\hbar^2}{L\pi} \int\limits_{-\infty}^{\infty} \left(\frac{1}{2} - \frac{1}{2}\cos(2kL)\right)\mathrm{d}k$$

$$= \frac{\hbar^2}{L\pi} \int\limits_{-\infty}^{\infty} \left(\frac{1}{2} - \frac{1}{2}\cos(2kL)\right)\mathrm{d}k = \frac{\hbar^2}{L\pi} \left[\frac{k}{2} - \frac{1}{4L}\sin(2kL)\right]_{-\infty}^{\infty}$$

$$= \frac{\hbar^2}{L\pi}\left(\frac{\infty}{2} - \frac{-\infty}{2} - 0 + 0\right) = \infty,$$

and so

$$\Delta p_x^2 = \langle p_x^2 \rangle - \langle p_x \rangle^2 = \infty.$$

The fact $\Delta p_x^2 = \infty$ may be understood as $|\psi(k)|^2$ not decreasing to zero fast enough in the limit $k \to \infty$.

The average value of position is

$$\langle x \rangle = \int_{-L}^{L} \psi(x) x \psi(x) \mathrm{d}x = \left(\frac{1}{\sqrt{2L}} \right)^2 \frac{1}{2} \left[x^2 \right]_{-L}^{L} = 0$$

and the value of $\langle x^2 \rangle$ is

$$\left\langle x^2 \right\rangle = \int \psi(x) x^2 \psi(x) \mathrm{d}x = \frac{1}{2L} \frac{1}{3} \left[x^3 \right]_{-L}^{L} = \frac{1}{3L} L^3 = \frac{L^2}{3},$$

giving a measure of the spread in measured values of

$$\Delta x^2 = \left\langle x^2 \right\rangle - \langle x \rangle^2 = \frac{L^2}{x} - 0.$$

It follows that $\Delta x = L/\sqrt{3}$, that is finite nonzero, so that

$$\Delta x \Delta p_x = \infty.$$

Solutions 4.6

(a) $[\hat{p}_x^2, \hat{p}_x] = -i\hbar \left(\frac{\mathrm{d}}{\mathrm{d}x} \frac{\mathrm{d}}{\mathrm{d}x} \frac{\mathrm{d}}{\mathrm{d}x} - \frac{\mathrm{d}}{\mathrm{d}x} \frac{\mathrm{d}}{\mathrm{d}x} \frac{\mathrm{d}}{\mathrm{d}x} \right) = 0.$

(b) $\left[\hat{V}(x), \hat{p}_x \right] = -i\hbar \left(V(x) \frac{\mathrm{d}}{\mathrm{d}x} - \frac{\mathrm{d}}{\mathrm{d}x} V(x) - V(x) \frac{\mathrm{d}}{\mathrm{d}x} \right) = i\hbar \frac{\mathrm{d}}{\mathrm{d}x} V(x).$

(c) Making use of the results in part (a) and (b), $\frac{\mathrm{d}}{\mathrm{d}t} \langle p_x \rangle = \frac{i}{\hbar} \left\langle \left[\hat{H}, \hat{p}_x \right] \right\rangle =$
$\frac{i}{\hbar} \left\langle \left[\frac{\hat{p}_x^2}{2m} + \hat{V}(x), \hat{p}_x \right] \right\rangle = \frac{i}{\hbar} \left\langle \left[\hat{V}(x), \hat{p}_x \right] \right\rangle = -\langle \frac{\partial}{\partial x} V(x) \rangle,$
which may be written as

$$\frac{\mathrm{d}}{\mathrm{d}t} \langle p_x \rangle = m \frac{\mathrm{d}^2}{\mathrm{d}t^2} \langle x \rangle = -\left\langle \frac{\mathrm{d}}{\mathrm{d}x} V(x) \right\rangle.$$

If the uncertainty $\Delta x = \left\langle (\hat{x} - \langle x \rangle)^2 \right\rangle^{1/2}$ is small so that the classical value of $x = \langle x \rangle$ *and* if the potential $V(x)$ is slowly varying such that

$$\frac{\mathrm{d}}{\mathrm{d}\langle x \rangle} V(\langle x \rangle) = \left\langle \frac{\mathrm{d}}{\mathrm{d}x} V(x) \right\rangle,$$

then Newton's classical result,

$$\frac{\mathrm{d}p_x}{\mathrm{d}t} = m \frac{\mathrm{d}^2 x}{\mathrm{d}t^2} = -\frac{\mathrm{d}}{\mathrm{d}x} V(x),$$

is obtained.

Solutions 4.7

The momentum operator is Hermitian so that matrix elements $A_{ij} = A_{ji}^*$ or, equivalently, $\left(\int \phi^* \hat{p} \psi \, \mathrm{d}^3 r \right)^* = \int \psi^* \hat{p} \phi \, \mathrm{d}^3 r = \left(\int \hat{p} \phi^* \psi \, \mathrm{d}^3 r \right)^*$. The x component of the momentum operator $\hat{p}_x = -\mathrm{i}\hbar(\mathrm{d}/\mathrm{d}x)$, and

$$\int \phi^* \hat{p}_x \psi \mathrm{d}x = \int \hat{p}_x \phi^* \psi \mathrm{d}x \text{ or } \langle \phi | \hat{p}_x \psi \rangle = \langle \hat{p}_x \phi | \psi \rangle.$$

The operator $\hat{p}_x$ can be seen to be Hermitian when integrating by parts, $\int U V' \mathrm{d}x = UV - \int U'V \, \mathrm{d}x$, so that

$$\int \phi^* \hat{p}_x \psi \, \mathrm{d}x = -\mathrm{i}\hbar \int \phi^* \frac{\mathrm{d}}{\mathrm{d}x} \psi \, \mathrm{d}x = \left[-\mathrm{i}\hbar \phi^* \psi \right]_{-\infty}^{\infty} + \mathrm{i}\hbar \int_{-\infty}^{\infty} \frac{\mathrm{d}\phi^*}{\mathrm{d}x} \psi \, \mathrm{d}x,$$

the term $\left[-\mathrm{i}\hbar \phi^* \psi \right]_{-\infty}^{\infty} = 0$, and $\mathrm{i}\hbar \int_{-\infty}^{\infty} \frac{\mathrm{d}\phi^*}{\mathrm{d}x} \psi \, \mathrm{d}x = \int \hat{p}_x \phi^* \psi \, \mathrm{d}x$. Thus $\langle \phi | \hat{p}\psi \rangle = \langle \hat{p}\phi | \psi \rangle$, provided that the wave function $\phi \to 0$ at $x \pm \infty$. If $\psi = \phi$, then $\langle p \rangle = \int \phi^* \hat{p} \phi \, \mathrm{d}r = \int \hat{p}^* \phi^* \phi \, \mathrm{d}r = \langle p \rangle^*$, which can only be true if $\langle p \rangle$ is real. Hence, the expectation value of the momentum operator is real.

To show that $-\mathrm{i}\hbar(\partial/\partial r)$ is not Hermitian in radial coordinates, the expectation value is found by integrating by parts in radial coordinates:

$$\int_{r=0}^{r=\infty} 4\pi r^2 \psi_m^* \left(-\mathrm{i}\hbar \frac{\partial}{\partial r} \right) \psi_n \mathrm{d}r = \left[4\pi r^2 \psi_m^* \left(-\mathrm{i}\hbar \frac{\partial}{\partial r} \psi_n \right) \right]_{r=0}^{r=\infty} - \int_{r=0}^{r=\infty} -\mathrm{i}\hbar \psi_n \frac{\partial}{\partial r} (4\pi r^2 \psi_m^*) \mathrm{d}r.$$

The first term on the right-hand side is zero, assuming that the eigenfunction vanishes at $r = \infty$, so that

$$\int_{r=0}^{r=\infty} 4\pi r^2 \psi_m^* \left(-\mathrm{i}\hbar \frac{\partial}{\partial r} \right) \psi_n \mathrm{d}r = \int \mathrm{i}\hbar 4\pi \psi_n \left(r^2 \frac{\partial}{\partial r} \psi_m^* - 2r\psi_m^* \right) \mathrm{d}r$$

$$= \int 4\pi r^2 \psi_n \left(\mathrm{i}\hbar \left(2r - \frac{\partial}{\partial r} \right) \psi_m \right)^* \mathrm{d}r.$$

Obviously, this does not satisfy $A_{ij}^* = A_{ji}$ for a Hermitian operator.

Solutions 4.8

A particle mass m is in a finite, one-dimensional rectangular potential well for which $V(x) = 0$ for $-L < x < L$ and $V(x) = V_0$ elsewhere, and for which the value of V_0 is a finite positive constant. The expectation value of kinetic energy is

$$\langle T \rangle = \int_{-\infty}^{\infty} \psi^*(x) \frac{\hat{p}^2}{2m} \psi(x) \mathrm{d}x = \frac{-\hbar^2}{2m} \int_{-\infty}^{\infty} \psi(x) \frac{\mathrm{d}}{\mathrm{d}x^2} \psi(x) \mathrm{d}x.$$

For even-parity *bound-state* solutions, including the ground state, the spatial wave functions in the well are of the form

$$\psi_n(x) = A_n \cos(k_n x),$$

and in the barrier they are of the form

$$\psi_n(x) = C_n e^{-\kappa_n x},$$

where the index n is an odd positive integer that labels the bound-state eigenvalue. For eigenenergy E_n,

$$k_n = \sqrt{2mE_n}/\hbar$$

and

$$\kappa_n = \sqrt{2m(V_0 - E_n)}/\hbar.$$

For the ground state $n = 1$, the expectation value for kinetic energy is

$$\langle T \rangle = \frac{\hbar^2 k_1^2 A^2}{m} \int_{x=0}^{x=L} \cos^2(k_1 x')\mathrm{d}x' - \frac{\hbar^2 k_1^2 C^2}{m} \int_{x=L}^{x=\infty} e^{-2\kappa_1 x'}\mathrm{d}x'.$$

This shows that the contribution to kinetic energy from the barrier region is negative.

Solutions 4.9

If $\langle \psi(t)|\psi(t)\rangle$ is constant in time then

$$\frac{\mathrm{d}}{\mathrm{d}t}\langle \psi(t)|\psi(t)\rangle = \left(\frac{\mathrm{d}}{\mathrm{d}t}\langle \psi(t)|\right)|\psi(t)\rangle + \langle \psi(t)|\left(\frac{\mathrm{d}}{\mathrm{d}t}|\psi(t)\rangle\right)$$

is zero. From the Schrödinger equation,

$$\frac{\partial}{\partial t}|\psi(t)\rangle = \frac{-i}{\hbar}\hat{H}|\psi(t)\rangle$$

and

$$\frac{\partial}{\partial t}\langle \psi(t)| = \frac{i}{\hbar}\langle \psi(t)|\hat{H}^\dagger = \frac{i}{\hbar}\langle \psi(t)|\hat{H}$$

from the hermiticity of the Hamiltonian. Hence,

$$\frac{\mathrm{d}}{\mathrm{d}t}\langle \psi(t)|\psi(t)\rangle = \left(\frac{i}{\hbar} - \frac{i}{\hbar}\right)\langle \psi(t)|\hat{H}|\psi(t)\rangle = 0,$$

and so the probability density, $\langle \psi(t)|\psi(t)\rangle$, does not evolve in time.

Solutions 4.10

For a source-free, linear, frequency-independent, lossless dielectric with $\varepsilon(\mathbf{r}) = \varepsilon_0(\mathbf{r})\varepsilon_r(\mathbf{r})$, and with relative magnetic permeability $\mu_r = 1$, Maxwell's equations lead directly to a wave equation for electromagnetic fields. The linearity allows separation of the time and space dependence into a set of harmonic solutions of the form $\mathbf{E}(\mathbf{r}, t) = \mathbf{E}_0 e^{i\mathbf{k} \cdot \mathbf{r}} e^{i\omega t}$ and $\mathbf{H}(\mathbf{r}, t) = \mathbf{H}_0 e^{i\mathbf{k} \cdot \mathbf{r}} e^{i\omega t}$, where complex numbers have been used for mathematical convenience (always taking the real part to obtain the *physical fields*). The $\mathbf{H}$ and $\mathbf{E}$ fields in Maxwell's equations are *real vector fields*. Because the dielectric is source-free, the divergence $\nabla \cdot \mathbf{D} = 0$ and $\nabla \cdot \mathbf{H} = 0$. There are no sources or sinks of displacement ($\mathbf{D}$) or magnetic fields ($\mathbf{H}$). Field configurations are built up out of plane waves that are transverse in such a way that $\mathbf{E}_0 \cdot \mathbf{k} = 0$ and $\mathbf{H}_0 \cdot \mathbf{k} = 0$.

Since the $\mathbf{H}$ and $\mathbf{E}$ fields are related through

$$\nabla \times \mathbf{E} = -\mu_0 \frac{\partial \mathbf{H}}{\partial t},$$

$$\nabla \times \mathbf{H} = \varepsilon(\mathbf{r}) \frac{\partial \mathbf{E}}{\partial t},$$

dividing the equation for $\nabla \times \mathbf{H}$ by $\varepsilon(\mathbf{r})$ and taking the curl gives

$$\nabla \times \left(\frac{1}{\varepsilon(\mathbf{r})} \nabla \times \mathbf{H}(\mathbf{r}) \right) = \nabla \times \frac{\partial \mathbf{E}(\mathbf{r})}{\partial t} = -\mu_0 \frac{\partial^2 \mathbf{H}(\mathbf{r})}{\partial t^2}.$$

And since $\mathbf{H}(\mathbf{r}, t) = \mathbf{H}(\mathbf{r}) e^{i\omega t}$,

$$\nabla \times \left(\frac{1}{\varepsilon_r(\mathbf{r})} \nabla \times \mathbf{H}(\mathbf{r}) \right) = \omega^2 \mu_0 \varepsilon_0 \frac{\partial^2 \mathbf{H}(\mathbf{r})}{\partial t^2} = \left(\frac{\omega}{c} \right)^2 \frac{\partial^2 \mathbf{H}(\mathbf{r})}{\partial t^2}.$$

After solving this wave equation for $\mathbf{H}(\mathbf{r})$, the transverse electric field may be obtained via

$$\mathbf{E}(\mathbf{r}) = \frac{1}{i\omega\, \varepsilon(\mathbf{r})} \nabla \times \mathbf{H}(\mathbf{r}).$$

Recognizing the wave equation for $\mathbf{H}(\mathbf{r})$ as an eigenvalue problem in which the differential operator $\tilde{H} = \nabla \times (1/\varepsilon_r(\mathbf{r})) \nabla$ is analogous to the Hamiltonian used in the Schrödinger equation gives

$$\tilde{H} \mathbf{H}(\mathbf{r}) = \left(\frac{\omega}{c} \right)^2 \mathbf{H}(\mathbf{r}).$$

The eigenvectors $\mathbf{H}(\mathbf{r})$ are the field patterns of the harmonic modes oscillating at frequency ω and the eigenvalues are $(\omega/c)^2$.

As with the Hamiltonian used in quantum mechanics, $\tilde{H}$ is a linear Hermitian operator with real eigenvalues. In fact, comparing the electrodynamics discussed above and quantum mechanics there are a number of similarities, but also important differences. This is expected because classical electrodynamics is just the macroscopic, incoherent, large-photon-number limit of the quantum treatment. In electrodynamics, $\mathbf{H}(\mathbf{r})$ is a *real vector field* with a simple time dependence $e^{i\omega t}$ that is used only as a mathematical convenience. To find a measurable value of $\mathbf{H}(\mathbf{r})$, the real part is evaluated. In quantum mechanics, Schrödinger's wave function $\psi(\mathbf{r})$ is a *complex scalar field*. To find the value of a measured

213

quantity, the real part of the wave function is not taken; rather, the expectation value of the measurable A associated with operator $\hat{A}$ is evaluated:

$$\langle A \rangle = \int \psi^*(\mathbf{r}, t) \hat{A} \psi(\mathbf{r}, t) \mathrm{d}^3 r.$$

The eigenvalues of the Schrödinger wave equation are related to the oscillation frequency through the energy $E = \hbar \omega$. In electrodynamics, eigenvalues are proportional to ω^2. In quantum mechanics, the Hamiltonian is separable if the potential is separable (e.g., $V(\mathbf{r}) = V(x)V(y)V(z)$). In electrodynamics, there is no such simplification as $\tilde{H}$ couples the different directions even if $\varepsilon(\mathbf{r})$ is separable. In electrodynamics, there is no *absolute* scale. Hence, in electrodynamics solutions from radio waves to visible light and beyond are based simply upon geometry and material parameters such as $\varepsilon(\mathbf{r})$. In quantum mechanics, there *is* an absolute scale, because $\hbar \neq 0$. Planck's constant $\hbar$ sets the scale. The corresponding length scale in atomic systems is set by the Bohr radius, $a_{\mathrm{B}} = 4\pi\varepsilon_0\hbar^2/m_0 e^2 = 0.529\,177 \times 10^{-10}$ m (or an *effective* Bohr radius in materials with $\varepsilon \neq \varepsilon_0$ and $m^* \neq m_0$). Typically, it is on this length scale (and corresponding energy and time scales) that quantum effects ($\hbar \neq 0$) are important.

Classical electromagnetic theory uses real magnetic and electric fields coupled via Maxwell's equations. The magnetic and electric fields each have physical meaning. Both fields are needed to describe the instantaneous state and time evolution of the system.

Time-reversal symmetry can occur in both quantum mechanics and classical electrodynamics. This time-reversal symmetry exists when the system under consideration is conservative.

4.9 Problems

Problem 4.1
An electron in a one-dimensional rectangular potential well of thickness L and infinite potential elsewhere is in the simple superposition state consisting of the ground and third excited state so that

$$\psi(x, t) = \frac{1}{\sqrt{2}}(\psi_1(x, t) + \psi_4(x, t)).$$

Find expressions for:
(a) The probability density, $|\psi(x, t)|^2$.
(b) The average particle position, $\langle x(t) \rangle$.
(c) The momentum probability density, $|\psi(p_x, t)|^2$.
(d) The average momentum, $\langle p_x(t) \rangle$.
(e) The current flux, $J_x(x, t)$.

Problem 4.2
(a) Show that the density of states for a free particle of mass m in two dimensions is

$$D_2(E) = \frac{m}{2\pi\hbar^2}.$$

(b) At low temperature, electrons in two electrodes occupy states up to the Fermi energy, E_{F}. The two closely spaced electrodes are connected by a two-dimensional conductance region. Derive an expression for the conductance of electrons flowing between the two electrodes as a function of applied voltage V_{bias}, assuming the transmission coefficient through the two-dimensional region is unity. Consider the two limiting cases $eV_{\text{bias}} \gg E_{\text{F}}$ and $eV_{\text{bias}} \ll E_{\text{F}}$.

Problem 4.3

Derive expressions for the two-dimensional $D_2^{\text{opt}}(\omega)$ and one-dimensional $D_1^{\text{opt}}(\omega)$ density of photon states in a homogeneous dielectric medium characterized by refractive index, n_{r}.

Problem 4.4

In a particular system the dispersion relation for electrons in one dimension is $E_k = \hbar\omega_k = 2t_{\text{hop},1}\cos(k_x L)$, where $t_{\text{hop},1}$ and L are constants and the wave vector in the x direction is $0 \leq k_x < \pi/L$. This dispersion relation can be derived using a nearest-neighbor tight binding model where $t_{\text{hop},1}$ is the overlap integral between atomic orbitals.

(a) Find the complex band structure for non-propagating electron states.

(b) Choosing one hundred equally spaced discrete values of k_x, plot the electron density of states,

$$N(E) = \sum_k \frac{\gamma/\pi}{(E - E_k)^2 + (\frac{\gamma}{2})^2},$$

using $\gamma = |t_{\text{hop},1}|/10$ and $t_{\text{hop},1} = -1$.

(c) If next nearest-neighbor interactions are included, the dispersion relation in (a) can, to within a scaling factor, be written $E_k = 2t_{\text{hop},1}\cos(k_x L) + 2t_{\text{hop},2}\cos(2k_x L)$. Write a computer program to plot the dispersion relation. Then calculate and plot the electron density of states using $\gamma = |t_{\text{hop},1}|/10$, $t_{\text{hop},1} = -1$, $t_{\text{hop},2} = -0.2$ and compare with results obtained in (a) including a comparison with the effective electron mass at the band edges.

(d) Calculate and plot the electron density of states for a square lattice with lattice constant L and a tight-binding nearest-neighbor dispersion relation given by $E_k = 2t_{\text{hop},1}(\cos(k_x L) + \cos(k_y L))$. Find an expression for the electron dispersion E_k when an additional next-nearest-neighbor hopping term $t_{\text{hop},2}$ is included and calculate the electron density of states. Use parameters $\gamma = |t_{\text{hop},1}|/10$, $t_{\text{hop},1} = -1$, $t_{\text{hop},2} = t_{\text{hop},1}/10$, and explain the differences in the density of states obtained for the two cases.

(e) Calculate and plot the electron density of states for a cubic lattice with lattice constant L and a tight-binding nearest-neighbor dispersion relation given by $E_k = 2t_{\text{hop},1}(\cos(k_x L) + \cos(k_y L) + \cos(k_z L))$. Find an expression for electron dispersion E_k when an additional next-nearest-neighbor hopping term $t_{\text{hop},2}$ is included and calculate the electron density of states. Use parameters $\gamma = |t_{\text{hop},1}|/10$, $t_{\text{hop},1} = -1$, $t_{\text{hop},2} = t_{\text{hop},1}/10$, and explain the differences in the density of states obtained for the two cases.

Submit the code used.

Problem 4.5

(a) Explain why a one-dimensional tight-binding model of N atoms on a lattice with nearest-neighbor spacing L, periodic boundary conditions, and nearest-neighbor hopping integral $t_{\text{hop},1}$ between atomic s-orbital sites at x_j and $x_i = x_j \pm nL$ has a symmetric electron density of states.

(b) In general, the electron dispersion relation, E_k, for a one-dimensional tight-binding model describing atoms on a lattice with nearest-neighbor spacing L and periodic boundary conditions is

$$E_k = -2 \sum_{n}^{N_{\text{atom}}} t_{\text{hop,n}} \cos{(nkL)},$$

where $t_{\text{hop,n}}$ is the overlap integral between atomic s-orbital sites at x_j and $x_i = x_j \pm nL$ with n a positive nonzero integer and N_{atom} is the number of atoms in the lattice. What values of electron hopping integral $t_{\text{hop,n}}$ result in a *linear* dispersion relation and what is the corresponding density of states? If configured as a conducting one-dimensional wire, what is the maximum conductance of the system?

(c) Show that *any* one-dimensional density of electron states is inversely proportional to electron group velocity, $v_{\text{g}} = \mathrm{d}\omega/\mathrm{d}k$. Then show that conductance is quantized in an electrically conducting one-dimensional wire, *independent* of its dispersion relation.

Problem 4.6

A hydrogen atom is in its ground state with electron wave function

$$\phi = \frac{2}{a_{\text{B}}^{3/2}} e^{-r/a_{\text{B}}} \left(\frac{1}{4\pi} \right)^{1/2}.$$

In this expression a_{B} is the Bohr radius and r is a radial coordinate.

Use spherical coordinates to find the expectation value of position r and momentum p_r for the electron in this state. In radial coordinates the Hermitian momentum operator is

$$\hat{p}_r = -i\hbar \frac{1}{r} \frac{\partial}{\partial r} r.$$

Problem 4.7

Using the fact that the Hamiltonian appearing in the Schrödinger equation

$$\frac{-i}{\hbar} \hat{H} |\psi(\mathbf{r}, t)\rangle = \left| \frac{\partial}{\partial t} \psi(\mathbf{r}, t) \right\rangle$$

is Hermitian, i.e., $\left\langle \psi | \hat{H} \psi \right\rangle = \left\langle \hat{H} \psi | \psi \right\rangle$, show that the time dependence of the average value of the observable A associated with the operator $\hat{A}$ is

$$\frac{\mathrm{d}}{\mathrm{d}t} \langle A \rangle = \frac{i}{\hbar} \left\langle \left[\hat{H}, \hat{A} \right] \right\rangle + \left\langle \frac{\partial}{\partial t} \hat{A} \right\rangle.$$

Problem 4.8

Show that:

(a) The position operator $\hat{x}$ acting on wavefunction $\psi(x)$ is Hermitian (i.e., $\hat{x}^{\dagger} = \hat{x}$).

(b) The operator $\mathrm{d}/\mathrm{d}x$ acting on the wavefunction $\psi(x)$ is anti-Hermitian (i.e., $(\mathrm{d}/\mathrm{d}x)^{\dagger} = -\mathrm{d}/\mathrm{d}x$).

(c) The momentum operator $-i\hbar(d/dx)$ acting on the wavefunction $\psi(x)$ is Hermitian.

Problem 4.9

A particle mass m is confined to motion in a one-dimensional potential $V(x)$. The Hamiltonian is

$$\hat{H} = -\frac{\hbar^2}{2m}\frac{d^2}{dx^2} + V(x)$$

and the momentum operator is

$$\hat{p} = -i\hbar\frac{d}{dx}.$$

(a) Find the commutator $\left[\hat{H}, \hat{p}\right]$.

(b) For what potentials, $V(x)$, are solutions of the time-independent Schrödinger equation also eigenstates of momentum?

Problem 4.10

Classical electromagnetic theory uses real magnetic and electric fields coupled via Maxwell's equations. The magnetic and electric fields each have physical meaning. Both fields are needed to describe the instantaneous state and time evolution of the system. Quantum mechanics uses one complex wave function to describe both the instantaneous state and time evolution of the system. It is also possible to describe quantum mechanics using two coupled real wave functions corresponding to the real and imaginary parts of the complex wave function. Why is this not done?

Problem 4.11

A classical bit of information has state 0 or 1, which in quantum mechanics corresponds to the orthonormal basis states $|0\rangle$ and $|1\rangle$. The difference between classical bits and quantum bits (qubits) is that a qubit can exist in a continuum of states between $|0\rangle$ and $|1\rangle$ as a superposition state $|\psi\rangle = a_0|0\rangle + a_1|1\rangle$ where $|a_0|^2 + |a_1|^2 = 1$.

(a) Two qubits have a normalized linear superposition state $|\psi\rangle = a_{00}|00\rangle + a_{01}|01\rangle + a_{10}|10\rangle + a_{11}|11\rangle$, where $|00\rangle$, $|01\rangle$, $|10\rangle$, and $|11\rangle$ are the basis states. Measuring just the first qubit gives eigenvalue 0 with probability $|a_{00}|^2 + |a_{11}|^2$. What is the normalized post-measurement state $|\psi'\rangle$?

(b) The first qubit in a two-qubit Bell state $|\psi\rangle = \frac{1}{\sqrt{2}}(|00\rangle + |11\rangle)$ is measured. It has probability $1/2$ that the post-state is $|\psi'\rangle = |11\rangle$. What is the result of measuring the second qubit when initially in state $|\psi\rangle$ and when initially in the state $|\psi'\rangle$?

Problem 4.12

A function $|f\rangle$ can be expressed as an expansion of complete orthonormal basis functions $|\phi_n\rangle$.

(a) Show that the identity operator $\hat{1} \equiv \sum_n |\phi_n\rangle\langle\phi_n|$ acting on the function $|f\rangle$ leaves it unchanged.

(b) The sum of diagonal elements of an operator $\hat{A}$ expressed as a matrix is called a trace operator, $\text{Tr}(\hat{A})$. Show that the trace operator is independent of the basis used.

(c) A unitary operator satisfies $\hat{U}^{-1} = \hat{U}^{\dagger}$. Show that the inner product of functions $|f_1\rangle$ and $|g_1\rangle$ is invariant under unitary transformation such that $|f_2\rangle = \hat{U}|f_1\rangle$ and $|g_2\rangle = \hat{U}|g_1\rangle$.

(d) Demonstrate that a unitary transformation can be used to change the representation of an operator from $\hat{A}$ to $\hat{B} = \hat{U}\hat{A}\hat{U}^{\dagger}$ by showing that the matrix elements satisfy $\langle g_1|\hat{A}|f_1\rangle = \langle g_2|\hat{B}|f_2\rangle$.

Problem 4.13

The nonzero state $|n, t\rangle$ evolves in time according to the Schrödinger equation $i\hbar\frac{\partial}{\partial t}|n, t\rangle = \hat{H}|n, t\rangle$, where $\hat{H}$ is the Hamiltonian. A unitary time-evolution operator $\hat{U}(t, t_0)$ evolves the state from time t_0 such that $|n, t\rangle = \hat{U}(t, t_0)|n, t\rangle$. For $\hat{H} \neq \hat{H}(t)$ show that

$$|n, t\rangle = e^{-i\hat{H}(t-t_0)/\hbar}|n, t_0\rangle,$$

and for $\hat{H} = \hat{H}(t)$, such that $\left[\hat{H}(t), \hat{H}(t')\right] = 0$ and $t \neq t'$, show that

$$|n, t\rangle = e^{\frac{-i}{\hbar}\int_{t'=t_0}^{t'=t}\hat{H}(t')dt'}|n, t_0\rangle.$$

Problem 4.14

Suppose the Hamiltonian $\hat{H}_\lambda$ describing a particle mass m constrained to motion in the x direction contains an adjustable parameter λ that may appear in the kinetic energy T, potential energy V, or both. The energy eigenvalue E_λ and eigenstate ψ_λ also depend on λ. For any λ the eigenvalue may be written $E_\lambda = \langle\psi_\lambda|\hat{H}_\lambda|\psi_\lambda\rangle$.

(a) Show that

$$\frac{dE_\lambda}{d\lambda} = \langle\psi_\lambda|\frac{d\hat{H}_\lambda}{d\lambda}|\psi_\lambda\rangle.$$

(b) Starting from the time-independent Schrödinger equation,

$$\left(-\frac{\hbar^2}{2m}\frac{d^2}{dx^2} + V(x)\right) = E\psi(x),$$

scale x such that $x \to \lambda x$ and apply the result in (a) to show that for any bound state the generalized virial theorem,

$$2\langle\psi|\frac{p^2}{2m}|\psi\rangle = \langle\psi|x\frac{d}{dx}V(x)|\psi\rangle,$$

is obtained when $\lambda \to 1$, so that if $V(x) \propto x^\gamma$ then $2\langle T\rangle = \gamma\langle V\rangle$.

Problem 4.15
(a) Show that

$$e^{\alpha \hat{A}} \hat{B} e^{-\alpha \hat{A}} = \hat{B} + \alpha \left[\hat{A}, \hat{B} \right] + \frac{\alpha^2}{2!} \left[\hat{A}, \left[\hat{A}, \hat{B} \right] \right] + \frac{\alpha^3}{3!} \left[\hat{A}, \left[\hat{A}, \left[\hat{A}, \hat{B} \right] \right] \right] + \cdots ,$$

where $\hat{A}$ and $\hat{B}$ are operators and α is a scalar.

(b) If $\left[\hat{A}, \left[\hat{A}, \hat{B} \right] \right] = \left[\hat{B}, \left[\hat{A}, \hat{B} \right] \right] = 0$, show that

$$e^{\hat{A}+\hat{B}} = e^{-\frac{1}{2}[\hat{A},\hat{B}]} e^{\hat{A}} e^{\hat{B}} = e^{\frac{1}{2}[\hat{B},\hat{A}]} e^{\hat{A}} e^{\hat{B}}$$

so if $\left[\hat{A}, \hat{B} \right] = 0$ then $e^{\hat{A}+\hat{B}} = e^{\hat{A}} e^{\hat{B}}$.

(c) If $\left[\hat{A}, \left[\hat{A}, \hat{B} \right] \right] = \left[\hat{B}, \left[\hat{A}, \hat{B} \right] \right] = 0$, rewrite $e^{\hat{A}} e^{\hat{B}} e^{\hat{A}}$ as a single exponential.

Problem 4.16
If the Hamiltonian $\hat{H}$ does not depend on time, the system is stationary and the Schrödinger equation $i\hbar \frac{\partial}{\partial t} |n, t\rangle$ can be integrated to give $|n, t\rangle = e^{-i\hat{H}t/\hbar} |n, t = 0\rangle$.

(a) Expand $e^{-i\hat{H}t/\hbar}$, take the time derivative, and show that
$i\hbar \frac{\partial}{\partial t} |n, t\rangle = \hat{H} e^{-i\hat{H}t/\hbar} |n, t = 0\rangle = \hat{H} |n, t\rangle$.

(b) The unitary operator $\hat{U}(t) = \hat{H} e^{-i\hat{H}t/\hbar}$ is the generator of time development and belongs to the unitary Lie group $U(1)$. If the system evolves for a short time interval, Δt, show that $|n, t\rangle = \hat{U}(\Delta t) |n, t = 0\rangle$, where $\hat{U}(\Delta t) = \hat{1} - \frac{i\hat{H}\Delta t}{\hbar} + 0(\Delta t)^2$.

(c) From the group property of the operator in (b), a finite time evolution may be built up from a product of N small time steps such that $t = N\Delta t$ where $\Delta t = t/N \to 0$. Making use of the fact that in the limit $N \to \infty$ the binomial theorem may be used to write $e^x = \left(1 + \frac{x}{N} \right)^N$, show that $\hat{U}(t) = e^{-i\hat{H}t/\hbar}$.

Problem 4.17
The active region of a single-walled carbon nanotube transistor is modeled as a conducting channel nanotube 2 nm in diameter that is coated in a 2 nm thick dielectric of relative permittivity ϵ_{r0} and a wrap-around metal gate of length $L = 100$ nm. Electrical conduction is via four quantized conductance channels (two electron paths each with spin up and down). What is the maximum characteristic frequency of operation of the device when configured to drive a fan-out of four identical transistors?

Problem 4.18
An electron of eigenenergy E in a one-dimensional potential $V(x)$ has stationary wave function

$$\psi(-L < x < L) = A \left(1 + \cos \left(\frac{\pi x}{L} \right) \right)$$

and

$$\psi(-L \geq x \geq L) = 0.$$

(a) Determine the normalization constant A.

(b) Find the position uncertainty Δx and the momentum uncertainty Δp and show that $\Delta x \Delta p \geq \frac{\hbar}{2}$.

(c) Determine the spatial extent of the classically allowed region.

(d) Find and plot an analytic expression for the potential $V(x)$ and calculate the numerical value of E expressed in eV when $L = 1\,\text{nm}$.

Problem 4.19

(a) Solve the Schrödinger equation and find the normalized bound-state wave functions and corresponding eigenenergies for a single particle of mass m trapped in a two-dimensional rectangular potential well with infinite barrier energy such that

$$V(0 < x < L_x, 0 < y < L_y) = 0$$

and

$$V(0 \geq x \geq L_x, 0 \geq y \geq L_y) = \infty.$$

(b) Assuming that $L_x = L_y = 1\,\text{nm}$ and ignoring degeneracy, plot the ground-state wave function $\psi_0(x, y)$ as well as the first three excited-state wave functions $\psi_1(x, y)$, $\psi_2(x, y)$, and $\psi_3(x, y)$ in the same figure by using MATLAB's "subplot" function. Do the same for the corresponding probability densities, functions $|\psi_0(x, y)|^2$, $|\psi_1(x, y)|^2$, $|\psi_2(x, y)|^2$, and $|\psi_3(x, y)|^2$. Use either "surf" or "imagesc" for the plots and make sure to label everything, including the corresponding bound-state energies in each subplot title (in units of $\hbar^2 \pi^2 / 2mL_x^2$).

(c) In units of $\hbar^2 \pi^2 / 2mL_x^2$, use MATLAB's "imagesc" function to plot the particle's energy as a function of the energy indices n_x and n_y, for the range $1 \leq n_x \leq 100$ and $1 \leq n_x \leq 100$. Generate three separate figures for the cases when $L_x = L_y$, $L_x = 2L_y$, and $L_x = 10L_y$. Comment on the results for each of these geometries.

Problem 4.20

N_{atom} atoms with s-orbital atomic states exist in a one-dimensional domain of length L. A physical model of the system includes periodic boundary conditions and a hopping integral whose value is $-t_{\text{hop}}/|x_i - x_j|^\alpha$ between an atom at position x_i and an atom at position x_j. Take $\alpha = 3$ and $t_{\text{hop}} = 1$.

(a) Write an optimization algorithm in which values of x_i are chosen at random to find an arrangement of $N_{\text{atom}} = 6$ atoms in a domain $0 \leq x < L = 12$ that gives a constant density of states in the energy range $-2 \times t_{\text{hop}} \leq E \leq 2 \times t_{\text{hop}}$. Use energy broadening $\gamma = 0.3 \times |t_{\text{hop}}|$ when evaluating the density of states.

(b) Repeat (a) but with $N_{\text{atom}} = 3$ and comment on what is learned.

(c) Explain the difference in *average* $\mathcal{L}^2$ convergence as a function of iteration number n in (a) and (b). Plot the data using logarithmic scales.

Problem 4.21

N_{atom} atoms with s-orbital atomic states exist in a one-dimensional chain with atom sites at positions x_j with $j = 1, 2, \ldots, N$. Nearest-neighbor electron hopping integral

$t_{hop,1} = -1$ and on-site real potential ν_j is zero except at the end-points of the chain located at x_1 and $x_{N_{atom}}$, respectively.

(a) For $N_{atom} = 2$, find the energy eigenvalues as a function of (i) $\nu_1 = \nu_2 = \nu$ and (ii) $\nu_1 = -\nu_2 = \nu$ with $0 \leq \nu \leq 3.5 \times t_{hop,1}$. Explain the difference in behavior between the results for (i) and (ii).

(b) For $N_{atom} = 7$, plot the real and imaginary parts of the energy eigenvalues as a function of $\nu_1 = \nu_{N_{atom}} = \nu$ with $0 \leq \nu \leq 3.5 \times t_{hop,1}$ in the Hermitian system and explain the emergence and symmetry of the degenerate edge-states.

(c) Compare the density of states for $\nu = 0$ and $\nu = 3.5 \times t_{hop,1}$ in (b) when each state is Lorentzian-broadened in energy by $\gamma = 0.1 \times |t_{hop,1}|$. Discuss why the edge-state with $\nu = 3.5 \times t_{hop,1}$ is expected to have a relatively long lifetime.

(d) Repeat (b) and (c) for the case $\nu = \nu_1 = -\nu_{N_{atom}}$.

(e) Repeat (d) for a non-Hermitian Hamiltonian that has pure imaginary on-site terms $\nu_1 = -\nu_{N_{atom}} = i\nu$ with real $\nu \geq 0$ and explain the exceptional-point behavior observed.

Problem 4.22

(a) Find the energy eigenvalues of a two-atom ($N_{atom} = 2$) non-Hermitian system with on-site terms $\nu_1 = -\nu_{N_{atom}} = i\nu$, interaction strength $0 \leq \nu \leq 3.5 \times t_{hop,1}$, and hopping integral $t_{hop} = -1$. State the value of the exceptional-point interaction strength, ν_{EP}, and show that there is a square-root dependence of energy eigenvalue on ν near ν_{EP}.

(b) Find the energy eigenvalues of a three-atom ($N_{atom} = 3$) non-Hermitian system with on-site terms $\nu_1 = -\nu_{N_{atom}} = i\nu$, interaction strength $0 \leq \nu \leq 3.5 \times t_{hop,1}$, and nearest-neighbor hopping integral $t_{hop,1} = -1$. State the value of ν_{EP} and find the dependence of energy eigenvalue on ν near ν_{EP}.

Problem 4.23

Following the work of Haydock and Kelly,[10] in 1998 Bender and Boettcher[11] suggested that the requirement of hermiticity of a Hamiltonian in quantum mechanics might be unnecessarily strict and could be replaced by a parity-time symmetry such that $\hat{H} = \hat{H}^{\mathcal{PT}}$ in which the parity operator is $\mathcal{P}: \hat{p} \rightarrow -\hat{p}, \hat{x} \rightarrow -\hat{x}$, and the time-reversal operator is $\mathcal{T}: \hat{p} \rightarrow -\hat{p}, i \rightarrow -i$, where $\hat{p}$ and $\hat{x}$ are the momentum and position operators, respectively, and i is the square root of minus one. A class of Hamiltonian with a distinct transition between purely real eigenvalues and real and complex eigenvalues includes

$$\hat{H} = \hat{p}^2 + e^{2ix}$$

and

$$\hat{H} = \hat{p}^2 - (ix)^\alpha,$$

where α is real and the potential is $V_\alpha(x) = -(ix)^\alpha$. For the latter case, symmetry is broken for values of $1 < \alpha < 2$ and there are a finite number of real positive eigenvalues and an infinite number of complex conjugate pairs of eigenvalues. For $\alpha \geq 2$, the eigenvalue spectrum is real and positive. The Hamiltonian reduces to that of a simple harmonic oscillator when $\alpha = 2$.

---

[10]R. Haydock and M. J. Kelly *J. Phys. C: Solid State Physics* **8**, L290 (1975).
[11]C. M. Bender and S. Boettcher, *Phys. Rev. Lett.* **80**, 5243 (1998).

(a) Show that $\mathcal{P} \colon \hat{p} \to -\hat{p}, \hat{x} \to -\hat{x}$, and $\mathcal{T} \colon \hat{p} \to -\hat{p}, \mathrm{i} \to -\mathrm{i}$.
(b) Show that the potential $V_\alpha(x)$ is $\mathcal{PT}$-symmetric.

Problem 4.24

An electron of eigenenergy E in a one-dimensional potential $V(x)$ has stationary continuous wave function

$$\psi(0 < x < L) = \sum_{n-0}^{N} a_n x^n$$

with the constraint

$$\psi(0 \geq x \geq L) = 0.$$

Find the potential $V(x)$ when $N = 2$ and calculate the numerical value of E expressed in eV when $L = 2\,\mathrm{nm}$.

Problem 4.25

Given a potential $V(x)$ with ground state ψ_0 of eigenenergy E_0, consider the shifted potential $V^{(-)}(x) = V(x) - E_0$ so that

$$\hat{H}^{(-)}\psi_0(x) = \left(-\frac{\hbar^2}{2m}\frac{\mathrm{d}^2}{\mathrm{d}x^2} + V^{(-)}(x) \right)\psi_0(x) = 0,$$

from which it follows that

$$V^{(-)}(x) = \frac{\hbar^2}{2m}\frac{1}{\psi_0(x)}\left(\frac{\mathrm{d}^2}{\mathrm{d}x^2}\psi_0(x) \right).$$

For any such potential, a corresponding *supersymmetric partner* $V^{(+)}(x)$ exists with Hamiltonian $\hat{H}^{(+)}$ that shares all the excited state ($n = 1, 2, \dots$) energy eigenvalues (and so is isospectral) with $\hat{H}^{(+)}$ *except* the ground state. It is difficult to distinguish two such potentials simply by measuring their energy eigenvalues. However, introducing a perturbation (such as an external electric field) can change the eigenvalues of $\hat{H}^{(-)}$ relative to $\hat{H}^{(+)}$, thereby making them distinguishable.
   (a) Show that $\hat{H}^{(-)} = \hat{A}^\dagger \hat{A}$ if

$$\hat{A} = \frac{\hbar}{\sqrt{2m}}\left(\frac{\mathrm{d}}{\mathrm{d}x} - \left(\frac{1}{\psi_0(x)}\frac{\mathrm{d}}{\mathrm{d}x}\psi_0(x) \right) \right).$$

   (b) Show that $\hat{H}^{(+)} \equiv \hat{A}\hat{A}^\dagger = -\frac{\hbar^2}{2m}\frac{\mathrm{d}^2}{\mathrm{d}x^2} + V^{(+)}(x)$ if

$$V^{(+)}(x) = -V^{(-)}(x) + \frac{\hbar^2}{m}\left(\frac{1}{\psi_0(x)}\frac{\mathrm{d}^2}{\mathrm{d}x^2}\psi_0(x) \right)^2.$$

(c) Show that if $\psi_{n>0}(x)$ is an eigenstate of $\hat{H}^{(-)}$ then $\hat{A}\psi_{n>0}$ is an eigenstate of $\hat{H}^{(+)}$ with the same eigenvalue. Show that this fails for $n = 0$. Similarly, show that if $\psi_n(x)$ is an eigenstate of $\hat{H}^{(+)}$ then $\hat{A}^\dagger\psi_n$ is an eigenstate of $\hat{H}^{(-)}$ with the same eigenvalue.

(d) An electron in the conduction band of GaAs has effective mass $m_e^* = 0.07 \times m_0$ and is confined by potential $V^{(-)}(0 < x < L) = 0$ with $L = 10\,\text{nm}$ and $V^{(-)}(0 \geq x \geq L) = \infty$. Find the first four lowest-energy eigenvalues and plot the corresponding eigenstates. Numerically determine and plot the supersymmetric partner potential $V^{(+)}$, check that it agrees with the analytic result, find the first four lowest energy eigenvalues, and plot the corresponding eigenstates.

(e) An applied uniform electric field $\mathbf{E}$ adds a perturbing potential $W = e|\mathbf{E}|x$. Plot the first four lowest energy eigenvalues of $\hat{H}^{(-)} + W$ and the first three lowest energy eigenvalues of $\hat{H}^{(+)} + W$ as a function of W. Explain the sensitivity of the solution to the perturbing potential, W.

Problem 4.26
Consider a one-dimensional lattice with lattice constant L that has integer N unit cells. Each unit cell contains two atoms labeled A and B, respectively, with corresponding atomic orbitals ϕ_A and ϕ_B. The tight-binding wave function can be written as

$$\phi_k(x) = \frac{1}{\sqrt{N}} \sum_{n=1}^{N} e^{iknL} (c_A \phi_A(x - nL) + c_B \phi_B(x - nL)),$$

where k is the Bloch wave vector, and c_A and c_B are constants. Multiplying the time-independent Schrödinger equation $\hat{H}|\psi_k\rangle = E|\psi_k\rangle$ from the left by each atomic orbital and integrating over space gives two equations:

$$\langle \phi_A|\hat{H}|\psi_k\rangle = E\langle\phi_A|\psi_k\rangle$$

and

$$\langle \phi_B|\hat{H}|\psi_k\rangle = E\langle\phi_B|\psi_k\rangle.$$

(a) Keeping only on-site and nearest-neighbor terms on the left-hand side of the equations and only the on-site term on the right-hand side of the equations, show that

$$E_A c_A - t_{\text{hop},1} c_B \left(1 + e^{-ikL}\right) = E c_A$$

and

$$-t_{\text{hop},1} c_A \left(1 + e^{ikL}\right) + E_B c_B = E c_B,$$

where on-site energies are $E_A = \langle\phi_A|\hat{H}|\phi_A\rangle$ and $E_B = \langle\phi_B|\hat{H}|\phi_B\rangle$, and overlap integral $t_{\text{hop},1} = -\langle\phi_A(x)|\hat{H}|\phi_B(x)\rangle = -\langle\phi_B(x - L)|\hat{H}|\phi_A(x)\rangle$.

(b) Write the two equations in (a) in matrix form and show that the characteristic polynomial is $E^2 - (E_A + E_B)E + E_A E_B - 2t_{\text{hop},1}^2(1 + \cos(kL)) = 0$ with solution

$$E = \frac{(E_A + E_B) \pm \sqrt{(E_A + E_B)^2 + 8t_{\text{hop},1}^2(1 + \cos(kL))}}{2}.$$

(c) Plot the real and imaginary contributions to the complex band structure in the reduced zone $0 \leq k < \pi/L$ when on-site energy for atom A is $E_A = 0\,\text{eV}$ and for atom B is $E_B = 1.5\,\text{eV}$ and nearest-neighbor hopping energy is $t_{\text{hop},1} = 1\,\text{eV}$.

Problem 4.27

Find an analytic expression for eigenenergies, E_k, of an isolated one-dimensional tight-binding chain of atoms with nearest-neighbor interaction energy $t_{\text{hop},1} = -1$ as a function of number of atoms, $1 \leq N_{\text{atom}} \leq 20$. Explain the results obtained.

5 The Harmonic Oscillator

5.1 The Harmonic Oscillator Potential

In classical mechanics, a particle of mass m subject to a restoring force linear in displacement, x, from a potential minimum such that $F(x) = -\kappa_0 x$, where κ_0 is the force constant, results in one-dimensional simple harmonic motion with an oscillation frequency $\omega = \sqrt{\kappa_0/m}$. The potential the particle moves in is quadratic, $V(x) = \kappa_0 x^2/2$, and so in this case the potential has a minimum at position $x = 0$. The idea that a quadratic potential may be used to describe a local minimum in an otherwise more complex potential is useful in both classical and quantum mechanics. In part this is because a local potential minimum often describes a point of stability and hence a resource for control of a system.

The positions of atoms that form a crystal are stabilized by the presence of a potential that has a local minimum at the location of each lattice site. To understand how the vibrational motion of atoms in a crystal determines properties such as the speed of sound and heat transfer, a model should be developed that describes the oscillatory motion of an atom about a local potential minimum. The same is true to understand the vibrational behavior of atoms in molecules.

As a starting point for the investigation of vibrational properties of atomic systems, a static potential is assumed, and expansion of the potential function taken in a power series about the classically stable minimum in potential experienced by one particular atom at position x_0. In one dimension,

$$V(x) = \sum_{n=0}^{\infty} \frac{1}{n!} \left(\frac{d^n}{dx^n} V(x_0) \right) (x - x_0)^n. \tag{5.1}$$

Assuming that higher-order terms in the polynomial expansion are of decreasing importance, only the first few terms are kept:

$$V(x) = V(x_0) + \left(\frac{d}{dx} V(x_0) \right) (x - x_0) + \frac{1}{2} \left(\frac{d^2}{dx^2} V(x_0) \right) (x - x_0)^2 + \cdots. \tag{5.2}$$

Because the atom position is stabilized by the potential, the potential has a local minimum, so the term in the first derivative of the series expansion about the position x_0 can be set to zero. This leaves

$$V(x) = V(x_0) + \frac{1}{2} \left(\frac{d^2}{dx^2} V(x_0) \right) (x - x_0)^2 + \cdots. \tag{5.3}$$

The first term on the right-hand side of the equation, $V(x_0)$, is a constant, and so it has no impact on the particle dynamics. The second term is just the quadratic potential of a one-dimensional harmonic oscillator for which the force constant is easily identified as a measure of the curvature of the potential about the equilibrium point:

$$\kappa_0 = \frac{\mathrm{d}^2}{\mathrm{d}x^2} V(x_0). \tag{5.4}$$

The importance of the harmonic oscillator in describing the dynamics of a particle in a local potential minimum may now be seen. Very often, a local minimum in potential energy can be approximated by the quadratic function describing a harmonic oscillator potential.

The classical conservative restoring force on a particle mass m attached to a light spring with spring constant κ_0 displaced by x from equilibrium is

$$F(x) = -\kappa_0 x, \tag{5.5}$$

and the scalar potential energy is

$$V(x) = -\int_{x'=0}^{x'=x} F(x')\mathrm{d}x' = \int_{x'=0}^{x'=x} \kappa_0 x'\mathrm{d}x' = \frac{\kappa_0}{2} x^2. \tag{5.6}$$

While it is convenient to visualize the harmonic oscillator in classical terms as illustrated in Fig. 5.1, when dealing with atomic-scale particles it is necessary to solve for the particle motion using quantum mechanics.

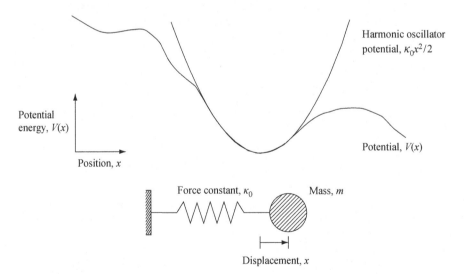

Fig. 5.1 Illustration of a one-dimensional potential with a local minimum that may be approximated by the parabolic potential of a harmonic oscillator. Also shown is a representation of a physical system that has a harmonic oscillator potential for small displacement from equilibrium. The classical system consists of a particle of mass m attached to a light spring with force constant κ_0. The one-dimensional displacement of the particle from its equilibrium position is x.

5.2 Creation and Annihilation Operators

In classical mechanics, a particle of mass m moving in the potential $V(x) = \kappa_0 x^2/2$ oscillates at frequency $\omega = \sqrt{\kappa_0/m}$, where κ_0 is the force constant. The Hamiltonian for this one-dimensional harmonic oscillator consists of kinetic energy and potential energy terms such that

$$H = T + V = \frac{p_x^2}{2m} + \frac{m\omega^2}{2}x^2, \tag{5.7}$$

where the x-directed particle momentum p_x is $m\,\mathrm{d}x/\mathrm{d}t$.

In quantum mechanics, the classical momentum p_x is replaced by the operator $\hat{p}_x = -\mathrm{i}\hbar\,\mathrm{d}/\mathrm{d}x$, so that the Hermitian Hamiltonian operator is

$$\hat{H} = \frac{\hat{p}_x^2}{2m} + \frac{m\omega^2}{2}\hat{x}^2. \tag{5.8}$$

Since the two operators $\hat{p}_x$ and $\hat{x}$ only appear as simple squares, the equation can be factored into two operators that are linear in $\hat{p}_x$ and $\hat{x}$. The new non-Hermitian operators are

$$\hat{b} = \sqrt{\frac{m\omega}{2\hbar}}\left(\hat{x} + \frac{\mathrm{i}\hat{p}_x}{m\omega}\right), \tag{5.9}$$

$$\hat{b}^\dagger = \sqrt{\frac{m\omega}{2\hbar}}\left(\hat{x} - \frac{\mathrm{i}\hat{p}_x}{m\omega}\right), \tag{5.10}$$

so that

$$\hat{x} = \sqrt{\frac{\hbar}{2m\omega}}\left(\hat{b} + \hat{b}^\dagger\right) \tag{5.11}$$

and

$$\hat{p}_x = \mathrm{i}\sqrt{\frac{\hbar m\omega}{2}}\left(\hat{b}^\dagger - \hat{b}\right). \tag{5.12}$$

The Hamiltonian (Eq. (5.8)) expressed in terms of the new operators is (see Problem 5.1)

$$\boxed{\hat{H} = \frac{\hbar\omega}{2}\left(\hat{b}\hat{b}^\dagger + \hat{b}^\dagger\hat{b}\right).} \tag{5.13}$$

The symmetry of this equation can both help simplify problem solving and provide new insight into the quantum-mechanical nature of the harmonic oscillator.

The commutation relations for the operators $\hat{b}^\dagger$ and $\hat{b}$ can be found by writing out the differential form and operating on a dummy wave function. However, this is not necessary if the operators $\hat{b}^\dagger$ and $\hat{b}$ are expressed in terms of the operators $\hat{x}$ and $\hat{p}_x$. For example, to find the commutation relation $\left[\hat{b}, \hat{b}^\dagger\right] = \hat{b}\hat{b}^\dagger - \hat{b}^\dagger\hat{b}$,

$$
\begin{aligned}
[\hat{b}, \hat{b}^\dagger] &= \hat{b}\hat{b}^\dagger - \hat{b}^\dagger\hat{b} \\
&= \left(\frac{m\omega}{2\hbar}\right)\left(\hat{x} + \frac{i\hat{p}_x}{m\omega}\right)\left(\hat{x} - \frac{i\hat{p}_x}{m\omega}\right) - \left(\frac{m\omega}{2\hbar}\right)\left(\hat{x} - \frac{i\hat{p}_x}{m\omega}\right)\left(\hat{x} + \frac{i\hat{p}_x}{m\omega}\right) \\
&= \left(\frac{im\omega}{\hbar}\right)\left(\frac{\hat{p}_x\hat{x}}{m\omega} - \frac{\hat{x}\hat{p}_x}{m\omega}\right) = \frac{i}{\hbar}(\hat{p}_x\hat{x} - \hat{x}\hat{p}_x) = \frac{i}{\hbar}[\hat{p}_x, \hat{x}] = i\frac{(-i\hbar)}{\hbar} = 1,
\end{aligned}
\tag{5.14}
$$

where use is made of the commutation relation $[\hat{p}_x, \hat{x}] = -i\hbar$.

The same process can be used to obtain all the commutation relations for the operators $\hat{b}$ and $\hat{b}^\dagger$. The result is

$$
\boxed{[\hat{b}, \hat{b}^\dagger] = \hat{b}\hat{b}^\dagger - \hat{b}^\dagger\hat{b} = 1,}
\tag{5.15}
$$

$$
\boxed{[\hat{b}^\dagger, \hat{b}] = \hat{b}^\dagger\hat{b} - \hat{b}\hat{b}^\dagger = -1,}
\tag{5.16}
$$

$$
\boxed{[\hat{b}, \hat{b}] = \hat{b}\hat{b} - \hat{b}\hat{b} = 0,}
\tag{5.17}
$$

$$
\boxed{[\hat{b}^\dagger, \hat{b}^\dagger] = \hat{b}^\dagger\hat{b}^\dagger - \hat{b}^\dagger\hat{b}^\dagger = 0.}
\tag{5.18}
$$

Thus, the Hamiltonian given by Eq. (5.13) may be rewritten as

$$
\hat{H} = \frac{\hbar\omega}{2}\left(\hat{b}\hat{b}^\dagger + \hat{b}^\dagger\hat{b}\right) = \frac{\hbar\omega}{2}\left(\hat{b}\hat{b}^\dagger - \hat{b}^\dagger\hat{b} + 2\hat{b}^\dagger\hat{b}\right) = \frac{\hbar\omega}{2}\left(1 + 2\hat{b}^\dagger\hat{b}\right).
\tag{5.19}
$$

Notice that use is made of the fact that $\left[\hat{b}, \hat{b}^\dagger\right] = \hat{b}\hat{b}^\dagger - \hat{b}^\dagger\hat{b} = 1$.

Hence, the Hamiltonian is

$$
\boxed{\hat{H} = \hbar\omega\left(\hat{b}^\dagger\hat{b} + \frac{1}{2}\right).}
\tag{5.20}
$$

The commutation relations, the Hamiltonian, and the constraint that a lowest-energy (ground) state exists *completely* specify the harmonic oscillator in terms of operators.

5.2.1 The Ground State of the Harmonic Oscillator

The ground-state wave function and energy of the one-dimensional harmonic oscillator obey the Schrödinger equation:

$$
\hat{H}\psi_n = \hbar\omega\left(\hat{b}^\dagger\hat{b} + \frac{1}{2}\right)\psi_n = E_n\psi_n.
\tag{5.21}
$$

Multiplying from the *left* by $\hat{b}$ gives

$$
\hbar\omega\left(\hat{b}\hat{b}^\dagger\hat{b} + \frac{\hat{b}}{2}\right)\psi_n = E_n\hat{b}\psi_n.
\tag{5.22}
$$

But $\left[\hat{b}, \hat{b}^\dagger\right] = \hat{b}\hat{b}^\dagger - \hat{b}^\dagger\hat{b} = 1$, so that $\hat{b}\hat{b}^\dagger = 1 + \hat{b}^\dagger\hat{b}$. Hence,

$$\hbar\omega\left(\left(1 + \hat{b}^\dagger\hat{b}\right)\hat{b} + \frac{\hat{b}}{2}\right)\psi_n = E_n\hat{b}\psi_n. \tag{5.23}$$

Factoring out the term $\hat{b}\psi_n$ on the left-hand side gives

$$\hbar\omega\left(\left(1 + \hat{b}^\dagger\hat{b}\right) + \frac{1}{2}\right)\left(\hat{b}\psi_n\right) = E_n\left(\hat{b}\psi_n\right). \tag{5.24}$$

Subtracting the term $\hbar\omega\left(\hat{b}\psi_n\right)$ from both sides results in

$$\hbar\omega\left(\hat{b}^\dagger\hat{b} + \frac{1}{2}\right)\left(\hat{b}\psi_n\right) = \left(E_n - \hbar\omega\right)\left(\hat{b}\psi_n\right), \tag{5.25}$$

$$\hbar\omega\left(\hat{b}^\dagger\hat{b} + \frac{1}{2}\right)\psi_{n-1} = E_{n-1}\psi_{n-1}, \tag{5.26}$$

which shows that $\psi_{n-1} = \left(\hat{b}\psi_n\right)$ is a new eigenfunction with energy eigenvalue $(E_n - \hbar\omega)$. In a similar way, it can be shown that the operator $\hat{b}^\dagger$ acting on eigenfunction ψ_n creates a new eigenfunction ψ_{n+1} with eigenenergy $(E_n + \hbar\omega)$.

Clearly, the operator $\hat{b}$ can only be used to reduce the energy eigenvalue of any eigenstate except the ground state. The existence of a ground state is *assumed*. Because there are, by definition, no eigenstates with energy less than the ground state (long-lived negative energy states are not allowed), the ground state must be defined by

$$\boxed{\hat{b}\psi_0 = 0.} \tag{5.27}$$

This, when combined with Eqs. (5.15)–(5.18) and Eq. (5.20), completes the definition of the harmonic oscillator in terms of the operators $\hat{b}^\dagger$ and $\hat{b}$.

The ground-state wave function requires $\hat{b}\psi_0 = 0$ so that

$$\left(\frac{m\omega}{2\hbar}\right)^{1/2}\left(x + \frac{\hbar}{m\omega}\frac{\mathrm{d}}{\mathrm{d}x}\right)\psi_0 = 0, \tag{5.28}$$

with solution of Gaussian form

$$\boxed{\psi_0 = A_0\mathrm{e}^{-x^2 m\omega/2\hbar},} \tag{5.29}$$

where the normalization constant A_0 is found from the requirement that $\int \psi_0^*\psi_0\mathrm{d}x = 1$. This gives

$$A_0 = \left(\frac{m\omega}{\hbar}\right)^{1/4}. \tag{5.30}$$

Notice that the ground-state wave function ψ_0 for the harmonic oscillator has the even-parity required by symmetry of the potential.

The eigenenergy of the ground state ψ_0 is found by substituting Eq. (5.29) into the Schrödinger equation for the one-dimensional harmonic oscillator:

$$\left(\frac{-\hbar^2}{2m}\frac{d^2}{dx^2} + \frac{m\omega^2}{2}x^2\right)\psi_0 = \left(\frac{-\hbar^2}{2m}\left(-\frac{2m\omega}{2\hbar} + 4x^2\left(\frac{m\omega}{2\hbar}\right)^2\right) + \frac{m\omega^2}{2}x^2\right)\psi_0$$

$$= \left(\frac{\hbar\omega}{2} - \frac{m\omega^2}{2}x^2 + \frac{m\omega^2}{2}x^2\right)\psi_0 = E_0\psi_0, \tag{5.31}$$

so that the value of the ground-state energy, E_0, is

$$\boxed{E_0 = \frac{\hbar\omega}{2}.} \tag{5.32}$$

5.2.1.1 Uncertainty in Position and Momentum for the Harmonic Oscillator in the Ground State

The ground-state wave function ψ_0 given by Eq. (5.29) is of even parity. This symmetry is helpful when evaluating integrals that give the expectation values for position and momentum.

The uncertainty (square root of the variance) in expectation value of position $\Delta x = \sqrt{\langle x^2\rangle - \langle x\rangle^2}$ requires evaluation of the expectation value of observable x associated with the position operator $\hat{x}$ and the expectation value of the observable x^2 associated with the operator $\hat{x}^2$. This is done by expressing the position operator and the position operator squared in terms of $\hat{b}^\dagger$ and $\hat{b}$:

$$\hat{x} = \sqrt{\frac{\hbar}{2m\omega}}\left(\hat{b} + \hat{b}^\dagger\right), \tag{5.33}$$

$$\hat{x}^2 = \left(\frac{\hbar}{2m\omega}\right)\left(\hat{b} + \hat{b}^\dagger\right)^2 = \left(\frac{\hbar}{2m\omega}\right)\left(\hat{b}\hat{b} + \hat{b}^\dagger\hat{b}^\dagger + \hat{b}\hat{b}^\dagger + \hat{b}^\dagger\hat{b}\right). \tag{5.34}$$

The expectation values $\langle x\rangle$ and $\langle x^2\rangle$ for the system in the ground state are easy to evaluate. We have that

$$\langle x\rangle = \int \psi_0^* x\psi_0 dx = 0. \tag{5.35}$$

The fact that $\langle x\rangle = 0$ follows directly from the observation that ψ_0 is an even function and x is an odd function, so the integral must, by symmetry, be zero.

The result for $\langle x^2\rangle$ is almost as straightforward to evaluate. The expectation value in integral form is

$$\langle x^2\rangle = \int \psi_0^* \hat{x}^2\psi_0 dx = \frac{\hbar}{2m\omega}\int \psi_0^*\left(\hat{b}\hat{b} + \hat{b}^\dagger\hat{b}^\dagger + \hat{b}\hat{b}^\dagger + \hat{b}^\dagger\hat{b}\right)\psi_0 dx = \frac{\hbar}{2m\omega}. \tag{5.36}$$

The terms involving $\hat{b}\psi_0$ must, by definition of the ground state, be zero. The term $\hat{b}^\dagger\hat{b}^\dagger\psi_0$ creates a state ψ_2 that is orthogonal to ψ_0^* and so must contribute zero to the integral. The remaining term $\hat{b}\hat{b}^\dagger\psi_0 = \psi_0$, from which $\int \psi_0^*\left(\hat{b}\hat{b}^\dagger\psi_0\right) dx = \int \psi_0^*\psi_0 dx = 1$.

The same approach may be used to evaluate the uncertainty in momentum $\Delta p_x = \sqrt{\langle p_x^2 \rangle - \langle p_x \rangle^2}$. As before, the momentum operator and the momentum operator squared are expressed in terms of $\hat{b}^\dagger$ and $\hat{b}$ so that

$$\hat{p}_x = i\sqrt{\frac{\hbar m\omega}{2}}\left(\hat{b}^\dagger - \hat{b}\right) \tag{5.37}$$

and

$$\hat{p}_x^2 = \left(\frac{\hbar m\omega}{2}\right)\left(-\hat{b}\hat{b} - \hat{b}^\dagger\hat{b}^\dagger + \hat{b}\hat{b}^\dagger + \hat{b}^\dagger\hat{b}\right). \tag{5.38}$$

It follows that

$$\langle p_x \rangle = 0 \tag{5.39}$$

and

$$\left\langle p_x^2 \right\rangle = \frac{\hbar m\omega}{2}. \tag{5.40}$$

Because $\langle x \rangle^2 = 0$ and $\langle p_x \rangle^2 = 0$, the uncertainty product has the value

$$\Delta x^2 \Delta p_x^2 = \left\langle x^2 \right\rangle \left\langle p_x^2 \right\rangle = \frac{\hbar^2 m\omega}{4m\omega} = \frac{\hbar^2}{4}. \tag{5.41}$$

Taking the square root of both sides gives the uncertainty product of standard deviation in position and momentum,

$$\Delta x \Delta p_x = \frac{\hbar}{2}, \tag{5.42}$$

which satisfies the uncertainty relation $\Delta p \Delta x \geq \hbar/2$.

The sum of kinetic and potential energy using Eq. (5.36) and Eq. (5.40) gives a ground-state energy of the harmonic oscillator that is $E_0 = \hbar\omega/2$. The important physical interpretation of this result is that, according to the uncertainty relation, the ground state energy represents a minimum uncertainty in the product of position and momentum.

In contrast, the lowest energy of a *classical* harmonic oscillator is *zero*. In the classical case, the minimum energy of a particle in the harmonic potential $V(x) = \kappa_0 x^2/2$ corresponds to both momentum and position simultaneously being zero. In quantum mechanics, this is impossible, since $\Delta x \Delta p_x \geq \hbar/2$.

5.2.2 Excited States of the Harmonic Oscillator and Eigenstate Normalization

If the ground state ψ_0 is known, then all other *excited states* ψ_n can be generated using the creation (or raising) operator $\hat{b}^\dagger$. To see that this is the case, the operator $\hat{b}^\dagger$ is applied to the Schrödinger equation describing the harmonic oscillator:

$$\hat{b}^\dagger \hat{H} \psi_n = \hbar\omega\left(\hat{b}^\dagger \hat{b}^\dagger \hat{b} + \frac{\hat{b}^\dagger}{2}\right)\psi_n = E_n \hat{b}^\dagger \psi_n. \tag{5.43}$$

Making use of the commutation relation $\hat{b}\hat{b}^\dagger - \hat{b}^\dagger\hat{b} = 1$ and substituting for $\hat{b}^\dagger\hat{b} = \hat{b}\hat{b}^\dagger - 1$,

$$\hbar\omega\left(\hat{b}^\dagger\left(\hat{b}\hat{b}^\dagger - 1\right) + \frac{\hat{b}^\dagger}{2}\right)\psi_n = \hbar\omega\left(\left(\hat{b}^\dagger\hat{b}\hat{b}^\dagger - \hat{b}^\dagger\right) + \frac{\hat{b}^\dagger}{2}\right)\psi_n = \hbar\omega\left(\left(\hat{b}^\dagger\hat{b} - 1\right) + \frac{1}{2}\right)\hat{b}^\dagger\psi_n$$

$$= E_n\hat{b}^\dagger\psi_n, \tag{5.44}$$

so that

$$\hbar\omega\left(\hat{b}^\dagger\hat{b} + \frac{1}{2}\right)\left(\hat{b}^\dagger\psi_n\right) = \left(E_n + \hbar\omega\right)\left(\hat{b}^\dagger\psi_n\right). \tag{5.45}$$

This shows that the operator $\hat{b}^\dagger$, acting on the eigenstate ψ_n, generates a new eigenstate $\left(\hat{b}^\dagger\psi_n\right)$ with energy eigenvalue $(E_n + \hbar\omega)$.

It is now clear that $\hat{b}^\dagger$, operating on ψ_n, increases the eigenenergy by an amount $\hbar\omega$, so that the eigenenergy for the nth state is

$$\boxed{E_n = \hbar\omega\left(n + \frac{1}{2}\right),} \tag{5.46}$$

where n is a positive integer $n = 0, 1, 2, \ldots$. Because the time-independent Schrödinger equation for the harmonic oscillator is

$$\hat{H}\psi_n = \hbar\omega\left(\hat{b}^\dagger\hat{b} + \frac{1}{2}\right)\psi_n = \hbar\omega\left(n + \frac{1}{2}\right)\psi_n = E_n\psi_n, \tag{5.47}$$

the number operator may be identified as

$$\hat{n} = \hat{b}^\dagger\hat{b}, \tag{5.48}$$

which, when operating on the eigenstate ψ_n, has eigenvalue n. See Section 5.2.2.2.

Here, $\hat{b}^\dagger$ and $\hat{b}$ are the creation (or raising) and annihilation (or lowering) operators, respectively, that act upon the state ψ_n in such a way that

$$\hat{b}^\dagger\psi_n = A_{n+1}\psi_{n+1} \tag{5.49}$$

and

$$\hat{b}\psi_n = A_{n-1}\psi_{n-1}, \tag{5.50}$$

where A_{n+1} and A_{n-1} are normalization constants. Assuming that the nth bound state is normalized,

$$\int \psi_n^*\psi_n dx = \langle n|n\rangle = 1. \tag{5.51}$$

Because the state $\hat{b}^\dagger\psi_n = |n+1\rangle$ is also required to be normalized,

$$|A_{n+1}|^2\left\langle \hat{b}^\dagger n\middle|\hat{b}^\dagger n\right\rangle = 1, \tag{5.52}$$

where A_{n+1} is the normalization constant to be found. Eq. (5.52) may be written

$$|A_{n+1}|^2 \left\langle n|\hat{b}\hat{b}^\dagger n \right\rangle = |A_{n+1}|^2 \left\langle n\left| \left(\hat{b}^\dagger \hat{b} + 1\right) n \right\rangle = |A_{n+1}|^2 (n+1)\langle n|n\rangle = 1, \tag{5.53}$$

where the commutation relation given by Eq. (5.16) (so that $\hat{b}\hat{b}^\dagger = \hat{b}^\dagger \hat{b} + 1$) is used and the fact that $|n\rangle$ is an eigenfunction of the number operator $\hat{n} = \hat{b}^\dagger \hat{b}$ with eigenvalue n. Hence,

$$|A_{n+1}|^2 = \frac{1}{n+1}. \tag{5.54}$$

Choosing A_{n+1} to be real, then

$$A_{n+1} = \frac{1}{\sqrt{n+1}} \tag{5.55}$$

or

$$|n+1\rangle = \frac{1}{\sqrt{n+1}} \left|\hat{b}^\dagger n \right\rangle. \tag{5.56}$$

Using this approach, it may be concluded that

$$\boxed{\left|\hat{b}^\dagger n\right\rangle = \sqrt{n+1}\,|n+1\rangle} \tag{5.57}$$

and

$$\boxed{\left|\hat{b}n\right\rangle = \sqrt{n}\,|n-1\rangle.} \tag{5.58}$$

If the ground state is normalized such that $\langle n = 0|n = 0\rangle = 1$, then a generating function for the normalized state $|n\rangle$ is

$$\boxed{|n\rangle = \frac{\left(\hat{b}^\dagger\right)^n}{\sqrt{n!}}|0\rangle} \tag{5.59}$$

and

$$\hat{b}^\dagger |n\rangle = \frac{\left(\hat{b}^\dagger\right)^{n+1}}{\sqrt{n!}}|0\rangle = \frac{\sqrt{n+1}\left(\hat{b}^\dagger\right)^{n+1}}{\sqrt{(n+1)!}}|0\rangle = \sqrt{n+1}\,|n+1\rangle. \tag{5.60}$$

5.2.2.1 Matrix Elements

The eigenstates $\psi_n = |n\rangle$ of the harmonic oscillator are orthonormal, so that

$$\langle n'|n\rangle = \delta_{n',n}. \tag{5.61}$$

In the notation used, $\langle n'|\hat{b}^\dagger|n\rangle = \int \psi_{n'}^* \hat{b}^\dagger \psi_n dx$ is a matrix element. It can be shown that the matrix elements involving $\hat{b}^\dagger$ and $\hat{b}$ only exist between adjacent states, so that

$$\langle n'|\hat{b}^\dagger|n\rangle = \sqrt{n+1}\, \delta_{n'=n+1} \tag{5.62}$$

and

$$\langle n'|\hat{b}|n\rangle = \sqrt{n}\, \delta_{n'=n-1}. \tag{5.63}$$

5.2.2.2 The Number Operator $\hat{n}$

It is convenient to define a *number* operator such that

$$\boxed{\hat{n} = \hat{b}^\dagger \hat{b}.} \tag{5.64}$$

The eigenvalue of the operator $\hat{n}$ applied to an eigenstate labeled by quantum number n is just n:

$$\hat{b}^\dagger \hat{b}|n\rangle = \hat{b}^\dagger \sqrt{n}\,|n-1\rangle = \sqrt{n}\,\hat{b}^\dagger|n-1\rangle = \sqrt{n}\sqrt{n-1+1}\,|n\rangle = n|n\rangle. \tag{5.65}$$

The operator $\hat{n} = \hat{b}^\dagger \hat{b}$ commutes with $\hat{b}$ and $\hat{b}^\dagger$ in the following way:

$$\left[\hat{n},\hat{b}\right] = \left[\hat{b}^\dagger\hat{b},\hat{b}\right] = \hat{b}^\dagger\left[\hat{b},\hat{b}\right] + \left[\hat{b}^\dagger,\hat{b}\right]\hat{b} \tag{5.66}$$

$$\left[\hat{n},\hat{b}^\dagger\right] = \left[\hat{b}^\dagger\hat{b},\hat{b}^\dagger\right] = \hat{b}^\dagger\left[\hat{b},\hat{b}^\dagger\right] + \left[\hat{b}^\dagger,\hat{b}^\dagger\right]\hat{b} \tag{5.67}$$

However, since $\left[\hat{b},\hat{b}\right] = 0$, $\left[\hat{b},\hat{b}^\dagger\right] = 1$, and $\left[\hat{b}^\dagger,\hat{b}\right] = -1$, it follows that

$$\left[\hat{n},\hat{b}\right] = -\hat{b} \tag{5.68}$$

and

$$\left[\hat{n},\hat{b}^\dagger\right] = \hat{b}^\dagger. \tag{5.69}$$

Since the Hamiltonian operator for the harmonic oscillator is $\hat{H} = \hbar\omega(\hat{n} + 1/2)$, the eigenfunctions of the Hamiltonian $\hat{H}$ are also eigenfunctions of the number operator $\hat{n}$.

In Fig. 5.2, the ground-state energy level and excited-state energy levels near the nth state of the one-dimensional harmonic oscillator are shown schematically. Transition between eigenstates of neighboring energy is achieved by applying the operators $\hat{b}^\dagger$ or $\hat{b}$ to a given eigenstate. The energy of an eigenstate is $\hbar\omega(n+1/2)$, and the value of n is found by applying the operator $\hat{b}^\dagger\hat{b} = \hat{n}$ to the eigenstate. From Eqs. (5.15) and (5.64) it follows that $\hat{b}\hat{b}^\dagger = \hat{n} + 1$. The ground state ψ_0 is defined by $\hat{b}\psi_0 = 0$.

Classical simple harmonic oscillation occurs in a single mode of frequency ω and the vibrational energy can be changed continuously by varying the oscillation amplitude. The quantum-mechanical oscillator also has a single oscillatory mode characterized by frequency ω but the vibrational energy is quantized in such a way that $E_n = \hbar\omega(n + 1/2)$.

The manipulation of operators is similar to ordinary algebra, with the obvious exception that the order of operators must be accurately maintained. There is another important

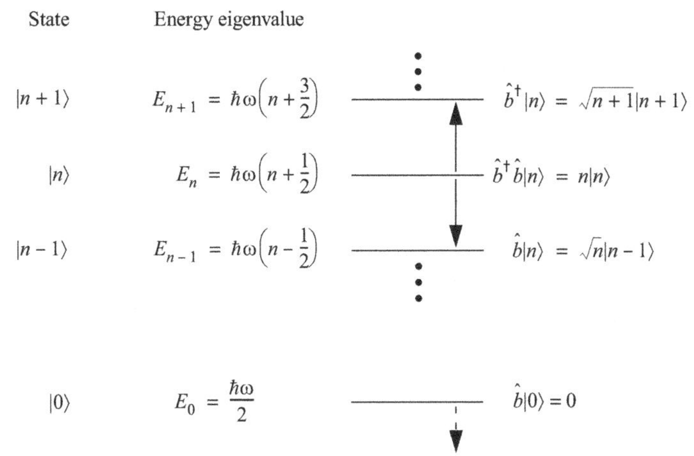

State	Energy eigenvalue	

$$\vdots$$

$|n+1\rangle$ $\quad E_{n+1} = \hbar\omega\left(n + \frac{3}{2}\right)$ ———— $\hat{b}^\dagger|n\rangle = \sqrt{n+1}|n+1\rangle$

$|n\rangle$ $\quad E_n = \hbar\omega\left(n + \frac{1}{2}\right)$ ———— $\hat{b}^\dagger \hat{b}|n\rangle = n|n\rangle$

$|n-1\rangle$ $\quad E_{n-1} = \hbar\omega\left(n - \frac{1}{2}\right)$ ———— $\hat{b}|n\rangle = \sqrt{n}|n-1\rangle$

$$\vdots$$

$|0\rangle$ $\quad E_0 = \dfrac{\hbar\omega}{2}$ ———— $\hat{b}|0\rangle = 0$

Fig. 5.2 Diagram showing the equally spaced energy levels of the one-dimensional harmonic oscillator. The raising or creation operator $\hat{b}^\dagger$ acts upon eigenstate $|n\rangle$ with eigenenergy E_n to form a new eigenstate $|n+1\rangle$ with eigenenergy E_{n+1}. In a similar way, the annihilation operator $\hat{b}$ acts upon eigenstate $|n\rangle$ with eigenenergy E_n to form a new eigenstate $|n-1\rangle$ with eigenenergy E_{n-1}. Energy levels are equally spaced in energy by $\hbar\omega$. The ground state $|0\rangle$ of the harmonic oscillator is the single state for which $\hat{b}|0\rangle = 0$. The ground-state energy is $\hbar\omega/2$.

rule. One must not divide by an operator $\hat{b}$. To show this, consider the state formed by $\psi = \hat{b}\psi_0$. If both sides are divided by $\hat{b}$, then

$$\frac{1}{\hat{b}}\psi = \psi_0. \qquad (5.70)$$

If ψ_0 ($\neq 0$) is the ground state then $\psi = \hat{b}\psi_0 = 0$ by the definition of the ground state (Eq. (5.27)). However, Eq. (5.70) has $(1/\hat{b})\psi = \psi_0 \neq 0$ and so is inconsistent with the definition of the ground state. While multiplication by $1/\hat{b}$ is not allowed, it is possible to apply an operator of the form $1/(c_b + \hat{b})$ where c_b is a constant and the operator may be expanded as a power series in $\hat{b}$.

5.2.2.3 Uncertainty in Occupation Number and Phase of the Harmonic Oscillator

For occupation number $n > 0$, a phase operator $\hat{\theta}$ with associated eigenstate $|\theta\rangle$ and eigenvalue θ can be defined such that

$$e^{i\hat{\theta}} = \frac{\hat{b}}{\sqrt{n}}. \qquad (5.71)$$

For $n > 0$, and making use of the fact that $\left[\hat{b}^\dagger, \hat{b}\right] = -1$, the commutator

$$\left[\hat{n}, e^{i\hat{\theta}}\right] = \left[\hat{n}, \frac{\hat{b}}{\sqrt{n}}\right] = \frac{1}{\sqrt{n}}\left[\hat{b}^\dagger \hat{b}, \hat{b}\right] = \frac{1}{\sqrt{n}}\left[\hat{b}^\dagger, \hat{b}\right]\hat{b} = \frac{-1}{\sqrt{n}}\hat{b} = -e^{i\hat{\theta}}, \qquad (5.72)$$

so that the operator

$$\hat{n} = i\frac{\partial}{\partial\hat{\theta}}, \tag{5.73}$$

for which

$$\left[\hat{n}, e^{i\hat{\theta}}\right] = \hat{n}e^{i\hat{\theta}} - e^{i\hat{\theta}}\hat{n} = -e^{i\hat{\theta}} + ie^{i\hat{\theta}}\frac{\partial}{\partial\hat{\theta}} - ie^{i\hat{\theta}}\frac{\partial}{\partial\hat{\theta}} = -e^{i\hat{\theta}}. \tag{5.74}$$

The commutation relation

$$\left[\hat{n}, \hat{\theta}\right] = i\frac{\partial}{\partial\hat{\theta}}\hat{\theta} - i\hat{\theta}\frac{\partial}{\partial\hat{\theta}} = i + i\hat{\theta}\frac{\partial}{\partial\hat{\theta}} - i\hat{\theta}\frac{\partial}{\partial\hat{\theta}} = i, \tag{5.75}$$

and using the generalized uncertainty relation $\Delta A \Delta B \geq \left|\frac{1}{2}\left\langle\left[\hat{A}, \hat{B}\right]\right\rangle\right|$ for operators $\hat{A}$ and $\hat{B}$ results in an uncertainty relation for occupation number and phase that is

$$\Delta n \Delta\theta \geq \left|\frac{1}{2}\left\langle\left[\hat{n}, \hat{\theta}\right]\right\rangle\right| = \frac{1}{2} \tag{5.76}$$

when $n > 0$.

5.3 The Harmonic Oscillator Wave Functions

Previously, expressions for the creation or raising operator $\hat{b}^{\dagger}$ and the normalized ground-state wave function ψ_0 for the one-dimensional harmonic oscillator were derived so that

$$\hat{b}^{\dagger} = \sqrt{\frac{m\omega}{2\hbar}}\left(x - \frac{\hbar}{m\omega}\frac{d}{dx}\right) \tag{5.77}$$

and

$$\psi_0(x) = \left(\frac{m\omega}{\hbar}\right)^{1/4}\exp\left(-x^2\frac{m\omega}{2\hbar}\right). \tag{5.78}$$

To simplify the notation, it is convenient to introduce a new spatial variable,

$$\xi = \sqrt{\frac{m\omega}{\hbar}}\, x. \tag{5.79}$$

Equation (5.77) may now be written as

$$\hat{b}^{\dagger} = \frac{1}{\sqrt{2}}\left(\sqrt{\frac{m\omega}{\hbar}}x - \sqrt{\frac{\hbar}{m\omega}}\frac{d}{dx}\right) = \frac{1}{\sqrt{2}}\left(\xi - \frac{d}{d\xi}\right), \tag{5.80}$$

and Eq. (5.78) for the normalized ground-state wave function becomes

$$\psi_0(\xi) = \left(\frac{1}{\pi}\right)^{1/4}\exp\left(-\xi^2/2\right). \tag{5.81}$$

The other higher-order states can be generated by using the operator $\hat{b}^\dagger$. Starting with the ground state and using Eq. (5.59) to ensure correct normalization, a natural sequence of wave functions is created:

$$\psi_0, \tag{5.82}$$

$$\psi_1 = \hat{b}^\dagger \psi_0, \tag{5.83}$$

$$\psi_2 = \frac{1}{\sqrt{2}} \hat{b}^\dagger \psi_1 = \frac{1}{\sqrt{2}} \left(\hat{b}^\dagger \right)^2 \psi_0, \tag{5.84}$$

$$\psi_3 = \frac{1}{\sqrt{3}} \hat{b}^\dagger \psi_2 = \frac{1}{\sqrt{2}\sqrt{3}} \left(\hat{b}^\dagger \right)^2 \psi_1 = \frac{1}{\sqrt{3!}} \left(\hat{b}^\dagger \right)^3 \psi_0, \tag{5.85}$$

$$\psi_n = \frac{1}{\sqrt{n!}} \left(\hat{b}^\dagger \right)^n \psi_0. \tag{5.86}$$

Because the ground-state wave function (Eq. (5.81)) is known, it is now possible to generate all the other excited states of the system. The first few states are

$$\psi_0 = \left(\frac{1}{\pi} \right)^{1/4} \exp(-\xi^2/2), \tag{5.87}$$

$$\psi_1 = \hat{b}^\dagger \psi_0 = \frac{1}{\sqrt{1!}} \frac{1}{\sqrt{2}} \left(\xi - \frac{d}{d\xi} \right) \psi_0 = \frac{1}{\sqrt{2}} 2\xi \psi_0, \tag{5.88}$$

$$\psi_2 = \hat{b}^\dagger \psi_1 = \frac{1}{\sqrt{2!}} \left(\hat{b}^\dagger \right)^2 \psi_0 = \frac{1}{\sqrt{2!}} \frac{1}{\sqrt{2}} \left(\xi - \frac{d}{d\xi} \right) \frac{1}{\sqrt{2}} 2\xi \psi_0 = \frac{1}{\sqrt{2}} \frac{1}{\sqrt{4}} \left(4\xi^2 - 2 \right) \psi_0, \tag{5.89}$$

$$\psi_3 = \hat{b}^\dagger \psi_2 = \frac{1}{\sqrt{3!}} \left(\hat{b}^\dagger \right)^3 \psi_0 = \frac{1}{\sqrt{6}} \frac{1}{\sqrt{8}} \left(8\xi^3 - 12\xi \right) \psi_0, \tag{5.90}$$

$$\psi_4 = \hat{b}^\dagger \psi_3 = \frac{1}{\sqrt{4!}} \left(\hat{b}^\dagger \right)^4 \psi_0 = \frac{1}{\sqrt{24}} \frac{1}{\sqrt{16}} \left(16\xi^4 - 48\xi^2 + 12 \right) \psi_0, \tag{5.91}$$

$$\psi_5 = \hat{b}^\dagger \psi_4 = \frac{1}{\sqrt{5!}} \left(\hat{b}^\dagger \right)^5 \psi_0 = \frac{1}{\sqrt{120}} \frac{1}{\sqrt{32}} \left(32\xi^5 - 160\xi^3 + 120\xi \right) \psi_0. \tag{5.92}$$

As required by symmetry of the potential, the wave functions are alternately even and odd functions.

It is clear from Eqs. (5.87)–(5.92) that there is a relationship between the wave functions that can be expressed as a Hermite polynomial $H_n(\xi)$ so that

$$\psi_n(\xi) = \hat{b}^\dagger \psi_{n-1}(\xi) = \frac{1}{\sqrt{2^n n!}} H_n(\xi) \psi_0(\xi). \tag{5.93}$$

The nth Hermite polynomial is related to the $n-1$ and $n-2$ polynomial via

$$H_n(\xi) = 2\xi H_{n-1}(\xi) - 2(n-1) H_{n-2}(\xi). \tag{5.94}$$

The Hermite polynomials themselves may be obtained from the generating function

$$\exp\left(-z^2 + 2z\xi \right) = \sum_{n=0}^{\infty} \frac{H_n(\xi)}{n!} z^n \tag{5.95}$$

or

$$H_n(\xi) = \frac{d^n}{dz^n} \exp^{-z^2 + 2z\xi}\bigg|_{z=0} = (-1)^n \exp\left(\xi^2\right) \frac{d^n}{d\xi^n} \exp\left(\xi^{-2}\right). \tag{5.96}$$

The first few Hermite polynomials are

$$H_0(\xi) = 1,$$
$$H_1(\xi) = 2\xi,$$
$$H_2(\xi) = 4\xi^2 - 2,$$
$$H_3(\xi) = 8\xi^3 - 12\xi,$$
$$H_4(\xi) = 16\xi^4 - 48\xi^2 + 12,$$
$$H_5(\xi) = 32\xi^5 - 160\xi^3 + 120\xi.$$

The Schrödinger equation for the one-dimensional harmonic oscillator can be written in terms of the variable ξ to give

$$\left(\frac{d^2}{d\xi^2} + \left(\frac{2E}{\hbar\omega} - \xi^2\right)\right)\psi_n(\xi) = 0. \tag{5.97}$$

The solutions are the Hermite–Gaussian functions

$$\psi_n(\xi) = \sqrt{\frac{1}{\sqrt{\pi}2^n n!}} H_n(\xi) \exp\left(-\xi^2/2\right) = \frac{1}{\sqrt{2^n n!}} H_n(\xi)\psi_0(\xi), \tag{5.98}$$

where $H_n(\xi)$ are Hermite polynomials. These satisfy the differential equation

$$\left(\frac{d^2}{d\xi^2} - 2\xi\frac{d}{d\xi} + 2n\right) H_n(\xi) = 0, \tag{5.99}$$

and n is related to the energy E_n by

$$E_n = \left(n + \frac{1}{2}\right)\hbar\omega, \tag{5.100}$$

where $n = 0, 1, 2, \ldots$, Alternately, if the two starting functions ψ_0 and ψ_1 are known, then the nth wave function can be generated by using

$$\psi_n(\xi) = \sqrt{\frac{2}{n}}\left(\xi\psi_{n-1}(\xi) - \sqrt{\frac{n-1}{2}}\psi_{n-2}(\xi)\right). \tag{5.101}$$

In Fig. 5.3, the wave function and probability function for the three lowest-energy states of the one-dimensional harmonic oscillator are plotted. The wave functions $\psi_n(\xi)$ form a complete orthogonal set. So it may be concluded that the eigenvalues given by Eq. (5.46) and the eigenfunctions $\psi_n(\xi)$ are the only solutions of the Hamiltonian describing the harmonic oscillator.

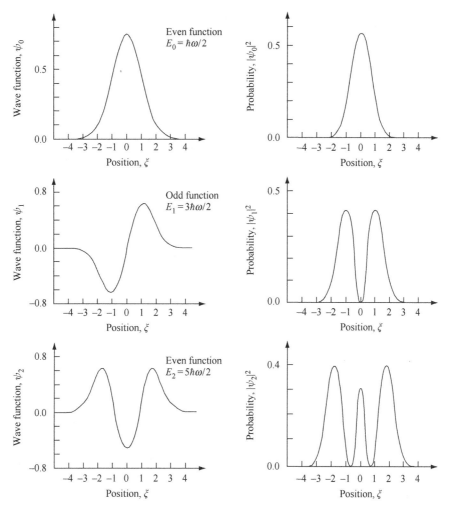

Fig. 5.3 Plot of wave function and probability function for the three lowest-energy states of the one-dimensional harmonic oscillator. Position is measured in normalized units of $\xi = x(m\omega/\hbar)^{1/2}$.

5.4 Time Dependence

The probability distribution of bound-state eigenfunctions is time independent and cannot carry flux or current. If the particle is in an eigenstate labeled by the positive integer quantum number n, then $\psi_n(x, t) = \psi_n(x)e^{-i\omega_n t}$, and no current flows because $|\psi_n(x, t)|^2 = |\psi_n(x)|^2$ is time independent. However, if the particle is in a linear superposition of eigenstates,

$$\psi(x, t) = \sum_n a_n \psi_n(x)e^{-i\omega_n t}, \tag{5.102}$$

239

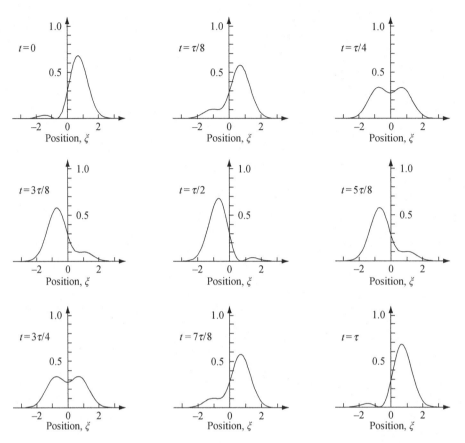

Fig. 5.4 Probability distribution at the indicated times t of a one-dimensional harmonic oscillator superposition state. In this particular case, the superposition state consists of the ground state and the first excited state, with equal weights and the same initial phase. The probability distribution oscillates with a period $\tau = 2\pi/\omega$. Position is normalized to units of $\xi = x\sqrt{m\omega/\hbar}$.

where a_n is a coefficient that weights each eigenfunction, then there can be a time dependence to the spatial probability distribution of the particle.

The total wave function describing the particle in a linear superposition of the ground state and first excited-state eigenfunctions, $\psi_0(x,t)$ and $\psi_1(x,t)$, respectively, is

$$\psi(x,t) = a_0\psi_0(x)e^{-i\omega t/2} + a_1\psi_1(x)e^{-i3\omega t/2}. \tag{5.103}$$

The energy of the nth eigenstate is $E_n = \hbar\omega(n+1/2)$, where $\hbar\omega$ is the energy eigenvalue difference between adjacent eigenstates. The coefficients a_0 and a_1 could be real or complex. Complex coefficients can be viewed as adding an initial phase to the eigenstate.

The probability density distribution for the superposition state $\psi(x,t)$ given by Eq. (5.103) is

$$|\psi(x,t)|^2 = a_0^2|\psi_0(x)|^2 + a_1^2\,|\psi_1(x)|^2 + 2a_0a_1\,\mathrm{Re}(\psi_0(x)\psi_1^*(x))\cos(\omega t), \tag{5.104}$$

where a_0 and a_1 have been assumed real and $\hbar\omega = E_1 - E_0$. For the special case in which $a_0 = a_1$, the expression for the probability density distribution is

$$|\psi(x,t)|^2 = a_0^2 \left(|\psi_0(x)|^2 + |\psi_1(x)|^2 + 2\mathrm{Re}\left(\psi_0(x)\psi_1^*(x)\right)\cos(\omega t) \right). \qquad (5.105)$$

The probability distribution oscillates with frequency ω, which is the difference in energy eigenvalue between the two eigenstates. There is a coherent superposition of ψ_0 and ψ_1, with equal weight in each state giving a total wave function probability distribution and an expectation value of position that oscillates at frequency ω. In general, unless the system is in a pure eigenstate, the probability distribution will be a function of time.

The time evolution of Eq. (5.105) is illustrated in Fig. 5.4. The superposition state $\psi(x,t)$ consists of the ground state and the first excited state, with equal weights and the same initial phase. In this case, the probability distribution $|\psi(x,t)|^2$ for the superposition state oscillates at frequency ω.

It is apparent from Eq. (5.105) and Fig. 5.4 that a superposition of harmonic oscillator eigenstates can be used to create a spatial oscillation in the probability distribution function. If the probability distribution function describes a charged particle, such as an electron, then the oscillation may give rise to a dipole moment and hence to a source of dipole radiation.

5.4.1 The Superposition Operator

It is possible to form an operator that creates a single particle superposition state. This is best shown by example. Suppose a superposition state consisting of the ground state of the harmonic oscillator and the first excited state is to be created so that the total wave function is

$$\psi = a_0\psi_0 + a_1\psi_1, \qquad (5.106)$$

where a_0 and a_1 are weights on each eigenfunction. This equation may always be written as

$$\psi = \left(a_0 + a_1\hat{b}^\dagger\right)\psi_0 = a_1\left(\frac{a_0}{a_1} + \hat{b}^\dagger\right)\psi_0 = a_1\left(c_\mathrm{b} + \hat{b}^\dagger\right)\psi_0. \qquad (5.107)$$

This shows that by addition of a number c_b to the operator $\hat{b}^\dagger$ a new operator of the form $\left(c_\mathrm{b} + \hat{b}^\dagger\right)$ results that acts to create a superposition state.

5.4.2 Superposition State, Measurement, and Correlations

A superposition state of the one-dimensional harmonic oscillator can dramatically alter the time dependence of probability density distributions and the time dependence of expectation values. If the particle described by the superposition state is charged, then measurable effects such as dipole radiation may result.

However, a single-particle superposition state cannot be measured *directly*. When a measurement is performed on the single-particle system, the only result possible is an eigenvalue corresponding to a single eigenfunction. These are the only long-lived (stationary) states of the system that can be measured. In the case of a state consisting of a linear superposition of eigenstates, there is no way to know beforehand what the result of measurement will be. After the measurement has been performed, the state of the system remains in the measured eigenstate. This phenomenon, the *collapse of the wave function*, is *noncausal* since the result of eigenstate measurement cannot be known beforehand. The process is probabilistic and fundamentally random. This mysterious behavior has to do with the nature of measurement, something that quantum mechanics does not describe very well. Measurement disturbs the system, forcing an initially coherent superposition of eigenstates into a definite stationary eigenstate. A probability distribution or an expectation value is created from a number of discrete events. Only after the measurement has been performed a large number of times, and each time on an identically prepared system, do the cumulative results asymptotically approach the predictions of the probability distribution or expectation value.

Superposition states do not always have a time dependence. Consider an isotropic two-dimensional harmonic oscillator with Hamiltonian $\hat{H} = \hbar\omega(\hat{n}_x + \hat{n}_y + 1)$ describing a particle of mass m with Cartesian position coordinates x and y. The number operators $\hat{n}_x$ and $\hat{n}_y$ have nonnegative integer eigenvalues n_x and n_y, respectively, with $n = n_x + n_y$. The eigenstates, $\psi_{n_x n_y}(x, y)$, are separable and may be written as a product of one-dimensional harmonic oscillator states such that $\psi_{n_x n_y}(x, y) = \phi_{n_x}(x)\phi_{n_y}(y)$. The energy eigenvalues of the pure state $\psi_{n_x n_y}(x, y)$ are $E_{n_x n_y}$ and have degeneracy $n+1$ so that when $n = 1$ then $E_{01} = E_{10}$. The $n = 1$ stationary eigenstate $\psi_{01}(x, y) = \phi_0(x)\phi_1(y)$ with eigenenergy E_{01} is plotted in Fig. 5.5(a). Because the system is separable and the eigenstate is a product of one-dimensional harmonic oscillator states $\phi_{n_x}(x)$ and $\phi_{n_y}(y)$, the results of measurement as a function of coordinate x are *independent* of y.

The situation is different when considering a superposition state such as the stationary state $\psi(x, y) = a_0 \left(\phi_0(x)\phi_1(y) + \frac{2}{3}\phi_1(x)\phi_0(y)\right)$, where a_0 is a normalization constant. This state, which is shown in Fig. 5.5(b), is not separable and so the results of measurement as a function of x are *not* independent of y.

For a non-separable superposition state, such as that illustrated in Fig. 5.5(b), there is a correlation between sequential measurement of particle position x followed by y.

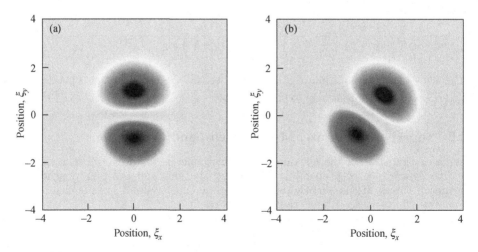

Fig. 5.5 (a) Separable $n = 1$ eigenstate $\psi_{01}(x, y) = \phi_0(x)\phi_1(y)$ of the isotropic two-dimensional harmonic oscillator. (b) A stationary superposition state $\psi(x, y) = a_0 \left(\phi_0(x)\phi_1(y) + \frac{2}{3}\phi_1(x)\phi_0(y)\right)$ that is not separable. Position is normalized to units of $\xi_x = x\sqrt{m\omega/\hbar}$ and $\xi_y = y\sqrt{m\omega/\hbar}$.

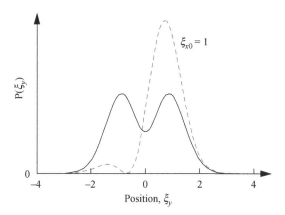

Fig. 5.6 Probability $P(y) = \int_\infty^\infty |\psi(x, y)|^2 \mathrm{d}x$ for the state shown in Fig. 5.5(b) before measuring the x position (solid curve). Position is normalized to units of $\xi_x = x\sqrt{m\omega/\hbar}$ and $\xi_y = y\sqrt{m\omega/\hbar}$. If the x position is measured first with the result $\xi_{x0} = 1$, where $\xi_{x0} = x_0\sqrt{m\omega/\hbar}$, then the probability of finding the particle at position y is $P(y)|_{x_0} = |\psi(x_0, y)|^2$ (dashed curve).

To illustrate this, Fig. 5.6 plots $P(y) = \int_\infty^\infty |\psi(x, y)|^2 \mathrm{d}x$ (solid curve), which is the probability of finding the particle at position y before measuring the particle at position x. Also shown is $P(y)|_{x_0} = |\psi(x_0, y)|^2$ (dashed curve), which is the probability of finding the particle at position y after measuring the x position of the particle and obtaining, in this particular case, a value $\xi_{x0} = 1$ where $\xi_{x0} = x_0\sqrt{m\omega/\hbar}$.

Since, in this example, the particle is found to be at position $\xi_{x0} = 1$ then, it turns out, it is most probable a subsequent measurement of the particle y position will result in a value near $\xi_y = 0.7$. In general, a non-separable superposition state creates correlations so that the result of a measurement of the x position changes the probability distribution of the particle's y position.

5.5 Time Dependence of Creation and Annihilation Operators

In quantum mechanics, the state of a system evolves *deterministically* according to the Schrödinger equation,

$$\hat{H}\psi(t) = i\hbar \frac{\partial}{\partial t}\psi(t). \tag{5.108}$$

So far, the case in which operators are time independent and the wave functions are time dependent has been considered. There is an alternative approach in which *operators are time dependent and the wave functions are time independent*. This is called the *Heisenberg representation*. To understand where this alternative view comes from, notice that Eq. (5.108) can always be written

$$\psi(t) = \mathrm{e}^{-i\hat{H}t/\hbar}\psi(0), \tag{5.109}$$

243

where $\psi(0)$ is the initial wave function at time $t = 0$. While Eq. (5.109) does not formally change anything mathematically, it does suggest a new viewpoint. Because quantum mechanics predicts the outcome of experiments, expectation values of the time-independent operator $\hat{A}$ are to be found where

$$\langle A \rangle = \langle \psi(t)|\hat{A}|\psi(t)\rangle. \tag{5.110}$$

Using Eq. (5.109), the expectation value may be written as

$$\langle A \rangle = \langle \psi(0)|e^{i\hat{H}t/\hbar}\hat{A}e^{-i\hat{H}t/\hbar}|\psi(0)\rangle = \langle \psi(0)|\tilde{A}(t)|\psi(0)\rangle, \tag{5.111}$$

where

$$\tilde{A}(t) = e^{i\hat{H}t/\hbar}\hat{A}e^{-i\hat{H}t/\hbar} \tag{5.112}$$

is a new time-dependent operator that acts on the initial wave function $\psi(0)$. These, the new operators in the Heisenberg representation, are time dependent, and the wave functions they operate on are time-independent initial states.

Differentiating Eq. (5.112) with respect to time using the product rule gives

$$\frac{\mathrm{d}}{\mathrm{d}t}\tilde{A}(t) = \frac{i}{\hbar}\hat{H}\tilde{A}(t) - \frac{i}{\hbar}\tilde{A}(t)\hat{H} + e^{i\hat{H}t/\hbar}\left(\frac{\partial}{\partial t}\hat{A}\right)e^{-i\hat{H}t/\hbar}, \tag{5.113}$$

and since $\hat{A}$ is a *time-independent* operator, $\frac{\partial}{\partial t}\hat{A} = 0$ and the expression becomes

$$\frac{\mathrm{d}}{\mathrm{d}t}\tilde{A}(t) = \frac{i}{\hbar}\left[\hat{H}, \tilde{A}\right]. \tag{5.114}$$

For the harmonic oscillator annihilation operator, $\hat{b}$, the corresponding time-dependent operator is $\tilde{b}(t) = e^{i\hat{H}t/\hbar}\hat{b}e^{-i\hat{H}t/\hbar}$ and

$$\frac{\mathrm{d}}{\mathrm{d}t}\tilde{b}(t) = \frac{i}{\hbar}\left[\hat{H}, \tilde{b}(t)\right] = i\omega e^{i\hat{H}t/\hbar}\left(\hat{b}^\dagger\hat{b}\hat{b} - \hat{b}\hat{b}^\dagger\hat{b}\right)e^{-i\hat{H}t/\hbar} = i\omega\left(\hat{b}^\dagger\hat{b} - \hat{b}\hat{b}^\dagger\right)\hat{b} = i\omega\left[\hat{b}^\dagger, \hat{b}\right]\hat{b}$$
$$= -i\omega\hat{b}, \tag{5.115}$$

where use is made of the fact that $\hat{H} = \hbar\omega\left(\hat{b}^\dagger\hat{b} + \frac{1}{2}\right)$ and $\left[\hat{b}^\dagger, \hat{b}\right] = -1$. The solution to the equation describing the time dependence of the annihilation operator,

$$\frac{\mathrm{d}}{\mathrm{d}t}\tilde{b}(t) = -i\omega\,\hat{b}, \tag{5.116}$$

found by integration is

$$\tilde{b}(t) = e^{-i\omega t}\hat{b}. \tag{5.117}$$

The adjoint of this expression is

$$\tilde{b}^\dagger(t) = e^{i\omega t}\hat{b}^\dagger. \tag{5.118}$$

244

Thus, the time development of the position operator

$$\hat{x} = \sqrt{\frac{\hbar}{2m\omega}} \left(\hat{b} + \hat{b}^{\dagger} \right) \tag{5.119}$$

becomes

$$\tilde{x}(t) = \sqrt{\frac{\hbar}{2m\omega}} \left(\hat{b}e^{-i\omega t} + \hat{b}^{\dagger}e^{i\omega t} \right). \tag{5.120}$$

5.5.1 Charged Particle in Harmonic Oscillator Potential Subject to Constant Electric Field E

A particle of mass m and charge e in a one-dimensional harmonic oscillator potential is subject to a constant applied electric field $\mathbf{E}$ in the positive x direction. The Hamiltonian for the system, which contains contributions from both the oscillator and the electric field, is

$$\hat{H} = \frac{\hat{p}^2}{2m} + \frac{m\omega^2}{2}\hat{x}^2 + e|\mathbf{E}|\hat{x}. \tag{5.121}$$

Equation (5.121) may be rewritten in terms of the operators $\hat{b}^{\dagger}$ and $\hat{b}$:

$$\hat{H} = \hbar\omega \left(\hat{b}^{\dagger}\hat{b} + \frac{1}{2} \right) + e|\mathbf{E}|\sqrt{\frac{\hbar}{2m\omega}} \left(\hat{b} + \hat{b}^{\dagger} \right). \tag{5.122}$$

The physics of the Hamiltonian describing the charged particle in a one-dimensional harmonic oscillator potential subject to a static uniform electric field is relatively straightforward. The particle oscillates at the same frequency, ω, established by the harmonic oscillator potential, but it is displaced by a distance

$$x_{\rm d} = \frac{e\mathbf{E}}{m\omega^2} \tag{5.123}$$

from the original position. The displacement in position due to application of the electric field causes the energy levels of the system to be changed by an amount

$$-\frac{e^2|\mathbf{E}|^2}{2m\omega^2} = -\frac{m\omega^2}{2}x_{\rm d}^2. \tag{5.124}$$

Figure 5.7 illustrates these ideas in graphical form.

The annihilation operator for states of the harmonic oscillator displaced by $x_{\rm d}$ is

$$\hat{b}_{\alpha} = \sqrt{\frac{m\omega}{2\hbar}} \left((\hat{x} - x_{\rm d}) + \frac{i\hat{p}}{m\omega} \right) = \hat{b} - \sqrt{\frac{m\omega}{2\hbar}}x_{\rm d}, \tag{5.125}$$

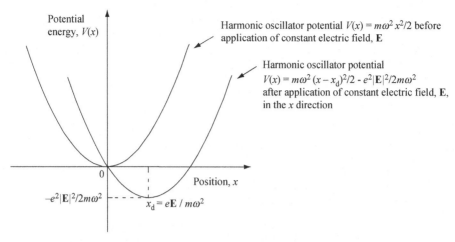

Fig. 5.7 Illustration of the potential of a particle of mass m and charge e in a one-dimensional harmonic potential before and after being subject to an applied constant electric field $\mathbf{E}$ in the x direction. The effect of the electric field is to shift the position of the oscillator potential by x_d and lower the potential energy by an amount $-e^2|\mathbf{E}|^2/2m\omega^2$. Extra energy $-e^2|\mathbf{E}|^2/2m\omega^2$ is stored in the potential due to displacement x_d.

and the ground-state eigenequation is

$$\hat{b}_\alpha|\alpha\rangle = \left(\hat{b} - \sqrt{\frac{m\omega}{2\hbar}}x_d\right)|\alpha\rangle = 0, \tag{5.126}$$

so that

$$\hat{b}|\alpha\rangle = \alpha|\alpha\rangle, \tag{5.127}$$

where eigenvalue $\alpha = \sqrt{\frac{m\omega}{2\hbar}}x_d$. The ground state, $|\alpha\rangle$, can be expressed as a coherent superposition of the harmonic oscillator states $|n\rangle$ before the displacement by x_d. In general, the wave-packet state is $|\alpha\rangle = \sum_{n=0}^{\infty} a_n|n\rangle$, where a_n are expansion coefficients. The larger the displacement value x_d, the greater the energy stored in the potential. If the particle displaced by x_d is released, e.g., by removing the electric field, it will oscillate at frequency ω.

5.6 Coherent States of the Harmonic Oscillator

The ground state of the harmonic oscillator minimizes uncertainty in position and momentum. Coherent states of the harmonic oscillator minimize *wave packet* position and momentum uncertainty. They are found by considering eigenstates of the (non-Hermitian) annihilation operator

$$\hat{b}|\alpha\rangle = \alpha|\alpha\rangle \tag{5.128}$$

and $\hat{b}^\dagger|\alpha\rangle = \alpha^*|\alpha\rangle$. In general, because operator $\hat{b}$ is non-Hermitian, the eigenvalue α can be complex. The family of eigenstates $|\alpha\rangle$ associated with operator $\hat{b}$ are over-complete. Expanding $|\alpha\rangle$ in terms of known oscillator excitation number (Fock) states gives

$$|\alpha\rangle = \sum_{n=0}^{\infty} a_n|n\rangle. \tag{5.129}$$

Substituting into the expression $\hat{b}|\alpha\rangle = \alpha|\alpha\rangle$,

$$\sum_{n=0}^{\infty} a_{n+1}\sqrt{n+1}\,|n\rangle = \alpha\sum_{n=0}^{\infty} a_n|n\rangle. \tag{5.130}$$

The first term $\hat{b}$ operates on in the sum is $|n = 1\rangle$ so that $\hat{b}|n+1\rangle = a_{n+1}\sqrt{n+1}|n\rangle$. This rearrangement of the sum avoids $\hat{b}$ operating on the vacuum for which $\hat{b}|0\rangle = 0$. Multiplying from the left by $\langle m|$ on both sides, only terms with $m = n$ survive to give

$$a_{n+1} = \frac{\alpha\,a_n}{\sqrt{n+1}}. \tag{5.131}$$

Assuming a ground-state coefficient a_0, it follows that $a_1 = \alpha a_0/\sqrt{1}$, and $a_2 = \alpha a_1/\sqrt{2} = \alpha^2 a_0/\sqrt{1}\sqrt{2}$, etc., so that

$$a_n = \frac{\alpha^n}{\sqrt{n!}}\,a_0. \tag{5.132}$$

Substitution into Eq. (5.129) gives

$$|\alpha\rangle = a_0\sum_{n=0}^{\infty} \frac{\alpha^n}{\sqrt{n!}}|n\rangle. \tag{5.133}$$

Normalization of $|\alpha\rangle$ can be used to find a_0 since[1]

$$1 = \langle\alpha|\alpha\rangle = |a_0|^2\sum_{n=0}^{\infty} \frac{|\alpha|^{2n}}{n!} = |a_0|^2 e^{|\alpha|^2}, \tag{5.134}$$

so that $a_0 = e^{-|\alpha|^2/2}$. If a_0 is real and positive then, after substituting into Eq. (5.133),

$$|\alpha\rangle = e^{-|\alpha|^2/2}\sum_{n=0}^{\infty} \frac{\alpha^n}{\sqrt{n!}}|n\rangle. \tag{5.135}$$

The $|\alpha\rangle$ states have interesting properties. For example, the average

$$\bar{n} = \langle\alpha|\hat{b}^\dagger\hat{b}|\alpha\rangle = |\alpha|^2 \tag{5.136}$$

(note the use of $\bar{n}$ when $\langle n\rangle$ could also be used).

---

[1] Use is made of $e^x = \sum_{n=0}^{\infty} \frac{x^n}{n!}$.

The probability P_n that $|n\rangle$ occurs in $|\alpha\rangle$ is

$$P_n = |\langle n|\alpha\rangle|^2 = \left|\langle n|e^{-|\alpha|^2/2}\sum_{n=0}^{\infty}\frac{\alpha^n}{\sqrt{n!}}|n\rangle\right|^2 = \frac{1}{n!}e^{-|\alpha|^2}|\alpha|^{2n}, \tag{5.137}$$

and since $\bar{n} = |\alpha|^2$, this may be rewritten as

$$\boxed{P_n = \frac{e^{-\bar{n}}\bar{n}^n}{n!},} \tag{5.138}$$

which is a Poisson distribution for energy spectrum $E_n = \hbar\omega(n+1/2)$. The states $|\alpha\rangle$ are *probability distribution amplitudes* for quanta of energy.

The average energy of the coherent state $|\alpha\rangle$ is

$$\left(\bar{n}+\frac{1}{2}\right)\hbar\omega = \left(|\alpha|^2 + \frac{1}{2}\right)\hbar\omega. \tag{5.139}$$

The expectation value of position for a particle mass m in a coherent harmonic oscillator state is

$$\langle x(t)\rangle = \langle\alpha|\tilde{x}(t)|\alpha\rangle = \sqrt{\frac{\hbar}{2m\omega}}\langle\alpha|\left(\hat{b}e^{-i\omega t} + \hat{b}^\dagger e^{i\omega t}\right)|\alpha\rangle = \sqrt{\frac{\hbar}{2m\omega}}\langle\alpha|\left(\alpha e^{-i\omega t} + \alpha^* e^{i\omega t}\right)|\alpha\rangle, \tag{5.140}$$

and so evolves sinusoidally in time (to within a constant arbitrary phase, ϕ) as

$$\langle x(t)\rangle = \sqrt{\frac{2\hbar}{m\omega}}|\alpha|\cos(\omega t + \phi), \tag{5.141}$$

where ω is the oscillation frequency and

$$\alpha = |\alpha|e^{-i\phi}. \tag{5.142}$$

To show that a coherent harmonic oscillator state $|\alpha\rangle$ minimizes uncertainty in the expectation value of position and momentum, it is necessary to evaluate the expectation value of variances Δx^2 and Δp^2, respectively. For example,

$$\Delta x^2 = \left\langle(\hat{x}-\langle x\rangle)^2\right\rangle = \left\langle\hat{x}^2 + \langle x\rangle^2 - 2\hat{x}\langle x\rangle\right\rangle = \langle x^2\rangle - \langle x\rangle^2, \tag{5.143}$$

so that

$$\Delta x^2 = \frac{\hbar}{2m\omega}\langle\alpha|\hat{b}\hat{b} + \hat{b}\hat{b}^\dagger + \hat{b}^\dagger\hat{b} + \hat{b}^\dagger\hat{b}^\dagger|\alpha\rangle - \frac{\hbar}{2m\omega}\langle\alpha|\hat{b}+\hat{b}^\dagger|\alpha\rangle^2$$

$$= \frac{\hbar}{2m\omega}\left(\langle\alpha|\hat{b}\hat{b}^\dagger - \hat{b}^\dagger\hat{b}|\alpha\rangle + (\alpha+\alpha^*)^2 - (\alpha+\alpha^*)^2\right), \tag{5.144}$$

where use was made of $\hat{b}|\alpha\rangle = \alpha|\alpha\rangle$ and $\hat{b}^\dagger|\alpha\rangle = \alpha^*|\alpha\rangle$. Noting that the commutation relation $\left[\hat{b},\hat{b}^\dagger\right] = \hat{b}\hat{b}^\dagger - \hat{b}^\dagger\hat{b} = 1$, then

$$\Delta x^2 = \frac{\hbar}{2m\omega}\left(\langle\alpha|\hat{b}\hat{b}^\dagger - \hat{b}^\dagger\hat{b}|\alpha\rangle\right) = \frac{\hbar}{2m\omega}\left[\hat{b},\hat{b}^\dagger\right] = \frac{\hbar}{2m\omega}. \tag{5.145}$$

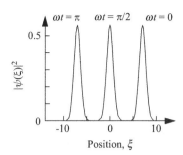

Fig. 5.8 Magnitude squared of the coherent state wave packet as a function of normalized position $\xi = x\sqrt{m\omega/\hbar}$ in a one-dimensional harmonic oscillator potential at time $t = 0$, $t = \pi/2\omega$, and $t = \pi/\omega$. The superposition state $\langle\xi|\alpha\rangle = \psi(\xi)$ is an eigenstate of the annihilation operator $\hat{b}$ such that $\hat{b}|\alpha\rangle = \alpha|\alpha\rangle$. The wave packet minimizes uncertainty in position and momentum and the expectation value of position oscillates sinusoidally with radial frequency ω. In the figure, eigenvalue $\alpha = 5$.

The variance in momentum follows, $\Delta p^2 = \frac{\hbar m\omega}{2}$, so that $\Delta x^2 \Delta p^2 = \frac{\hbar^2}{4}$ and

$$\Delta x \Delta p = \frac{\hbar}{2}. \tag{5.146}$$

Hence, the coherent harmonic oscillator superposition state $|\alpha\rangle$ defined in Eq. (5.135) minimizes uncertainty in the expectation values of position and momentum. Time evolution of the magnitude squared of the coherent state with $\alpha = 5$ is illustrated in Fig. 5.8. This quantum mechanical wave packet superposition state is closest to the behavior of a classical harmonic oscillator.

5.7 Quantization of Electromagnetic Fields

In free space the total classical electromagnetic energy density can be written as

$$U = \frac{1}{2}(\mathbf{E} \cdot \mathbf{D} + \mathbf{B} \cdot \mathbf{H}) = \frac{\varepsilon_0}{2}|\mathbf{E}(\mathbf{r}, t)|^2 + \frac{1}{2\mu_0}|\mathbf{B}(\mathbf{r}, t)|^2. \tag{5.147}$$

In free space there is no charge density, so that $\nabla V(\mathbf{r}, t) = 0$ and the electric and magnetic fields can be written in terms of the vector potential $\mathbf{A}(\mathbf{r}, t)$ in such a way that $\mathbf{E}(\mathbf{r}, t) = -\frac{\partial}{\partial t}(\mathbf{A}(\mathbf{r}, t))$ and $\mathbf{B}(\mathbf{r}, t) = \nabla \times \mathbf{A}(\mathbf{r}, t)$. To make connection with the harmonic oscillator, the vector potential is expressed in terms of its Fourier components,

$$\mathbf{A}(\mathbf{r}, t) = \frac{1}{(2\pi)^{3/2}} \int \mathbf{A}(\mathbf{k}, t) e^{i\mathbf{k}\cdot\mathbf{r}} d\mathbf{k}. \tag{5.148}$$

Substituting into the expression for energy density gives

$$U = \frac{\varepsilon_0}{2}\left|-\frac{\partial}{\partial t}(\mathbf{A}(\mathbf{k}, t))\right|^2 + \frac{k^2}{2\mu_0}|\mathbf{A}(\mathbf{k}, t)|^2. \tag{5.149}$$

Comparing this with the energy of a particle of mass m in a harmonic oscillator potential,

$$E = \frac{\hat{p}^2}{2m} + \frac{\kappa_0}{2}\hat{x}^2, \qquad (5.150)$$

which oscillates at frequency $\omega = \sqrt{\kappa_0/m}$, it is clear that the electromagnetic field may be viewed as a number of harmonic oscillators of amplitude $\mathbf{A}(\mathbf{k}, t)$ and frequency $\omega = \sqrt{k^2/\mu_0\varepsilon_0} = kc$, where $c = 1/\sqrt{\mu_0\varepsilon_0}$ is the speed of light.

A rather surprising conclusion may now be drawn. Because the electromagnetic field must be quantized in the same way as a harmonic oscillator, the energy density of the electromagnetic field in vacuum is never zero. This result is a direct consequence of the fact that each electromagnetic mode of wave vector $\mathbf{k}$ and amplitude $\mathbf{A}(\mathbf{k}, t)$ has nonzero ground-state energy.

The quanta of the electromagnetic field is the *photon*. Each photon particle carries energy $\hbar\omega$ and integer spin quantum number $\sigma_s = \pm 1$ carrying angular momentum $\pm\hbar$. Photon spin corresponds to left- or right-circular polarization of plane waves in classical electromagnetism. All integer spin particles such as photons are classified as *bosons*.

5.7.1 Laser Light

Continuous single-mode laser light emission is dominated by a single frequency of oscillation, ω. Each component of the electromagnetic field describes simple harmonic motion. Quantization of the electromagnetic field follows naturally so that each photon contributing to the field has energy $\hbar\omega$. The total energy due to all photons in the single mode is $E = \hbar\omega(n + 1/2)$, where n is the number of photons in the field. The lasing photon field may be described as a *coherent state* of a harmonic oscillator of frequency ω. The uncertainty relation between photon number and phase is $\Delta n\Delta\theta \geq 1/2$. Every time an additional photon is added to the system, so is additional energy, $\hbar\omega$. Applying the operator $\hat{b}^\dagger$ to a state function ψ, which describes coherent laser light, increases the energy by $\hbar\omega$. Thus, $\hat{b}^\dagger$ operates to create a photon. In the same way, $\hat{b}$ annihilates a photon. The operators $\hat{b}^\dagger$ and $\hat{b}$ are creation and annihilation operators for photon particles in the photon field and ψ_n is a multi-particle state function because it contains n photons.

Since $\hat{b}\psi_0 = 0$ describes the ground state, ψ_0, the vacuum contains no real photons for $\hat{b}$ to annihilate. The ground-state energy $\hbar\omega/2$ is the energy in the electromagnetic field *before* the laser has been turned on. This is called vacuum energy. In this picture, the energy level $E_n = \hbar\omega(n + 1/2)$ describes the situation when n photons have been added to the vacuum.

5.7.2 Quantization of an Electrical Resonator

A series LC circuit has resonant frequency such that $\omega = 1/\sqrt{LC}$ where L is an inductor and C is a capacitor. Current flow in the circuit oscillates in time as $I(t) = I_0 e^{i\omega t}$, and electromagnetic energy is stored in the capacitor and inductor. There are, of course, many practical applications in which such a circuit may be used, including the production of electromagnetic radiation at the resonant frequency ω. The circuit behaves as a harmonic oscillator with electromagnetic energy quantized as photons in such a way that $E_n = \hbar\omega(n + 1/2)$ where $\hbar\omega = \hbar/\sqrt{LC}$.

The electromagnetic energy stored in the LC circuit is the sum of the energy in the inductor $LI^2/2$ and the energy in the capacitor $CV^2/2$. Since magnetic flux $\phi_B = LI$, charge $Q = CV$, and, on resonance, $C = 1/L\omega^2$, the stored electromagnetic energy may be written as

$$E = \frac{\phi_B^2}{2L} + \frac{L\omega^2 Q^2}{2}. \tag{5.151}$$

Comparing this with the energy of a particle of mass m in a harmonic oscillator potential (Eq. (5.150)) the "mass" of the resonator is L, the "spring constant" is $L\omega^2$, and the coordinates $\hat{p} = \hat{\phi}_B$ and $\hat{x} = \hat{Q}$. The operators $\hat{\phi}_B$ and $\hat{Q}$ form a conjugate pair so that $\hat{\phi}_B = -i\hbar d/d\hat{Q}$ and $\left[\hat{\phi}_B, \hat{Q}\right] = -i\hbar$.

Quantum mechanics predicts that charge and magnetic flux obey the uncertainty relation $\Delta Q \Delta \phi_B \geq \hbar/2$ and so cannot be measured simultaneously to arbitrary accuracy. Quantum mechanics also predicts that the minimum electromagnetic energy in the resonant circuit is $E_0 = \hbar\omega/2 = \hbar/\sqrt{4LC}$. If inductance, L, and capacitance, C, are constants, then it may be concluded that quantum mechanics does not allow current and charge in the resonator to be simultaneously measured to arbitrary accuracy. Thus $\Delta(CV)\Delta(LI) \geq \hbar/2$, so that $\Delta V \Delta I \geq \hbar/2LC$ or, equivalently, $\Delta V \Delta I \geq \hbar\omega^2/2$.

5.8 Quantization of Lattice Vibrations

A linear monatomic chain consisting of N particles, each of mass m and average nearest-neighbor lattice spacing L, is modeled in the harmonic nearest-neighbor interaction approximation with periodic boundary conditions. Displacement from the lattice position of the jth particle is x_j. A vibrational normal mode of the chain is characterized by frequency $\omega(q)$ and wave vector $q = 2\pi/\lambda$. The Hamiltonian for the linear chain in quantum form makes use of the momentum operator $\hat{p}$ and the displacement operator $\hat{x}$ to give

$$\hat{H} = \sum_{j}^{N} \frac{\hat{p}_j^2}{2m} + \frac{\kappa_0}{2} \sum_{j}^{N} \left(2\hat{x}_j^2 - \hat{x}_j\hat{x}_{j+1} - \hat{x}_j\hat{x}_{j-1}\right), \tag{5.152}$$

where the sum is over all N particles in the chain. To convert this into a more convenient diagonal form, it is necessary to perform a canonical transformation. To do this, new operators are defined in terms of a linear combination of displacements and the momenta of each particle so that

$$\hat{b}_q = \frac{1}{\sqrt{N}} \sum_{j}^{N} \sqrt{\frac{m\omega_q}{2\hbar}} \, e^{-iqjL} \left(\hat{x}_j + \frac{i\hat{p}_j}{m\omega_q}\right), \tag{5.153}$$

$$\hat{b}_q^\dagger = \frac{1}{\sqrt{N}} \sum_{j}^{N} \sqrt{\frac{m\omega_q}{2\hbar}} \, e^{iqjL} \left(\hat{x}_j - \frac{i\hat{p}_j}{m\omega_q}\right), \tag{5.154}$$

where e^{iqjL} is a Bloch phase factor and ω_q is chosen to diagonalize the Hamiltonian. The new operators, $\hat{b}_q^\dagger$ and $\hat{b}_q$, which obey the usual commutation relations (Eqs. (5.15)–(5.18)), may be used in linear combination to give

$$\hat{x}_j = \frac{1}{\sqrt{N}} \sum_q^N \sqrt{\frac{\hbar}{2m\omega_q}}\, e^{iqjL} \left(\hat{b}_q + \hat{b}_{-q}^\dagger \right), \tag{5.155}$$

$$\hat{p}_j = \frac{-i}{\sqrt{N}} \sum_q^N \sqrt{\frac{m\hbar\omega_q}{2}}\, e^{iqjL} \left(\hat{b}_q - \hat{b}_{-q}^\dagger \right). \tag{5.156}$$

Substitution of these new expressions for $\hat{x}_j$ and $\hat{p}_j$ into Eq. (5.152) results in a Hamiltonian,

$$\hat{H} = -\frac{1}{4} \sum_q \hbar\omega_q \left(\hat{b}_q - \hat{b}_{-q}^\dagger \right) \left(\hat{b}_{-q} - \hat{b}_q^\dagger \right)$$
$$+ \frac{1}{4} \sum_q \frac{\hbar}{\omega_q} \frac{\kappa_0}{m} \left(\hat{b}_q + \hat{b}_{-q}^\dagger \right) \left(\hat{b}_{-q} + \hat{b}_q^\dagger \right) \left(2 - e^{iqL} - e^{-iqL} \right). \tag{5.157}$$

Recognizing that if

$$\omega_q^2 = \frac{\kappa_0}{m} \left(2 - e^{iqL} - e^{-iqL} \right), \tag{5.158}$$

the Hamiltonian takes on the familiar diagonal form (Eqs. (5.13) and (5.20)):

$$\hat{H} = \sum_q \frac{\hbar\omega_q}{2} \left(\hat{b}_q \hat{b}_q^\dagger + \hat{b}_q^\dagger \hat{b}_q \right) = \sum_q \hbar\omega_q \left(\hat{b}_q^\dagger \hat{b}_q + \frac{1}{2} \right) = \sum_q \hbar\omega_q \left(n_q + \frac{1}{2} \right). \tag{5.159}$$

This is the sum of independent linear oscillators of frequency $\omega(q)$. Modes of vibrational frequency ω_q and wave vector q are described by the dispersion relation $\omega = \omega(q)$, which, for the case considered, is the same as the classical result. Each quantized vibrational mode of the linear chain is made up of N individual particles of mass m oscillating about their lattice site coupled via the interaction potential. A lattice vibration of wave vector $q = 2\pi/\lambda$ is quantized with energy $E_n = \hbar\omega(q)(n_q + 1/2)$ and contains n_q *phonons*. Phonons have zero integer spin and, like photons, are bosons.

5.9 Quantization of Mechanical Vibrations

Lattice vibrations in a crystal are quantized and so are oscillations of small mechanical structures. An example of a small mechanical structure is the cantilever beam shown schematically in Fig. 5.9.

The lowest-frequency classical vibrational mode of a long, thin cantilever beam oscillates at radial frequency

$$\omega = 3.52 \frac{d}{l^2} \sqrt{\frac{E_{\text{Young}}}{12\rho}}, \tag{5.160}$$

where l is the length, d is the thickness, ρ is the density of the beam, and E_{Young} is Young's modulus. Classical mechanics predicts that the vibrational energy of a cantilever with width w and free-end displacement amplitude A is

$$E_{\text{classical}} = \frac{wd^3 A^2 E_{\text{Young}}}{6l^3}. \tag{5.161}$$

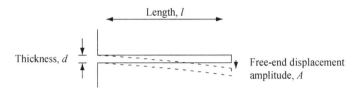

Fig. 5.9 Cross-section of a cantilever beam of length l and thickness d with a free-end displacement amplitude A.

If energy $E_{\text{classical}} = m\omega^2 A^2/2$ is equated with the characteristic energy of a quantized one-dimensional harmonic oscillator $\hbar\omega$, then an estimate of the amplitude of oscillation of the free end is

$$A = \sqrt{\frac{21.12 \times \hbar \times l}{w \times d^2 \sqrt{12 \times \rho \times E_{\text{Young}}}}}. \tag{5.162}$$

If the cantilever is made of silicon by a micro-electro-mechanical systems (MEMS) process then the bulk values for density, $\rho = 2.328 \times 10^3$ kg m^{-3}, and Young's modulus, $E_{\text{Young}} = 1.96 \times 10^{11}$ N m^{-2}, may be used to estimate oscillation frequency and amplitude. Equations (5.160) and (5.162) predict that a small cantilever structure with dimensions $l = 10\,\mu$m, $w = 50$ nm, and $d = 10$ nm has a free-end oscillation frequency $\nu = 148$ kHz and an amplitude $A = 0.0078$ nm. To make sure that low-energy vibrational motion dominates, temperature $T \leq \hbar\omega/k_{\text{B}} = 7$ K.

This combination of very small vibrational amplitude and low temperature makes direct measurement of quantized mechanical motion in a MEMS structure quite challenging.[2] Nevertheless, it is a prediction of quantum mechanics that the motion of small mechanical structures is quantized in such a way that vibrational energy is $E_n = \hbar\omega(n + 1/2)$.

5.10 Example Exercises

Exercise 5.1
A one-dimensional harmonic oscillator is in the $n = 1$ state, for which

$$\psi_{n=1}(x) = 2 \left(2\pi^{1/2} x_0\right)^{-1/2} (x/x_0) e^{-0.5(x/x_0)^2}$$

with $x_0 = (\hbar/m\omega)^{1/2}$. Calculate the probability of finding the particle in the interval x to $x + dx$. Show that, according to classical mechanics, the probability is

$$P_{\text{classical}}(x)dx = \frac{1}{\pi} \left(A_0^2 - x^2\right)^{-1/2} dx$$

---

[2] R. G. Knobel and A. N. Cleland, *Nature* **424**, 291 (2003); M. D. LaHaye, O. Buu, B. Camarota, and K. C. Schwab, *Science* **304** (5667), 74 (2004). For a review, see K. L. Ekinci and M. L. Roukes, *Rev. Sci. Inst.* **76**, 061101 (2005).

for $-A_0 < x < A_0$ and zero elsewhere (A_0 is the classical amplitude). With the aid of sketches, compare this probability distribution with the quantum mechanical one. Locate the maxima for the probability distribution for the $n = 1$ quantum state relative to the classical turning point. Assume that the classical amplitude, A_0, is such that the total energy is identical in both cases.

Exercise 5.2
A two-dimensional potential experienced by a particle of mass m is of the form

$$V(x, y) = m\omega^2 \left(x^2 + xy + y^2 \right).$$

Write the potential as a 2×2 matrix and find new coordinates u and v that diagonalize the matrix. Find the energy levels of the particle.

Exercise 5.3
The purpose of this exercise is to show that a *coherent superposition* of high quantum number energy eigenstates of a harmonic oscillator with a spread in energies that is small compared with their mean energy *behaves as a classical oscillator*. Consider a one-dimensional harmonic oscillator characterized by mass m and angular frequency ω. Let $|n\rangle$ be an eigenstate of the Hamiltonian corresponding to the quantum number n. The state $\psi(t)$ of the oscillator at time $t = 0$ is

$$\psi(t = 0) = \frac{1}{\sqrt{2\Delta N}} \sum_{n=N-\Delta N}^{n=N+\Delta N} |n\rangle,$$

where $1 \ll \Delta N \ll N$.

(a) Find the expectation value of position as a function of time.

(b) Compare the result in (a) to the predictions of a classical harmonic oscillator.

(c) Write a computer program that plots the nth eigenstate $|n\rangle$ and probability $\langle n|n\rangle$ of the one-dimensional harmonic oscillator. Use Eq. (5.101) to generate the wave function, and plot the $|n = 18\rangle$ eigenstate and its probability function.

(d) In general, a superposition wave function of the one-dimensional harmonic oscillator can be formed so that

$$\psi(t) = \sum a_n |n\rangle e^{-i\omega_n t},$$

where a_n is a weighting factor that contains amplitude and phase information for each eigenstate $|n\rangle$ and $\omega_n = \omega(n + 1/2)$. However, if equal weights and a contiguous sum are assumed, then at time $t = 0$ the superposition wave function,

$$\psi(t = 0) = \frac{1}{\sqrt{2\Delta N}} \sum_{n=N-\Delta N}^{n=N+\Delta N} |n\rangle,$$

represents a particle at an extreme of its motion. Use (c) and write a computer program that plots the superposition wave function and particle probability function for the specific case in which $N = 18$ and $\Delta N = 2$. Compare the peak in probability with the classical turning point for the $|n = 18\rangle$ eigenstate.

Exercise 5.4
A particle of mass m in a one-dimensional harmonic potential $V(x) = m\omega^2 x^2/2$ is in an eigenstate $\psi_n(\xi)$ with eigenvalue n and eigenenergy $E_n = \hbar\omega(n + 1/2)$. The probability of the particle being found in the nonclassical region of the harmonic oscillator is given by

$$P_n^{\text{nonclassical}} = \int_{|\xi_n| > 1} \psi_n^*(\xi)\psi_n(\xi)\mathrm{d}\xi,$$

where $\xi = (m\omega/\hbar)^{1/2}x$ and the classical turning point is $\xi_n = (2n + 1)^{1/2}$.

Find the values of $P_n^{\text{nonclassical}}$ for the first excited state $n = 1$ and for the second excited state $n = 2$.

Exercise 5.5
In Section 5.2 it was shown that the creation operator $\hat{b}^\dagger$ and the annihilation operator $\hat{b}$ for the one-dimensional harmonic oscillator with Hamiltonian $\hat{H}$ are related to each other through the commutation relation $[\hat{b}\hat{b}^\dagger] = (\hat{b}\hat{b}^\dagger - \hat{b}^\dagger\hat{b}) = 1$.

Verify that the expressions for the commutation relations $\left[\hat{b}, \hat{b}^\dagger\right]$, $\left[\hat{b}, \hat{b}\right]$, and $\left[\hat{b}^\dagger, \hat{b}^\dagger\right]$ given in Section 5.2 are correct.

If a new operator $\hat{n} = \hat{b}^\dagger\hat{b}$ is defined, show that

$$\hat{H} = \hbar\omega\left(\hat{n} + \frac{1}{2}\right)$$

and derive expressions for the commutation relations $\left[\hat{n}, \hat{b}\right]$ and $\left[\hat{n}, \hat{b}^\dagger\right]$.

Exercise 5.6
What are the degeneracies of the three lowest levels of a symmetric three-dimensional harmonic oscillator? Find a general expression for the degeneracy of the nth level.

Exercise 5.7
The ground-state wave function of a particle of mass m in a harmonic potential is $\psi_0 = A_0 e^{-x^2/4\sigma_x^2}$, where $\sigma_x^2 = \hbar/2m\omega$. Derive the standard deviation in position and momentum, and show that they satisfy the uncertainty relation.

Exercise 5.8
The ground state and the *second* excited state of a charged particle of mass m in a one-dimensional harmonic oscillator potential are both excited. What is the expectation value of the particle position x as a function of time? What happens to the expectation value if the potential is subject to a constant electric field $\mathbf{E}$ in the x direction?

Exercise 5.9

Using the method outlined in Section 2.4, write a computer program to solve the Schrödinger wave equation for the first four eigenvalues and eigenstates of an electron with effective mass $m_e^* = 0.07 \times m_0$ confined to a parabolic potential well in such a way that $V(x) = (x - L/2)^2/(L/2)^2$ eV and $L = 100$ nm.

Solutions

Solutions 5.1

Starting with a one-dimensional harmonic oscillator in the $n = 1$ state for which

$$\psi_{n=1}(x) = 2(2x_0\sqrt{\pi})^{-1/2}(x/x_0)e^{-0.5(x/x_0)^2}$$

with $x_0 = (\hbar/m\omega)^{1/2}$, the quantum mechanical probability of finding the particle in the interval x to $x + dx$ is

$$P_{\text{quantum}}(x)dx = |\psi_1(x)|^2 dx = \frac{2}{x_0\sqrt{\pi}}(x/x_0)^2 e^{-(x/x_0)^2} dx.$$

The maxima in this distribution occurs when

$$0 = \frac{d}{dx}P_{\text{quantum}}(x) = \frac{2}{x_0^3\sqrt{\pi}}2xe^{-(x/x_0)^2} - \frac{2}{x_0^3\sqrt{\pi}}2x\left(\frac{x}{x_0}\right)^2 e^{-(x/x_0)^2},$$

$$0 = 1 - \left(\frac{x}{x_0}\right)^2,$$

so that $x_{\text{max}} = \pm x_0$ is the peak in the quantum-mechanical probability distribution.

Classically, $x(t) = A_0 \sin(\omega t)$, where A_0 is the classical amplitude of oscillation. The energy of the classical harmonic oscillator is found from the solution to the equation of motion,

$$\kappa_0 x + m\frac{d^2 x}{dt^2} = 0,$$

or the Hamiltonian

$$H = T + V = \frac{1}{2}m\left(\frac{dx}{dt}\right)^2 + \frac{1}{2}\kappa_0 x^2,$$

where κ_0 is the force constant and m is the particle mass. The potential energy is

$$V = \frac{1}{2}\kappa_0 A^2 \cos^2(\omega_0 t + \phi) \quad \text{and the kinetic energy is } T = \frac{1}{2}m\omega_0^2 A^2 \sin^2(\omega_0 t + \phi),$$

where ϕ is an arbitrary phase.

Hence, since $\sin^2(\theta) + \cos^2(\theta) = 1$ and $\kappa_0 = m\omega_0^2$, the total energy

$$H = \frac{1}{2}m\omega_0^2 A^2 = \frac{1}{2}\kappa_0 A^2.$$

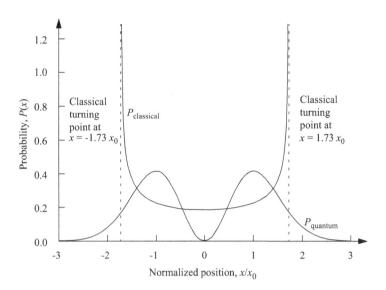

Fig. 5.E10

Equating the total energy for the classical and quantum-mechanical case gives

$$E_{\text{total}} = \frac{1}{2}m\omega^2 A_0^2 = \frac{3}{2}\hbar\omega.$$

Therefore,

$$A_0 = \sqrt{\frac{3\hbar\omega}{m\omega^2}} = \sqrt{\frac{\hbar}{m\omega}}\sqrt{3} = x_0\sqrt{3} = 1.73x_0.$$

Classically, $x(t) = A_0\sin(\omega t)$, where A_0 is the classical amplitude of oscillation and $\tau = 2\pi/\omega$ is the oscillation period. For any time interval during the period of oscillation for which $0 < t < \tau = 2\pi/\omega$, the oscillator will be between x and $x + dx$ at the time

$$t = \frac{1}{\omega}\sin^{-1}\left(\frac{x}{A_0}\right).$$

Hence, the probability of finding the classical particle in the interval x to $x + dx$ is

$$P_{\text{classical}}(x)dx = \frac{2dt}{\tau} = \frac{2dt}{2\pi/\omega} = \frac{\omega dt}{\pi}.$$

The factor of 2 arises because the particle passes position x twice during one oscillation period. Substituting the expression for dt results in

$$dt = \frac{dt}{dx}dx = \frac{d}{dx}\left(\frac{1}{\omega}\sin^{-1}(x/A_0)\right)dx = \frac{1}{\omega}\frac{1/A_0}{\sqrt{1-(x/A_0)^2}}dx = \frac{1}{\omega}\left(A_0^2 - x^2\right)^{-1/2}dx,$$

257

so that

$$P_{\text{classical}}(x)\mathrm{d}x = \frac{1}{\pi}\left(A_0^2 - x^2\right)^{-1/2}\mathrm{d}x.$$

Solutions 5.2

The symmetric potential,

$$V(x,y) = m\omega^2 \left(x^2 + xy + y^2\right),$$

can be written as a 2×2 matrix

$$V(x,y) = \frac{m}{2}\omega^2 [x\,y] \begin{bmatrix} 2 & 1 \\ 1 & 2 \end{bmatrix} \begin{bmatrix} x \\ y \end{bmatrix}.$$

Because the eigenvalues of the 2×2 matrix are 3 and 1, there must be coordinates u and v that diagonalize the matrix to give

$$V(u,v) = \frac{m}{2}\omega^2 [u\,v] \begin{bmatrix} 3 & 0 \\ 0 & 1 \end{bmatrix} \begin{bmatrix} u \\ v \end{bmatrix}.$$

Notice that the symmetric 2×2 matrix has been diagonalized by an orthogonal change of variables,

$$u = \frac{1}{\sqrt{2}}(x + y) \text{ and } v = \frac{1}{\sqrt{2}}(x - y).$$

Multiplying out the matrix, the potential is now given by

$$V(u,v) = \frac{m}{2}\omega^2 \left(3u^2 + v^2\right),$$

and so the eigenenergy values of the particle are

$$E_{n_u,n_v} = \sqrt{3}\hbar\omega \left(n_u + \frac{1}{2}\right) + \hbar\omega \left(n_v + \frac{1}{2}\right),$$

where the quantum numbers n_u and n_v are positive integers $0, 1, 2, \ldots$.

Solutions 5.3

(a) The state function at time $t = 0$ of the one-dimensional harmonic oscillator is given as

$$\psi(0) = \frac{1}{\sqrt{2\Delta N}} \sum_{n=N-\Delta N}^{n=N+\Delta N} |n\rangle,$$

where $1 \ll \Delta N \ll N$. This is a coherent superposition of high quantum number eigenfunctions the relative phase of which is specified at time $t = 0$. It follows that the wave function evolves in time as

$$\psi(t) = \frac{1}{\sqrt{2\Delta N}} \sum_{n=N-\Delta N}^{n=N+\Delta N} |n\rangle e^{-iE_n t/\hbar} = \frac{1}{\sqrt{2\Delta N}} \sum_{n=N-\Delta N}^{n=N+\Delta N} |n\rangle e^{-i\left(n+\frac{1}{2}\right)\omega t},$$

since the energy of the nth eigenfunction is $E_n = \left(n + \frac{1}{2}\right)\hbar\omega$. The position operator,

$$\hat{x} = \sqrt{\frac{\hbar}{2m\omega}}\left(\hat{b} + \hat{b}^\dagger\right),$$

has an associated expectation value given by

$$\langle x \rangle = \sqrt{\frac{\hbar}{2m\omega}}\langle\psi(t)|\hat{b} + \hat{b}^\dagger|\psi(t)\rangle = \sqrt{\frac{\hbar}{2m\omega}}\sum_{nn'}\langle n'|\hat{b} + \hat{b}^\dagger|n\rangle\exp(\mathrm{i}(n' - n)\omega t).$$

The relations $\hat{b}|n\rangle = \sqrt{n}\,|n-1\rangle$ and $\hat{b}^\dagger|n\rangle = \sqrt{n+1}\,|n+1\rangle$ show that the matrix element is zero unless $n' = n \pm 1$. Hence,

$$\langle x \rangle = \sqrt{\frac{\hbar}{2m\omega}}\frac{1}{2\Delta N}\sum_{n=N-\Delta N}^{n=N+\Delta N}\left(\sqrt{n}e^{-\mathrm{i}\omega t} + \sqrt{n+1}e^{\mathrm{i}\omega t}\right).$$

The summation over n is from $n = N - \Delta N$ to $n = N + \Delta N$. However, since $1 \ll \Delta N \ll N$, it is possible to approximate $\sqrt{n} \cong \sqrt{n+1} \cong \sqrt{N}$. Since the sum is from $n = N - \Delta N$ to $n = N + \Delta N$ there are $2\Delta N + 1 \cong 2\Delta N$ equal terms in the summation, and thus

$$\sum_n \sqrt{n}e^{-\mathrm{i}\omega t} + \sqrt{n+1}e^{\mathrm{i}\omega t} \approx 2 \times 2\Delta N\sqrt{N}\cos(\omega t),$$

giving an expectation value for position

$$\langle x \rangle = \sqrt{\frac{\hbar}{2m\omega}}\cos(\omega t).$$

(b) Classically, $m\left(\mathrm{d}^2x/\mathrm{d}t^2\right) + \kappa_0 x = 0$, where $\kappa_0 = m\omega^2$ with solution for position coordinate $x = A_0\cos(\omega t)$ and total energy $E = \kappa_0 x^2/2 + m(\mathrm{d}x/\mathrm{d}t)^2/2 = m\omega^2 A_0^2/2$. In the quantum calculation, the energy is $E \approx (N+1/2)\hbar\omega \approx N\hbar\omega$. Therefore, the quantum amplitude $(2\hbar N/m\omega)^{1/2} \approx (2E/m\omega^2)^{1/2} = A_0$. Hence, the coherent combination of a large number of energy eigenstates with a spread in energies that is small compared with their mean energy behaves as a classical oscillator.

(c) Figure 5.E11 plots the wave function $\psi(\xi)$ and the modulus of the wave function squared $|\psi(\xi)|^2$ for the $n = 18$ state of the one-dimensional harmonic oscillator. The normalized spatial coordinate $\xi = (m\omega/\hbar)x$, where m is the particle mass and ω is the oscillation frequency. The eigenenergy of the state $\psi_{18}(\xi)$ is $E_{n=18} = (n + 1/2)\hbar\omega = 18.5 \times \hbar\omega$, and the classical turning point occurs at $\xi_{n=18} = \pm(2n + 1)^{1/2} = \pm 6.083$.

(d) Figure 5.E12 plots the superposition wave function,

$$\psi(t = 0) = \frac{1}{\sqrt{2\Delta N}}\sum_{n=N-\Delta N}^{n=N+\Delta N}|n\rangle,$$

and the modulus of the wave function squared for the specific situation when $N = 18$ and $\Delta N = 2$. The total wave function is a coherent sum of five states centered on the $|n = 18\rangle$ state. In this case, for which time $t = 0$, the particle may be viewed as at an extreme of its

259

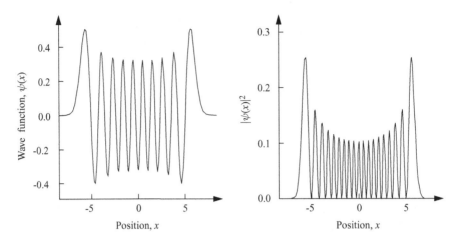

Fig. 5.E11

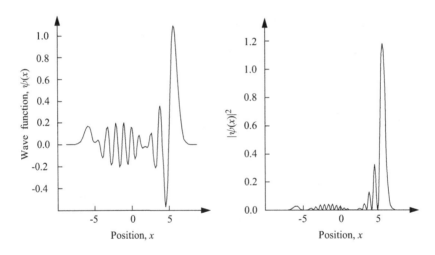

Fig. 5.E12

motion with the peak probability occurring at position $\xi = 5.7$, which is almost the same as the classical turning point $\xi_{n=18} = 6.1$ for the $|n = 18\rangle$ eigenstate.

Solutions 5.4

$$P_n^{\text{nonclassical}} = \int\limits_{|\xi| > \xi_n} \psi_n^*(\xi)\psi_n(\xi)\mathrm{d}\xi,$$

where $\xi = (m\omega/\hbar)^{1/2}x$ and the classical turning point is $\xi_n = (2n+1)^{1/2}$. For the first excited state and the second excited state,

$$\psi_1(\xi) = \left(\frac{4}{\pi}\right)^{1/4} \xi e^{-\xi^2/2},$$

$$\psi_2(\xi) = \left(\frac{1}{64\pi}\right)^{1/4} (4\xi^2 - 2)e^{-\xi^2/2}.$$

The probability of finding the particle in the nonclassical region is

$$P_1^{\text{nonclassical}} = \int\limits_{|\xi|>\sqrt{3}} \psi_1^*(\xi)\psi_1(\xi)d\xi = 0.112,$$

$$P_2^{\text{nonclassical}} = \int\limits_{|\xi|>\sqrt{5}} \psi_2^*(\xi)\psi_2(\xi)d\xi = 0.095.$$

With increasing quantum number n, the probability of finding the particle in the nonclassical region decreases. With increasing energy eigenvalue the slope of the potential at the classical turning point becomes greater, which has the effect of decreasing tunneling.

Solutions 5.5

To find $[\hat{n}, \hat{b}]$ and $[\hat{n}, \hat{b}^\dagger]$, the expressions using $\hat{n} = \hat{b}^\dagger\hat{b}$ are expanded and rearranged in terms of commutator relations,

$$\left[\hat{n}, \hat{b}\right] = \left[\hat{b}^\dagger\hat{b}, \hat{b}\right] = \hat{b}^\dagger\hat{b}\hat{b} + \left(-\hat{b}^\dagger\hat{b}\hat{b} + \hat{b}^\dagger\hat{b}\hat{b}\right) + \hat{b}\hat{b}^\dagger\hat{b} = \hat{b}^\dagger \left[\hat{b}, \hat{b}\right] + \left[\hat{b}^\dagger, \hat{b}\right]\hat{b} = -\hat{b}$$

and

$$\left[\hat{n}, \hat{b}^\dagger\right] = \left[\hat{b}^\dagger\hat{b}, \hat{b}^\dagger\right] = \hat{b}^\dagger\hat{b}\hat{b}^\dagger + \left(-\hat{b}^\dagger\hat{b}^\dagger\hat{b} + \hat{b}^\dagger\hat{b}^\dagger\hat{b}\right) + \hat{b}^\dagger\hat{b}^\dagger\hat{b} = \hat{b}^\dagger \left[\hat{b}, \hat{b}^\dagger\right] + \left[\hat{b}^\dagger, \hat{b}^\dagger\right]\hat{b} = \hat{b}^\dagger,$$

where use was made of the fact that $\left[\hat{b}, \hat{b}\right] = 0$, $\left[\hat{b}^\dagger, \hat{b}\right] = -1$, $\left[\hat{b}^\dagger, \hat{b}^\dagger\right] = 0$, and $\left[\hat{b}, \hat{b}^\dagger\right] = 1$.

Solutions 5.6

The nth energy level of the symmetric three-dimensional harmonic oscillator has energy $E_n = (n+3/2)\hbar\omega$ for which $n = n_x + n_y + n_z$ or $n - n_x = n_y + n_z$. To find the degeneracy of the nth state, consider n_x fixed. There are now $(n - n_x + 1)$ possibilities for the pair $\{n_y, n_z\}$, which are

$$\{0, n - n_x\}, \{1, n - n_x - 1\}, \ldots, \{n - n_x, 0\}.$$

The degeneracy of the nth energy level is therefore

$$g_n = \sum_{n_x=0}^{n} (n - n_x + 1) = (n+1)\sum_{n_x=0}^{n} 1 - \sum_{n_x=0}^{n} n_x = \frac{(n+1)(n+2)}{2},$$

261

or

$$g_n = \frac{(n+2)!}{n!2!} = \frac{(n+2)(n+1)n!}{2n!} = \frac{1}{2}(n+2)(n+1).$$

Solutions 5.7

For the operator $\hat{A}$,

$$(\Delta A)^2 = \left\langle \left(\hat{A} - \langle A \rangle\right)^2 \right\rangle = \left\langle \hat{A}^2 + \langle A \rangle^2 - 2\hat{A}\langle A \rangle \right\rangle = \left\langle A^2 \right\rangle + \langle A \rangle^2 - 2\langle A \rangle\langle A \rangle.$$

Hence,

$$(\Delta A)^2 = \left\langle A^2 \right\rangle - \langle A \rangle^2$$

or

$$\Delta A = \sqrt{\langle A^2 \rangle - \langle A \rangle^2}.$$

There are two possible approaches to solving the problem: (i) calculate the matrix elements by performing the integrals directly, or (ii) use the properties of the raising and lowering operators of the harmonic oscillator to obtain the solution.

First approach to a solution:

The variance $\Delta x^2 = \langle x^2 \rangle - \langle x \rangle^2$ and $\langle x \rangle = 0$ from symmetry, so only $\langle x^2 \rangle$ need be found. To do so, the integral

$$\langle x^2 \rangle = \frac{\int\limits_{-\infty}^{\infty} x^2 e^{-x^2/2\sigma_x^2} dx}{\int\limits_{-\infty}^{\infty} |\psi_0|^2 dx}$$

must be solved. Using the standard integrals,

$$\int\limits_0^{\infty} x^2 e^{-ax^2} dx = \frac{1}{4a}\sqrt{\frac{\pi}{a}} \quad \text{and} \quad \int\limits_0^{\infty} e^{-ax^2} dx = \frac{1}{2}\sqrt{\frac{\pi}{a}},$$

it follows that

$$\langle x^2 \rangle = \Delta x^2 = \frac{2\sigma_x^2}{2} \frac{\sigma_x\sqrt{2\pi}}{\sigma_x\sqrt{2\pi}} = \sigma_x^2 = \frac{\hbar}{2m\omega},$$

so

$$A_0 = (2\pi)^{-1/4}\sigma_x^{-1/2} = \left(\frac{m\omega}{\hbar}\right)^{1/4}.$$

The uncertainty in momentum is found using

$$\Delta p_x^2 = \left\langle p_x^2 \right\rangle - \langle p_x \rangle^2,$$

where $\hat{p} = -i\hbar\, d/dx$, and

$$\langle p_x \rangle = A_0^2 \int_{-\infty}^{\infty} e^{-x^2/4\sigma_x^2} \hbar \frac{d}{dx} e^{-x^2/4\sigma_x^2} dx = A_0^2 \int_{-\infty}^{\infty} e^{-x^2/2\sigma_x^2} \frac{-2x\hbar}{4\sigma_x^2} dx = 0$$

from symmetry. Hence,

$$\Delta p_x^2 = \langle p_x^2 \rangle = \frac{-\hbar^2}{\sigma_x \sqrt{2\pi}} \int_{-\infty}^{\infty} e^{-x^2/4\sigma_x^2} \frac{d^2}{dx^2} e^{-x^2/4\sigma_x^2} dx.$$

Note that

$$\frac{d^2}{dx^2} e^{-x^2/4\sigma_x^2} = \frac{-2}{4\sigma_x^2} e^{-x^2/2\sigma_x^2} + \frac{4x^2}{16\sigma_x^4} e^{-x^2/2\sigma_x^2},$$

so

$$\Delta p_x^2 = \frac{-\hbar^2}{\sigma_x \sqrt{2\pi}} \left(\frac{-1}{\sigma_x} \int_{-\infty}^{\infty} e^{-x^2/2\sigma_x^2} dx + \frac{1}{2\sigma_x^4} \int_{-\infty}^{\infty} x^2 e^{-x^2/2\sigma_x^2} dx \right)$$

$$= \frac{-\hbar^2}{\sigma_x \sqrt{2\pi}} \left(\frac{-1}{2\sigma_x^2} \sigma_x \sqrt{2\pi} + \frac{1}{2\sigma_x^4} \frac{\sigma_x^2}{2} \sigma_x \sqrt{2\pi} \right) = \frac{-\hbar^2}{\sigma_x \sqrt{2\pi}} \left(\frac{-1}{4\sigma_x^2} \sigma_x \sqrt{2\pi} \right).$$

This gives

$$\Delta p_x^2 = \frac{\hbar^2}{4\sigma_x^2} = \frac{\hbar\omega m}{2},$$

and so

$$\Delta p_x^2 \Delta x^2 = \frac{\hbar^2}{4\sigma_x^2} \sigma_x^2 = \frac{\hbar^2}{4}$$

or

$$\Delta p_x \Delta x = \frac{\hbar}{2},$$

which is the uncertainty relation.

The *second* approach to a solution begins by noting that the ground state of the harmonic oscillator is defined by $\hat{b}|0\rangle = 0$ and that $\langle j|k \rangle = \delta_{jk}$.

Since

$$\hat{x} = \sqrt{\frac{\hbar}{2m\omega}} \left(\hat{b} + \hat{b}^\dagger \right),$$

it follows that

$$\langle x \rangle = \sqrt{\frac{\hbar}{2m\omega}} \left\langle 0 \left| \left(\hat{b} + \hat{b}^\dagger \right) \right| 0 \right\rangle = \sqrt{\frac{\hbar}{2m\omega}} \left(\langle 0|\hat{b}|0 \rangle + \langle 0|1 \rangle \right) = 0,$$

$$\langle x^2 \rangle = \sqrt{\frac{\hbar}{2m\omega}} \left\langle 0 \left| \left(\hat{b}\hat{b} + \hat{b}^\dagger\hat{b}^\dagger + \hat{b}\hat{b}^\dagger + \hat{b}^\dagger\hat{b} \right) \right| 0 \right\rangle = \left(\frac{\hbar}{2m\omega} \right),$$

for the ground state and for the general state $|n\rangle$,

$$\left\langle x^2 \right\rangle = \left(\frac{\hbar}{2m\omega} \right)(1 + 2n).$$

Similarly, $\langle p_x \rangle = 0$, $\langle p_x^2 \rangle = (\hbar m\omega/2)$, and $\langle p_x^2 \rangle = (\hbar m\omega/2)(1+2n)$ for general state $|n\rangle$. Hence,

$$\Delta x^2 \Delta p_x^2 = \left\langle x^2 \right\rangle \left\langle p_x^2 \right\rangle = \hbar^2 m\omega/4m\omega = \hbar^2/4$$

for the ground state, or

$$\Delta x^2 \Delta p_x^2 = \left\langle x^2 \right\rangle \left\langle p_x^2 \right\rangle = \frac{\hbar^2}{4}(1 + 2n)^2$$

for the general state $|n\rangle$. For the ground state,

$$\Delta x \Delta p_x = \frac{\hbar}{2},$$

which is the minimum value given by the uncertainty relation. For the general state $|n\rangle$, the result is

$$\Delta x \Delta p_x = \frac{\hbar(1 + 2n)}{2}.$$

Solutions 5.8
The ground state and the *second* excited state of a charged particle of mass m in a one-dimensional harmonic oscillator potential are both excited. To find the expectation value of the particle position x as a function of time, it should be noted that $|0\rangle$ and $|2\rangle$ have no overlap when operated on by the position operator $\hat{x}$. Hence,

$$\langle x(t) \rangle = \frac{\hbar}{2m\omega} \langle a_0 \langle 0| + a_2 a_0 \langle 2|\hat{b} + \hat{b}^\dagger a_0|0\rangle + a_2|2\rangle \rangle = 0,$$

and so there is no time dependence for the position expectation operator. $|0, t\rangle = \phi_0(x)e^{-i\hbar\omega t/2}$ and $|2, t\rangle = \phi_2(x)e^{-i5\hbar\omega t/2}$, where $\phi_0(x)$ and $\phi_2(x)$ are the spatial wave functions for the ground state and the second excited state, respectively.

In the presence of a constant electric field $\mathbf{E}$ in the x direction, the particle of charge e in the harmonic oscillator potential experiences a constant x-directed force $e|\mathbf{E}|$, which shifts the expectation value of position from $x = 0$ to x_d.

The new Hamiltonian is

$$\hat{H} = \frac{\hat{p}^2}{2m} + \frac{\kappa_0 \hat{x}^2}{2} + e|\mathbf{E}|\hat{x},$$

and the amount of the shift is given by $e|\mathbf{E}| = \kappa_0 x_d$, where $\kappa_0 = m\omega^2$ is the oscillator force constant. The potential energy is

$$V = \frac{1}{2}m\omega^2 x^2 - \frac{1}{2}\frac{e^2|\mathbf{E}|^2}{\kappa_0}.$$

The oscillator frequency remains the same, but all of the energy levels are uniformly reduced by a fixed amount

$$\frac{-e^2|\mathbf{E}|^2}{2m\omega^2}.$$

Solutions 5.9

The main MATLAB computer program for this exercise calls solve_schM, which was used in the solution of Exercise 2.6.

In this exercise, the first four energy eigenvalues are $E_0 = 0.0147\,\text{eV}$, $E_1 = 0.0443\,\text{eV}$, $E_2 = 0.0738\,\text{eV}$, and $E_3 = 0.1032\,\text{eV}$. The potential and first four wave functions (eigenfunctions) are shown in Fig. 5.E13. As expected for a harmonic oscillator, the separation in energy between adjacent states is independent of eigenvalue, in this case $\hbar\omega = 0.0295\,\text{eV}$.

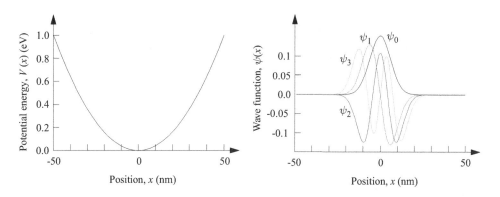

Fig. 5.E13

5.11 Problems

Problem 5.1

(a) Write down the Hamiltonian for a particle of mass m in a one-dimensional harmonic oscillator potential in terms of momentum $\hat{p}_x$ and position $\hat{x}$ operators.

(b) If operators

$$\hat{b} = \sqrt{\frac{m\omega}{2\hbar}}\left(\hat{x} + \frac{i\hat{p}_x}{m\omega}\right),$$

$$\hat{b}^\dagger = \sqrt{\frac{m\omega}{2\hbar}}\left(\hat{x} - \frac{i\hat{p}_x}{m\omega}\right),$$

are defined show that the Hamiltonian can be expressed as

$$\hat{H} = \frac{\hbar\omega}{2}\left(\hat{b}\hat{b}^\dagger + \hat{b}^\dagger\hat{b}\right).$$

265

(c) Derive the commutation relation $\left[\hat{b}, \hat{b}^{\dagger}\right]$ by writing out the differential form of $\hat{b}$ and $\hat{b}^{\dagger}$ and operating on a dummy wave function.

(d) Using the result from (c), show that the Hamiltonian is

$$\hat{H} = \hbar\omega \left(\hat{b}^{\dagger}\hat{b} + \frac{1}{2}\right).$$

Problem 5.2

(a) Find the expectation value of position and momentum for the first excited state for a particle of mass m in a one-dimensional harmonic oscillator potential.

(b) Find the value of the product in uncertainty in position Δx and momentum Δp_x for the first excited state of a particle of mass m in a one-dimensional harmonic oscillator potential.

Problem 5.3

Often an operator $\hat{A}$ is time-independent but the corresponding numerical value of the observable A has a spread in values ΔA about an average value $\langle A(t)\rangle$ and varies with time because the system is described by a wave function $\psi(x, t)$ that is not an eigenstate. The change in $\langle A(t)\rangle$ in time interval Δt is the slope $(\mathrm{d}/\mathrm{d}t)\langle A(t)\rangle$ multiplied by Δt. Hence, the exact time t at which the numerical value of the observable A passes through a specific value will actually have a spread in values Δt such that

$$\Delta t = \frac{\Delta A}{\left|\frac{\mathrm{d}}{\mathrm{d}t}\langle A\rangle\right|}.$$

(a) Use the generalized uncertainty relation

$$\Delta A \Delta B \geq \left|\frac{1}{2}\left\langle\left[\hat{A}, \hat{B}\right]\right\rangle\right|$$

for time independent operators $\hat{A}$ and $\hat{B}$ to show that

$$\Delta E \Delta t \geq \frac{\hbar}{2}.$$

(b) Show that the spread in photon number Δn and phase $\Delta\theta$ for light of frequency ω is

$$\Delta n \Delta\theta \geq \frac{1}{2},$$

and that for a Poisson distribution of such photons,

$$\Delta\theta \geq \frac{1}{2\sqrt{\langle n\rangle}}.$$

(c) Apply the results in (b) and determine Δn and $\Delta\theta$ for a 100 ps pulse of $\lambda = 1\,500\,\mathrm{nm}$ wavelength light from a 10 μW source.

Problem 5.4

A particle of charge e, mass m, and momentum p oscillates in a one-dimensional harmonic potential $V(x) = m\omega_0^2 x^2/2$ and is subject to an oscillating electric field $|\mathbf{E}|\cos(\omega t)$.

(a) Write down the Hamiltonian of the system.

(b) Find $\frac{d}{dt}\langle x \rangle$.

(c) Find $\frac{d}{dt}\langle p \rangle$ and show that

$$\frac{d}{dt}\langle p \rangle = -\left\langle \frac{d}{dx}V(x)\right\rangle.$$

Describe conditions when the quantum mechanical result

$$m\frac{d^2}{dt^2}\langle x \rangle = -\left\langle \frac{d}{dx}V(x)\right\rangle$$

is the same as Newton's second law in which the force on a particle is

$$F = m\frac{d^2 x}{dt^2} = -\frac{d}{dx}V(x).$$

(d) Find $\frac{d}{dt}\langle H \rangle$.

(e) Use the results in (b) and (c) to find the time dependence of the expectation value of position. What happens to the maximum value of $\langle x \rangle$ as a function of time when $\omega = \omega_0$ and when ω is close in value to ω_0?

Problem 5.5

Express the total ground-state energy of a one-dimensional harmonic oscillator as the sum of potential and kinetic energy terms involving displacement Δx and momentum Δp_x. Assume the minimum uncertainty relation $\Delta x \Delta p_x = \hbar/2$ and find the ground-state energy of the system.

Problem 5.6

(a) What is the minimum energy E_0 stored in a resonant LC circuit?

(b) Find an expression for the value of capacitance C if the charging energy associated with the coulomb blockade for the capacitor is the same as E_0.

(c) If the inductor has value $L = 10^{-8}$ H, what is the resonant oscillation frequency of the circuit and what is the value of the capacitance C?

(d) If the current in the circuit can be measured to an accuracy of one electron per oscillation, how accurately can the voltage of the circuit be determined?

Problem 5.7

An electron is confined by a one-dimensional harmonic potential.

(a) What is the value of the mean electric-field dipole moment when the electron is in eigenstate $|\phi_n\rangle$?

(b) What is the value of the mean electric-field dipole moment in the presence of a uniform static electric field $\mathbf{E}$ in the x direction?

(c) What is the static electric susceptibility and permittivity of the system?

(d) Estimate the frequency-dependent electric susceptibility of the system.

Problem 5.8

The annihilation operator for a particle of mass m in a one-dimensional harmonic oscillator potential is

$$\hat{b} = \sqrt{\frac{m\omega}{2\hbar}}\left(\hat{x} + \frac{i\hat{p}_x}{m\omega}\right),$$

where $\hat{x}$ is the position operator and $\hat{p}_x$ is the momentum operator.

(a) Show that $\langle n|\hat{b}^\dagger|m\rangle = \langle m|\hat{b}|n\rangle^*$ and $\langle \hat{b}^\dagger n|\hat{b}^\dagger n\rangle = \langle n|\hat{b}\hat{b}^\dagger n\rangle$.

(b) Show that $\langle m|\hat{b}^\dagger|n\rangle \neq \langle n|\hat{b}^\dagger|m\rangle^*$ and, hence, that $\hat{b}^\dagger$ is not a Hermitian operator.

(c) Show that the number operator $\hat{n} = \hat{b}^\dagger\hat{b}$ is Hermitian.

(d) Show that the position operator $\hat{x}$ and the momentum operator $\hat{p}_x$ are Hermitian operators.

Problem 5.9

(a) Numerically evaluate and plot the time evolution of expectation value of position $\langle x(t)\rangle$, probability $|\psi(x,t)|^2$, and current density $J(x,t)$, for a superposition of the ground state, ψ_0, and first excited state, ψ_1, of an electron confined to motion in a one-dimensional rectangular potential well of thickness L and infinite potential elsewhere. The superposition state is

$$\psi = \frac{1}{\sqrt{N+1}}\sum_{n=0}^{N}\psi_n,$$

where $N = 1$. Repeat the calculation but now for a superposition state in which $N = 9$, and explain the differences in the results of the two calculations.

What is the value of the guaranteed full revival time and what is the smallest possible time between full revivals?

(b) Numerically evaluate the time evolution of $\langle x(t)\rangle$ and $|\psi(x,t)|^2$ for a superposition of the ground state and first excited state of the harmonic oscillator.

(c) Numerically evaluate the time evolution of $\langle x(t)\rangle$ and $|\psi(x,t)|^2$ for $N = 18$ and $\Delta N = 2$ in which the superposition state of a harmonic oscillator with eigenfunctions $\psi_n(x,t)$ is

$$\psi(x,t) = \frac{1}{\sqrt{2\Delta N}}\sum_{n=N-\Delta N}^{n=N+\Delta N}\psi_n(x,t).$$

(d) The coherent quantum superposition of harmonic oscillator eigenstates $\psi_n(x)$ that best describes the classical harmonic oscillator is

$$\psi_\alpha(x,t) = e^{\frac{-|\alpha|^2}{2}}\sum_{n=0}^{\infty}\frac{\alpha^n}{\sqrt{n!}}\psi_n(x)e^{-i\omega_n t},$$

where $\omega_n = \omega\left(n + \frac{1}{2}\right)$, and α is an eigenvalue of the annihilation operator $\hat{b}$ acting on the state $|\alpha\rangle$. Numerically evaluate the time evolution of $\langle x(t)\rangle$ and $|\psi_\alpha(x,t)|^2$ for $\alpha = 1$, $\alpha = 2$, and $\alpha = 9$. Find analytic expressions for the time evolution of expectation value of position and momentum.

Problem 5.10

Pure exponential decay $e^{-t/\tau}$ starting from a constant value at time $t = 0$ is forbidden in a closed unitary system evolving due to a Hermitian time-independent Hamiltonian $\hat{H}$.

(a) The probability amplitude that an initial state $|j\rangle$ is observed in state $|k\rangle$ at time t is $A_{kj}(t) = \langle k|\hat{U}(t)|j\rangle$, where $\hat{U}(t)$ is the unitary time evolution operator. Analytically show that the probability of measuring and observing state $|j\rangle$ is symmetric in time.

(b) Show analytically that any measurable state of the system cannot evolve as a simple exponential, $e^{-t/\tau}$, for either short times ($t \ll \tau$) or long times ($t \gg \tau$), thereby proving that exponential decay is incompatible with unitary evolution.

(c) Show numerically that the initial change in expectation value of position for the closed unitary system described in Problem 5.9(a) with $N \gg 2$ does not evolve as a simple exponential.

Problem 5.11

(a) Consider the wave function in Problem 5.9(a) with $N = 4$. Calculate numerically the real part of ψ as a function of time at position $x_0 = L/5$ and find the value of the full revival time (the time it takes for the wave function to return to its original state). Show that peaks in the FFT spectrum are energy eigenvalues of the system. Show that this result generalizes to states of an arbitrary potential, $V(x)$, so long as x_0 is not an eigenfunction node.

(b) Demonstrate how to use the information in (a) to find the eigenfunctions ψ_n. Show that the numerical approach generalizes and so may be used to find the eigenstates of an arbitrary potential, $V(x)$.

(c) Consider the wave function in Problem 5.9(d) with $\alpha = 2$. Calculate numerically the real part of ψ as a function of time at position $x_0 = 0$. Show that peaks in the FFT spectrum are energy eigenvalues of the system. Comment on anything learned.

Problem 5.12

Find the eigenenergies, eigenfunctions, and degeneracy of an isotropic two-dimensional harmonic oscillator by separation into Cartesian coordinates.

Problem 5.13

Numerical methods exist to solve the dynamics of a classical particle of mass m with position x and momentum p described by a Hamiltonian of the form $H = T(p) + V(x)$. For the simple harmonic oscillator with spring constant κ_0, the kinetic energy $T = p^2/2m$ and the potential energy $V = m\omega^2 x^2/2$, where angular frequency $\omega = \sqrt{\kappa_0/m}$.

(a) Defining canonical relations $-\frac{dH}{dx} = \frac{dp}{dt}$ and $\frac{dH}{dp} = \frac{dx}{dt}$, show that

$$\dot{x} = v \equiv \frac{dx}{dt} \text{ and } \dot{v} = -\omega^2 x,$$

and that the analytic solution for position is $x(t) = x(0)\cos(\omega t) + (v(0)/\omega)\sin(\omega t)$.

(b) A phase-space plot of p as a function of x is an ellipse whose area is constant because energy is conserved. Rewrite equations for $\dot{x}$ and $\dot{v}$ in (a) using discretization of space and

269

time where $dx = \Delta x$ and $dt = \Delta t$, such that $x_n = n\Delta x$ and $t_n = n\Delta t$, to obtain a set of equations for x_{n+1} and v_{n+1} in terms of x_n, v_n, Δt, and ω. Write the set of linear equations in matrix form $z_{n+1} = \mathbf{A}z_n$, such that $z_n = \mathbf{A}z_{n-1} = \mathbf{A}^n z_0$.

(c) Numerical stability implies that the phase-space vector norm $\|z_n\|$ remains bounded for all n. For any consistent matrix norm,

$$\rho(\mathbf{A}) = \max_i(|\lambda_i|) = \lim_{n\to\infty} \|\mathbf{A}^n\|^{1/n},$$

where λ_i are the eigenvalues of $\mathbf{A}$. The spectral radius theorem states that given a matrix $\mathbf{A}$ over the complex numbers, the iterations $z_n = \mathbf{A}^n z_0$ are bounded if $\rho(\mathbf{A}) \leq 1$. Show that under Euler discretization in part (b), energy is not conserved by explicitly demonstrating that $\rho(\mathbf{A}) > 1$. Thus the phase-space area is not conserved over time.

(d) In addition to the utility of the Baker–Campbell–Hausdorff formula in quantum mechanics (see Problem 4.15(b)), the identity can be exploited to integrate Hamiltonians of the form $H = T(p) + V(x)$. By constructing explicit and time-reversible symplectic integrators of higher order that maintain the structure of the Hamiltonian, it is possible to suppress numerical error stemming from the energy non-conserving discretization of sets of coupled equations. If $e^{\hat{A}+\hat{B}} = e^{-\frac{1}{2}[\hat{A},\hat{B}]}e^{\hat{A}}e^{\hat{B}}$ is true, show that

$$e^{2\hat{A}+\hat{B}} = e^{\hat{A}}e^{\hat{B}}e^{\hat{A}}.$$

(e) For Hamiltonians of the form $H = T(p) + V(x) = p^2/2m + V(x)$, show that

$$e^{t\frac{d}{dt}} = e^{t(P+X)},$$

where

$$P = -\frac{dV}{dx}\frac{d}{dp}$$

and

$$X = \frac{dT}{dp}\frac{d}{dx}.$$

(f) Since $[\hat{A}, [\hat{A}, \hat{B}]] \neq 0$ in general, show that to order $O(\Delta t^2)$, the symmetric symplectic integrator can be written as

$$U(\Delta t)^{t/\Delta t} = e^{t(P+X)} + O\left(\Delta t^2\right),$$

where

$$U(\Delta t)^{t/\Delta t} = \left(e^{\Delta t P/2}e^{\Delta t X}e^{\Delta t P/2}\right)^{t/\Delta t}.$$

What does this result indicate about the energy conserving properties of a symplectic integrator?

Problem 5.14

(a) Show that the Hamiltonian for a parallel LC circuit with current flow I and voltage V may be written as

$$\hat{H} = \frac{\hat{\phi}_B^2}{2L} + \frac{\hat{Q}^2}{2C},$$

for which the commutator for charge, $\hat{Q}$, and magnetic flux, $\hat{\phi}_B$, operators is $\left[\hat{Q}, \hat{\phi}_B\right] = i\hbar$ and resonant frequency $\omega_0 = 1/\sqrt{LC}$.

(b) Define the operator

$$\hat{b} = \left(\frac{1}{2\hbar}\sqrt{\frac{C}{L}}\right)^{1/2} \hat{\phi}_B + i\left(\frac{1}{2\hbar}\sqrt{\frac{L}{C}}\right)^{1/2} \hat{Q},$$

and show that

$$\hat{\phi}_B = \left(\frac{\hbar}{2}\sqrt{\frac{L}{C}}\right)^{1/2} \left(\hat{b} + \hat{b}^\dagger\right),$$

$$\hat{Q} = \left(\frac{\hbar}{2}\sqrt{\frac{C}{L}}\right)^{1/2} \left(\hat{b}^\dagger - \hat{b}\right),$$

and

$$\hat{H} = \hbar\omega_0 \left(\hat{b}^\dagger \hat{b} + \frac{1}{2}\right).$$

(c) For number states $|n\rangle$ and deviations in observable A such that $(\Delta A)^2 = \langle A^2\rangle - \langle A\rangle$, show and explain why

$$\langle\phi_B\rangle = \langle Q\rangle = 0,$$

$$(\Delta\phi_B)^2 = \hbar\sqrt{\frac{L}{C}}\left(n + \frac{1}{2}\right),$$

$$(\Delta Q)^2 = \hbar\sqrt{\frac{C}{L}}\left(n + \frac{1}{2}\right),$$

and

$$\Delta\phi_B\Delta Q = \hbar\left(n + \frac{1}{2}\right).$$

(d) Using results from (c), show that $\langle I\rangle = \langle V\rangle = 0$,

$$(\Delta I)^2 = \frac{\hbar\omega_0}{L}\left(n + \frac{1}{2}\right),$$

$$(\Delta V)^2 = \frac{\hbar\omega_0}{C}\left(n + \frac{1}{2}\right),$$

and

$$\Delta I \Delta V = \frac{\hbar \omega_0^2}{L} \left(n + \frac{1}{2} \right).$$

(e) If the current in a resonant circuit with inductance $L = 1\,\text{nH}$ and capacitance $C = 100\,\text{fF}$ can be measured to an accuracy of one electron per oscillation, how accurately can the voltage be determined? Compare the result with capacitor thermal voltage noise $\sqrt{k_B T / C}$ at room temperature ($T = 300\,\text{K}$).

Problem 5.15

The eigenstate representation uses matrix elements $\langle m|\hat{A}|n\rangle$ for operator $\hat{A}$ where $|n\rangle$ and $|m\rangle$ are eigenstates. For a particle mass m_0 in a one-dimensional harmonic oscillator potential:

(a) Show that $\langle m|\hat{H}|n\rangle = \hbar\omega \left(n + \frac{1}{2} \right) \delta_{m,n}$ and form the matrix $\mathbf{H}$ associated with the Hamiltonian whose rows and columns are labeled by m and n, respectively.

(b) Find the matrix $\mathbf{b}^\dagger$ associated with the creation operator.

(c) Find the matrix $\mathbf{b}$ associated with the annihilation operator.

(d) Find the matrix $\mathbf{x}$ associated with the position operator.

(e) Find the matrix $\mathbf{p}$ associated with the momentum operator.

(f) By multiplying the matrix $\mathbf{x}$ and $\mathbf{p}$ find the matrix $\mathbf{xp} - \mathbf{px}$ associated with the commutator for position and momentum.

(g) What is the expectation value $\langle x \rangle$ and $\langle x^2 \rangle$ of a pure state $|n\rangle$?

(h) What is the expectation value $\langle x \rangle$ and $\langle x^2 \rangle$ of a mixed state $\frac{|n\rangle + |m\rangle}{\sqrt{2}}$?

Problem 5.16

(a) The ground state, $|n = 0\rangle$, of an electron mass m_0 moving in a one-dimensional potential $V(x) = \xi x^2$ is defined by $\hat{b}|n = 0\rangle = 0$, where the annihilation operator is

$$\hat{b} = \left(\frac{m_0 \xi}{2\hbar^2} \right)^{\frac{1}{4}} \left(\hat{x} + \frac{i\hat{p}}{\sqrt{2m_0\xi}} \right),$$

$\hat{x}$ is the position operator, $\hat{p}$ is the momentum operator, and ξ is a constant. Find the normalized wave function and numerical value of the energy eigenvalue for the ground state $|n = 0\rangle$ when $\xi = 4.6 \times 10^{-3}\,\text{kg}\,\text{rad}^2\,\text{s}^{-2}$.

(b) State the value of the lowest three energy eigenvalues and sketch the corresponding wave functions if the potential in (a) is *modified* so that $V(x < 0) = \infty$.

(c) Sketch the wave function for the ground state with energy eigenvalue E_0 and first excited state with energy eigenvalue E_1 if the potential in (b) is modified as shown in Fig 5.P14.

In answering part (a) of this question, use may be made of the standard integral

$$\int_{-\infty}^{\infty} e^{-ax^2}\,dx = \sqrt{\frac{\pi}{a}}.$$

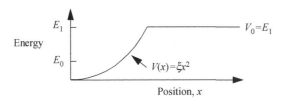

Fig. 5.P14

Problem 5.17

The Hamiltonian of a particle mass m moving in a one-dimensional harmonic oscillator potential may be written

$$\hat{H} = \hbar\omega\left(\hat{b}^\dagger\hat{b} + \frac{1}{2}\right),$$

where ω is the angular frequency of oscillation and the operator

$$\hat{b} = \sqrt{\frac{m\omega}{2\hbar}}\left(\hat{x} + \frac{i\hat{p}}{m\omega}\right)$$

satisfies the commutation relations $\left[\hat{b}, \hat{b}^\dagger\right]$ and $\left[\hat{b}, \hat{b}\right] = \left[\hat{b}^\dagger, \hat{b}^\dagger\right] = 0$.

(a) Show that ground-state wave function $\psi_{n=0}$ is defined by $\hat{b}|0\rangle$.

(b) Find the normalized ground state and first excited state wave function.

(c) The operator

$$\frac{1}{\sqrt{2}}\left(1 + \frac{\hat{b}^\dagger}{\sqrt{n+1}}\right)$$

acts on the state ψ_n and creates a new wave function $\psi(x,t)$. Find the probability density $|\psi(x,t)|^2$, the expectation value of energy, the expectation value of position $\langle x(t)\rangle$, the value of $\langle x^2\rangle$, the uncertainty in position $\Delta x(t)$, and the minimum and maximum values of the position-momentum uncertainty product $\Delta x\Delta p$.

Problem 5.18

Expressed in terms of orthonormal Fock states $|n\rangle$, the normalized coherent state is

$$|\alpha\rangle = e^{-|\alpha|^2/2}\sum_{n=0}^{\infty}\frac{\alpha^n}{\sqrt{n!}}|n\rangle.$$

(a) Show that coherent states α_1 and α_2 are not orthogonal and have overlap such that

$$|\langle\alpha_1|\alpha_2\rangle|^2 = e^{-|\alpha_1-\alpha_2|^2},$$

so that orthogonality occurs as $|\alpha_1 - \alpha_2| \to \infty$.

(b) Show that coherent states are complete by integrating $\frac{1}{\pi} \int d\alpha \, |\alpha\rangle\langle\alpha| = 1$ over the complex plane.

(c) Show that coherent states form an over-complete set of states.

Problem 5.19

A stationary superposition non-separable state of the two-dimensional isotropic harmonic oscillator is $\psi(x, y) = a_0 \left(\phi_2(x)\phi_3(y) + \frac{2}{3}\phi_3(x)\phi_2(y) \right)$ where x and y are Cartesian coordinates.

(a) Plot $\psi(\xi_x, \xi_y)$, where position is normalized to units of $\xi_x = x\sqrt{m\omega/\hbar}$ and $\xi_y = y\sqrt{m\omega/\hbar}$.

(b) Plot probability $P(\xi_y) = \int_{\infty}^{\infty} |\psi(\xi_x, \xi_y)|^2 d\xi_x$ for the state in (a) before measuring position ξ_x. If ξ_x is measured first with the result $\xi_{x0} = 2$, plot the probability $P(\xi_y)|_{\xi_{x0}} = |\psi(\xi_{x0}, \xi_y)|^2$.

6 Fermions and Bosons

6.1 Introduction

The Hamiltonian for N particles subject to mutual two-body interactions is

$$\hat{H} = \sum_{j}^{N} \frac{\hat{p}_j^2}{2m_j} + \sum_{j}^{N} \hat{V}_j(x_j) + \sum_{j,k}^{j>k} \hat{V}_{j,k}(x_j - x_k), \tag{6.1}$$

where $j > k$ in the sum avoids double counting. The corresponding multi-particle wave function obeys the Schrödinger equation,

$$\hat{H}\psi(x_1, x_2, x_3, \ldots, x_N, t) = i\hbar \frac{\partial}{\partial t}\psi(x_1, x_2, x_3, \ldots, x_N, t), \tag{6.2}$$

where (if the particles are confined to a finite region of space) the quantity $|\psi(x_1, x_2, x_3, \ldots, x_N, t)|^2 dx_1 dx_2 dx_3 \ldots dx_N$ is the probability of finding particle 1 in the interval x_1 to x_1+dx_1, particle 2 in the interval x_2 to x_2+dx_2, and so on. The key idea is that there is a single multi-particle wave function that describes the state of the N-particle system.

As it stands, this is a complex multi-particle, or *many-body* problem that is difficult to solve. However, removing the mutual two-body interactions in Eq. (6.1) simplifies the Hamiltonian to:

$$\hat{H} = \sum_{j}^{N} \left(\frac{\hat{p}_j^2}{2m_j} + \hat{V}_j(x_j) \right) = \sum_{j}^{N} \hat{H}_j. \tag{6.3}$$

If the one-body potential $V_j(x_j)$ is time independent, then the multi-particle wave function is a product,

$$\psi(x_1, x_2, x_3, \ldots, x_N) = \psi_1(x_1)\psi_2(x_2)\psi_3(x_3)\cdots\psi_N(x_N) = \prod_{j}^{N} \psi_j(x_j), \tag{6.4}$$

that satisfies the time-independent Schrödinger equation

$$\hat{H}_j\psi_j(x_j) = E_j\psi_j(x_j), \tag{6.5}$$

and the system may be described as a sum of N single-particle Hamiltonians.

To see this, note that if a Hamiltonian is of separable form, $\hat{H} = \hat{H}_1(x_1) + \hat{H}_2(x_2)$, then, since there are no terms depending on both x_1 and x_2, the wave function can be factorized such that $\psi(x_1, x_2) = \psi(x_1)\psi(x_2)$.

6.1.1 The Symmetry of Indistinguishable Particles

Unlike in classical mechanics, where labels can be used to distinguish mechanically identical objects, in quantum mechanics elementary particles are indistinguishable. This fact introduces a symmetry to a multi-particle wave function describing a system containing identical particles. One consequence is that the statistically most likely energy distribution of identical indistinguishable noninteracting particles in thermal equilibrium falls into one of two classes. Identical integer-spin particles behave as *bosons* and half-odd-integer-spin particles behave as *fermions*.

The Bose–Einstein energy distribution applies to *boson* particles such as photons and phonons. Photons have spin quantum number ± 1, and phonons have spin zero.

The Fermi–Dirac energy distribution applies to many identical indistinguishable non-interacting half-odd-integer-spin particles, an example of which are electrons that have spin quantum number $\sigma_s = \pm 1/2$. Such *fermion* particles obey the Pauli exclusion principle, which states that no identical indistinguishable noninteracting half-odd-integer-spin particles may occupy the same state. An almost equivalent statement is that in a system of N fermion particles, the total eigenfunction must be antisymmetric (must change sign) upon the permutation of any two particles.

To illustrate permutation symmetry, consider two noninteracting identical particles at positions x_1 and x_2, respectively, which, for the purpose of reasoning, are labeled (1) and (2). The noninteracting particles satisfy their respective Hamiltonians so that

$$\hat{H}_1 \psi_1(x_1) = E_1 \psi_1(x_1), \tag{6.6}$$

$$\hat{H}_2 \psi_2(x_2) = E_2 \psi_2(x_2), \tag{6.7}$$

where E_1 and E_2 are energy eigenvalues. The total Hamiltonian is just the sum of the individual Hamiltonians,

$$\hat{H} = \hat{H}_1 + \hat{H}_2, \tag{6.8}$$

with the solution characterized by

$$\hat{H}\psi = (E_1 + E_2)\psi, \tag{6.9}$$

where $E_1 + E_2$ is the total energy in the system and ψ is the multiparticle wave function.

Because the particles are identical, it is possible to interchange (or permute) them without affecting the total energy eigenvalue of the Hamiltonian, $\hat{H}$. A *permutation operator* $\hat{\mathcal{P}}_{12}$ that interchanges the identical particles (1) and (2) must commute with the Hamiltonian $\hat{H}$ so that $\hat{\mathcal{P}}_{12}\hat{H} = \hat{H}\hat{\mathcal{P}}_{12}$, or

$$\left[\hat{\mathcal{P}}_{12}, \hat{H}\right] = 0. \tag{6.10}$$

It is therefore possible to choose common eigenfunctions for the two operators $\hat{H}$ and $\hat{\mathcal{P}}_{12}$. The permutation operator has eigenvalue λ such that

$$\hat{\mathcal{P}}_{12}\psi = \lambda\psi, \tag{6.11}$$

$$\hat{\mathcal{P}}_{12}\hat{\mathcal{P}}_{12}\psi = \lambda^2\psi = \psi, \tag{6.12}$$

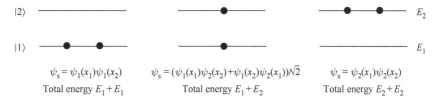

$\psi_{\mathrm{s}} = \psi_1(x_1)\psi_1(x_2)$
Total energy $E_1 + E_1$

$\psi_{\mathrm{s}} = (\psi_1(x_1)\psi_2(x_2) + \psi_1(x_2)\psi_2(x_1))/\sqrt{2}$
Total energy $E_1 + E_2$

$\psi_{\mathrm{s}} = \psi_2(x_1)\psi_2(x_2)$
Total energy $E_2 + E_2$

Fig. 6.1 Ways of arranging two identical indistinguishable particles of integer spin between two eigenfunctions and their associated energy eigenvalues.

and so $\lambda^2 = 1$ or $\lambda = \pm 1$. This means that the wave function ψ can only be symmetric or antisymmetric under particle exchange due to symmetry built into the system.[1] The symmetric wave functions for the two-particle system under consideration are

$$\psi_{\mathrm{s}} = \psi_1(x_1)\psi_1(x_2), \tag{6.13}$$
$$\psi_{\mathrm{s}} = \psi_2(x_1)\psi_2(x_2), \tag{6.14}$$
$$\psi_{\mathrm{s}} = \frac{1}{\sqrt{2}}(\psi_1(x_1)\psi_2(x_2) + \psi_1(x_2)\psi_2(x_1)). \tag{6.15}$$

The subscript on ψ labels the eigenstate. Figure 6.1 illustrates the different ways of maintaining a symmetric multi-particle wave function while distributing two identical indistinguishable particles between two eigenfunctions $|1\rangle$ and $|2\rangle$ with eigenenergies E_1 and E_2, respectively. A symmetric multi-particle wave function allows particles to occupy the same integer spin state.

The antisymmetric wave function ψ_{a} may be found from the determinant of the 2×2 matrix that describes state and particle occupation,

$$\psi_{\mathrm{a}} = \frac{1}{\sqrt{2}} \begin{vmatrix} \psi_1(x_1) & \psi_1(x_2) \\ \psi_2(x_1) & \psi_2(x_2) \end{vmatrix}. \tag{6.16}$$

Here, *rows label the state* and *columns label the particle*. Equation (6.16) is an example of a *Slater determinant*, which, in this case, gives the wave function

$$\psi_{\mathrm{a}} = \frac{1}{\sqrt{2}}(\psi_1(x_1)\psi_2(x_2) - \psi_1(x_2)\psi_2(x_1)). \tag{6.17}$$

Figure 6.2 illustrates the only way of arranging two identical indistinguishable half-odd integer spin particles between two eigenfunctions and their associated energy eigenvalues.

For more than two identical noninteracting fermion particles, the antisymmetric state can be found using the Slater determinant of a larger matrix. For N particles, this gives a ground-state wave function,

$$\psi_{\mathrm{a}}(x_1, x_2, \ldots, x_N) = \frac{1}{\sqrt{N!}} \begin{vmatrix} \psi_1(x_1) & \psi_1(x_2) & \cdots & \psi_1(x_N) \\ \psi_2(x_1) & \psi_2(x_2) & \cdots & \psi_2(x_N) \\ \vdots & & & \vdots \\ \psi_N(x_1) & & \cdots & \psi_N(x_N) \end{vmatrix}. \tag{6.18}$$

[1] The discrete permutation symmetry for identical indistinguishable particles is always obeyed. However, some quasiparticles in two dimensions exhibit any phase, called a Berry phase, upon adiabatic interchange. These unusual particles are called anyons.

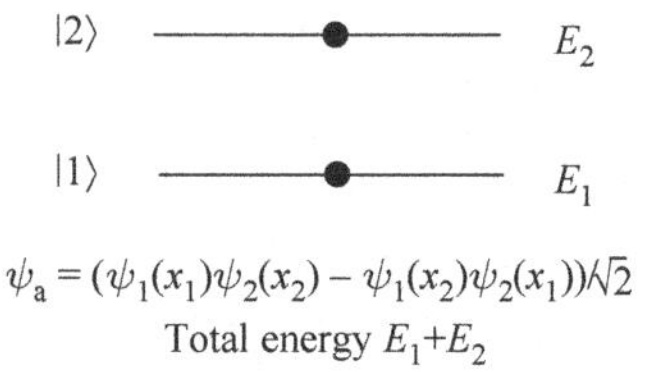

$$\psi_a = (\psi_1(x_1)\psi_2(x_2) - \psi_1(x_2)\psi_2(x_1))/\sqrt{2}$$
Total energy $E_1 + E_2$

Fig. 6.2 The only way of arranging two identical indistinguishable half-odd-integer-spin particles between two eigenfunctions and their associated energy eigenvalues.

The interchange of any two particles causes the sign of the antisymmetric multi-particle wave function ψ_a to change, since it involves the interchange of two columns. Expansion of the determinant has $N!$ terms, which take into account all possible permutations of the particles among N states. If any two single-particle eigenfunctions are the same, then those two particles are in the same state, and it follows that the multi-particle wave function $\psi_a = 0$, since the determinant will vanish. This fact is known as the Pauli exclusion principle. The Slater determinant ensures that no two noninteracting identical fermion particles can possess the same quantum numbers.

In Dirac notation, the state of a single particle in its Hilbert space $\mathcal{H}$ is $|1\rangle$. Addition of a noninteracting identical particle placed in state $|2\rangle$ creates a total state $|1\rangle \otimes |2\rangle$ and a new Hilbert space $\mathcal{H} \otimes \mathcal{H} = \mathcal{H}^2$. The wave function for the two particles is

$$\langle x_1, x_2| (|1\rangle \otimes |2\rangle) = \langle x_1|1\rangle\langle x_2|2\rangle = \psi_1(x_1)\psi_2(x_2).$$

Swapping the positions of the identical particles gives

$$\langle x_1, x_2| (|2\rangle \otimes |1\rangle) = \langle x_1|2\rangle\langle x_2|1\rangle = \psi_2(x_1)\psi_1(x_2),$$

so that the total two-particle state is

$$|\psi\rangle = a|1\rangle \otimes |2\rangle + b|2\rangle \otimes |1\rangle.$$

Fermion states are antisymmetric so that $a = -b$ and the normalized total two-particle state is

$$|\psi\rangle = \frac{1}{\sqrt{2}} (|1\rangle \otimes |2\rangle - |2\rangle \otimes |1\rangle).$$

The wave function is

$$\langle x_1, x_2|\psi\rangle = \frac{1}{\sqrt{2}} (\psi_1(x_1)\psi_2(x_2) - \psi_2(x_1)\psi_1(x_2))$$

$$= \frac{1}{\sqrt{2}} \begin{vmatrix} \psi_1(x_1) & \psi_1(x_2) \\ \psi_2(x_1) & \psi_2(x_2) \end{vmatrix},$$

which is the Slater determinant.

6.1.1.1 Fermion Creation and Annihilation Operators

The concept of creation and annihilation operators can be applied to multi-particle fermion systems. It is, however, necessary to take into account the Pauli exclusion principle and the antisymmetry of the wave functions. Unlike bosons, where the number of particles in a single mode of frequency ω can be increased to an arbitrary value, the number of fermion particles in a given state is limited to unity by the Pauli exclusion principle.

Suppose the occupation of a fermion state $|i\rangle$ is zero, so that $|n_i = 0\rangle$ where n_i is the occupation number of the state. Application of the fermion creation operator $\hat{c}_i^\dagger$ will create one fermion in that state, so that $|n_i = 1\rangle$. However, creation of an additional fermion in the same state is not allowed because of the Pauli exclusion principle. Hence, it is required that

$$\hat{c}_i^\dagger \hat{c}_i^\dagger |\psi\rangle = \left(\hat{c}_i^\dagger\right)^2 |\psi\rangle = 0, \tag{6.19}$$

where $|\psi\rangle$ is any state of the multi-particle system. The same must be true of the fermion annihilation operator $\hat{c}_i$, so that

$$\hat{c}_i \hat{c}_i |\psi\rangle = (\hat{c}_i)^2 |\psi\rangle = 0. \tag{6.20}$$

Consider the sequence of operations $\left(\hat{c}_i^\dagger \hat{c}_i + \hat{c}_i \hat{c}_i^\dagger\right)$. If the state $|i\rangle$ is *empty*, the first term is zero because $\hat{c}_i |n_i = 0\rangle = 0$. The second term creates a fermion and then annihilates it, so that $\hat{c}_i \hat{c}_i^\dagger |n_i = 0\rangle = |n_i = 0\rangle$, and from this it may be concluded that $\hat{c}_i \hat{c}_i^\dagger = 1$. (This result also follows from Eqs. (5.57) and (5.58) in Chapter 5.) *If* the state $|i\rangle$ is *occupied*, then the first term is $\hat{c}_i^\dagger \hat{c}_i = 1$ and the second term $\hat{c}_i \hat{c}_i^\dagger = 0$ because of the Pauli exclusion principle. These results suggest the existence of *anti-commutation* relations between fermion creation and annihilation operators. The anti-commutation relations for the state $|i\rangle$ are $\left\{\hat{c}_i, \hat{c}_i^\dagger\right\} = \hat{c}_i \hat{c}_i^\dagger + \hat{c}_i^\dagger \hat{c}_i = 1$, $\left\{\hat{c}_i^\dagger, \hat{c}_i\right\} = \hat{c}_i^\dagger \hat{c}_i + \hat{c}_i \hat{c}_i^\dagger = 1$, $\{\hat{c}_i, \hat{c}_i\} = \hat{c}_i \hat{c}_i + \hat{c}_i \hat{c}_i = 0$, and $\left\{\hat{c}_i^\dagger, \hat{c}_i^\dagger\right\} = \hat{c}_i^\dagger \hat{c}_i^\dagger + \hat{c}_i^\dagger \hat{c}_i^\dagger = 0$.

Applying these ideas to two states $|j\rangle$ and $|i\rangle$, gives

$$\left\{\hat{c}_j, \hat{c}_i^\dagger\right\} = \hat{c}_j \hat{c}_i^\dagger + \hat{c}_i^\dagger \hat{c}_j = \delta_{ji}, \tag{6.21}$$

$$\{\hat{c}_j, \hat{c}_i\} = 0, \tag{6.22}$$

$$\{\hat{c}_j^\dagger, \hat{c}_i^\dagger\} = 0. \tag{6.23}$$

The curly brackets are used to distinguish the anti-commutation relations for fermions from the commutation relations for bosons. The anti-commutation relations are identical to the commutation relations with the exception that the minus signs in Eqs. (5.15), (5.16), (5.17), and (5.18) are replaced with plus signs. This small change in the equations has a dramatic effect on the quantum mechanical behavior of particles in the system. In particular, it forces the multi-particle wave functions to be antisymmetric.

Since the multi-particle antisymmetric wave function ψ_{a} is characterized by specifying the number of particles in each state, it seems natural to adopt a particle number representation with basis vectors,

$$|n_1, n_2, n_3, \ldots\rangle, \tag{6.24}$$

where n_i is the number of particles in state ψ_i. In mathematics, the particle number representation is said to exist in *Fock space*.

If there are no particles in any of the states, then $|0, 0, 0, \ldots, 0\rangle$. If there is one particle in one state, then any single 0 may be replaced by a 1. Likewise, if two states are occupied, then any two 0s may be replaced with 1s, and so on.

For a total of N particles in the system, the antisymmetric ground-state wave function ψ_a (projected out from the Fock state) is

$$\psi_a(x_1, x_2, x_3, \ldots, x_N) = \langle x_1, x_2, x_3, \ldots, x_N | n_1, n_2, n_3, \ldots, n_N \rangle. \tag{6.25}$$

To make use of the particle number representation, it must be possible to apply either the fermion creation operator $\hat{c}_i^\dagger$ or the annihilation operator $\hat{c}_i$ to change the occupation of state $|n_i\rangle$.

The existence of the vacuum state $|0\rangle = |n_1 = 0, \ n_2 = 0, \ n_3 = 0, \ldots\rangle$ requires

$$\hat{c}_i | n_i = 0\rangle = 0 \tag{6.26}$$

for all values of i. It follows that the many-electron state can be written as

$$|n_1, n_2, n_3, \ldots\rangle = \prod_i \left(\hat{c}_i^\dagger\right)^{n_i} |n_i = 0\rangle, \tag{6.27}$$

where the values of n_i can only be 1 or 0 because $\left(\hat{c}_i^\dagger\right)^2 = 0$ and $\left(\hat{c}_i^\dagger\right)^0 = 1$. The energy of the state is

$$E = \sum_i E_i n_i, \tag{6.28}$$

and the total number of fermions is found by using the number operator $\hat{n}_i = \hat{c}_i^\dagger \hat{c}_i$ with eigenvalues n_i such that

$$N = \sum_i n_i. \tag{6.29}$$

The annihilation operator $\hat{c}_i$, acting on the many-electron state, gives

$$\hat{c}_i | n_1, \ldots, n_i = 1, \ldots\rangle = (-1)^{\sum_{j<i} n_j} | n_1, \ldots, n_i = 0, \ldots\rangle \tag{6.30}$$

when $n_i = 1$ and

$$\hat{c}_i | n_1, \ldots, n_i = 0, \ldots\rangle = 0 \tag{6.31}$$

when $n_i = 0$. The term $(-1)^{\sum_{j<i} n_j}$ in Eq. (6.30) gives a factor of -1 for each occupied state to the left of i. The origin of the term may be seen by considering the removal of the ψ_i row in the Slater determinant (Eq. (6.18)). The required interchange of rows introduces a factor $(-1)^{i-1}$ and the total number of electrons is reduced by one.

Likewise, the creation operator $\hat{c}_i^\dagger$, acting on the many-electron state, gives

$$\hat{c}_i^\dagger | n_1, \ldots, n_i = 1, \ldots\rangle = 0 \tag{6.32}$$

when $n_i = 1$ and

$$\hat{c}_i^\dagger |n_1, \ldots, n_i = 0, \ldots\rangle = (-1)^{\sum\limits_{j<i} n_j} |n_1, \ldots, n_i = 1, \ldots\rangle \tag{6.33}$$

when $n_i = 0$.

The Hamiltonian and other ordinary quantum mechanical operators can be expressed in terms of Fermi creation and annihilation operators. This *second quantization* method is quite a powerful way of dealing with many-particle systems and, importantly, is the common starting point for the quantum field theory description of solids.[2]

Second quantization arises because in quantum mechanics forces between particles are caused by other particles that themselves are quantized. In this way classical force fields are replaced by quantized fields. For example, an electron described by Schrödinger's equation is a first quantization. However, the force exerted between two electrons due to the coulomb interaction is mediated via exchange of photons and so is an example of second quantization. Likewise, the collision of an electron with a lattice vibration (a phonon) followed by a collision of another electron with the same phonon results in an effective force between electrons via the phonon field. In Chapter 5, creation and annihilation operators were used to describe the quantized phonon field. In the same way, fermion creation and annihilation operators can be used to describe the electron field.

Operators expressed in terms of $\hat{c}^\dagger$ and $\hat{c}$ must have matrix elements calculated using the occupation number representation that are the same as those calculated using Slater determinant wave functions. If the occupation number operator is $\hat{A}_{\hat{c}^\dagger \hat{c}}$ and the Slater determinant operator is $\hat{A}$, then, for a single particle,

$$\langle 0, 0, \ldots, n_j = 1, \ldots |\hat{A}_{\hat{c}^\dagger \hat{c}}|, \ldots, n_i = 1, \ldots\rangle = \langle \psi_j |\hat{A}|\psi_i\rangle = \int \psi_j^* \hat{A} \psi_i \mathrm{d}x = A_{ji} \tag{6.34}$$

and, since

$$\langle 0, 0, \ldots, n_j = 1, \ldots, |\hat{A}_{\hat{c}^\dagger \hat{c}}|, \ldots, n_i = 1, \ldots\rangle \tag{6.35}$$
$$= \sum_{ml} A_{ml} \langle \ldots, n_j = 1, \ldots, |\hat{c}_m^\dagger \hat{c}_l|, \ldots, n_i = 1, \ldots\rangle = \sum_{ml} A_{ml} \delta_{jm} \delta_{li} = A_{ji},$$

then

$$\hat{A}_{\hat{c}^\dagger \hat{c}} = \sum_{ml} A_{ml} \hat{c}_m^\dagger \hat{c}_l. \tag{6.36}$$

This result is also true for a system of N particles.

To illustrate the usefulness of the occupation number representation, consider a finite linear chain of identical atoms with nearest-neighbor separation L and periodic boundary conditions. Each lattice site x_j is occupied by a single atom. In the absence of any mutual interaction, an s-electron in each atom of the chain will have exactly the same eigenenergy, E_0. This energy may be set to zero so $E_0 = 0$. The eigenenergies of the system due to the presence of interaction strength $V_{i,j}$ between lattice sites may be found using the Hamiltonian

---

[2] For an introduction, see H. Haken, *Quantum Field Theory of Solids*, Amsterdam, North-Holland, 1988 (ISBN 0 444 86737 6).

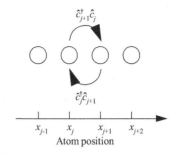

Fig. 6.3 Nearest-neighbor hopping of an electron between adjacent lattice sites at position x_j and x_{j+1} in which $L = x_{j+1} - x_j$.

$$\hat{H}_{\hat{c}^\dagger \hat{c}} = \sum_i E_0 \hat{c}_i^\dagger \hat{c}_i + \sum_{i,j} V_{i,j} \hat{c}_i^\dagger \hat{c}_j = \sum_{i,j}^{i>j} V_{i,j} \left(\hat{c}_i^\dagger \hat{c}_j + \hat{c}_j^\dagger \hat{c}_i \right), \tag{6.37}$$

where $i > j$ in the sum avoids double counting and use is made of the fact $E_0 = 0$. The matrix element $V_{i,j} = \langle i|V|j \rangle = \int \psi_i^*(x) V(x) \psi_j(x) \mathrm{d}x$. If the interaction is restricted to act on nearest neighbors then the sum in Eq. (6.37) can only include nearest-neighbor matrix elements $V_{j,j+1} = -t_{\mathrm{hop},1}$ so that

$$\hat{H}_{\hat{c}^\dagger \hat{c}} = -t_{\mathrm{hop},1} \sum_j \left(\hat{c}_j^\dagger \hat{c}_{j+1} + \hat{c}_{j+1}^\dagger \hat{c}_j \right). \tag{6.38}$$

This Hamiltonian describes the behavior of each s-electron in the linear chain that is able to hop from its lattice site to its nearest-neighbor site in real space. See Fig. 6.3. It is convention that the hopping matrix element has value $-t_{\mathrm{hop},1}$ for s-electrons.

Expressing the creation and annihilation operators in terms of their discrete Fourier transform,

$$\hat{c}_j^\dagger = \frac{1}{\sqrt{L}} \sum_{k'} e^{-ik'x_j} \hat{c}_{k'}^\dagger, \tag{6.39}$$

$$\hat{c}_{j+1} = \frac{1}{\sqrt{L}} \sum_k e^{ik(x_j+L)} \hat{c}_k, \tag{6.40}$$

and substituting into the Hamiltonian gives

$$\hat{H}_{\hat{c}^\dagger \hat{c}} = \frac{-t_{\mathrm{hop},1}}{L} \sum_{j,k,k'} \left(e^{-ik'x_j} e^{ik(x_j+L)} \hat{c}_{k'}^\dagger \hat{c}_k + e^{ik'x_j} e^{-ik(x_j+L)} \hat{c}_k^\dagger \hat{c}_{k'} \right). \tag{6.41}$$

Making use of the fact that

$$\frac{1}{L} \sum_j e^{-i(k'-k)x_j} = \delta_{k,k'} \tag{6.42}$$

gives

$$\hat{H}_{\hat{c}^\dagger\hat{c}} = -t_{\text{hop},1} \sum_{k,k'} \left(\delta_{kk'} e^{ikL} \hat{c}_{k'}^\dagger \hat{c}_k + \delta_{kk'} e^{-ikL} \hat{c}_k^\dagger \hat{c}_{k'} \right) = -t_{\text{hop},1} \sum_k \left(e^{ikL} + e^{-ikL} \right) \hat{c}_k^\dagger \hat{c}_k$$

$$= -2t_{\text{hop},1} \sum_k \cos(kL) \hat{c}_k^\dagger \hat{c}_k, \tag{6.43}$$

which is a diagonal Hamiltonian matrix with matrix elements $-2t_{\text{hop},1} \cos(k_j L)$. Hence, the dispersion relation for the continuum system $(N \to \infty)$ is given by the eigenvalues

$$E_k = -2t_{\text{hop},1} \cos(kL). \tag{6.44}$$

This is just the dispersion relation for the tight binding model with nearest-neighbor interactions discussed in Section 3.6.6.

6.2 Fermi–Dirac Distribution and Chemical Potential

Understanding the probability of occupying electron or photon states as a function of energy is of great practical significance. For example, it plays a crucial role in determining the behavior of a semiconductor laser diode. The distribution in energy of electrons in the conduction band and holes in the valence band of a direct band-gap semiconductor such as GaAs determines the presence or absence of optical gain.

A large number of identical indistinguishable noninteracting half-odd-integer-spin particles of energy $E_\mathbf{k}$ in thermal equilibrium characterized by absolute temperature T have a Fermi–Dirac probability distribution,

$$\boxed{f_\mathbf{k}(E_\mathbf{k}) = \frac{1}{e^{(E_\mathbf{k}-\mu)/k_\mathrm{B}T} + 1},} \tag{6.45}$$

where k_B is the Boltzmann constant and μ is the chemical potential. The appearance of the chemical potential in Eq. (6.45) is due to the fact that particle number is assumed conserved. In classical thermodynamics, the chemical potential is defined as the change in total energy needed to place an extra particle in the system of N particles at constant entropy S and volume V so that $\mu = (\partial E/\partial N)_{S,V}$.

The fact that half-odd-integer-spin particles are quantized according to Fermi–Dirac statistics can be justified using relativistic quantum field theory along with the assumption that the system has a lowest-energy state.[3,4] The same theory shows that integer-spin particles are quantized according to Bose–Einstein statistics.

The total number of noninteracting identical indistinguishable spin one-half electrons in homogeneous isotropic three-dimensional free space is just the integral over k-states multiplied by the distribution function $f_\mathbf{k}(E_\mathbf{k})$. For electrons, the electron density in the particle-number conserved system is

[3] W. Pauli, *Phys. Rev.* **58**, 716 (1940).
[4] For an introduction, see J. J. Sakurai *Advanced Quantum Mechanics*, Reading, Massachusetts, Addison Wesley, 1967 (ISBN 0 201 06710 2).

$$no = \int \frac{d^3k}{(2\pi)^3} 2f_{\mathbf{k}}(E_{\mathbf{k}}),$$

(6.46)

where the factor 2 appears in the integral because each electron may be in a state of either $+\hbar/2$ or $-\hbar/2$ spin (corresponding to spin quantum number $\sigma_s = \pm 1/2$).

In the low-temperature limit $T \rightarrow 0\,\mathrm{K}$ and $f(E_{\mathbf{k}}) = 1$ for $E_{\mathbf{k}} \le \mu$ and $f(E_{\mathbf{k}}) = 0$ for $E_{\mathbf{k}} > \mu$. This is an important limit in which

$$\mu(T = 0\,\mathrm{K}) \equiv E_{\mathrm{F}}$$

(6.47)

is defined as the Fermi energy, or

$$E_{\mathrm{F}} = \frac{\hbar^2 k_{\mathrm{F}}^2}{2m_{\mathrm{e}}^*},$$

(6.48)

where k_{F} is the Fermi wave vector for electrons and the effective electron mass is m_{e}^*. For electron density in three dimensions, and taking the low-temperature limit, Eq. (6.46) may be used to find a relationship between n_0 and the Fermi wave vector k_{F}:

$$n_0(T = 0\,\mathrm{K}) = \int \frac{2}{(2\pi)^3} d^3k = \frac{2}{(2\pi)^3} \int_0^{k_{\mathrm{F}}} 4\pi k^2 \, \mathrm{d}k = \frac{2}{(2\pi)^3} \frac{4\pi}{3} k_{\mathrm{F}}^3 = \frac{k_{\mathrm{F}}^3}{3\pi^2}.$$

(6.49)

Hence, the Fermi wave vector in three dimensions is independent of m_{e}^* and is given by

$$k_{\mathrm{F}} = (3\pi^2 n_0)^{1/3},$$

(6.50)

where $k_{\mathrm{F}} = 2\pi/\lambda_{\mathrm{F}}$. In this case, λ_{F} is the de Broglie wavelength associated with an electron of energy E_{F}.

Table 6.1 lists values of Fermi energy for different three-dimensional carrier concentrations and two representative values of effective electron mass, in which $m_{\mathrm{e}}^* = 0.07 \times m_0$ is the effective electron mass in the conduction band of GaAs, and $m_{\mathrm{hh}}^* = 0.50 \times m_0$ is the effective heavy-hole mass in the valence band of GaAs.

The three-dimensional density of states at the Fermi energy can also be calculated since

$$D_3(k_{\mathrm{F}}) = 2\frac{4\pi k_{\mathrm{F}}^2}{(2\pi)^3} = \frac{k_{\mathrm{F}}^2}{\pi^2}.$$

(6.51)

This is just Eq. (4.102) evaluated at k_{F} and multiplied by a factor 2 to account for electron spin. However, for electrons of mass m_{e}^*, the Fermi energy $E_{\mathrm{F}} = \hbar^2 k_{\mathrm{F}}^2/2m_{\mathrm{e}}^*$, so that $\mathrm{d}E_{\mathrm{F}} = \hbar^2 k_{\mathrm{F}} \mathrm{d}k_{\mathrm{F}}/m_{\mathrm{e}}^*$, which means that the density of states at energy E_{F} is

$$D_3(E_{\mathrm{F}}) = \frac{k_{\mathrm{F}}^2}{\pi^2} \frac{m_{\mathrm{e}}^*}{\hbar^2 k_{\mathrm{F}}} = \frac{m_{\mathrm{e}}^* k_{\mathrm{F}}}{\hbar^2 \pi^2}.$$

(6.52)

Table 6.1 Fermi energy for different three-dimensional carrier concentrations

Carrier concentration $n_0\ (\mathrm{cm}^{-3})$	Fermi wave vector, $k_F\ (\times 10^8\ \mathrm{m}^{-1})$	Fermi wavelength, λ_F (nm)	Fermi energy, E_F(meV) $(m_e^* = 0.07 \times m_0)$	Fermi energy, E_F(meV)$(m_{hh}^* = 0.50 \times m_0)$
1×10^{19}	6.66	9.4	241.6	33.8
1×10^{18}	3.09	20.3	52.1	7.3
1×10^{17}	1.44	43.8	11.2	1.6
1×10^{16}	0.67	94.3	2.4	0.3

For finite temperatures such that $k_B T \ll E_F$, a Taylor expansion of Eq. (6.46) to second order in temperature about the energy E_F gives

$$n_0 \sim \int_0^{E_F} D_3(E)\mathrm{d}E + \left((\mu - E_F)D_3(E_F) + \frac{\pi^2}{6}(k_B T)^2 \frac{\mathrm{d}D_3(E_F)}{\mathrm{d}E_F} \right). \tag{6.53}$$

Since the first term on the right-hand side is the carrier density n_0 at zero temperature and n_0 is assumed to be independent of temperature, then

$$0 \sim (\mu - E_F)D_3(E_F) + \frac{\pi^2}{6}(k_B T)^2 \frac{\mathrm{d}D_3(E_F)}{\mathrm{d}E_F}. \tag{6.54}$$

Hence, the chemical potential is

$$\mu \sim E_F - \frac{\pi^2}{6} \frac{(k_B T)^2}{D_3(E_F)} \frac{\mathrm{d}D_3(E_F)}{\mathrm{d}E_F}. \tag{6.55}$$

In three dimensions, the chemical potential may be approximated to second order in temperature as

$$\mu \sim E_F - \frac{\pi^2}{12} \frac{(k_B T)^2}{E_F} - O\left(\frac{(k_B T)^4}{E_F} \right). \tag{6.56}$$

It may be shown that to fourth order in the temperature, the chemical potential in three dimensions is

$$\mu \sim E_F - \frac{\pi^2}{12} \frac{(k_B T)^2}{E_F} - \frac{7\pi^4}{960} \frac{(k_B T)^4}{E_F^3} - O\left(\frac{(k_B T)^6}{E_F^5} \right). \tag{6.57}$$

Figure 6.4 plots the numerical, second-order, and fourth-order approximations of chemical potential energy as a function of thermal energy. The axes are normalized to Fermi energy, E_F. Both approximations become less accurate as $k_B T$ approaches E_F. Because metals such as Au and Cu have Fermi energy in the range 5–7 eV, the second- and fourth-order expansions are accurate at room temperature since thermal energy $k_B T \ll E_F$. The situation is very different for semiconductors such as Si and GaAs with carrier concentrations such that $E_F \sim k_B T$. In this case, accurate numerical calculations of chemical potential are required.

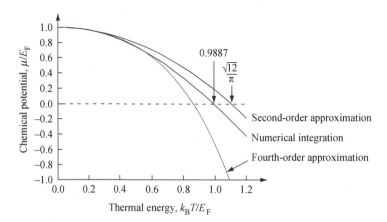

Fig. 6.4 Chemical potential for a three-dimensional gas of electrons as a function of thermal energy calculated using numerical integration and compared with results from second-order and fourth-order approximations. The chemical potential μ and thermal energy $k_B T$ are normalized to the Fermi energy, E_F.

The temperature at which the chemical potential is zero estimated using Eq. (6.56) is $T = \sqrt{12}E_F/\pi k_B$, while more accurate calculations (see Exercise 6.4) give a temperature $T = 0.9887 \times E_F/k_B$. The temperature at which the chemical potential is zero estimated using the fourth-order approximation is *less* accurate than the second-order result.

6.2.1 Writing a Computer Program to Calculate the Chemical Potential

The expression for carrier density in Eq. (6.46) may be written as

$$n_0 = \int_0^\infty D_3(E)\, 2f(E)\mathrm{d}E = \frac{1}{2\pi^2}\left(\frac{2m}{\hbar^2}\right)^{3/2} \int_{E_{\min}=0}^{E_{\max}=\infty} \sqrt{E}\,\frac{1}{\mathrm{e}^{(E-\mu)/k_B T}+1}\mathrm{d}E, \tag{6.58}$$

where $D_3(E)$ is the three-dimensional density of states for electrons of mass m_e^* and μ is the chemical potential.

The maximum possible value for the chemical potential is given by the Fermi energy

$$\mu_{\max} = E_F = \frac{\hbar^2 k_F^2}{2m}, \tag{6.59}$$

where $k_F = (3\pi^2 n_0)^{1/3}$ for a three-dimensional carrier density n_0 (Eq. (6.50)). The minimum possible value of the chemical potential, $\mu_{\min}$, is given by the high-temperature limit $(T \to \infty)$ that, for fixed particle density n_0, is $\mu/k_B T \to -\infty$. In this limit, the Fermi–Dirac distribution function becomes

$$f(E)|_{T\to\infty} = \frac{1}{\mathrm{e}^{(E-\mu)/k_B T}+1}\bigg|_{T\to\infty} = \mathrm{e}^{(\mu-E)/k_B T}, \tag{6.60}$$

which is the Boltzmann distribution. In the limit $T \to \infty$, occupation probability at energy $E = 0$ takes on the value $\mathrm{e}^{\mu/k_B T}$.

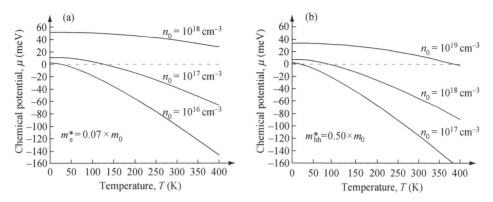

Fig. 6.5 Calculated chemical potential for electrons of carrier concentration n_0 with (a) effective mass $m_e^* = 0.07\, m_0$ and (b) effective mass $m_{hh}^* = 0.50\, m_0$ as a function of temperature T. The dashed line corresponds to chemical potential $\mu = 0\,\text{meV}$.

From classical thermodynamics, it is known that a three-dimensional electron gas in this high-temperature limit has chemical potential[5]

$$\mu_{\min} = k_B T \ln\left(\frac{n_0}{2}\left(\frac{2\pi\hbar^2}{mk_B T}\right)^{3/2}\right). \tag{6.61}$$

A computer program may now be used to calculate a carrier density n' for given temperature T by using an initial estimate for the chemical potential $\mu' = \mu_{\min} + (\mu_{\max} - \mu_{\min})/2$ and numerically integrating Eq. (6.58). In practice, choosing a cut-off value for $E_{\max} = E_F + 15k_B T$ works well for most cases of interest.

If the value of n_0' calculated using μ' is less than the actual value n_0, then the new best estimate for $\mu_{\min} = \mu'$. If $n_0' \geq n_0$, then $\mu_{\max} = \mu'$. A new value of μ' can now be calculated and the integration to calculate a new value of n_0' performed again. In this way, it is possible to iterate to the desired level of accuracy in μ. See Exercise 6.2.

Figure 6.5(a) and (b) plots the temperature dependence of chemical potential $\mu(T)$ for the indicated carrier concentration and effective electron mass. As expected, in the limit of low temperature $(T \to 0\,\text{K})$ the chemical potential approaches the Fermi energy, E_F. With increasing temperature, the chemical potential monotonically decreases in value, eventually taking on a negative value. Equation (6.58) can be used to find the value of the temperature at which the chemical potential is zero (Exercise 6.4).

Figure 6.6 is a three-dimensional plot of chemical potential, μ, as a function of temperature, T, and carrier concentrations, n_0. The effective electron mass is $m_e^* = 0.07 \times m_0$.

[5] L. D. Landau and E. M. Lifshitz, *Statistical Physics*, Oxford, Pergamon Press, 1985 (ISBN 0 08 023039 3).

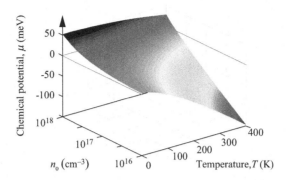

Fig. 6.6 Calculated chemical potential as a function of electron carrier concentration, n_0, and temperature, T. Effective electron mass is $m_e^* = 0.07 \times m_0$

6.2.2 Plotting the Fermi–Dirac Distribution

At zero temperature with increasing electron energy there is a step-like distribution going from 1 to 0 at energy $E = E_F$. For the situation shown in Fig. 6.7, the Fermi energy is $E_F = 52$ meV above the conduction-band minimum. At finite temperatures, the step function is smeared out in energy. The broadening of the step transition is controlled by the value of the chemical potential μ and the temperature T in Eq. (6.45). The same equation always has value 0.5 when $E = \mu$.

In the limit in which the chemical potential tends to a large negative value, the distribution function tends to a Boltzmann function. To learn more about this high-energy tail of the distribution, it is convenient to plot the occupation probability on a natural logarithmic scale. As may be seen in Fig. 6.8, when the occupation probability is displayed in this way, at high electron energy it is linear with increasing energy. This means that high-energy occupation probability scales with increasing energy as a Boltzmann factor $e^{-E/k_B T}$.

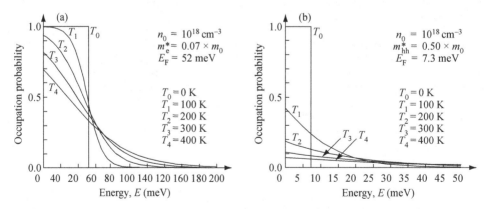

Fig. 6.7 The Fermi–Dirac distribution as a function of electron energy for carrier concentration $n = 10^{18}$ cm^{-3} with (a) effective electron mass $m_e^* = 0.07 \times m_0$ and (b) effective electron mass $m_{hh}^* = 0.5 \times m_0$. The distribution functions are calculated for the indicated temperatures. Occupation probability is 0.5 when $E = \mu$

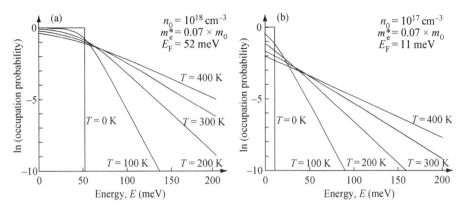

Fig. 6.8 Natural logarithm of Fermi–Dirac distribution as a function of electron energy for carrier concentration with (a) $n_0 = 10^{18}$ cm^{-3} and (b) $n_0 = 10^{17}$ cm^{-3} for the indicated temperatures. The effective electron mass is $m_e^* = 0.07 \times m_0$.

This may be confirmed by comparing Fig. 6.8(a) with Fig. 6.8(b). Here, it can be seen that, while for given temperature T carrier density differs by a factor of ten and Fermi energy by a factor of almost five, the slope of the high-energy tail in the distribution is the same and has the value $-1/k_B T$.

6.2.3 Fermi–Dirac Distribution Function and Thermal Equilibrium Statistics

There are a number of ways to explore the origin of the Fermi–Dirac statistical distributions. In the following, the statistically most likely arrangement of a *large* number of particle states is sought when the particles are in thermodynamic equilibrium. *Thermodynamic equilibrium exists when there is no net macroscopic flux of particles or energy within or between systems.* Further:

In equilibrium, any two microscopically distinguishable arrangements of a system with the same total energy are equally likely.

This means that the probability of finding the system with n_1 particles in the energy range E to $E+\Delta E$ is proportional to the number of microscopically distinguishable arrangements that correspond to the same macroscopic arrangement.

To find the distribution function for identical indistinguishable particles such as electrons that have half-odd-integer spin, consider an energy level labeled by E_j for which each energy level has a degeneracy n_j. A total of N electrons are placed into the system using the following rules:

(i) Electrons are indistinguishable, and each state can only accommodate one electron (the Pauli exclusion principle).

(ii) The total number of electrons is fixed by the sum rule, and $N = \sum_j N_j$ is a constant, where N_j is the number of electrons in an energy level E_j.

(iii) The total energy of the system is fixed so that $E_{\text{total}} = \sum_j E_j N_j$.

At equilibrium there is a distribution function that describes the most probable arrangement of electrons as a function of energy. This is the Fermi–Dirac function by which the probability of finding the particle at energy E_j is

$$f(E_j) = \frac{N_j}{n_j}. \tag{6.62}$$

To find this probability distribution, the number of ways in which N_j indistinguishable electrons in the E_j energy level can be placed in n_j states must be found. Because of the Pauli exclusion principle, $n_j \geq N_j$. Starting with the first particle, there are n_j states from which to choose. For the second particle there are $(n_j - 1)$ states available, and so on, giving

$$n_j(n_j - 1)\cdots(n_j - N_j + 1). \tag{6.63}$$

The number of possible permutations of N_j particles among themselves is $N!$. Because the particles are indistinguishable, these permutations do not lead to distinguishable arrangements, so the total number of distinct arrangements with N_j particles is

$$\frac{n_j(n_j - 1)\cdots(n_j - N_j + 1)}{N_j!} = \frac{n_j(n_j - 1)\cdots(n_j - N_j + 1)}{N_j!}\frac{(n_j - N_j)}{(n_j - N_j)}, \tag{6.64}$$

$$\frac{n_j(n_j - 1)\cdots(n_j - N_j + 1)}{N_j!} = \frac{n_j!}{(n_j - N_j)!N_j!}. \tag{6.65}$$

The total number of ways of arranging N electrons in the multilevel system is

$$\mathcal{P} = \prod_j \frac{n_j!}{(n_j - N_j)!N_j!}. \tag{6.66}$$

To find the most probable set of values for N_j, an extreme value of $\mathcal{P}$ must be found so that $d\mathcal{P} = 0$. Taking the logarithm of both sides,

$$\ln(\mathcal{P}) = \sum_j (\ln(n_j!) - \ln((n_j - N_j)!) - \ln(N_j!)). \tag{6.67}$$

For *large* values of N_j and n_j, Sterling's approximation may be used so that $\ln(x!) = x\ln(x) - x$. This gives

$$\ln(\mathcal{P}) \sim \sum_j (n_j \ln(n_j) - n_j - (n_j - N_j)\ln(n_j - N_j) + (n_j - N_j) - N_j \ln(N_j) + N_j). \tag{6.68}$$

And, because $dn_j = 0$, the derivative is

$$d(\ln(\mathcal{P})) = \sum_j \frac{\partial}{\partial N_j}\ln(P)dN_j = \sum_j (\ln(n_j - N_j) + 1 - \ln(N_j) - 1)dN_j$$

$$= \sum_j \left(\ln\left(\frac{n_j - N_j}{N_j}\right)\right)dN_j. \tag{6.69}$$

Finding the extremes of this function under the constraint of total energy and particle conservation requires that $\sum_j dN_j = 0$ and $\sum_j E_j dN_j = 0$, so that

$$\sum_j \left(\ln\left(\frac{n_j - N_j}{N_j}\right) - \alpha - \beta E_j \right) dN_j = 0, \tag{6.70}$$

where α and β are Lagrange multipliers. The sum vanishes if

$$\ln\left(\frac{n_j - N_j}{N_j}\right) - \alpha - \beta E_j = 0, \tag{6.71}$$

for all j. The Fermi–Dirac distribution requires $f(E_j) = N_j/n_j$ be found. Rearranging Eq. (6.71) and taking the exponential of both sides gives $n_j/N_j = 1 + e^{(\alpha + \beta E_j)}$; Eq. (6.62) may be written as

$$\frac{N_j}{n_j} = f(E_j) = \frac{1}{e^{(\alpha + \beta E_j)} + 1}. \tag{6.72}$$

For a continuum of energy levels, $E_j \rightarrow E$ and

$$f_{\mathbf{k}}(E_{\mathbf{k}}) = \frac{1}{e^{(\alpha + \beta E_{\mathbf{k}})} + 1}. \tag{6.73}$$

The coefficients α and β are found from classical thermodynamics to be[6] $\alpha = -\mu/k_B T$ and $\beta = 1/k_B T$, where μ is the chemical potential, k_B is the Boltzmann constant, and T is the absolute temperature. There is, therefore, agreement with the Fermi–Dirac distribution given by Eq. (6.45).

In the limit $T \rightarrow \infty$, the Fermi–Dirac distribution function $f_{\mathbf{k}}(E_{\mathbf{k}})|_{T \rightarrow \infty} = e^{-E_{\mathbf{k}}/k_B T}$, which is the classical Maxwell–Boltzmann ratio (see Exercise 6.3).

6.3 The Bose–Einstein Distribution Function

The Bose–Einstein probability distribution for indistinguishable integer-spin particles of energy $\hbar\omega$ in thermal equilibrium characterized by absolute temperature T is

$$\boxed{g_{BE}(\hbar\omega) = \frac{1}{e^{(\hbar\omega - \mu)/k_B T} - 1}.} \tag{6.74}$$

However, integer spin particles such as phonons and photons are not typically conserved, and in such circumstances the chemical potential $\mu = 0$, so that Eq. (6.74) reduces to the Bose distribution,

$$\boxed{g(\hbar\omega) = \frac{1}{e^{\hbar\omega/k_B T} - 1}.} \tag{6.75}$$

---

[6] See any text on statistical physics such as L. D. Landau and E. M. Lifshitz, *Statistical Physics*, Oxford, Pergamon Press, 1985 (ISBN 0 08 023039); F. Reif, *Fundamentals of Statistical and Thermal Physics*, Boston, Massachusetts, McGraw Hill, 1965 (ISBN 07 051800 9).

As an example of the use of Eq. (6.75), recall from Section 5.7 that an electromagnetic field may be described in terms of a number of harmonic oscillators. Each photon contributing to the electromagnetic field is quantized in energy in such a way that $E_n = \hbar\omega(n + 1/2)$, where the factor $1/2$ comes from the contribution of zero-point energy. Assuming that the oscillators are excited thermally and are in equilibrium characterized by an absolute temperature T, the probability of excitation into the nth state $P_{\text{prob}}(n)$ is given by a Boltzmann factor

$$P_{\text{prob}}(n) = \frac{e^{-E_n/k_{\text{B}}T}}{\sum_n e^{-E_n/k_{\text{B}}T}} = \frac{e^{-n\hbar\omega/k_{\text{B}}T}}{\sum_n e^{-n\hbar\omega/k_{\text{B}}T}}. \tag{6.76}$$

Notice that in the expression for energy E_n, the zero-point energy term cancels out. The sum in the denominator may be written

$$\sum_{n=0}^{n=\infty} e^{-n\hbar\omega/k_{\text{B}}T} = \frac{1}{1 - e^{-\hbar\omega/k_{\text{B}}T}}. \tag{6.77}$$

The probability of excitation in the nth state becomes

$$P_{\text{prob}}(n) = \left(1 - e^{-\hbar\omega/k_{\text{B}}T}\right) e^{-n\hbar\omega/k_{\text{B}}T}, \tag{6.78}$$

and the average number of photons excited in the nth field mode at temperature T is

$$g(\hbar\omega) = \sum_{n=0}^{n=\infty} nP_{\text{prob}}(n) = \left(1 - e^{-\hbar\omega/k_{\text{B}}T}\right) \sum_{n=0}^{n=\infty} ne^{-n\hbar\omega/k_{\text{B}}T}$$

$$= \left(1 - e^{-\hbar\omega/k_{\text{B}}T}\right) e^{-\hbar\omega/k_{\text{B}}T} \frac{\partial}{\partial \left(e^{-\hbar\omega/k_{\text{B}}T}\right)} \sum_{n=0}^{n=\infty} e^{-n\hbar\omega/k_{\text{B}}T} = \frac{e^{-\hbar\omega/k_{\text{B}}T}}{1 - e^{-\hbar\omega/k_{\text{B}}T}}. \tag{6.79}$$

Thus, finally,

$$g(\hbar\omega) = \frac{1}{e^{\hbar\omega/k_{\text{B}}T} - 1}, \tag{6.80}$$

which is in agreement with Eq. (6.75).

It follows that the average energy in excess of the zero-point energy for the electromagnetic field in thermal equilibrium is just $\hbar\omega \times g(\hbar\omega)$. The radiative energy density is this average energy multiplied by the density of optical field modes in the frequency range ω to $\omega + d\omega$. In three-dimensional free space, and allowing for the fact that the electromagnetic field can have one of two orthogonal polarizations, the radiative energy density is

$$U_{\text{S}}(\omega) = 2g(\hbar\omega)\hbar\omega D_3^{\text{opt}}(\omega) = \frac{2\hbar\omega}{e^{\hbar\omega/k_{\text{B}}T} - 1} \frac{\omega^2}{\pi^2 c^3} = \frac{\hbar\omega^3}{\pi^2 c^3} \frac{1}{e^{\hbar\omega/k_{\text{B}}T} - 1}. \tag{6.81}$$

This is Planck's radiative energy density spectrum for thermal light discussed in Section 1.1.2. $U_{\text{S}}(\omega)$ is measured in units of J s m^{-3}.

6.4 Example Exercises

Exercise 6.1
Calculate k_F in two dimensions for a GaAs quantum well with electron density $n = 10^{12} \, \text{cm}^{-2}$. What is the de Broglie wavelength for an electron at the Fermi energy?

Exercise 6.2
Write a computer program to calculate the chemical potential for n electrons per unit volume at temperature T. The electrons have effective mass of $m_e^* = 0.067 \times m_0$.

(a) Calculate the value of the chemical potential when $n_0 = 10^{18} \, \text{cm}^{-3}$ and $T = 300 \, \text{K}$. Compare the results of the calculation with the expressions for μ using second-order and fourth-order expansions. How do the results compare with the value of chemical potential when $T = 0 \, \text{K}$?

(b) Repeat the calculations of (a) for the case in which $n_0 = 10^{14} \, \text{cm}^{-3}$. Explain why the results from the second-order and fourth-order expansions are inaccurate in this case.

Exercise 6.3
Use the method of Section 6.2.3 to determine the classical Maxwell–Botzmann distribution function $g_{MB}(E)$ and the ratio for distinguishable identical particles.

Exercise 6.4
At what temperature is the chemical potential of a three-dimensional electron gas zero?

Solutions

Solutions 6.1
For carrier density n_0 electrons per unit area in two dimensions,

$$n_0(T = 0) = \int \frac{2}{(2\pi)^2} d^2k = \frac{2}{(2\pi)^2} \int^{k_F} 2\pi k dk = \frac{2}{(2\pi)^2} \pi k_F^2 = \frac{1}{2\pi} k_F^2,$$

where the factor 2 accounts for electron spin. Hence, in two dimensions,

$$k_F = \sqrt{2\pi n_0}.$$

For $n_0 = 10^{12} \, \text{cm}^{-2}$, $k_F = 2\pi/\lambda_{F_e} = 2.5 \times 10^6 \, \text{cm}^{-1}$ and $\lambda_{F_e} = 4 \, \text{nm}$.

Solutions 6.2
(a) The carrier density n_0 of a three-dimensional electron gas in equilibrium at absolute temperature T is

$$n_0 = \int_0^\infty D_3(E) \, 2f(E) dE = \frac{1}{2\pi^2} \left(\frac{2m_e^*}{\hbar^2} \right)^{3/2} \int_{E_{min}=0}^{E_{max}=\infty} \sqrt{E} \frac{1}{e^{(E-\mu)/k_B T} + 1} dE,$$

where $D_3(E)$ is the three-dimensional density of states, $f(E)$ is the Fermi–Dirac distribution, and m_e^* is the effective electron mass.

The maximum possible value of chemical potential is given by the Fermi energy

$$\mu_{max} = E_F = \frac{\hbar^2 k_F^2}{2m},$$

where $k_F = (3\pi^2 n_0)^{1/3}$ for a three-dimensional carrier density n_0 (Eq. (6.50)). The minimum possible value of the chemical potential, μ_{min}, is given by the high-temperature limit $(T \to \infty)$, which, for fixed particle density n_0, is $\mu/k_B T \to -\infty$. In this limit, the Fermi–Dirac distribution function becomes

$$f(E)|_{T \to \infty} = \frac{1}{e^{(E-\mu)/k_B T} + 1}\bigg|_{T \to \infty} = e^{(\mu - E)/k_B T},$$

which is the Maxwell–Boltzmann distribution. Classical thermodynamics may be used to show that for a three-dimensional electron gas in this limit,

$$\mu_{min} = k_B T \ln \left(\frac{n_0}{2} \left(\frac{2\pi \hbar^2}{m_e^* k_B T} \right)^{3/2} \right).$$

The carrier density n_0' for given temperature T is found using an initial estimate for the chemical potential $\mu' = \mu_{min} + (\mu_{max} - \mu_{min})/2$ and numerically integrating Eq. (6.58). Notice that a cut-off for E_{max} in Eq. (6.58) must be chosen. In practice, adopting a value $E_{max} = E_F + 15 k_B T$ works well for most cases of interest.

If the value of n' calculated using μ' is less than the actual value n, then the new best estimate for $\mu_{min} = \mu'$. If $n' \geq n$, then $\mu_{max} = \mu'$. A new value of μ' can now be calculated and the integration to calculate a new value of n' can be performed again. In this way it is possible to iterate to a desired level of accuracy in μ.

For temperature $T = 300$ K, a computer program gives $\mu_{T=300\,K} = 41.9$ meV, which may be compared with the second-order approximation

$$\mu \sim E_F - \frac{\pi^2}{12} \frac{(k_B T)^2}{E_F},$$

giving $\mu_{T=300\,K} = 44.3$ meV. The agreement is quite good because thermal energy at temperature $T = 300$ K is $k_B T = 25.8$ meV and this value is sufficiently smaller than the Fermi energy, $E_F = 54$ meV, to make the expansion quite accurate. The fourth-order approximation,

$$\mu \sim E_F - \frac{\pi^2}{12} \frac{(k_B T)^2}{E_F} - \frac{7\pi^4}{960} \frac{(k_B T)^4}{E_F^3} - O\left(\frac{(k_B T)^6}{E_F^5} \right),$$

gives a slightly more accurate result, $\mu_{T=300\,K} = 42.4$ meV.

(b) When $n_0 = 10^{14}$ cm^{-3} the computer program gives $\mu_{T=300\,K} = -217$ meV. The results from the second-order and fourth-order expansion are inaccurate, because now $k_B T$ is much greater than the Fermi energy, $E_F = 0.12$ meV.

Solutions 6.3

The number of microscopically distinguishable arrangements of n_1 distinguishable particles among a total of N particles may be found by choosing the first particle from a total of N particles, the second from $(N - 1)$ and so on, so that the total number of choices is

$$N(N - 1)(N - 2) \cdots (N - n_1 + 1) = N(N - 1)(N - 2) \cdots (N - n_1 + 1)\frac{(N - n_1)!}{(N - n_1)!}$$

$$= \frac{N!}{(N - n_1)!}.$$

Dividing by the number of ways of arranging n_1 distinguishable particles among themselves gives the number of microscopic arrangements of n_1 particles in the energy range E to $E + \Delta E$. Hence,

$$\mathcal{P}_1 = \frac{N!}{n_1!(N - n_1)!}, \quad \mathcal{P}_2 = \frac{(N - n_1)!}{n_2!(N - n_1 - n_2)!}, \quad \mathcal{P}_3 = \frac{(N - n_1 - n_2)!}{n_3!(N - n_1 - n_2 - n_3)!},$$

and so on. The total number of arrangements is

$$\mathcal{P}_1(n_j) = \mathcal{P}_1\mathcal{P}_2\mathcal{P}_3 \cdots \mathcal{P}_j = \frac{N!}{n_1!(N - n_1)!}\frac{(N - n_1)!}{n_2!(N - n_1 - n_2)!}\frac{(N - n_1 - n_2)!}{n_3!(N - n_1 - n_2 - n_3)!} \cdots$$

$$= N! \prod_{j=1}^{j=\infty} \frac{1}{n_j!}.$$

Since, at equilibrium, all microscopically distinguishable distributions with a fixed number of particles and the same total energy are equally likely, it follows that the most probable macroscopic distribution is one in which the number of microscopically distinguishable arrangements $\mathcal{P}(n_1, n_2, \ldots, n_j)$ is a *maximum* subject to the *constraints* of particle conservation,

$$f = \left(\sum_{j=1}^{j=\infty} n_j\right) - N = 0,$$

and conservation of total energy,

$$g = \left(\sum_{j=1}^{j=\infty} E_j n_j\right) - E_{\text{total}} = 0.$$

It is easier to maximize a new function $G = \ln(\mathcal{P}(n_j))$ instead of $\mathcal{P}(n_j)$ itself. Maximization of a function subject to constraints has been worked on a great deal in the past, especially in the context of classical mechanics. One approach uses the Lagrange method of undetermined multipliers. To maximize the function

$$F(n_1, n_2, \ldots \alpha, \beta) = G(n_1, n_2, \ldots) - \alpha f(n_1, n_2, \ldots) - \beta g(n_1, n_2, \ldots)$$

and solve for $n_1, n_2, \ldots, \alpha$, and β in such a way that G is a maximum is done by requiring $\partial F/\partial n_j = 0$ for all j, $\partial F/\partial \alpha = 0$, and $\partial F/\partial \beta = 0$.

Since

$$F = \ln(\mathcal{P}) - \alpha \left(\left(\sum_j n_j \right) - N \right) - \beta \left(\left(\sum_j E_j n_j \right) - E \right),$$

then

$$\ln(\mathcal{P}) = \ln \left(N! \prod_{j=1}^{j=\infty} \frac{1}{n_j} \right) = \ln(N!) + \sum_{j=1}^{j=\infty} (\ln(1) - \ln(n_j!)) = \ln(N!) - \sum_{j=1}^{j=\infty} \ln(n_j!)$$

because $\ln(1) = 0$. For *large* n_j, Sterling's formula $\ln(n!) \sim (n \ln(n) - n)$ may be used so that

$$\ln(\mathcal{P}) = \ln(N!) - \sum_{j=1}^{j=\infty} (n_j \ln(n_j) - n_j).$$

Fixing j and taking the derivative of F with respect to n_j gives

$$\frac{\partial F}{\partial n_j} = \frac{\partial}{\partial n_j} \left(\ln(N!) - \sum_{j=1}^{j=\infty} (n_j \ln(n_j) - n_j) - \alpha \left(\left(\sum_j n_j \right) - N \right) - \beta \left(\left(\sum_j E_j n_j \right) - E \right) \right),$$

and since $\partial F/\partial n_j = 0$, one has

$$0 = 0 - \ln(n_j) - n_j \frac{\partial}{\partial n_j} \ln(n_j) + 1 - \alpha - \beta E_j = \ln(n_j) - \frac{n_j}{n_j} + 1 - \alpha - \beta E_j$$

$$= -\ln(n_j) - \alpha - \beta E_j.$$

Rewriting $-\ln(n_j) = \alpha + \beta E_j$ so that $n_j = \frac{1}{\exp(\alpha + \beta E_j)}$, the Maxwell–Boltzmann distribution function

$$g_{\mathrm{MB}}(E) = \frac{1}{\exp((E - \mu)/k_{\mathrm{B}}T)}$$

results, where $\alpha = \mu/k_{\mathrm{B}}T$ and $\beta = 1/k_{\mathrm{B}}T$ may be obtained from the classical theory of gases. This distribution specifies the probability that an available state of energy E is occupied under equilibrium conditions.

The Maxwell–Boltzmann ratio (the Boltzmann factor) $n_j/n_k = 1/e^{(E_j - E_k)\beta}$ is $e^{-E/k_{\mathrm{B}}T}$.

Solutions 6.4

To find the temperature at which the chemical potential is zero in a three-dimensional electron gas, Eq. (6.58) is written:

$$n_0 = \frac{1}{2\pi^2} \left(\frac{2m_{\mathrm{e}}^*}{\hbar^2} \right)^{3/2} \int_{E_{\min}=0}^{E_{\max}=\infty} \sqrt{E} \frac{1}{\exp((E - \mu)/k_{\mathrm{B}}T) + 1} \, dE.$$

Setting the chemical potential to zero, $\mu = 0$, and normalizing energy to the Fermi energy, E_F, gives

$$n_0 = \frac{1}{2\pi^2} \left(\frac{2m_e^*}{\hbar^2}\right)^{3/2} (E_F)^{3/2} \int_{E_{min}=0}^{E_{max}=\infty} \left(\frac{E}{E_F}\right)^{1/2} \frac{1}{\exp((E/E_F)(E_F/k_BT)) + 1} d\left(\frac{E}{E_F}\right).$$

Introducing the variable $x = E/E_F$ and the value $r = E_F/k_BT$, and making use of the fact that $E_F = \hbar^2 k_F^2/2m_e^*$ (Eq. (6.48)), where $k_F = (3\pi^2 n_0)^{1/3}$ (Eq. (6.50)), then

$$n_0 = \frac{1}{2\pi^2} \left(\frac{2m_e^*}{\hbar^2}\right)^{3/2} \left(\frac{\hbar^2}{2m_e^*}\right)^{3/2} 3\pi^2 n_0 \int_0^\infty x^{1/2} \frac{1}{e^{rx} + 1} dx.$$

The carrier density n_0 cancels, and after some rearrangement the expression

$$\frac{2}{3} = \int_0^\infty \frac{x^{1/2}}{e^{rx} + 1} dx = \int_0^\infty \frac{x^{p-1}}{e^{rx} - q} dx = \frac{1}{qr^p} \Gamma(p) \sum_{k=1}^\infty \frac{q^k}{k^p}$$

may be obtained. The integral is of known form[7] in which $\Gamma(p)$ is the gamma function,[8] $p > 0$, $r > 0$, and $-1 < q < 1$. In this particular case, $p = 3/2$, $r = E_F/k_BT$, and $q = -1$, giving

$$\frac{k_BT}{E_F} \left(\frac{-2}{3 \times \Gamma(1.5) \sum_{k=1}^\infty \frac{-1^k}{k^{3/2}}}\right)^{2/3}.$$

Putting in the numbers, $\Gamma(1.5) = \pi^{1/2}/2 = 0.886\,227$ and the sum $\sum_{k=1}^\infty \frac{-1^k}{k^{3/2}} = -0.765\,147$. Hence, the chemical potential of a three-dimensional electron gas is always zero when temperature T is such that

$$\frac{k_BT}{E_F} = \left(\frac{2}{3 \times 0.886227 \times 0.765147}\right)^{2/3} = 0.9887.$$

6.5 Problems

Problem 6.1
(a) Numerically calculate the chemical potential for n_0 identical indistinguishable non-interacting electrons per unit volume in thermal equilibrium at temperature T.

(b) Calculate the value of the chemical potential for electrons of effective mass $m_e^* = 0.07 \times m_0$ and carrier density $n_0 = 1.5 \times 10^{18}$ cm^{-3} at temperature $T = 300$ K.

---

[7] I. S. Gradshteyn and I. M. Ryzhik, *Table of Integrals, Series, and Products*, San Diego, Academic Press, 1980, p. 326 (ISBN 0 12 294760 6).
[8] M. Abramowitz and I. A. Stegun, *Handbook of Mathematical Functions*, New York, Dover Publications, 1970, pp. 267–273 (ISBN 0 486 61272 4).

(c) Repeat (b) for electrons with effective electron mass $m_e^* = 0.50 \times m_0$.

(d) Plot the Fermi–Dirac distribution function for case (b) and (c).

(e) Repeat (b), (c), and (d) for the case when temperature $T = 77\,\mathrm{K}$.

Problem 6.2

(a) Show that

$$\frac{1}{\exp\left((E - \mu)/k_\mathrm{B}T\right) + 1} = 1 - \frac{1}{\exp\left((\mu - E)/k_\mathrm{B}T\right) + 1}.$$

(b) A semiconductor has valence band electron energy dispersion relation $E_\mathrm{VB} = E(\mathbf{k})$ and conduction band electron energy dispersion relation $E_\mathrm{CB} = E_0 - E(\mathbf{k})$, where E_0 is a constant such that the conduction band and valence band are separated by an energy band gap, E_g. Show that when particle number is conserved, the chemical potential is in the middle of the band gap with value $\mu = E_0/2$ and is independent of temperature.

Problem 6.3

(a) Calculate the average energy of an electron in a three-dimensional electron gas at temperature T. Show that $\langle E_\mathrm{3D}(T \to 0)\rangle = \frac{3}{5}E_\mathrm{F}$ in the limit of low absolute temperature and that $\langle E_\mathrm{3D}(T \to \infty)\rangle = \frac{3}{2}k_\mathrm{B}T$ in the high temperature limit.

(b) Calculate the average energy of an electron in a two-dimensional gas of electrons at temperature T. Show that $\langle E_\mathrm{2D}(T \to 0)\rangle = \frac{1}{2}E_\mathrm{F}$ in the low temperature limit and that $\langle E_\mathrm{2D}(T \to \infty)\rangle = k_\mathrm{B}T$ in the high temperature limit.

Problem 6.4

The antisymmetric wave function that describes two identical indistinguishable non-interacting particles is given by the Slater determinant,

$$\psi_\mathrm{a} = \frac{1}{\sqrt{2}}\begin{vmatrix} \psi_1(x_1) & \psi_1(x_2) \\ \psi_2(x_1) & \psi_2(x_2) \end{vmatrix},$$

where rows label the single-particle state and columns label the particle. The position coordinate for particle 1 is x_1 and for particle 2 it is x_2.

(a) Plot $\psi_\mathrm{a}(x_1, x_2)$ for the case when ψ_1 is the single-particle ground state of a one-dimensional rectangular potential well with infinite barrier energy and ψ_2 is the first excited state. Comment on the value of $|\psi_\mathrm{a}(x_1, x_2)|^2$ when $x_1 = x_2$.

(b) Repeat (a) but now for the case when ψ_1 is the single-particle first excited state and ψ_2 is the second excited state. Comment on the results.

(c) Repeat (a) and (b) for symmetric wave functions $\psi_s(x_1, x_2)$. Comment on the results.

Problem 6.5

(a) Consider the conduction-band minimum potential profile shown in Fig. 6.P9. It consists of two GaAs contact layers and an AlGaAs potential energy barrier region. The contacts have the same n-type impurity concentration and the AlGaAs is intrinsic. The current due to a single electron in state $|\mathbf{k}\rangle$ with energy $E_\mathbf{k}$ is

$$J_e = e\frac{\hbar k_\perp}{m_e^*}T(E_\perp),$$

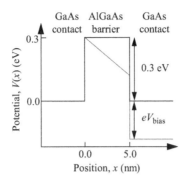

Fig. 6.P9

where $k_\perp$ is the component of $\mathbf{k}$ perpendicular to the layer interface and $T(E_\perp)$ is the transmission coefficient.

The total current flowing left-to-right involves all electron states in the left contact and so requires integration over both $k_\perp$ and $k_\parallel$, where $k_\parallel = \sqrt{2m_e^* E_\parallel}/\hbar$ is the component of $\mathbf{k}$ parallel to the contact–barrier interface, and $E_\mathbf{k} = E_\perp + E_\parallel$. If the probability of an electron mass m_e^* and charge e occupying state $|\mathbf{k}\rangle$ is given by the Fermi–Dirac function $f(E_\mathbf{k}, \mu)$, the current due to the left contact is

$$J_\mathrm{L} = e \int T(E_\perp) \frac{\hbar k_\perp}{m_e^*} \frac{\mathrm{d}k_\perp}{2\pi} \int f(E_\mathbf{k}, \mu) \frac{\mathrm{d}^2 k_\parallel}{(2\pi)^2},$$

where μ is the chemical potential in the contact. The total current is the difference between the left contact and the right contact current. A positive bias voltage, V_bias, lowers the chemical potential energy of the right-hand contact by eV_bias, and the total current is

$$J = e \int T(E_\perp) \frac{\hbar k_\perp}{m_e^*} \frac{\mathrm{d}k_\perp}{2\pi} \int (f(E_\mathbf{k}, \mu) - f(E_\mathbf{k}, \mu - eV_\mathrm{bias})) \frac{\mathrm{d}^2 k_\parallel}{(2\pi)^2}.$$

Starting from this expression, use the one- and two-dimensional densities of states to convert the integrals to energy and evaluate the integration over $k_\parallel$ to show that

$$J = \frac{em_e^* k_\mathrm{B} T}{\pi^2 \hbar^3} \int_0^\infty T(E_\perp) \ln \left(\frac{1 + e^{(\mu - E_\perp)/k_\mathrm{B}T}}{1 + e^{(\mu - E_\perp - eV_\mathrm{bias})/k_\mathrm{B}T}} \right) \mathrm{d}E_\perp.$$

(b) Calculate the current density through the potential barrier using an effective electron mass $m_e^* = 0.07 \times m_0$ and a voltage bias range of $0 \leq V_\mathrm{bias} \leq 0.3\,\mathrm{V}$. As in the figure, assume that the potential change due to V_bias is linear with position across the barrier. Plot the calculated current density using both linear and log scales. Explain the dependence of current density on the voltage bias when the impurity concentration has value $n_0 = 10^{18}\,\mathrm{cm}^{-3}$ and $n_0 = 10^{16}\,\mathrm{cm}^{-3}$ and when the temperature has value $T = 300\,\mathrm{K}$ and $T = 4.2\,\mathrm{K}$. Explain what is observed when the voltage bias range is extended to $0 \leq V_\mathrm{bias} \leq 2.3\,\mathrm{V}$.

299

Problem 6.6

The minimum value of chemical potential, $\mu_{\min}$, at finite absolute temperature T may be found by assuming a particle distribution function that is obtained in the $T \to \infty$ limit.

(a) Show that in this case a three-dimensional electron gas of fixed density n_{3D} has minimum chemical potential

$$\mu_{\min}^{3D} = k_B T \ln \left(\frac{n_{3D}}{2} \left(\frac{2\pi \hbar^2}{m_0 k_B T} \right)^{\frac{3}{2}} \right),$$

where m_0 is the electron mass.

(b) Find the expression for $\mu_{\min}^{2D}$ of a two-dimensional electron gas.

(c) Find $\mu_{\min}^{1D}$ of a one-dimensional electron gas.

(d) Plot the exact μ and minimum $\mu_{\min}$ from part (a), (b), and (c), for normalized chemical potential μ/E_F as a function of normalized thermal energy $k_B T/E_F$, where E_F is the Fermi energy. Explain the differences observed.

Problem 6.7

Three identical indistinguishable noninteracting particles in a one-dimensional harmonic oscillator potential obey the Pauli exclusion principle. Assuming that any two microscopically distinguishable arrangements of the system with the same total energy are equally likely (the ergodic theorem), plot the probability of occupation as a function of energy for:

(a) the lowest energy state of the system,

(b) when the total energy is $E_{total} = 13.5 \times \hbar\omega$, and

(c) when total energy is $E_{total} = 48.5 \times \hbar\omega$.

(d) Repeat (a)–(c) for the case when there are five particles. Comment on anything learned.

Problem 6.8

The Hubbard model for a one-dimensional (1D) lattice with only two sites has Hamiltonian for electrons of spin $\sigma_s = \uparrow$ or $\downarrow$ that is

$$\hat{H} = -t_{\text{hop},1} \sum_{i,j,\sigma_s} (\hat{c}_{2,\sigma_s}^\dagger \hat{c}_{1,\sigma_s} + \hat{c}_{1,\sigma_s}^\dagger \hat{c}_{2,\sigma_s}) + U \sum_i \hat{n}_{i,\uparrow} \hat{n}_{i,\downarrow},$$

where $t_{\text{hop},1}$ is the nearest-neighbor hopping matrix element, U is the additional energy associated with two electrons of opposite spin occupying the same lattice site, i, the electron creation and annihilation operators are $\hat{c}^\dagger$ and $\hat{c}$, respectively, and $\hat{n}$ is the electron number operator.

(a) If there are two electrons in the system ($N = 2$), write down all possible states in the Fock particle number basis.

(b) Using the Fock basis and $\hat{H}$, calculate all matrix elements and derive an expression for the eigenenergies in terms of the coupling strengths U and $t_{\text{hop},1}$ by finding the eigenvalues for the matrix. Plot the eigenenergies as a function of the ratio $U/t_{\text{hop},1}$.

(c) Repeat (a) and (b) for the same one-dimensional lattice with two sites, but now for the cases when $N = 1, 3, 4$ and comment on what is learned.

Problem 6.9
Explain why the statement "The wave function $\psi(\mathbf{r})$ is a function of three-dimensional real-space" is incorrect.

Problem 6.10
An electron gas obeys Fermi–Dirac statistics at temperature T and is constrained to motion in the x direction. A potential barrier of infinite energy is placed at position $x = 0$ such that potential $V(x > 0) = \infty$ and $V(x \leq 0) = 0$.

(a) By considering a region that confines electrons of effective mass m_e^* from $x = 0$ to $x = L < 0$, find the normalized electron energy eigenstates and explain how taking the limit $L \to -\infty$ creates a continuum of states.

(b) At high temperatures, the Fermi–Dirac distribution may be approximated by a Maxwell–Boltzmann distribution. Find an analytic expression for the electron density $n = n(x, T)$ in terms of a characteristic length scale, $\lambda_e = \hbar/\sqrt{2m_e^* k_B T}$. Explain why n has a smaller value near $x = 0$ than when $x \ll -\lambda_e$. Plot the density, $n(x, T)$, at temperatures $T = 300\,\mathrm{K}$, $T = 100\,\mathrm{K}$, and $T = 10\,\mathrm{K}$ for electrons with effective mass $m_e^* = 0.07 \times m_0$.

(c) When temperature $T < E_F/k_B$, the Fermi–Dirac distribution may not be approximated by a Maxwell–Boltzmann distribution. Find and explain the spatial dependence of carrier concentration $n(x, T)$ for temperatures $T = 300\,\mathrm{K}$, $T = 100\,\mathrm{K}$, and $T = 10\,\mathrm{K}$.

Note: $\int_0^\infty e^{-a\xi^2} d\xi = \frac{1}{2}\sqrt{\frac{\pi}{a}}$ and $\int_{\xi=0}^{\xi=\infty} e^{-a\xi^2} \cos(b\xi) d\xi = \sqrt{\frac{\pi}{4a}} e^{\frac{-b^2}{4a}}$ for $\mathrm{Re}(a) > 0$.

Problem 6.11
A fermion system has chemical potential $\mu(T = 3\,\mathrm{K}) = 52.1\,\mathrm{meV}$ and $\mu(T = 300\,\mathrm{K}) = 38.9\,\mathrm{meV}$. Plot the Fermi–Dirac distribution $f(E)$ *and* the Bose distribution $g(E)$ as a function of energy for $0 \leq E \leq 200\,\mathrm{meV}$. Explain the difference in the curves obtained. What is expected to happen to the distribution functions in the high-temperature limit?

7 Time-Dependent Perturbation

7.1 Introduction

Engineers who design transistors, lasers, and other semiconductor components want to understand and control the cause of resistance to current flow so that they may better optimize device performance. A detailed microscopic understanding of electron motion from one part of a semiconductor to another requires the explicit calculation of electron scattering probability. One would like to know how to predict electron scattering from one state to another – something quantum mechanics can do.

In addition to understanding electron motion in a semiconductor, knowledge about how to make devices that emit or absorb light is also of practical importance. In Chapter 5 it was shown that a superposition of two harmonic oscillator eigenstates could give rise to dipole radiation and emission of a photon. For electron bound states, the creation of a photon is only possible if a superposition state exists between appropriate eigenfunctions. This leads directly to the concept of rules determining pairs of electron eigenstates that can give rise to photon emission. Such selection rules are a useful tool to help understand the emission and absorption of light by matter. However, the real challenge is to use what is known to make practical devices that operate using emission and absorption of photons. This usually requires imposing some control over atomic-scale physical processes that, of course, can only be understood using quantum mechanics.

In this chapter, electronic transitions due to an abrupt time-dependent change in potential is considered first. Following this, important results from first-order time-dependent perturbation theory are derived and used to calculate excitation of a charged particle in a harmonic potential due to a transient electric field pulse. The *golden rule* is then derived. As an example, it will be used to calculate the elastic scattering rate from ionized impurities for electrons in the conduction band of n-type GaAs. Such calculations are of practical importance for the design of high-performance transistors and laser diodes. The study results in a number of predictions, such as the temperature dependence of conductivity and the fact that the response of many mobile electrons to the presence of a scattering site must be taken into account. It is also shown that, by controlling the position of scattering sites on an atomic scale, the probability of elastic scattering can be dramatically altered.

As a starting point, and by way of example, control of an electronic transition from a ground state to an excited state in a quantum mechanical system is demonstrated. The key idea is application of a time-dependent potential to change the distribution of occupied states. In principle, the change in potential could take place smoothly or abruptly in time. To explore the influence of a time-varying potential in a quantum system, an abrupt change in potential is considered first.

7.1.1 An Abrupt Change in Potential

An electron of mass m_0 is in a one-dimensional rectangular potential well where $V(x) = 0$ for $0 < x < L$ and $V(x) = \infty$ elsewhere. The energy eigenvalues are $E_n = \hbar^2 k_n^2 / 2m_0$, and the bound-state eigenfunctions are $\psi_n = \sqrt{2/L} \sin(k_n x)$, where $k_n = n\pi/L$ and the quantum number n is a positive nonzero integer.

The particle is initially prepared in the ground state ψ_1 with eigenenergy E_1. Then, as illustrated in Fig. 7.1, at time $t = 0$ the potential is very rapidly changed (*quenched*) in such a way that the original wave function remains the same but $V(x) = 0$ for $0 < x < 2L$ and $V(x) = \infty$ elsewhere. The task is to determine the effect such a change in potential has on the expectation value of particle energy and the probability that the particle is in an excited state of the system at time $t > 0$. The expectation value of particle energy before

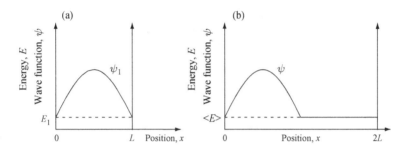

Fig. 7.1 (a) Sketch of a one-dimensional rectangular potential well with infinite barrier energy showing the lowest-energy eigenvalue E_1 and its associated ground-state wave function ψ_1. (b) The potential barrier at position L is suddenly moved to position $2L$, resulting in a new wave function ψ. The energy expectation value of the new state is $\langle E \rangle = E_1$.

the potential well is abruptly increased in thickness is known to be $\langle E \rangle = E_1$ for $t < 0$. Since the wave function after the change in potential ψ is the same as the original wave function ψ_1 but with the addition of a constant zero value for $L < x < 2L$, the expectation value in energy is $\langle E \rangle = E_1$ for time $t \geq 0$. It is important to check that the kink in the wave function at position $x = L$ does not contribute $\Delta \langle E \rangle = -\hbar^2 \psi(x = L) \Delta \psi / 2m$ to the energy (see Section 2.1.1). Clearly, since $\psi(x = L) = 0$, the kink does not make a contribution, and so $\langle E \rangle = E_1$ after the change in potential because the curvature of the wave function determining particle energy does not change.

When time $t \geq 0$, the new state ψ is not an eigenfunction of the system. The *new* eigenfunctions of a rectangular potential well of thickness $2L$ with infinite barrier energy are $\psi_m = \sqrt{1/L} \sin(k_m x)$, where $k_m = m\pi/2L$ and the index $m = 1, 2, 3, \ldots$. Since the state ψ is *not* an eigenfunction, it may be expressed as a sum of the new eigenfunctions, so that

$$\psi = \sum_m a_m \psi_m. \tag{7.1}$$

The coefficients a_m are found by multiplying both sides by ψ_m^* and integrating over all space. This overlap integral gives the coefficients

$$a_m = \int \psi_m^* \psi \, dx. \tag{7.2}$$

303

The effect of the overlap integral is to *project out* the components of the new eigenstates that contribute to the wave function ψ. The value of $|a_m|^2$ is the probability of finding the particle in the eigenstate ψ_m.

To illustrate how to find the contribution of the new eigenstates to the wave function ψ, the probability that the particle is in the new ground state $\psi_{m=1}$ when $t \geq 0$ is calculated. This is given by the square of the magnitude of the overlap integral, a_m, with $m = 1$:

$$a_1 = \int_{x=0}^{x=L} \sqrt{\frac{1}{L}} \sin\left(\frac{\pi x'}{2L}\right) \sqrt{\frac{2}{L}} \sin\left(\frac{\pi x'}{L}\right) dx'. \tag{7.3}$$

Using $2\sin(x)\sin(y) = \cos(x-y) - \cos(x+y)$, then

$$a_1 = \frac{\sqrt{2}}{L} \int_{x=0}^{x=L} \left(\frac{1}{2}\cos\left(\frac{\pi x'}{2L}\right) - \frac{1}{2}\cos\left(\frac{3\pi x'}{2L}\right)\right) dx'$$

$$= \frac{\sqrt{2}}{L}\left[\frac{1}{2}\frac{2L}{\pi}\sin\left(\frac{\pi x'}{2L}\right) - \frac{1}{2}\frac{2L}{3\pi}\sin\left(\frac{3\pi x'}{2L}\right)\right]_0^L = \frac{\sqrt{2}}{L}\left(\frac{L}{\pi} + \frac{L}{3\pi}\right) = \frac{4\sqrt{2}}{3\pi}. \tag{7.4}$$

Hence, the probability of finding the particle in the new ground state of the system is

$$|a_1|^2 = \frac{32}{9\pi^2} = 0.36025. \tag{7.5}$$

In the preceding, the solutions for the new eigenfunctions are known. When the new solutions are not known, it is possible to perform an expansion in the initial state basis. In the preceding, it was also assumed the potential energy changed much faster than the particle's response time. Classically, this means that the potential barrier moved faster than the velocity of a particle with energy E. However, if the potential barrier velocity is comparable to that of a particle with energy E, then a different method of calculating transition probabilities is needed. This is analyzed next.

7.1.2 Time-Dependent Change in Potential

Consider a quantum-mechanical system described by Hamiltonian $\hat{H}_0$ and for which the solutions to the time-independent Schrödinger equation,

$$\hat{H}_0|n\rangle = E_n|n\rangle, \tag{7.6}$$

are known. The time-independent eigenvalues are $E_n = \hbar\omega_n$, and the orthonormal bound-state eigenfunctions are $|n\rangle$. The eigenfunction $|n\rangle$ evolves in time according to

$$|n,t\rangle = |n\rangle e^{-i\omega_n t} \tag{7.7}$$

(so that $\langle x|n\rangle e^{-i\omega_n t} = \phi_n(x)e^{-i\omega_n t}$) and satisfies

$$i\hbar\frac{\partial}{\partial t}|n\rangle e^{-i\omega_n t} = \hat{H}_0|n\rangle e^{-i\omega_n t}. \tag{7.8}$$

At time $t = 0$, a time-dependent change in potential $\hat{W}(t)$ is applied the effect of which is to create a new Hamiltonian,

$$\hat{H} = \hat{H}_0 + \hat{W}(t), \tag{7.9}$$

and state $\psi(t)$, which evolves in time according to

$$i\hbar \frac{\partial}{\partial t}\psi(t) = \left(\hat{H}_0 + \hat{W}(t)\right)\psi(t). \tag{7.10}$$

The time-dependent change in potential energy $\hat{W}(t)$ might, for example, be a step function or an oscillatory function.

Solutions are sought to the time-dependent Schrödinger equation, which includes the change in potential (Eq. (7.10)), in the form of a sum over the *known* eigenstates of the unperturbed system,

$$\psi(t) = \sum_n^\infty a_n(t)|n\rangle e^{-i\omega_n t}, \tag{7.11}$$

where $a_n(t)$ are time-dependent coefficients.

Substituting Eq. (7.11) into Eq. (7.10) gives

$$i\hbar \frac{d}{dt}\sum_n a_n(t)|n\rangle e^{-i\omega_n t} = \left(\hat{H}_0 + \hat{W}(t)\right)\sum_n a_n(t)|n\rangle e^{-i\omega_n t}. \tag{7.12}$$

Using the product rule for differentiation $((fg)' = (f'g + fg'))$, the left-hand side may be written as

$$i\hbar \sum_n \left(\left(\frac{\partial}{\partial t}a_n(t)\right)|n\rangle e^{-i\omega_n t} + a_n(t)\left(\frac{\partial}{\partial t}|n\rangle e^{-i\omega_n t}\right)\right) = \left(\hat{H}_0 + \hat{W}(t)\right)\sum_n a_n(t)|n\rangle e^{-i\omega_n t}. \tag{7.13}$$

Making use of Eq. (7.8), the terms

$$i\hbar \sum_n a_n(t)\frac{\partial}{\partial t}|n\rangle e^{-i\omega_n t} = \sum_n a_n(t)\hat{H}_0|n\rangle e^{-i\omega_n t} \tag{7.14}$$

may be removed from Eq. (7.13) to leave

$$i\hbar \sum_n |n\rangle e^{-i\omega_n t}\frac{\partial}{\partial t}a_n(t) = \sum_n a_n(t)\hat{W}(t)|n\rangle e^{-i\omega_n t}. \tag{7.15}$$

Multiplying both sides by $\langle m|e^{i\omega_n t}$ and using the orthonormal relationship $\langle m|n\rangle = \delta_{mn}$ gives

$$i\hbar \frac{d}{dt}a_m(t) = \sum_n a_n(t)\langle m|\hat{W}(t)|n\rangle e^{i\omega_{mn} t}. \tag{7.16}$$

However, since

$$\int \phi_m^*(x) e^{i\omega_m t} \hat{W} \phi_n(x) e^{-i\omega_n t} dx = W_{mn} e^{i\omega_{mn} t}, \tag{7.17}$$

where $\hbar\omega_{mn} = E_m - E_n$, and W_{mn} is defined as the matrix element $\int \phi_m^*(x)\hat{W}\phi_n(x)\,dx = \langle m|W|n\rangle = W_{mn}$, then

$$i\hbar\frac{d}{dt} a_m(t) = \sum_n a_n(t) W_{mn} e^{i\omega_{mn} t}. \tag{7.18}$$

The intrinsic time-dependence of each state can be factored out by introducing different time-dependent coefficients so that $c_n(t) = a_n(t)e^{i\omega_{mn} t}$. This is called the *interaction picture*, in contrast to the a-coefficient formalism that is the *Schrödinger picture*.

The probability that the system can be found in a stationary state $|n\rangle$ after the perturbation is turned off at time t is

$$P_n(t) = |a_n(t)|^2 = |c_n(t)|^2. \tag{7.19}$$

It is important to recognize that Eq. (7.18) is an *exact* result. However, the right-hand side contains the time-dependent coefficients that are sought. At first sight, not a great deal of progress seems to have been made. To make a little more headway, it helps to be quite specific.

Suppose the system is initially in the ground state $|0\rangle$ so that $a_n(t < 0) = c_n(t < 0) = \delta_{n,0}$. It is now assumed that there is a constant step change in potential $W(t)$ that is turned on at time $t = 0$ for duration τ so that $\hat{W}(0 \le t \le \tau) \ne 0$, $\partial\hat{W}(0 \le t \le \tau)/\partial t = 0$, and $\hat{W}(t < 0, t > \tau) = 0$. Figure 7.2 is an illustration of this time-dependent potential.

The following may be stated about the time evolution of the system subject to the time-dependent potential, $\hat{W}(t)$.

Before application of $\hat{W}(t)$: Time-independent, stationary-state solutions satisfy $\hat{H}_0|n\rangle = E_n|n\rangle$. The probability of finding the state, in this case $|0\rangle$, is assumed to be unity.

During application of $\hat{W}(t)$: The state of the system evolves according to

$$\psi(t) = \sum_n a_n(t)|n\rangle e^{-iE_n t/\hbar}. \tag{7.20}$$

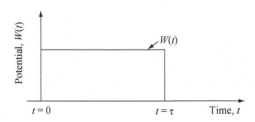

Fig. 7.2 Plot of the time-dependent potential $W(t)$ discussed in the text.

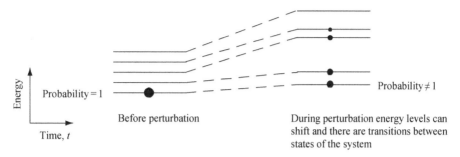

Fig. 7.3 Illustration of energy level shifts and changes in population of energy levels during perturbation.

The time-dependent change in potential causes *transitions between eigenstates* of the initial system (*off-diagonal* matrix elements) and can *shift energy levels* of the initial eigenstates (*diagonal* matrix elements).

After application of $\hat{W}(t)$: Stationary-state solutions satisfy $\hat{H}_0|n\rangle = E_n|n\rangle$. Any time after $\hat{W}(t)$ is turned off at time $t = \tau$ there is a probability of finding the state in any of the known initial time-independent stationary solutions. The final state of the system may be a *superposition* of these eigenstates,

$$\psi(t) = \sum_n a_n(t)|n\rangle e^{-iE_n t/\hbar}. \tag{7.21}$$

For the above example, the perturbation may be visualized as changing the population of eigenstates and shifting energy eigenvalues. This is illustrated in Fig. 7.3.

7.2 First-Order Time-Dependent Perturbation

While Eq. (7.18) is an *exact* result, it is often necessary to make approximations to calculate actual probabilities. One way to proceed is to assume $a_n(t \le 0) = 1$ for the eigenstate $|n\rangle$ and $a_j(t \le 0) = 0$ for $j \ne n$, and then *approximate* the value of $a_n(t > 0) = 1$. Such an approximation is called first-order perturbation theory. Typically, this approach is valid when the time-dependent change in potential $\hat{W}(t)$ is small and thus may be considered a perturbation to the initial system described by the Hamiltonian $\hat{H}_0$.

Consider the case in which the system is in an eigenstate $|n\rangle$ of $\hat{H}_0$ at $t \le 0$, so $a_n(t = 0) = 1$ and $a_j(t = 0) = 0$ for $j \ne n$. There is now only one term on the right-hand side of Eq. (7.18), which, for $m \ne n$ and $t > 0$, becomes

$$i\hbar \frac{d}{dt} a_m(t) = W_{mn} e^{i\omega_{mn}t} \tag{7.22}$$

since all coefficients $a_m(t = 0) = 0$ except for $a_n(t = 0) = 1$. This means that the matrix element W_{mn} couples $|n\rangle$ to $|m\rangle$ and creates the coefficient $a_m(t)$ for times $t > 0$ and $m \ne n$.

To find how $a_m(t)$ evolves in time from $t = 0$, Eq. (7.22) is rewritten as an integral,

$$a_m(t) = \frac{1}{i\hbar} \int_{t'=0}^{t'=t} W_{mn} e^{i\omega_{mn}t'} dt'.$$
(7.23)

It is assumed that the state with eigenenergy E_n is not degenerate, so that the perturbed wave function can be expressed as a sum of unperturbed states weighted by coefficients $a_k(t)$,

$$\psi(x, t) = \sum_k a_k(t) e^{-i\omega_k t} \phi_k(x).$$
(7.24)

If $|a_m|^2 \ll 1$, then the approximation that the coefficient $a_n(t) = 1$ can be used. In this case, Eq. (7.24) may be written as

$$\psi(x, t) = \phi_n(x) e^{-i\omega_n t} + \sum_{m \neq n} \frac{1}{i\hbar} \int_{t'=0}^{t'=t} W_{mn} e^{i\omega_{mn}t'} dt' e^{-i\omega_m t} \phi_m(x).$$
(7.25)

There is an obvious problem with normalization of the scattered state given by Eq. (7.25). Clearly, in a complete theory correction for the normalization error must be performed self-consistently.

7.2.1 Higher-Order Terms in Time-Dependent Perturbation

Another way to develop time-dependent perturbation theory starts with the exact result Eq. (7.16),

$$i\hbar \frac{d}{dt} a_m(t) = \sum_n a_n(t) \langle m|\hat{W}(t)|n\rangle e^{i\omega_{mn}t},$$
(7.26)

for which states $|n\rangle$ and eigenvalues $\hbar\omega_n = E_n$ are known and $\omega_{mn} = \omega_m - \omega_n$. Perturbation $W(t)$ is also known and applied at time $t > 0$.

If the coefficients $a_n(t)$ can be expanded in a power series of the perturbing potential, then

$$a_n(t) = a_n^{(0)} + \lambda a_n^{(1)}(t) + \lambda^2 a_n^{(2)}(t) + \cdots$$
(7.27)

and

$$W(t) = \lambda W(t),$$
(7.28)

where λ is a dummy variable to keep track of the order of the terms in the power series. At the end of the calculation, the dummy variable is set to $\lambda = 1$. Substituting into the exact result (Eq. (7.26)),

308

$$i\hbar \frac{d}{dt} \left(a_m^{(0)} + \lambda a_m^{(1)}(t) + \lambda^2 a_m^{(2)}(t) + \cdots \right)$$

$$= \sum_n \left(a_n^{(0)} + \lambda a_n^{(1)}(t) + \lambda^2 a_n^{(2)}(t) + \cdots \right) \langle m|\lambda \hat{W}(t)|n\rangle e^{i\omega_{mn}t}$$

$$= \sum_n \left(\lambda a_n^{(0)} + \lambda^2 a_n^{(1)}(t) + \lambda^3 a_n^{(2)}(t) + \cdots \right) \langle m|\hat{W}(t)|n\rangle e^{i\omega_{mn}t}. \tag{7.29}$$

Equating zeroth-order terms in λ gives

$$i\hbar \frac{d}{dt} a_m^{(0)}(t) = 0, \tag{7.30}$$

and so $a_m^{(0)}(t) = a_m^{(0)}$, there is no time dependence, $W(t)$ has no effect, and system is unchanged.

Equating the first-order terms in λ gives

$$i\hbar \frac{d}{dt} a_m^{(1)}(t) = \sum_n a_n^{(0)} \langle m|\hat{W}(t)|n\rangle e^{i\omega_{mn}t}, \tag{7.31}$$

where $a_n^{(0)}$ are constant coefficients describing the known starting state of the system at time $t = 0$. The starting state of the system is, in general, a weighted sum of known eigenstates (and not the single eigenstate considered previously, Eq. (7.22)). Because the initial state of the system is known, the coefficients $a_m^{(1)}(t)$ can be found by integration and so an approximate solution for the expansion coefficients to first order, $a_m(t) \approx a_m^{(0)} + a_m^{(1)}(t)$, is obtained. To go beyond first-order time-dependent perturbation theory, it is possible to iterate to higher orders using

$$i\hbar \frac{d}{dt} a_m^{(k+1)}(t) = \sum_n a_n^{(k)} \langle m|\hat{W}(t)|n\rangle e^{i\omega_{mn}t}. \tag{7.32}$$

If this power series expansion converges, then the accuracy of the solution can be improved by using the next higher-order correction.

7.2.2 Charged Particle in a Harmonic Oscillator Potential

As an application of first-order time-dependent perturbation theory, consider a particle of mass m_0 and charge e moving in a one-dimensional harmonic oscillator potential. The potential might have been created in a molecule. The system is perturbed by application of a macroscopic external pulse of electric field that has a Gaussian time dependence. While use of a macroscopic field to manipulate a molecular-sized entity may seem a little crude, it is in fact quite a powerful way to control single electrons.

Assume that the charged particle is initially (at time $t = -\infty$) in the ground state of the one-dimensional harmonic oscillator potential $V(x)$ of the molecule. The bound state of the charged particle is to be controlled by applying a pulse of electric field $\mathbf{E}(t) = -\mathbf{E}_0 e^{-t^2/\tau^2}$, where $\mathbf{E}_0$ and τ are constants, $|\mathbf{E}_0|$ is the maximum strength of the applied electric field and $2\tau\sqrt{\ln(2)}$ is the full-width-half-maximum (FWHM) of the electric field pulse. The potential seen by the particle with charge e is $\hat{W}(t) = e|\mathbf{E}(t)|x$. The value of τ that gives the *maximum* probability of the system being in an excited state a long time after application of the pulse is of interest.

Starting from the exact result, Eq. (7.18),

$$i\hbar \frac{d}{dt}a_m(t) = \sum_n a_n(t)W_{mn}e^{i\omega_{mn}t}, \tag{7.33}$$

where $a_n(t)$ is approximated by its initial value $a_n(t = -\infty)$. So, if the system is initially in an eigenstate $|n\rangle$ of the Hamiltonian $\hat{H}_0$, then $a_n(t = -\infty) = 1$ and $a_m(t = -\infty) = 0$ for $m \neq n$. There is now only one term on the right-hand side of the equation, so that

$$i\hbar \frac{d}{dt}a_m(t) = a_n(t)W_{mn}e^{i\omega_{mn}t} = W_{mn}e^{i\omega_{mn}t}. \tag{7.34}$$

Since $a_n(t)$ is approximated by its initial value, $a_n(t) = 1$. Integration gives

$$a_m(t) = \frac{1}{i\hbar} \int\limits_{t=-\infty}^{t=\infty} W_{mn}e^{i\omega_{mn}t'}dt'. \tag{7.35}$$

The charged particle starts in the ground state $|n = 0\rangle$ of the harmonic potential. The probability of the system being in an excited state after the electrical pulse has gone $(t \to \infty)$ is given by the sum

$$P_{t\to\infty} = \left| \sum_{m \neq n} a_m(t = \infty) \right|^2. \tag{7.36}$$

If each final state $|m\rangle$ is an *independent* parallel *channel* for the scattering process, then

$$P_{t\to\infty} = \sum_{m \neq n} |a_m(t = \infty)|^2. \tag{7.37}$$

The matrix element for transitions from the ground state of the harmonic oscillator in the presence of a uniform electric field is

$$W_{m,n=0} = e|\mathbf{E}_0|e^{-t^2/\tau^2}\langle m|\hat{x}|n = 0\rangle. \tag{7.38}$$

The only matrix element that contributes to the sum for the probability P couples the ground state to the first excited state separated in energy by $\hbar\omega$. Substituting Eq. (5.11) into Eq. (7.38) gives matrix element

$$W_{10} = e|\mathbf{E}_0|e^{-t^2/\tau^2}\left(\frac{\hbar}{2m_0\omega}\right)^{1/2}\langle 1|\left(\hat{b} + \hat{b}^\dagger\right)|0\rangle = e|\mathbf{E}_0|e^{-t^2/\tau^2}\left(\frac{\hbar}{2m_0\omega}\right)^{1/2}. \tag{7.39}$$

Hence, the probability of the system being in the first excited state after the electrical pulse has gone $(t \to \infty)$ is

$$P_{t\to\infty} = \frac{1}{\hbar^2}\left| \int\limits_{t=-\infty}^{t=\infty} W_{10}e^{i\omega_{10}t'}dt' \right|^2 = \left(\frac{\hbar}{2m_0\omega}\right)\frac{e^2|\mathbf{E}_0|^2}{\hbar^2}\left| \int\limits_{t=-\infty}^{t=\infty} e^{-t'^2/\tau^2}e^{i\omega t'}dt' \right|^2, \tag{7.40}$$

where the frequency $\omega_{mn} = \omega_{10} = \omega$. Completing the square in the exponent in such a way that $(-t'^2/\tau^2 + i\omega t') = -(t'/\tau - i\omega\tau/2)^2 - \omega^2\tau^2/4$, then

$$P_{t\to\infty} = \left(\frac{e^2|\mathbf{E}_0|^2}{2m_0\hbar\omega}\right) e^{-\omega^2\tau^2/2} \left| \int_{t=-\infty}^{t=\infty} e^{-(t'/\tau - i\omega\tau/2)^2} dt' \right|^2. \tag{7.41}$$

Fortunately, the integral is standard, with the solution[1]

$$\int_{t=-\infty}^{t=\infty} e^{-(t'/\tau - i\omega\tau/2)^2} dt' = \tau\sqrt{\pi}. \tag{7.42}$$

Hence, Eq. (7.41) may be written

$$P_{t\to\infty} = \left(\frac{\pi e^2|\mathbf{E}_0|^2}{2m_0\hbar\omega}\right) \tau^2 e^{-\omega^2\tau^2/2}. \tag{7.43}$$

The physics of how the transition is induced is illustrated in Fig. 7.4. The electric field pulse exerts a force on the charged particle through a change in the potential energy. At any given instant at which the electric field has value $\mathbf{E}$, the parabolic harmonic oscillator potential energy $V(x)$ is shifted in position by $x_d = e|\mathbf{E}|/2m_0\omega^2$ and reduced in energy by $-e^2|\mathbf{E}|^2/2m_0\omega^2$ (see Chapter 5). Transitions from the ground state are induced by this change in potential energy.

The maximum transition probability occurs when the derivative of the probability function with respect to the variable τ is zero (that this is a maximum may be confirmed by taking the second derivative):

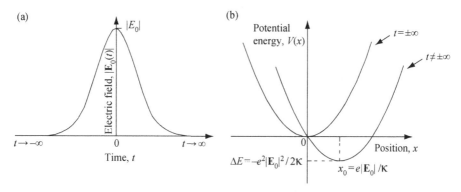

Fig. 7.4 (a) The electric field pulse is Gaussian in shape and centered at time $t = 0$. (b) The harmonic oscillator potential, $V(x, t)$, evolves in time. Maximum displacement is x_0 when electric field has value E_0 and energy shift is ΔE. The spring constant $\kappa = m\omega^2$

---

[1] To see this, change variables to $y = t'/\tau - i\omega\tau/2$ so $dy = dt'/\tau$. The integral becomes $\int_{-\infty}^{\infty} e^{-y^2} \tau \, dy = \tau\sqrt{\pi}$.

$$0 = \frac{d}{d\tau} P_{t\to\infty} = 2\tau e^{-\omega^2\tau^2/2} - \frac{2\omega^2\tau}{2}\tau^2 e^{-\omega^2\tau^2/2} = 2\tau - \omega^2\tau^3. \tag{7.44}$$

Hence, the maximum transition probability occurs when the value of τ is

$$\tau = \frac{\sqrt{2}}{\omega}. \tag{7.45}$$

According to Eq. (7.45), if $\omega = 10^{12}\,\mathrm{rad\,s^{-1}}$ then the maximum transition probability occurs for $\tau = 1.41\,\mathrm{ps}$, corresponding to an electric field pulse with an FWHM of 2.35 ps.

The value of the maximum transition probability is found by substituting Eq. (7.45) into Eq. (7.43). This gives

$$P_{\max} = \left(\frac{\pi e^2|\mathbf{E}_0|^2}{m_0\hbar\omega^3}\right)e^{-1}. \tag{7.46}$$

There is an obvious difficulty with this result, since when $\omega \to 0$ or $|\mathbf{E}_0| \to \infty$ the maximum transition probability $P_{\max} > 1$. According to Eq. (7.46), $P_{\max} = 1$ when $\omega = (\pi e^2|\mathbf{E}_0|^2 e^{-1}/m_0\hbar)^{1/3}$.

The resolution of this inconsistency is that the perturbation theory being used only applies when the time-dependent change in potential $W(t)$ is small. The assumption that the perturbation is weak means that the probability of scattering out of the initial state $|n\rangle$ is small, so that $|a_m|^2 \ll 1$. This condition is simply $P_{\max} \ll 1$, which constrains the value of ω and $|\mathbf{E}_0|$ so that

$$\left(\frac{\pi e^2|\mathbf{E}_0|^2\hbar^2}{m_0}\right)e^{-1} \ll \hbar^3\omega^3. \tag{7.47}$$

For the case considered, with $\omega = 10^{12}\,\mathrm{rad\,s^{-1}}$, this results in $|\mathbf{E}_0| \ll 5.7 \times 10^4\,\mathrm{V\,m^{-1}}$. Clearly, while perturbation theory can be used to calculate transition rates, it is always important to identify and understand the limitations of the calculation.

7.3 The Golden Rule

A perturbing potential $\hat{W}(t)$ can be used to control transitions between eigenstates of the initially unperturbed Hamiltonian. If the system is first prepared in state $|n\rangle$ and $\hat{W}(t) = 0$ for $t \le 0$, then up to time $t = 0$ the system remains in state $|n\rangle$ with energy $E_n = \hbar\omega_n$. For time $t > 0$, $\hat{W}(t) \ne 0$ is allowed. Hence, when $t > 0$ it is possible that the system is in a different state $|m\rangle$ with energy $E_m = \hbar\omega_m$.

The transition probability may be calculated by starting with Eq. (7.23),

$$a_m(t) = \frac{1}{i\hbar}\int_{t'=0}^{t'=t} W_{mn}e^{i\omega_{mn}t'}dt'. \tag{7.48}$$

If the perturbation $\hat{W}(t > 0) \ne 0$ does not depend on time ($\frac{\partial}{\partial t}\hat{W}(t > 0) = 0$) then the probability of scattering out of state $|n\rangle$ into state $|m\rangle$ as a function of time is

$$P_n(t) = |a_m(t)|^2 = \frac{1}{\hbar^2} \left| \int_{t'=0}^{t'=t} W_{mn} e^{i\omega_{mn}t'} dt' \right|^2 = \frac{|W_{mn}|^2}{\hbar^2} \left| \frac{e^{i\omega_{mn}t} - 1}{\omega_{mn}} \right|^2, \tag{7.49}$$

where $\hbar\omega_{mn} = E_m - E_n$ and $W_{mn} = \langle m|\hat{W}|n\rangle$. Since

$$\left| e^{ix} - 1 \right|^2 = 4\sin^2(x/2), \tag{7.50}$$

and making the substitution $x \to \omega_{mn}t$, Eq. (7.49) may be written as

$$P_n(t) = \frac{|W_{mn}|^2}{\hbar^2} \frac{\sin^2((\omega_{mn}t)/2)}{(\omega_{mn}/2)^2}. \tag{7.51}$$

Equation (7.51) may be simplified by considering the long time limit so that $t \to \infty$ and using the relationship

$$\left. \frac{\sin^2(xt)}{\pi t x^2} \right|_{t\to\infty} = \delta(x), \tag{7.52}$$

so that

$$P_n(t) = \frac{|W_{mn}|^2}{\hbar^2} \pi t \delta(\omega_{mn}/2). \tag{7.53}$$

Making use of the fact

$$\delta(ax) = \delta(x)/|a|, \tag{7.54}$$

and setting $x = \omega_{mn}/2$ and $a = 2\hbar$, then $\delta(\omega_{mn}/2) = 2\hbar\delta(\hbar\omega_{mn})$ and

$$P_n(t) = \frac{2\pi t}{\hbar} |W_{mn}|^2 \delta(\hbar\omega_{mn}) = \frac{2\pi t}{\hbar} |W_{mn}|^2 \delta(E_m - E_n). \tag{7.55}$$

Notice that the probability of a transition is *linearly* proportional to time. The reason for this is embedded in the approximations used to obtain this result. Also note that the delta function $\delta(E_m - E_n)$ in Eq. (7.55) ensures energy conservation. For a constant perturbation $\hat{W}$ in the long time limit, the energy of the final state is the same as the energy of the initial state. This is, therefore, an example of elastic scattering, in which, a long time after the scattering event, the incident particle in state $|n\rangle$ neither gains nor loses energy from the perturbation as it scatters into state $|m\rangle$.

If, rather than a discrete final state $|m\rangle$, there is a continuum of final states described by a density of states $D(E_m)$ in the energy interval E_m to $E_m + dE_m$ then the transition probability is given by

$$P_n(t) = \frac{2\pi t}{\hbar} |W_{mn}|^2 \int D(E)\delta(E_m - E_n) \, dE, \tag{7.56}$$

where it is assumed that each scattering process is an *independent* parallel *channel* and the magnitude of the matrix element squared, $|W_{mn}|^2$, varies slowly over the energy interval of interest.

The *transition rate* is the time derivative of the probability $P_n(t)$,

$$\frac{d}{dt} P_n(t) = \frac{2\pi}{\hbar} |W_{mn}|^2 D(E_m). \tag{7.57}$$

Recognizing $dP_n(t)/dt$ as the inverse probability lifetime τ_n of the state $|n\rangle$, then the golden rule may be written as

$$\boxed{\frac{1}{\tau_n} = \frac{2\pi}{\hbar} |W_{mn}|^2 D(E)\delta(E_m - E_n),} \tag{7.58}$$

in which the inverse lifetime $1/\tau_n$ of the initial state $|n\rangle$ only depends upon the magnitude of the matrix element squared coupling the initial state to any scattered state $|m\rangle$ multiplied by the final density of scattered states $D(E)$. The delta function in Eq. (7.58) ensures energy conservation.

In the case where the perturbation is harmonic in time so that, for example,

$$\hat{W}(t > 0) = \left(\hat{b}\, e^{i\omega t} + \hat{b}^\dagger e^{-i\omega t} \right), \tag{7.59}$$

then the transition rate for scattering out of state $|n\rangle$ into state $|m\rangle$ in the $t \to \infty$ limit is (see Problem 7.9)

$$\frac{1}{\tau_n} = \frac{2\pi}{\hbar} \left| \langle m|\hat{b}|n\rangle \right|^2 \delta(E_m - E_n + \hbar\omega) + \frac{2\pi}{\hbar} \left| \langle m|\hat{b}^\dagger|n\rangle \right|^2 \delta(E_m - E_n - \hbar\omega). \tag{7.60}$$

Again, the delta functions ensure energy conservation. For the delta function $\delta(E_m - E_n + \hbar\omega)$, the final energy is $E_m = E_n - \hbar\omega$ indicating an inelastic scattering event in which the system was initially in an excited state and, after the perturbation, *emitted* a quanta of energy $\hbar\omega$. This process is called stimulated emission and is important, for example, when describing the operation of a laser diode. For the delta function $\delta(E_m - E_n - \hbar\omega)$, the final energy is $E_m = E_n + \hbar\omega$ indicating that the system *absorbed* a quanta of energy $\hbar\omega$.

If, rather than a discrete final state $|m\rangle$, there is a continuum of final states described by a density of states $D(E_m)$ then the golden rule for the stimulated emission rate is given by

$$\frac{1}{\tau_n^{\text{emi}}} = \frac{2\pi}{\hbar} \left| \langle m|\hat{b}|n\rangle \right|^2 D(E)\delta(E_m - E_n + \hbar\omega), \tag{7.61}$$

and the absorption rate is

$$\frac{1}{\tau_n^{\text{abs}}} = \frac{2\pi}{\hbar} \left| \langle m|\hat{b}^\dagger|n\rangle \right|^2 D(E)\delta(E_m - E_n - \hbar\omega). \tag{7.62}$$

Notice that because the perturbation $\hat{W}$ is Hermitian, the matrix elements in Eq. (7.61) and Eq. (7.62) are related via

$$\langle m|\hat{b}|n\rangle = \langle n|\hat{b}^\dagger|m\rangle^*, \tag{7.63}$$

so that

$$\left| \langle m|\hat{b}|n\rangle \right|^2 = \left| \langle n|\hat{b}^\dagger|m\rangle \right|^2. \tag{7.64}$$

It follows from Eq. (7.64) that Eq. (7.61) and Eq. (7.62) satisfy

$$\tau_{mn}^{emi} D(E_m = E_n - \hbar\omega) = \tau_{nm}^{abs} D(E_m = E_n + \hbar\omega), \tag{7.65}$$

which is known as the principle of detailed balance.

7.3.1 The Golden Rule for Unbound States

So far, in the development of time-dependent perturbation theory, the perturbation was taken to act on orthonormal bound states of a system. However, the same theory can be applied to unbound (scattering) states.

In a pictorial way, depicted in Fig. 7.5, the scattering event may be visualized using arrows to indicate initial and final unbound states. The lengths of the arrows are a measure of $|a_m|^2$. The initial state is $|n\rangle$ with $|a_n|^2 = 1$ for time $t \leq 0$. After the scattering event, a number of states are excited in such a way that $|a_m|^2 \neq 0$ for $m \neq n$. However, because the perturbation is assumed weak, the final states are dominated by the state $|n\rangle$ with $|a_n|^2 = 1$. The final states after scattering by a weak perturbation have probability $|a_m|^2 \geq 0$, which may be found using

$$a_m(t) = \frac{1}{i\hbar} \int_{t'=0}^{t'=t} W_{mn} e^{i\omega_{mn} t'} dt', \tag{7.66}$$

where t is the time the perturbation is applied and W_{mn} is the matrix element for scattering out of the initial state $|n\rangle$. The assumption that the perturbation is weak means that the probability of scattering out of the initial state is small. This means that $|a_m|^2 \ll 1$.

The derivation of the golden rule involved a number of approximations that can limit its validity in some applications. For example, use of the $t \to \infty$ limit implies that the collision is completed. Hence, it is important to check that the perturbing potential is small so that collisions do not overlap in space or time. It is also assumed that the probability of scattering out of the initial state $|n\rangle$ is so small that $|a_n|^2 = 1$ and conservation of the number of particles can be ignored. Also, *if*, as will often happen, a plane wave initial state characterized by wave vector $\mathbf{k}$ and a final plane wave state characterized by wave vector $\mathbf{k'}$ is assumed, then the actual collision *is localized in real space*, so that the use of Fourier components is justifiable. All of these assumptions can be violated in modern semiconductor devices and other situations in which scattering can be quite strong and nonlocal effects can become important.

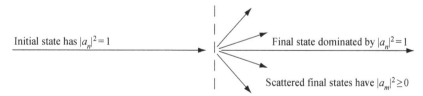

Initial state has $|a_n|^2 = 1$

Final state dominated by $|a_n|^2 = 1$

Scattered final states have $|a_m|^2 \geq 0$

Fig. 7.5 Illustration of initial unbound plane-wave state with probability amplitude such that $|a_n|^2 = 1$ and final state after scattering (to the right of the vertical dashed line) being either the same state or a scattered state with probability amplitude such that $|a_m|^2 \geq 0$. The arrows represent the plane-wave k-state direction or the direction of motion of the particle. The length of the lines is related to the probability amplitude.

The golden rule can be used to calculate the average distance (the mean free path) an electron travels in an unbound state before scattering in a bulk semiconductor that has ionized substitutional impurity concentration n. The microscopic model developed enables estimation of the electrical mobility and electrical conductivity of the material.

The golden rule can also be used to calculate the probability of inducing stimulated optical transitions between electronic states. Stimulated emission of light is a key ingredient determining the operation of lasers.

7.4 Elastic Scattering from Ionized Impurities

Establishing a method to control electrical conductivity in semiconductors is essential for many practical device applications. The performance of transistors and lasers depends critically upon flow of current through specific regions of a semiconductor. An important way to control the electrical conductivity of a semiconductor is by a technique called *substitutional doping* that involves introducing a small number of impurity atoms into the semiconductor crystal. It is energetically favorable for an impurity atom to replace an atom on a lattice site of the original semiconductor crystal. For example, a density n of Si donor impurity atoms can occupy Ga sites in a GaAs crystal. At each impurity site, three of the four chemically active Si electrons are used to replace Ga valence electrons. At low temperatures, the remaining Si electron is bound by the positive charge of the Si donor impurity ion. In a GaAs crystal, this extra electron has an effective electron mass of $m_e^* = 0.07 \times m_0$ and is in a hydrogenic s-like electronic state. The coulomb potential seen by the electron is screened by the presence of the semiconductor dielectric, which is characterized by dielectric permittivity ε. The screened coulomb potential and the low effective electron mass in the conduction band of GaAs increases the Bohr radius of a hydrogenic state from

$$a_B = \frac{4\pi\varepsilon_0\hbar^2}{m_0 e^2} = 0.0529\,\text{nm} \tag{7.67}$$

to an effective Bohr radius given by

$$a_B^* = \frac{4\pi\varepsilon_0\varepsilon_{r0}\hbar^2}{m_e^* e^2} = a_B \frac{\varepsilon_{r0}}{m_e^*} = 10\,\text{nm}, \tag{7.68}$$

where the use of the low-frequency relative dielectric (permittivity) constant, $\varepsilon_{r0} = 13.2$, implies consideration of low-frequency processes. For GaAs, the ratio $a_B^*/a_B = 189$.

In addition, the hydrogenic binding energy is reduced from its value of a Rydberg in atomic hydrogen,

$$\text{Ry} = \frac{m_0}{2}\frac{e^4}{(4\pi\varepsilon_0)^2\hbar^2} = 13.6058\,\text{eV}, \tag{7.69}$$

to a new value (an effective Rydberg constant, Ry^*), which is given by

$$\text{Ry}^* = E_{\text{CBmin}} - E_{\text{donor}} = \frac{m_e^*}{2}\frac{e^4}{(4\pi\varepsilon)^2\hbar^2} = \text{Ry}\left(\frac{m_e^*/m_0}{\varepsilon_{r0}^2}\right) \tag{7.70}$$

$$= 13.6\frac{0.07}{(13.2)^2} = 5.5\,\text{meV}.$$

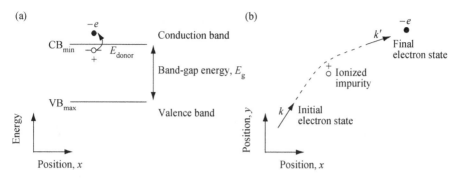

Fig. 7.6 (a) Conduction-band minimum ($\mathrm{CB_{min}}$) and valence-band maximum ($\mathrm{VB_{max}}$) of a semiconductor with band-gap energy E_g. The energy level of a donor impurity is shown. When an electron is excited from the donor level into the conduction band, a positively charged ion core is left at the donor site. This is called an ionized impurity. (b) An electron moving in the conduction band in an initial unbound plane-wave state $\mathbf{k}$ can be elastically scattered into the final state $\mathbf{k}'$ by the coulomb potential of an ionized impurity. The sketch is supposed to represent the trajectory of the wave vector associated with the electron.

The small value of $\mathrm{Ry^*}$ relative to ambient thermal energy suggests that the donor electron is only loosely bound to the donor ion in GaAs. The reasons for this are the small effective electron mass and the value of the low frequency relative permittivity ε_{r0} in the semiconductor.

At finite temperatures, lattice vibrations or interaction with freely moving electrons can easily excite the donor electron from its bound state into unbound states in the conduction band. This ionization process is shown schematically in Fig. 7.6(a).

For temperatures $k_BT > |E_{\mathrm{CB_{min}}} - E_{\mathrm{donor}}|$ and low n-type substitutional Si-impurity concentrations in GaAs, the loosely bound donor electron has a high probability of being excited into the conduction band, leaving behind a positive Si$^+$ ion core. For high-impurity concentrations in which there is a significant overlap between donor wave functions, electrons can also move freely through the conduction band. In either case, the coulomb potential due to the positive ion core acts as a scattering potential for electrons moving in the conduction band. If a density n of Si impurities is introduced into the semiconductor to increase the number of electrons, then the number of ionized impurity sites in the crystal that can scatter these electrons is also increased.

There is then the interesting question of why it is possible to increase the conductivity of a semiconductor by increasing the number of electrons in the conduction band through substitutional impurity doping. It is, after all, not yet obvious that such a strategy would work, as the role of electron scattering from the increased number of ionized impurity sites must be considered.

Anticipating the solution, it will be assumed that electrons in the conduction band can be described by well-defined unbound $|\mathbf{k}\rangle$ states that scatter into final states $|\mathbf{k}'\rangle$, transferring momentum $\hbar\mathbf{q}$ in such a way that $\mathbf{k} = \mathbf{k}' - q$.[2] This is illustrated in Fig. 7.6(b), where an electron moving past an ionized impurity is scattered from an initial state $|\mathbf{k}\rangle$ to a final state $|\mathbf{k}'\rangle$. It is assumed that there is a dilute number of impurity sites, weak scattering,

---

[2] It is possible to choose $\mathbf{k} = \mathbf{k}' + \mathbf{q}$ as opposed to $\mathbf{k} = \mathbf{k}' - \mathbf{q}$ to indicate that momentum $\hbar\mathbf{q}$ is transferred from the incident particle. D. Pines and P. Nozières, *Theory of Quantum Liquids*, New York, Benjamin, 1966, p. 86.

and a plane-wave description for initial and final unbound states. On average, the distance between scattering events is l_k. This distance, called the *mean free path*, is assumed to be longer than the electron wavelength. Also assumed is no energy transfer, so that only elastic scattering is considered.

A natural question concerns the justification for using time-dependent perturbation theory for an electron scattering from a static potential $v(r)$ that has no explicit time dependence. The answer to this question is that an expansion of a state function belonging to one time-independent Hamiltonian in the eigenfunctions of another time-independent Hamiltonian have time-dependent expansion coefficients. In the case being considered, the incident particle is described by a plane-wave state far away from the scattering center. The final state is also described by a plane-wave state far away from the scattering center. However, near the scattering center the particle definitely cannot be described as a plane wave. As indicated in Problem 7.5, it is straightforward to show that in the limit of a static scattering potential the matrix element coupling initial and final states is related to the transition probability per unit time in the same way as Eq. (7.57) and hence the golden rule applies.

An electron that scatters from a plane-wave state $|\mathbf{k}\rangle$ (for which $\psi_{\mathbf{k}} = Ae^{i\mathbf{k}\cdot\mathbf{r}}$) of energy $E(\mathbf{k})$ to a final plane-wave state $|\mathbf{k}'\rangle$ with the same energy will be considered. The golden rule requires evaluation of the matrix element $\langle\mathbf{k}'|v(r)|\mathbf{k}\rangle$, where $v(r)$ is the spatially symmetric coulomb potential. Simple substitution of initial and final plane-wave states into this matrix element reveals that $\langle\mathbf{k}'|v(r)|\mathbf{k}\rangle = v(\mathbf{q})$, where $v(\mathbf{q})$ is the Fourier transform of the coulomb potential in real space. It is clear then, that an expression for the coulomb potential in momentum (wave vector) space is sought.

7.4.1 The Coulomb Potential

The *bare* coulomb potential energy in real space between charge e and $-e$ separated by a distance $r = |\mathbf{r}|$ is

$$v(r) = \frac{-e^2}{4\pi\varepsilon_0 r}. \tag{7.71}$$

The potential energy in wave vector space is found by taking the Fourier transform. The result for the bare coulomb potential (see Exercise 7.2) is

$$v(q) = \frac{-e^2}{\varepsilon_0 q^2}. \tag{7.72}$$

In a uniform dielectric such as a homogeneous isotropic semiconductor characterized by a relative dielectric (permittivity) function ε_r, the term ε_0 in Eq. (7.72) is replaced with $\varepsilon = \varepsilon_0\varepsilon_r$. In this case, ε is constant over real space, but the value of ε_r depends more generally upon wave vector $\mathbf{q}$, so that $\varepsilon = \varepsilon(\mathbf{q}) = \varepsilon_0\varepsilon_r(\mathbf{q})$. Such a dependence is expected because many electrons can respond to the long-range electric field components of the real-space coulomb potential. This effect, called *screening*, reduces the coulomb potential for small values of $\mathbf{q}$. Because screening is embedded in ε there is a screened *dielectric response function*, $\varepsilon(\mathbf{q})$. In general, the dielectric response should also have a frequency dependence, so that $\varepsilon = \varepsilon(\mathbf{q}, \omega)$. However, this is ignored at present. In the following, a constant value of ε is replaced with $\varepsilon(\mathbf{q})$. Later, expressions for the functional form of $\varepsilon(\mathbf{q})$ will be found.

The coulomb potential energy in wave vector space in a homogeneous isotropic medium characterized by permittivity $\varepsilon(\mathbf{q})$ is

$$v(\mathbf{q}) = \frac{-e^2}{\varepsilon(\mathbf{q})q^2},$$

(7.73)

and in real space it is

$$v_{\mathbf{q}}(r) = \frac{-e^2}{4\pi\varepsilon_{\mathbf{q}}r},$$

(7.74)

where the subscript $\mathbf{q}$ is used because $v(r)$ and ε depend upon the scattered wave vector.

The potential energy at position $\mathbf{r}$ due to the ion charge at position $\mathbf{R}_j$ is

$$v_{\mathbf{q}}(\mathbf{r} - \mathbf{R}_j) = \frac{-e^2}{4\pi\varepsilon_{\mathbf{q}}|\mathbf{r} - \mathbf{R}_j|},$$

(7.75)

which only depends upon the separation of charges.

To obtain $v(\mathbf{q})$, the Fourier transform is taken,

$$v(\mathbf{q}) = \int d^3r\, v(\mathbf{r} - \mathbf{R}_j)e^{-i\mathbf{q}\cdot(\mathbf{r}-\mathbf{R}_j)},$$

(7.76)

where the integral is over all space. Clearly, in a homogeneous isotropic medium, if $v(\mathbf{q})$ is determined at one position in space, then it has been found for all space.

7.4.1.1 Elastic Scattering of Electrons by Ionized Impurities in GaAs

Suppose $\mathbf{R}_j$ is the position of the jth dopant atom in GaAs and, as shown in Fig. 7.7, an electron is at position $\mathbf{r}$. The interaction potential in real space is the sum of the contributions from the n individual ions per cm^3. Thus, the total potential is

$$V(\mathbf{r}) = \sum_{j=1}^{n} v(\mathbf{r} - \mathbf{R}_j).$$

(7.77)

In wave vector space the sum of Fourier transforms of $v(\mathbf{r} - \mathbf{R}_j)$ is

$$V(\mathbf{q}) = \sum_{j=1}^{n} \int d^3r\, v(\mathbf{r} - \mathbf{R}_j)e^{-i\mathbf{q}\cdot\mathbf{r}} = \sum_{j=1}^{n} \int d^3r\, v(\mathbf{r} - \mathbf{R}_j)e^{-i\mathbf{q}\cdot\mathbf{r}}e^{i\mathbf{q}\cdot\mathbf{R}_j}e^{-i\mathbf{q}\cdot\mathbf{R}_j}$$

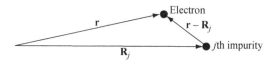

Fig. 7.7 Diagram illustrating the relative position $\mathbf{r} - \mathbf{R}_j$ of an electron at position $\mathbf{r}$ and the jth ionized impurity at position $\mathbf{R}_j$.

$$= \sum_{j=1}^{n} \int d^3 r\, v(\mathbf{r} - \mathbf{R}_j) e^{-i\mathbf{q}\cdot(\mathbf{r}-\mathbf{R}_j)} e^{-i\mathbf{q}\cdot\mathbf{R}_j} = v(\mathbf{q}) \sum_{j=1}^{n} e^{-i\mathbf{q}\cdot\mathbf{R}_j}. \tag{7.78}$$

Hence, the total potential seen by the electron in the presence of n ionized impurities per unit volume is

$$\boxed{V(\mathbf{q}) = v(\mathbf{q}) \sum_{j=1}^{n} e^{-i\mathbf{q}\cdot\mathbf{R}_j}.} \tag{7.79}$$

For elastic scattering, transitions between an unbound state $|\mathbf{k}\rangle$ (for which $\psi_\mathbf{k} = A e^{i\mathbf{k}\cdot\mathbf{r}}$) of energy $E(\mathbf{k})$ and an unbound final state $|\mathbf{k}'\rangle$ with the same energy are considered. The golden rule (the first term in the Born series) involves evaluating the matrix element $\langle \mathbf{k}'|v(r)|\mathbf{k}\rangle$. Since $|\mathbf{k}\rangle$ and $\langle \mathbf{k}'|$ are plane-wave states of the form $e^{i\mathbf{k}\cdot\mathbf{r}}$ (assuming that $kl_k \gg 1$, so the mean free path l_k is many electron wavelengths long),

$$\langle \mathbf{k}'|v(r)|\mathbf{k}\rangle = \int d^3 r\, e^{-i\mathbf{k}'\cdot\mathbf{r}} v(r) e^{i\mathbf{k}\cdot\mathbf{r}} = \int d^3 r\, v(r) e^{i(\mathbf{k}-\mathbf{k}')\cdot\mathbf{r}}, \tag{7.80}$$

which is just the Fourier transform of the coulomb potential in real space,

$$\langle \mathbf{k}'|v(r)|\mathbf{k}\rangle = \int d^3 r\, v(r) e^{-i\mathbf{q}\cdot\mathbf{r}} = v(\mathbf{q}). \tag{7.81}$$

In this expression $\mathbf{q} = \mathbf{k}' - \mathbf{k}$, since momentum conservation requires $\mathbf{k}' = \mathbf{k} + \mathbf{q}$. As illustrated in Fig. 7.8, the scattering angle θ for elastic scattering (no energy loss) is such that $k\sin(\theta/2) = q/2$.

The probability of elastic scattering between the two states is

$$1/\tau_{\mathbf{k}\mathbf{k}'} = \frac{2\pi}{\hbar}|v(\mathbf{q})|^2 \delta(E(\mathbf{k}) - E(\mathbf{k}+\mathbf{q})), \tag{7.82}$$

where the δ function ensures that no energy is exchanged. The total scattering rate is a sum over all transitions, so that for a *single impurity*,

$$\boxed{1/\tau_{\text{el}} = \frac{2\pi}{\hbar} \int \frac{d^3 q}{(2\pi)^3} |v(\mathbf{q})|^2 \delta(E(\mathbf{k}) - E(\mathbf{k}+\mathbf{q})).} \tag{7.83}$$

(a)   (b)

Fig. 7.8 (a) Diagram illustrating initial unbound state wave vector $\mathbf{k}$, final wave vector $\mathbf{k}'$, and transferred momentum $\hbar\mathbf{q} = \hbar(\mathbf{k}' - \mathbf{k})$. (b) For elastic scattering in a homogeneous isotropic bulk medium, the scattered angle θ is related to q by $k\sin(\theta/2) = q/2$.

Elastic scattering from n impurities can now be calculated using

$$|V(\mathbf{q})|^2 = |v(\mathbf{q})|^2 \sum_{j=1}^{n} e^{-i\mathbf{q}\cdot\mathbf{R}_j} \sum_{k=1}^{n} e^{i\mathbf{q}\cdot\mathbf{R}_k} = |v(\mathbf{q})|^2 s(\mathbf{q}), \tag{7.84}$$

where

$$s(\mathbf{q}) = \sum_{j=1}^{n} e^{-i\mathbf{q}\cdot\mathbf{R}_j} \sum_{k=1}^{n} e^{i\mathbf{q}\cdot\mathbf{R}_k}. \tag{7.85}$$

Here, $s(\mathbf{q})$ is a *structure factor* that contains phase information on the scattered wave from site $\mathbf{R}_j$. For n large and random $\mathbf{R}_j$, the sum over n random phases is $\sqrt{n}$, and so the sum squared is n. It follows that if there are n spatially *uncorrelated* scattering sites corresponding to random impurity positions, then $s(\mathbf{q}) = n$. To show that this is so, see Exercise 7.3. For a large number of spatially random impurity-site positions, the magnitude of the matrix element squared given by Eq. (7.84) becomes

$$|V(\mathbf{q})|^2 = n|v(\mathbf{q})|^2, \tag{7.86}$$

and the *total elastic scattering rate* from n impurities per unit volume is

$$1/\tau_{\mathrm{el}} = \frac{2\pi}{\hbar} n \int \frac{\mathrm{d}^3 q}{(2\pi)^3} \left| \frac{e^2}{\varepsilon(\mathbf{q})q^2} \right|^2 \delta(E(\mathbf{k}) - E(\mathbf{k}+\mathbf{q})), \tag{7.87}$$

where the integral over $\mathrm{d}^3 q$ is the final density of states.

Physically, each impurity is viewed as contributing independently, so that the scattering rate is n times the scattering rate from a *single impurity atom*. It is possible to also see why increasing the impurity concentration n does not necessarily result in a linear increase in scattering rate $1/\tau_{\mathrm{el}}$. The integral contains the magnitude of a matrix element squared, $|e^2/\varepsilon(\mathbf{q})q^2|^2$, which can influence the scattering rate. The $1/q^2$ term reflects the fact that ionized impurity coulomb scattering is weighted toward final states with small q transfer. This means that electrons moving in a given direction are mainly scattered by small angles without too much deviation from the forward direction. The dielectric function $\varepsilon(\mathbf{q})$ also has an influence on scattering rate, in part because of its $\mathbf{q}$ dependence but also because the function depends upon carrier concentration, n.

7.4.1.2 Correlation Effects Due to Spatial Position of Dopant Atoms

The constraint that *substitutional* impurity atoms in a crystal *occupy crystal lattice sites* gives rise to a *correlation effect because double occupancy of a site is not allowed*. Suppose a fraction f_{sites} of sites are occupied. In this case, there is no longer a truly random distribution, and, for small q and small f_{sites}, the scattering rate will be reduced by $s(\mathbf{q}) = n(1 - f_{\mathrm{sites}})$. The factor $(1 - f_{\mathrm{sites}})$ reflects the fact that not allowing double occupancy of a site is a correlation effect.

Other spatial correlation effects are possible and can, in principle, dramatically alter scattering rates.[3]

---

[3] A. F. J. Levi, S. L. McCall, and P. M. Platzman, *Appl. Phys. Lett.* **54**, 940 (1989).

7.4.1.3 Calculating Electron Mean Free Path

To calculate the mean free path of a conduction band electron in a homogeneous isotropic semiconductor that has been doped with a density of n randomly positioned substitutional impurities, it is convenient to start from the expression for the total elastic scattering rate of an electron mass m_e^* and charge e given by Eq. (7.87). To evaluate the total elastic scattering rate as a function of the incoming electron energy E, it is necessary to express the volume element d^3q in terms of energy $E = \hbar^2 k^2 / 2m_e^*$ and scattering angle θ. It can be shown (Exercise 7.4) that

$$1/\tau_{el} = \frac{2\pi m_e^*}{\hbar^3 k^3} n \left(\frac{e^2}{4\pi\varepsilon_0}\right)^2 \int_{\eta=0}^{\eta=1} \frac{d\eta}{(\varepsilon_r(q))^2 \eta^3}, \tag{7.88}$$

or, as a function of energy,

$$1/\tau_{el}(E) = \frac{\pi}{(2m_e^*)^{1/2}} n \left(\frac{e^2}{4\pi\varepsilon_0}\right)^2 E^{-3/2} \int_{\eta=0}^{\eta=1} \frac{d\eta}{(\varepsilon_r(2k\eta))^2 \eta^3}, \tag{7.89}$$

where $\eta = \sin(\theta/2)$, scattered wave vector $q = 2k\eta$, and dielectric function $\varepsilon(q) = \varepsilon_0 \varepsilon_r(q)$.

Using parameters for GaAs with $n = 10^{18}$ cm^{-3}, a conduction band effective electron mass $m_e^* = 0.07 \times m_0$, and Fermi wave vector in three dimensions $k_F = (3\pi^2 n)^{1/3} = 3 \times 10^6$ cm^{-1}, the prefactor in Eq. (7.88) is

$$\frac{2\pi}{\hbar^3} n \frac{e^4 m_e^*}{(4\pi\varepsilon_0)^2 k_F^3} = \frac{2\pi n c (m_e^*/m_0)}{3\pi^2 n} \frac{m_0 e^2}{4\pi\varepsilon_0 \hbar^2} \frac{e^2}{4\pi\varepsilon_0 \hbar c} = \frac{2c(m_e^*/m_0)}{3\pi a_B \alpha_f^{-1}}. \tag{7.90}$$

Putting in the numbers,

$$\frac{2c(m_e^*/m_0)}{3\pi a_B \alpha_f^{-1}} = \frac{2 \times 3 \times 10^8 \times 0.07}{3\pi \times 0.53 \times 10^{-10} \times 137} = 6.14 \times 10^{14}\, \mathrm{s}^{-1}. \tag{7.91}$$

Notice that when evaluating the prefactor, known *physical values* and *dimensionless units* are used as much as possible. This helps to avoid mistakes.

Assuming $\varepsilon_r(q) \sim \varepsilon_{r0} \sim 10$, the integral in Eq. (7.88) may be estimated as $1/\varepsilon_{r0}^2 = 1/100$, so that $1/\tau_{el} \sim 6.14 \times 10^{14}/\varepsilon_{r0}^2 = 6.14 \times 10^{12}\,\mathrm{s}^{-1}$.

The mean free path is the characteristic length l_k between electron scattering events. For elastic scattering at the Fermi energy, the mean free path is

$$l_{k_F} = v_F \tau_{el}, \tag{7.92}$$

where the Fermi velocity $v_F = \hbar k_F / m_e^*$. Hence, in GaAs with $n = 10^{18}$ cm^{-3}, the mean free path is $l_{k_F} = 5 \times 10^7 / 6 \times 10^{12} = 83$ nm. This length can be compared with the average spacing between impurity sites, which is only 10 nm for an impurity concentration $n = 10^{18}$ cm^{-3}. Obviously, this impurity concentration is *not* the *dilute* limit that was previously assumed, since the electron wavelength $\lambda_F = 2\pi/k_F = 20$ nm is similar to the average spacing between impurities. However, $l_{k_F} > \lambda_F$, so that $k_F l_{k_F} \gg 1$, justifying the assumption of weak scattering. The average spacing between impurity sites, which is 10 nm (many times the GaAs lattice constant $L = 0.565\,33\,\text{nm}$), may be compared with the effective Bohr radius for a hydrogenic impurity which, using a value $\varepsilon_{r0} = 13.2$ for the *low-frequency* dielectric constant, gives $a_B^* = 10\,\text{nm}$. Because a_B^* is similar to the average spacing between impurity sites, there should be a significant overlap between donor electron wave functions giving rise to metallic behavior. More formally, the parameter r_s that is the radius of a sphere occupied, on average, by one electron, assuming a uniform electron density n, divided by the effective Bohr radius, a_B^* is introduced,

$$r_s = \left(\frac{3}{4\pi n}\right)^{1/3} \frac{1}{a_B^*}.$$
(7.93)

For GaAs with an impurity concentration $n = 10^{18}\,\text{cm}^{-3}$, this gives $r_s = 0.63$. Again, because $r_s < 1$, metallic behavior is expected. This is indeed the case, and the calculation of scattering rate $1/\tau$ can be used to estimate electron mobility and conductivity.

7.4.1.4 Calculating Mobility and Conductivity

Electron mobility is defined as

$$\boxed{\mu = e\tau_{el}^*/m_e^*,}$$
(7.94)

where $1/\tau_{el}^*$ is an appropriate elastic scattering rate and m_e^* is the effective electron mass. It is not uncommon for mobility to be quoted in CGS units of $\text{cm}^2\,\text{V}^{-1}\,\text{s}^{-1}$. If the mobility of electrons that are characterized on average by a Fermi wave vector k_F and a mean free path l_{k_F}, where it is assumed $\tau_{el}^* = \tau_{el}$, then the mobility is

$$\mu = e\tau_{el}/m_e^* = el_{k_F}/\hbar k_F.$$
(7.95)

Conductivity is proportional to mobility and is defined as

$$\boxed{\sigma = ne\mu = \frac{ne^2\tau_{el}^*}{m_e^*}.}$$
(7.96)

Assuming $\tau_{el}^* = \tau_{el}$, then the conductivity of the material is

$$\sigma = ne\mu = \frac{ne^2\tau_{el}}{m_e^*} = \frac{ne^2 l_{k_F}}{\hbar k_F}.$$
(7.97)

Putting in numbers for mobility of GaAs doped to $n = 10^{18}\,\text{cm}^{-3}$, the value for τ_{el} in Eq. (7.94) gives, in CGS units,

$$\mu_n = e\tau_{el}/m_e^* = 1.6 \times 10^{-12}/6 \times 10^{12} \times 0.07 \times 9.1 \times 10^{-28}$$
$$= 4.1 \times 10^3\,\text{cm}^2\,\text{V}^{-1}\,\text{s}^{-1}.$$
(7.98)

As shown in Fig. 7.9, the experimentally measured value of electron mobility in bulk GaAs with carrier concentration $n = 10^{18}\,\text{cm}^{-3}$ at temperature $T = 300\,\text{K}$ is $\mu_n = 2\text{--}3\times10^3\,\text{cm}^2\,\text{V}^{-1}\,\text{s}^{-1}$. At the lower temperature of $T = 77\,\text{K}$, the mobility is measured to be $\mu_n = 3\text{--}4 \times 10^3\,\text{cm}^2\,\text{V}^{-1}\,\text{s}^{-1}$.

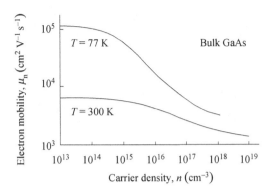

Fig. 7.9 Experimentally determined electron mobility of bulk GaAs as a function of carrier density n on a logarithmic scale for the indicated temperatures, T.

As a next step to *improve* the calculation of elastic scattering rate, $1/\tau_{el}$, modification due to the presence of mobile electron charge in the system is considered. The coulomb potential of each ionized impurity is changed, or *screened*, by mobile charge carriers. Because the coulomb potential is modified, the total elastic scattering rate and its angular dependence is also also changed.

7.4.2 Linear Screening of the Coulomb Potential

Previously, it has been assumed that a doped semiconductor has an ionized impurity distribution that is random. The static ionized charge distribution is $\rho_i(\mathbf{r})$, with net charge $Q_i = e \int d^3r\, \rho_i(\mathbf{r})$. The total mobile charge attracted is exactly $-Q_i$. The mobile charge (which is considered a nearly free electron gas) is a *screening charge* and has its own distribution in space given by $\rho_s(\mathbf{r})$.

The screened potential energy from the static impurity charge and the mobile screening charge is exactly

$$V(r) = \int d^3 r' \frac{-e^2(\rho_i(\mathbf{r}') + \rho_s(\mathbf{r}'))}{4\pi\varepsilon_0|\mathbf{r} - \mathbf{r}'|}. \tag{7.99}$$

As shown schematically in Fig. 7.10, electron charge density is expected to increase around the positive impurity ion and in this way the mobile electron charge density *screens* the coulomb potential due to the impurity.

7.4.2.1 Calculating the Screened Potential in Real Space

To calculate the screened coulomb potential, consider a homogeneous isotropic three-dimensional free-electron gas with equilibrium time-averaged electron particle density n_0. Each electron has an effective mass m_e^*. Suppose a test charge (the impurity ion) is placed at position $\mathbf{r} = 0$ into this electron gas. The test charge will try to create a coulomb potential $\phi_{ex}(r) = -e/4\pi\varepsilon r$. However, the response of the electron gas to the presence of the test charge is to create a new electron particle density $n(r)$, which will create a new screened potential $\phi(r)$.

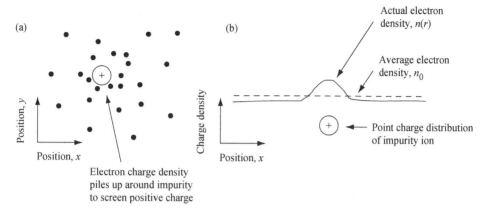

Fig. 7.10 (a) Illustration to represent the response of an electron gas to the presence of a positive charge distribution due to an impurity ion. On average, electrons spend more time in the vicinity of the impurity. (b) The electron density is greater than average near the positively charged impurity ion. The impurity ion may be modeled as a static point charge.

A simplifying assumption is that the relationship between the energy of an electron at position $\mathbf{r}$ and its wave vector is only modified from its free-electron value by the *local potential*, so that

$$E(k) = \frac{\hbar^2 k^2}{2m} - e\phi(r). \tag{7.100}$$

The assumption of a local potential can only be true for electrons localized in space and so described (semiclassically) as wave packets. However, the electron wave packets are spread out in real space by a characteristic distance (at least $1/k_F$ for a low-temperature degenerate electron gas). To ensure that use of a local potential is an accurate approximation, it is required that $\phi(r)$ vary slowly on the scale of the wave packet size.

The new screened potential may be written in the form

$$\phi(r) = \frac{1}{r} f_{\mathrm{r}}(r), \tag{7.101}$$

where $f_{\mathrm{r}}(r)$ is a function that will be determined using Poisson's equation $\nabla^2 \phi = -\rho(r)/\varepsilon$, which relates the *local* charge density to the *local* potential. The change in equilibrium charge density is

$$\rho(r) = -e(n(r) - n_0), \tag{7.102}$$

where the averaged electron particle density is

$$n_0 = \int \frac{d^3 k}{(2\pi)^3} 2 f_k = \int \frac{d^3 k}{(2\pi)^3} 2 \frac{1}{e^{(E_k - \mu)/k_B T} + 1}, \tag{7.103}$$

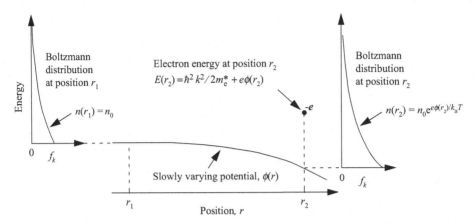

Fig. 7.11 Electron energy $E(r)$ changes when subject to a slowly varying potential, $\phi(r)$. For a Boltzmann distribution of electron occupation probability there is an exponential dependence on local potential, $n(r) = n_0 e^{e\phi(r)/k_B T}$.

and the *local* particle density at position r is

$$n(r) = \int \frac{\mathrm{d}^3 k}{(2\pi)^3} \, 2 f_k = \int \frac{\mathrm{d}^3 k}{(2\pi)^3} \, 2 \, \frac{1}{e^{(E_k - e\phi(r) - \mu)/k_B T} + 1}. \tag{7.104}$$

In the expressions for n_0 and $n(r)$, notice that f_k is the equilibrium Fermi–Dirac distribution function and that the factor 2 in the integral accounts for electron spin $\pm \hbar/2$. Eqs. (7.103) and (7.104) allow Eq. (7.102) to be rewritten as

$$\rho(r) = -e(n_0(\mu + e\phi(r)) - n_0(\mu)). \tag{7.105}$$

If the potential ϕ is small, then Eq. (7.105) may be expanded to first order to give

$$\rho(r) = -e \left(n_0(\mu) + \frac{\partial n_0}{\partial \mu} e\phi(r) - n_0(\mu) \right) = -e^2 \frac{\partial n_0}{\partial \mu} \phi(r), \tag{7.106}$$

which shows how the change in equilibrium charge density is related to the screened potential.

For a homogeneous isotropic, *nondegenerate, three-dimensional electron gas at equilibrium*, the local change in number of carriers due to a local change in potential is given by a Boltzmann factor, so

$$n = n_0 e^{e\phi/k_B T}, \tag{7.107}$$

and, as illustrated in Fig. 7.11, there is an exponential dependence on local charge density with potential $\phi(r)$. Poisson's equation becomes

$$\nabla^2 \phi(r) = \frac{-\rho(r)}{\varepsilon} = \frac{e(n - n_0)}{\varepsilon} = \frac{e n_0}{\varepsilon} \left(e^{e\phi/k_B T} - 1 \right). \tag{7.108}$$

This is a *nonlinear* differential equation for $\phi(r)$. To simplify the equation, it is assumed that the test charge is small so that the induced potential is also expected to be small. *If* this is true, then the exponential can be expanded to give $e^{e\phi/k_BT} \sim 1 + e\phi/k_BT + \cdots$, so that

$$\nabla^2 \phi(r) \approx \frac{e^2 n_0}{\varepsilon k_B T} \phi(r), \tag{7.109}$$

which is a *linear* differential equation for $\phi(r)$. Substituting $\phi(r) = f_r(r)/r$ gives

$$\nabla^2 \phi(r) = \frac{e^2 n_0}{\varepsilon k_B T} \frac{1}{r} f_r(r). \tag{7.110}$$

In spherical coordinates, the left-hand side of Poisson's equation can be written

$$\nabla^2 \phi(r) = \nabla^2 \left(\frac{1}{r} f_r(r) \right) = \frac{1}{r^2} \frac{\partial}{\partial r} \left(r^2 \frac{\partial}{\partial r} \frac{1}{r} f_r(r) \right)$$

$$= \frac{1}{r^2} \frac{\partial}{\partial r} \left(r^2 \left(\frac{-1}{r^2} f_r(r) + \frac{1}{r} \frac{\partial f_r(r)}{\partial r} \right) \right) = \frac{1}{r^2} \frac{\partial}{\partial r} \left(r \frac{\partial f_r(r)}{\partial r} - f_r(r) \right)$$

$$= \frac{1}{r^2} \left(\frac{\partial f_r(r)}{\partial r} + r \frac{\partial^2 f_r(r)}{\partial r^2} - \frac{\partial f_r(r)}{\partial r} \right) = \frac{1}{r} \frac{\partial^2 f_r(r)}{\partial r^2}, \tag{7.111}$$

so that, making use of Eq. (7.110),

$$\frac{\partial^2 f_r(r)}{\partial r^2} = \frac{e^2 n_0}{\varepsilon k_B T} f_r(r). \tag{7.112}$$

Hence, the solution for $f_r(r)$ is proportional to $e^{-q_D \cdot r}$, where

$$q_D^2 = \frac{n_0 e^2}{\varepsilon k_B T}. \tag{7.113}$$

The value $1/q_D$ is the Debye screening length. It applies to the *equilibrium nondegenerate electron gas* and scales with carrier density as $\sqrt{1/n_0}$ and with thermal energy as $\sqrt{k_B T}$. In this case, the screened coulomb potential energy in real space becomes

$$\phi(r) = \frac{1}{r} f_r(r) = \frac{-e^2}{4\pi\varepsilon} \frac{e^{-q_D \cdot r}}{r}, \tag{7.114}$$

where $\varepsilon = \varepsilon_0 \varepsilon_{r0}$.

For a three-dimensional, *degenerate electron system* at low temperature the Thomas–Fermi screening length is obtained by simply identifying the Fermi energy as the characteristic energy of the system so that E_F substitutes for thermal energy $3k_B T/2$. Substitution into the previous expression for q_D gives

$$q_{TF}^2 = \frac{3n_0 e^2}{2\varepsilon E_F}. \tag{7.115}$$

Since $E_F = \hbar^2 k_F^2 / 2m_e^*$ and $n_0 = k_F^3/3\pi^2$, this expression may be rewritten as

$$\boxed{q_{TF}^2 = \frac{k_F m_e^* e^2}{\varepsilon \pi^2 \hbar^2}.} \tag{7.116}$$

The low-temperature ($k_B T \ll \mu$) Thomas–Fermi screening length $1/q_{TF}$ scales with the characteristic Fermi wave number as $\sqrt{1/k_F}$. In this case, the screened coulomb potential energy in real space becomes

$$\phi(r) = \frac{-e^2}{4\pi\varepsilon}\frac{e^{-q_{TF}\cdot r}}{r}. \qquad (7.117)$$

7.4.2.2 Calculating the Screened Potential and Dielectric Function in Wave Vector Space

A single ion at position $\mathbf{R}_j = 0$ in a homogeneous isotropic dielectric medium has coulomb potential energy

$$v_q(r) = \frac{-e^2}{4\pi\varepsilon_0\varepsilon_{r0}}\frac{1}{r}. \qquad (7.118)$$

This is a long-range interaction that will be screened by mobile electron charge. Suppose there is a characteristic screening length r_{scr}. Then the screened potential energy is approximated as a function similar to Eq. (7.114) or Eq. (7.117), so that

$$v_q(r) = \frac{-e^2}{4\pi\varepsilon_0\varepsilon_{r0}}\frac{e^{-r/r_{scr}}}{r}. \qquad (7.119)$$

To find $v(q)$ for this static screened potential energy, the Fourier transform is taken:

$$v(q) = \int d^3r\; v(r)\, e^{-i q \cdot r}. \qquad (7.120)$$

Leaving the integration to Exercise 7.5, the solution is

$$v(q) = \frac{-e^2}{\varepsilon_0\varepsilon_{r0}\left(q^2 + 1/r_{scr}^2\right)}. \qquad (7.121)$$

This may now be compared with the previous expression in terms of a dielectric function

$$v(q) = \frac{-e^2}{\varepsilon(q)q^2} = \frac{-e^2}{\varepsilon_0\varepsilon_{r0}q^2\left(1 + 1/q^2 r_{scr}^2\right)}, \qquad (7.122)$$

where r_{scr} is a characteristic screening length. It is apparent that the effect of screening is to modify the dielectric function in such a way that $\varepsilon = \varepsilon(q)$. For a degenerate three-dimensional electron gas in the low-temperature limit, the Thomas–Fermi screening length may be used such that $1/r_{scr}^2 = q_{TF}^2$. In this case,

$$\varepsilon(q) = \varepsilon_0\varepsilon_{r0}\left(1 + \frac{q_{TF}^2}{q^2}\right). \qquad (7.123)$$

This is the Thomas–Fermi dielectric function that, *if* valid for all q, describes a statically screened, real-space potential energy of the Yukawa type,

$$v_q(r) = \frac{-e^2}{4\pi\varepsilon_0\varepsilon_{\mathrm{r}0}}\frac{e^{-q_{\mathrm{TF}}\cdot r}}{r}, \tag{7.124}$$

where

$$q_{\mathrm{TF}}^2 = \frac{k_{\mathrm{F}}m_{\mathrm{e}}^* e^2}{\varepsilon\pi^2\hbar^2} = \frac{k_{\mathrm{F}}m_{\mathrm{e}}^* e^2}{\varepsilon_0\varepsilon_{\mathrm{r}0}\pi^2\hbar^2} \tag{7.125}$$

is the Thomas–Fermi wave number. The inverse of the Thomas–Fermi wave number defines the length scale for screening.

In bulk, single-crystal GaAs with impurity concentration $n = 10^{18}\,\mathrm{cm}^{-3}$ and a conduction-band effective electron mass of $m_{\mathrm{e}}^* = 0.07 \times m_0$ then $q_{\mathrm{TF}} = 2 \times 10^6\,\mathrm{cm}^{-1}$. This may be compared with the Fermi wave vector, which has a value $k_{\mathrm{F}} = (3\pi^2 n)^{1/3} = 3 \times 10^6\,\mathrm{cm}^{-1}$. The fact that $1/q_{\mathrm{TF}} = 5\,\mathrm{nm}$ and $1/k_{\mathrm{F}} = 3\,\mathrm{nm}$ have comparable values is not unexpected, since they are both a measure of highest spatial frequency that can be used by the electrons to screen the coulomb interaction.

In Fig. 7.12(a), the Thomas–Fermi dielectric function is plotted as a function of wave vector normalized to q_{TF}. In Fig. 7.12(b), the Thomas–Fermi statically screened real-space potential energy is shown, along with the bare-coulomb potential energy as a function of position, r.

Large values of r in the real-space potential correspond to long-wavelength excitations or, equivalently, small q scattering in wave vector space. At large values of r, there are many conduction electrons between the impurity and the test charge. Hence, many conduction band electrons can respond to and effectively screen the impurity potential. Short-wavelength or high-spatial-frequency components of the potential correspond to small values of r in the real-space potential. In this case, there are few electrons that can respond to screen the impurity potential. High q scattering from a real-space potential involves the incident electron getting close to the ionized impurity. When this happens, there are fewer electrons available to screen the ion.

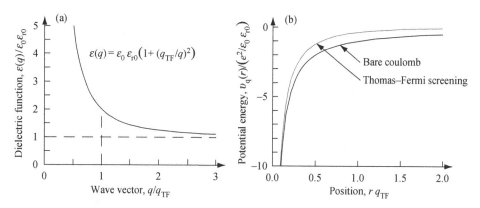

Fig. 7.12 (a) The Thomas–Fermi screened dielectric function as a function of q/q_{TF} and (b) the bare-coulomb and screened-coulomb potential as a function of position $r\,q_{\mathrm{TF}}$.

In general, the dielectric function will have a dynamic (or frequency-dependent) part, so $\varepsilon = \varepsilon(q, \omega)$. However, energy exchange processes (which change electron energy by $\hbar\omega$) can be ignored as only elastic electron scattering is considered. There are other limitations to this type of model dielectric function. For example, Debye and Thomas–Fermi screening adopted a semiclassical approximation that required the screened potential to vary slowly. This approximation is not valid in the limit of $r \to 0$ (or, equivalently, large q). Substitution of the Thomas–Fermi screened-coulomb potential into Poisson's equation predicts a screened charge density proportional to $\left(q_{TF}^2 e^{-q_{TF} \cdot r}\right)/r$, which diverges as $r \to 0$. This deficiency may be overcome by using a different model dielectric function that does not require the screened potential to vary slowly. In one such approach, called the random phase approximation (RPA), due to Lindhard[4],[5] an approximation that the induced charge density contributes linearly to the total potential is exploited. The Schrödinger equation is then used to calculate the electronic wave functions self-consistently in the presence of the new potential. However, for most calculations of practical interest (see Exercise 7.7), differences between the RPA and Thomas–Fermi results are relatively small, and so the Thomas–Fermi dielectric function can often be used to calculate elastic, ionized-impurity electron scattering rates in semiconductors.

7.4.2.3 Using the Thomas–Fermi Dielectric Function to Calculate Elastic, Ionized-Impurity, Electron Scattering in GaAs

From Eq. (7.88), the elastic scattering rate is

$$1/\tau_{el} = \frac{2\pi m_e^*}{\hbar^3 k^3} n \left(\frac{e^2}{4\pi\varepsilon_0}\right)^2 \int_{\eta=0}^{\eta=1} \frac{d\eta}{(\varepsilon_r(q))^2 \eta^3}, \tag{7.126}$$

where $\eta = \sin(\theta/2)$ and $q = 2k\eta$. For the Thomas–Fermi dielectric function the scattering rate is given by the expression

$$1/\tau_{el} = \frac{2\pi m_e^*}{\hbar^3 k^3} n \left(\frac{e^2}{4\pi\varepsilon_0\varepsilon_{r0}}\right)^2 \int_{\eta=0}^{\eta=1} \frac{d\eta}{\left(1 + \dfrac{q_{TF}^2}{q^2}\right)^2 \eta^3}. \tag{7.127}$$

Figure 7.13(a) shows results when using Eq. (7.127) to calculate the total elastic scattering rate in GaAs for the indicated values of impurity concentration. The conduction-band effective electron mass is taken to be $m_e^* = 0.07 \times m_0$, and the value of ε_{r0} is 13.2. The elastic scattering rate for an electron of energy $E = 200$ meV in GaAs with $n = 10^{17}$ cm^{-3} is about 2×10^{12} s^{-1}, corresponding to a scattering time of $\tau = 0.5$ ps.

Figure 7.13(b) shows the calculated elastic scattering rate as a function of scattered angle for an electron of energy $E = 100$ meV and $E = 300$ meV in the conduction band of GaAs with $n = 10^{17}$ cm^{-3}. It is clear that the coulomb potential favors small-angle scattering. This is particularly true when the electron has a large value of energy, E.

[4] J. Lindhard, *Kgl. Danske Videnskab. Selskab Mat.-Fys. Medd.* **28** no. 8 (1954).
[5] A. F. J. Levi, *Essential Electron Transport for Device Physics*, Melville, New York, AIP Publishing, 2020 (ISBN 978 0 7354 2158 5).

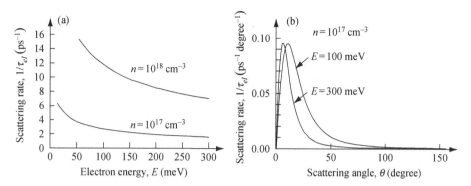

Fig. 7.13 (a) Calculated total elastic scattering rate for an electron of energy E in the conduction band of GaAs due to the presence of the indicated random ionized impurity density. The calculation uses the Thomas–Fermi dielectric function and $T = 0\,\mathrm{K}$. (b) Elastic scattering rate as a function of scattered angle for an electron of energy $E = 100\,\mathrm{meV}$ and $E = 300\,\mathrm{meV}$ in the conduction band of GaAs with $n = 10^{17}\,\mathrm{cm}^{-3}$, $\varepsilon_{r0} = 13.2$, and $m_{\mathrm{e}}^{*} = 0.07 \times m_0$.

Figure 7.14 illustrates the difference in calculated elastic scattering rate as a function of scattered angle with and without Thomas–Fermi screening for the indicated electron energies in the conduction band of GaAs with $n = 10^{18}\,\mathrm{cm}^{-3}$.

The effect of screening is to increase the dielectric constant for small scattered wave vector q, since

$$\varepsilon(q) = \varepsilon_0 \varepsilon_{r0} \left(1 + \frac{q_{\mathrm{TF}}^2}{q^2} \right). \tag{7.128}$$

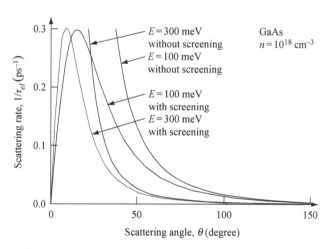

Fig. 7.14 Calculated total elastic scattering rate from random ionized impurities as a function of angle θ with and without Thomas–Fermi screening of the coulomb potential. The parameters used in the calculation are those for GaAs with impurity concentration $n = 10^{18}\,\mathrm{cm}^{-1}$, low-frequency relative permittivity $\varepsilon_{r0} = 13.2$, and effective electron mass $m_{\mathrm{e}}^{*} = 0.07 \times m_0$.

331

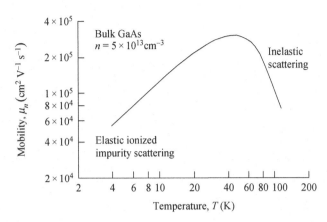

Fig. 7.15 Measured temperature dependence of mobility in bulk GaAs with an impurity concentration of $n = 5 \times 10^{13}$ cm^{-3}. The axes use a logarithmic scale. At low temperatures, elastic ionized impurity scattering dominates mobility, and mobility increases with increasing temperature. At temperatures above $T = 50$ K, inelastic scattering from lattice vibrations dominates, causing a decrease in mobility with increasing temperature.

This reduces the value of the integral when q is small. Small q corresponds to small-angle scattering since $q = 2k\sin(\theta/2)$. It is small-angle scattering (or long-wavelength *excitations*) that are suppressed.

Figure 7.13(a) shows that a high-energy electron scatters less than an electron of low energy. This is characteristic behavior for coulomb scattering, the origin of which in this case can be traced back to the $E^{-3/2}$ term in Eq. (7.89). Using this energy dependence, it is straightforward to show that, when elastic scattering from ionized impurities dominates electron dynamics, calculations predict that mobility has a $T^{3/2}$ temperature dependence (see Exercise 7.6). Typically, elastic scattering from ionized impurities is most significant at low temperatures, and so, as shown in Fig. 7.15, mobility increases with increasing temperature for $T < 50$ K. However, at temperatures above $T = 50$ K, *inelastic* scattering from lattice vibrations can dominate, causing a decrease in mobility with increasing temperature.

7.5 Photon Emission due to Electronic Transitions

To understand and control the emission of photons from atoms or solids, it is helpful to know about the density of optical modes, light intensity, the background energy density in thermal equilibrium, the golden rule for optical transitions, the occupation factor for thermally distributed photons, and the Einstein $\mathcal{A}$ and $\mathcal{B}$ coefficients.

7.5.1 Density of Optical Modes in Three Dimensions

For plane-wave optical modes characterized by wave vector $\mathbf{k}$, the density of states in three dimensions is

$$D_3^{\text{opt}}(k)\mathrm{d}k = 2 \times 4\pi k^2 \frac{\mathrm{d}k}{(2\pi)^3} = \frac{k^2}{\pi^2}\mathrm{d}k, \tag{7.129}$$

where the factor 2 is from the two orthogonal polarizations. This is the density of modes per unit volume in k-space. However, in a three-dimensional homogeneous isotropic nondispersive medium with refractive index n_r, the wave vector $k = n_r\omega/c$ and $dk = n_r d\omega/c$. Hence, the mode density as a function of angular frequency is

$$D_3^{\text{opt}}(\omega)d\omega = 2 \times 4\pi k^2 \frac{1}{(2\pi)^3} \frac{dk}{d\omega}d\omega = 2 \times 4\pi \frac{\omega^2 n_r^3}{c^3} \frac{d\omega}{(2\pi)^3}, \tag{7.130}$$

$$\boxed{D_3^{\text{opt}}(\omega)d\omega = \frac{\omega^2 n_r^3}{\pi^2 c^3}d\omega.} \tag{7.131}$$

Notice that the density of optical modes in a medium with refractive index $n_r > 1$ is larger than that of free space (where $n_r = 1$).

7.5.2 Energy Density of Light

The Poynting vector $\mathbf{S} = \mathbf{E} \times \mathbf{H}$ can be used to determine the energy flux density of a sinusoidally varying electromagnetic field. The magnitude of *average* power flux is given by

$$|\mathbf{S}_{\text{av}}| = \frac{1}{2}|\mathbf{E} \times \mathbf{H}| = \frac{1}{2}|\mathbf{E}_0||\mathbf{H}_0| = \frac{1}{2}|\mathbf{E}_0|\frac{\omega\varepsilon_0}{k}|\mathbf{E}_0| = \frac{1}{2}c\varepsilon_0|\mathbf{E}_0|^2, \tag{7.132}$$

where the factor $1/2$ comes from taking the average. This is proportional to the average light intensity of a sinusoidally oscillating electromagnetic field in free space. It follows that the energy density for photons per unit frequency interval in free space is

$$U(\omega) = \frac{1}{2}\varepsilon_0|\mathbf{E}_0|^2. \tag{7.133}$$

7.5.3 Background Photon Energy Density at Thermal Equilibrium

The average value of radiation energy density at frequency ω is the product of the density of states, the occupation factor, and the energy per photon:

$$U(\omega) = D_3^{\text{opt}}(\omega)g(\omega)\hbar\omega. \tag{7.134}$$

The occupation factor for a system in thermal equilibrium is given by the Bose–Einstein distribution function $g(\omega)$, so, using $D_3^{\text{opt}}(\omega)$ from Eq. (7.131) with $n_r = 1$,

$$U(\omega) = \frac{\omega^2}{\pi^2 c^3}\frac{1}{e^{\hbar\omega/k_B T} - 1}\hbar\omega = \frac{\hbar\omega^3}{\pi^2 c^3}\frac{1}{e^{\hbar\omega/k_B T} - 1}. \tag{7.135}$$

Here, $U(\omega)$ is the background radiative photon energy per unit volume per unit frequency at thermal equilibrium. In a homogeneous isotropic medium with refractive index n_r, Eq. (7.135) is modified to

$$\boxed{U(\omega) = \frac{\hbar\omega^3 n_r^3}{\pi^2 c^3}\frac{1}{e^{\hbar\omega/k_B T} - 1}.} \tag{7.136}$$

7.5.4   The Golden Rule for Stimulated Optical Transitions

When deriving the golden rule (Eq. (7.23)),

$$a_m(t) = \frac{1}{i\hbar} \int\limits_{t'=0}^{t'=t} W_{mn} e^{i\omega_{mn}t'} dt', \tag{7.137}$$

where $\omega_{mn} = \omega_m - \omega_n$. The simplest description of interaction between an atomic dipole and an oscillating electric field is by way of the dipole matrix element,

$$W_{mn} = \mathbf{d}_{mn} \cdot \mathbf{E} = er_{mn}|\mathbf{E}_0| \cos(\omega t), \tag{7.138}$$

where $d_{mn} = e\langle m|\hat{\mathbf{r}}|n\rangle = er_{mn}$ is the dipole and $\mathbf{E} = \mathbf{E}_0 \cos(\omega t) = \mathbf{E}_0(e^{i\omega t} + e^{-i\omega t})/2$ is the oscillating classical electric field.

The golden rule can be used to calculate the transition rate between two states of an atomic system due to the presence of a sinusoidally oscillating electric field. For convenience, an electric field in the z direction is considered so that the dipole matrix element between state $|n\rangle$ and $|m\rangle$ changes from r_{mn} to z_{mn}. Equation (7.137) may now be written as

$$\begin{aligned}
a_m(t) &= \frac{1}{i\hbar} \frac{e|\mathbf{E}_0|}{2} z_{mn} \int\limits_{t'=0}^{t'=t} \left(e^{i\omega t'} + e^{-i\omega t'} \right) e^{i\omega_{mn}t'} dt' \\
&= \frac{1}{\hbar} \frac{e|\mathbf{E}_0|}{2} z_{mn} \left(\frac{e^{i(\omega+\omega_{mn})t} - 1}{\omega + \omega_{mn}} + \frac{e^{-i(\omega-\omega_{mn})t} - 1}{\omega - \omega_{mn}} \right).
\end{aligned} \tag{7.139}$$

Since $\omega + \omega_{mn} \gg \omega - \omega_{mn}$ for ω near ω_{mn}, the first term can be set to zero:

$$a_m(t) \sim \frac{1}{\hbar} \frac{e|\mathbf{E}_0|}{2} z_{mn} \left(\frac{e^{-i(\omega-\omega_{mn})t} - 1}{\omega - \omega_{mn}} \right). \tag{7.140}$$

With this approximation,

$$\begin{aligned}
a_m(t) &= \frac{1}{\hbar} \frac{e|\mathbf{E}_0|}{2} z_{mn} \frac{e^{-i(\omega-\omega_{mn})t/2}}{\omega - \omega_{mn}} \left(e^{-i(\omega-\omega_{mn})t/2} - e^{i(\omega-\omega_{mn})t/2} \right) \\
&= \frac{1}{\hbar} e|\mathbf{E}_0| z_{mn} \frac{e^{-i(\omega-\omega_{mn})t/2}}{\omega - \omega_{mn}} \sin((\omega - \omega_{mn})t/2).
\end{aligned} \tag{7.141}$$

Hence, the transition probability is

$$|a_m(t)|^2 = \left(\frac{e|\mathbf{E}_0|}{\hbar} \right)^2 |z_{mn}|^2 \frac{\sin^2((\omega - \omega_{mn})t/2)}{(\omega - \omega_{mn})^2}. \tag{7.142}$$

The probability of a transition to the continuum or over the complete line shape is found by integrating over all frequencies ω, so that

$$|a_m(t)|^2 = \left(\frac{e|\mathbf{E}_0|}{\hbar} \right)^2 |z_{mn}|^2 \int\limits_{-\infty}^{\infty} \frac{\sin^2((\omega - \omega_{mn})t/2)}{(\omega - \omega_{mn})^2} d\omega. \tag{7.143}$$

But

$$\int_{-\infty}^{\infty} \frac{\sin^2(ax)}{x^2} dx = a\pi. \tag{7.144}$$

So, changing variables such that $a = t/2$ and $x = (\omega - \omega_{mn})$, the expression becomes

$$|a_m(t)|^2 = \left(\frac{e|\mathbf{E_0}|}{\hbar}\right)^2 |z_{mn}|^2 \frac{\pi t}{2} = \frac{\pi e^2}{\varepsilon_0 \hbar^2} |z_{mn}|^2 U(\omega) t, \tag{7.145}$$

where Eq. (7.133) is used to eliminate $|\mathbf{E_0}|^2$. For an isotropic energy density $U(\omega)$, averaging over electric field directions to correspond to unpolarized radiation, and differentiating with respect to time, gives the transition rate

$$\frac{\mathrm{d}}{\mathrm{d}t} |a_m(t)|^2 = \frac{\pi e^2}{3\varepsilon_0 \hbar^2} |r_{mn}|^2 U(\omega) = \mathcal{B} U(\omega). \tag{7.146}$$

The factor $1/3$ comes from the average over electric field directions, and $\mathcal{B}U(\omega)$ is called the stimulated emission rate. The probability per unit time that an atom in state $|n\rangle$ makes a transition to any possible state $|m\rangle$ stimulated by electromagnetic radiation is

$$\mathcal{B}_{mn} U(\omega) = \frac{\pi e^2}{3\varepsilon_0 \hbar^2} |\langle m|\hat{\mathbf{r}}|n\rangle|^2 U(\omega). \tag{7.147}$$

7.5.5 The Einstein $\mathcal{A}$ and $\mathcal{B}$ Coefficients

Electromagnetic radiation can stimulate transitions between electronic states. In addition to stimulated transitions, spontaneous transitions from a high-energy state to a lower-energy state are possible. The existence of spontaneous emission from an excited state is *required* as a mechanism to drive the system back to thermal equilibrium. For a system in *thermal equilibrium*, stimulated and spontaneous transition rates are related to each other.[6]

Consider a two-energy-level atom system illustrated in Fig. 7.16 in which it is assumed an electromagnetic field can also be used to describe a photon field. Optical transitions take place between states $|1\rangle$ and $|2\rangle$. There are N_1 atoms in state $|1\rangle$ and N_2 atoms in state $|2\rangle$. The total stimulated transition rate is proportional to Eq. (7.147) multiplied by the number of occupied initial states. Under conditions of thermal equilibrium, the net rate of transition from $|2\rangle$ to $|1\rangle$ must equal that from $|1\rangle$ to $|2\rangle$, so that

$$N_2(\mathcal{B}_{21} U(\omega) + \mathcal{A}) = N_1 \mathcal{B}_{12} U(\omega), \tag{7.148}$$

where a spontaneous transition rate $\mathcal{A}$ has been introduced. Spontaneous transitions occur from $|2\rangle$ to $|1\rangle$ due to vacuum fluctuations in photon density.

In thermal equilibrium, the ratio of occupation of levels, N_1/N_2, is given by a Boltzmann factor, so that

$$\frac{\mathcal{B}_{21}}{\mathcal{B}_{12}} + \frac{\mathcal{A}}{\mathcal{B}_{12} U(\omega)} = \frac{N_1}{N_2} = \mathrm{e}^{\hbar\omega/k_\mathrm{B}T}. \tag{7.149}$$

---

[6] A. Einstein, *Phys. Z.* **18**, 121 (1917).

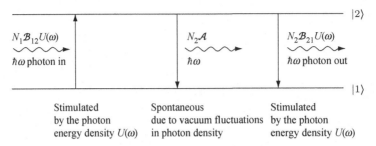

Fig. 7.16 Energy level diagram of a two-level atom system in which N_1 atoms are in state $|1\rangle$ and N_2 atoms are in state $|2\rangle$.

The Boltzmann factor is the ratio of the number of atoms, N_1/N_2. Substituting the expression for electromagnetic energy density $U(\omega) = \frac{\hbar\omega^3}{\pi^2 c^3} \frac{1}{e^{\hbar\omega/k_B T} - 1}$ for black-body radiation gives

$$\frac{\mathcal{B}_{21}}{\mathcal{B}_{12}} + \frac{\mathcal{A}}{\mathcal{B}_{12}} \frac{\pi^2 c^3}{\hbar\omega^3} \left(e^{\hbar\omega/k_B T} - 1\right) = e^{\hbar\omega/k_B T}. \tag{7.150}$$

Separating out the temperature-dependent and temperature-independent terms,

$$\frac{\mathcal{A}}{\mathcal{B}_{12}} \frac{\pi^2 c^3}{\hbar\omega^3} = 1 \tag{7.151}$$

and

$$\frac{\mathcal{B}_{21}}{\mathcal{B}_{12}} = 1. \tag{7.152}$$

Hence, the *Einstein relations* between spontaneous and stimulated transition probability are obtained,

$$\boxed{\begin{aligned} \mathcal{A} &= \frac{\hbar\omega^3}{\pi^2 c^3} \mathcal{B}_{12} \\ \mathcal{B}_{12} &= \mathcal{B}_{21}. \end{aligned}} \tag{7.153}$$

Using Eq. (7.147),

$$\mathcal{B} = \frac{\pi e^2}{3\varepsilon_0 \hbar^2} |\langle j|\hat{\mathbf{r}}|k\rangle|^2, \tag{7.154}$$

and it follows that the spontaneous emission rate, $1/\tau_{\text{sp}} = \mathcal{A}$, is just

$$\mathcal{A} = \frac{e^2 \omega^3}{3\pi\varepsilon_0 \hbar c^3} |\langle j|\hat{\mathbf{r}}|k\rangle|^2. \tag{7.155}$$

It can be shown (see Exercise 7.9) that if the time dependence of spontaneous light emission intensity from a number of excited atoms is $I(t) = I(t = 0)e^{-\mathcal{A}t} = I(t =$

$0)e^{-t/\tau_{\mathrm{sp}}}$, the associated spectral line has a Lorentzian line shape with FWHM $1/\tau_{\mathrm{sp}} = \mathcal{A}$ measured in units of rad s^{-1}.

If light emission occurs in a homogeneous isotropic medium characterized by refractive index $n_{\mathrm{r}} > 1$ then the density of optical modes contributing to $U(\omega)$ is given by Eq. (7.131), and Eq. (7.155) is modified to

$$\mathcal{A} = \frac{e^2 \omega^3 n_{\mathrm{r}}^3}{3\pi\varepsilon_0 \hbar c^3} \, |\langle j|\hat{\mathbf{r}}|k\rangle|^2 . \tag{7.156}$$

7.5.5.1 Estimation of the Spontaneous Emission Coefficient $\mathcal{A}$ for the Hydrogen $|2p\rangle \to |1s\rangle$ Transition

If a hydrogen atom in free space is in an excited electron state with $n = 2$, then quantum mechanics predicts the atom will relax to the $n = 1$ ground state by spontaneous emission of a photon. This physical process limits the excited-state lifetime, on average, to a time characterized by the spontaneous emission lifetime $\tau_{\mathrm{sp}} = 1/\mathcal{A}$. To estimate this value, the expression for the spontaneous emission coefficient given by Eq. (7.155) is used. For the transition considered, the emission wavelength is $\lambda_{\mathrm{photon}} = 122$ nm and the optical frequency is found using $ck = \omega = c2\pi/\lambda_{\mathrm{photon}}$. The dipole matrix element can be estimated as $\langle 1s|\hat{\mathbf{r}}|2p\rangle \sim a_{\mathrm{B}} = 0.053$ nm, where a_{B} is the Bohr radius of the electron in a hydrogen atom. Putting in the numbers gives

$$\mathcal{A} = \frac{(2\pi)^3}{\lambda_{\mathrm{photon}}^3} \frac{e^2}{3\pi\varepsilon_0 \hbar} a_{\mathrm{B}}^2 = 1.12 \times 10^9 \, \mathrm{s}^{-1} = \frac{1}{\tau_{\mathrm{sp}}}. \tag{7.157}$$

Hence, an estimate for the spontaneous emission time is $\tau_{\mathrm{sp}} = 0.89$ ns. A more detailed calculation gives $|\langle 1s|\hat{\mathbf{r}}|2p\rangle|^2 = \frac{1}{3}\left(\frac{32}{27}\right)^3 a_{\mathrm{B}}^2$ for the $|2p\rangle \to |1s\rangle$ transition in hydrogen, so that $\tau_{\mathrm{sp}} = 1.6$ ns (see Problem 10.3).

The photon field produced by the transition takes energy from the excited electron state and converts it to photon energy. Typically, the field intensity decays as $e^{-t/\tau_{\mathrm{sp}}}$, so that the *length* of a photon in free space when $\tau_{\mathrm{sp}} \sim 1$ ns is about 0.3 m.

As the electron makes its transition, the superposition of the $|2p\rangle$ and $|1s\rangle$ states causes the hydrogen electron probability density cloud to oscillate at difference energy $\hbar\omega = \Delta E = E_2 - E_1 = 10.2$ eV. The oscillation in expectation value of electron position creates a dipole moment, and a photon field is emitted at wavelength $\lambda_{\mathrm{photon}} = 2\pi c/\omega = 0.122\,\mu$m, carrying away angular momentum of magnitude $\pm\hbar$ (quantum number ± 1).

7.5.5.2 Dipole Selection Rules for Optical Transitions

The dipole matrix element $d_{jk} = e\langle j|\hat{\mathbf{r}}|k\rangle$ gives rise to a set of rules for optical transitions at frequency ω between initial eigenstate $|k\rangle$ and final eigenstate $|j\rangle$. Dipole radiation requires a parity difference (even-to-odd or odd-to-even) between initial and final states to ensure oscillation in the mean position of charge. Without oscillation in the mean position of charge, there can be no dipole radiation. Clearly, the dipole matrix element $\langle even(odd)|r|odd(even)\rangle \neq 0$, whereas $\langle even(odd)|r|even(odd)\rangle = 0$ from symmetry. Hence, for quantum numbers that sequentially alternate between odd and even parity it is expected that $\langle j|r|k\rangle \neq 0$ for $j - k = odd$. This type of condition is often called a dipole selection rule. Other rules also apply. For example, energy conservation requires that the separation in energy between initial and final states is the energy of the photon, $\hbar\omega$.

7.6 Example Exercises

Exercise 7.1

(a) A particle of mass m is in a one-dimensional, rectangular potential well for which $V(x) = 0$ for $0 < x < L$ and $V(x) = \infty$ elsewhere. The particle is initially prepared in the ground state ψ_1 with eigenenergy E_1. Then, at time $t = 0$, the potential is very rapidly changed so that the original wave function remains the same but $V(x) = 0$ for $0 < x < 2L$ and $V(x) = \infty$ elsewhere. Find the probability that the particle is in the first, second, third, and fourth excited state of the system when $t \geq 0$.

(b) Consider the same situation as (a) but for the case in which at time $t = 0$ the potential is very rapidly changed so that the original wave function remains the same but $V(x) = 0$ for $0 < x < \gamma L$, where $1 < \gamma < 5$, and $V(x) = \infty$ elsewhere. Write a computer program that plots the probability of finding the particle in the ground, first, second, third and fourth excited states of the system as a function of the parameter γ.

Exercise 7.2

The coulomb potential energy in real space between charges e and $-e$ is

$$v(r) = \frac{-e^2}{4\pi\varepsilon_0\varepsilon_{r0}r},$$

where ε_{r0} is the low-frequency dielectric constant of the three-dimensional homogeneous isotropic medium. By taking the Fourier transform of $v(r)$, show that the coulomb potential energy in wave vector space is

$$v(q) = \frac{-e^2}{\varepsilon_0\varepsilon_{r0}q^2}.$$

Exercise 7.3

Show how the structure factor can be given by

$$s(\mathbf{q}) = \sum_{j=1}^{n} e^{-i\mathbf{q}\cdot\mathbf{R}_j} \sum_{k=1}^{n} e^{i\mathbf{q}\cdot\mathbf{R}_k} = n$$

for a density of n sites at random positions $\mathbf{R}_j$.

Exercise 7.4

Starting from the expression for total elastic scattering rate of a particle of mass m and charge e from a density of n ionized impurities given by

$$1/\tau_{\mathrm{el}} = \frac{2\pi}{\hbar}n\int \frac{d^3q}{(2\pi)^3} \left|\frac{e^2}{\varepsilon(\mathbf{q})q^2}\right|^2 \delta(E(\mathbf{k}) - E(\mathbf{k}+\mathbf{q})),$$

show that this may be rewritten as

$$1/\tau_{\mathrm{el}} = \frac{2\pi m}{\hbar^3 k^3} n \left(\frac{e^2}{4\pi\varepsilon_0}\right)^2 \int_{\eta=0}^{\eta=1} \frac{\mathrm{d}\eta}{(\varepsilon_{\mathrm{r}}(q))^2 \eta^3}$$

for a three-dimensional homogeneous isotropic semiconductor with a density of n randomly positioned ionized impurities. In this expression, $\varepsilon(q) = \varepsilon_0 \varepsilon_{\mathrm{r}}(q)$ and $\eta = \sin(\theta/2)$, where θ is the angle between the initial wave vector $\mathbf{k}$ and final wave vector $(\mathbf{k}+\mathbf{q})$ of the charged particle. Assume that the kinetic energy of the particle is given by $E = \hbar^2 k^2/2m$.

Exercise 7.5
A *screened* coulomb potential energy in a homogeneous isotropic medium is

$$v_q(r) = \frac{-e^2}{4\pi\varepsilon_0\varepsilon_{\mathrm{r}0}} \frac{\mathrm{e}^{-r/r_{\mathrm{scr}}}}{r}.$$

Calculate $v(q)$ by taking the Fourier transform, and show that

$$v(q) = \frac{-e^2}{\varepsilon_0\varepsilon_{\mathrm{r}0}(q^2 + 1/r_{\mathrm{scr}}^2)}.$$

Exercise 7.6
Analyze the influence on mobility of conduction-band electrons scattering elastically off ionized impurities in a bulk n-doped semiconductor crystal when:

(a) The ionized substitutional impurities are randomly positioned throughout the bulk crystal. The temperature of the crystal is increased. What is the expected behavior of electron mobility as a function of temperature?

(b) The positions of n ionized substitutional impurities per unit area are placed in a single crystal plane and are spatially correlated. Consider the case of spatial correlations arising from a net repulsion between ionized impurity sites occurring during crystal growth and the case of net attraction (resulting in spatial clustering of ionized impurities in the crystal).

Exercise 7.7
Write a computer program to calculate the total elastic scattering rate $1/\tau_{\mathrm{el}}(E)$ of a single conduction-band electron from randomly positioned ionized impurities in a bulk n-doped semiconductor in the low-temperature limit. Specifically, use the parameters for GaAs (a) with an impurity concentration $n = 1 \times 10^{18}\,\mathrm{cm}^{-3}$ and (b) with an impurity concentration $n = 1 \times 10^{14}\,\mathrm{cm}^{-3}$. The effective electron mass near the conduction-band minimum of GaAs is $m_{\mathrm{e}}^* = 0.07 \times m_0$ (where m_0 is the bare electron mass), and the low-frequency dielectric constant is $\varepsilon_{\mathrm{r}0} = 13.2$. Use the Thomas–Fermi dielectric function, and compare the results with the RPA dielectric function, which is

$$\varepsilon_{\mathrm{r}}(q) = \varepsilon_{\mathrm{r}0} + \frac{r_{\mathrm{s}}^0}{x^3}\xi\left(x + \left(1 - \frac{x^2}{4}\right)\ln\left|\frac{x+2}{x-2}\right|\right),$$

where

$$r_s^0 = \left(\frac{3}{4\pi n}\right)^{1/3} \left(\frac{m_e^* e^2}{4\pi\varepsilon_0\hbar^2}\right),$$

$$\xi = \frac{1}{\pi^2}\left(\frac{32\pi^2}{9}\right)^{1/3},$$

$$x = \frac{q}{k_F}.$$

In the equation, q is the scattered wave number and $k_F = (3\pi^2 n)^{1/3}$ is the Fermi wave number. Calculate the scattering rate for electron energy $E = 300\,\text{meV}$ and $E = 100\,\text{meV}$ above the conduction-band minimum as a function of scattering angle θ for both cases and discuss the significance and meaning of the results. What are the changes to both the results and interpretation when the angular integral is weighted by a factor $(1 - \cos(\theta))$?

Exercise 7.8
A time-varying Hamiltonian $\hat{H}(t')$ induces transitions from state $|k\rangle$ at time $t' = 0$ to a state $|j\rangle$ at time $t' = t$, with probability $P_{k\rightarrow j}(t)$. Use first-order time-dependent perturbation theory to show that if $P_{j\rightarrow k}(t)$ is the probability that the same Hamiltonian brings about the transition from state $|j\rangle$ to state $|k\rangle$ in the same time interval, then $P_{k\rightarrow j}(t) = P_{j\rightarrow k}(t)$.

Exercise 7.9
A single excited state of an atom decays radiatively to the ground state. Derive the time evolution of radiated power, $P(t)$, for N_0 atoms. Show that $P(t) = N_0\hbar\omega_0 A \exp(-\gamma t)$, where $\hbar\omega_0$ is the average photon energy and $1/\gamma$ is the radiative lifetime. The electric field is $\mathbf{E}(t) = \mathbf{E}_0 e^{-\gamma t/2}\cos(\omega_0 t)$ for $t \geq 0$ and $\mathbf{E}(t) = 0$ for $t < 0$. Derive the expression for the homogeneously broadened spectral intensity, $|\mathbf{S}(\omega)|$. In the limit $\gamma \ll \omega_0$, find an expression for $|\mathbf{S}(\omega)|$ near $\omega = \omega_0$. If there are different isotopes of the atom in the gas or Doppler shifts, how is the appearance of the line shape, $|\mathbf{S}(\omega)|$, expected to change?

Exercise 7.10
(a) What determines the selection rules for optical transitions at radial frequency ω between states $|k\rangle$ and $|j\rangle$?
   (b) Show that the inverse of the Einstein spontaneous emission coefficient, τ,

$$\frac{1}{A} = \frac{3\pi\varepsilon_0\hbar c^3}{e^2\omega^3|\langle j|\hat{x}|k\rangle|^2} = \tau,$$

can be rewritten for light emission of wavelength λ from electronic transitions in a harmonic oscillator potential as

$$\frac{1}{A} = 45 \times \lambda^2(\mu\text{m}) = \tau(\text{ns}),$$

where wavelength is measured in μm and time τ is measured in ns.

(c) Calculate the spontaneous emission lifetime and spectral line width for an electron making a transition from the first excited state to the ground state of a harmonic oscillator potential characterized by force constant $\kappa = 3.59 \times 10^{-3}$ kg s^{-2}.

Solutions

Solutions 7.1

(a) Following the solution given at the beginning of this chapter, the probability of finding the particle in an excited state when $t \geq 0$ is given by the square of the overlap integral between the ground state ψ_1 when $t < 0$ and the excited state when $t \geq 0$. After the change in potential, the new state ψ is not an eigenfunction of the system. The new eigenfunctions of a rectangular potential well of thickness $2L$ with infinite barrier energy are $\psi_m = \sqrt{1/L}\sin(k_m x)$, where $k_m = m\pi/2L$ for $m = 1, 2, 3, \ldots$.

Since the state ψ is not an eigenfunction, it may be expressed as a sum of the new eigenfunctions, so that

$$\psi = \sum_m a_m \psi_m.$$

The coefficients a_m are found by multiplying both sides by ψ_m^* and integrating over all space. This overlap integral gives the coefficients

$$a_m = \int \psi_m^* \psi \, dx.$$

The overlap integral projects out the components of the new eigenstates that contribute to the wave function ψ. The value of $|a_m|^2$ is the probability of finding the particle in the eigenstate ψ_m.

The probability that the particle is in the new first excited state when $t \geq 0$ is given by the square of the overlap integral, a_2:

$$a_2 = \int\limits_{x=0}^{x=L} \sqrt{\frac{1}{L}} \sin\left(\frac{2\pi x'}{2L}\right) \sqrt{\frac{2}{L}} \sin\left(\frac{\pi x'}{L}\right) dx'$$

$$= \frac{\sqrt{2}}{L} \int\limits_{x=0}^{x=L} \left(\frac{1}{2} - \frac{1}{2}\cos\left(\frac{2\pi x'}{L}\right)\right) dx' = \frac{\sqrt{2}}{L}\left[\frac{x}{2} + \frac{1}{2}\frac{L}{2\pi}\sin\left(\frac{2\pi x'}{L}\right)\right]_0^L$$

$$= \frac{\sqrt{2}}{L}\left(\frac{L}{2} + 0\right) = \frac{1}{\sqrt{2}}.$$

The probability of finding the particle in the first excited state $|\psi_{m=2}\rangle$ is

$$|a_2|^2 = \frac{1}{2} = 0.50.$$

The probability of excitation of the next few states are

$$|a_3|^2 = \frac{32}{25\pi^2} = 0.129\,69,$$

$$|a_4|^2 = 0.0,$$

$$|a_5|^2 = \frac{32}{21^2 \pi^2} = 0.007\ 352\ 1.$$

That $|a_2|^2$ has the highest probability and $|a_4|^2$ is zero is a direct consequence of symmetry imposed by the fact that in the problem the thickness of the potential well exactly doubled.

(b) In this part of the exercise, the new eigenfunctions of a rectangular potential well of thickness γL with infinite barrier energy are $\psi_m = \sqrt{2/\gamma L}\,\sin(k_m x)$, where $k_m = m\pi/\gamma L$ for $m = 1, 2, 3, \ldots$, and γ takes on values $1 < \gamma \leq 5$.

As in (a), the probability of finding the particle in the eigenstate ψ_m when $t \geq 0$ is given by the square of the overlap integral, a_m:

$$a_m = \frac{2}{L}\sqrt{\frac{1}{\gamma}} \int\limits_{x=0}^{x=L} \sin\left(\frac{m\pi x'}{\gamma L}\right) \sin\left(\frac{\pi x'}{L}\right)\, dx'.$$

Using $2\sin(x)\sin(y) = \cos(x-y) - \cos(x+y)$ to rewrite the integrand gives

$$a_m = \frac{1}{L}\sqrt{\frac{1}{\gamma}} \int\limits_{x=0}^{x=L} \cos\left(\left(\frac{m}{\gamma}-1\right)\pi x/L\right) - \cos\left(\left(\frac{m}{\gamma}+1\right)\pi x/L\right)\, dx'$$

$$= \sqrt{\frac{1}{\gamma}}\left(\frac{\sin\left(\left(\frac{m}{\gamma}-1\right)\pi\right)}{\left(\frac{m}{\gamma}-1\right)\pi} - \frac{\sin\left(\left(\frac{m}{\gamma}+1\right)\pi\right)}{\left(\frac{m}{\gamma}+1\right)\pi}\right).$$

The probability of finding the particle in the state $|\psi_m\rangle$ is just $|a_m|^2$. See Fig. 7.E17.

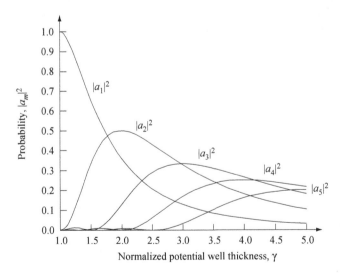

Fig. 7.E17

Solutions 7.2

The bare coulomb potential energy in real space is

$$v(r) = \frac{-e^2}{4\pi\varepsilon_0 r},$$

where e is the electron charge. The Fourier transform is taken to obtain the potential energy in wave-vector space. However, in general, for the Fourier transform of a function $f(x)$ to be well behaved requires the function converge to zero *fast enough* in the limit $x \to \infty$. Unfortunately, the $1/x$ behavior of the coulomb potential does not converge fast enough in this limit. One way to regularize the Fourier transform of the coulomb potential energy is to multiply $f(x)$ by $e^{-\gamma x}$, perform the Fourier transform, and then take the limit $\gamma \to 0$.

Ignoring factors of $-e^2$ and $4\pi\varepsilon_0$ for the moment, the Fourier transform of the *regularized* coulomb (Yukawa-type) potential,

$$v(r) = \frac{e^{-\gamma r}}{r},$$

is sought. Hence,

$$v(q) = \int_0^{2\pi} d\phi \int_0^\pi \sin(\theta) d\theta \int_0^\infty r^2 \frac{e^{-\gamma r}}{r} e^{-i\mathbf{q} \cdot \mathbf{r}} dr = 2\pi \int_0^\pi d\theta \int_0^\infty r^2 \frac{e^{-\gamma r}}{r} e^{-iqr \cos(\theta)} \sin(\theta) dr$$

$$= 2\pi \int_0^\infty r^2 \frac{e^{-\gamma r}}{iqr^2} \left[e^{-iqr \cos(\theta)} \right]_0^\pi dr = 2\pi \int_0^\infty \frac{e^{-\gamma r}}{iq} \left[e^{iqr} - e^{-iqr} \right] dr$$

$$= \frac{4\pi}{q} \int_0^\infty e^{-\gamma r} \sin(qr) dr.$$

Writing $\sin(qr) = \mathrm{Im}\left(e^{iqr}\right)$,

$$v(q) = \frac{4\pi}{q} \mathrm{Im} \int_0^\infty e^{(iq-\gamma)r} dr = \frac{4\pi}{q} \mathrm{Im} \left[\frac{e^{(iq-\gamma)r}}{iq - \gamma} \right]_0^\infty = \frac{4\pi}{q} \mathrm{Im} \left[\frac{1}{\gamma - iq} \right] = \frac{4\pi}{q} \mathrm{Im} \left[\frac{\gamma + iq}{\gamma^2 + q^2} \right]$$

so that

$$v(q) = \frac{4\pi}{\gamma^2 + q^2}.$$

Now, taking the limit $\gamma \to 0$ and inserting the factors of factors of $-e^2$ and $4\pi\varepsilon_0$ results in

$$v(q) = \frac{-e^2}{\varepsilon_0 q^2},$$

which is the usual expression for the coulomb potential in k-space.

The Yukawa-type potential that has been used implies that, while taking the limit, the photons that create the coulomb potential have mass m_γ. The finite mass condition may be eliminated if the electron charge giving rise to the coulomb potential is uniformly distributed in a sphere of radius

$$r_e = \frac{1}{4\pi\epsilon_0} \frac{e^2}{m_0 c^2} = 2.818 \times 10^{-15} \text{ m}.$$

In this case $m_\gamma = 0$ and the need for a Yukawa-type potential is removed.

However, there is no experimental evidence for the existence of a classical electron radius, r_e. Electrons seem to be point particles and so photons may indeed have mass. Experiments using data from the cosmological background radiation indicate that if photon mass exists it must be such that $m_\gamma < 2 \times 10^{-54}$ kg and the associated photon lifetime is $\tau_\gamma > 1 \times 10^8$ s.[7,8,9]

Assuming $E = m_\gamma c^2$, the inequality $\Delta E \Delta t \geq \hbar/2$, and the age of the universe is 10^{10} years, then a lower limit of the photon mass is

$$\Delta m_\gamma \geq \frac{\hbar}{2\Delta t c^2} = \frac{1.05 \times 10^{-34}}{2 \times 3.16 \times 10^{17} \times 9 \times 10^{16}} = 1.8 \times 10^{-69} \text{ kg},$$

which is a very small number.

Solutions 7.3

The structure factor for an impurity density n of ions at positions $\mathbf{R}_j$ may be written as

$$s(\mathbf{q}) = \sum_{j=1}^{n} e^{-i\mathbf{q}\cdot\mathbf{R}_j} \sum_{k=1}^{n} e^{i\mathbf{q}\cdot\mathbf{R}_k} = \sum_{j=k}^{n} 1 + \sum_{j\neq k}^{n} e^{-i\mathbf{q}\cdot(\mathbf{R}_j - \mathbf{R}_k)}.$$

The second term on the right-hand side can be written in terms of sine and cosine functions,

$$s(\mathbf{q}) = n + \sum_{j\neq k}^{n} (\cos(\mathbf{q}\cdot(\mathbf{R}_j - \mathbf{R}_k)) - i\sin(\mathbf{q}\cdot(\mathbf{R}_j - \mathbf{R}_k))).$$

For a *large number* of impurities *randomly positioned* on lattice sites, the average value of the sine and cosine terms can be assumed to tend to zero so that, in this approximation,

$$s(\mathbf{q}) = n.$$

It is important to remember that in small systems containing relatively few *randomly positioned* impurity sites, the average over the sine and cosine terms cannot, in general, be assumed to tend to zero.

Solutions 7.4

It is given that the total elastic scattering rate of an electron of mass m and charge e from a density of n ionized impurities is

$$1/\tau_{el} = \frac{2\pi}{\hbar} n \int \frac{d^3q}{(2\pi)^3} \left| \frac{e^2}{\varepsilon(\mathbf{q})q^2} \right|^2 \delta(E(\mathbf{k}) - E(\mathbf{k}+\mathbf{q})).$$

[7] A. S. Goldhaber and M. M. Nieto, *Rev. Mod. Phys.* **82**, 939 (2010).
[8] J. Heeck, *Phys. Rev. Lett.* **111**, 021801 (2013).
[9] M. Reece, *J. High Energy Phys.* **2019**, 181 (2019).

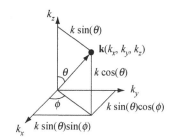

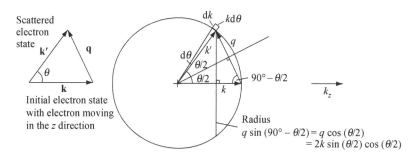

Fig. 7.E18

The task is to evaluate the total elastic scattering rate of an electron with energy E, wave vector k, and mass m, in the energy interval dE. To do this, the volume element d^3q should be expressed in terms of energy E and scattering angle θ. A good way to proceed is to draw a diagram of the volume element in k-space (see Fig. 7.E18).

For elastic scattering in an isotropic material, $q = 2k\sin(\theta/2)$. The energy of the electron is $E = \hbar^2 k^2/2m$, and momentum conservation requires $\mathbf{k} = \mathbf{k'} - \mathbf{q}$. The volume element d^3q is

$$d^3q = k\sin(\theta)d\phi\, k\, d\theta\, dk.$$

Noting that $\sin(\theta) = 2\sin(\theta/2)\cos(\theta/2)$, then

$$d^3q = 2k\sin(\theta/2)\cos(\theta/2)d\phi\, k\, d\theta\, dk.$$

Over the infinitesimal energy dE, this becomes

$$d^3q = 2k\sin(\theta/2)\cos(\theta/2)d\phi\, k\, d\theta \left(\frac{dk}{dE}\right) dE.$$

Since energy $E = \hbar^2 k^2/2m$, then $dk/dE = m/\hbar^2 k$. Substituting into the expression for d^3q and integrating over ϕ gives

$$4\pi\frac{km}{\hbar^2}\sin(\theta/2)\cos(\theta/2)d\theta\, dE.$$

345

Substituting this into the equation for the total elastic scattering rate gives

$$1/\tau_{el} = \frac{2\pi}{\hbar} n \int\limits_{\theta=0}^{\theta=\pi} \frac{4\pi \, d\theta}{(2\pi)^3} \left| \frac{e^2}{\varepsilon(q)4k^2 \sin^2(\theta/2)} \right|^2 \frac{km}{\hbar^2} \sin(\theta/2)\cos(\theta/2)$$

$$= \frac{\pi m}{\hbar^3 k^3} n \frac{e^4}{16\pi^2} \int\limits_{\theta=0}^{\theta=\pi} d\theta \, \frac{\sin(\theta/2)\cos(\theta/2)}{\varepsilon(q)^2 \sin^4(\theta/2)},$$

so that

$$1/\tau_{el} = \frac{2\pi m}{\hbar^3 k^3} n \left(\frac{e^2}{4\pi\varepsilon_0} \right)^2 \int\limits_{\theta=0}^{\theta=\pi} d(\theta/2) \frac{\cos(\theta/2)}{\varepsilon_r(q)^2 \sin^3(\theta/2)},$$

where the fact $\varepsilon(q) = \varepsilon_0 \varepsilon_r(q)$ has been used. Letting $\eta = \sin(\theta/2)$, so that $d\eta = \cos(\theta/2)d(\theta/2)$, results in

$$1/\tau_{el} = \frac{2\pi m}{\hbar^3 k^3} n \left(\frac{e^2}{4\pi\varepsilon_0} \right)^2 \int\limits_{\eta=0}^{\eta=1} \frac{d\eta}{(\varepsilon_r(q))^2 \eta^3},$$

which is the same as Eq. (7.88).

Since $k = \sqrt{2mE/\hbar^2}$, then

$$1/\tau_{el}(E) = \frac{2\pi m}{\hbar^3} n \left(\frac{e^2}{4\pi\varepsilon_0} \right)^2 \left(\frac{\hbar^2}{2mE} \right)^{3/2} \int\limits_{\eta=0}^{\eta=1} \frac{d\eta}{(\varepsilon_r(2k\eta))^2 \eta^3},$$

which may be written as

$$1/\tau_{el}(E) = \frac{\pi}{(2m)^{1/2}} n \left(\frac{e^2}{4\pi\varepsilon_0} \right)^2 E^{-3/2} \int\limits_{\eta=0}^{\eta=1} \frac{d\eta}{(\varepsilon_r(2k\eta))^2 \eta^3}.$$

This is just Eq. (7.89).

Solutions 7.5
It is given that the screened coulomb potential energy is

$$v_q(r) = \frac{-e^2}{4\pi\varepsilon_0\varepsilon_{r0}} \frac{e^{-r/r_{scr}}}{r}.$$

Notice that the subscript q appearing in $v_q(r)$ indicates that the potential in wave-vector space is q-dependent. To calculate the potential energy in wave-vector space the Fourier transform for the real-space potential energy is taken. This gives

$$v(q) = \int d^3r \; v(r) \; e^{-i\mathbf{q} \cdot \mathbf{r}} = \int_{\phi=0}^{\phi=2\pi} d\phi \int_{\theta=0}^{\theta=\pi} d\theta \sin(\theta) \int_{r=0}^{r=\infty} dr \; r^2 \; v(r) \; e^{-iqr \cos \theta}$$

$$= 2\pi \int_{r=0}^{r=\infty} dr \; r^2 v(r) \int_{\theta=0}^{\theta=\pi} d\theta \sin(\theta) e^{-iqr \cos \theta}.$$

Changing variables to $x = \cos(\theta)$, so that $dx = -\sin(\theta)$, and remembering that $\cos(\pi) = -1$ and $\cos(0) = 1$, gives

$$v(q) = 2\pi \int_{r=0}^{r=\infty} dr \; r^2 v(r) \int_{x=-1}^{x=1} dx \; e^{-iqrx} = 2\pi \int_{r=0}^{r=\infty} dr \; r^2 v(r) \left(\frac{1}{iqr} e^{iqr} - \frac{1}{iqr} e^{-iqr} \right).$$

Substituting in the function

$$v_q(r) = \frac{-e^2}{4\pi\varepsilon_0\varepsilon_{r0}} \frac{e^{-r/r_{scr}}}{r}$$

allows the integral to be rewritten as

$$v(q) = \frac{-e^2}{2\varepsilon_0\varepsilon_{r0}} \frac{1}{iq} \int_{r=0}^{r=\infty} \left(e^{-\left(\frac{1}{r_{scr}} - iq\right)r} - e^{-\left(\frac{1}{r_{scr}} + iq\right)r} \right) dr = \frac{-e^2}{2\varepsilon_0\varepsilon_{r0}} \frac{1}{iq} \left(\frac{1}{\frac{1}{r_{scr}} - iq} - \frac{1}{\frac{1}{r_{scr}} + iq} \right)$$

$$= \frac{-e^2}{2\varepsilon_0\varepsilon_{r0}} \frac{1}{iq} \left(\frac{2iq}{\frac{1}{r_{scr}^2} + q^2} \right) = \frac{-e^2}{\varepsilon_0\varepsilon_{r0}} \left(\frac{1}{\frac{1}{r_{scr}^2} + q^2} \right) = \frac{-e^2}{\varepsilon_0\varepsilon_{r0}q^2(1 + 1/q^2r_{scr}^2)}.$$

An alternative way to perform the integral is as follows. Starting with

$$v(q) = 2\pi \int_{r=0}^{r=\infty} dr \; r^2 v(r) \left(\frac{1}{iqr} e^{iqr} - \frac{1}{iqr} e^{-iqr} \right) = 2\pi \int_{r=0}^{r=\infty} dr \; r^2 v(r) \frac{2}{qr} \sin(qr)$$

and substituting in the function

$$v_q(r) = \frac{-e^2}{4\pi\varepsilon_0\varepsilon_{r0}} \frac{e^{-r/r_{scr}}}{r}$$

gives

$$v(q) = \frac{-e^2}{\varepsilon_0\varepsilon_{r0}q} \int_{r=0}^{r=\infty} dr \; e^{-r/r_{scr}} \sin(qr) = \frac{-e^2}{\varepsilon_0\varepsilon_{r0}q} I,$$

where the integral I has been separated out:

$$I = \int_{r=0}^{r=\infty} dr \; e^{-r/r_{scr}} \sin(qr).$$

Performing this integral by parts, $\int UV' = UV - \int U'V$, where $U = \sin(qr)$,

$U' = q\cos(qr)$, $V' = e^{-r/r_{\text{scr}}}$, and $V = -r_{\text{scr}}e^{-r/r_{\text{scr}}}$:

$$I = \left[-r_{\text{scr}}\sin(qr)e^{-r/r_{\text{scr}}}\right]_{r=0}^{r=\infty} + \int_{r=0}^{r=\infty} dr\, e^{-r/r_{\text{scr}}}q^2 r_0^2 \cos(qr).$$

The term in the square brackets is zero. Performing the remaining integral by parts using $U = \cos(qr)$, $U' = -q\sin(qr)$, $V' = e^{-r/r_{\text{scr}}}$, and $V = -r_{\text{scr}}e^{-r/r_{\text{scr}}}$ gives

$$I = \left[-qr_{\text{scr}}^2\cos(qr)e^{-r/r_{\text{scr}}}\right]_{r=0}^{r=\infty} - \int_{r=0}^{r=\infty} dr\, e^{-r/r_{\text{scr}}}q^2 r_{\text{scr}}^2 \cos(qr) = qr_{\text{scr}}^2 - q^2 r_{\text{scr}}^2 I,$$

$$I(q^2 r_{\text{scr}}^2 + 1) = qr_{\text{scr}}^2,$$

and

$$I = \frac{qr_{\text{scr}}^2}{(q^2 r_{\text{scr}}^2 + 1)} = \frac{q}{(q^2 + 1/r_{\text{scr}}^2)}.$$

Substituting in the expression for $v(q)$ gives

$$v(q) = \frac{-e^2}{\varepsilon_0 \varepsilon_{\text{r}0} q}\frac{q}{(q^2 + 1/r_{\text{scr}}^2)} = \frac{-e^2}{\varepsilon_0 \varepsilon_{\text{r}0}(q^2 + 1/r_{\text{scr}}^2)} = \frac{-e^2}{\varepsilon_0 \varepsilon_{\text{r}0}q^2(1 + 1/q^2 r_{\text{scr}}^2)}.$$

Solutions 7.6

Mobility $\mu = e\tau/m^*$ is a measure of an electron scattering time τ. Conductivity is related to the mobility via the relation $\sigma = ne^2\tau/m^* = en\mu$.

(a) If elastic ionized impurity scattering dominates mobility in a bulk n-type semiconductor then it is expected that $\tau \to \tau_{\text{el}}$ where τ_{el} is the elastic scattering time given by

$$1/\tau_{\text{el}}(E) = \frac{\pi}{(2m)^{1/2}}n\left(\frac{e^2}{4\pi\varepsilon_0}\right)^2 E^{-3/2}\int_{\eta=0}^{\eta=1}\frac{d\eta}{(\varepsilon_{\text{r}}(2k\eta))^2\eta^3}.$$

Hence, $\tau_{\text{el}}E^{3/2}$. Since, for a nondegenerate electron gas, the average energy of an electron in thermal equilibrium is proportional to the thermal energy $k_{\text{B}}T$, it follows that $\tau_{\text{el}} \sim E^{3/2} \sim (k_{\text{B}}T)^{3/2}$. When elastic ionized impurity scattering dominates mobility, it is possible to surmise that the mobility will *increase* with increasing temperature as $T^{3/2}$.

(b) Coulomb scattering is weighted to small angle (small-q) scattering. Elastic ionized impurity scattering is screened. Small-q scattering is screened most effectively. If there are spatial correlations in the positions of ionized impurities, this can influence the structure factor for a given q.

A net repulsion between impurities will tend to give rise to a long-range correlation in impurity position. Long-range order will move some spectral weight in the structure factor to points in the Brillouin zone of the resulting sub-lattice. Such order can result in a suppression in small-q scattering.

A net attraction between impurities will tend to give rise to clusters of impurity atom positions. In this case, there is not any long-range order, and scattering strength can be enhanced for high-q scattering because clusters can have large effective coulomb scattering cross-sections.

Solutions 7.7

Thomas–Fermi (TF) screening predicts lower scattering rate (lower of each curve for given energy in Fig. 7.E19) compared with the RPA model. This occurs because screening is overestimated in the Thomas–Fermi model. In fact, in this model screened charge density is proportional to $\left(q_{\mathrm{TF}}^2\, e^{-q_{\mathrm{TF}}\cdot r}\right)/r$, which diverges as $r \to 0$. Notice that in Fig. 7.E19(a) there is very little difference between the two models when carrier concentration is high and screening is very effective in reducing scattering in both models. The differences between Thomas–Fermi and RPA are enhanced when the carrier concentration is low (see Fig. 7.E19(b)). Weighting the angular integral by $(1-\cos(\theta))$ has the effect of suppressing small-angle scattering. Such $(1-\cos(\theta))$ weighting is used as a way to estimate the influence scattering angle has on electrical conductivity $\sigma = ne^2\tau_{\mathrm{el}}/m_e^*$. The intuitively obvious fact that back scattering, corresponding to $\theta = 180°$, has a much larger effect in reducing conductivity than small-angle scattering is quantified by using the $(1 - \cos(\theta))$ weighting term when calculating elastic scattering time, τ_{el}.

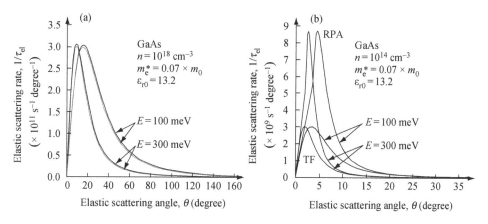

Fig. 7.E19

Solutions 7.8

The probability of a transition from state $|k\rangle$ to state $|j\rangle$ at time t is

$$P_{k\to j}(t) = |c_{k\to j}(t)|^2,$$

and the first-order expression for $c_{k\to j}(t)$ is

$$c_{k\to j}(t) = \frac{1}{i\hbar} \int_{t'=0}^{t'=t} \langle j|\hat{H}(t')|k\rangle \, e^{i\omega_{jk}t'} \, \mathrm{d}t'.$$

349

The coefficient $c_{j\to k}(t)$ for the reverse transition is given by the same expression with the indices k and j interchanged:

$$c_{j\to k}(t) = \frac{1}{i\hbar} \int_{t'=0}^{t'=t} \langle k|\hat{H}(t')|j\rangle \, e^{i\omega_{kj}t'} \, dt'.$$

Since the Hamiltonian $\hat{H}$ is a Hermitian operator, it follows that $\langle k|\hat{H}(t')|j\rangle = \langle j|\hat{H}(t')|k\rangle^*$. Also, the change in energy due to the transition is $\hbar\omega_{kj} = E_k - E_j = -\hbar\omega_{jk}$. Hence,

$$P_{j\to k}(t) = |c_{j\to k}(t)|^2 = P_{k\to j}(t).$$

The probability of transition between two states due to external stimulus is the same for transitions in either direction. This result is known as the principle of *detailed balance*.

Solutions 7.9

There are N_0 excited atoms that can radiatively decay to the ground state with an average radiative lifetime $1/\gamma$. If the distrubution of lifetimes is strongly peaked near $1/\gamma$ then the rate equation for the population of excited atoms is $dN/dt = -\gamma N$, with solution $N(t) = N_0 e^{-\gamma t}$. Power radiated is the number of photons emitted per unit time multiplied by the energy per photon, so that $P(t) = N_0 \hbar\omega_0 \gamma e^{-\gamma t}$. Given the electric field as a function of time is $\mathbf{E}(t) = \mathbf{E}_0 e^{-\gamma t/2} \cos(\omega_0 t)$ for $t \geq 0$, $\mathbf{E}(t) = 0$ for $t < 0$. The spectral function, $|\mathbf{S}(\omega)| = \mathbf{E}^*(\omega) \cdot \mathbf{E}(\omega)/Z_0$, is the Fourier transform of the temporal electric field intensity.

The frequency dependence of the electric field is

$$\mathbf{E}(\omega) = \int_0^\infty \mathbf{E}(t) e^{-i\omega t} \, dt.$$

Substituting in the expression for $\mathbf{E}(t)$ and noting that $\cos(\omega_0 t) = \frac{1}{2}(e^{i\omega_0 t} + e^{-i\omega_0 t})$,

$$\mathbf{E}(\omega) = \frac{\mathbf{E}_0}{2} \int_0^\infty e^{-\gamma t/2} e^{-i\omega t} (e^{i\omega_0 t} + e^{-i\omega_0 t}) \, dt$$

$$= \frac{\mathbf{E}_0}{2} \int_0^\infty (e^{-(\gamma/2+i\omega)t} e^{i\omega_0 t} + e^{-(\gamma/2+i\omega)t} e^{-i\omega_0 t}) \, dt$$

$$= \frac{\mathbf{E}_0}{2} \int_0^\infty (e^{-(\gamma/2+i(\omega-\omega_0))t} + e^{-(\gamma/2+i(\omega+\omega_0))t}) \, dt,$$

$$\mathbf{E}(\omega) = \frac{\mathbf{E}_0}{2} \left[\frac{e^{-(\gamma/2+i(\omega-\omega_0))t}}{-(\gamma/2+i(\omega-\omega_0))} + \frac{e^{-(\gamma/2+i(\omega+\omega_0))t}}{-(\gamma/2+i(\omega+\omega_0))} \right]_0^\infty$$

$$= \frac{\mathbf{E}_0}{2} \left(\frac{1}{\gamma/2+i(\omega-\omega_0)} + \frac{1}{\gamma/2+i(\omega+\omega_0)} \right)$$

$$= \frac{\mathbf{E}_0}{2} \left(\frac{\gamma + 2i\omega}{\omega_0^2 - \omega^2 + (\gamma/2)^2 + \frac{i\gamma\omega}{2} + \frac{i\gamma\omega_0}{2} + \frac{i\gamma\omega}{2} - \frac{i\gamma\omega_0}{2}} \right)$$

$$= \frac{\mathbf{E}_0}{2} \left(\frac{\gamma + 2i\omega}{\omega_0^2 - \omega^2 + (\gamma/2)^2 + i\gamma\omega} \right) = \mathbf{E}_0 \left(\frac{\gamma/2 + i\omega}{\omega_0^2 - \omega^2 + (\gamma/2)^2 + i\gamma\omega} \right),$$

$$|\mathbf{S}(\omega)| = \frac{\mathbf{E}^*(\omega) \times \mathbf{E}(\omega)}{Z_0} = \frac{|\mathbf{E}_0|^2}{Z_0} \frac{(\gamma/2)^2 + \omega^2}{(\omega_0^2 - \omega^2 + (\gamma/2)^2)^2 + (\gamma\omega)^2},$$

where $Z_0 = \sqrt{\mu_0/\varepsilon_0} = 376.73\,\Omega$ is the impedance of free space.

The Lorentzian line shape approximation may be assumed if $\gamma \ll \omega_0$. Such a small value of γ gives peaked $|\mathbf{S}(\omega)|$, so $\gamma^2 \to 0$ and $(\omega + \omega_0) \cong 2\omega_0$ for ω near ω_0. Hence,

$$|\mathbf{S}(\omega)| = \frac{|\mathbf{E}_0|^2}{Z_0} \frac{\omega_0^2}{(\omega_0 - \omega)^2(\omega_0 + \omega)^2 + (\gamma\omega)^2} = \frac{|\mathbf{E}_0|^2}{Z_0} \frac{\omega_0^2}{4\omega_0^2((\omega_0 - \omega)^2 + (\gamma/2)^2)}$$

$$= \frac{|\mathbf{E}_0|^2}{4Z_0} \frac{1}{(\omega_0 - \omega)^2 + (\gamma/2)^2}.$$

This is in the form of a Lorentzian function. The Lorentzian function is symmetric in frequency about ω_0, but the exact solution is not. Different isotopes (distinguishable particles) or Doppler shifts give different contributions to peaks in the spectrum and an asymmetric line shape.

In Fig. 7. E20, a comparison between an exact and Lorentzian line shape is made. The figures show a comparison of exact and Lorentzian line shape for different values of γ. Note that there is very little difference between the curves when $\omega_0 = 10$, $\gamma = 0.5$, as shown in Fig. 7.E20(a). The value γ is the full-width-half-maximum (FWHM), and γ determines the peak value $1/\gamma^2$ at ω_0. The differences between the exact and Lorentzian approximation are more apparent when $\omega_0 = 10$ and $\gamma = 5$, as shown in Fig. 7.E20(b). For this large γ, heavy damping case, the exact result shows that the peak in the spectral response is shifted to a frequency above $\omega_0 = 10$, the peak value is greater than $1/\gamma^2$, the spectral line shape is asymmetric, but the FWHM is still close in value to γ.

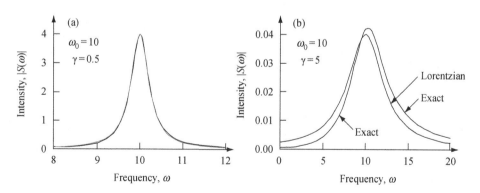

Fig. 7.E20

Solutions 7.10

(a) Dipole radiation requires a parity difference between the initial and final states to ensure oscillation in mean position and charge. This can most easily be seen by considering the dipole matrix element $\langle even(odd)|\hat{\mathbf{r}}|odd(even)\rangle \neq 0$, whereas $\langle even(odd)|\hat{\mathbf{r}}|even(odd)\rangle = 0$ from symmetry. Hence, it is expected that $\langle j|\hat{\mathbf{r}}|k\rangle \neq 0$ for $j - k = odd$. Conservation of spin requires that the photon, which obeys boson statistics, must take spin quantum number ± 1 away from the system. Energy conservation requires that the separation in energy between the initial and final states is the energy of the photon, $\hbar\omega$.

(b) For the harmonic oscillator, the matrix element is found using the position operator

$$\hat{x} = \left(\frac{\hbar}{2m_0\omega}\right)^{1/2}\left(\hat{b} + \hat{b}^\dagger\right).$$

Hence,

$$|\langle j|\hat{x}|k\rangle|^2 = \frac{\hbar}{2m_0\omega}\left|\langle j|\hat{b} + \hat{b}^\dagger|k\rangle\right|^2.$$

For spontaneous emission, only the transition from the first excited state to the ground state will be considered. In this situation, the magnitude of the matrix element squared is

$$|\langle j|\hat{x}|k\rangle|^2 = \frac{\hbar}{2m_0\omega}.$$

Substituting into the expression for the spontaneous emission lifetime,

$$\frac{1}{A} = \tau = \frac{3\pi\varepsilon_0\hbar c^3}{e^2\omega^3|\langle j|r|k\rangle|^2} = \frac{3\pi\varepsilon_0\hbar c^3 2m_0\omega}{e^2\hbar\omega^3} = \frac{6\pi\varepsilon_0 c^3 m_0}{e^2\omega^2} = \frac{3\varepsilon_0\lambda^2 c m_0}{2\pi e^2},$$

so that

$$\tau = \frac{3 \times 8.8 \times 10^{-12} \times \lambda^2 \times 3 \times 10^8 \times 9.1 \times 10^{-31}}{2\pi(1.6 \times 10^{-19})^2} = 45 \times 10^3 \times \lambda^2,$$

which, if wavelength is measured in µm and time in ns, becomes

$$\tau(\text{ns}) = 45 \times \lambda^2(\text{µm}).$$

(c) The force constant κ is related to oscillator frequency ω and particle mass m_0 through $\kappa = \omega^2 m_0$. Given that $\kappa = 3.59 \times 10^{-3}\,\text{kg s}^{-2}$ and m_0 is the bare electron mass, the oscillator frequency is $10\,\text{THz}$ ($\hbar\omega = 41.36\,\text{meV}$), and the emission wavelength is $\lambda = 30\,\text{µm}$. This gives a spontaneous emission lifetime $\tau(\text{ns}) = 45 \times 900\,\text{ns} = 40\,\text{µs}$.

7.7 Problems

Problem 7.1

(a) Starting from the exact result

$$i\hbar\frac{\mathrm{d}}{\mathrm{d}t}a_m(t) = \sum_n a_n(t)\langle m|\hat{W}(x,t)|n\rangle e^{i\omega_{mn}t},$$

for which eigenstates $|n\rangle$ and eigenvalues $\hbar\omega_n = E_n$ are known prior to application of perturbation $\hat{W}(x,t)$ at time $t \geq 0$, time-dependent coefficients $a_n(t)$ describe the time-dependent state

$$|\psi(t \geq 0)\rangle = \sum_n a_n(t)|n\rangle e^{i\omega_n t}.$$

If $a_n(t)$ can be expressed as a power series in the perturbing potential then

$$\hat{W}(x,t) = \lambda\hat{W}(x,t)$$

and

$$a_n(t) = a_m^{(0)}(t) + \lambda a_m^{(1)}(t) + \lambda^2 a_m^{(2)}(t) + \lambda^3 a_m^{(3)}(t) + \ldots,$$

where λ is a dummy variable used to keep track of the order of terms in the power series and is set to unity at the end of the calculation. Substitute into the expression for $i\hbar\frac{d}{dt}a_m(t)$ and evaluate zeroth-order, first-order, and kth order terms.

(b) A particle initially in eigenstate $|n\rangle$ of the unperturbed Hamiltonian for all time $t < 0$ can for time $t \geq 0$ scatter into state $|m\rangle$. Show using first-order time-dependent perturbation theory that the scattering amplitude is

$$a_m(t) = \frac{a}{i\hbar}\int\limits_{t'=0}^{t'=t} W_{mn}e^{i\omega_{mn}t'}\,dt',$$

where the matrix element $W_{mn} = \langle m|\hat{W}(x,t)|n\rangle$ and $\hbar\omega_{mn} = E_m - E_n$ is the difference in eigenenergies of the states $|m\rangle$ and $|n\rangle$.

(c) A particle of mass m_0 is initially in the ground state of a one-dimensional harmonic oscillator. At time $t = 0$, a perturbation $\hat{W}(x,t) = V_0 x^3 e^{-t/\tau}$ is applied where V_0 and τ are constants. Using the result in part (a), calculate the probability of transition to each excited state of the system in the long time limit, $t \to \infty$.

Problem 7.2

An electron is in the ground state of a one-dimensional rectangular potential well for which $V(x) = 0$ in the range $0 < x < L$ and $V(x) = \infty$ elsewhere. It is decided to control the state of the electron by applying a pulse of electric field $\mathbf{E}(t) = \mathbf{E}_0\, e^{-t^2/\tau^2}$ in the x direction starting at time $t = -\infty$, where τ is a constant and $|\mathbf{E}_0|$ is the maximum strength of the applied electric field.

(a) Calculate the probability P_{12} that the particle will be found in the first excited state in the long time limit, $t \to \infty$.

(b) If the electron is in a semiconductor and has an effective mass $m^* = 0.07 \times m_0$, where m_0 is the bare electron mass, and the potential well is of thickness $L = 10\,\mathrm{nm}$, calculate the minimum value of $|\mathbf{E}_0|$ for which $P_{12} = 1$. Comment on the result.

Note the standard integral,

$$\int\limits_{t'=-\infty}^{t'=\infty} e^{-ax^2}\,dx = \sqrt{\frac{\pi}{a}}.$$

Problem 7.3

An electron is initially in the ground state of a one-dimensional rectangular potential well for which $V(x) = 0$ in the range $0 < x < L$ and $V(x) = \infty$ elsewhere. The ground state energy is E_1 and the first excited state energy is E_2. At time $t = 0$, the system is subject to a perturbation $\hat{W}(x, t) = V_0 x^2 e^{-t/\tau}$. Calculate analytically and then use a computer program to plot the probability of finding the particle in the first excited state as a function of time for $t \geq 0$. In the plot, normalize time to units of τ and consider the three values of $\omega_{21} = 1/2\pi$, $\omega_{21} = 1$, and $\omega_{21} = 2\pi$, where $\hbar\omega_{21} = E_2 - E_1$. Explain the results.

Problem 7.4

Use the Einstein spontaneous emission coefficient,

$$
A = \frac{4\omega^3 e^2}{3\hbar c^3 4\pi\varepsilon_0} |\langle j\, |\hat{\mathbf{r}}|k\rangle|^2,
$$

to estimate the numerical value of the spontaneous emission lifetime of the 2p excited state of atomic hydrogen. Use the results to estimate the spontaneous emission lifetime of the 2p transition of atomic He^+ ions. Describe the expected spontaneous emission spectral line-shape. Does the He^+ ion 2p spontaneous emission line spectrum have a larger or smaller full-width at half maximum compared to atomic hydrogen?

Problem 7.5

(a) A particle in a continuum system described by Hamiltonian $\hat{H}_0$ is prepared in eigenstate $|n\rangle$ with eigenvalue $E_n = \hbar\omega_n$. Consider the effect of a perturbation turned on at time $t = 0$ that is harmonic in time such that

$$
\hat{W}(x, t) = V(x)\cos(\omega t),
$$

where $V(x)$ is the spatial part of the potential and ω is the frequency of oscillation. Start by writing down the Schrödinger equation for the complete system including the perturbation and then go on to show that the scattering rate in the static limit ($\omega \to 0$) is given by the golden rule

$$
\frac{1}{\tau_n} = \frac{2\pi}{\hbar} |W_{mn}|^2 D(E)\delta(E_m - E_n),
$$

where the matrix element $W_{mn} = \langle m|\hat{W}(x, t)|n\rangle$ couples state $|n\rangle$ to state $|m\rangle$ via the static potential $V(x)$, the density of final continuum states is $D(E)$, and $\delta(E_m - E_n)$ ensures energy conservation.

(b) An electron of energy E moving in the x direction in the conduction band of a semiconductor has effective electron mass m_e^* and is incident on two identical ionized impurities, one at position $x = 0\,\mathrm{nm}$ and the other at $x = 10\,\mathrm{nm}$. Calculate and plot the structure factor $s(\theta)$ as a function of electron energy $0 \leq E \leq 0.15\,\mathrm{eV}$, where θ is the elastic scattering angle. Explain the significance of the results.

Problem 7.6

(a) In a uniform dielectric, the dielectric function is a constant over space but depends on wave vector so that $\varepsilon = \varepsilon(\mathbf{q})$. Given that an impurity potential at position $\mathbf{r}$ due to a charge e at position $\mathbf{R}_i$ is

$$v_q(\mathbf{r} - \mathbf{R}_j) = \frac{-e^2}{4\pi\varepsilon_q |\mathbf{r} - \mathbf{R}_i|},$$

derive an expression for $v(\mathbf{q})$.

(b) Use the expression for $v(\mathbf{q})$ and the golden rule to evaluate the total elastic scattering rate for an electron of initial energy $E(\mathbf{k})$ due to a single impurity in a dielectric with dielectric function $\varepsilon = \varepsilon(\mathbf{q})$. Describe any assumptions that have been made. Outline how to extend the calculations to include elastic scattering from n ionized impurities in a substitutionally doped crystalline semiconductor.

(c) What differences in scattering rate are expected in (b) for the case of randomly positioned impurities and for the case of strongly correlated impurity positions?

Problem 7.7

(a) Using the method outlined in Exercise 2.6 as a starting point, calculate numerically the dipole matrix elements between the ground state and the first three excited states for an electron with effective mass $m_e^* = 0.07 \times m_0$ confined to the asymmetric potential well sketched in the figure and bounded by barriers of infinite energy at $x \le 0\,\text{nm}$ and $x \ge 50\,\text{nm}$. The value of the step change in potential energy in Fig. 7.P21 is $V_{\text{step}} = 0.2\,\text{eV}$.

(b) Explain the difference in matrix elements obtained.

(c) Calculate the spontaneous emission rate associated with each transition in a medium with refractive index $n_r = 3.3$.

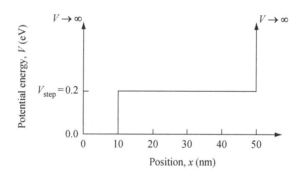

Fig. 7.P21

Problem 7.8

An electron is initially in the ground state of a one-dimensional harmonic oscillator characterized by frequency ω_0. At time $t = 0$, a uniform electric field $\mathbf{E}$ is applied in the x direction for time τ. Calculate the probability of transition to the first excited state of the oscillator and plot the result as a function of the electric field pulse duration τ.

Problem 7.9

A lattice vibration at frequency ω that lasts for a time t can result in an electron making a transition from an initial state $|\psi_i\rangle$ with eigenenergy E_i to a final state $|\psi_f\rangle$ with eigenenergy $E_f = E_i + \hbar\omega$ or $E_f = E_i - \hbar\omega$. The lattice vibration can be viewed as a harmonic perturbation of the form

$$\hat{W}(t') = W_0 \left(\hat{b} e^{i\omega t'} + \hat{b}^\dagger e^{-i\omega t'} \right),$$

where W_0 is a constant. According to first-order time-dependent perturbation theory and assuming that each scattering process is an independent parallel channel, the transition probability is

$$P(t) = \frac{1}{\hbar^2} \sum_f \left| \int_{t'=0}^{t'=t} \langle \psi_f | \hat{W}(t') | \psi_i \rangle e^{i\omega_{fi} t'} dt' \right|^2$$

$$= \frac{W_0^2}{\hbar^2} |\langle \psi_f | \hat{b} | \psi_i \rangle|^2 \left| \int_{t'=0}^{t'=t} e^{i(\omega_{fi}+\omega)t'} dt' \right|^2 + \frac{W_0^2}{\hbar^2} |\langle \psi_f | \hat{b}^\dagger | \psi_i \rangle|^2 \left| \int_{t'=0}^{t'=t} e^{i(\omega_{fi}-\omega)t'} dt' \right|^2,$$

where $\hbar\omega_{fi} = (E_f - E_i)$.

(a) Show that

$$P(t) = \frac{4W_0^2}{\hbar^2} \left| \langle \psi_f | \hat{b} | \psi_i \rangle \right|^2 \frac{\sin^2((\omega_{fi}+\omega)t/2)}{(\omega_{fi}+\omega)^2 t} t + \frac{4W_0^2}{\hbar^2} \left| \langle \psi_f | \hat{b}^\dagger | \psi_i \rangle \right|^2 \frac{\sin^2((\omega_{fi}-\omega)t/2)}{(\omega_{fi}-\omega)^2 t} t.$$

(b) In the limit $t \to \infty$, show that the transition rate is

$$\frac{1}{\tau} = \frac{2\pi}{\hbar} W_0^2 \left| \langle \psi_f | \hat{b} | \psi_i \rangle \right|^2 \delta(E_f - E_i + \hbar\omega) + \frac{2\pi}{\hbar} W_0^2 \left| \langle \psi_f | \hat{b}^\dagger | \psi_i \rangle \right|^2 \delta(E_f - E_i - \hbar\omega)$$

and explain the physical meaning of the result.

Note the relations

$$|e^{ix} - 1|^2 = 4\sin^2 \left(\frac{x}{2} \right),$$

$$\delta(x) = \frac{1}{\pi} \lim_{\eta \to \infty} \frac{\sin^2(\eta x)}{\eta x^2},$$

and

$$\delta(ax) = \frac{1}{|a|} \delta(x),$$

so that if $x = \omega/2$ and $a = 2\hbar$ then $2\hbar\delta(\hbar\omega) = \delta\left(\frac{\omega}{2} \right)$.

Problem 7.10

(a) A hydrogen atom excited in a 2p state is placed inside a cavity. At what temperature of the cavity are the spontaneous and induced photon emission rates equal?

(b) An electron in a GaAs quantum dot is modeled as a particle of mass $m_e^* = 0.07 \times m_0$ embedded in a medium with refractive index $n_r = 3.3$ and confined by a potential that is infinite everywhere except for the region $0 < x < L$, $0 < y < L$, and $0 < z < L$, where the potential is zero. The value of $L = 20\,\text{nm}$. The electron is in the first excited state of the quantum dot. At what temperature are the spontaneous and induced photon emission rates equal? Comment on any assumptions made.

Problem 7.11

A two-level atom described by Hamiltonian $\hat{H}_0$ has eigenstates $|1\rangle$ and $|2\rangle$ with energy separation $\hbar\omega_{21} = E_2 - E_1$. The atom is initially in its ground state $|1\rangle$ and at time $t \geq 0$ it is illuminated with an electric field $\mathbf{E} = \mathbf{E}_0(e^{i\omega t} + e^{-i\omega t})$ in the x direction. The electric field oscillates at frequency ω and has amplitude $|\mathbf{E}_0|$.

(a) Write down the Hamiltonian for time $t \geq 0$ in terms of $\hat{H}_0$ and a perturbation $\hat{W}$.

(b) The solution at time $t \geq 0$ is of the form $|x, t\rangle = a_1(t)e^{-i\omega_1 t}|1\rangle + a_2(t)e^{-i\omega_2 t}|2\rangle$ where $E_1 = \hbar\omega_1$ and $E_2 = \hbar\omega_2$. Substitute this into the time-dependent Schrödinger equation and show that

$$i\hbar\left(\frac{\mathrm{d}}{\mathrm{d}t}a_1(t)\right)e^{-i\omega_1 t}|1\rangle + i\hbar\left(\frac{\mathrm{d}}{\mathrm{d}t}a_2(t)\right)e^{-i\omega_2 t}|2\rangle = a_1(t)e^{-i\omega_1 t}\hat{W}|1\rangle + a_2(t)e^{-i\omega_2 t}\hat{W}|2\rangle.$$

(c) If $|\mathbf{E}_0|$ is small and $\omega = \omega_{21}$, show that the probability that the atom will be in state $|2\rangle$ at time $t > 0$ is $|a_2(t)|^2 = \sin^2(W_{21}t/\hbar)$, where $W_{21} = |\mathbf{E}_0|\langle 2|\hat{x}|1\rangle$.

(d) How is the result in (c) modified if ω is slightly detuned from ω_{21}?

Problem 7.12

A conduction band electron is initially in the lowest energy state of a GaAs quantum well that has thickness $L = 10\,\text{nm}$. The effective electron mass is $m_e^* = 0.07 \times m_0$, where m_0 is the bare electron mass and the ground-state energy eigenvalue is E_1. An electric field pulse $\mathbf{E}(t) = \mathbf{E}_0 e^{-t^2/\tau^2}$ is applied in the x direction across the quantum well starting at time $t = -\infty$, where τ is a constant and $|\mathbf{E}_0| = 1.17 \times 10^6\,\text{Vm}^{-1}$ is the maximum strength of the applied electric field.

(a) Modeling the quantum well as a rectangular potential well for which $V(x) = 0$ in the range $0 < x < L$ and $V(x) = \infty$ elsewhere, find the value of τ that maximizes the probability that the electron will be found in the first excited state in the long time limit, $t \to \infty$.

(b) The quantum well has an area $10 \times 10\,\mu\text{m}^2$ and electron density $10^{12}\,\text{cm}^{-2}$. Using the value of τ calculated in (a) and assuming any electron in the first excited state flows as current in an external circuit, how many electrons contribute to the current?

(c) Is it possible to operate this particular device at room temperature?

Note the standard integral,

$$\int_0^\infty e^{-ax^2}\,\mathrm{d}x = \frac{1}{2}\sqrt{\frac{\pi}{a}}.$$

Problem 7.13

Consider a time-dependent Hamiltonian $\hat{H}(t)$ with state $|\psi(t)\rangle$ evolving from time t_0 that satisfies the Schrödinger equation

$$i\hbar\frac{\partial}{\partial t}|\psi(t)\rangle = \hat{H}(t)|\psi(t)\rangle. \tag{7P.1}$$

If at any given instant in time, the eigenstates $|\phi_n(t)\rangle$ and energy eigenvalues $E_n(t) = \hbar\omega_n(t)$ are known then they can be used as a basis to expand the time-evolving state function

$$|\psi(t)\rangle = \sum_n a_n(t)|\phi(t)\rangle e^{i\theta_n(t)}, \tag{7P.2}$$

where

$$\theta_n(t) = \int_{t'=t_0}^{t'=t} \omega_n(t')dt',$$

so that $\frac{d}{dt}\theta_n(t) \equiv \dot{\theta}_n(t) = \omega_n(t)$ and $a_n(t)$ is the occupation amplitude of the nth state at time t.

(a) Substitute Eq. (7.P2) into Eq. (7.P1) and show that

$$i\hbar \sum_n e^{i\theta_n(t)} \left(a_n(t)|\phi_n(t)\rangle - i\dot{\theta}_n(t)a_n(t)|\phi_n(t)\rangle + a_n(t)|\dot{\phi}_n(t)\rangle \right)$$
$$= \sum_n E_n(t)a_n(t)|\phi_n(t)\rangle e^{i\theta_n(t)}.$$

(b) Multiply both sides of (a) by $\langle\phi_m(t)|$ and show that

$$\dot{a}_m(t)e^{i\theta_m(t)} = -\sum_n a_n(t)\langle\phi_m(t)|\dot{\phi}_n(t)\rangle e^{i\theta_n(t)}.$$

(c) To find $\langle\phi_m(t)|\dot{\phi}_n(t)\rangle$, differentiate $\hat{H}(t)|\phi_n(t)\rangle = E_n(t)|\phi_n(t)\rangle$ with respect to time and then show that

$$\dot{a}_m(t) = -a_m\langle\phi_m|\dot{\phi}_m\rangle + \sum_{n\neq m} a_n \frac{\langle\phi_m|\dot{H}(t)|\phi_n\rangle}{E_m - E_n} e^{i(\theta_n - \theta_m)}.$$

8 The Semiconductor Laser

8.1 Introduction

The history of the laser dates back to at least 1951 and an idea of Townes. He wanted to use ammonia molecules to amplify microwave radiation. Townes and two students completed a prototype device in late 1953 and gave it the name *maser* or *microwave amplification by stimulated emission of radiation*.[1] In 1958, Townes and Schawlow published the results of a study showing that a similar device could be made to amplify light.[2] The device was named a *laser*, which is an acronym for *light amplification by stimulated emission of radiation*. In principle, a large flux of essentially single-wavelength electromagnetic radiation could be produced by a laser. Independently, Prokhorov and Basov proposed related ideas. The first laser used a rod of ruby and was constructed in 1960 by Maiman.

In late 1962, lasing action in a current-driven GaAs p-n diode maintained at liquid nitrogen temperature (77 K) was reported.[3] Room-temperature operation and other improvements followed. It became apparent that, with appropriate design, a p-n diode driven by a direct current could produce controlled coherent electromagnetic radiation with an oscillation frequency of several hundred terahertz.

Soon, telephone companies recognized the potential of such components for use in communication systems. However, it took some time before useful devices and suitable glass-fiber transmission media became available. The first fiber-optic telephone installation was put in place in 1977 and consisted of a 2.4 km long link under downtown Chicago.

Another type of laser diode suitable for use in data communication applications was inspired by the work of Iga published in 1977.[4] By the late 1990s, these vertical-cavity surface-emitting lasers (VCSELs) had appeared in volume-manufactured commercial products.

In the 1990s, the largest volume of semiconductor lasers were produced for compact disk (CD) and digital versatile disk (DVD) video applications. Today, laser diodes are volume-manufactured for fiber-optic communication products, cell phones, laser printers, and laser copiers. There is additional low-volume production of laser diodes for numerous specialty markets.

Typically, the successful design of a semiconductor laser diode that addresses a particular market requirement involves using models that efficiently capture the behavior of multiple physical processes.[5] Depending on the application, device specification and control requirements, this can be quite involved.

---

[1] J. P. Gordon, H. J. Zeiger, and C. H. Townes, *Phys. Rev.* **95**, 282(1954).

[2] A. L. Schawlow and C. H. Townes, *Phys. Rev.* **112**, 1940 (1958).

[3] R. Hall, G. E. Fenner, J. Kingsley, T. J. Soltys, and R. O. Carlson, *Phys. Rev. Lett.* **9**, 366 (1962).

[4] H. Soda, K. Iga, C. Kitahara, and Y. Suematsu, *Jap. J. Appl. Phys.* **18**, 2329 (1977).

[5] A. F. J. Levi, *Essential Semiconductor Laser Device Physics*, San Raphael, Morgan and Claypool Publishers, 2018 (ISBN 978 1 6432 7029 6).

8.2 Spontaneous and Stimulated Emission

Figure 8.1(a) is a schematic energy-level diagram showing stimulated and spontaneous optical transitions between two electronic states $|1\rangle$ and $|2\rangle$ with energy eigenvalues E_1 and E_2, respectively. In the figure, $P(E)$ is the spectral density of electromagnetic modes measured in units of *number* of photons per unit volume per unit energy interval and $f_1(f_2)$ is the Fermi–Dirac occupation factor for particles in state $|1\rangle$ ($|2\rangle$). When describing the semiconductor laser, it is convention to use *photon density per unit energy interval* $P(E)$ (instead of the energy density distribution per unit frequency interval).

In a semiconductor laser diode, optical transitions take place between conduction-band states and valence-band states. The active region where these transitions occur is typically a *direct band-gap* semiconductor, examples of which include GaAs, InP, and InGaAs. In such a semiconductor the energy minimum of the conduction band lines up with the maximum energy of the valence band in k-space. This fact is of particular importance for direct interaction of semiconductor electronic states with light of wave vector k_{opt} because the dispersion relation of light, $\omega = ck_{\text{opt}}$, is almost vertical compared with the dispersion relation for electrons in a given band, $\omega = \hbar k^2/2m^*$, where m^* is an effective electron mass. Conservation of momentum during a transition from occupied to unoccupied electronic states via the emission or absorption of a photon requires, therefore, an almost vertical transition in k-space.

The simplest model of a direct band-gap semiconductor typically includes a heavy-hole valence band hh with effective hole mass m^*_{hh} and conduction band e with effective electron mass m^*_{e}. Figure 8.1(b) shows schematically spontaneous emission of a photon of energy $\hbar\omega$ accompanied by an electronic transition of a state characterized by wave vector $\mathbf{k}$ and energy $E_{\mathbf{k}}$ in the conduction band to wave vector state $\mathbf{k}$ energy $E_{\mathbf{k}} - \hbar\omega$ in the valence band. Because the initial and final electronic states have the same value of $\mathbf{k}$, it is assumed that a vertical transition in k-space occurs and that crystal momentum is conserved.

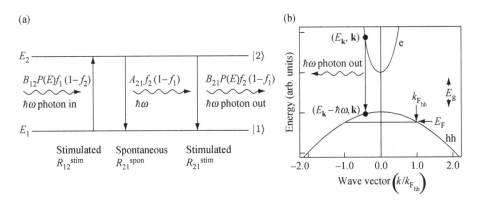

Fig. 8.1 (a) Schematic energy-level diagram showing stimulated and spontaneous optical transitions between two electronic states $|1\rangle$ and $|2\rangle$ with energy eigenvalues E_1 and E_2, respectively. (b) Band structure of a direct band-gap semiconductor showing valence heavy-hole band hh, conduction band e, and band-gap energy E_{g}. The semiconductor is doped p-type, and at low temperature the Fermi energy is E_{F} and the Fermi wave vector is $k_{\text{F}_{\text{hh}}}$. Electrons can make a transition from a state characterized by wave vector $\mathbf{k}$ and energy $E_{\mathbf{k}}$ in the conduction band to wave vector state $\mathbf{k}$ energy $E_{\mathbf{k}} - \hbar\omega$ in the valence band by emitting a photon of energy $\hbar\omega$.

If energy is measured from the top of the valence band, then the energy of an electron in the conduction band with effective electron mass m_e^* is

$$E_2 = E_g + \frac{\hbar^2 k^2}{2m_e^*}, \tag{8.1}$$

where E_g is the band-gap energy. The energy of an electron in the valence band with effective electron mass m_{hh}^* is

$$E_1 = -\frac{\hbar^2 k^2}{2m_{hh}^*}. \tag{8.2}$$

The energy due to an electronic transition from the conduction band to the valence band is

$$\hbar\omega = E_2 - E_1 = \frac{\hbar^2 k^2}{2} \left(\frac{1}{m_e^*} + \frac{1}{m_{hh}^*} \right) + E_g. \tag{8.3}$$

A reduced effective electron mass, m_r^*, may be defined such that

$$\frac{1}{m_r^*} = \frac{1}{m_e^*} + \frac{1}{m_{hh}^*}. \tag{8.4}$$

For GaAs, $m_e^* = 0.07 \times m_0$ and $m_{hh}^* = 0.5 \times m_0$, giving $m_r^* = 0.06 \times m_0$. Equation (8.4) allows Eq. (8.3) to be written as

$$\hbar\omega = \frac{\hbar^2 k^2}{2m_r^*} + E_g. \tag{8.5}$$

It follows that the three-dimensional density of electronic states coupled to vertical optical transitions of energy $\hbar\omega$ is

$$D_3(\hbar\omega) = \frac{1}{2\pi^2} \left(\frac{2m_r^*}{\hbar^2} \right)^{3/2} (\hbar\omega - E_g)^{1/2}. \tag{8.6}$$

If Fig. 8.1(b) is representative of the physical processes involved in an optically active semiconductor, then the *golden rule* might be used to calculate transition probability between electronic states.

First, consider a discrete two-level system inside an optical cavity that is a cube of side L, volume V_{vol}, and temperature T. Possible transitions are shown in Fig. 8.1(a). For periodic boundary conditions, allowed optical modes are k-states $k_{n_j} = 2\pi n_j/L$, where $j = x, y, z$ and n is an integer. The spectral density $P(E)$ of electromagnetic modes at a specific energy is found by multiplying the density of photon states, $D_3^{opt}(E)$, by the Bose–Einstein occupation factor for photons, $g(E)$. Since

$$D_3^{opt}(k_{opt}) dk_{opt} = \frac{1}{V_{vol}} 2 \left(\frac{L}{2\pi} \right)^3 4\pi k_{opt}^2 dk_{opt} = \left(\frac{k_{opt}}{\pi} \right)^2 dk_{opt}, \tag{8.7}$$

and noting that $E = \hbar\omega = \hbar c k_{opt}$ in free space and $dE = \hbar c\, dk_{opt}$, then

$$D_3^{opt}(E) = \frac{E^2}{\pi^2 \hbar^3 c^3}, \tag{8.8}$$

giving a spectral density measured in units of *number of photons per unit volume per unit energy interval*:

$$P(E) = D_3^{\text{opt}}(E)g(E) = \frac{E^2}{\pi^2 \hbar^3 c^3} g(E) = \frac{E^2}{\pi^2 \hbar^3 c^3} \frac{1}{e^{E/k_B T} - 1}. \tag{8.9}$$

If the electromagnetic modes exist in a homogeneous dielectric medium characterized by refractive index n_r at frequency ω, then $E = \hbar\omega = \hbar c k_{\text{opt}}/n_r$. If $n_r = n_r(\omega)$, then $dk_{\text{opt}} = (1/c)(n_r + \omega (dn_r/d\omega))d\omega$. Ignoring dispersion in the refractive index ($\omega dn_r/d\omega = 0$) gives

$$P(E) = \frac{E^2 n_r^3}{\pi^2 \hbar^3 c^3} \frac{1}{e^{E/k_B T} - 1}. \tag{8.10}$$

Assuming that the golden rule may be used to calculate transition rates between states $|1\rangle$ and $|2\rangle$, it makes sense to define

$$B_{21} \equiv \frac{2\pi}{\hbar} |W_{21}|^2, \tag{8.11}$$

where $|W_{21}|^2$ is the magnitude of the matrix element squared coupling the initial and final states. The stimulated and spontaneous rates for photons of energy $E_{21} = E_2 - E_1$ become

$$R_{12}^{\text{stim}} = B_{12} P(E_{21}) f_1 (1 - f_2), \tag{8.12}$$
$$R_{21}^{\text{stim}} = B_{21} P(E_{21}) f_2 (1 - f_1), \tag{8.13}$$
$$R_{21}^{\text{spon}} = A_{21} f_2 (1 - f_1), \tag{8.14}$$

where the Fermi–Dirac distribution function gives the probability of electron occupation f_1 at energy E_1 and the probability of an unoccupied electron state $(1 - f_2)$ at energy E_2.

When the system is in thermal equilibrium, the rates must balance, and there is only one chemical potential, so $\mu_1 = \mu_2 = \mu$. If thermal equilibrium exists between the two-level system and the electromagnetic modes of the cavity, then

$$R_{12}^{\text{stim}} = R_{21}^{\text{stim}} + R_{21}^{\text{spon}}, \tag{8.15}$$
$$B_{12} P(E_{21}) f_1 (1 - f_2) = B_{21} P(E_{21}) f_2 (1 - f_1) + A_{21} f_2 (1 - f_1), \tag{8.16}$$
$$P(E_{21})(B_{12} f_1 (1 - f_2) - B_{21} f_2 (1 - f_1)) = A_{21} f_2 (1 - f_1), \tag{8.17}$$

$$P(E_{21}) = \frac{A_{21} f_2 (1 - f_1)}{B_{12} f_1 (1 - f_2) - B_{21} f_2 (1 - f_1)} = \frac{A_{21}(f_2 - f_1 f_2)}{B_{12}(f_1 - f_1 f_2) - B_{21}(f_2 - f_1 f_2)}$$

$$= \frac{A_{21}\left(\dfrac{1}{f_1} - 1\right)}{B_{12}\left(\dfrac{1}{f_2} - 1\right) - B_{21}\left(\dfrac{1}{f_1} - 1\right)} = \frac{A_{21}}{B_{12}\left(\dfrac{1/f_2 - 1}{1/f_1 - 1}\right) - B_{21}}$$

$$= \frac{A_{21}}{B_{12} e^{E_{21}/k_B T}\left(\dfrac{e^{-\mu_2/k_B T}}{e^{-\mu_1/k_B T}}\right) - B_{21}}. \tag{8.18}$$

362

Since the system is in equilibrium, $\mu_1 = \mu_2$. Making use of Eq. (8.10), the number of photons per unit volume per unit energy interval may be written as

$$P(E_{21}) = \frac{A_{21}}{B_{12}e^{E_{21}/k_B T} - B_{21}} = D_3^{\text{opt}}(E_{21})\frac{1}{e^{E_{21}/k_B T} - 1}, \tag{8.19}$$

which must hold for any temperature when the system is in equilibrium, so that

$$\boxed{B_{12} = B_{21},} \tag{8.20}$$

$$\boxed{\frac{A_{21}}{B_{12}} = D_3^{\text{opt}}(E_{21}) = \frac{E_{21}^2 n_r^3}{\pi^2 \hbar^3 c^3}.} \tag{8.21}$$

Note that it is assumed that the electron system is in thermal equilibrium with the electromagnetic modes of the cavity. *Thermal equilibrium* means that the complete system may be characterized by a single temperature T and there is no net flow of particles or energy from outside or within the system. Equations (8.20) and (8.21) are the Einstein relations introduced in Chapter 7.

8.2.1 Absorption and Its Relation to Spontaneous Emission

Photons of energy $E = \hbar\omega$ incident on a two-level system can cause transitions between two states $|1\rangle$ and $|2\rangle$ with energy eigenvalues $E_1 < E_2$. Absorption α may be defined as the *ratio of the number of absorbed photons per second per unit volume to the number of incident photons per second per unit area*. Hence,

$$\boxed{\alpha = \frac{R_{\text{stim}}^{\text{net}}}{S/\hbar\omega} = \frac{R_{12}^{\text{stim}} - R_{21}^{\text{stim}}}{S/\hbar\omega},} \tag{8.22}$$

where S is the magnitude of the Poynting vector and $S/\hbar\omega$ is the number of incident photons per second per unit area. It is usual for α to be measured in units of cm^{-1}. Since the absorption coefficient multiplied by the photon flux is the net stimulated rate, then

$$\alpha = \frac{B_{12}P(E_{21})f_1(1 - f_2) - B_{21}P(E_{21})f_2(1 - f_1)}{P(E_{21})\dfrac{c}{n_r}} = \frac{n_r}{c}B_{12}(f_1 - f_2). \tag{8.23}$$

The ratio of spontaneous emission and absorption is

$$\frac{R_{21}^{\text{spon}}}{\alpha} = \frac{A_{21}f_2(1 - f_1)}{\dfrac{n_r}{c}B_{12}(f_1 - f_2)} = \frac{A_{21}(1 - f_1)}{\dfrac{n_r}{c}B_{12}\left(\dfrac{f_1}{f_2} - 1\right)} = \frac{A_{21}\left(\dfrac{1}{f_1} - 1\right)}{\dfrac{n_r}{c}B_{12}\left(\dfrac{1}{f_2} - \dfrac{1}{f_1}\right)}$$

$$= \frac{A_{21}e^{(E_1 - \mu_1)/k_B T}}{\dfrac{n_r}{c}B_{12}\left(e^{(E_2 - \mu_2)/k_B T} - e^{(E_1 - \mu_1)/k_B T}\right)} = \frac{A_{21}}{\dfrac{n_r}{c}B_{12}\left(\dfrac{e^{(E_2 - \mu_2)/k_B T}}{e^{(E_1 - \mu_1)/k_B T}} - 1\right)}$$

$$= \frac{A_{21}}{\dfrac{n_r}{c}B_{12}\left(e^{E_{21}/k_B T}e^{-(\mu_2 - \mu_1)/k_B T} - 1\right)}. \tag{8.24}$$

Substituting for the ratio A_{21}/B_{12} using Eq. (8.21) gives

$$\frac{R_{21}^{\text{spon}}}{\alpha} = \frac{c}{n_r} D_3^{\text{opt}}(E_{21}) \frac{1}{e^{(E_{21} - (\mu_2 - \mu_1))/k_B T} - 1} = \frac{E_{21}^2 n_r^2}{\pi^2 c^2 \hbar^3} \frac{1}{e^{(E_{21} - \Delta\mu)/k_B T} - 1}, \qquad (8.25)$$

where $\Delta\mu = \mu_2 - \mu_1$ is the difference in quasi-chemical potentials used to describe the distribution of electronic states at energy E_2 and E_1, respectively. The approximation made is that Eq. (8.21) (which was derived for the equilibrium condition $\Delta\mu = 0$) remains valid for a driven system when $\Delta\mu \neq 0$. This is likely to be true when $\Delta\mu < k_B T$.

The relationship between absorption and spontaneous emission for a (nonequilibrium) system characterized by temperature T and difference in chemical potential $\Delta\mu$ given by Eq. (8.25) may be rewritten in a convenient form as

$$\boxed{\alpha = \frac{\pi^2 c^2 \hbar^3}{E_{21}^2 n_r^2} R_{21}^{\text{spon}} \left(e^{(E_{21} - \Delta\mu)/k_B T} - 1 \right).} \qquad (8.26)$$

Net optical gain exists when absorption α is negative. Since spontaneous emission is always positive, the only way the value of absorption α can change sign is if the right-hand side term in parenthesis in Eq. (8.26) changes sign. An easy way to see this is by noticing that Eq. (8.24) may be rewritten as

$$\alpha = \frac{R_{21}^{\text{spon}} \frac{n_r}{c} B_{12}(f_1 - f_2)}{(n_r/c) A_{21} f_2 (1 - f_1)}. \qquad (8.27)$$

The denominator in Eq. (8.27) is always positive, since $0 < f_1 < 1$ and $0 < f_2 < 1$. The numerator is positive, giving positive absorption α if $f_1 > f_2$ and giving negative absorption (or optical gain $g_{\text{opt}} \equiv -\alpha$) if $f_1 < f_2$. The condition for optical gain is $f_2 - f_1 > 0$, or

$$\Delta\mu > E_{21}. \qquad (8.28)$$

This expresses the fact that the separation in quasi chemical potentials must be greater than the photon energy for net optical gain to exist. Equation (8.28) is called the Bernard–Duraffourg condition.[6]

In a semiconductor there are not just two energy levels E_1 and E_2 to be considered, but rather a continuum of energy levels in the conduction band and valence band. This is illustrated in Fig. 8.2. Electrons in the conduction-band have a Fermi–Dirac distribution f_2, and in the valence band they have a distribution f_1. As was shown in Chapter 6, for equal carrier concentrations injected into the conduction band and valence band at fixed temperature, the Fermi–Dirac distribution functions f_2 and f_1 are different since, in general, the quasi chemical potentials in the driven system are different. This occurs because the effective electron mass in each band is different, giving a different density of states, and hence different quasi chemical potentials for carrier concentration n and temperature T.

For convenience, the way electron energy in the conduction and valence bands is measured is modified. The energy of holes (absence of electrons) in the valence band will be measured from the valence-band maxima *down*. The energy of holes is negative,

---

[6] M. G. A. Bernard and G. Duraffourg, *Phys. Stat. Solidi* **1**, 699 (1961).

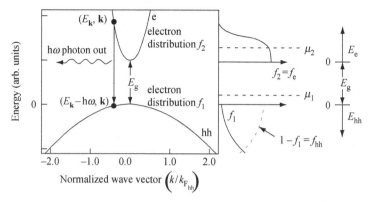

Fig. 8.2 A schematic diagram showing an electronic transition from a state of energy $E_{\mathbf{k}}$ and wave vector $\mathbf{k}$ to state of energy $E_{\mathbf{k}} - \hbar\omega$ and wave vector $\mathbf{k}$, resulting in the emission of a photon of energy $\hbar\omega$. A conduction-band electron has an effective electron mass m_e^* and a valence-band electron has an effective electron mass m_{hh}^*. The Fermi–Dirac distribution of electron states in the conduction-band is f_2, and in the valence band it is f_1. The quasi chemical potentials are μ_2 and μ_1, respectively. Sometimes it is convenient to measure electron energy E_e from the conduction-band minimum and hole energy E_{hh} from the valence-band maximum.

and the distribution of holes is $f_h = (1 - f_1)$. The energy of electrons in the conduction band will be measured from the conduction band minimum *up*. The energy of electrons is positive, and the distribution of electrons is $f_e = f_2$. When calculating photon energy, $\hbar\omega$, for an inter-band transition the band-gap energy, E_g, must be added. Also, the condition for optical gain previously given by Eq. (8.28) becomes

$$\Delta\mu_{e-hh} > 0. \tag{8.29}$$

The total spontaneous emission, $r_{\text{spon}}(\hbar\omega)$, for photons of energy $E = \hbar\omega$ is the sum of all energy levels separated by vertical transitions of energy E. Substituting Eq. (8.21) into Eq. (8.14), using the definition of $|W_{21}|^2$ given by Eq. (8.11), and performing the sum over allowed vertical k-state transitions gives

$$r_{\text{spon}}(\hbar\omega) = \frac{2\pi}{\hbar} \frac{E_{21}^2 n_r^3}{\pi^2 \hbar^3 c^3} \sum_{k_2, k_1} |W_{21}|^2 f_2 (1 - f_1) \delta(E_{21} - \hbar\omega), \tag{8.30}$$

where the delta function ensures energy conservation. If the matrix element W_{21} is slowly varying as a function of E_{21}, it may be treated as a constant and brought outside the sum. Converting the sum to an integral and substituting $f_e = f_2$ and $f_{hh} = (1 - f_1)$ gives

$$r_{\text{spon}}(\hbar\omega) = \frac{2\pi}{\hbar} \frac{E_{21}^2 n_r^3}{\pi^2 \hbar^3 c^3} |W_{21}|^2 \int \frac{d^3k}{(2\pi)^3} f_e f_{hh} \delta\left(E_g + \frac{\hbar^2 k^2}{2m_r} - \hbar\omega\right), \tag{8.31}$$

which may be written as

$$r_{\text{spon}}(\hbar\omega) = \frac{2\pi}{\hbar} \frac{\hbar^2 \omega^2 n_r^3}{\pi^2 \hbar^3 c^3} |W_{21}|^2 \frac{1}{2\pi^2} \left(\frac{2m_r^*}{\hbar^2}\right)^{3/2} (\hbar\omega - E_g)^{1/2} f_e f_{hh}. \tag{8.32}$$

365

In this expression, $(\hbar\omega - E_g)^{1/2}$ is the energy dependence of the reduced three-dimensional density of electronic states in Eq. (8.6).

It follows from Eq. (8.26) that optical gain, $g_{opt}(\hbar\omega)$, is related to $r_{spon}(\hbar\omega)$ through

$$g_{opt}(\hbar\omega) = -\alpha(\hbar\omega) = \hbar \left(\frac{c\pi}{n_r\omega}\right)^2 r_{spon}(\hbar\omega)\left(1 - e^{(\hbar\omega - E_g - \Delta\mu_{e-hh})/k_B T}\right). \tag{8.33}$$

This is a useful relationship since the absorption function may be obtained if the spontaneous emission is known. Note that if $\Delta\mu \neq 0$ then the electron hole system is out of thermal equilibrium and the assumptions made in deriving this relationship are no longer valid. For example, n_r will depend upon $\Delta\mu$.

Optical gain may also be found directly by substituting Eq. (8.11) into Eq. (8.23) and performing the integral over allowed initial and final electronic density of states (Eq. (8.6)). This gives

$$g_{opt}(\hbar\omega) = \frac{2\pi n_r}{c\hbar}|W_{21}|^2 \frac{1}{2\pi^2}\left(\frac{2m_r^*}{\hbar^2}\right)^{3/2}(\hbar\omega - E_g)^{1/2}(f_e + f_{hh} - 1). \tag{8.34}$$

8.3 Optical Transitions Using the Golden Rule

The matrix element W_{21} appearing in Eqs. (8.32) and (8.34) remains to be evaluated. This may be done by applying the golden rule.

Consider a semiconductor illuminated with light. The interaction between a classical optical electric field of the form

$$\mathbf{E}_{opt} = \mathbf{E}_0 e^{i(\mathbf{k}_{opt}\cdot\mathbf{x} - \omega t)}, \tag{8.35}$$

and an electron with motion in the x direction is to be described by the perturbation

$$\hat{W} = -e|\mathbf{E}_0|x e^{i(\mathbf{k}_{opt}\cdot\mathbf{x} - \omega t)}. \tag{8.36}$$

Electron states in a crystal are Bloch functions, so the dipole matrix element coupling a conduction-band initial state $\psi_e(x) = U_{ek}(x)e^{i\mathbf{k}\cdot\mathbf{x}}$ and a heavy-hole valence-band final state $\psi_{hh}(x) = U_{hhk'}(x)e^{i\mathbf{k'}\cdot\mathbf{x}}$ is

$$W_{ehh} = \langle\psi_{hh}|\hat{W}|\psi_e\rangle = -e|\mathbf{E}_0|\int U_{hhk'}^*(x)U_{ek}(x)x e^{i(\mathbf{k'}-\mathbf{k}-\mathbf{k}_{opt})\cdot\mathbf{x}}dx. \tag{8.37}$$

The term $e^{i(\mathbf{k'}-\mathbf{k}-\mathbf{k}_{opt})\cdot\mathbf{x}}$ in the integral rapidly oscillates, resulting in $W_{ehh} \to 0$ *except* when $\mathbf{k'}-\mathbf{k} = \mathbf{k}_{opt}$. Since electronic states have $|\mathbf{k}| \sim 3\times10^6$ cm$^{-1} \gg |\mathbf{k}_{opt}|$ (see Table 6.1), it is reasonable to set $|\mathbf{k}_{opt}| = 0$, so that $\mathbf{k'} = \mathbf{k}$. As discussed in Section 8.2, transitions between initial and final electron states conserve crystal momentum and so have the same k-vector.

Finding the value of the matrix element in Eq. (8.37) requires detailed knowledge of the Bloch wave functions involved in the transition. The calculations can be quite involved.[7]

---

[7] For an introduction, see S. L. Chuang, *Physics of Optoelectronic Devices*, New York, Wiley, 1995 (ISBN 0 471 10939 8).

With a pragmatic approach, the coefficients, including the magnitude of the matrix element squared in Eqs. (8.32) and (8.34), are taken to be constants. The $\hbar^2\omega^2$ term in Eq. (8.32) is slowly varying and may be treated as a constant, since an energy range of approximately $\Delta\mu_{\text{e-hh}}$ around E_{g} is of interest and typically $\Delta\mu_{\text{e-hh}}/E_{\text{g}} \ll 1$. It is then reasonable to write spontaneous emission (Eq. (8.32)) in a bulk direct band-gap semiconductor as

$$r_{\text{spon}}(\hbar\omega) = r_0(\hbar\omega - E_{\text{g}})^{1/2} f_{\text{e}} f_{\text{hh}}, \tag{8.38}$$

where r_0 is a material-dependent constant. Optical gain becomes

$$g_{\text{opt}}(\hbar\omega) = g_0(\hbar\omega - E_{\text{g}})^{1/2}(f_{\text{e}} + f_{\text{hh}} - 1), \tag{8.39}$$

where g_0 is also a material-dependent constant. The ratio $g_0/r_0 = \pi^2 c^2/\omega^2 n_{\text{r}}^2$. In this simple model, the range in energy over which optical gain exists is given by the difference in chemical potential, $\Delta\mu_{\text{e-hh}}$.

Obviously, the use of constants r_0 and g_0 is a quite crude approximation. However, it does allow estimation of trends, such as temperature dependence of gain. In fact, it turns out that to create a model of optical gain in a semiconductor that is qualitatively more advanced than that described in this chapter is a very challenging task and the subject of ongoing research.

Figure 8.3 shows the result of calculating $g_{\text{opt}}(\hbar\omega)$ using Eq. (8.39) and parameters appropriate for the direct band-gap semiconductor GaAs.

According to the results shown in Fig. 8.3, there is no optical gain when carrier density $n = 1 \times 10^{18}\,\text{cm}^{-3}$. As carrier density increases in the driven nonequilibrium system, optical gain first appears near the band-gap energy, $E_{\text{g}} = 1.4\,\text{eV}$. When carrier density $n = 2 \times 10^{18}\,\text{cm}^{-13}$, a peak optical gain of $330\,\text{cm}^{-1}$ occurs for photon energy near $\hbar\omega = 1.415\,\text{eV}$, and the gain bandwidth is $\Delta\mu_{\text{e-hh}} = 39\,\text{meV}$.

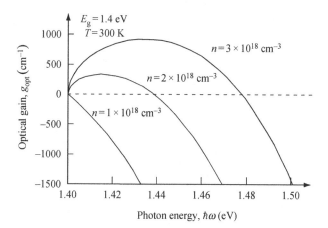

Fig. 8.3 Calculated optical gain in a bulk semiconductor for the indicated carrier densities, band-gap energy $E_{\text{g}} = 1.4\,\text{eV}$, temperature $T = 300\,\text{K}$, effective electron mass $m_{\text{e}}^* = 0.07 \times m_0$, effective hole mass $m_{\text{hh}}^* = 0.5 \times m_0$, and $g_0 = 2.64 \times 10^4\,\text{cm}^{-1}\,\text{eV}^{-1/2}$.

8.3.1 Optical Gain in the Presence of Electron Scattering

Inelastic electron scattering has the effect of broadening the energy of electron states. Because a typical inelastic electron scattering rate $1/\tau_{\text{in}}$ can be tens of ps^{-1}, corresponding to several meV energy broadening, this effect can be significant and on the same scale as the difference in chemical potential.

A Lorentzian broadening function has an energy full-width-half-maximum (FWHM) $\gamma_{\mathbf{k}} = \hbar/\tau_{\text{in}}$, so that if $\tau_{\text{in}} = 25$ fs then $\gamma_{\mathbf{k}} = 26$ meV. There is a subscript $\mathbf{k}$ in $\gamma_{\mathbf{k}}$ because, in general, scattering rate depends upon electron crystal momentum $\hbar\mathbf{k}$. However, in practice this fact is usually ignored and $\gamma_{\mathbf{k}}$ is treated as a constant.

To calculate optical gain in the presence of electron scattering, spontaneous emission using the Lorentzian broadening function might be used so that Eq. (8.38) is modified to

$$r_{\text{spon}}(\hbar\omega) = r_0 \int\limits_0^\infty E^{1/2} f_e f_{\text{hh}} \frac{\gamma_{\mathbf{k}}/2\pi}{(E_g + E - \hbar\omega)^2 + \left(\frac{\gamma_{\mathbf{k}}}{2}\right)^2} \, dE, \tag{8.40}$$

where the factor 2π ensures proper normalization of the Lorentzian function.

Optical gain, g_{opt}, as a function of photon energy, $\hbar\omega$, is then calculated using Eq. (8.33). This ensures that optical transparency in the semiconductor occurs at a photon energy of $\Delta\mu_{\text{e-hh}} + E_g$, where $\Delta\mu_{\text{e-hh}}$ is the difference in chemical potential and E_g is the band-gap energy. Optical transparency occurring at a different energy violates basic concepts of energy conservation in thermodynamics.

Unfortunately, even this elementary consideration is often ignored in conventional theories, which put Lorentzian broadening directly in the gain function (Eq. (8.34)).[8] As illustrated in Fig. 8.4, not only does this result in optical transparency at an energy less than $\Delta\mu_{\text{e-hh}} + E_g$, but it also predicts substantial absorption of sub-band-gap energy photons[9] – something that is not observed experimentally.

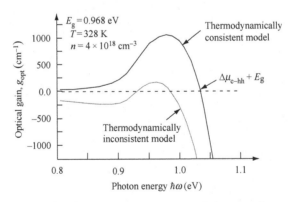

Fig. 8.4 Calculated optical modal gain, including effects of electron–electron scattering in bulk InGaAsP with room-temperature band-gap energy $E_g = 0.968$ eV and an inelastic scattering broadening factor $\gamma_{\mathbf{k}} = 25$ meV.

---

[8] For example, see Peter Zory (ed.), *Quantum Well Lasers*, San Diego, Academic Press, 1993 (ISBN 0 12 781890 1). Contributions from Corzine, Yang, Coldren, Asada, Kapon, Englemann, Shieh, and Shu all explicitly and incorrectly put Lorentzian broadening directly in the gain function.

[9] For example, see S. L. Chuang, J. O'Gorman, and A. F. J. Levi, *IEEE J. Quantum Electron.* **QE-29**, 1631 (1993).

Correct calculation of gain spectra is important to ensure lasers can be designed optimally. At a minimum, a model should be able to accurately predict both peak gain and gain bandwidth. Details matter. For example, simple exponential Markovian decay of an electronic state with time dependence $e^{-t/\tau_{\text{in}}}$ dominated by a single time constant $\tau_{\text{in}} = \hbar/\gamma_{\mathbf{k}}$ has spectrally long tails that are not present in measured gain spectra. Replacing the Lorentzian function in Eq. (8.40) with a sech function,

$$\frac{1}{\pi \gamma_{\mathbf{k}}} \text{sech} \left(\frac{E - \hbar\omega}{\gamma_{\mathbf{k}}} \right),$$

suppresses long spectral tails and allows a better fit to experimentally measured spectra. A self-consistent physical model of band-edge spectral tails (*Urbach* tails) should include correlations in electron scattering processes in the *complex* band structure – something that is not well described using a single dominant scattering time.

8.4 Designing a Laser Diode

The existence of optical gain in a direct band-gap semiconductor such as GaAs or InGaAsP can be exploited to make a laser diode. When a p-n diode is forward-biased to pass a current, I_{inj}, electrons are injected into the conduction band and holes into the valence band. The optically active region of the semiconductor is where the electrons and holes overlap in real space, so that vertical optical transitions can take place in k-space. If the density of carriers injected into the active region is great enough, then Eq. (8.28) is satisfied and optical gain exists for light at some wavelength in the semiconductor. There is, however, more to designing a useful device. Among other things, it is important to ensure that a high intensity of lasing light emission occurs at a specific wavelength.

Because a typical value of gain for an optical mode in a semiconductor laser diode is not very large ($\sim 500 \, \text{cm}^{-1}$), and in order to precisely control emission wavelength, the active semiconductor is placed in a high-Q optical cavity. The optical cavity has the effect of storing light at a particular wavelength, allowing it to interact with the gain medium for a longer time. In this way, relatively modest optical gain may be used to build up high light intensity in a given optical mode. Electrons contributing to injection current I_{inj} are converted into lasing photons that have a single mode and wavelength. The efficiency of this conversion process is enhanced if only one high-Q optical-cavity resonance is in the same wavelength range as semiconductor optical gain.

8.4.1 The Optical Cavity

Figure 8.5(a), (b), and (c) illustrate optical cavities into which an optically active semiconductor may be placed to form a Fabry–Perot laser, VCSEL,[10] and microdisk laser,[11] respectively.

The Fabry–Perot optical cavity is, at least superficially, quite easy to understand. Assuming that photons travel normally to the two mirror planes and in the z direction, the task is to find expressions for the corresponding longitudinal optical resonances.

---

[10]K. Iga, M. Oikawa, S. Misawa, J. Banno, and Y. Kokubun, *Appl. Opt.* **21**, 3456 (1982).
[11]S. L. McCall, A. F. J. Levi, R. E. Slusher, S. J. Pearton, and R. A. Logan, *Appl. Phys. Lett.* **60**, 289 (1992).

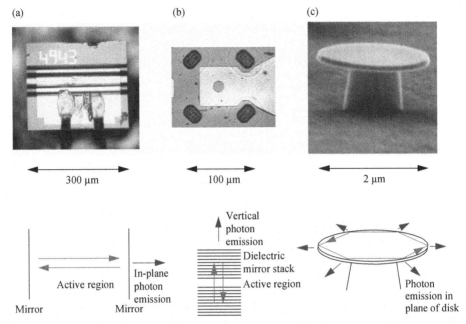

Fig. 8.5 (a) Photograph of top view of a Fabry–Perot, edge-emitting, semiconductor laser diode showing horizontal gold metal stripe used to make electrical contact to the p-type contact of the diode. The n-type contact is made via the substrate. Two gold wire bonds attach to the large gold pad in the lower half of the picture. This particular device has a multiple-quantum-well InGaAsP active region, lasing emission wavelength $\lambda_0 = 1\,310$ nm, and a laser threshold current of 3 mA. The sketch shows the side view of the 300 μm long optical cavity formed by reflection at the cleaved semiconductor–air interface. (b) Photograph of top view of a VCSEL showing gold metallization used to make electrical contact to the p-type contact of the diode. Lasing light is emitted from the small aperture in the center of the device. This VCSEL has a multiple quantum-well GaAs active region, lasing emission wavelength $\lambda_0 = 850$ nm, and a laser threshold current of 1 mA. (c) Scanning electron microscope image of a microdisk laser. The semiconductor disk is 2 μm in diameter and 0.1 μm thick. This particular device has a single quantum-well InGaAs active region, lasing emission wavelength $\lambda_0 = 1\,550$ nm, and an external incident optical laser threshold pump power at 980 nm wavelength of 300 μW.

8.4.1.1 Longitudinal Resonances in the z Direction

The Fabry–Perot laser consists of an index-guided active gain region placed within a Fabry–Perot optical resonator.

Index guiding helps maintain the z-oriented trajectory of photons travelling perpendicular to the mirror plane. Index guiding is achieved in a buried heterostructure laser diode by surrounding the semiconductor active region with a semiconductor of lower refractive index. Usually this involves etching the semiconductor wafer to define a narrow, z-oriented active-region stripe and then planarizing the etched regions by epitaxial growth of nonactive, lower refractive index, wider band-gap semiconductor.

Optical loss for a photon inside the Fabry–Perot cavity is minimized at cavity resonances. Figure 8.6 shows a schematic diagram of a Fabry–Perot optical resonator consisting of a semiconductor active-gain medium and two mirrors with electric field reflectivity r_1 and r_2, respectively, forming an optical cavity of length L_C. The photon round-trip time in this cavity is $t_{\mathrm{round-trip}}$.

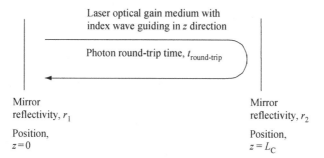

Fig. 8.6 Schematic diagram of a Fabry–Perot optical resonator consisting of a semiconductor active-gain medium and two mirrors with reflectivity r_1 and r_2, respectively, forming an optical cavity of length L_C.

The Fabry–Perot cavity has an optical mode spacing given by $kL_C = \pi m$, where $m = 1, 2, 3, \ldots$ and $k = \omega n_r/c$. The refractive index of the dielectric is n_r, and c is the speed of light in vacuum. Adjacent modes are spaced in angular frequency according to

$$\Delta\omega = \frac{c(k_{m+1} - k_m)}{n_r} = \frac{c\pi(m + 1 - m)}{L_C n_r} = \Delta\omega = \frac{c\pi}{L_C n_r} = 2\pi\Delta\nu. \tag{8.41}$$

This mode spacing is also called the free spectral range of the cavity.

Measured in units of Hz,

$$\Delta\nu = \frac{c}{2L_C n_r} = \frac{1}{t_{\text{round-trip}}}, \tag{8.42}$$

where $t_{\text{round-trip}}$ is the round-trip time for a photon in the cavity and $\nu = \omega/2\pi$. The mode spacing as a function of wavelength is $\Delta\lambda = \lambda^2/(2L_C n_r)$.

In a Fabry-Perot cavity consisting of two parallel plane mirrors, the round-trip attenuation factor for light amplitude in the cavity is $r = r_1 r_2$.[12] The roundtrip phase accumulated by a plane-wave electromagnetic field propagating in the z direction, normal to the mirror, with wave vector $k = 2\pi/\lambda$ is

$$\phi = 2kL_C = \frac{4\pi}{\lambda}L_C = 4\pi L_C \frac{\nu}{c}. \tag{8.43}$$

On resonance, the electromagnetic field must accumulate an integer multiple of 2π so that $k_0 L_C = n\pi$, where $n = 1, 2, 3, \ldots$. The possibility of resonant zero or negative phase accumulation is usually ignored. In general, if the initial value of the electric field amplitude is E_0 and the fractional loss of amplitude remaining after one round trip in the cavity is $r < 1$ then the electric field after one round trip in the cavity is

$$\mathbf{E}_1 = \mathbf{E}_0 r e^{-i\phi}. \tag{8.44}$$

---

[12]For a lossless dielectric Fabry-Perot etalon with refractive index n_r, the mirror reflectivity at a cleaved dielectric-to-air interface is $r_{1,2} = |(1 - n_r)/(1 + n_r)|^2$.

If the electric field is allowed to propagate indefinitely in the resonator, then the total electric field is an infinite sum of terms,

$$\mathbf{E} = \mathbf{E}_0(1 + re^{-i\phi} + r^2e^{-2i\phi} + \cdots) = \frac{\mathbf{E}_0}{1 - re^{-i\phi}}, \tag{8.45}$$

since

$$\sum_{n=0}^{n=\infty} ax^n = a\frac{1}{1 - x}$$

for $x < 1$ and constant a.

The total electric field intensity in the resonator is

$$I_{opt} = |\mathbf{E}|^2 = \left|\frac{\mathbf{E}_0}{1 - re^{-i\phi}}\right|^2 = \frac{I_0}{1 + r^2 - 2r\cos(\phi)}, \tag{8.46}$$

where intensity $I_0 = |\mathbf{E}_0|^2$. Adding and subtracting $2r$ in the denominator gives

$$1 + r^2 - 2r + 2r - 2r\cos(\phi) = (1 - r)^2 + 4r\left(1 - \sin^2\left(\frac{\phi}{2}\right)\right),$$

and so the equation for $I_{opt} = |\mathbf{E}|^2$ may be written

$$I_{opt} = \frac{I_0}{(1 - r)^2 - 4r\sin^2\left(\frac{\phi}{2}\right)}. \tag{8.47}$$

Introducing finesse of the optical cavity,

$$\mathcal{F} \equiv \frac{\pi r^{1/2}}{1 - r}, \tag{8.48}$$

and

$$I_{max} = \frac{I_0}{(1 - r)^2}, \tag{8.49}$$

the spectral intensity as a function of frequency inside a Fabry–Perot cavity pumped by light of initial intensity I_0 in the cavity may be written as[13]

$$I_{opt} = \frac{I_{max}}{1 + \left(\frac{2\mathcal{F}}{\pi}\right)^2\sin^2\left(\frac{\phi}{2}\right)} = \frac{I_{max}}{1 + \left(\frac{2\mathcal{F}}{\pi}\right)^2\sin^2\left(\frac{\pi\nu}{\Delta\nu}\right)}. \tag{8.50}$$

When finesse is large ($\mathcal{F} \gg 1$), the optical line width γ_{opt} is much smaller than $\Delta\nu$ and $\mathcal{F} = \Delta\nu/\gamma_{opt}$. In this limit of $\mathcal{F} \gg 1$, the expression for the FWHM of the resonance becomes $\gamma_{opt} = \Delta\nu/\mathcal{F}$ and the optical-Q associated with the cavity is the frequency ν_0 of the resonance divided by γ_{opt} so that $Q = \nu_0/\gamma_{opt}$.

---

[13]For example, see B. E. A. Saleh and M. C. Teich, *Fundamentals of Photonics*, New York, John Wiley and Sons, 1992 (ISBN 0 471 83965 5).

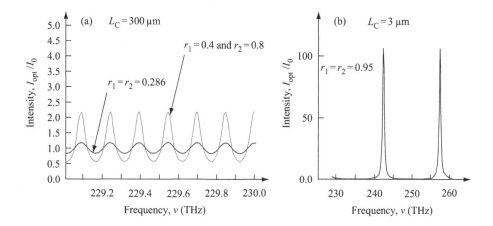

Fig. 8.7 (a) Spectral intensity as a function of frequency when $r_1 = r_2 = 0.286$ for photons inside a Fabry–Perot resonant cavity of length $L_C = 300\,\mu$m and refractive index $n_r = 3.3$. Optical resonances are spaced by $\Delta\nu = 151\,$GHz. The vertical axis is normalized in such a way that $I_0 = 1$ in the calculation. Also shown is the case in which $r_1 = 0.4$ and $r_2 = 0.8$. In this situation, finesse $\mathcal{F} = 2.613$ and $\gamma_{\rm opt} = 57.8\,$GHz. At an optical frequency of $\nu = 229\,$THz, this gives an optical of $Q = \nu/\gamma_{\rm opt} = 3\,963$. (b) Spectral intensity as a function of frequency when $r_1 = r_2 = 0.95$ for photons inside a Fabry–Perot resonant cavity of length $L_C = 3\,\mu$m and refractive index $n_r = 3.3$. Optical resonances are spaced by $\Delta\nu = 15.1\,$THz. In this case, the finesse $\mathcal{F} = \Delta\nu/\gamma_{\rm opt} = 30.6$ and $\gamma_{\rm opt} = 495\,$GHz. At an optical frequency of $\nu = 242\,$THz, this gives an optical $Q = \nu/\gamma_{\rm opt} = 489$.

Consider a Fabry–Perot laser diode with cavity length $L_C = 300\,\mu$m, an effective refractive index $n_r = 3.3$, and emission wavelength near $\lambda = 1\,310\,$nm, corresponding to a frequency $\nu_0 = 229\,$THz. The spectral intensity as a function of frequency when $r_1 = r_2 = 0.286$ is shown in Fig. 8.7(a). Finesse is $\mathcal{F} = 0.978$. Also shown is the case in which $r_1 = 0.4$ and $r_2 = 0.8$, which has a slightly improved finesse of $\mathcal{F} = 2.613$. In this case $Q = \nu_0/\gamma_{\rm opt} = 3\,963$ and $\gamma_{\rm opt} = 58\,$GHz. Figure 8.7(b) shows the spectral intensity of a Fabry–Perot optical cavity of length $L_C = 3\,\mu$m, effective refractive index $n_r = 3.3$, mirror reflectivity $r_1 = r_2 = 0.95$, and emission wavelength near $\lambda_0 = 1\,240\,$nm corresponding to a frequency $\nu_0 = 242\,$THz.

8.4.1.2 Mode Profile in an Index-Guided Slab Waveguide

The optical mode profile in the x and y directions of an index-guided slab geometry structure in which the refractive index of the active region, n_a, is greater than the refractive index, n_c, of the surrounding material can act to guide light close to the active region. Confining light to the active region is important because only the fraction Γ of light that overlaps with the active region can experience optical gain and be amplified.

The optical confinement factor Γ may be found by using the classical time-independent electromagnetic wave equation for light traveling in the z direction:

$$\nabla^2\mathbf{E} + \varepsilon(x,y)k_0^2\mathbf{E} = 0, \tag{8.51}$$

where $k_0 = \omega/c$ is the propagation constant in free space and $\varepsilon(x,y)$ is the spatially varying dielectric function in the slab waveguide geometry. This has solution

$$\mathbf{E} = \mathbf{e}^{\sim} \phi(y) \psi(x) e^{i\beta z}, \tag{8.52}$$

where $\mathbf{e}^{\sim}$ is the electric-field unit vector and β is the propagation constant in the dielectric.

The optical confinement factor,

$$\Gamma = \frac{\int\limits_{y_1}^{y_2} \phi^2(y)\,dy}{\int\limits_{-\infty}^{\infty} \phi^2(y)\,dy}, \tag{8.53}$$

is found by assuming that $\varepsilon(x, y)$ varies slowly in the x direction compared with the y direction and by adopting an effective index approximation. The solution is found for the simple slab waveguide geometry depicted in Fig. 8.8(a), in which the thickness of the active region is $t_c = y_2 - y_1$.

Figure 8.8(b) shows the results of calculating the optical confinement factor of TE and TM modes in a slab waveguide as a function of bulk active-layer thickness. The parameters used are typical for a laser diode with emission wavelength $\lambda_0 = 1\,310\,\text{nm}$ and an InGaAsP active region. For a given active region thickness, the optical confinement factor for TE polarization is greater than that for TM polarization. TE-polarized light propagating in the z direction has its electric field parallel to the x direction and so it is in the plane of the active-region layer.

For most Fabry–Perot laser designs that use index guiding, the ratio of active region thickness t_c to emission wavelength λ_0 is small, and the confinement factor for TE-polarized light may be found using the approximation[14]

$$\Gamma_{\text{TE}} = 2(n_a^2 - n_c^2)\left(\frac{\pi t_c}{\lambda_0}\right)^2. \tag{8.54}$$

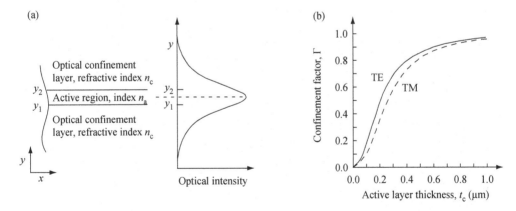

Fig. 8.8 (a) Slab waveguide geometry showing active region of thickness, $t_c = y_2 - y_1$, and optical confinement layers. The propagation of light is in the z direction, which is into the page. Optical intensity peaks in the active region, which has refractive index n_a. The refractive index of the optical confinement layer is n_c. (b) Calculated optical confinement factor Γ for TE and TM modes in a slab waveguide as a function of bulk active layer thickness, t_c. The lasing emission wavelength is $\lambda_0 = 1\,310\,\text{nm}$, the InGaAsP active layer has refractive index $n_a = n_{\text{InGaAsP}} = 3.51$, and the InP optical confinement layers have refractive index $n_c = n_{\text{InP}} = 3.22$.

[14]W. P. Dumpke, *IEEE J. Quantum Electron.* **QE-11**, 400 (1975).

In the case of TM-polarized light, the approximation for the confinement factor is

$$\Gamma_{TM} = 2(n_a^2 - n_c^2) \left(\frac{\pi n_c t_c}{n_a \lambda_0} \right)^2.$$

(8.55)

8.4.2 Mirror Loss and Photon Lifetime

A simple rate equation analysis shows that photon density grows exponentially in the presence of optical gain, $g_{opt} = -\alpha$, so that $S = S_0 e^{-2\alpha z}$. Note the factor 2 because S is an intensity.

To convert optical gain to a rate, the rate of increase in optical intensity, $G = 2g_{opt}c/n_r$, is introduced, where n_r is the effective refractive index and c is the velocity of light in vacuum. For steady-state emission in a Fabry–Perot cavity, the photons reflected back to the start in a single round-trip time $t_{\text{round-trip}} = 2Lcn_r/c$ must have the same density. So, if S_0 photons start out from mirror r_1 of the Fabry–Perot cavity, then $r_2 S_0 e^{(G \cdot t_{\text{round-trip}})/2}$ are reflected back from mirror r_2 to grow with another pass down the laser, and $r_1 r_2 S_0 e^{G \cdot t_{\text{round-trip}}}$ is reflected from mirror r_1. Hence, in steady-state,

$$S_0 = r_1 r_2 S_0 e^{G \cdot t_{\text{round-trip}}},$$

(8.56)

which, by taking the logarithm of both sides, may be rewritten as

$$G = \frac{1}{t_{\text{round-trip}}} \ln\left(\frac{1}{r_1 r_2} \right) = \frac{c}{2Lcn_r} \ln\left(\frac{1}{r_1 r_2} \right).$$

(8.57)

Ignoring spontaneous emission, a rate equation for the photon density is

$$\frac{dS}{dt} = \left(G - \frac{1}{\tau_{\text{photon}}} \right) S = 0,$$

(8.58)

giving $G\tau_{\text{photon}} = 1$, so that the photon loss rate due to the mirrors is

$$\boxed{\frac{1}{\tau_{\text{mirror}}} = \frac{c}{2Lcn_r} \ln\left(\frac{1}{r_1 r_2} \right).}$$

(8.59)

To include the possibility of additional loss mechanisms, including elastic scattering of light out of the cavity due to imperfections, an extra photon loss term, $1/\tau_{\text{internal}}$, can be introduced. Combining the loss rates together, the total photon loss rate is

$$\frac{1}{\tau_{\text{photon}}} = \frac{1}{\tau_{\text{internal}}} + \frac{1}{\tau_{\text{mirror}}},$$

(8.60)

or, equivalently,

$$\boxed{\kappa = \alpha_i + \alpha_m,}$$

(8.61)

where α_i is the internal photon loss rate, α_m is the photon mirror loss, and κ is the total optical loss rate.

8.4.3 The Fabry–Perot Laser Diode

Figure 8.9 is a sketch of a semiconductor, buried-heterostructure, Fabry–Perot laser diode. The diagram shows the bulk-active or quantum-well region exposed at one of the two cleaved-mirror faces. Carriers are injected into the region from the n-type substrate and the p-type epitaxially grown layers from below and above the p-n junction. Electrical contact to the diode is achieved by depositing a metal film and subsequent alloying into a surface layer of the semiconductor.

Typically, a bulk active InGaAsP has a composition such that its band gap is $\lambda_g = 1\,280$ nm. Under lasing conditions, various physical effects cause the lasing emission wavelength to increase to longer wavelength, so that the device lases at $\lambda_0 = 1\,310$ nm. In a typical device, the bulk or multiple quantum-well active region is 0.12 µm thick and 0.8 µm wide. The wafer is thinned before cleaving to form the two mirror facets. Thinning the wafer to about 120 µm thickness helps to ensure that stress-induced irregularities are avoided on the cleaved mirror faces. The buried heterostructure is achieved using an etching and semiconductor regrowth process. Index guiding of the $\lambda_0 = 1\,310$ nm lasing mode occurs because of the refractive index difference between the InGaAsP active layer with $n_{\text{InGaAsP}} = 3.51$ and the InP optical confinement layers with refractive index $n_{\text{InP}} = 3.22$. For an index-guided buried heterostructure, this ensures a single transverse mode and a high optical confinement of around $\Gamma = 0.25$. The Fabry–Perot cavity length is $L_C = 300$ µm. A multi-layer dielectric mirror coating is used to increase reflectivity to 0.4 on one mirror and 0.8 on the other. This reduces optical loss and reduces laser threshold current to a value that is typically around 3 mA.

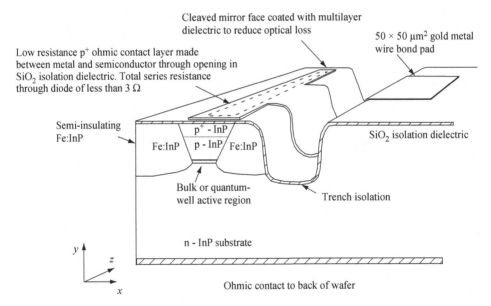

Fig. 8.9 Schematic diagram of a semiconductor buried-heterostructure Fabry–Perot laser diode. The difference in refractive index between the active region and the surrounding dielectric causes index guiding of light propagating in the z direction.

8.4.4 Semiconductor Laser Diode Rate Equations

To understand the operation of the semiconductor buried-heterostructure Fabry–Perot laser diode shown schematically in Figs. 8.9 and 8.10, a model must be developed. The simplest approach is to use rate equations to describe the behavior of the laser diode that is driven from thermodynamic equilibrium by an externally applied injection current, I_{inj}.

It is assumed that lasing occurs in only one optical mode of frequency ω_s. The calculations will not incorporate any variation in optical gain, optical loss, or carrier density along the longitudinal (z) axis. This is equivalent to a lumped-element model. Simple approximations for gain, spontaneous emission, and nonradiative recombination will be used. It is also assumed that the conduction-band and valence-band electrons have the same density and that they are thermalized so that they may be characterized by a single temperature.

When considering the rate equations, care must be taken to define the parameters used. The current, I_{inj}, injected into the diode is measured in amps, and the volume of the active region into which electrons flow is V_{vol}. The carrier density in the active region is n and is measured in either m^{-3} or cm^{-3}. The photon density in the optical mode of frequency ω_s is S. The two mirrors used to form the optical cavity for photons in the device have reflectivity r_1 and r_2, respectively. Optical loss at frequency ω_s in the cavity is κ and is measured in m^{-1} or cm^{-1}. The fraction of spontaneous emission, β, that feeds into the lasing mode at frequency ω_s must be defined. It turns out that this is somewhat difficult to do in an active device. In a large Fabry–Perot laser with $L_C = 300\,\mu m$, a typical value for β is in the range $10^{-4} < \beta < 10^{-5}$. In devices with smaller L_C, the value of β becomes larger. Theoretically, the maximum possible value of β is near unity.

Rate equations are used to describe measurable flux into and out of a region of interest. The physical quantities monitored are carrier density n, photon density S in a single optical mode of frequency ω_s, and injection current I_{inj} as a function of time, t. Rate equations keep track of current, carrier, and photon flow in and out of the device.

To illustrate the approach, a bucket may be drawn to represent the active region of the semiconductor (see Fig. 8.11). Charge carriers supplied by a current I_{inj} are poured into

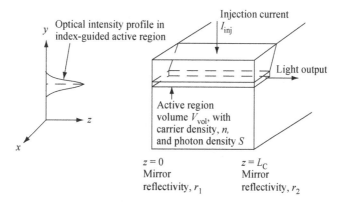

Fig. 8.10 Section through a buried-heterostructure Fabry–Perot laser diode. Index guiding ensures optical intensity is tightly confined near the active region of the device. Fabry–Perot cavity length is L_C, and mirror reflectivity is r_1 at $z = 0$ and r_2 at $z = L_C$. When current I_{inj} is injected into the diode, the active region of volume V_{vol} has carrier density n and photon density S.

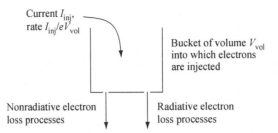

Current I_{inj},
rate $I_{\text{inj}}/eV_{\text{vol}}$

Bucket of volume V_{vol}
into which electrons
are injected

Nonradiative electron
loss processes

Radiative electron
loss processes

Fig. 8.11 Bucket model for electron rates into and out of a semiconductor active region.

the bucket at a rate so that the *density* of electrons per second increases as $I_{\text{inj}}/eV_{\text{vol}}$, where e is the electron charge and V_{vol} is the volume of the active region. There are losses or leaks in the bucket that represent mechanisms for removing electrons from the system. Electrons can be removed by emitting a photon or by some nonradiative process.

Because n and S are of interest, there are two coupled rate equations that must be solved. The first equation describes the rate of change in carrier density, dn/dt, in the device. Carrier density will increase as more current is injected, so dn/dt is expected to have a term proportional to injection current, I_{inj}. Carriers are removed from the active region of the device by nonradiative and spontaneous photon-emission carrier-recombination processes. This carrier loss is characterized by carrier lifetime τ_n, so that dn/dt should be proportional to $-n/\tau_n$. The negative sign reflects the fact that carriers are removed from the system. There are also carrier losses due to stimulated photon emission, in which an electron in the conduction band and a hole in the valence band are removed to create a photon in the lasing mode. This stimulated emission process influences the number of carriers via the rate $-GS$, where G is the optical gain and S is the photon density in the lasing mode.

The second equation describes the rate of change in photon density, dS/dt, in the device. Photon density increases due to the presence of optical gain, so a term proportional to GS is expected. There is also optical loss that can be described by a total optical loss rate κ, giving a term $-\kappa S$. Finally, there is a fraction β of total spontaneous emission r_{spon} feeding into the lasing mode that makes a contribution βr_{spon}.

These considerations allow the basic coupled rate equations that describe the behavior of the laser diode to be written as:

$$\frac{dn}{dt} = \frac{I_{\text{inj}}}{eV_{\text{vol}}} - \frac{n}{\tau_n} - GS, \tag{8.62}$$

$$\frac{dS}{dt} = (G - \kappa)S + \beta r_{\text{spon}}. \tag{8.63}$$

Equations (8.62) and (8.63) are the single-mode rate equations. In Eq. (8.62), $1/\tau_n$ is an experimentally determined carrier recombination rate, where $n/\tau_n = A_{\text{nr}}n + Bn^2 + Cn^3$, for which A_{nr} is the nonradiative recombination rate, B is the spontaneous emission coefficient, and C is a higher-order term. The total spontaneous emission rate in the device is $r_{\text{spon}} = Bn^2$, and the function for optical gain at mode frequency ω_s in a bulk–active region device is

$$G_{\text{bulk}} = \Gamma G_{\text{slope}}(n - n_{\text{ot}})(1 - \varepsilon_{\text{bulk}}S), \tag{8.64}$$

where G_{slope} is the differential optical gain with respect to carrier density. The carrier density needed to achieve optical transparency at frequency ω_s is n_{ot}.

For a quantum-well active region, the optical gain function at mode frequency ω_s may be phenomenologically approximated by

$$G_{\text{QW}} = \Gamma G_{\text{const}} \left(\ln \left(\frac{n}{n_{\text{ot}}} \right) \right) (1 - \varepsilon_{\text{QW}} S).$$ (8.65)

Note that, for convenience, the optical confinement factor Γ is included in the expressions for gain.

Both $\varepsilon_{\text{bulk}}$ and ε_{QW} are gain saturation terms, which become important at high optical intensities in the cavity. In practice, the gain functions used represent the variation of peak gain with increasing carrier density fairly well.

To explain the steady-state carrier density and photon density characteristics of a laser diode as a function of injected current, Eq. (8.62) for the steady-state case becomes

$$\frac{dn}{dt} = \frac{I_{\text{inj}}}{eV_{\text{vol}}} - \frac{n}{\tau_n} - GS = 0.$$ (8.66)

This equation shows that in steady state, the rate of electron density injected into the active region is exactly balanced by the removal of electrons via the carrier recombination n/τ_n and the optically stimulated recombination GS.

In steady state, Eq. (8.63) is

$$\frac{dS}{dt} = (G - \kappa)S + \beta r_{\text{spon}} = 0,$$ (8.67)

which may be rewritten as

$$S = \frac{\beta r_{\text{spon}}}{(\kappa - G)}.$$ (8.68)

It is this last equation that may be used as a starting point to explain how a laser works. The numerator βr_{spon} on the right-hand side of Eq. (8.68) shows that the optical output L_{out} of the laser amplifier is fed by a small fraction of the total spontaneous emission in the device. Because spontaneous emission is a stochastic (random) quantum-mechanical process, the laser may be viewed as amplifying noise. The denominator $(\kappa - G)$ on the right-hand side of Eq. (8.68) is the term responsible for optical amplification and lasing emission. As electrons are injected into the device, optical gain increases, and $(\kappa - G)$ approaches zero, the amplification of spontaneous emission increases. This increase in photon density in the device is so great that the stimulated carrier recombination term $-GS$ in Eq. (8.66) becomes large. As $(\kappa - G)$ continues to approach zero, the net optical amplification of spontaneous emission $1/(\kappa - G)$ becomes very large and the stimulated recombination $-GS$ dominates Eq. (8.66). At this point, every additional electron injected by current I_{inj} into the active region recombines very rapidly to create an additional photon in the lasing mode at frequency ω_s.

Figure 8.12 illustrates the situation as optical gain approaches a minimum in Fabry–Perot optical-cavity loss in such a way that $(G - \kappa) \to 0$. The lasing mode will coincide with that of the Fabry–Perot resonance of frequency ω_s nearest to peak optical gain. Because

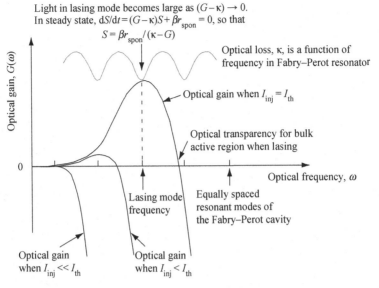

Fig. 8.12 Schematic plot of optical gain as a function of optical frequency. Lasing will occur when optical gain approaches optical loss. This will usually happen at a resonance of the Fabry–Perot resonator, the optical loss of which is also shown.

in this case only one high-Q optical-cavity resonance is in the same frequency range as semiconductor optical gain, lasing occurs in one optical mode at frequency ω_s.

As stimulated recombination begins to dominate, light emission at frequency ω_s rapidly increases. The point at which this occurs is called laser threshold. Associated with laser diode threshold is a threshold current, I_{th}, and a threshold carrier density, n_{th}. For currents above the laser threshold carrier density, n does not increase very much, because the

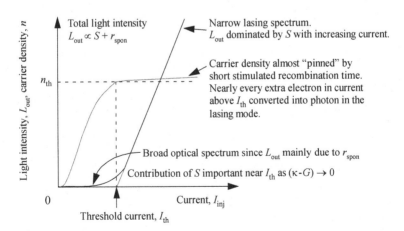

Fig. 8.13 Total light output intensity L_{out} and carrier density n as a function of injected current I_{inj}. Carrier density is pinned to a value of approximately n_{th} when current is greater than the threshold current I_{th}.

stimulated recombination $-GS$ dominates the carrier dynamics described by Eq. (8.66). The carrier density is said to be *pinned* above threshold. Such carrier pinning results in a rapid *linear* increase in laser light output intensity with increasing injection current because almost every extra injected electron is converted to a lasing photon.

Figure 8.13 shows total light output intensity L_{out} and carrier density n as a function of injected current I_{inj}. Carrier density is pinned to a value of approximately n_{th} when current is greater than the threshold current I_{th}.

8.5 Large-Signal Transient Response

While the single-mode laser diode rate equations can be used to predict the steady-state behavior of a laser diode, there is also interest in the high-speed, large-signal response of the device. Data transmission via an optical fiber medium requires a photon source in which lasing light intensity is modulated. A simple way to change lasing light emission is to change the injection current. Modulating in a one-bit digital fashion, a high level of light might correspond to binary 1 and a low level of light might correspond to binary 0. To efficiently pass data in a fiber-optic link, it is helpful to know how fast a laser can switch from a 1 state to a 0 state. To answer such basic questions concerning laser performance it is necessary to resort to numerical methods that are capable of predicting the large-signal

Table 8.1 Fabry–Perot laser diode rate equation parameters

Description	Parameter	GaAs 850 nm wavelength	InGaAsP 1310 nm wavelength	InGaAsP 1550 nm wavelength
Refractive index	n_r	3.3	4	4
Cavity length	L_C (cm)	250×10^{-4}	300×10^{-4}	500×10^{-4}
Active layer thickness	t_C (cm)	0.14×10^{-4}	0.14×10^{-4}	0.14×10^{-4}
Active layer width	w_C (cm)	0.8×10^{-4}	0.8×10^{-4}	0.8×10^{-4}
Integration time increment	t_{inc} (s)	1×10^{-12}	1×10^{-12}	1×10^{-12}
Nonradiative recombination coefficient	A_{nr} (s^{-1})	2×10^8	2×10^8	1×10^8
Radiative recombination coefficient	B(cm^3 s^{-1})	1×10^{-10}	1×10^{-10}	1×10^{-10}
Nonlinear recombination coefficient	C(cm^6 s^{-1})	1×10^{-29}	1×10^{-29}	5×10^{-29}
Transparency carrier density	n_{ot}(cm^{-3})	1×10^{18}	1×10^{18}	1×10^{18}
Optical gain-slope coefficient	G_{slope} (cm^2)	3.3×10^{-16}	2.5×10^{-16}	2.0×10^{-16}
Gain saturation coefficient	ε_{bulk} (cm^3)	2×10^{-18}	3×10^{-18}	5×10^{-18}
Spontaneous emission coefficient	β	1×10^{-4}	5×10^{-5}	1×10^{-5}
Optical confinement factor	Γ	0.25	0.25	0.25
Mirror 1 reflectivity	r_1	0.3	0.32	0.32
Mirror 2 reflectivity	r_2	0.3	0.32	0.32
Internal optical loss	α_i(cm^{-1})	20	40	50

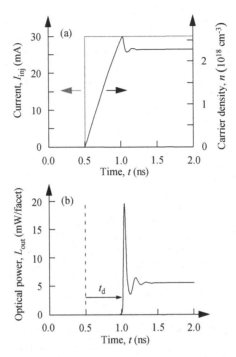

Fig. 8.14 (a) Input step current as a function of time. Current (left) is initially zero, increasing to a value $I_{\text{step}} = 30\,\text{mA}$ at time $t = 0.5\,\text{ns}$. Laser threshold is $I_{\text{th}} = 5.7\,\text{mA}$. Conduction-band and valence-band carrier density (right) as a function of time. The calculation uses parameters for an InGaAsP device as given in Table 8.1. (b) Laser diode light output into the lasing mode per mirror facet as a function of time, showing turn-on delay, t_{d}, and overshoot in optical light output. Mirror reflectivity is the same for each facet.

dynamic response of a laser diode's carrier density and laser light emission in response to rapid changes in injection current.

The coupled rate equations (Eqs. (8.62) and (8.63)) can be integrated using the fourth-order Runge–Kutta method.

When writing a computer program to model the behavior of a laser diode with a bulk active gain region, it is necessary to define the functions in the rate equations (Eqs. (8.62) and (8.63)), including assigning numerical values. Table 8.1 gives some typical values for the parameters introduced in Section 8.4.4.

In Fig. 8.14, the result of calculating the large-signal response of a diode laser to a 30 mA step change in current is shown. The model uses parameters given in Table 8.1 for a laser with emission wavelength near $\lambda_0 = 1\,310\,\text{nm}$.

Because the device is turned on from a zero-current state, there is a significant *turn-on delay*, t_d, associated with the fact that it takes time to inject enough carriers to bring the device to a lasing state.

As expected with coupled rate equations, there is a time delay between carrier density, n, and photon density, S. The electrons lead the photon density. This gives rise to substantial photon density overshoot and *relaxation oscillations* in both the optical output, L_{out}, and the carrier density, n, as the system tries to establish steady-state conditions. This is typical of a response to a large step change in injection current, especially if the injection current, I_{inj}, passes through the threshold value, I_{th}. Relaxation oscillations limit the useful switching speed of laser diodes.

8.5.1 Scaling with Spontaneous Emission Factor β

The fraction of spontaneous emission, β, feeding into the laser mode depends on the size of the active region, details of the cavity design, and laser operating conditions. Figure 8.15 shows the results of using continuum mean-field rate equations to calculate light output power, L_{out}, and carrier density, n, as a function of drive current for a typical Fabry–Perot laser diode. In Fig. 8.15(a), the laser diode threshold current, I_{th}, is determined by linearly extrapolating the high-slope portion of the curve to $L_{out} = 0$. The $n - I_{inj}$ plot shows that carrier density is pinned above threshold with a value that does not increase significantly with increasing current, I_{inj}. The log-log $L_{out} - I_{inj}$ curve shown in Fig. 8.15(b) is helpful in showing the behavior of below-threshold light level. For the parameters used, laser threshold is associated with a very rapid change in optical output power near $I_{inj} = I_{th}$.

The spontaneous emission factor β can be used to predict behavior as the laser diode geometry is scaled. Assuming the laser diode optical cavity may be reduced in size such that β increases but *no* other parameters change, Fig. 8.16 shows predicted (a) $L_{out} - I_{inj}$ and (b) $n - I_{inj}$ behavior on a log-log plot for different values of β. The laser threshold becomes less well defined as the device is reduced in size resulting in an increase in the spontaneous emission coefficient, β. When $\beta = 1$, all spontaneous emission feeds into the laser mode and there is no longer any evidence in the $L - I_{inj}$ characteristic of a definite transition from non-lasing to lasing. Similarly, carrier pinning above threshold disappears as β approaches unity. If the existence of a laser threshold implies a second-order nonequilibrium phase transition (from disordered light below threshold to ordered light above threshold) then, as β approaches unity, this transition is no longer well defined in the $L_{out} - I_{inj}$ or $n - I_{inj}$ behavior.

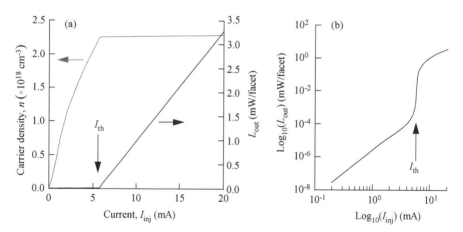

Fig. 8.15 (a) Calculated light emission into the lasing mode and carrier density in a Fabry–Perot laser diode plotted using linear scale. Laser threshold is $I_{th} = 5.7$ mA. (b) Light output power into the lasing mode versus injection current plotted using logarithmic scales. The calculation uses parameters for an InGaAsP device given in Table 8.1 with emission wavelength $\lambda_0 = 1\,310$ nm.

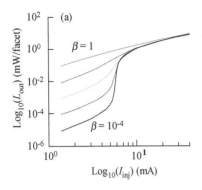
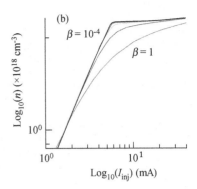

Fig. 8.16 (a) Calculated light emission into the lasing mode and (b) carrier density in a Fabry–Perot laser diode for values of spontaneous emission factor $\beta = 10^{-4}$, $\beta = 10^{-3}$, $\beta = 10^{-2}$, $\beta = 10^{-1}$, and $\beta = 1$ plotted using logarithmic scales. The calculation uses parameters for an InGaAsP device given in Table 8.1 with emission at $1\,310$ nm wavelength.

8.5.2 Critical Slowing

A time delay, t_d, exists between the onset of a step change in diode forward current from zero current at time $t = 0$ and laser light output L_{out} to reach half of the steady-state value. Figure 8.17 (a) plots calculated time delay, t_d, as a function of step current, I_{step}, for different values of spontaneous emission factor β. For small values of β there is an increase in delay for values of current near the threshold current, I_{th}. This characteristic behavior is called critical slowing and is associated with the existence of a phase transition. Laser threshold occurs at a second-order nonequilibrium phase transition in which the photon field acts as the order parameter[15] and critical slowing is due to long-lived fluctuations.[16] Formally, a laser threshold associated with a nonequilibrium phase transition can *only* exist in the large particle number limit (the thermodynamic limit) for which β is necessarily small. Increasing the value of β damps the critical slowing phenomena and, for a given step increase in injection current, reduces the off-on time delay, t_d. However, this does *not* mean that a laser diode with $\beta \sim 1$ has a faster small-signal on-on modulation response compared to a conventional device with $\beta \ll 1$. This is because as β becomes large, carriers are no longer pinned and so the recombination rate $-GS$ never dominates.

A plot of inverse delay time as a function of current is given in Fig. 8.17(b) and shows that $I_{step} \propto 1/t_d$ for current $I_{inj} > I_{th}$. The minimum value of the dip in $1/t_d$ at I_{th} is proportional to an energy gap that separates the sustained lasing and non-lasing states of the system.

[15]V. DeGiorgio and M. O. Scully, *Phys. Rev. A* **2**, 1170 (1970); R. Graham and H. Haken, *Z. Physik.* **237**, 31 (1970); S. Grossmann and P. H. Richter, *Z. Physik.* **242**, 458 (1971); M. Corti and V. DeGiorgio, *Phys. Rev. Lett.* **36**, 1173 (1976); V. B. Pakhalov and A. S. Chirkin, *Sov. J. Quantum Electron.* **7**, 715 (1977); R. Salomaa and S. Stenholm, *Appl. Phys.*, **14** 355 (1977); J. O. Gorman, A. F. J. Levi, S. Schmitt-Rink, T. Tanbun-Ek, D. L. Coblentz, and R. A. Logan, *Appl. Phys. Lett.* **60**, 157 (1992).
[16]H. Haken, *Rev. Mod. Phys.* **47**, 67 (1975).

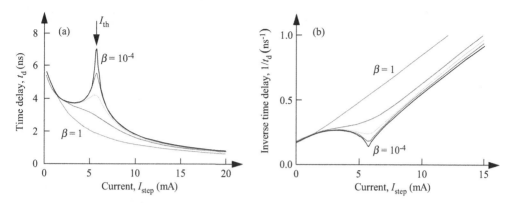

Fig. 8.17 (a) Time delay, t_d, as a function of step current, I_{step}, for values of spontaneous emission factor $\beta = 10^{-4}$, $\beta = 10^{-3}$, $\beta = 10^{-2}$, $\beta = 10^{-1}$, and $\beta = 1$. Critical slowing occurs near threshold current, $I_{th} = 5.7\,\text{mA}$. (b) Inverse of time delay, $1/t_d$, as a function of step current for the same values of spontaneous emission factor, β, used in (a). The calculation uses parameters for an InGaAsP device given in Table 8.1 with emission wavelength $\lambda_0 = 1\,310\,\text{nm}$.

8.5.3 Cavity Formation

Consider a photon inside a Fabry–Perot laser diode. The photon cannot know it is in a Fabry–Perot resonator or on resonance until it interacts with the mirrors. If there are no mirrors, the device is merely a light-emitting diode (LED). Hence, lasing emission into an optical cavity resonance requires that the photon experience at least one round trip within the resonator. The device cannot behave as a laser until the photon cavity has formed and this takes *at least* one photon round-trip time, which, in a conventional Fabry–Perot laser diode, is about 10 ps.

Predicting the effect multiple photon round trips have on the time evolution of lasing light emission intensity and spectra is not particularly easy to do because usually *photon cavity formation* is obscured by the nonlinear coupling of the optical field with the optical gain medium. Under normal conditions, it is difficult to measure the effect of multiple round-trips on the evolution of lasing light intensity and lasing spectra due to the short cavity round-trip time and charge carrier lifetime. However, by adiabatically decoupling the cavity formation (by making a large external cavity) from other processes such as charge carrier dynamics, experiments can be performed that explore this issue.[17] The results can also be predicted using time-delayed, single-mode, or multi-mode rate equations. Importantly, the experiments show how a laser uses cavity formation to drive the device from an LED to a laser. Surprisingly, approximately 200 photon round trips are needed to approach steady-state laser characteristics that are *independent* of the laser injection current (see Fig. 8.18). Obtaining pure steady-state spectral behavior requires even more photon round trips.

Continuum mean-field single-mode time-delayed rate equations can be used to model transient optical intensity build up in a laser with a passive external cavity. The inset in Fig. 8.19(a) shows the device configuration consisting of a semiconductor diode gain

---

[17]J. O'Gorman, A. F. J. Levi, D. Coblentz, T. Tanbun-Ek, and R. A. Logan, *Appl. Phys. Lett.* **61**, 889 (1992).

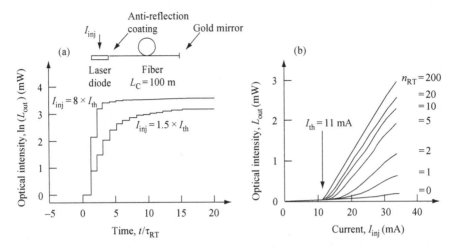

Fig. 8.18 (a) Natural logarithm of laser emission intensity as a function of time, t/τ_{RT}. Time is measured in units of the external cavity round-trip time $\tau_{\mathrm{RT}} = 0.995\,\mu\mathrm{s}$. Inset shows the experimental arrangement. The laser diode is the active region of a 100 m long Fabry–Perot resonator formed using a glass fiber. The diode is pumped by a step change in current, I_{inj}, and the optical output, L_{out}, is monitored as a function of cavity round-trip time using a high-speed photodetector. (b) Optical output power as a function of injection current for the indicated number of photon round trips, n_{RT}. When $n_{\mathrm{RT}} = 0$, the device behaves as an LED. When $n_{\mathrm{RT}} = 200$, the current–light output characteristics approach the steady-state behavior of a laser with threshold current $I_{\mathrm{th}} = 11\,\mathrm{mA}$.

medium of length L and refractive index n_{r} coupled to an external fiber cavity with photon round-trip time of $\tau_{\mathrm{RT}} = 10\,\mathrm{ns}$. At time $t = 0$, there is a step change in injection current from $I_{\mathrm{inj}} = 0\,\mathrm{mA}$ to $I_{\mathrm{inj}} = 30\,\mathrm{mA}$. For time $t > \tau_{\mathrm{RT}}$, a fraction of the photons emitted into the external cavity will be reintroduced into the semiconductor gain medium and some will be reflected from mirror 2 and remain in the external cavity. To take into account photons delayed by an integer number of round-trips, n_{RT}, before being reinjected into the active gain medium, the single mode rate equations can be modified to

$$\frac{\mathrm{d}S}{\mathrm{d}t} = (G - \kappa)S + \beta r_{\mathrm{spon}} + \kappa_2 \left(\sum_{j=1}^{n_{\mathrm{RT}}} (1 - r_2) r_2^{(j-1)} r_3^j\, S(t - j\tau_{\mathrm{RT}}) \right) \Theta(t - \tau_{\mathrm{RT}}), \qquad (8.69)$$

where S is the photon density in the semiconductor gain medium and the rate of photon loss from mirror 2 is given by

$$\kappa_2 = \frac{1 - r_2}{2 - r_1 - r_2} \left(\frac{c}{2Ln_{\mathrm{r}}} \right) \ln \left(\frac{1}{r_1 r_2} \right). \qquad (8.70)$$

In this expression, reflectivity of mirrors 1 and 2 are r_1 and r_2, respectively. If it is assumed that reflectivity at mirror 2 is very small, i.e., $r_2 \ll 1$, then only the first term in the sum in Eq. (8.69) need be retained and the delayed rate equation for photon density becomes

$$\frac{\mathrm{d}S}{\mathrm{d}t} = (G - \kappa)S + \beta r_{\mathrm{spon}} + (1 - r_2) r_3 \kappa_2 S(t - \tau_{\mathrm{RT}}). \qquad (8.71)$$

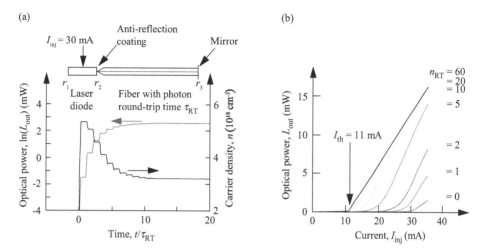

Fig. 8.19 (a) Natural logarithm of calculated laser emission intensity, L_{out}, from the mirror with reflectivity r_1 as a function of time, t/τ_{RT}. Time is measured in units of the external cavity round-trip time $\tau_{RT} = 10$ ns. Light output power, L_{out}, is monitored as a function of the cavity round-trip time using a high-speed photodetector. Calculated carrier density, n, as a function of time is also shown in the figure. Inset shows the device configuration and laser diode with step change in injection current of $I_{inj} = 30$ mA. (b) Calculated light output power as a function of injection current for the indicated number of photon round-trip trips, n_{RT}. When $n_{RT} = 0$, the device behaves as an LED. When $n_{RT} = 60$, the current-light output characteristics approach the steady-state behavior of a laser diode with threshold current $I_{th} = 11$ mA. Parameters are those of the laser diode with emission wavelength $\lambda_0 = 1\,310$ nm given in Table 8.1 with mirror reflectivity $r_1 = 0.32$, $r_2 = 2 \times 10^{-6}$, and $r_3 = 1$.

Figure 8.19(a) shows calculated laser emission intensity as a function of time, t/τ_{RT}, using this approximation. Also shown is the calculated carrier density as a function of time. Both optical emission and carrier density have a time dependence that is determined by the external cavity round-trip time $\tau_{RT} = 10$ ns.

Figure 8.19(b) shows results of calculating optical output power as a function of injection current, I_{inj}, for the indicated number of photon round-trip trips, n_{RT}. When $n_{RT} = 0$, the device behaves as an LED. When $n_{RT} = 60$, the current-light output characteristics approach the steady-state behavior of a laser diode with threshold current $I_{th} = 11$ mA. The qualitative agreement between the calculations shown in Fig. 8.19(b) and the experimental results shown in Fig. 8.18 serves to validate the approach taken to model the device.

8.6 Noise in Laser Diode Light Emission

A laser works by amplifying spontaneous emission. Spontaneous emission is a *fundamentally quantum-mechanical* and *random process*. The total energy due to all photons in the single mode is $E = \hbar(n + \frac{1}{2})$, where n is the number of photons in the field. The photon field emission produced when the laser diode injection current is above the lasing threshold value may be described as a coherent state of a harmonic oscillator of frequency ω. Every time an additional photon is added to the system so is an additional quanta of

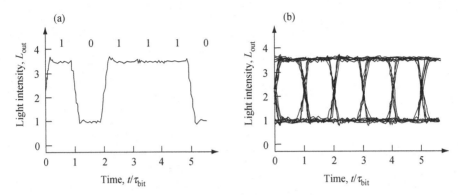

Fig. 8.20 (a) Intensity-modulated laser light transmits bits of information in which a high level of light is binary 1 and a low level of light is binary 0. Laser intensity noise is a source of bit errors. (b) A continuously sampled datastream creates an eye diagram in which the eye width and height define the area in which the receiver circuitry can minimize the probability of bit errors.

energy, $\hbar\omega$. The uncertainty relation between photon number and photon field phase is $\Delta n \Delta \theta = \frac{1}{2}$. The frequency, intensity, and phase purity of single-mode laser diode light emission may be exploited for fiber-optic communication systems as well as many other applications.

Because, in its most primitive form, fiber-optic communication systems use intensity modulation of laser light, there is interest in developing accurate models of intensity noise. Figure 8.20(a) shows how intensity-modulated laser light can be used to transmit bits of information. In this case, a high level of light is taken to represent binary 1 and a low level of light is binary 0. Because the receiver circuitry needs time to decide between high and low light levels, data is transmitted at a well-defined rate called the bit rate, $1/\tau_{\mathrm{bit}}$. A typical bit rate in fiber-optic communication systems is $10\,\mathrm{Gb\,s^{-1}}$. In this case, $\tau_{\mathrm{bit}} = 100\,\mathrm{ps}$ and 10^{10} bits of information can be transmitted through the system in one second. With such large information capacity, it is important to minimize the chance of bit errors due to laser intensity noise.

If a binary data stream is randomly and continuously sampled in time while synchronized to the bit rate, the result is the *eye diagram* of the type shown in Fig. 8.20(b). The opening in the center of the eye diagram is where the receiver circuitry decides between high and low light levels. The greater the amount of intensity noise in the laser signal, the smaller the region in which the receiver can make its decision and the greater the chance of errors.

8.6.1 Relative Intensity Noise (RIN)

Noise in the intensity of light output from a laser diode is characterized by relative intensity noise (RIN). The optical intensity S contains noise, so that

$$S(t) = \langle S \rangle + \delta S(t), \tag{8.72}$$

where $\langle S \rangle$ is the time-averaged optical intensity and $\delta S(t)$ is the deviation from the average value at any given instant in time, t. The time average in the fluctuation $\delta S(t)$ is $\langle \delta S(t) \rangle = 0$. The optical noise may be characterized in the time domain by the autocorrelation function,

$$g_s(\tau) = \langle \delta S(t) \delta S(t - \tau) \rangle, \tag{8.73}$$

or in the frequency domain by the noise spectral density

$$\left\langle |\delta S(\omega)|^2 \right\rangle = \int_{\tau=-\infty}^{\tau=\infty} g_s(\tau) e^{-i\omega\tau} d\tau = \frac{1}{t'}\bigg|_{t'\to\infty} \left| \int_0^{t'} \delta S(t) e^{-i\omega t} dt \right|^2. \tag{8.74}$$

Optical relative intensity noise is defined as the noise spectral density at frequency ω divided by the average value of optical intensity, $\langle S \rangle = S_0$, squared, so that

$$\mathrm{RIN}(\omega) = \frac{\left\langle |\delta S(\omega)|^2 \right\rangle}{\langle S \rangle^2} = \frac{\left\langle |\delta S(\omega)|^2 \right\rangle}{S_0^2}. \tag{8.75}$$

Relative intensity noise is measured either in units of $\mathrm{dB\,Hz}^{-1}$ or Hz^{-1}.

A high-speed photodetector converts photons to electrical current, $i(t)$. Photon flux is proportional to electric current and so the fluctuation $\Delta S(t) \propto i(t)$. Because electrical *power* across load resistor R_{L} is $|i(t)|^2 R_{\mathrm{L}}$, it follows that the average noise power spectral density per unit frequency in the photodetector electrical current is

$$\delta P(\omega) = \langle |i(\omega)|^2 \rangle R_{\mathrm{L}} \propto \langle |\delta S(\omega)|^2 \rangle \tag{8.76}$$

and the average power in the electrical current is

$$\langle P \rangle = P_0 \propto \langle S \rangle^2. \tag{8.77}$$

Hence, RIN measured using a perfect photodetector and a perfect electrical spectrum analyzer is simply

$$\mathrm{RIN}(\omega) = \frac{\delta P(\omega)}{P_0}. \tag{8.78}$$

8.6.2 Shot-Noise Limit to RIN

The relative intensity noise due to the discrete quantum nature of the photon gives a shot-noise contribution to RIN. A time-average optical power, P_0, detected by the photodetector consists of a flux of photons of energy $\hbar\omega$ with shot-noise RIN of

$$\mathrm{RIN}_{\mathrm{shot\text{-}noise}} = \frac{2\hbar\omega}{P_0}. \tag{8.79}$$

The factor of two takes into account both positive and negative frequency contributions because it is assumed that the photodetector is attached to a spectrum analyzer that measures a single-sided frequency spectrum.

As an example, $P_0 = 1\,\mathrm{mW}$ laser light of emission wavelength $\lambda_0 = 1310\,\mathrm{nm}$ has $\mathrm{RIN}_{\mathrm{shot\text{-}noise}} = 3 \times 10^{-16}\,\mathrm{Hz}^{-1} = -155\,\mathrm{dB\,Hz}^{-1}$.[18] Increasing P_0 reduces $\mathrm{RIN}_{\mathrm{shot\text{-}noise}}$. Assuming a broad (white) spectrum for shot noise, $\mathrm{RIN}_{\mathrm{shot\text{-}noise}}$ is, in practice, a noise-floor.

---

[18]$10 \times \log_{10}\left(3 \times 10^{-16}\right) = -155.$

8.6.3 Langevin Intensity Rate Equations

Relative noise intensity may be investigated theoretically using a slight modification of the single-mode rate equations (Eqs. (8.62) and (8.63)). The *Langevin* rate equations are

$$\frac{dn}{dt} = \frac{I_{inj}}{eV_{vol}} - \frac{n}{\tau_n} - GS + F_e(t) \tag{8.80}$$

and

$$\frac{dS}{dt} = (G - \kappa)S + \beta r_{spon} + F_s(t), \tag{8.81}$$

where S and n are the photon and carrier density in the cavity, respectively, G is optical gain, κ is optical loss, β is the spontaneous emission factor, n/τ_n is the carrier recombination rate, e is the charge of an electron, r_{spon} accounts for spontaneous emission into all optical modes, and V_{vol} is the volume of the semiconductor active region. A source of random noise in the rate equations is included through the terms $F_s(t)$ and $F_e(t)$. These are called the Langevin noise terms. Equations (8.80) and (8.81) are the simplest way to include noise in a model of a laser diode. A slightly more detailed approach also takes into account optical phase noise.[19] In the Markovian approximation, corresponding to instantaneous changes in $F_s(t)$ and $F_e(t)$, the autocorrelation and cross-correlation functions are given by

$$\langle F_e(t)F_e(t')\rangle = \left(\frac{I_{inj}}{eV_{vol}} + GS + \frac{n}{\tau_n}\right)\delta(t - t'), \tag{8.82}$$

$$\langle F_s(t)F_s(t')\rangle = ((G + \kappa)S + \beta r_{spon})\delta(t - t'), \tag{8.83}$$

$$\langle F_s(t)F_e(t')\rangle = -(GS + \beta r_{spon})\delta(t - t'). \tag{8.84}$$

The Markovian approximation is guaranteed by use of $\delta(t - t')$, $\langle F_e(t)F_e(t')\rangle$ is the square of Gaussian fluctuations around the mean value of n given by the rate equation, Eq. (8.62), and $\langle F_s(t)F_s(t')\rangle$ is just the square of Gaussian fluctuations around the mean value of S given by the rate equation, Eq. (8.63). The cross-correlation term $\langle F_s(t)F_e(t')\rangle$ shows that the rate equations for S and n are coupled and, hence, correlated. The negative sign in Eq. (8.84) indicates that $F_s(t)$ and $F_e(t)$ are anti-correlated.

A peak in RIN occurs near the relaxation oscillation frequency and is due to fluctuations in carrier density and photon density working in phase to amplify the response to a noise fluctuation.

8.7 Why the Model Works

It is truly remarkable that the simple model used throughout this chapter to describe the behavior of semiconductor lasers gives useful results. Engineers can use this, and slightly improved versions of the model, to successfully design and optimize the performance of laser diodes. The fact that this is so is largely accidental. A great deal more needs to be included in the model to understand detailed physical processes in a device.

---

[19]For a somewhat more complete model, see M. Ahmed, M. Yamada, and M. Saito, *IEEE J. Quantum Electron.* **QE-37**, 1600 (2001).

For example, the low-temperature energy dependence of the low-carrier density optical absorption in GaAs is very different from the simple square-root behavior predicted by Eq. (8.34). An electron in the conduction band is attracted by way of the coulomb interaction to a hole in the valence band and can form a correlated state called an exciton.[20] Absorption due to excitons can give rise to a spectrally sharp absorption peak for photons with energy *less* than the semiconductor band gap. At room temperature, this peak is broadened by thermal processes, but still makes a contribution to absorption for photon energies near the band-gap energy.

For carrier densities at levels necessary to achieve useful optical gain in a laser diode, electron scattering broadens electron energy levels in the semiconductor on an energy scale that is comparable to the measure of optical gain bandwidth, $\Delta\mu_{e-hh}$. In addition, the relatively high carrier density screens the coulomb interaction. This reduces the ability of the system to form excitons. High carrier density can also reduce the value of the band-gap energy.

While these and other high-carrier-density effects make it very hard to create a detailed physical model of optical gain in a semiconductor, it is also responsible for the success of a naive approach. Associated with high carrier density and room-temperature operation are energy broadening effects that quite accidently and serendipitously conspire to turn the simple model into something that is useful. The fact remains, however, that while the model gives results that may be used to design lasers, the model itself is physically incorrect.

Relying too heavily on a simple model can result in misunderstanding and misinterpretation of device behavior.

8.8 Example Exercises

Exercise 8.1
Write a computer program to calculate spontaneous emission and optical gain in a bulk direct band-gap semiconductor as a function of photon energy $\hbar\omega$ using Eqs. (8.38) and (8.39). Plot results for GaAs with carrier concentration $n = j \times 10^{18} \, \mathrm{cm}^{-3}$, where $j = 1, 2, \ldots, 10$, temperature $T = 300 \, \mathrm{K}$, band-gap energy $E_g = 1.4 \, \mathrm{eV}$, and constant $g_0 = 2.64 \times 10^4 \, \mathrm{cm}^{-1} \, \mathrm{eV}^{-1/2}$.

Exercise 8.2
Repeat the calculation of Exercise 8.1, but now plot difference in chemical potential, $\Delta\mu_{e-hh}$, peak optical gain, g_{peak}, and total spontaneous emission, $r_{spon-total}$ as functions of carrier density in the range $n = 10^{18} \, \mathrm{cm}^{-3}$ to $n = 10^{19} \, \mathrm{cm}^{-3}$. What happens to these functions if the active region is a two-dimensional quantum well?

Exercise 8.3
Plot spontaneous emission using the Lorentzian broadening function (Eq. (8.40)) to simulate the effect of electron–electron scattering. Use the parameters $n = 2 \times 10^{18} \, \mathrm{cm}^{-3}$, $T = 300 \, \mathrm{K}$, $E_g = 1.4 \, \mathrm{eV}$, $n_g = 3.3$, $\gamma_k = 15 \, \mathrm{meV}$, and the relationship

[20]This is just one of a number of different optically induced excitations that contribute to interesting optical processes in semiconductors. For a review of such phenomena, see the contribution by D. S. Chemla in the series *Semiconductors and Semimetals* **58**, edited by R. K. Willardson and E. R. Weber, New York, Academic Press, 1999 (ISBN 0 12 752167 5).

$$g_{opt}(\hbar\omega) = -\alpha_{opt}(\hbar\omega) = \hbar \left(\frac{c\pi}{n_r\omega}\right)^2 r_{spon}(\hbar\omega) \left(1 - e^{(\hbar\omega - E_g - \Delta\mu_{e-hh})/k_B T}\right)$$

to calculate optical gain as a function of photon energy $\hbar\omega$. Assume that $g_0 = 2.64 \times 10^4 \, \text{cm}^{-1} \, \text{eV}^{-1/2}$.

Exercise 8.4
A Fabry–Perot laser diode has a bulk active region 300 μm long, 0.8 μm wide, and 0.14 μm thick. The laser has emission wavelength $\lambda_0 = 1310$ nm, internal optical loss of $40 \, \text{cm}^{-1}$, an optical confinement factor of $\Gamma = 0.25$, a mirror reflectivity of 0.32, and a spontaneous emission factor of $\beta = 10^{-4}$. Optical transparency occurs at carrier density $n_{ot} = 10^{18} \, \text{cm}^{-3}$, the refractive index of the semiconductor is $n_r = 4.0$, and the peak optical gain at carrier density n is $g_{opt} = g_{slope}(n - n_{ot})(1 - \varepsilon S)$, where $g_{slope} = 2.5 \times 10^{-16} \, \text{cm}^2 \, \text{s}^{-1}$, $\varepsilon = 5 \times 10^{-18} \, \text{cm}^3$, and S is the photon density.

Write a computer program that uses the fourth-order Runge–Kutta method to solve the rate equations for the device. Then plot light output, L_{out}, as a function of time; carrier density, n, as a function of time; and output power as a function of carrier density for a step current of 20 mA. Assume a nonradiative carrier recombination rate $A_{nr} = 2 \times 10^8 \, \text{s}^{-1}$, a radiative carrier recombination rate coefficient $B = 1 \times 10^{-10} \, \text{cm}^3 \, \text{s}^{-1}$, and a nonlinear carrier recombination rate coefficient $C = 1 \times 10^{-29} \, \text{cm}^6 \, \text{s}^{-1}$.

Solutions

Solutions 8.1
Figure 8.E21 shows the result of calculating spontaneous emission and optical gain in a bulk direct band-gap semiconductor as a function of photon energy $\hbar\omega$ using Eqs. (8.38) and (8.39). In the figure, carrier concentration $n = j \times 10^{18} \, \text{cm}^{-3}$, where $j =$

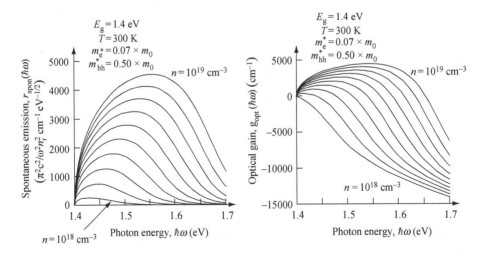

Fig. 8.E21

$1, 2, \ldots, 10$, temperature $T = 300\,\mathrm{K}$, and band-gap energy $E_\mathrm{g} = 1.4\,\mathrm{eV}$. Optical gain g_opt $(\hbar\omega)$ is calculated using the constant $g_0 = 2.64 \times 10^4\,\mathrm{cm}^{-1}\,\mathrm{eV}^{-1/2}$. Because $g_0/r_0 = \pi^2 c^2/\omega^2 n_\mathrm{r}^2$, it is not necessary to specify r_0.

The MATLAB computer program to calculate the figures calls the algorithm (in this case a function called mu, which in turn uses the function fermi) developed in Exercise 6.2 to determine the chemical potential for electrons and holes.

Solutions 8.2

The calculation of Exercise 8.1 is repeated, but now difference in chemical potential, $\Delta\mu_\mathrm{e-hh}$, peak optical gain, g_peak, and total spontaneous emission, $r_\mathrm{spon-total}$, is plotted as functions of carrier density in the range $n = 10^{18}\,\mathrm{cm}^{-3}$ to $n = 10^{19}\,\mathrm{cm}^{-3}$. Figure 8.E22 shows the results of performing the calculations. Notice how peak optical gain increases essentially linearly with increasing carrier concentration, n. This justifies the use of the linear approximation for optical gain given by Eq. (8.64). The total spontaneous emission increases faster than linear with increasing carrier concentration, supporting the use of the approximation $r_\mathrm{spon-total} = Bn^2$.

If the active region is a two-dimensional quantum well then the energy dependence of the electronic density of states per sub-band is a constant. This changes how the difference

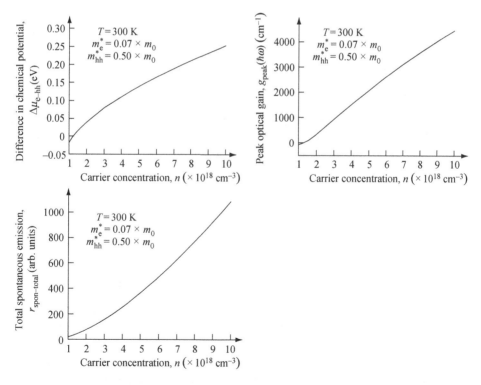

Fig. 8.E22

in chemical potential, $\Delta\mu_{e-hh}$, peak optical gain, g_{peak}, and total spontaneous emission, $r_{spon-total}$, depend upon carrier density. Peak gain as a function of n can be approximated by the logarithmic function given by Eq. (8.65).

Solutions 8.3

In this exercise a Lorentzian broadening function is used to simulate the effect of electron–electron scattering and spontaneous emission (Eq. (8.40)) is plotted for the parameters $n = 2 \times 10^{18}$ cm^{-3}, $T = 300$ K, $E_g = 1.4$ eV, $n_g = 3.3$, and $\gamma_k = 15$ meV. The relationship

$$g_{opt}(\hbar\omega) = -\alpha_{opt}(\hbar\omega) = \hbar\left(\frac{c\pi}{n_r\omega}\right)^2 r_{spon}(\hbar\omega)\left(1 - e^{(\hbar\omega - E_g - \Delta\mu_{e-hh})/k_BT}\right)$$

is used to calculate optical gain as a function of photon energy $\hbar\omega$, assuming that $g_0 = 2.64 \times 10^4$ cm^{-1} eV$^{-1/2}$.

The computer program used to do this exploits what was developed in Exercise 8.1. The additional complication is the inclusion of the Lorentzian broadening function. The results shown in Fig. 8.E23 are indicative of what happens to the spontaneous emission and gain spectrum when γ_k is included. The peak values of both r_{spon} and g_{opt} decrease and there is a low-energy tail that extends into the band-gap region.

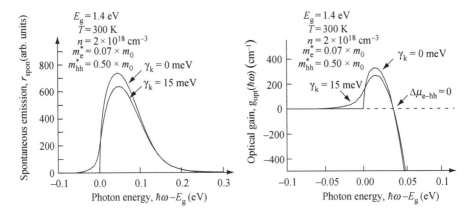

Fig. 8.E23

Solutions 8.4

In this exercise, the task is to write a computer program that simulates the large-signal response of a Fabry–Perot laser diode using the parameters provided.

Figure 8.E24 plots (clockwise from top left): step injection current I_{inj}; light output, L_{out}, from one mirror facet as a function of time; carrier density, n, as a function of time, t; and light output power as a function of carrier density for a step-current of 20 mA. Notice the relaxation oscillations in both the L_{out}–t and n–t plots (top-right and bottom-right plots, respectively). These oscillations arise from the response of the system to a large-signal step change in current. There is a phase lag between the photons and the carriers during the transient that can be seen in the plots for L_{out}–n (bottom left).

394

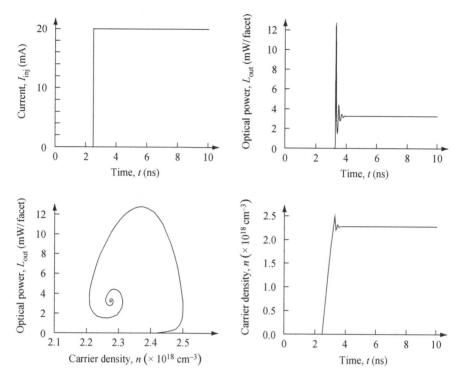

Fig. 8.E24

The MATLAB computer program is written in two parts so that the Runge–Kutta integrator can be called from the main program.

8.9 Problems

Problem 8.1
A new type of nanowire laser is to be designed with optical emission wavelength $\lambda_0 = 1\,550$ nm. The laser consists of an InGaAs semiconductor active region with refractive index $n_{\mathrm{InGaAs}} = 4$ inside a Fabry–Perot cavity of length L_C. The InGaAs is of width 40 nm and of thickness 40 nm. The sides of the wire are embedded in InP which has a refractive index of $n_{\mathrm{InP}} = 3.22$. One mirror end of the wire is faced with gold of reflectivity 0.99 and the other end is bonded to sapphire (Al$_2$O$_3$), which has refractive index $n_{\mathrm{Al_2O_3}} = 1.78$.

(a) Estimate the optical confinement factor of the structure.

(b) Find the reflectivity of the mirror formed at the semiconductor–sapphire interface.

(c) Assume a spontaneous emission coefficient $\beta = 10^{-4}$, optical confinement factor $\Gamma = 0.1$, nonradiative recombination rate $A_{\mathrm{nr}} = 10^8\,\mathrm{s}^{-1}$, radiative recombination coefficient $B = 10^{-10}\,\mathrm{cm}^3\,\mathrm{s}^{-1}$, nonlinear recombination coefficient $C = 5 \times 10^{-29}\,\mathrm{cm}^6\,\mathrm{s}^{-1}$, a bulk gain model with optical transparency at carrier density $n_{\mathrm{ot}} = 10^{18}\,\mathrm{cm}^{-3}$, optical gain

395

slope coefficient $G_{slope} = 2 \times 10^{-16}\,cm^2\,s^{-1}$, optical gain saturation coefficient $\varepsilon_{bulk} = 5 \times 10^{-18}\,cm^3$, and internal optical loss $\alpha_i = 10\,cm^{-1}$. Calculate the $L_{out} - I_{inj}$ characteristics and threshold current for a wire of length $L_C = 10\,\mu m$, $L_C = 30\,\mu m$, and $L_C = 100\,\mu m$.

(d) Discuss how to decrease the laser threshold current.

Problem 8.2
Driving a laser with a short electrical pulse can generate an optical pulse.

(a) Using the laser diode described in Exercise 8.4, plot the FWHM of laser optical pulse output as a function of electrical pulse width. The electrical pulse has a rectangular shape with maximum value $I_{max} = 50\,mA$, minimum value $I_{min} = 0\,mA$, and width τ.

(b) What happens to the minimum optical pulse width if I_{min} is allowed to increase such that $I_{max} > I_{min} > 0\,mA$?

(c) Find values of I_{min}, I_{max}, and τ that, on average, result in emission of one photon per pulse.

Problem 8.3
(a) Plot the noise-free transient and steady-state behavior for the laser diode of Exercise 8.4 but with each mirror having reflectivity 0.999, active region volume $2 \times 1 \times 0.2\,\mu m^3$ with cavity length $2\,\mu m$, and a step current of $100\,\mu A$.

(b) Repeat (a) only now in the presence of uncorrelated Gaussian photon and carrier number noise, i.e., use Eq. (8.82) and Eq. (8.83), but ignore Eq. (8.84).

(c) Use (b) to numerically find the mean and standard deviation of photon number for times long after the transient associated with the step current being turned on.

Problem 8.4
Modify the computer program used in Exercise 8.4 to demonstrate cavity formation, as illustrated in Fig. 8.18. Apply delayed rate equations using the same device parameters as in Exercise 8.4, but with a 1 ns round-trip photon delay from a 100% reflecting external cavity mirror as indicated in Fig. 8.P25.

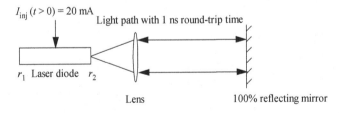

Fig. 8.P25

Problem 8.5
The probability of s discrete photons and n electrons in a laser diode is $P_{n,s}$.

(a) Write down an equation for the time evolution of $P_{n,s}$.

(b) Write a computer program to find $P_{n,s}$ as a function of time and sketch how you expect $P_{n,s}$ to evolve as the injection current in a laser diode is increased.

Problem 8.6

In a bulk semiconductor, optical gain medium electron scattering times are short compared to the spontaneous emission time and carriers relax to a quasi-equilibrium described by the Fermi–Dirac distribution, $f_{\mathbf{k}}$. When carrier density, n, is small (and/or temperature, T, is high) such that the chemical potential $\mu \ll E_F$, where E_F is the Fermi energy, then $f_{\mathbf{k}}$ may be approximated by a Maxwell–Boltzmann distribution. Assuming a reduced electron mass m_r^* such that $1/m_r^* = 1/m_e^* + 1/m_{hh}^*$, where m_e^* is the conduction band electron effective mass and m_{hh}^* is the heavy hole effective mass, show that the total photon spontaneous emission rate is proportional to n^2 and has temperature dependence $T^{3/2}$. How is the n^2 dependence of *total* photon spontaneous emission rate modified when n becomes large?

Problem 8.7

Exercise 8.3 illustrates the influence of a Lorentzian broadening function used to model the effect of finite electron–electron scattering time on the optical spontaneous emission and gain spectrum. The spectral broadening $\gamma_{\mathbf{k}}$ is included in the spontaneous emission (Eq. (8.38)) and the transformation to optical gain (Eq. (8.33)) guarantees that optical transparency occurs when photon energy equals the difference in chemical potential $\Delta\mu$. The parameters are those for GaAs with electron effective mass $m_e^* = 0.07 \times m_0$, heavy hole effective mass $m_{hh}^* = 0.5 \times m_0$, carrier density $n = 2 \times 10^{18} \text{ cm}^{-3}$, absolute temperature $T = 300 \text{ K}$, band-gap energy $E_g = 1.4 \text{ eV}$, $n_r = 3.3$, $\gamma_{\mathbf{k}} = 15 \text{ meV}$, and $g_0 = 2.64 \times 10^4 \text{ cm}^{-1} \text{ eV}^{-1/2}$.

(a) Reproduce the solution to Exercise 8.3 and on the same scale add a plot of optical gain using Eq. (8.39) with Lorentzian spectral broadening and $\gamma_{\mathbf{k}} = 15 \text{ meV}$. Comment on the value of optical gain for photon energies $\hbar\omega < E_g$ and when $\hbar\omega = \Delta\mu$.

(b) Add to (a) a plot of optical gain using Eq. (8.33) with a broadening function

$$\frac{1}{\pi\gamma_{\mathbf{k}}} \text{sech}\left(\frac{E - \hbar\omega}{\gamma_{\mathbf{k}}}\right).$$

Comment on the value of optical gain for photon energies $\hbar\omega < E_g$ and when $\hbar\omega = \Delta\mu$.

Problem 8.8

The fourth-order Runge–Kutta method with uniform time step, t_{step}, can be used to integrate the Langevin single-mode laser diode rate-equations,

$$\frac{dn}{dt} = \frac{I_{\text{inj}}}{eV_{\text{vol}}} - \frac{n}{\tau_n} - GS + F_e(t)$$

and

$$\frac{dS}{dt} = (G - \kappa)S + \beta r_{\text{spon}} + F_s(t),$$

where a normalized Gaussian random number generator (randn in MATLAB) gives

$$F_e(t) = \text{randn} \times \sqrt{\frac{\frac{I_{\text{inj}}}{eV_{\text{vol}}} + GS + \frac{n}{\tau_n}}{t_{\text{step}}}},$$

397

$$F_s(t) = \text{randn} \times \sqrt{\frac{(G + \kappa)S + \beta r_{\text{spon}}}{t_{\text{step}}}}.$$

(a) For a laser diode with the parameters given in Exercise 8.4 and injection current $I_{\text{inj}} = 20\,\text{mA}$, plot relative intensity noise,

$$\text{RIN}(\omega) = \frac{\delta P(\omega)}{P_0},$$

as a function of frequency in the range $0.1\,\text{GHz} < f < 10\,\text{GHz}$. In the expression for RIN, P_0 is the time-average optical power and $\delta P(\omega)$ is the average noise power spectral density per unit frequency in the photodetector electrical current.

(b) What is the shot-noise limit of RIN for a laser diode with emission wavelength $\lambda_0 = 1310\,\text{nm}$ and $P_0 = 1\,\text{mW}$?

9 Time-Independent Perturbation

9.1 Introduction

Often there are situations in which the solutions to the time-independent Schrödinger equation are known for a particular potential but not for a similar but different potential. Time-independent perturbation theory provides a means of finding approximate solutions using an expansion in the known eigenfunctions.

As an example, consider the one-dimensional, rectangular potential well with infinite barrier energy. The thickness of the well is L, and the potential energy is $V(0 < x < L) = 0$ and infinity elsewhere. The time-independent Schrödinger equation for an electron with effective mass m_e^* in the potential $V(x)$ is $\hat{H}\psi_n(x) = E_n\psi_n(x)$. Eigenfunction solutions are $\psi_n(x) = \sqrt{2/L}\sin(k_n x)$ and eigenenergies are $E_n = \hbar^2 k_n^2/2m_e^*$ where $k_n = n\pi/L$ and $n = 1, 2, 3, \ldots$.

Suppose that the potential $V(x)$ is deformed by the presence of an additional term $W(x)$. In Fig. 9.1, $W(x)$ is shown as a small curved bump. The challenge is to find the new eigenstates and eigenvalues of the system. One approach is to use time-independent perturbation theory.

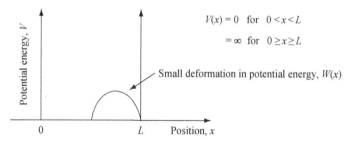

$$V(x) = 0 \quad \text{for} \quad 0 < x < L$$
$$= \infty \quad \text{for} \quad 0 \geq x \geq L$$

Small deformation in potential energy, $W(x)$

Fig. 9.1 Sketch of a one-dimensional, rectangular potential well with infinite barrier energy. The thickness of the well is L, and the potential energy is $V(0 < x < L) = 0$ and $V(x \leq 0, x \geq L) = \infty$. The known eigenfunctions for this potential can be used to obtain approximate solutions in the presence of a "small" deformation in the potential.

9.2 Time-Independent Nondegenerate Perturbation

To develop time-independent nondegenerate perturbation theory, a Hamiltonian of the form

$$\boxed{\hat{H} = \hat{H}^{(0)} + \hat{W}} \tag{9.1}$$

399

is assumed, where $\hat{H}^{(0)}$ is the unperturbed Hamiltonian and $\hat{W}$ is the perturbation. Formally, the solution for the *total* Hamiltonian $\hat{H}$ has eigenfunctions ψ_n and eigenenergies E_n for which

$$\hat{H}\psi_n = E_n\psi_n. \tag{9.2}$$

The solution for the *unperturbed* Hamiltonian $\hat{H}^{(0)}$ has eigenfunctions $\psi_m^{(0)}$ and eigenenergies $E_m^{(0)}$ such that

$$\hat{H}^{(0)}\psi_m^{(0)} = E_m^{(0)}\psi_m^{(0)}. \tag{9.3}$$

For the situations considered, $\hat{H}^{(0)}$, $\psi_m^{(0)}$, and $E_m^{(0)}$ are known. Because the perturbation $\hat{W}$ is taken to be small, it should be possible to expand ψ_n and E_n as a power series in $\hat{W}$.

The eigenfunctions and eigenvalues of the perturbed system are written

$$\psi = \psi^{(0)} + \lambda\psi^{(1)} + \lambda^2\psi^{(2)} + \cdots,$$
$$E = E^{(0)} + \lambda E^{(1)} + \lambda^2 E^{(2)} + \cdots, \tag{9.4}$$

where $\lambda = 1$ and is a dummy variable employed to keep track of the order of the terms in the power series used. Hence,

$$\left(\hat{H}^{(0)} + \lambda\hat{W}\right)\left(\psi^{(0)} + \lambda\psi^{(1)} + \cdots\right) = \left(E^{(0)} + \lambda E^{(1)} + \cdots\right)\left(\psi^{(0)} + \lambda\psi^{(1)} + \cdots\right). \tag{9.5}$$

Equating equal powers of λ gives

$$\left(\hat{H}^{(0)} - E^{(0)}\right)\psi^{(0)} = 0, \tag{9.6}$$

$$\left(\hat{H}^{(0)} - E^{(0)}\right)\psi^{(1)} = \left(E^{(1)} - \hat{W}\right)\psi^{(0)}, \tag{9.7}$$

$$\left(\hat{H}^{(0)} - E^{(0)}\right)\psi^{(2)} = \left(E^{(1)} - \hat{W}\right)\psi^{(1)} + E^{(2)}\psi^{(0)}, \tag{9.8}$$

$$\left(\hat{H}^{(0)} - E^{(0)}\right)\psi^{(3)} = \left(E^{(1)} - \hat{W}\right)\psi^{(2)} + E^{(2)}\psi^{(1)} + E^{(3)}\psi^{(0)}. \tag{9.9}$$

Equation (9.6) is the zero-order solution. Equation (9.7) has the first-order correction. Equation (9.8) is to second order and so on.

This theory has an important flaw. There is no self-consistent method to decide when to terminate the power series. Often, this is not a limitation because the high spatial frequency components of a potential usually contribute little to the total potential. However, this is not always the case, and so it is necessary to proceed with caution.

9.2.1 The First-Order Correction

Because $\psi^{(0)}$ forms a complete set, it is possible to expand each correction term in $\psi^{(0)}$. For $\psi^{(1)}$, this gives

$$\psi^{(1)} = \sum_n a_n^{(1)} \psi_n^{(0)}, \tag{9.10}$$

where it is worth noting that the unperturbed solution is Eq. (9.3).

The first-order corrected solution for the perturbed mth eigenvalue and eigenfunction is (Eq. (9.7)):

$$\left(\hat{H}^{(0)} - E_m^{(0)} \right) \psi^{(1)} = \left(E^{(1)} - \hat{W} \right) \psi_m^{(0)}. \tag{9.11}$$

Substituting for $\psi^{(1)}$ gives

$$\left(\hat{H}^{(0)} - E_m^{(0)} \right) \sum_n a_n^{(1)} \psi_n^{(0)} = \left(E^{(1)} - \hat{W} \right) \psi_m^{(0)}. \tag{9.12}$$

Multiplying both sides by $\psi_k^{(0)*}$, integrating, and making use of the orthonormal property of eigenfunctions so that $\langle i|j \rangle = \delta_{ij}$, results in

$$\int d^3 r \psi_k^{(0)*} \hat{H}^{(0)} \sum_n a_n^{(1)} \psi_n^{(0)} - E_m^{(0)} \int d^3 r \psi_k^{(0)*} \sum_n a_n^{(1)} \psi_n^{(0)} = E^{(1)} \delta_{km} - W_{km}, \tag{9.13}$$

$$\boxed{a_k^{(1)} \left(E_k^{(0)} - E_m^{(0)} \right) = E^{(1)} \delta_{km} - W_{km},} \tag{9.14}$$

where W_{km} is the matrix element $\int \psi_k^{(0)*} \hat{W} \psi_m^{(0)} d^3 r = \langle k | \hat{W} | m \rangle$. Thus, for $k \neq m$,

$$\boxed{a_k^{(1)} = \frac{W_{km}}{E_m^{(0)} - E_k^{(0)}},} \tag{9.15}$$

and for $k = m$,

$$\boxed{E^{(1)} = W_{mm},} \tag{9.16}$$

where $E^{(1)}$ is the first-order correction to eigenvalue E_m under the perturbation $\hat{W}$. Notice that in Eq. (9.15) the denominator $\left(E_m^{(0)} - E_k^{(0)} \right) \neq 0$ for $k \neq m$ since the method is restricted to nondegenerate eigenstates.

Evaluation of $a_m^{(1)}$ is achieved by requiring that $\psi = \psi^{(0)} + \psi^{(1)}$, the first-order corrected wave function, is normalized to unity:

$$\int \psi^* \psi \, d^3 r = \int \left(\psi_m^{(0)} + \lambda \psi^{(1)} \right) \left(\psi_m^{(0)} + \lambda \psi^{(1)} \right) d^3 r$$

$$= \int \left(\psi_m^{(0)} + \lambda \sum_i a_i^{(1)} \psi_i^{(0)} \right)^* \left(\psi_m^{(0)} + \lambda \sum_j a_j^{(1)} \psi_j^{(0)} \right) d^3 r$$

$$= 1 + \lambda a_m^{(1)} + \lambda a_m^{(1)*} + \lambda^2 \sum_i a_i^{(1)} a_i^{(1)*} = 1. \tag{9.17}$$

Neglecting the second-order term gives $a_m^{(1)} = 0$ as a solution ($a_m^{(1)}$ pure imaginary is also a solution). Hence, it may be concluded that the eigenfunction and eigenvalue to first order are

401

$$
\boxed{
\begin{aligned}
\psi &= \psi_m^{(0)} + \sum_{k \neq m} \frac{W_{km}}{E_m^{(0)} - E_k^{(0)}} \psi_k^{(0)}, \\
E &= E_m^{(0)} + W_{mm}.
\end{aligned}
}
\tag{9.18}
$$

9.2.2 The Second-Order Correction

The second-order correction to the eigenfunction $\psi^{(2)}$ may be expanded as

$$
\psi^{(2)} = \sum_n a_n^{(2)} \psi_n^{(0)}.
\tag{9.19}
$$

The second-order solution for the perturbed mth eigenvalue and eigenfunction is (Eq. (9.8))

$$
\left(\hat{H}^{(0)} - E_m^{(0)} \right) \psi^{(2)} = \left(E^{(1)} - \hat{W} \right) \psi^{(1)} + E^{(2)} \psi_m^{(0)}.
\tag{9.20}
$$

Expanding $\psi^{(1)}$ (Eq. (9.10)) and $\psi^{(2)}$ (Eq. (9.19)) in terms of $\psi^{(0)}$ results in

$$
\hat{H}^{(0)} \sum_n a_n^{(2)} \psi_n^{(0)} - E_m^{(0)} \sum_n a_n^{(2)} \psi_n^{(0)} = E^{(1)} \sum_n a_n^{(1)} \psi_n^{(0)} - \hat{W} \sum_n a_n^{(1)} \psi_n^{(0)} + E^{(2)} \psi_m^{(0)}.
\tag{9.21}
$$

Multiplying both sides by $\psi_k^{(0)*}$ and integrating gives

$$
\int \mathrm{d}^3 r \psi_k^{(0)*} \hat{H}^{(0)} \sum_n a_n^{(2)} \psi_n^{(0)} - E_m^{(0)} \int \mathrm{d}^3 r \psi_k^{(0)*} \sum_n a_n^{(2)} \psi_n^{(0)}
\tag{9.22}
$$

$$
= E^{(1)} \int \mathrm{d}^3 r \psi_k^{(0)*} \sum_n a_n^{(1)} \psi_n^{(0)} - \int \mathrm{d}^3 r \psi_k^{(0)*} \hat{W} \sum_n a_n^{(1)} \psi_n^{(0)} + E^{(2)} \int \mathrm{d}^3 r \psi_k^{(0)*} \psi_m^{(0)}.
$$

Hence,

$$
\boxed{
a_k^{(2)} \left(E_k^{(0)} - E_m^{(0)} \right) = a_k^{(1)} E^{(1)} - \sum_n a_n^{(1)} W_{kn} + E^{(2)} \delta_{mk}.
}
\tag{9.23}
$$

9.2.2.1 Second-Order Correction to Eigenvalues ($k = m$)

For $k = m$, Eq. (9.23) may be written

$$
E^{(2)} = \sum_n a_n^{(1)} W_{mn} - a_m^{(1)} E^{(1)} = \sum_{n \neq m} a_n^{(1)} W_{mn} + a_m^{(1)} W_{mm} - a_m^{(1)} E^{(1)}.
\tag{9.24}
$$

But from first-order perturbation $E^{(1)} = W_{mm}$, so the two right-hand terms cancel and

$$
E^{(2)} = \sum_{n \neq m} a_n^{(1)} W_{mn}.
\tag{9.25}
$$

Substituting for $a_n^{(1)}$ from first-order perturbation results in

$$E^{(2)} = \sum_{n \neq m} \frac{|W_{mn}|^2}{E_m^{(0)} - E_n^{(0)}}. \tag{9.26}$$

9.2.2.2 Second-Order Coefficients $a_k^{(2)}$

There are two situations to be considered – the values of $a_k^{(2)}$ when $k \neq m$ and the value when $k = m$. For the case in which $k \neq m$, the previous results, Eq. (9.15) and Eq. (9.16), may be substituted into Eq. (9.23) to give

$$a_k^{(2)} \left(E_m^{(0)} - E_k^{(0)} \right) = \sum_n a_n^{(1)} W_{kn} - a_k^{(1)} E^{(1)} = \sum_n \frac{W_{nm} W_{kn}}{E_m^{(0)} - E_n^{(0)}} - \frac{E^{(1)} W_{km}}{E_m^{(0)} - E_k^{(0)}}, \tag{9.27}$$

so that

$$a_k^{(2)} = \sum_n \frac{W_{nm} W_{kn}}{\left(E_m^{(0)} - E_n^{(0)} \right) \left(E_m^{(0)} - E_k^{(0)} \right)} - \frac{W_{mm} W_{km}}{\left(E_m^{(0)} - E_k^{(0)} \right)^2}, \qquad k \neq m. \tag{9.28}$$

For the case in which $k = m$, the value of $a_m^{(2)}$ is found by using normalization of the corrected wave function and is

$$a_m^{(2)} = -\frac{1}{2} \sum_{n \neq m} \frac{|W_{mn}|^2}{\left(E_m^{(0)} - E_n^{(0)} \right)^2}, \qquad k = m. \tag{9.29}$$

The eigenvalues and eigenfunctions of the perturbed system to second order are

$$E = E^{(0)} + E^{(1)} + E^{(2)}, \tag{9.30}$$

$$E = E_m^{(0)} + W_{mm} + \sum_{n \neq m} \frac{|W_{mn}|^2}{E_m^{(0)} - E_n^{(0)}}, \tag{9.31}$$

and

$$\psi = \psi^{(0)} + \psi^{(1)} + \psi^{(2)}$$
$$= \psi_m^{(0)} + \sum_k a_k^{(1)} \psi_k^{(0)} + \sum_k a_k^{(2)} \psi_k^{(0)}, \tag{9.32}$$

$$\psi = \psi_m^{(0)} + \sum_{k \neq m} \frac{W_{mk}}{E_m^{(0)} - E_k^{(0)}} \psi_k^{(0)}$$

$$+ \sum_{k \neq m} \left(\sum_{n \neq m} \frac{W_{kn} W_{mn}}{\left(E_m^{(0)} - E_n^{(0)} \right) \left(E_m^{(0)} - E_k^{(0)} \right)} - \frac{W_{mm} W_{km}}{\left(E_m^{(0)} - E_k^{(0)} \right)^2} \right) \psi_k^{(0)} \tag{9.33}$$

$$- \sum_{n \neq m} \frac{1}{2} \frac{|W_{mn}|^2}{\left(E_m^{(0)} - E_n^{(0)} \right)^2} \psi_m^{(0)}.$$

9.2.3 Harmonic Oscillator Subject to Perturbing Potential in x

Consider an electron with effective mass m_e^* moving in a one-dimensional harmonic oscillator potential that is subject to a constant small electric field, $\mathbf{E}$, in the x direction.

To find the new energy eigenvalues and eigenfunctions for the perturbed system, the Hamiltonian is first written as

$$\hat{H} = \frac{\hat{p}^2}{2m_e^*} + \frac{m_e^* \omega^2}{2} \hat{x}^2 + \hat{W}, \tag{9.34}$$

where the perturbation is $\hat{W} = e|\mathbf{E}|\hat{x}$. This perturbation involves the position operator $\hat{x}$, and so the matrix elements $W_{nm} = e|\mathbf{E}|x_{nm}$, where $x_{nm} = \langle n|\hat{x}|m \rangle$, must be found.

The position operator may be written as a linear combination of creation and annihilation operators in such a way that $\hat{x} = (\hbar/2m_e^*\omega)^{1/2} \left(\hat{b}^\dagger + \hat{b} \right)$. Since the matrix elements $\langle n|\hat{b}^\dagger|m \rangle = n^{1/2}\delta_{n=m+1}$ and $\langle n|\hat{b}|m \rangle = (n+1)^{1/2}\delta_{n=m-1}$, it follows that matrix element $W_{nm} = e|\mathbf{E}|x_{nm}$ has only two nonzero values:

$$W_{n,n+1} = e|\mathbf{E}| \left(\frac{\hbar}{2m_e^*\omega} \right)^{1/2} (n+1)^{1/2} \tag{9.35}$$

and

$$W_{n,n-1} = e|\mathbf{E}| \left(\frac{\hbar}{2m_e^*\omega} \right)^{1/2} n^{1/2}. \tag{9.36}$$

The unperturbed energy levels of the harmonic oscillator are

$$E_n^{(0)} = \hbar\omega \left(n + \frac{1}{2} \right). \tag{9.37}$$

The first-order correction due to the presence of the perturbation is

$$E^{(1)} = W_{mm} = e|\mathbf{E}|x_{mm} = 0. \tag{9.38}$$

This is zero because the matrix element $\langle m|\hat{x}|m \rangle = x_{mm} = 0$. However, the second-order correction is

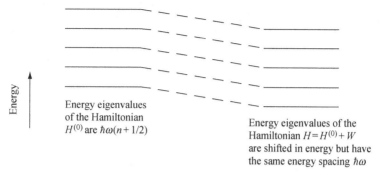

Fig. 9.2 Illustration of energy eigenvalues of the one-dimensional harmonic oscillator with (left) Hamiltonian $H^{(0)}$ and (right) subject to perturbation $W = -e|\mathbf{E}|x$ in the potential. In the presence of the perturbation, the energy levels are shifted but the energy-level spacing remains the same.

$$E^{(2)} = \sum_{n \neq m} \frac{|W_{mn}|^2}{E_m^{(0)} - E_n^{(0)}} = \frac{e^2|\mathbf{E}|^2 \hbar}{2m_e^* \omega} \left(\frac{x_{n,n+1}^2}{-\hbar\omega} + \frac{x_{n,n-1}^2}{\hbar\omega} \right)$$

$$= \frac{e^2|\mathbf{E}|^2}{2m_e^* \omega^2} \left(-(n+1) + n \right) = -\frac{e^2|\mathbf{E}|^2}{2m_e^* \omega^2}. \tag{9.39}$$

The new energy levels of the oscillator are, to second order,

$$E = \hbar\omega \left(n + \frac{1}{2} \right) - \frac{e^2|\mathbf{E}|^2}{2m_e^* \omega^2}, \tag{9.40}$$

which is the same as the exact result. Physically, the particle oscillates at the same frequency, ω, as the unperturbed case, but it is displaced a distance of magnitude $e|\mathbf{E}|/m\omega^2$, and the new energy levels are shifted in value by $-e^2|\mathbf{E}|^2/2m_e^*\omega^2$ (see Fig. 9.2).

The fact that perturbation theory gives the same result previously obtained by an exact calculation is a validation of the approach.

9.2.4 Harmonic Oscillator Subject to Perturbing Potential in x^2

Following success calculating the energy levels for a one-dimensional harmonic oscillator subject to a perturbation linear in x, the case of a perturbation in x^2 is now considered.

As usual, the Hamiltonian for the complete system is written down first:

$$\hat{H} = \frac{\hat{p}^2}{2m_e^*} + \frac{m_e^* \omega^2}{2} \hat{x}^2 + \hat{W}, \tag{9.41}$$

where the perturbation is $\hat{W} = \kappa\xi x^2/2$.

In this case, the effect of the perturbation is to change the value of κ. If $\omega'^2 = \omega^2(1 + \xi)$, then

$$\hat{H} = \frac{\hat{p}^2}{2m_e^*} + \frac{m_e^* \omega^2}{2} \hat{x}^2 + \frac{m_e^* \omega^2}{2} \xi \hat{x}^2 = \frac{\hat{p}^2}{2m_e^*} + \frac{m_e^* \omega^2}{2} (1 + \xi) \hat{x}^2 = \frac{\hat{p}^2}{2m_e^*} + \frac{m_e^* \omega'^2}{2} \hat{x}^2. \tag{9.42}$$

The right-hand side of Eq. (9.42) is another harmonic oscillator. Therefore, this can be solved exactly. The eigenvalues are

$$E_n = \left(n + \frac{1}{2}\right)\hbar\omega' = \left(n + \frac{1}{2}\right)\hbar\omega\sqrt{1 + \xi} = \left(n + \frac{1}{2}\right)\hbar\omega\left(1 + \frac{\xi}{2} - \frac{\xi^2}{8} + \cdots\right), \quad (9.43)$$

where the expansion $(1 + x)^n = 1 + nx + n(n - 1)x^2/2! + \cdots$ has been used. The same result may be found using perturbation theory.

The perturbation $\hat{W}$ may be written as

$$\hat{W} = \frac{m_e^*\omega^2}{2}\xi\hat{x}^2 = \frac{\xi}{4}\hbar\omega\left(\hat{b}^\dagger + \hat{b}\right)^2 = \frac{\xi}{4}\hbar\omega\left(\hat{b}^{\dagger 2} + \hat{b}^2 + \hat{b}\hat{b}^\dagger + b^\dagger b\right), \quad (9.44)$$

and, since $\hat{b}\hat{b}^\dagger = \hat{b}^\dagger\hat{b} + 1$,

$$\hat{W} = \frac{\xi}{4}\hbar\omega\left(\hat{b}^{\dagger 2} + \hat{b}^2 + 2\hat{b}^\dagger\hat{b} + 1\right). \quad (9.45)$$

As usual, the nonzero matrix elements are easy to find. They are

$$\langle\phi_n|\hat{W}|\phi_n\rangle = \frac{1}{2}\xi\left(n + \frac{1}{2}\right)\hbar\omega, \quad (9.46)$$

$$\langle\phi_{n+2}|\hat{W}|\phi_n\rangle = \frac{1}{4}\xi\sqrt{(n + 1)(n + 2)}\hbar\omega, \quad (9.47)$$

$$\langle\phi_{n-2}|\hat{W}|\phi_n\rangle = \frac{1}{4}\xi\sqrt{n(n - 1)}\hbar\omega. \quad (9.48)$$

These results may now be used to evaluate the energy terms up to second order:

$$E = E^{(0)} + E^{(1)} + E^{(2)}, \quad (9.49)$$

$$E_n = E_n^{(0)} + W_{nn} + \sum_{m \neq n}\frac{|W_{nm}|^2}{E_n^{(0)} - E_m^{(0)}}$$

$$= E_n^{(0)} + \frac{\xi}{2}\left(n + \frac{1}{2}\right)\hbar\omega - \frac{\xi^2}{16}(n + 1)(n + 2)\frac{\hbar\omega}{2} + \frac{\xi^2}{16}n(n - 1)\frac{\hbar\omega}{2} + \cdots$$

$$= E_n^{(0)} + \left(n + \frac{1}{2}\right)\hbar\omega\frac{\xi}{2} - \left(n + \frac{1}{2}\right)\hbar\omega\frac{\xi^2}{8} + \cdots$$

$$= \left(n + \frac{1}{2}\right)\hbar\omega\left(1 + \frac{\xi}{2} - \frac{\xi^2}{8} + \cdots\right), \quad (9.50)$$

which is the same as the previous result. The energy-level diagram in Fig. 9.3 illustrates the effect of the perturbation.

Again, perturbation theory is in good agreement with an alternative exact approach, thereby justifying the method.

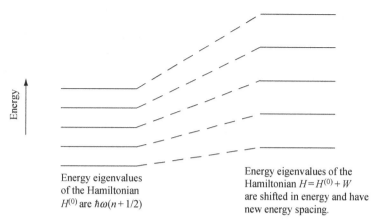

Energy eigenvalues
of the Hamiltonian
$H^{(0)}$ are $\hbar\omega(n+1/2)$

Energy eigenvalues of the
Hamiltonian $H = H^{(0)} + W$
are shifted in energy and have
new energy spacing.

Fig. 9.3 Illustration of energy eigenvalues of the one-dimensional harmonic oscillator with (left) Hamiltonian $H^{(0)}$ and (right) subject to perturbation $W = \kappa\xi x^2/2$ in the potential. In the presence of the perturbation, both the energy levels and the energy-level spacing are changed. In the figure it is assumed that $\xi < 1$.

9.2.5 Harmonic Oscillator Subject to Perturbing Potential in x^3

Consider a one-dimensional harmonic oscillator subject to a perturbation in x^3. As before, the way to proceed is by writing down the Hamiltonian for the complete system:

$$\hat{H} = \frac{\hat{p}^2}{2m_e^*} + \frac{\kappa}{2}\hat{x}^2 + \hat{W}, \tag{9.51}$$

where the perturbation is $\hat{W} = \xi x^3 \hbar\omega(m_e^*\omega/\hbar)^{3/2}$.

Because $\sqrt{\hbar/m_e^*\omega}$ has been factored out, the position operator may be written as $\hat{x} = \left(\hat{b}^\dagger + \hat{b}\right)/\sqrt{2}$, and since the perturbation is in x^3, an expression for $\hat{x}^3 \propto \left(\hat{b}^\dagger + \hat{b}\right)^3$ must be found:

$$\left(\hat{b}^\dagger + \hat{b}\right)^3 = \left(\hat{b}^{\dagger 2} + \hat{b}\hat{b}^\dagger + \hat{b}^\dagger\hat{b} + \hat{b}^2\right)\left(\hat{b}^\dagger + \hat{b}\right)$$

$$= \left(b^{\dagger 3} + \hat{b}\hat{b}^\dagger\hat{b}^\dagger + \hat{b}^\dagger\hat{b}\hat{b}^\dagger + \hat{b}\hat{b}\hat{b}^\dagger + \hat{b}^\dagger\hat{b}^\dagger\hat{b} + \hat{b}\hat{b}^\dagger\hat{b} + \hat{b}^\dagger\hat{b}\hat{b} + \hat{b}^3\right). \tag{9.52}$$

Making use of the operators $\hat{n} = \hat{b}^\dagger\hat{b}$ and $\hat{n} + 1 = \hat{b}\hat{b}^\dagger$, and substituting into the equation gives

$$\left(\hat{b}^\dagger + \hat{b}\right)^3 = \left(\hat{b}^{\dagger 3} + (\hat{n} + 1)\hat{b}^\dagger + \hat{n}\hat{b}^\dagger + \hat{b}(\hat{n} + 1) + \hat{b}^\dagger\hat{n} + (\hat{n} + 1)\hat{b} + \hat{n}\hat{b} + \hat{b}^3\right). \tag{9.53}$$

Using the commutation relations $\left[\hat{n}, \hat{b}\right] = \hat{n}\hat{b} - \hat{b}\hat{n} = -\hat{b}$ and $\left[\hat{n}, \hat{b}^\dagger\right] = \hat{n}\hat{b}^\dagger - \hat{b}^\dagger\hat{n} = \hat{b}^\dagger$ the following result is found:

$$\left(\hat{b}^\dagger + \hat{b}\right)^3 = \left(\hat{b}^{\dagger 3} + \hat{n}\hat{b}^\dagger + \hat{b}^\dagger + \hat{n}\hat{b}^\dagger + (\hat{n} + 1)\hat{b} + \hat{b} + \hat{n}\hat{b}^\dagger - \hat{b}^\dagger + (\hat{n} + 1)\hat{b} \right.$$
$$\left. + (\hat{n} + 1)\hat{b} - \hat{b} + \hat{b}^3\right)$$

$$= \left(\hat{b}^{\dagger 3} + 3\hat{n}\hat{b}^{\dagger} + 3(\hat{n}+1)\hat{b} + \hat{b}^3 \right). \tag{9.54}$$

Therefore, the perturbation is

$$\hat{W} = \frac{\xi \hbar \omega}{2^{3/2}} \left(\hat{b}^{\dagger 3} + 3\hat{n}\hat{b}^{\dagger} + 3(\hat{n}+1)\hat{b} + \hat{b}^3 \right). \tag{9.55}$$

Hence, the only nonzero matrix elements of the perturbation $\hat{W}$ have the effect of mixing $|\phi_n\rangle$ with states $|\phi_{n+1}\rangle$, $|\phi_{n-1}\rangle$, $|\phi_{n+3}\rangle$, and $|\phi_{n-3}\rangle$. The matrix elements are

$$\langle \phi_{n+3}|\hat{W}|\phi_n \rangle = \xi \left(\frac{(n+3)(n+2)(n+1)}{8} \right)^{1/2} \hbar \omega, \tag{9.56}$$

$$\langle \phi_{n-3}|\hat{W}|\phi_n \rangle = \xi \left(\frac{n(n-1)(n-2)}{8} \right)^{1/2} \hbar \omega, \tag{9.57}$$

$$\langle \phi_{n+1}|\hat{W}|\phi_n \rangle = 3\xi \left(\frac{n+1}{2} \right)^{3/2} \hbar \omega, \tag{9.58}$$

$$\langle \phi_{n-1}|\hat{W}|\phi_n \rangle = 3\xi \left(\frac{n}{2} \right)^{3/2} \hbar \omega. \tag{9.59}$$

Using these results to evaluate the energy terms up to second order gives

$$E_n = E_n^{(0)} + W_{nn} + \sum_{m \neq n} \frac{|W_{nm}|^2}{E_n^{(0)} - E_m^{(0)}} = \left(n + \frac{1}{2} \right) \hbar \omega + 0 + \sum_{m \neq n} \frac{|W_{nm}|^2}{E_n^{(0)} - E_m^{(0)}}$$

$$= \left(n + \frac{1}{2} \right) \hbar \omega - \frac{1}{8} \xi^2 \left(30n^2 + 30n + 11 \right) \hbar \omega + \cdots . \tag{9.60}$$

As shown in Fig. 9.4, the effect of $\hat{W}$ is to *lower* the unperturbed energy levels. The energy levels are lowered *whatever the sign of* ξ. In addition, the difference between adjacent energy eigenvalues decreases with increasing energy.

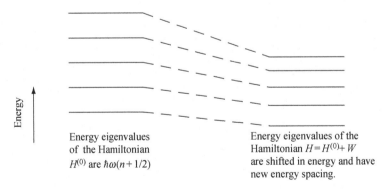

Fig. 9.4 Illustration of energy eigenvalues of the one-dimensional harmonic oscillator with (left) Hamiltonian $H^{(0)}$ and (right) subject to perturbation $W = \xi x^3 \hbar \omega (m_e^* \omega / \hbar)^{3/2}$ in the potential. In the presence of the perturbation, both the energy levels and the energy-level spacing are changed. The effect of W is to lower the unperturbed energy levels whatever the sign of ξ. In addition, the difference between adjacent energy eigenvalues decreases with increasing energy.

9.3 Time-Independent Degenerate Perturbation

Often, the potential in which a particle moves contains symmetry that results in eigenvalues that are degenerate. If the symmetry producing this degeneracy is destroyed by the perturbation $\hat{W}$, the degenerate state separates, or splits, into distinct energy levels. This is illustrated schematically in Fig. 9.5.

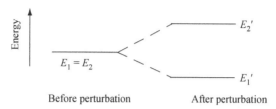

Before perturbation   After perturbation

Fig. 9.5 Energy-level diagram illustrating degenerate energy eigenvalues split into separate energy levels by the presence of a perturbing potential.

It is assumed that the Hamiltonian describing the system is of the form

$$\hat{H} = \hat{H}^{(0)} + \hat{W}, \tag{9.61}$$

where $\hat{H}^{(0)}$ is the unperturbed Hamiltonian with degenerate energy levels and $\hat{W}$ is the perturbation. Formally, the solution for the Schrödinger equation with total Hamiltonian $\hat{H}$ has eigenfunctions ψ_n and eigenenergies E_n, so that

$$\hat{H}\psi_n = E_n\psi_n. \tag{9.62}$$

The solution for the Schrödinger equation using the unperturbed Hamiltonian $\hat{H}^{(0)}$ has eigenfunctions $\psi_m^{(0)}$ and degenerate eigenenergies $E_m^{(0)}$ so that

$$\hat{H}^{(0)}\psi_m^{(0)} = E_m^{(0)}\psi_m^{(0)}. \tag{9.63}$$

9.3.1 A Two-Fold Degeneracy Split by Time-Independent Perturbation

If the *first-order* corrected wave functions are expanded in terms of the unperturbed wave functions, then

$$\psi_m^{(1)} = \sum_n a_{mn}^{(1)}\psi_n^{(0)} \tag{9.64}$$

and

$$a_{mn}^{(1)} = \frac{W_{mn}}{E_m^{(0)} - E_n^{(0)}}. \tag{9.65}$$

If degeneracy exists such that $E_m^{(0)} - E_n^{(0)} = 0$ then a_{mn} is infinite. This problem may be circumvented by diagonalizing (finding solutions) for the Hamiltonian $\hat{H}$.

9.3.2 Matrix Method

In the matrix method, the Hamiltonian is treated as a finite-sized matrix equation. Diagonalization of this $N \times N$ matrix gives the solution to how a group of initially unperturbed states interact via the perturbation *with one another*.

The matrix form of the Schrödinger equation $\hat{H}|a\rangle = E|a\rangle$ describing the perturbed system may be written as

$$\mathbf{Ha} = \begin{bmatrix} H_{11} & H_{12} & H_{13} & \cdot & \cdot & \cdot \\ H_{21} & H_{22} & H_{23} & \cdot & \cdot & \cdot \\ H_{31} & H_{32} & H_{33} & \cdot & \cdot & \cdot \\ \cdot & \cdot & \cdot & \cdot & \cdot & \cdot \\ \cdot & \cdot & \cdot & \cdot & \cdot & \cdot \\ \cdot & \cdot & \cdot & \cdot & \cdot & H_{NN} \end{bmatrix} \begin{bmatrix} a_1 \\ a_2 \\ a_3 \\ \cdot \\ \cdot \\ a_N \end{bmatrix} = E \begin{bmatrix} a_1 \\ a_2 \\ a_3 \\ \cdot \\ \cdot \\ a_N \end{bmatrix} = E\mathbf{a}. \tag{9.66}$$

In Eq. (9.66), $\mathbf{H}$ is the Hamiltonian matrix with matrix elements $H_{mn} = \langle m|\hat{H}|n\rangle$.

The approximation in the matrix method arises because *only N terms are used*. This method is good for the problem of degenerate energy levels split by a perturbation. Hence, it is sometimes called degenerate perturbation theory.

This approximation works because, first, the larger the energy separation between states the weaker the effect of the perturbation and, second, the smaller the matrix element $W_{mn} = \langle m|\hat{W}|n\rangle$ the weaker the effect of the perturbation.

The matrix equation (Eq. (9.66)) may be rewritten as

$$(\mathbf{H} - E\mathbf{1})\mathbf{a} = \sum_{m=1,n=1}^{N} [H_{mn} - E\delta_{mn}]a_n = 0, \tag{9.67}$$

which has a nontrivial solution if the characteristic determinant vanishes, giving the *secular equation*

$$\begin{vmatrix} H_{11} - E & H_{12} & \cdot & \cdot & \cdot \\ H_{21} & H_{22} - E & \cdot & \cdot & \cdot \\ \cdot & \cdot & \cdot & \cdot & \cdot \\ \cdot & \cdot & \cdot & \cdot & \cdot \\ \cdot & \cdot & \cdot & \cdot & H_{NN} - E \end{vmatrix} = 0. \tag{9.68}$$

9.3.2.1 Matrix Method for two States

As a simple example of the matrix method, consider two states. The secular equation is

$$\begin{vmatrix} H_{11} - E & H_{12} \\ H_{21} & H_{22} - E \end{vmatrix} = 0, \tag{9.69}$$

$$\begin{aligned} 0 &= (H_{11} - E)(H_{22} - E) - H_{12}H_{21} \\ &= E^2 - EH_{22} - EH_{11} + H_{11}H_{22} - H_{12}H_{21} \\ &= E^2 - (H_{22} + H_{11})E + H_{11}H_{22} - H_{12}H_{21}. \end{aligned} \tag{9.70}$$

This equation is of the form $ax^2 + bx + c = 0$, and so it has a solution $x = (-b \pm \sqrt{b^2 - 4ac})/2a$. Hence, the two new energy eigenvalues are

$$E = \frac{(H_{11} + H_{22}) \pm \sqrt{(H_{11} + H_{22})^2 - 4(H_{11}H_{22} - H_{12}H_{21})}}{2}$$

$$= \frac{(H_{11} + H_{22})}{2} \pm \frac{\sqrt{H_{11}^2 + H_{22}^2 + 2H_{11}H_{22} - 4H_{11}H_{22} + 4H_{12}H_{21}}}{2}, \tag{9.71}$$

$$\boxed{E_\pm = \frac{(H_{11} + H_{22})}{2} \pm \sqrt{\frac{1}{4}(H_{11} - H_{22})^2 + H_{12}H_{21}},} \tag{9.72}$$

or

$$E_\pm = \frac{(H_{11} + H_{22})}{2} \pm \sqrt{\frac{1}{4}(H_{11} - H_{22})^2 + |H_{12}|^2}. \tag{9.73}$$

The coefficients a_1 and a_2 may be determined by rewriting Eq. (9.66) as

$$\begin{bmatrix} H_{11} - E & H_{12} \\ H_{21} & H_{22} - E \end{bmatrix} \begin{bmatrix} a_1 \\ a_2 \end{bmatrix} = 0. \tag{9.74}$$

Multiplying out the matrix gives two equations:

$$(H_{11} - E)a_1 + H_{12}a_2 = 0 \tag{9.75}$$

and

$$H_{21}a_1 + (H_{22} - E)a_2 = 0. \tag{9.76}$$

In addition to these two equations, there is the constraint that ψ is normalized, requiring

$$|a_1|^2 + |a_2|^2 = 1. \tag{9.77}$$

Solving Eq. (9.75) is achieved by writing

$$(H_{11} - E)a_1 = -H_{12}a_2 \tag{9.78}$$

and then squaring both sides to give

$$(H_{11} - E)^2 |a_1|^2 = H_{12}^2 |a_2|^2. \tag{9.79}$$

Using the fact that $|a_1|^2 = 1 - |a_2|^2$ (Eq. (9.77)), it follows that

$$(H_{11} - E)^2 |a_1|^2 = H_{12}^2 \left(1 - |a_1|^2\right), \tag{9.80}$$

$$\left((H_{11} - E)^2 + H_{12}^2\right) |a_1|^2 = H_{12}^2. \tag{9.81}$$

Hence,

$$|a_1|^2 = \frac{|H_{12}|^2}{(H_{11} - E)^2 + |H_{12}|^2}. \tag{9.82}$$

Letting

$$b_0^2 = |H_{12}|^2 + (H_{11} - E)^2, \tag{9.83}$$

then

$$a_1 = \frac{H_{12}}{b_0}, \tag{9.84}$$

and substituting this into Eq. (9.75) gives

$$a_2 = \frac{-(H_{11} - E)a_1}{H_{12}} = \frac{E - H_{11}}{b_0}. \tag{9.85}$$

9.3.3 The Two-Dimensional Harmonic Oscillator Perturbed in xy

The unperturbed Hamiltonian for an electron with effective mass m_e^* in an isotropic two-dimensional harmonic oscillator potential is

$$\hat{H}^{(0)} = \frac{\hat{p}_x^2 + \hat{p}_y^2}{2m_e^*} + \frac{m_e^* \omega^2}{2} \left(\hat{x}^2 + \hat{y}^2 \right). \tag{9.86}$$

Figure 9.6 illustrates the potential, $V(x, y)$.

The unperturbed Hamiltonian given by Eq. (9.86) can be rewritten in terms of creation and annihilation operators:

$$\hat{H}^{(0)} = \hbar\omega \left(\hat{b}_x^\dagger \hat{b}_x + \frac{1}{2} + \hat{b}_y^\dagger \hat{b}_y + \frac{1}{2} \right) = \hbar\omega \left(\hat{b}_x^\dagger \hat{b}_x + \hat{b}_y^\dagger \hat{b}_y + 1 \right), \tag{9.87}$$

where $\hat{x} = \left(\frac{\hbar}{2m_e^* \omega} \right)^{1/2} \left(\hat{b}_x + \hat{b}_x^\dagger \right)$ and $\hat{y} = \left(\frac{\hbar}{2m_e^* \omega} \right)^{1/2} \left(\hat{b}_y + \hat{b}_y^\dagger \right)$.

The eigenstates of $H^{(0)}$ are of the form

$$\phi_{nm} = \phi_n(x)\phi_m(y) = |nm\rangle \tag{9.88}$$

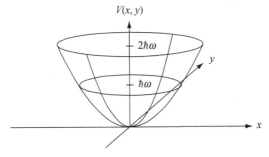

Fig. 9.6 Illustration of the harmonic oscillator potential in two dimensions, showing quantization energy $\hbar\omega$ and $2\hbar\omega$.

and the eigenenergy is

$$E_{nm} = \hbar\omega(n + m + 1) = \hbar\omega\left(n + \frac{1}{2} + m + \frac{1}{2}\right), \tag{9.89}$$

which is $(n + m + 1)$-fold degenerate. For example, states with energy $2\hbar\omega$ are two-fold degenerate, since $E_{10} = E_{01} = 2\hbar\omega$. The corresponding eigenstates are $|10\rangle$ and $|01\rangle$.

The task is to consider the effect of the perturbing potential, $\hat{W} = \kappa'\hat{x}\hat{y}$, on degenerate states $|10\rangle$ and $|01\rangle$ and find the two new wave functions and eigenvalues that diagonalize $\hat{W}$. The new wave functions are linear combinations of the unperturbed wave functions, so that

$$\psi_1 = a_1\phi_{10} + a_2\phi_{01} \tag{9.90}$$

and

$$\psi_2 = a_1\phi_{10} - a_2\phi_{01}. \tag{9.91}$$

The submatrix $\mathbf{W}$ in the basis $\{\phi_{10}, \phi_{01}\}$ is

$$\mathbf{W} = \kappa'\begin{bmatrix} \langle 10|xy|10\rangle & \langle 10|xy|01\rangle \\ \langle 01|xy|10\rangle & \langle 01|xy|01\rangle \end{bmatrix} = \kappa'\begin{bmatrix} W_{11} & W_{12} \\ W_{21} & W_{22} \end{bmatrix}. \tag{9.92}$$

Evaluation of the matrix elements gives

$$W_{12} = \langle 10|xy|01\rangle = \frac{\hbar}{2m_{\rm e}^*\omega}\langle 10|\left(\hat{b}_x + \hat{b}_x^\dagger\right)\left(\hat{b}_y + \hat{b}_y^\dagger\right)|01\rangle$$

$$= \frac{\hbar}{2m_{\rm e}^*\omega}\langle 10|\hat{b}_x\hat{b}_y + \hat{b}_x^\dagger\hat{b}_y + \hat{b}_x\hat{b}_y^\dagger + \hat{b}_x^\dagger\hat{b}_y^\dagger|01\rangle = \frac{\hbar}{2m_{\rm e}^*\omega}\langle 10|\hat{b}_x^\dagger\hat{b}_y|01\rangle = \frac{\hbar}{2m_{\rm e}^*\omega} \tag{9.93}$$

and

$$W_{11} = \langle 10|xy|10\rangle = \frac{\hbar}{2m_{\rm e}^*\omega}\langle 10|\hat{b}_x\hat{b}_y + \hat{b}_x^\dagger\hat{b}_y + \hat{b}_x\hat{b}_y^\dagger + \hat{b}_x^\dagger\hat{b}_y^\dagger|10\rangle = 0. \tag{9.94}$$

Hence, substituting these results into Eq. (9.92) results in

$$\mathbf{W} = \frac{\hbar\kappa'}{2m_{\rm e}^*\omega}\begin{bmatrix} 0 & 1 \\ 1 & 0 \end{bmatrix}, \tag{9.95}$$

and the secular equation

$$\begin{vmatrix} H_{11} - E & H_{12} \\ H_{21} & H_{22} - E \end{vmatrix} = 0 \tag{9.96}$$

may be written as

$$\begin{vmatrix} -E & \dfrac{\hbar\kappa'}{2m_{\rm e}^*\omega} \\ \dfrac{\hbar\kappa'}{2m_{\rm e}^*\omega} & -E \end{vmatrix} = 0, \tag{9.97}$$

413

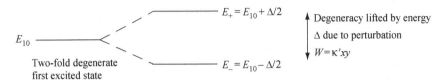

Fig. 9.7 The two-fold degeneracy of the first excited state of the two-dimensional harmonic oscillator is split into two energy eigenvalues by the perturbation $W = \kappa' xy$.

which has the solutions

$$E = \pm \frac{\hbar \kappa'}{2 m_e^* \omega} = \pm \frac{\Delta}{2}. \tag{9.98}$$

The perturbation is found to separate the first excited state by an amount

$$\frac{2\hbar \kappa'}{2 m_e^* \omega} = \frac{\hbar \kappa'}{m \omega} = \Delta. \tag{9.99}$$

The lifting of the first excited-state degeneracy by the perturbation $\hat{W} = \kappa' \hat{x} \hat{y}$ is illustrated in Fig. 9.7.

The new wave functions are obtained by substituting these values into the matrix equation

$$\sum_{m=1, n=1}^{N} [H_{nm} - E\delta_{nm}] a_n = 0. \tag{9.100}$$

For the case being considered,

$$\begin{bmatrix} -E & \dfrac{\Delta}{2} \\ \dfrac{\Delta}{2} & -E \end{bmatrix} \begin{bmatrix} a_1 \\ a_2 \end{bmatrix} = 0. \tag{9.101}$$

The symmetric state,

$$\psi_1 = \frac{1}{\sqrt{2}} (\phi_{10} + \phi_{01}), \tag{9.102}$$

has eigenenergy

$$E_- = -\frac{\Delta}{2} \tag{9.103}$$

and the antisymmetric state,

$$\psi_2 = \frac{1}{\sqrt{2}} (\phi_{10} - \phi_{01}), \tag{9.104}$$

has the eigenenergy

$$E_+ = \frac{\Delta}{2}. \tag{9.105}$$

9.3.4 Perturbation of Two-Dimensional Potential with Infinite Barrier Energy

Consider an electron with effective mass m_e^* confined to a two-dimensional potential $V(x, y)$ that is infinite except in a region $0 < x < L$ and $0 < y < L$ where $V(x, y) = 0$. The time-independent Schrödinger equation for the unperturbed system is

$$\hat{H}^{(0)} \psi_n^{(0)}(x, y) = E_n^{(0)} \psi_n^{(0)}(x, y), \qquad (9.106)$$

where

$$\hat{H}^{(0)} = \frac{-\hbar^2}{2m_e^*} \left(\frac{d^2}{dx^2} + \frac{d^2}{dy^2} \right) \qquad (9.107)$$

in the region $0 < x < L$ and $0 < y < L$. Each eigenstate $\psi_n^{(0)}(x, y)$ is separable such that $\psi_n^{(0)}(x, y) = \phi_{n_x}(x)\phi_{n_y}(y)$. The functions $\phi_{n_x}(x)$ and $\phi_{n_y}(y)$ are of the form

$$\phi_{n_x}(x) = \sqrt{\frac{2}{L}} \sin(k_{n_x} x), \qquad (9.108)$$

where $k_{n_x} = \frac{n_x \pi}{L}$ and n_x is a nonzero positive integer. Hence, the eigenfunctions for the unperturbed system are

$$\psi_{n_x, n_y}^{(0)}(x, y) = \frac{2}{L} \sin(k_{n_x} x) \sin(k_{n_y} y).$$

Substitution into the time-independent Schrödinger equation gives eigenenergies

$$E_{n_x, n_y}^{(0)} = \frac{\pi^2 \hbar^2}{2m_e^* L^2} \left(n_x^2 + n_y^2 \right). \qquad (9.109)$$

Notice that the ground state is not degenerate and the first excited state is $\psi_{12}^{(0)}$ or $\psi_{21}^{(0)}$, has eigenenergy $E_{12}^{(0)} = \frac{5\pi^2 \hbar^2}{2m_e^* L^2}$, and degeneracy 2.

In matrix notation, the general form of the time-independent Schrödinger equation for the system is

$$\sum_{m=1, n=1}^{N} [H_{nm}^{(0)} - E\delta_{nm}]a_n^{(0)} \qquad (9.110)$$

$$= \begin{bmatrix} H_{11}^{(0)} - E & H_{12}^{(0)} & H_{13}^{(0)} & \cdot & \cdot \\ H_{21}^{(0)} & H_{22}^{(0)} - E & H_{23}^{(0)} & \cdot & \cdot \\ H_{31}^{(0)} & H_{32}^{(0)} & H_{33}^{(0)} - E & \cdot & \cdot \\ \cdot & & & & \\ \cdot & \cdot & \cdot & \cdot & H_{NN}^{(0)} - E \end{bmatrix} \begin{bmatrix} a_1^{(0)} \\ a_2^{(0)} \\ a_3^{(0)} \\ \cdot \\ a_N^{(0)} \end{bmatrix} = 0.$$

For the system considered, all the matrix elements $H_{nm}^{(0)} = 0$ so that

$$\sum_{m=1,n=1}^{N} \left[H_{nm}^{(0)} - E\delta_{nm}\right] a_n^{(0)} = \begin{bmatrix} -E & 0 & 0 & \cdot & \cdot \\ 0 & -E & 0 & \cdot & \cdot \\ 0 & 0 & -E & \cdot & \cdot \\ \cdot & \cdot & \cdot & \cdot & \cdot \\ \cdot & \cdot & \cdot & \cdot & -E \end{bmatrix} \begin{bmatrix} a_1^{(0)} \\ a_2^{(0)} \\ a_3^{(0)} \\ \cdot \\ a_N^{(0)} \end{bmatrix} = 0. \quad (9.111)$$

Consider the effect on the ground state and the first excited state of a perturbation potential such that $\hat{W} = V_0$ in the region for which $0 < x < \frac{L}{2}$, and $0 < y < \frac{L}{2}$ and $\hat{W} = 0$ elsewhere. The value V_0 is a constant.

The correction to ground state energy after the perturbation is applied is, to first order, given by the diagonal matrix element of *nondegenerate* perturbation theory. Hence,

$$E_{11}^{(1)} = \left\langle \psi_{11}^{(0)} \left| \hat{W} \right| \psi_{11}^{(0)} \right\rangle \quad (9.112)$$

$$= \left(\frac{2}{L}\right)^2 V_0 \int\limits_0^{L/2} \sin^2\left(\frac{\pi x}{L}\right) dx \int\limits_0^{L/2} \sin^2\left(\frac{\pi y}{L}\right) dy = \left(\frac{2}{L}\right)^2 V_0 \left(\frac{L}{4}\right)^2 = \frac{V_0}{4}.$$

The first excited state is two-fold degenerate and so *degenerate* perturbation theory must be used. The 2×2 sub-matrix in the basis $\left\{\psi_{12}^{(0)}, \psi_{21}^{(0)}\right\}$ is

$$\begin{bmatrix} W_{11} & W_{12} \\ W_{21} & W_{22} \end{bmatrix} = \begin{bmatrix} \left\langle \psi_{12}^{(0)} \left| \hat{W} \right| \psi_{12}^{(0)} \right\rangle & \left\langle \psi_{12}^{(0)} \left| \hat{W} \right| \psi_{21}^{(0)} \right\rangle \\ \left\langle \psi_{21}^{(0)} \left| \hat{W} \right| \psi_{12}^{(0)} \right\rangle & \left\langle \psi_{21}^{(0)} \left| \hat{W} \right| \psi_{21}^{(0)} \right\rangle \end{bmatrix}. \quad (9.113)$$

The diagonal matrix elements are

$$W_{11} = W_{22} = \left\langle \psi_{12}^{(0)} \left| \hat{W} \right| \psi_{12}^{(0)} \right\rangle = \left(\frac{2}{L}\right)^2 V_0 \int\limits_0^{L/2} \sin^2\left(\frac{\pi x}{L}\right) dx \int\limits_0^{L/2} \sin^2\left(\frac{2\pi y}{L}\right) dy \quad (9.114)$$

$$= \left(\frac{2}{L}\right)^2 V_0 \left(\frac{L}{4}\right)^2 = \frac{V_0}{4}.$$

The off-diagonal elements are

$$W_{12} = W_{21} = \left\langle \psi_{12}^{(0)} \left| \hat{W} \right| \psi_{21}^{(0)} \right\rangle \quad (9.115)$$

$$= \left(\frac{2}{L}\right)^2 V_0 \int\limits_0^{L/2} \sin\left(\frac{\pi x}{L}\right) \sin\left(\frac{2\pi x}{L}\right) dx \int\limits_0^{L/2} \sin\left(\frac{2\pi y}{L}\right) \sin\left(\frac{\pi y}{L}\right) dy,$$

which may be rewritten as

$$W_{12} = \left(\frac{4V_0}{L^2}\right)\frac{1}{2}\int_0^{L/2}\left(\cos\left(\frac{-\pi x}{L}\right) - \cos\left(\frac{3\pi x}{L}\right)\right)dx\frac{1}{2}\int_0^{L/2}\left(\cos\left(\frac{\pi y}{L}\right) - \cos\left(\frac{3\pi y}{L}\right)\right)dy.$$

(9.116)

Performing the integral,

$$W_{12} = \left(\frac{4V_0}{L^2}\right)\frac{1}{4}\left[\frac{L}{\pi}\sin\left(\frac{\pi x}{L}\right) - \frac{L}{3\pi}\sin\left(\frac{3\pi x}{L}\right)\right]_0^{L/2}\left[\frac{L}{\pi}\sin\left(\frac{\pi x}{L}\right) - \frac{L}{3\pi}\sin\left(\frac{3\pi x}{L}\right)\right]_0^{L/2}$$

(9.117)

$$= \frac{V_0}{L^2}\left(\frac{L}{\pi} + \frac{L}{3\pi}\right)\left(\frac{L}{\pi} + \frac{L}{3\pi}\right) = \frac{V_0}{L^2}\left(\frac{4L}{3\pi}\right)\left(\frac{4L}{3\pi}\right) = \frac{16V_0}{9\pi^2}.$$

Hence,

$$\begin{bmatrix} W_{11} & W_{12} \\ W_{21} & W_{22} \end{bmatrix} = \begin{bmatrix} \dfrac{V_0}{4} & \dfrac{16V_0}{9\pi^2} \\ \dfrac{16V_0}{9\pi^2} & \dfrac{V_0}{4} \end{bmatrix} = \frac{V_0}{4}\begin{bmatrix} 1 & \dfrac{64}{9\pi^2} \\ \dfrac{64}{9\pi^2} & 1 \end{bmatrix}.$$

(9.118)

Solutions to

$$(\mathbf{H} - E'\mathbf{1})\begin{bmatrix} a_1 \\ a_2 \end{bmatrix} = \left(\mathbf{H}^{(0)} + \mathbf{W} - E'\mathbf{1}\right)\begin{bmatrix} a_1 \\ a_2 \end{bmatrix}$$

$$= \begin{bmatrix} (E_{12}^{(0)} + \dfrac{V_0}{4} - E) & \dfrac{16V_0}{9\pi^2} \\ \dfrac{16V_0}{9\pi^2} & (E_{12}^{(0)} + \dfrac{V_0}{4} - E) \end{bmatrix}\begin{bmatrix} a_1 \\ a_2 \end{bmatrix} = 0$$

(9.119)

are sought and found using the characteristic equation

$$\left(E_{12}^{(0)} + \frac{V_0}{4} - E\right)\left(E_{12}^{(0)} + \frac{V_0}{4} - E\right) - \Delta^2 = 0$$

(9.120)

with roots

$$E_{\pm} = E_{12}^{(0)} + \frac{V_0}{4} \pm \Delta,$$

(9.121)

where $\Delta = \dfrac{16V_0}{9\pi^2}$. The new energy levels are

$$E_+ = E_{12}^{(0)} + \frac{V_0}{4}\left(1 + \frac{64}{9\pi^2}\right),$$

(9.122)

and

$$E_- = E_{12}^{(0)} + \frac{V_0}{4}\left(1 - \frac{64}{9\pi^2}\right),$$

(9.123)

417

and the new eigenstates are

$$\psi_+ = \frac{1}{\sqrt{2}}\left(\psi_{12}^{(0)} - \psi_{21}^{(0)}\right), \tag{9.124}$$

$$\psi_- = \frac{1}{\sqrt{2}}\left(\psi_{12}^{(0)} + \psi_{21}^{(0)}\right). \tag{9.125}$$

The originally degenerate energy levels split into new energy levels $E_\pm$ because when the perturbation is turned on, the off-diagonal matrix elements W_{12} and W_{12} are no longer zero.

9.4 Example Exercises

Exercise 9.1
A particle of mass m moves in a one-dimensional, infinitely deep potential well having a parabolic bottom, $V(x) = \infty$ for $|x| \geq L$ and $V(x) = \xi x^2/L^2$ for $-L < x < L$, where ξ is small compared with the ground-state energy. Treat the term $\xi x^2/L^2$ as a perturbation on a rectangular potential well (denoting the unperturbed states as ϕ_0, ϕ_1, $\phi_2, \ldots$ in order of increasing energy), and calculate, to first order in ξ only, the energy and the amplitudes A_0, A_1, A_2, A_3 of the first four perturbed states.

Exercise 9.2
Calculate the energy levels of an anharmonic oscillator with potential of the form

$$V(x) = \frac{\kappa}{2}x^2 + \xi x^3 \hbar\omega,$$

where the constant κ sets the energy scale for a harmonic oscillator potential. Find the difference between two adjacent perturbed levels, $E_n - E_{n-1}$. A heterodiatomic molecule can absorb or emit electromagnetic waves, the frequency of which coincides with the vibrational frequency of anharmonic oscillations of the molecule about its equilibrium position. For a molecule initially in the ground state, what is the expected absorption spectrum of the molecule?

Exercise 9.3
The potential function of a one-dimensional oscillator of mass m and angular frequency ω is $V(x) = \kappa x^2/2 + \xi x^4$, where $\kappa = m\omega^2$ and the second term is small compared with the first.

(a) Show that, to first order, the effect of the anharmonic term is to change the energy of the ground state by $3\xi(\hbar/2m\omega)^2$.

(b) What would be the first-order effect of an additional x^3 term in the potential?

Exercise 9.4
An electron with effective mass m_e^* is confined to motion in the x direction and experiences a harmonic oscillator potential characterized by angular frequency ω.

(a) Show, using perturbation theory, that the effect of an applied uniform electric field $\mathbf{E}$ in the x direction is to lower all the energy levels by $e^2|\mathbf{E}|^2/2m_e^*\omega^2$.

(b) Compare this with the classical result.

(c) Use perturbation theory to calculate the new ground-state wave function.

Exercise 9.5

The potential seen by an electron with effective mass m_e^* in a GaAs quantum well of thickness $2L$ is approximated as $V(0 < x < 2L) = 0$ and $V(x) = \infty$ elsewhere.

(a) Find the eigenvalues E_n, eigenfunctions ψ_n, and the parity of ψ_n.

(b) The system is subject to a perturbation in the potential energy so that $\hat{W}(x) = e|\mathbf{E}|\hat{x}$, where $\mathbf{E}$ is a constant electric field in the x direction. Find the value of the new energy eigenvalues to first order (the linear Stark effect) for a quantum well of thickness 10 nm subject to an electric field of 10^5 V cm^{-1}. Compare the change in energy value with thermal energy at room temperature.

(c) Find the expression for the second-order correction to the energy eigenvalues for the perturbation in (b).

Exercise 9.6

How does the three-fold degenerate energy $E = 3\hbar\omega$ and the four-fold degenerate energy $E = 4\hbar\omega$ of the isotropic two-dimensional harmonic oscillator separate due to the perturbation $\hat{W} = \kappa'\hat{x}\hat{y}$?

Solutions

Solutions 9.1

The eigenfunctions for a rectangular potential well of thickness $2L$ centered at $x = 0$ and with infinite barrier energy may be expressed in terms of sine functions. Hence,

$$\phi_n^{(0)} = \frac{1}{\sqrt{L}}\sin\left(\frac{(n+1)\pi(x+L)}{2L}\right) = \frac{1}{\sqrt{L}}\sin(k_n(x+L)),$$

where the index $n = 0, 1, 2, \ldots$ labels the eigenstate $k_n = (n+1)\pi/2L$, and the eigenvalues are

$$E_n^{(0)} = \frac{\hbar^2 k_n^2}{2m} = \frac{\hbar^2\pi^2(n+1)^2}{8mL^2}.$$

In the presence of the perturbation $\hat{W}(x) = \xi x^2/L^2$, the energy eigenvalues and eigenfunctions are to first order given by

$$E = E_m^{(0)} + W_{mm},$$

$$\psi = \psi_m^{(0)} + \sum_{k \neq m} \frac{W_{km}}{E_m^{(0)} - E_k^{(0)}} \psi_k^{(0)},$$

where the first-order correction to energy eigenvalues are the diagonal matrix elements

$$E_m^{(1)} = W_{mm} = \langle \phi_m | \xi \frac{x^2}{L^2} | \phi_m \rangle,$$

$$W_{mm} = \frac{\xi}{L^3} \int_{-L}^{L} \sin \left(\frac{(m+1)\pi(x+L)}{2L} \right) x^2 \sin \left(\frac{(m+1)\pi(x+L)}{2L} \right) dx.$$

Using $2 \sin(x) \sin(y) = \cos(x - y) - \cos(x + y)$, this integral may be written as

$$W_{mm} = \frac{\xi}{2L^3} \int_{-L}^{L} x^2 \left(1 - \cos \left(\frac{(m+1)\pi(x+L)}{L} \right) \right) dx$$

$$= \frac{\xi}{2L^3} \left(\frac{2L^3}{3} - \int_{-L}^{L} x^2 \cos \left(\frac{(m+1)\pi(x+L)}{L} \right) dx \right).$$

Solving the integral by parts using $\int UV' dx = UV - \int U'V dx$, with

$$U = x^2,$$
$$U' = 2x,$$
$$V' = \cos \left(\frac{(m+1)\pi(x+L)}{L} \right),$$
$$V = \frac{L}{(m+1)\pi} \sin \left(\frac{(m+1)\pi(x+L)}{L} \right), \text{gives}$$

$$\int UV' dx = \frac{Lx^2}{(m+1)\pi} \sin \left(\frac{(m+1)\pi(x+L)}{L} \right) \Bigg|_{-L}^{L}$$

$$- \frac{2L}{(m+1)\pi} \int_{-L}^{L} x \sin \left(\frac{(m+1)\pi(x+L)}{L} \right) dx.$$

The first term on the right-hand side is zero for all integer values of m. Solving the integral by parts again:

$$U = \frac{2Lx}{(m+1)\pi},$$
$$U' = \frac{2L}{(m+1)\pi},$$
$$V' = \sin \left(\frac{(m+1)\pi(x+L)}{L} \right),$$
$$V = \frac{-L}{(m+1)\pi} \cos \left(\frac{(m+1)\pi(x+L)}{L} \right),$$

$$-\int UV' \, \mathrm{d}x = \frac{2L^2 x}{(m+1)^2 \pi^2} \cos\left(\frac{(m+1)\pi(x+L)}{L}\right)\Big|_{-L}^{L}$$

$$+\frac{2L^2}{(m+1)^2 \pi^2} \int_{-L}^{L} \cos\left(\frac{(m+1)\pi(x+L)}{L}\right) \mathrm{d}x = \frac{4L^3}{(m+1)^2 \pi^2} + 0,$$

so that

$$W_{mm} = \frac{\xi}{2L^3}\left(\frac{2L^3}{3} - \frac{4L^3}{(m+1)^2 \pi^2}\right)$$

and the first-order corrected eigenvalues are

$$E_m = E_m^{(0)} + W_{mm} = \frac{\hbar^2 \pi^2 (n+1)^2}{8mL^2} + \xi\left(\frac{1}{3} - \frac{2}{(m+1)^2 \pi^2}\right).$$

Notice that in the limit $m \to \infty$ the first-order correction to energy eigenvalues is $W_{mm} \to \xi/3$. This limit is easy to understand, since for states with large m the probability of finding the particle somewhere in the range $-L < x < L$ is uniform. In this case, the energy shift is given by the average value of the perturbation in the potential:

$$\langle V(x) \rangle = \frac{1}{L^2} \int_{-L}^{L} \xi x^2 \, \mathrm{d}x \Big/ \int_{-L}^{L} \mathrm{d}x = \frac{1}{3}\xi.$$

The term $-2\xi/(m+1)^2 \pi^2$ is significant for states with small values of m. This decrease in energy shift with increasing quantum number m compared with the average value of the potential is due to the fact that the probability of finding the particle somewhere in the range $-L < x < L$ is nonuniform. For low values of m, particle probability tends to be greater near to $x = 0$ compared with $|x| = L$. Because the perturbing potential is zero at $x = 0$, the first-order energy shift for states with small values of m is always smaller than for states with large values of m.

To find the eigenfunctions in the presence of the perturbation it is necessary to evaluate the matrix elements:

$$W_{km} = \langle \phi_k | \xi \frac{x^2}{L^2} | \phi_m \rangle = \frac{\xi}{L^3} \int_{-L}^{L} \sin\left(\frac{(k+1)\pi(x+L)}{2L}\right) x^2 \sin\left(\frac{(m+1)\pi(x+L)}{2L}\right) \mathrm{d}x$$

$$= \frac{\xi}{2L^3} \int_{-L}^{L} x^2 \left(\cos\left(\frac{(k-m)\pi(x+L)}{2L}\right) - \cos\left(\frac{(k+m)\pi(x+L)}{2L}\right)\right) \mathrm{d}x$$

$$= \frac{8\xi}{\hbar^2}\left(\frac{1}{(m-n)^2} - \frac{1}{(m+n)^2}\right),$$

$$\psi_1 = \phi_1 + \sum_{m \neq 1} \frac{W_{m1}}{E_1^{(0)} - E_m^{(0)}} \phi_m = \phi_1 + \sum_{m(\mathrm{odd}) \neq 1} \frac{\frac{8\xi}{\pi^2}\left(\frac{1}{(m-n)^2} - \frac{1}{(m+n)^2}\right)}{\frac{\hbar^2 \pi^2}{8mL^2}(1 - m^2)} \phi_m,$$

421

$$\psi_2 = \phi_2 + \sum_{m \neq 2} \frac{W_{m2}}{E_2^{(0)} - E_m^{(0)}} \phi_m = \phi_2 + \sum_{m(\text{even}) \neq 2} \frac{\frac{8\xi}{\pi^2}\left(\frac{1}{(m-n)^2} - \frac{1}{(m+n)^2}\right)}{\frac{\hbar^2 \pi^2}{8mL^2}(4 - m^2)} \phi_m,$$

$$\psi_3 = \phi_3 + \sum_{m \neq 3} \frac{W_{m3}}{E_3^{(0)} - E_m^{(0)}} \phi_m = \phi_3 + \sum_{m(\text{odd}) \neq 3} \frac{\frac{8\xi}{\pi^2}\left(\frac{1}{(m-n)^2} - \frac{1}{(m+n)^2}\right)}{\frac{\hbar^2 \pi^2}{8mL^2}(9 - m^2)} \phi_m.$$

Solutions 9.2

The Hamiltonian for the one-dimensional harmonic oscillator subject to perturbation $\hat{W}$ is

$$\hat{H} = \frac{\hat{p}^2}{2m} + \frac{\kappa}{2}\hat{x}^2 + \hat{W},$$

where $\kappa = m\omega^2$. From Section 9.2.5, for the perturbation

$$\hat{W} = \xi \hat{x}^3 \hbar\omega(m\omega/\hbar)^{3/2},$$

$$E_n = \left(n + \frac{1}{2}\right)\hbar\omega - \frac{1}{8}\xi^2\left(30n^2 + 30n + 11\right)\hbar\omega,$$

Hence,

$$E_n - E_{n-1} = \hbar\omega - \frac{1}{8}\xi^2 60 n \hbar\omega,$$

Because, in this exercise, $\hat{W}$ does not explicitly contain the factor $(m\omega/\hbar)^{3/2}$, this needs to be put back in. Since ξ appears as a squared term, the inverse of the factor $(m\omega/\hbar)^{3/2}$ is also squared to give

$$E_n - E_{n-1} = \hbar\omega\left(1 - \frac{1}{8}\xi^2\left(\frac{\hbar}{m\omega}\right)^3 60n\right).$$

The absorption spectrum of an anharmonic diatomic molecule initially in the ground state will consist of a series of absorption lines with energy separation between adjacent lines that decreases with increasing energy. The absorption lines will be at energy

$$E_n - E_0 = n\hbar\omega - \frac{1}{8}\xi^2(30n^2 + 30n)\hbar\omega,$$

and the photon wavelength is given by $\lambda_{\text{phot}} = 2\pi\hbar c/(E_n - E_0)$.

Solutions 9.3

(a) Using the position operator $\hat{x} = (\hbar/2m\omega)^{1/2}\left(\hat{b}^\dagger + \hat{b}\right)$, the perturbation may be expressed as

$$\hat{W} = \xi\hat{x}^4 = \xi\left(\frac{\hbar}{2m\omega}\right)^2\left(\hat{b}^\dagger + \hat{b}\right)^4$$

$$= \xi\left(\frac{\hbar}{2m\omega}\right)^2\left(\hat{b}^\dagger + \hat{b}\right)\left(\hat{b}^{\dagger3} + \hat{b}\hat{b}^\dagger\hat{b}^\dagger + \hat{b}^\dagger\hat{b}\hat{b}^\dagger + \hat{b}\hat{b}\hat{b}^\dagger + \hat{b}^\dagger\hat{b}^\dagger\hat{b} + \hat{b}\hat{b}^\dagger\hat{b} + \hat{b}^\dagger\hat{b}\hat{b} + \hat{b}^3\right)$$

$$= \xi \left(\frac{\hbar}{2m\omega} \right)^2 \left(\hat{b}^{\dagger 4} + \hat{b}^\dagger \hat{b} \hat{b}^\dagger \hat{b}^\dagger + \hat{b}^\dagger \hat{b}^\dagger \hat{b} \hat{b}^\dagger + \hat{b}^\dagger \hat{b} \hat{b} \hat{b}^\dagger + \hat{b}^\dagger \hat{b}^\dagger \hat{b}^\dagger \hat{b} + \hat{b}^\dagger \hat{b} \hat{b}^\dagger \hat{b} + \hat{b}^\dagger \hat{b}^\dagger \hat{b} \hat{b} \right.$$
$$\left. + \hat{b}^\dagger \hat{b}^3 + \hat{b} \hat{b}^{\dagger 3} + \hat{b} \hat{b} \hat{b}^\dagger \hat{b}^\dagger + \hat{b} \hat{b}^\dagger \hat{b} \hat{b}^\dagger + \hat{b} \hat{b} \hat{b} \hat{b}^\dagger + \hat{b} \hat{b}^\dagger \hat{b}^\dagger \hat{b} + \hat{b} \hat{b} \hat{b}^\dagger \hat{b} + \hat{b} \hat{b}^\dagger \hat{b} \hat{b} + \hat{b}^4 \right).$$

Energy eigenvalues in first-order perturbation theory couple the same state, so only symmetric terms with two $\hat{b}^\dagger$s and two $\hat{b}$s will contribute:

$$W_{nn} = \xi \left(\frac{\hbar}{2m\omega} \right)^2 \left((n+1)(n+2) + (n+1)^2 + n^2 + n(n-1) + 2n(n+1) \right)$$
$$= \xi \left(\frac{\hbar}{2m\omega} \right)^2 \left(6n^2 + 6n + 3 \right),$$

so, to first order,

$$E_n = \hbar\omega \left(n + \frac{1}{2} \right) + \xi \left(\frac{\hbar}{2m\omega} \right)^2 \left(6n^2 + 6n + 3 \right),$$

and the new ground state eigenenergy is

$$E_0 = \frac{\hbar\omega}{2} + 3\xi \left(\frac{\hbar}{2m\omega} \right)^2.$$

(b) There is no first-order correction for a perturbation in x^3 because it cannot couple to the same state. (There are always an odd number of operators $\hat{b}^\dagger$ or $\hat{b}$.)

Solutions 9.4

(a) The solution follows that already given in this chapter.

(b) The new energy levels of the oscillator are, to second order,

$$E = \hbar\omega \left(n + \frac{1}{2} \right) - \frac{e^2 |\mathbf{E}|^2}{2m_e^* \omega^2},$$

which is the same as the exact result. Physically, the particle oscillates at the same frequency, ω, as the unperturbed case, but it is displaced a distance of magnitude $e|\mathbf{E}|/m_e^* \omega^2$, and the new energy levels are shifted by $-e^2 |\mathbf{E}|^2 / 2m_e^* \omega^2$.

(c) The ground-state wave function of the unperturbed harmonic oscillator is

$$\psi_0(x) = \left(\frac{m_e^* \omega}{\pi \hbar} \right)^{1/4} e^{-x^2 m_e^* \omega / 2\hbar}.$$

After the perturbation, it is

$$\psi_0(x) = \left(\frac{m_e^* \omega}{\pi \hbar} \right)^{1/4} e^{-\frac{m_e^* \omega}{2\hbar} \left(x - \frac{e|\mathbf{E}|}{m_e^* \omega^2} \right)^2}.$$

The result using second-order perturbation theory is

$$\psi = \psi_m^{(0)} + \sum_{k \neq m} \frac{W_{mk}}{E_m - E_k} \psi_k^{(0)}$$

$$+ \sum_{k \neq m} \left(\sum_{n \neq m} \frac{W_{kn} W_{mn}}{\left(E_m^{(0)} - E_n^{(0)}\right)\left(E_m^{(0)} - E_k^{(0)}\right)} - \frac{W_{mm} W_{km}}{\left(E_m^{(0)} - E_k^{(0)}\right)^2} \right) \psi_k^{(0)}$$

$$- \sum_{n \neq m} \frac{1}{2} \frac{|W_{mn}|^2}{\left(E_m^{(0)} - E_n^{(0)}\right)^2} \psi_m^{(0)}.$$

Solutions 9.5

(a) The eigenfunctions and eigenvalues for the electron in a rectangular well of thickness $2L$ and infinite barrier energy are found by solving the time-independent Schrödinger equation

$$\hat{H}^{(0)} \psi_n^{(0)} = E_n^{(0)} \psi_n^{(0)},$$

where the Hamiltonian for the electron in the potential is

$$\hat{H}^{(0)} = \frac{\hat{p}^2}{2m_e^*} + V(x).$$

The solution for the eigenfunctions is

$$\psi_n^{(0)} = \frac{1}{\sqrt{L}} \sin\left(\frac{n\pi x}{2L}\right), \qquad \text{where } n = 1, 2, \ldots,$$

and the parity of the eigenfunctions is even for odd-integer and odd for even-integer values of n.

The solution for the eigenvalues is

$$E_n^{(0)} = \frac{\hbar^2 k_n^2}{2m_e^*} = \frac{\hbar^2 n^2 \pi^2}{8m_e^* L^2}, \qquad \text{where } k_n = \frac{n2\pi}{2 \times 2L} = \frac{n\pi}{2L}.$$

(b) First-order correction to energy eigenvalues is

$$E_n^{(1)} = \langle n|\hat{W}|n \rangle = \langle n|e|\mathbf{E}|\hat{x}|n \rangle = W_{nn} = \frac{e|\mathbf{E}|}{L} \int_{x=0}^{x=2L} x \sin^2\left(\frac{n\pi x}{2L}\right) dx.$$

Using the relation $2\sin(x)\sin(y) = \cos(x-y) - \cos(x+y)$ with $x = y = n\pi x/2L$ allows the integrand to be rewritten, giving

$$W_{nn} = \frac{e|\mathbf{E}|}{2L} \int_{x=0}^{x=2L} x \left(1 - \cos\left(\frac{n\pi x}{L}\right)\right) dx = \frac{e|\mathbf{E}|}{2L}\left[\frac{x^2}{2}\right]_{x=0}^{x=2L} = e|\mathbf{E}|L,$$

which is the *linear Stark effect*. The energy-level shift for an electric field of magnitude $|\mathbf{E}| = 10^5 \, \mathrm{V \, cm^{-1}}$ in the x direction across a well of thickness $2L = 10 \, \mathrm{nm}$ is

$$\Delta = 10^7 \times 5 \times 10^{-9} = 50 \, \mathrm{meV}.$$

At temperature $T = 300 \, \mathrm{K}$ and $\Delta > k_{\mathrm{B}}T = 25 \, \mathrm{meV}$, which indicates that the Stark effect produces a large-enough change in energy eigenvalue to be of potential use in a room-temperature device.

(c) The new energy levels to second order are found using

$$E_n = E_n^{(0)} + E_n^{(1)} + E_n^{(2)},$$

where $E_k^{(1)} = W_{kk}$ and $E_k^{(2)} = \sum_{j \neq k} W_{kj} W_{jk} / \left(E_k^{(0)} - E_j^{(0)} \right)$.

The second-order matrix elements are the off-diagonal terms

$$W_{kj} = \langle k | \hat{W} | j \rangle = \langle k | e | \mathbf{E} | \hat{x} | j \rangle = \frac{e|\mathbf{E}|}{L} \int\limits_{x=0}^{x=2L} x \sin\left(\frac{k\pi x}{2L} \right) \sin\left(\frac{j\pi x}{2L} \right) dx.$$

Using the relation $2 \sin(x) \sin(y) = \cos(x - y) - \cos(x + y)$ with $x = k\pi x/2L$ and $y = j\pi x/2L$ allows the integrand to be rewritten, giving

$$W_{kj} = \frac{e|\mathbf{E}|}{2L} \int\limits_{x=0}^{x=2L} x \left(\cos\left(\frac{(k - j)\pi x}{2L} \right) - \cos\left(\frac{(k + j)\pi x}{2L} \right) \right) dx.$$

For $(k \pm j)$ odd and integrating by parts, using $\int UV' \, dx = UV - \int U'V \, dx$ with $U = x$ and $V' = \cos((k \pm j)\pi x/2L)$, gives

$$W_{kj} = \frac{e|\mathbf{E}|}{2L} \left(\frac{4L^2(\cos((k - j)\pi) - 1)}{\pi^2(k - j)^2} - \frac{4L^2(\cos((k + j)\pi) - 1)}{\pi^2(k + j)^2} \right)$$

$$= \frac{-4e|\mathbf{E}|L}{\pi^2} \left(\frac{1}{(k + j)^2} - \frac{1}{(k - j)^2} \right) = \frac{-16eE_x L}{\pi^2} \frac{kj}{(k^2 - j^2)^2}.$$

For $(k \pm j)$ even, symmetry requires that

$$W_{kj} = 0.$$

So, the perturbing potential only mixes states of different parity.

Solutions 9.6

A two-dimensional harmonic oscillator with motion in the x–y plane is subject to perturbation $\hat{W} = \kappa' \hat{x}\hat{y}$. The task is to find how the three-fold degenerate energy, $E = 3\hbar\omega$, and the four-fold degenerate energy, $E = 4\hbar\omega$, separate due to this perturbation. Separation of variables x and y allows the unperturbed Hamiltonian to be written as

$$\hat{H}^{(0)} = \frac{(\hat{p}_x^2 + \hat{p}_y^2)}{2m} + \frac{\kappa}{2}\left(\hat{x}^2 + \hat{y}^2 \right) = \hbar\omega \left(\hat{b}_x^\dagger \hat{b}_x + \hat{b}_y^\dagger \hat{b}_y + 1 \right),$$

425

where $\hat{x} = \sqrt{\hbar/2m\omega}(\hat{b}_x + \hat{b}_x^\dagger)$ and $\hat{y} = \sqrt{\hbar/2m\omega}(\hat{b}_y + \hat{b}_y^\dagger)$. The eigenstates are of the form $\psi_{nm}^{(0)} = \phi_n^{(0)}(x)\phi_m^{(0)}(y) = |nm\rangle$, and the energy eigenvalues are

$$E_{nm} = \hbar\omega(n + m + 1),$$

where n and m are positive integers. States with eigenenergy $3\hbar\omega$ are E_{02}, E_{11}, and E_{20}, and so they are three-fold degenerate. To find the effect of the perturbation $\hat{W} = \kappa'\hat{x}\hat{y}$ the total Hamiltonian is written as

$$\hat{H} = \hat{H}^{(0)} + \hat{W} = \hbar\omega\left(\hat{b}_x^\dagger\hat{b}_x + \hat{b}_y^\dagger\hat{b}_y + 1\right) + \frac{\hbar\kappa'}{2m\omega}\left(\hat{b}_x + \hat{b}_x^\dagger\right)\left(\hat{b}_y + \hat{b}_y^\dagger\right).$$

This has eigenfunction solutions that are linear combinations of the unperturbed eigenstates so that

$$\psi_j = a_1(j)\psi_{11}^{(0)} + a_2(j)\psi_{20}^{(0)} + a_3(j)\psi_{02}^{(0)}.$$

The coefficients a_n may be found by writing the Schrödinger equation in matrix form,

$$\begin{bmatrix} \langle 02|\hat{H}|02\rangle & \langle 02|\hat{H}|11\rangle & \langle 02|\hat{H}|20\rangle \\ \langle 11|\hat{H}|02\rangle & \langle 11|\hat{H}|11\rangle & \langle 11|\hat{H}|20\rangle \\ \langle 20|\hat{H}|02\rangle & \langle 20|\hat{H}|11\rangle & \langle 20|\hat{H}|20\rangle \end{bmatrix} \begin{bmatrix} a_1 \\ a_2 \\ a_3 \end{bmatrix} = E \begin{bmatrix} a_1 \\ a_2 \\ a_3 \end{bmatrix}.$$

The diagonal terms have a value that is the unperturbed eigenvalue $3\hbar\omega$. To show this, consider

$$\langle 02|\hat{H}|02\rangle$$

$$= \hbar\omega\langle 02|\hat{b}_x^\dagger\hat{b}_x + \hat{b}_y^\dagger\hat{b}_y + 1|02\rangle + \frac{\hbar\kappa'}{2m\omega}\langle 02|\hat{b}_x\hat{b}_y + \hat{b}_x^\dagger\hat{b}_y + \hat{b}_y^\dagger\hat{b}_x + \hat{b}_x^\dagger\hat{b}_y^\dagger|02\rangle.$$

The first term on the right-hand side has value $3\hbar\omega$, and the second term of the right-hand side is zero. Because the perturbation W is linear in x and y, only the off-diagonal terms adjacent to the diagonal are finite. For example,

$$\langle 11|\hat{H}|02\rangle = \hbar\omega\langle 11|\hat{b}_x^\dagger\hat{b}_x + \hat{b}_y^\dagger\hat{b}_y + 1|02\rangle + \frac{\hbar\kappa'}{2m\omega}\langle 11|\left(\hat{b}_x + \hat{b}_x^\dagger\right)\left(\hat{b}_y + \hat{b}_y^\dagger\right)|02\rangle.$$

The first term on the right-hand side is zero, leaving

$$\langle 11|\hat{H}|02\rangle = \frac{\hbar\kappa'}{2m\omega}\langle 11|\hat{b}_x\hat{b}_y + \hat{b}_x^\dagger\hat{b}_y + \hat{b}_x\hat{b}_y^\dagger + \hat{b}_x^\dagger\hat{b}_y^\dagger|02\rangle = \frac{\hbar\kappa'}{2m\omega}\langle 11|\hat{b}_x^\dagger\hat{b}_y|02\rangle.$$

Recalling that $|\hat{b}^\dagger n\rangle = (n+1)^{1/2}|n+1\rangle$ and $|\hat{b}n\rangle = n^{1/2}|n-1\rangle$ then

$$\langle 11|\hat{H}|02\rangle = \frac{\hbar\kappa'}{2m\omega}\langle 11|\hat{b}_x^\dagger\hat{b}_y|02\rangle = \frac{\sqrt{2}\hbar\kappa'}{2m\omega}\langle 11|\hat{b}_x^\dagger|01\rangle = \frac{\sqrt{2}\hbar\kappa'}{2m\omega}\langle 11|11\rangle = \frac{\sqrt{2}\hbar\kappa'}{2m\omega}.$$

It follows that

$$
\begin{bmatrix}
3\hbar\omega & \dfrac{\sqrt{2}\hbar\kappa'}{2m\omega} & 0 \\[2mm]
\dfrac{\sqrt{2}\hbar\kappa'}{2m\omega} & 3\hbar\omega & \dfrac{\sqrt{2}\hbar\kappa'}{2m\omega} \\[2mm]
0 & \dfrac{\sqrt{2}\hbar\kappa'}{2m\omega} & 3\hbar\omega
\end{bmatrix}
\begin{bmatrix} a_1 \\ a_2 \\ a_3 \end{bmatrix}
= E
\begin{bmatrix} a_1 \\ a_2 \\ a_3 \end{bmatrix} .
$$

The eigenvalues of the matrix are

$$E_1 = 3\hbar\omega,$$

$$E_2 = \frac{3m\hbar\omega^2 - \hbar\kappa'}{m\omega},$$

$$E_3 = \frac{3m\hbar\omega^2 + \hbar\kappa'}{m\omega}.$$

The eigenfunctions are given by the coefficients

$$
\begin{bmatrix} a_1 \\ a_2 \\ a_3 \end{bmatrix} = \frac{1}{\sqrt{2}} \begin{bmatrix} -1 \\ 0 \\ 1 \end{bmatrix},
$$

$$
\begin{bmatrix} a_1 \\ a_2 \\ a_3 \end{bmatrix} = \frac{1}{2} \begin{bmatrix} 1 \\ -\sqrt{2} \\ 1 \end{bmatrix},
$$

$$
\begin{bmatrix} a_1 \\ a_2 \\ a_3 \end{bmatrix} = \frac{1}{2} \begin{bmatrix} 1 \\ \sqrt{2} \\ 1 \end{bmatrix}.
$$

A similar procedure is followed to find how the four-fold degenerate levels of a two-dimensional harmonic oscillator with motion in the x–y plane subject to perturbation $\hat{W} = \kappa' \, \hat{x}\hat{y}$ change. The perturbed Schrödinger equation matrix is

$$
\begin{bmatrix}
4\hbar\omega & \dfrac{\sqrt{3}\hbar\kappa'}{2m\omega} & 0 & 0 \\[2mm]
\dfrac{\sqrt{3}\hbar\kappa'}{2m\omega} & 4\hbar\omega & \dfrac{\hbar\kappa'}{m\omega} & 0 \\[2mm]
0 & \dfrac{\hbar\kappa'}{m\omega} & 4\hbar\omega & \dfrac{\sqrt{3}\hbar\kappa'}{2m\omega} \\[2mm]
0 & 0 & \dfrac{\sqrt{3}\hbar\kappa'}{2m\omega} & 4\hbar\omega
\end{bmatrix}
\begin{bmatrix} a_1 \\ a_2 \\ a_3 \\ a_4 \end{bmatrix}
= E
\begin{bmatrix} a_1 \\ a_2 \\ a_3 \\ a_4 \end{bmatrix},
$$

which has eigenvalues

$$E_1 = \frac{8m\hbar\omega^2 - 3\hbar\kappa'}{2m\omega},$$

$$E_2 = \frac{8m\hbar\omega^2 - \hbar\kappa'}{2m\omega},$$

$$E_3 = \frac{8m\hbar\omega^2 + \hbar\kappa'}{2m\omega},$$

$$E_4 = \frac{8m\hbar\omega^2 + 3\hbar\kappa'}{2m\omega}.$$

The eigenfunctions are given by the coefficients

$$\begin{bmatrix} a_1 \\ a_2 \\ a_3 \\ a_4 \end{bmatrix} = \frac{1}{2\sqrt{2}} \begin{bmatrix} -1 \\ \sqrt{3} \\ -\sqrt{3} \\ 1 \end{bmatrix},$$

$$\begin{bmatrix} a_1 \\ a_2 \\ a_3 \\ a_4 \end{bmatrix} = \frac{\sqrt{3}}{2\sqrt{2}} \begin{bmatrix} 1 \\ -1/\sqrt{3} \\ -1/\sqrt{3} \\ 1 \end{bmatrix},$$

$$\begin{bmatrix} a_1 \\ a_2 \\ a_3 \\ a_4 \end{bmatrix} = \frac{\sqrt{3}}{2\sqrt{2}} \begin{bmatrix} -1 \\ -1/\sqrt{3} \\ 1/\sqrt{3} \\ 1 \end{bmatrix},$$

$$\begin{bmatrix} a_1 \\ a_2 \\ a_3 \\ a_4 \end{bmatrix} = \frac{1}{2\sqrt{2}} \begin{bmatrix} 1 \\ \sqrt{3} \\ \sqrt{3} \\ 1 \end{bmatrix}.$$

9.5 Problems

Problem 9.1
Consider a three-dimensional potential, $V(x, y, z)$, that is infinite except in a region $0 < x < L, 0 < y < L$, and $0 < z < L$, where $V(x, y, z) = 0$.

(a) Write down the time-independent Schrödinger equation for a particle mass m confined to motion in the potential and solve for the eigenfunctions.

(b) Show that the eigenenergies are $E^{(0)}_{n_x,n_y,n_z} = \frac{\pi^2\hbar^2}{2mL^2}\left(n_x^2 + n_y^2 + n_z^2\right)$, where n_x, n_y, and n_z are nonzero positive integers. What is the degeneracy of the ground state and what is the degeneracy of the first excited state?

(c) The system is perturbed by introducing a potential $\hat{W} = V_0$ in a region for which $0 < x < \frac{L}{2}, 0 < y < \frac{L}{2}$, and $0 < z < L$. The perturbation $\hat{W} = 0$ elsewhere and V_0 is a constant. Use first-order perturbation theory to find the new ground state energy.

(d) What are the new eigenenergies and eigenfunctions of the first excited state?

Problem 9.2
A particular unperturbed Hamiltonian expressed in matrix form is

$$\mathbf{H}^{(0)} = \begin{bmatrix} 1 & 0 & 0 \\ 0 & 3 & 0 \\ 0 & 0 & 2 \end{bmatrix}.$$

The system is subject to the perturbation

$$\mathbf{W} = \begin{bmatrix} 0 & \Delta & 0 \\ \Delta & 0 & 0 \\ 0 & 0 & \Delta \end{bmatrix},$$

where $\Delta \ll 1$.
(a) Find the exact eigenvalues of $\mathbf{H} = \mathbf{H}^{(0)} + \mathbf{W}$.
(b) Find the eigenvalues to second order using time-independent nondegenerate perturbation theory.
(c) Compare the results obtained in (a) and (b).

Problem 9.3
(a) An electron moves in a one-dimensional box of length X. Apply periodic boundary conditions and find the electron eigenfunction and eigenvalues.
(b) Now apply a weak periodic potential $V(x) = V(x + L)$ to the system, where $X = NL$ and N is a large positive integer. Using nondegenerate perturbation theory, find the first-order correction to the wave functions and the second-order correction to the eigenenergies.
(c) When the wave vector k is close to $n\pi/L$, where n is an integer, the result in (b) is no longer valid. Use two-state degenerate perturbation theory to find the corrected energy values for $k = n\pi((1 + \Delta)/L)$ and $k' = n\pi((1 - \Delta)/L)$, where Δ is small compared with π/L.
(d) Use the results of (b) and (c) to draw the electron dispersion relation, $E(k)$.
(e) If the lowest-frequency Fourier component of the perturbative periodic potential is chosen in part (b), then $V(x) = V_1 \cos(\pi x/L)$. Repeat (b), (c), and (d) using this potential.

Hint: $V(x) = V_0 + \sum_{n \neq 0} V_n e^{i2\pi nx/L}$, and choose $V_0 = 0$.

Problem 9.4
A semiconductor quantum dot is modeled as a three-dimensional box of side L and infinite barrier energy. An electron in the quantum dot has energy $E = \frac{3\pi^2 \hbar^2}{m_e^* L^2}$, where m_e^* is the effective electron mass.
(a) Calculate the first-order correction to the electron energy when a uniform electric field $\mathbf{E}$ is applied in the z direction.
(b) If $L = 20\,\text{nm}$, the effective electron mass is $m_e^* = 0.07 \times m_0$, and the strength of the applied electric field is $|\mathbf{E}| = 10^4\,\text{V cm}^{-1}$, what is the value of the new electron energy level?
(c) Explain the degeneracy of the system after the perturbation is applied.

Problem 9.5

The first four lowest energy states of a one-dimensional harmonic oscillator with characteristic frequency ω_0 are subject to the perturbation

$$\mathbf{W} = \begin{bmatrix} W_{00} & W_{01} & W_{02} & W_{03} \\ W_{10} & W_{11} & W_{12} & W_{13} \\ W_{20} & W_{21} & W_{22} & W_{23} \\ W_{30} & W_{31} & W_{32} & W_{33} \end{bmatrix} = \Delta\hbar\omega_0 \begin{bmatrix} 1 & 0 & \frac{-1}{\sqrt{2}} & 0 \\ 0 & 0 & 0 & 0 \\ \frac{-1}{\sqrt{2}} & 0 & \frac{1}{2} & 0 \\ 0 & 0 & 0 & 0 \end{bmatrix},$$

where $\Delta \ll 1$.
 (a) Find the new eigenenergies to first order in time-independent perturbation theory.
 (b) Find the new eigenenergies to second order in time-independent perturbation theory.

Problem 9.6

An electron is confined in a one-dimensional rectangular potential well of thickness $2L$ such that $V(x) = 0$ for $0 < x < 2L$ and $V(x) = \infty$ elsewhere. The system is subject to a constant electric field $\mathbf{E}$ in the x direction.
 (a) Write down an analytic expression for the new eigenfunctions and energy eigenvalues evaluated to first order in time-independent perturbation theory.
 (b) For an electron with effective electron mass $m_{\mathrm{e}}^* = 0.07 \times m_0$, where m_0 is the bare electron mass, the well thickness is $2L = 10\,\mathrm{nm}$ and the electric field $2 \times 10^5\,\mathrm{V}\ \mathrm{cm}^{-1}$, use the result from (a) to find the new eigenfunctions and energy eigenvalues.
 (c) Sketch and explain how, according to first-order time-independent perturbation theory, the unperturbed ground-state wave function is modified under the influence of the perturbation. Under what circumstances are the results of first-order time-independent perturbation theory expected to be valid?

Problem 9.7

An electron mass m_0 confined to motion in a one-dimensional harmonic potential with characteristic frequency ω is subject to a perturbing potential $\hat{W} = \xi x^3 \hbar\omega (m_0\omega/\hbar)^{3/2}$.
 (a) Write down the Hamiltonian for the system.
 (b) Calculate to second order the eigenenergies of the perturbed system.
 (c) Calculate to first order the eigenstates for the perturbed system.

Problem 9.8

The unperturbed Hamiltonian for an electron with effective mass m_{e}^* and kinetic energy $\hat{T} = (\hat{p}_x^2 + \hat{p}_y^2)/2m_{\mathrm{e}}^*$ in a two-dimensional harmonic oscillator potential $\hat{V} = \kappa(\hat{x}^2 + \hat{y}^2)/2$ is $\hat{H}^{(0)} = \hat{T} + \hat{V}$ where $\kappa = m_{\mathrm{e}}^*\omega^2$. The eigenstates associated with $\hat{H}^{(0)}\phi_{nm} = E_{nm}\phi_{nm}$ are of the form $\phi_{nm} = \phi_n(x)\phi_m(y) = |nm\rangle$ with $n, m = 0, 1, 2, \ldots$. The eigenstates are $(n + m + 1)$-fold degenerate with eigenenergies $E_{nm} = \hbar(n + m + 1)$.
 (a) Find the position of minimum potential and the amount by which any perturbing potential $\hat{W} = \kappa'\hat{x}/2$ or $\hat{W} = \kappa'(\hat{x} + \hat{y})/2$ shifts eigenenergy values and show that the perturbation does not break the degeneracy of the states.
 (b) Create a contour plot of the potential $V(x, y)$ in the range $-2\,\mathrm{nm} < x < 2\,\mathrm{nm}$ and $-2\,\mathrm{nm} < y < 2\,\mathrm{nm}$ for $\kappa = 2\,\mathrm{eV}\,\mathrm{nm}^{-2}$, and overlay a contour plot of $\hat{V}(x, y) + \hat{W}(x) =$

$\kappa(\hat{x}^2+\hat{y}^2)/2+\kappa'\hat{x}/2$ for $\kappa' = 1\,\text{eV nm}$. Use the contour plots to explain why the perturbing potential fails to break the degeneracy of the states.

(c) Create a contour plot of the potential $V(x,y)$ in the range $-2\,\text{nm} < x < 2\,\text{nm}$ and $-2\,\text{nm} < y < 2\,\text{nm}$ for $\kappa = 2\,\text{eV nm}^{-2}$, and overlay a contour plot of $\hat{V}(x,y)+\hat{W}(x,y) = \kappa(\hat{x}^2+\hat{y}^2)/2 + \kappa'\hat{x}\hat{y}/2$ for $\kappa' = 1\,\text{eV nm}^{-2}$. Use the contour plots to explain why the perturbing potential breaks the degeneracy of the states in this case.

Problem 9.9

Consider a system such that $\hat{H}|\psi\rangle = (\hat{H}_0+\lambda\hat{W})|\psi\rangle = E|\psi\rangle$ for which $(\hat{H}^{(0)} + \lambda\hat{W})(\phi_n^{(0)} + \lambda\phi_n^{(1)}+\cdots) = (E^{(0)}+\lambda E^{(1)}+\cdots)(\phi_n^{(0)}+\lambda\phi_n^{(1)}+\cdots)$, where the unperturbed Hamiltonian is $\hat{H}^{(0)}$ with known eigenstates $\phi_n^{(0)}$, the perturbation is $\hat{W}$, and λ is a dummy variable of unit magnitude used to keep track of the order of terms in the perturbative expansion. The energy eigenstate to first order in time-independent nondegenerate perturbation theory is normalized to unity so that $\int \psi_n^* \psi_n \; \mathrm{d}^3 r = 1 + \lambda a_m^{(1)} + \lambda a_m^{(1)*} + O(\lambda^2) = 1$ with $\psi_n = \phi_n^{(0)} + \phi_n^{(1)}$ and $a_m^{(1)} = 0$ as one possible solution. Another solution is $a_m^{(1)}$ pure imaginary so that $a_m^{(1)} = ia = ae^{i\pi/2}$ with a real.

(a) Show that for $a_m^{(1)}$ pure imaginary the terms $\phi_m^{(0)} + a_m^{(1)}\phi_m^{(0)}$ may be written $e^{ia}\psi_m^{(0)}$.

(b) The phase term e^{ia} in (a) may be chosen arbitrarily. Choose the value of a so that $\langle\phi_m^{(0)}|\phi_m^{(1)}\rangle = 0$ and comment on the result.

Problem 9.10

A particle of mass m with motion confined to one-dimension in a potential $V(x)$ has lowest energy (ground state) wave function

$$\psi(x) = \left(\frac{1}{\pi\sigma^2}\right)^{1/4} e^{-x^2/2\sigma^2}.$$

(a) If $\sigma^2 = \hbar/m\omega$, find an expression for the potential $V(x)$ and the value of the ground-state energy.

(b) Calculate to first order the change in ground-state energy when the system is subject to a perturbation

$$\hat{W} = \frac{\lambda}{x^2 + \gamma^2}$$

when $\gamma \ll \sigma$.

(c) Repeat the calculation in (b) for the case when $\gamma \gg \sigma$.

Note that

$$\int_{-\infty}^{\infty} \frac{1}{x^2 + \gamma^2}\mathrm{d}x = \frac{\pi}{\gamma}$$

and

$$\int_{-\infty}^{\infty} e^{ax^2}\mathrm{d}x = \sqrt{\frac{\pi}{a}}.$$

10 Angular Momentum and the Hydrogenic Atom

10.1 Angular Momentum

In classical mechanics, the constants of motion of an isolated system are energy, linear momentum, and angular momentum. So far in this book, angular momentum has not been considered. This chapter starts by defining classical angular momentum and then proceeds to find the corresponding quantum operators. Following this, a hydrogenic atom is studied as a prototype application.

10.1.1 Classical Angular Momentum

Suppose, as illustrated in Fig. 10.1, a point particle mass m_0 in a rigid body rotating about a fixed point has linear velocity $\mathbf{v}$ at instant $t = 0$, and momentum $\mathbf{p} = m_0\mathbf{v}$. Angular momentum $\mathbf{L}$ about the fixed point $\mathbf{r} = \mathbf{0}$ is defined as[1]

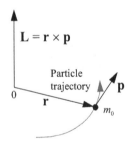

Fig. 10.1 The angular momentum of an electron point particle mass m_0 at position $\mathbf{r}$ moving with momentum $\mathbf{p}$ is defined as $\mathbf{L} = \mathbf{r} \times \mathbf{p}$ where the principal axis of rotation passes through $\mathbf{r} = \mathbf{0}$.

$$\mathbf{L} = \mathbf{r} \times \mathbf{p}, \tag{10.1}$$

where $\mathbf{r}$ is the position of particle. Angular momentum is measured in units of J s.

This equation can be written in terms of angular frequency vector $\underline{\omega}$, normal to the plane of rotation, to give, for the case of an electron of mass m_0,

$$\mathbf{L} = m_0(\mathbf{r} \times \mathbf{v}) = m_0(\mathbf{r} \times (\underline{\omega} \times \mathbf{r})). \tag{10.2}$$

---

[1] For example, H. Goldstein, *Classical Mechanics*, Massachusetts, Addison-Wesley, 3rd ed. 2002 (ISBN 0 20 165702 3); E. T. Whittaker, *A Treatise on Analytical Dynamics of Particles and Rigid Bodies*, Cambridge, Cambridge University Press, 4th ed. 1937, reprinted 1988 (ISBN 0 52 135883 3).

Making use of the vector relationship $\mathbf{a} \times (\mathbf{b} \times \mathbf{c}) = (\mathbf{a} \cdot \mathbf{c})\mathbf{b} - (\mathbf{a} \cdot \mathbf{b})\mathbf{c}$, the expression for $\mathbf{L}$ may be rewritten as

$$\mathbf{L} = m_0 \left(r^2 \underline{\omega} - \mathbf{r}(\mathbf{r} \cdot \underline{\omega}) \right) = \mathbf{I}_{\mathrm{L}}\underline{\omega}, \tag{10.3}$$

where the inertia tensor is

$$\mathbf{I}_{\mathrm{L}} = m_0 \left(r^2 \mathbf{1} - \mathbf{rr} \right). \tag{10.4}$$

In Eq. (10.4), $\mathbf{1}$ is the identity (unit) matrix. If $\underline{\omega} = \omega\mathbf{n}$, where $\mathbf{n}$ is the unit vector in the direction of $\underline{\omega}$, then the scalar

$$I_{\mathrm{L}} = m_0 \left(r^2 - (\mathbf{r} \cdot \mathbf{n})^2 \right) \tag{10.5}$$

is the moment of inertia for an electron of mass m_0 about the axis of rotation. From this definition it is clear that the moment of inertia of a point particle, such as an electron with mass m_0, at distance r measured from a principal axis of rotation is just

$$I_{\mathrm{L}} = m_0 r^2. \tag{10.6}$$

The kinetic energy of a rigid body with moment of inertia I_{L} and rotating at angular frequency ω is

$$E = \frac{1}{2} I_{\mathrm{L}} \omega^2 = \frac{L^2}{2I_{\mathrm{L}}}. \tag{10.7}$$

The classical Cartesian components of total angular momentum $\mathbf{L}$ are

$$L_x = yp_z - zp_y, \tag{10.8}$$
$$L_y = zp_x - xp_z, \tag{10.9}$$
$$L_z = xp_y - yp_x. \tag{10.10}$$

Note the cyclic permutation in x, y, z results, for example, in only the p_x and p_y components of linear momentum contributing to the cross-product that gives the z-component of angular momentum, L_z. Figure 10.2 is a diagram that illustrates the contributions of linear momentum in the $x-y$ plane that give rise to angular momentum L_z.

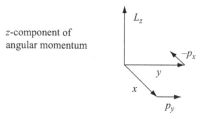

Fig. 10.2 Illustration, in Cartesian coordinates, of the z-component of angular momentum, L_z, which has contributions from linear momentum components p_y and p_x.

10.2 The Angular Momentum Operator

The quantum mechanical angular momentum operator is found by replacing the classical momentum $\mathbf{p}$ with operator $\hat{\mathbf{p}} = -i\hbar\nabla$. Hence, the components of angular momentum may be written in terms of products of position operators, $\hat{x}, \hat{y}, \hat{z}$, and linear momentum operators, $\hat{p}_x, \hat{p}_y, \hat{p}_z$, where $\hat{p}_x = -i\hbar\dfrac{\partial}{\partial x}, \hat{p}_y = -i\hbar\dfrac{\partial}{\partial y}$, and $\hat{p}_z = -i\hbar\dfrac{\partial}{\partial z}$, so that

$$
\boxed{
\begin{aligned}
\hat{L}_x &= \hat{y}\hat{p}_z - \hat{z}\hat{p}_y = -i\hbar\left(y\frac{\partial}{\partial z} - z\frac{\partial}{\partial y}\right), \\
\hat{L}_y &= \hat{z}\hat{p}_x - \hat{x}\hat{p}_z = -i\hbar\left(z\frac{\partial}{\partial x} - x\frac{\partial}{\partial z}\right), \\
\hat{L}_z &= \hat{x}\hat{p}_y - \hat{y}\hat{p}_x = -i\hbar\left(x\frac{\partial}{\partial y} - y\frac{\partial}{\partial x}\right),
\end{aligned}
}
\tag{10.11}
$$

or, in general, the total momentum operator is

$$
\boxed{\hat{\mathbf{L}} = -i\hbar(\hat{\mathbf{r}} \times \nabla).}
\tag{10.12}
$$

Planck's constant has dimensions of angular momentum so it is reasonable to anticipate that angular momentum is quantized in units of $\hbar$.

The commutator relations for these operators (Eqs. (10.11) and (10.12)) can now be found. As an example, consider the commutator between $\hat{L}_x$ and $\hat{L}_y$:

$$
\begin{aligned}
\left[\hat{L}_x, \hat{L}_y\right] = \hat{L}_x\hat{L}_y - \hat{L}_y\hat{L}_x &= [(\hat{y}\hat{p}_z - \hat{z}\hat{p}_y), (\hat{z}\hat{p}_x - \hat{x}\hat{p}_z)] \\
&= (\hat{y}\hat{p}_z - \hat{z}\hat{p}_y)(\hat{z}\hat{p}_x - \hat{x}\hat{p}_z) - (\hat{z}\hat{p}_x - \hat{x}\hat{p}_z)(\hat{y}\hat{p}_z - \hat{z}\hat{p}_y) \\
&= [\hat{y}\hat{p}_z, \hat{z}\hat{p}_x] - [\hat{y}\hat{p}_z, \hat{x}\hat{p}_z] - [\hat{z}\hat{p}_y, \hat{z}\hat{p}_x] + [\hat{z}\hat{p}_y, \hat{x}\hat{p}_z] \\
&= [\hat{y}\hat{p}_z, \hat{z}\hat{p}_x] - 0 - 0 + [\hat{z}\hat{p}_y, \hat{x}\hat{p}_z] \\
&= (\hat{y}\hat{p}_z\hat{z}\hat{p}_x - \hat{z}\hat{p}_x\hat{y}\hat{p}_z + \hat{z}\hat{p}_y\hat{x}\hat{p}_z - \hat{x}\hat{p}_z\hat{z}\hat{p}_y).
\end{aligned}
\tag{10.13}
$$

This expression may be rearranged as

$$
\begin{aligned}
\left[\hat{L}_x, \hat{L}_y\right] &= (\hat{p}_z\hat{z}\hat{y}\hat{p}_x - \hat{z}\hat{p}_z\hat{y}\hat{p}_x + \hat{z}\hat{p}_z\hat{p}_y\hat{x} - \hat{p}_z\hat{z}\hat{x}\hat{p}_y) \\
&= (\hat{p}_z\hat{z} - \hat{z}\hat{p}_z)\hat{y}\hat{p}_x - (\hat{p}_z\hat{z} - \hat{z}\hat{p}_z)\hat{x}\hat{p}_y = (\hat{p}_z\hat{z} - \hat{z}\hat{p}_z)(\hat{y}\hat{p}_x - \hat{x}\hat{p}_y).
\end{aligned}
\tag{10.14}
$$

Substituting in the commutator of position and linear momentum, $[\hat{p}_z, \hat{z}] = \hat{p}_z\hat{z} - \hat{z}\hat{p}_z = -i\hbar$, gives

$$
\left[\hat{L}_x, \hat{L}_y\right] = i\hbar(\hat{x}\hat{p}_y - \hat{y}\hat{p}_x) = i\hbar\hat{L}_z,
\tag{10.15}
$$

where the fact $\hat{L}_z = \hat{x}\hat{p}_y - \hat{y}\hat{p}_x$ has been used. Hence,

$$
\boxed{\left[\hat{L}_x, \hat{L}_y\right] = i\hbar\hat{L}_z,}
\tag{10.16}
$$

$$\left[\hat{L}_y, \hat{L}_z\right] = i\hbar\hat{L}_x,$$ (10.17)

$$\left[\hat{L}_z, \hat{L}_x\right] = i\hbar\hat{L}_y.$$ (10.18)

These commutation relations have the physical consequence that no two components of angular momentum can be measured to arbitrary accuracy.

The total angular momentum operator is

$$\hat{L} = \hat{L}_x + \hat{L}_y + \hat{L}_z,$$ (10.19)

so that

$$\hat{L}^2 = \hat{L}_x^2 + \hat{L}_y^2 + \hat{L}_z^2.$$ (10.20)

To show that there are states of a system in which $\hat{L}_z$ and $\hat{L}^2$ can be simultaneously specified and measured to arbitrary accuracy, it is necessary to show that $\hat{L}_z$ and $\hat{L}^2$ commute:

$$
\begin{aligned}
\left[\hat{L}_z, \hat{L}^2\right] &= \left[\hat{L}_z, \hat{L}_x^2 + \hat{L}_y^2 + \hat{L}_z^2\right] = \left[\hat{L}_z, \hat{L}_x^2\right] + \left[\hat{L}_z, \hat{L}_y^2\right] + 0 \\
&= \hat{L}_x\left[\hat{L}_z, \hat{L}_x\right] + \left[\hat{L}_z, \hat{L}_x\right]\hat{L}_x + \hat{L}_y\left[\hat{L}_z, \hat{L}_y\right] + \left[\hat{L}_z, \hat{L}_y\right]\hat{L}_y \\
&= i\hbar\left[\hat{L}_x\hat{L}_y + \hat{L}_y\hat{L}_x - \hat{L}_y\hat{L}_x - \hat{L}_x\hat{L}_y\right],
\end{aligned}
$$ (10.21)

$$\left[\hat{L}_z, \hat{L}^2\right] = 0,$$ (10.22)

where the fact that $\left[\hat{L}_y, \hat{L}_z\right] = i\hbar\hat{L}_x$, etc., has been used.

Because no special significance is given to the z direction, it may be infered that

$$\left[\hat{L}_x, \hat{L}^2\right] = \left[\hat{L}_y, \hat{L}^2\right] = \left[\hat{L}_z, \hat{L}^2\right] = 0$$ (10.23)

and

$$\left[\hat{L}, \hat{L}^2\right] = 0.$$ (10.24)

Because $\hat{L}^2$ and $\hat{L}_z$ commute, it follows that they have common eigenfunctions $\psi_{\lambda m}$. It is possible to simultaneously know the eigenvalues of the z-component of the angular momentum operator $\hat{L}_z$ and the angular momentum squared operator $\hat{L}^2$. However, since $\hat{L}_z$ does not commute with $\hat{L}_x$ or $\hat{L}_y$, it is not possible to simultaneously know the eigenvalue of the z-component of angular momentum *and* the eigenvalues for the x- and y-components of angular momentum. It is conventional to choose the $\hat{L}_z$ and $\hat{L}^2$ pair of commuting operators in the study of quantized angular momentum. Of course, there is nothing special about using $\hat{L}_z$, either the $\hat{L}_x$ or $\hat{L}_y$ components could equally well have been chosen.

10.2.1 Eigenvalues of Angular Momentum Operators $\hat{L}_z$ and $\hat{L}^2$

To find the eigenvalue equations for the operators $\hat{L}_z$ and $\hat{L}^2$, creation and annihilation operators can be introduced in much the same way as was done for the harmonic oscillator in Chapter 5. The operator $\hat{L}_\pm$ is defined as

$$\hat{L}_\pm = \hat{L}_x \pm i\hat{L}_y, \tag{10.25}$$

so that

$$\hat{L}_x = \frac{1}{2}\left(\hat{L}_+ + \hat{L}_-\right), \tag{10.26}$$

$$\hat{L}_y = \frac{1}{2i}\left(\hat{L}_+ - \hat{L}_-\right), \tag{10.27}$$

and

$$\hat{L}_x^2 = \frac{1}{4}\left(\hat{L}_+^2 + \hat{L}_+\hat{L}_- + \hat{L}_-\hat{L}_+ + \hat{L}_-^2\right), \tag{10.28}$$

$$\hat{L}_y^2 = \frac{-1}{4}\left(\hat{L}_+^2 - \hat{L}_+\hat{L}_- - \hat{L}_-\hat{L}_+ + \hat{L}_-^2\right). \tag{10.29}$$

The commutation relations of Eqs. (10.16), (10.17), and (10.18) can be used to show that (Exercise 10.1):

$$\left[\hat{L}^2, \hat{L}_\pm\right] = 0, \tag{10.30}$$

$$\left[\hat{L}_+, \hat{L}_-\right] = 2\hbar\hat{L}_z, \tag{10.31}$$

$$\left[\hat{L}_z, \hat{L}_\pm\right] = \pm\hbar\hat{L}_\pm. \tag{10.32}$$

In addition,

$$\hat{L}_+\hat{L}_- = \hat{L}_x^2 + \hat{L}_y^2 + \hbar\hat{L}_z = \hat{L}^2 - \hat{L}_z^2 + \hbar\hat{L}_z, \tag{10.33}$$

$$\hat{L}_-\hat{L}_+ = \hat{L}_x^2 + \hat{L}_y^2 - \hbar\hat{L}_z = \hat{L}^2 - \hat{L}_z^2 - \hbar\hat{L}_z, \tag{10.34}$$

and

$$\hat{L}^2 = \hat{L}_\pm\hat{L}_\mp + \hat{L}_z^2 \mp \hbar\hat{L}_z, \tag{10.35}$$

$$\hat{L}^2 = \frac{1}{2}\left(\hat{L}_+\hat{L}_- + \hat{L}_-\hat{L}_+\right) + \hat{L}_z^2. \tag{10.36}$$

Associated with operators $\hat{L}^2$ and $\hat{L}_z$ are the eigenvalues $\hbar^2\lambda$ and $\hbar m$, respectively, such that

$$\hat{L}^2\psi_{\lambda m} = \hbar^2\lambda\psi_{\lambda m} \tag{10.37}$$

and

$$\hat{L}_z\psi_{\lambda m} = \hbar m\psi_{\lambda m}, \tag{10.38}$$

where λ and m are quantum numbers and $\psi_{\lambda m}$ is an eigenstate.

Since, according to Eq. (10.32), the creation and annihilation operators $\hat{L}_+$ or $\hat{L}_-$ do not commute with $\hat{L}_z$, the state $\psi_{\lambda m}$ is not an eigenstate of either $\hat{L}_+$ or $\hat{L}_-$. Using Eqs. (10.32) and (10.38) results in

$$\hat{L}_z \left(\hat{L}_\pm \psi_{\lambda m} \right) = \left(\hat{L}_\pm \hat{L}_z \pm \hbar \hat{L}_\pm \right) \psi_{\lambda m} = \hbar (m \pm 1) \left(\hat{L}_\pm \psi_{\lambda m} \right), \tag{10.39}$$

so $\left(\hat{L}_\pm \psi_{\lambda m} \right)$ is an eigenstate of $\hat{L}_z$ with eigenvalue $\hbar(m \pm 1)$. Because, according to Eq. (10.22), $\hat{L}_z$ commutes with $\hat{L}^2$, the state $\left(\hat{L}_\pm \psi_{\lambda m} \right)$ must also be an eigenstate of $\hat{L}^2$ with eigenvalue $\hbar^2 \lambda$. Making use of Eqs. (10.30) and (10.37) gives

$$\hat{L}^2 \left(\hat{L}_\pm \psi_{\lambda m} \right) = \hat{L}_\pm \hat{L}^2 \psi_{\lambda m} = \hbar^2 \lambda \left(\hat{L}_\pm \psi_{\lambda m} \right). \tag{10.40}$$

The operator $\hat{L}_\pm$ acting on the state $\psi_{\lambda m}$ does not change quantum number λ, but increases or decreases the quantum number m by one unit so that

$$\hat{L}_\pm \psi_{\lambda m} = A_{\lambda m}^\pm \psi_{\lambda m \pm 1}, \tag{10.41}$$

where $A_{\lambda m}^\pm = \hbar \sqrt{\lambda - m(m \pm 1)}$ is found by considering $\langle \psi_{\lambda m} | \hat{L}_\mp \hat{L}_\pm | \psi_{\lambda m} \rangle$, using Eqs. (10.33), (10.34), (10.37), (10.38), and the bound-state normalization properties of $\psi_{\lambda m}$.

Because, from Eq. (10.20), $\hat{L}^2 - \hat{L}_z^2 = \hat{L}_x^2 + \hat{L}_y^2 \geq 0$, then

$$\langle \psi_{\lambda m} | \left(\hat{L}^2 - \hat{L}_z^2 \right) | \psi_{\lambda m} \rangle = \hbar^2 (\lambda - m^2) \geq 0, \tag{10.42}$$

and so $\lambda \geq m^2$. Hence, for a given value of quantum number λ there is a maximum value of the quantum number $m_{\max}$ such that

$$\hat{L}_+ \psi_{\lambda m_{\max}} = 0. \tag{10.43}$$

Making use of this fact and Eq. (10.34), it follows that

$$\hat{L}_- \hat{L}_+ \psi_{\lambda m_{\max}} = \left(\hat{L}^2 - \hat{L}_z^2 - \hbar \hat{L}_z \right) \psi_{\lambda m_{\max}} = \hbar^2 \left(\lambda - m_{\max}^2 - m_{\max} \right) \psi_{\lambda m_{\max}} = 0, \tag{10.44}$$

so that

$$\lambda = m_{\max}(m_{\max} + 1). \tag{10.45}$$

Likewise, there is a minimum value of the quantum number $m_{\min}$ such that

$$\hat{L}_- \psi_{\lambda m_{\min}} = 0. \tag{10.46}$$

Making use of Eq. (10.33) results in

$$\lambda = m_{\min}(m_{\min} - 1). \tag{10.47}$$

Equations (10.45) and (10.47) require

$$m_{\max} = -m_{\min}. \tag{10.48}$$

The orbital angular momentum quantum number is defined as $l = m_{max}$ and using Eqs. (10.37), (10.38), and (10.45), it can be shown that the quantum numbers l and m are related to the eigenvalues of $\hat{L}^2$ and $\hat{L}_z$ by the eigenvalue equations

$$\hat{L}^2 \psi_{lm} = \hbar^2 l(l+1)\psi_{lm}, \tag{10.49}$$

$$\hat{L}_z \psi_{lm} = \hbar m \psi_{lm}, \tag{10.50}$$

where the orbital angular momentum quantum number,

$$l = 0, 1, 2, 3, \ldots, \tag{10.51}$$

and the quantum number,

$$m = -l, -(l-1), \ldots, (l-1), l, \tag{10.52}$$

such that

$$-l \leq m \leq l. \tag{10.53}$$

10.2.2 Geometrical Representation

Because the Hamiltonian, $\hat{L}_z$, and $\hat{L}^2$ commute it follows that they have common eigenfunctions and it is possible to simultaneously know the eigenvalues of $\hat{H}, \hat{L}_z$, and $\hat{L}^2$. Since $\hat{L}_z$ and $\hat{L}^2$ commute, the quantized angular momentum states can be described in a geometrical representation. For example, angular momentum states with quantum numbers $l = 2$ and $m = 2, 1, 0, -1, -2$ can be represented as lines of latitude on a sphere of radius $L = \hbar\sqrt{l(l+1)}$. This is illustrated in Fig. 10.3. Each line of latitude defines a plane that bisects the z-axis with value $m\hbar$. Only the radius of the sphere L and the bisection point value $m\hbar$ are quantized. The values of angular momentum L_y and L_x are not known precisely because, according to Eqs. (10.16), (10.17), and (10.18), $\hat{L}_z$ does not commute with $\hat{L}_x$ or $\hat{L}_y$. For this reason, the state has *indeterminate* values of L_x and L_y on the circumference of a circle defined by the bisection of the (L_x, L_y) plane that passes through the quantized value $L_z = m\hbar$ with the sphere of quantized radius $L = \hbar\sqrt{l(l+1)}$.

A measure in the uncertainty in L_x and L_y can be found by substituting $\hat{L}_x$ and $\hat{L}_y$ into the generalized uncertainty relation for operators $\hat{A}$ and $\hat{B}$ (see Chapter 4), $\Delta A \Delta B \geq \left|\frac{1}{2}\left\langle\left[\hat{A}, \hat{B}\right]\right\rangle\right|$. Making use of the commutator $\left[\hat{L}_x, \hat{L}_y\right] = i\hbar\hat{L}_z$ (Eq. (10.16)) and the eigenvalue equation for $\hat{L}_z$ (Eq. (10.50)) gives

$$\Delta L_x \Delta L_y \geq \frac{m\hbar^2}{2}. \tag{10.54}$$

It follows that the only common state for which all components of L can be simultaneously known corresponds to $|l = 0, m = 0\rangle$.

Notice that it is not possible to have the value of the z-component of angular momentum, L_z, equal to the radius of the sphere L. Angular momentum cannot point exactly along

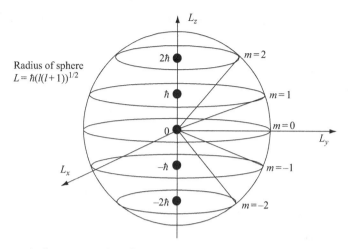

Fig. 10.3 Geometrical representation of quantized angular momentum for a state with $l = 2$. The z-component of angular momentum is quantized such that $L_z = m\hbar$ where integer m has values $-l \le m \le l$. The state exists with indeterminate values of L_x and L_y at the bisection of the (L_x, L_y) plane that passes through the quantized value $L_z = m\hbar$ with the sphere of quantized radius $L = \hbar(l(l+1))^{1/2}$.

the z direction since to achieve this requires knowing the precise values of L_x, L_y, and L_z, and this is not possible because the operators $\hat{L}_x$, $\hat{L}_y$, and $\hat{L}_z$ do not commute.

10.2.3 Spherical Coordinates and Spherical Harmonics

Because rotational symmetry is inherent to angular momentum, it is convenient to express angular momentum in terms of spherical coordinates. The relationship between Cartesian coordinates (x, y, z) and spherical coordinates (r, θ, ϕ) is illustrated in Fig. 10.4 and may be expressed via the transformations

$$x = r\sin(\theta)\cos(\phi),$$
$$y = r\sin(\theta)\sin(\phi), \qquad (10.55)$$
$$z = r\cos(\theta).$$

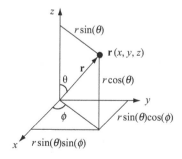

Fig. 10.4 Illustration of a spherical coordinate system. Position vector $\mathbf{r}$ is described using the radial coordinate r, the polar angle θ, and the azimuthal angle ϕ.

439

Substituting into the expressions for $\hat{L}_x$, $\hat{L}_y$, and $\hat{L}_z$ results in (see Exercise 10.1)

$$\hat{L}_x = i\hbar \left(\sin(\phi)\frac{\partial}{\partial \theta} + \cot(\theta)\cos(\phi)\frac{\partial}{\partial \phi} \right), \tag{10.56}$$

$$\hat{L}_y = i\hbar \left(-\cos(\phi)\frac{\partial}{\partial \theta} + \cot(\theta)\sin(\phi)\frac{\partial}{\partial \phi} \right), \tag{10.57}$$

$$\hat{L}_z = -i\hbar \frac{\partial}{\partial \phi}, \tag{10.58}$$

$$\hat{L}^2 = -\hbar^2 \left(\frac{1}{\sin(\theta)}\frac{\partial}{\partial \theta}\left(\sin(\theta)\frac{\partial}{\partial \theta} \right) + \frac{1}{\sin^2(\theta)}\frac{\partial^2}{\partial \phi^2} \right). \tag{10.59}$$

The eigenfunction solutions for these operators are spherical harmonics, Y_l^m. Substituting into the eigenvalue equations (Eqs. (10.49) and (10.50)) gives

$$\hat{L}^2 Y_l^m = \hbar^2 l(l+1)Y_l^m, \tag{10.60}$$

$$\hat{L}_z Y_l^m = \hbar m Y_l^m, \tag{10.61}$$

and because $\hat{L}_z = -i\hbar\frac{\partial}{\partial \phi}$, the last equation may be written as

$$-i\hbar \frac{\partial}{\partial \phi} Y_l^m = \hbar m Y_l^m \tag{10.62}$$

or

$$\frac{\partial}{\partial \phi} Y_l^m = i m Y_l^m. \tag{10.63}$$

Since this equation only determines the ϕ (azimuthal angle) dependence of Y_l^m, it is possible to separate variables such that

$$Y_l^m(\theta, \phi) = \Phi_m(\phi)\Theta_m^l(\theta). \tag{10.64}$$

Hence,

$$\frac{\partial}{\partial \phi} \Phi_m(\phi) = i m \Phi_m(\phi). \tag{10.65}$$

This equation has solution

$$\Phi_m(\phi) = \frac{1}{\sqrt{2\pi}} e^{im\phi}, \tag{10.66}$$

where Φ_m has been normalized so that

$$\int_0^{2\pi} |\Phi_m(\phi)|^2 d\phi = 1. \tag{10.67}$$

If, as is physically reasonable, single valueness of $\Phi(\phi)$ is required, then

$$\Phi_m(\phi) = \Phi_m(\phi + 2\pi), \tag{10.68}$$

so that

$$e^{im\phi} = e^{im(\phi+2\pi)}. \tag{10.69}$$

Hence, the condition of single valueness is equivalent to

$$e^{im(2\pi)} = 1, \tag{10.70}$$

which is only satisfied for integer values $m = 0, \pm 1, \pm 2, \pm 3, \ldots$.

Using Eq. (10.66) and substituting Eq. (10.64) into Eq. (10.62), it is apparent that solutions to Eq. (10.60) are sought of the form

$$Y_l^m = \frac{1}{\sqrt{2\pi}} e^{im\phi} \Theta_l^m(\theta). \tag{10.71}$$

Substitution of Eq. (10.71) into Eq. (10.60), and using Eq. (10.59) gives

$$\frac{1}{\sin(\theta)} \frac{d}{d\theta} \left(\sin(\theta) \frac{d}{d\theta} \Theta_l^m(\theta) \right) + \left(l(l+1) - \frac{m^2}{\sin^2(\theta)} \right) \Theta_l^m(\theta) = 0. \tag{10.72}$$

A change of variable to $\mu = \cos(\theta)$ gives

$$\frac{d}{d\mu} \left(1 - \mu^2 \frac{d}{d\mu} \Theta_l^m(\theta) \right) + \left(l(l+1) - \frac{m^2}{1 - \mu^2} \right) \Theta_l^m(\theta) = 0. \tag{10.73}$$

For *integer* values of m and l, this equation has solutions

$$\boxed{P_l^m(\mu) = (-1)^m \left(1 - \mu^2 \right)^{m/2} \frac{d^m}{d\mu^m} P_l(\mu),} \tag{10.74}$$

where $P_l^m(\mu)$ are called *associated Legendre polynomials.*[2] In this expression, the *Legendre polynomials* of degree l satisfy

$$P_l(\mu) = \frac{1}{2^l l!} \frac{d^l}{d\mu^l} \left(\mu^2 - 1 \right)^l. \tag{10.75}$$

Legendre polynomials can be written as

$$P_l(\mu) = \sum_{k=0}^{l} \frac{(-1)^k (l+k)!}{(1-k)!(k!)^2 2^{k+1}} \left((1-\mu)^k + (-1)^l (1+\mu)^k \right). \tag{10.76}$$

The Legendre polynomials $P_l(\mu)$ for $l = 0$ to $l = 5$ are given in Table 10.1.

Spherical harmonics are normalized so that the square of the modulus of the function Y_l^m integrated over all angles is unity. This means that

$$\int |Y_l^m|^2 \sin(\theta) d\theta d\phi = \int_0^{2\pi} d\phi \left| \frac{e^{im\phi}}{\sqrt{2\pi}} \right|^2 \int_{-1}^{1} d\mu |\Theta_l^m(\mu)|^2 = 1 \tag{10.77}$$

---

[2] These formulae can be found in I. S. Gradshteyn and I. M. Ryzhik, *Table of Integrals, Series, and Products*, San Diego, Academic Press, 1980, p. 1025 (ISBN 0 12 294760 6) for the Legendre polynomials and p. 1014 for the associated Legendre polynomials.

Table 10.1 Legendre polynomials

$P_0(\mu) = 1$
$P_1(\mu) = \mu$
$P_2(\mu) = \frac{1}{2}\left(3\mu^2 - 1\right)$
$P_3(\mu) = \frac{1}{2}\left(5\mu^3 - 3\mu\right)$
$P_4(\mu) = \frac{1}{8}\left(35\mu^4 - 30\mu^2 + 3\right)$
$P_5(\mu) = \frac{1}{8}\left(63\mu^5 - 70\mu^3 + 15\mu\right)$

or

$$\int_{-1}^{1} d\mu \, |\Theta_l^m(\mu)|^2 = 1, \tag{10.78}$$

and results in

$$\Theta_l^m(\mu) = \left(\frac{2l+1}{2}\frac{(1-m)!}{(l+m)!}\right)^{1/2} P_l^m(\mu). \tag{10.79}$$

Hence, combining Eqs. (10.64), (10.66), and (10.79), the spherical harmonics are

$$\boxed{Y_l^m(\theta, \phi) = \left(\frac{2l+1}{4\pi}\frac{(l-m)!}{(l+m)!}\right)^{1/2} P_l^m(\cos(\theta)) e^{im\phi},} \tag{10.80}$$

where $Y_l^m(\theta, \phi)$ is of the form given by Eq. (10.71) and $|Y_l^m(\theta, \phi)|^2$ is independent of ϕ. The spherical harmonics given by Eq. (10.80) are normalized, orthogonal, and complete. The orthogonality of spherical harmonics may be expressed as

$$\int_{\theta=0}^{\theta=\pi} \int_{\phi=0}^{\phi=2\pi} Y_l^{m*} Y_{l'}^{m'} \sin(\theta) d\theta \, d\phi = \delta_{ll'} \delta_{mm'}. \tag{10.81}$$

In addition, it is a property of spherical harmonics that

$$\sum_{m=-l}^{m=l} |Y_l^m(\theta, \phi)|^2 = \frac{2l+1}{4\pi}. \tag{10.82}$$

The first few spherical harmonics are

$$Y_0^0 = \left(\frac{1}{4\pi}\right)^{1/2}, \tag{10.83}$$

442

$$Y_1^1 = -\frac{1}{2} \left(\frac{3}{2\pi} \right)^{1/2} \sin(\theta) \, e^{i\phi}, \tag{10.84}$$

$$Y_1^0 = \frac{1}{2} \left(\frac{3}{\pi} \right)^{1/2} \cos(\theta), \tag{10.85}$$

$$Y_1^{-1} = \frac{1}{2} \left(\frac{3}{2\pi} \right)^{1/2} \sin(\theta) \, e^{-i\phi}, \tag{10.86}$$

$$Y_2^2 = \frac{1}{4} \left(\frac{15}{2\pi} \right)^{1/2} \sin^2(\theta) \, e^{2i\phi}, \tag{10.87}$$

$$Y_2^1 = -\frac{1}{2} \left(\frac{15}{2\pi} \right)^{1/2} \sin(\theta) \cos(\theta) \, e^{i\phi}, \tag{10.88}$$

$$Y_2^0 = \frac{1}{4} \left(\frac{5}{\pi} \right)^{1/2} \left(3 \, \cos^2(\theta) - 1 \right), \tag{10.89}$$

$$Y_2^{-1} = \frac{1}{2} \left(\frac{15}{2\pi} \right)^{1/2} \sin(\theta) \cos(\theta) \, e^{-i\phi}, \tag{10.90}$$

$$Y_2^{-2} = \frac{1}{4} \left(\frac{15}{2\pi} \right)^{1/2} \sin^2(\theta) \, e^{-2i\phi}, \tag{10.91}$$

$$Y_3^3 = -\frac{1}{8} \left(\frac{35}{\pi} \right)^{1/2} \sin^3(\theta) \, e^{3i\phi}, \tag{10.92}$$

$$Y_3^2 = \frac{1}{4} \left(\frac{105}{2\pi} \right)^{1/2} \sin^2(\theta) \cos(\theta) \, e^{2i\phi}, \tag{10.93}$$

$$Y_3^1 = -\frac{1}{8} \left(\frac{21}{\pi} \right)^{1/2} \sin(\theta) \left(5 \, \cos^2(\theta) - 1 \right) e^{i\phi}, \tag{10.94}$$

$$Y_3^0 = \frac{1}{4} \left(\frac{7}{\pi} \right)^{1/2} \left(5 \, \cos^3(\theta) - 3 \, \cos(\theta) \right), \tag{10.95}$$

$$Y_3^{-1} = \frac{1}{8} \left(\frac{21}{\pi} \right)^{1/2} \sin(\theta) \left(5 \, \cos^2(\theta) - 1 \right) e^{-i\phi}, \tag{10.96}$$

$$Y_3^{-2} = \frac{1}{4} \left(\frac{105}{2\pi} \right)^{1/2} \sin^2(\theta) \cos(\theta) \, e^{-2i\phi}, \tag{10.97}$$

$$Y_3^{-3} = \frac{1}{8} \left(\frac{35}{\pi} \right)^{1/2} \sin^3(\theta) \, e^{-3i\phi}. \tag{10.98}$$

Figure 10.5 is a plot of the spherical harmonic Y_0^0 on the unit sphere.

Because, in general, Y_l^m is a complex function, it may be necessary to plot the absolute value, real part, and imaginary part. This is illustrated for $Y_{l=1}^{m=1}$ in Fig. 10.6, for $Y_{l=2}^{m=1}$ in Fig. 10.7, and for $Y_{l=2}^{m=0}$ in Fig. 10.8. It should be easy to relate these figures to Y_l^m plotted in Fig. 10.9 as a function of polar angle θ in a plane through the z axis.

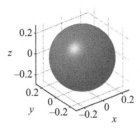

Fig. 10.5 Plot of the spherical harmonic Y_l^m with $l = m = 0$. The Cartesian coordinates x, y, and z are indicated.

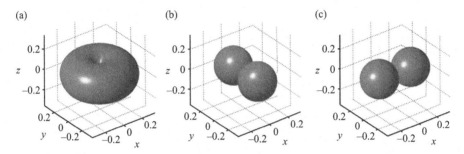

Fig. 10.6 The (a) absolute magnitude, (b) real part, and (c) imaginary part of the $l = 1$ and $m = 1$ spherical harmonic Y_l^m.

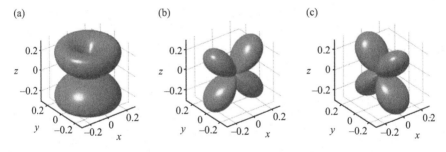

Fig. 10.7 The (a) absolute magnitude, (b) real part, and (c) imaginary part of the $l = 2$ and $m = 1$ spherical harmonic Y_l^m.

As illustrated in Fig. 10.9, it is also possible to plot $Y_l^m(\theta, \phi)$ as a function of θ in any plane through the z-axis.

Using spherical harmonics and quantization of angular momentum, it is possible to solve the eigenvalue and eigenfunctions of many atomic and molecular systems. For these systems it is angular momentum that plays an essential role in determining the potential seen by electrons.

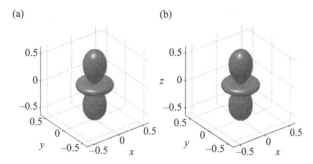

Fig. 10.8 The (a) absolute magnitude and (b) real part of the $l = 2$ and $m = 0$ spherical harmonic Y_l^m.

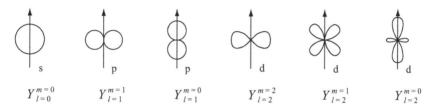

Fig. 10.9 Illustration of the first few spherical harmonics by plotting the value of the absolute value of Y_l^m as a function of polar angle θ in a plane through the z axis.

10.2.4 The Rigid Rotator

A rigid rotator system consisting of two identical atoms each of mass m_1 separated by a fixed distance $2a$ is shown schematically in Fig. 10.10. Longitudinal motion along the axis connecting the two particles is ignored by assuming the system is in a ground state for longitudinal vibrational motion. The moment of inertia about the midpoint of the rotator along the z-axis is

$$I_L = \sum_i m_i r_i^2 = 2m_1 a^2. \tag{10.99}$$

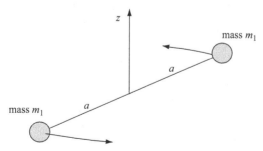

Fig. 10.10 Illustration of a rigid rotator consisting of two particles of equal mass with fixed midpoint.

Table 10.2 Quantized angular momentum of rigid rotator

Orbital angular momentum state	Energy	Quantum number l	Quantum number m	Degeneracy $2l + 1$
s	0	0	0	1
p	$2\hbar^2/2I_L$	1	$-1, 0, 1$	3
d	$6\hbar^2/2I_L$	2	$-2, -1, 0, 1, 2$	5
f	$12\hbar^2/2I_L$	3	$-3, -2, -1, 0, 1, 2, 3$	7

In classical mechanics, the kinetic energy for rotation about the z-axis is just

$$E = \frac{1}{2}I_L\omega^2 = \frac{L^2}{2I_L}. \tag{10.100}$$

In quantum mechanics, L^2 is replaced with the operator $\hat{L}^2$ so that the time-independent Schrödinger equation describing the system becomes

$$\hat{H}\psi_{lm} = \frac{\hat{L}^2}{2I_L}\psi_{lm} = E_l\psi_{lm} = \frac{\hbar^2 l(l+1)}{2I_L}\psi_{lm}. \tag{10.101}$$

The solutions are $\psi_{lm} = Y_l^m(\theta, \phi)$ and the energy eigenvalues are $(2l+1)$-fold degenerate. See Table 10.2. This degeneracy means that a diatomic molecule prepared in a state characterized by quantum number l may not be in a pure state because there are $2l + 1$ values of quantum number m with which to create a linear superposition state with the same energy E_l.

10.3 The Hydrogen Atom

All materials are made up of atoms whose behavior is governed by quantum mechanics. It is common to believe that the simplest atom is the hydrogen atom since it consists of only one electron and one proton. This is a two particle system whose relative motion and reduced mass may be used to simplify a description to an effective one-particle problem. The electron at position **r** moves in the *radially symmetric coulomb potential* of the proton. In the slightly more general case, a *hydrogenic* atom has one electron and a nucleus of charge Ze with Z positive integer number of protons. The coulomb potential seen by the electron is

$$V(r) = \frac{-e^2 Z}{4\pi\varepsilon_0\varepsilon_{r0}r}. \tag{10.102}$$

In this expression, the low frequency relative permittivity $\varepsilon_{r0} = 1$ for an isolated hydrogenic atom or a dilute gas of hydrogenic atoms.

For the hydrogen atom, the reduced mass of the electron and proton system is m_r and is determined using $1/m_r = 1/m_0 + 1/m_p$ where m_0 is the bare electron mass and m_p is the proton mass. The coulomb potential given by Eq. (10.102) suggests that a continuum of electron states exists for electron energies $E > 0$ and bound electron states with discrete energy eigenvalues exists for $E < 0$.

10.3.1 Eigenstates and Eigenvalues of the Hydrogen Atom

The hydrogen atom is a two-particle (electron (e) and proton (p)) system with particle motion in three dimensions described by Hamiltonian

$$\hat{H} = -\frac{\hbar^2}{2m_0}\nabla_e^2 - \frac{\hbar^2}{2m_p}\nabla_p^2 + V(\mathbf{r}_e - \mathbf{r}_p). \tag{10.103}$$

This can be rewritten in terms of a center of mass coordinate $\mathbf{R}$ and relative coordinate $\mathbf{r} = \mathbf{r}_e - \mathbf{r}_p$, total mass $M = m_0 + m_p$, and reduced mass m_r.
Hence,

$$\hat{H} = -\frac{\hbar^2}{2M}\nabla_R^2 - \frac{\hbar^2}{2m_r}\nabla_r^2 + V(\mathbf{r}), \tag{10.104}$$

where ∇_R and ∇_r are gradients with respect to $\mathbf{R}$ and $\mathbf{r}$, respectively. The time-independent Schrödinger equation,

$$\hat{H}\psi(\mathbf{r}_e, \mathbf{r}_p) = E_{\text{total}}\psi(\mathbf{r}_e, \mathbf{r}_p), \tag{10.105}$$

can be separated into parts so that

$$\psi(\mathbf{r}_e, \mathbf{r}_p) = f(\mathbf{R})\psi(\mathbf{r}), \tag{10.106}$$

where the center of mass coordinate has no contribution to energy from the relative position of $\mathbf{r}$,

$$-\frac{\hbar^2}{2M}\nabla_R^2 f(\mathbf{R}) = E_K f(\mathbf{R}), \tag{10.107}$$

and for the relative motion,

$$\left(-\frac{\hbar^2}{2m_r}\nabla_r^2 + V(r)\right)\psi(\mathbf{r}) = E_n\psi(\mathbf{r}). \tag{10.108}$$

The total energy is

$$E_{\text{total}} = E_K + E_n, \tag{10.109}$$

where E_K is the eigenenergy appearing in Eq. (10.107) and E_n is the eigenenergy for the relative motion appearing in Eq. (10.108).
The solution to Eq. (10.107) is that of a free particle of mass M moving in volume V_{vol} where

$$f(\mathbf{R}) = \frac{1}{\sqrt{V_{\text{vol}}}}e^{i(\mathbf{K}\cdot\mathbf{R})} \tag{10.110}$$

and

$$E_K = \frac{\hbar^2 K^2}{2M}. \tag{10.111}$$

The definition of ∇_r^2 used in Eq. (10.108) is

$$\nabla_r^2 = \frac{1}{r^2}\frac{\partial}{\partial r}\left(r^2\frac{\partial}{\partial r}\right) + \frac{1}{r^2\sin(\theta)}\frac{\partial}{\partial\theta}\left(\sin(\theta)\frac{\partial}{\partial\theta}\right) + \frac{1}{r^2\sin^2(\theta)}\frac{\partial^2}{\partial\phi^2}. \tag{10.112}$$

The solution to Eq. (10.108) with the free-space coulomb potential energy is obtained by separation of variables,

$$\psi_{nlm}(r,\theta,\phi) = R_{nl}(r)\Theta_m^l(\theta)\Phi_m(\phi), \tag{10.113}$$

$$\boxed{\psi_{nlm}(r,\theta,\phi) = R_{nl}(r)Y_l^m(\theta,\phi),} \tag{10.114}$$

where $Y_l^m(\theta,\phi)$ are the normalized spherical harmonics of Section 10.2.3 and n, l, and m are quantum numbers.

Substituting Eq. (10.114) into the time-independent Schrödinger equation for the relative motion (Eq. (10.108)) gives

$$\frac{-\hbar^2}{2m_r}\left(\frac{Y_l^m(\theta,\phi)}{r^2}\frac{\partial}{\partial r}\left(r^2\frac{\partial}{\partial r}R_{nl}(r)\right) + \frac{R_{nl}(r)}{r^2\sin(\theta)}\frac{\partial}{\partial\theta}\left(\sin(\theta)\frac{\partial}{\partial\theta}Y_l^m(\theta,\phi)\right)\right.$$

$$\left.+\frac{R_{nl}(r)}{r^2\sin^2(\theta)}\frac{\partial^2}{\partial\phi^2}Y_l^m(\theta,\phi)\right) + V(r)R_{nl}(r)Y_l^m(\theta,\phi) = E_n R_{nl}(r)Y_l^m(\theta,\phi). \tag{10.115}$$

It is now helpful to divide by $R_{nl}(r)Y_l^m(\theta,\phi)$ and multiply by $-2r^2m_r/\hbar^2$, resulting in

$$\left(\frac{1}{R_{nl}(r)}\frac{d}{dr}\left(r^2\frac{d}{dr}R_{nl}(r)\right) - \frac{2r^2m_r}{\hbar^2}(V(r)-E_n)\right)$$

$$+\frac{1}{Y_l^m(\theta,\phi)}\left(\frac{1}{\sin(\theta)}\frac{\partial}{\partial\theta}\left(\sin(\theta)\frac{\partial}{\partial\theta}Y_l^m(\theta,\phi)\right) + \frac{1}{\sin^2(\theta)}\frac{\partial^2}{\partial\phi^2}Y_l^m(\theta,\phi)\right) = 0. \tag{10.116}$$

Terms in the first bracket only depend on r and terms in the second bracket only depend on θ and ϕ. Hence, separation of variables may be achieved by writing down two equations equal to a separation constant, λ. For the radial part,

$$\frac{1}{R_{nl}(r)}\frac{d}{dr}\left(r^2\frac{d}{dr}R_{nl}(r)\right) - \frac{2r^2m_r}{\hbar^2}(V(r)-E_n) = \lambda, \tag{10.117}$$

and for the angular part,

$$\frac{1}{Y_l^m(\theta,\phi)}\left(\frac{1}{\sin(\theta)}\frac{\partial}{\partial\theta}\left(\sin(\theta)\frac{\partial}{\partial\theta}Y_l^m(\theta,\phi)\right) + \frac{1}{\sin^2(\theta)}\frac{\partial^2}{\partial\phi^2}Y_l^m(\theta,\phi)\right) = -\lambda. \tag{10.118}$$

From Eq. (10.59) the operator $\hat{L}^2/\hbar^2$ can be identified, and from Eq. (10.49) the eigenvalue associated with this operator is $\lambda = l(l+1)$. Hence, the angular dependence of the wave function in Eq. (10.114) satisfies

$$\left(\frac{1}{\sin(\theta)}\frac{\partial}{\partial\theta}\left(\sin(\theta)\frac{\partial}{\partial\theta}\right) + \frac{1}{\sin^2(\theta)}\frac{\partial^2}{\partial\phi^2}\right)Y_l^m(\theta,\phi) = -l(l+1)Y_l^m(\theta,\phi), \tag{10.119}$$

or

$$\hat{L}^2 Y_l^m(\theta, \phi) = -\hbar^2 \left(\frac{1}{\sin(\theta)} \frac{\partial}{\partial \theta} \left(\sin(\theta) \frac{\partial}{\partial \theta} \right) + \frac{1}{\sin^2(\theta)} \frac{\partial^2}{\partial \phi^2} \right) Y_l^m(\theta, \phi) \tag{10.120}$$

$$= \hbar^2 l(l+1) Y_l^m(\theta, \phi),$$

which shows that $Y_l^m(\theta, \phi)$ is an eigenfunction of $\hat{L}^2$ with eigenvalue $\hbar^2 l(l+1)$.

Multiplying the radial equation, Eq. (10.117), by $\frac{\hbar^2}{2m_r} \frac{R_{nl}(r)}{r^2}$ and substituting $\lambda = l(l+1)$ results in

$$\boxed{\frac{-\hbar^2}{2m_r} \frac{1}{r^2} \frac{\mathrm{d}}{\mathrm{d}r} \left(r^2 \frac{\mathrm{d}}{\mathrm{d}r} R_{nl}(r) \right) + \left(\frac{\hbar^2}{2m_r} \frac{l(l+1)}{r^2} + V(r) \right) R_{nl}(r) = E_n R_{nl}(r).} \tag{10.121}$$

This radial equation is a one-dimensional time-independent Schrödinger equation with an effective potential,

$$V_{\mathrm{eff}} = \frac{\hbar^2}{2m_r} \frac{l(l+1)}{r^2} + V(r), \tag{10.122}$$

that contains the coulomb potential $V(r)$ and an additional centrifugal term due to the angular momentum. Because the orbital angular momentum quantum numbers can have integer value $l = 0, 1, 2, \ldots$ (Eq. (10.51)), the effective potential seen by an electron depends on l. The contributions to the effective potential are illustrated in Fig. 10.11.

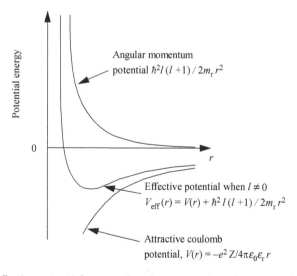

Fig. 10.11 The effective potential energy for electron motion about a positive charge eZ has contributions from the bare coulomb potential and the angular momentum (or centrifugal) potential.

For a hydrogen atom in free space, $Z = 1, \varepsilon_{r0} = 1$, and the coulomb potential

$$V(r) = \frac{-e^2}{4\pi\varepsilon_0 r},\tag{10.123}$$

so that Eq. (10.121) becomes

$$\frac{-\hbar^2}{2m_r}\frac{1}{r^2}\frac{d}{dr}\left(r^2\frac{d}{dr}R_{nl}(r)\right) + \left(\frac{\hbar^2}{2m_r}\frac{l(l+1)}{r^2} - \frac{e^2}{4\pi\varepsilon_0 r}\right)R_{nl}(r) = E_n R_{nl}(r).\tag{10.124}$$

Following the standard approach,[3]

$$\alpha^2 = -2m_r E_n/\hbar^2,\tag{10.125}$$

$$\lambda = m_r e^2/4\pi\varepsilon_0\hbar^2\alpha,\tag{10.126}$$

and

$$\rho = 2\alpha r\tag{10.127}$$

are substituted into Eq. (10.124) resulting in

$$\frac{1}{\rho^2}\frac{d}{d\rho}\left(\rho^2\frac{d}{d\rho}R_{nl}(\rho)\right) + \left(\frac{\lambda}{\rho} - \frac{1}{4} - \frac{l(l+1)}{\rho^2}\right)R_{nl}(\rho) = 0.\tag{10.128}$$

Noting the product rule for differentiation gives

$$\frac{1}{\rho^2}\frac{d}{d\rho}\left(\rho^2\frac{d}{d\rho}\right) = \frac{d^2}{d\rho^2} + \frac{2}{\rho}\frac{d}{d\rho},\tag{10.129}$$

so that in the limit $\rho \to \infty$, only $\frac{d^2}{d\rho^2}$ in the equation survives, and Eq. (10.128) becomes

$$\left(\frac{d^2}{d\rho^2} - \frac{1}{4}\right)R_{nl}(\rho \to \infty) = 0,\tag{10.130}$$

with solution

$$R_{nl}(\rho \to \infty) = e^{-\rho/2}.\tag{10.131}$$

The other solution, $R_{nl}(\rho \to \infty) = e^{\rho/2}$, is not allowed because the wave function must be finite everywhere.

It is now assumed that the solution for all values of ρ is of the form

$$R_{nl}(\rho) = e^{-\rho/2}F(\rho),\tag{10.132}$$

where $F(\rho)$ is a function found by substituting Eq. (10.132) into Eq. (10.128) and dividing by $e^{-\rho/2}$ to give

---

[3] For example, L. D. Landau and E. M. Lifshitz, *Quantum Mechanics*, Oxford, Pergamon Press, 1977 (ISBN 0 08 020940 8); L. Pauling and E. B. Wilson, *Introduction to Quantum Mechanics*, New York, Dover, 1985 (ISBN 0 486 64871 0); L. I. Schiff, *Quantum Mechanics*, New York, McGraw-Hill Companies, 1968 (ISBN 0 07 055287 8).

$$\frac{d^2 F}{d\rho^2} + \left(\frac{2}{\rho} - 1\right)\frac{dF}{d\rho} + \left(\frac{\lambda}{\rho} - \frac{l(l+1)}{\rho^2} - \frac{1}{\rho}\right)F = 0. \tag{10.133}$$

To remove the singularity that occurs when $\rho = 0$ it is required that

$$F(\rho) = \rho^l L(\rho), \tag{10.134}$$

where $L(\rho)$ is a power series such that

$$L(\rho) = \sum_v a_v \rho^v = a_0 + a_1\rho + a_2\rho^2 + \cdots. \tag{10.135}$$

Substitution of Eq. (10.134) into Eq. (10.133) gives

$$\rho\frac{d^2 L}{d\rho^2} + (2(l+1) - \rho)\frac{dL}{d\rho} + (\lambda - l - 1)L = 0. \tag{10.136}$$

When Eq. (10.135) is substituted into Eq. (10.136), the sum of all the coefficients of each power of ρ must be zero; hence the coefficients for the first few powers of ρ are

$$(\lambda - l - 1)a_0 + 2(l+1)a_1 = 0, \tag{10.137}$$
$$(\lambda - l - 1 - 1)a_1 + (4(l+1) + 2)a_2 = 0, \tag{10.138}$$
$$(\lambda - l - 1 - 2)a_2 + (6(l+1) + 6)a_3 = 0. \tag{10.139}$$

This leads to a recursion relation for the coefficients:

$$a_{v+1} = \frac{(v + l + 1 - \lambda)}{(v+1)(v + 2l + 2)}a_v. \tag{10.140}$$

To ensure the wave function goes to zero as $\rho \to \infty$, the series given by Eq. (10.140) requires

$$\lambda - l - 1 - n' = 0. \tag{10.141}$$

The principal quantum number is defined as $n = \lambda$ so that

$$n = n' + l + 1. \tag{10.142}$$

The radial quantum number is n', and the orbital quantum number l can have a positive integer number or zero value (Eq. (10.51)).

The polynomials $L(\rho)$ defined by the recursion relation in Eq. (10.140) are the *associated Laguerre polynomials*.[4] These can be written as

$$L_{n+l}^{2l+1}(\rho) = \sum_{k=0}^{n-l-1}(-1)^{k+2l+1}\frac{((n+l)!)^2(\rho)^k}{(n-l-1-k)!(2l+1+k)!k!}, \tag{10.143}$$

or, in differential form,

$$L_n^m(z) = \frac{1}{n!}e^z z^{-m}\frac{d^n}{dz^n}(e^{-z}z^{n+m}). \tag{10.144}$$

[4] I. S. Gradshteyn and I. M. Ryzhik, *Table of Integrals, Series, and Products*, San Diego, Academic Press, 1980, p. 1037 (ISBN 0 12 294760 6).

Now, using $n = \lambda$ and substituting Eq. (10.126) into Eq. (10.125), the eigenenergies are

$$E_n = \frac{-\hbar^2 \alpha^2}{2m_r} = \frac{-m_r e^4}{2(4\pi\varepsilon_0)^2 \hbar^2 n^2}, \tag{10.145}$$

and the characteristic inverse length scale is

$$\alpha = \frac{m_r e^2}{4\pi\varepsilon_0 \hbar^2 n}. \tag{10.146}$$

Making the approximation $m_r = m_0$ and defining

$$1/\alpha(n = 1) = a_B \tag{10.147}$$

gives

$$a_B = \frac{4\pi\varepsilon_0 \hbar^2}{m_0 e^2} = 0.529\,177 \times 10^{-10} \text{m}. \tag{10.148}$$

The characteristic length scale for the hydrogen atom, a_B, is called the *Bohr radius*. Choosing $m_r = m_0$ is the same as assuming an infinite proton mass for the hydrogen atom. If the reduced mass, m_r, is chosen, then the radius has a value $0.528\,889 \times 10^{-10}$ m.

If $k_{a_B} = 1/a_B$ then the *Rydberg constant* for the hydrogen atom is

$$Ry = \frac{\hbar^2 k_{a_B}^2}{2m_0} = \frac{m_0 e^4}{2(4\pi\varepsilon_0)^2 \hbar^2} = 13.6058 \text{ eV}, \tag{10.149}$$

which is the characteristic energy given by Eq. (10.145) with $n = 1$. Also, from Eq. (10.145) it is apparent that the radial solutions of a hydrogen wave function give energy eigenvalues that depend on quantum number n such that

$$E_n = \frac{-m_0 e^4}{2(4\pi\varepsilon_0)^2 \hbar^2} \left(\frac{1}{n}\right)^2 = \frac{-Ry}{n^2}. \tag{10.150}$$

It is the form of the coulomb potential (Eq. (10.102)) that gives rise to eigenenergies that only depend on quantum number n rather than n and l. The nonzero positive integer n is the principal quantum number of the hydrogen atom, the orbital angular momentum quantum number is an integer $0 \le l \le n - 1$, and the quantum number m is an integer such that $-l \le m \le l$.

Equation (10.150) may be used to explain the optical emission spectrum of the hydrogen atom. Allowed values of l and m for $n = 1, 2, 3$ along with the spectroscopic notation of the states are given in Table 10.3.

Figure 10.12 illustrates the n^2 degenerate states corresponding to principal quantum number n. Horizontal lines represent quantum states ψ_{nlm}. Optical emission involves transitions between electronic states (see Section 10.3.3).

Combining Eqs. (10.132), (10.134), and (10.143), it is clear that, to within a normalization factor, the radial wave function is $e^{-\rho/2} \rho^l \, L_{n+l}^{2l+1}(\rho)$. Normalization of $R_{nl}(\rho)$ requires

Table 10.3 Quantum numbers of hydrogen

Principal quantum number, n	1	2		3		
Orbital angular momentum quantum number, l	0	0	1	0	1	2
Spectroscopic notation of state	1s	2s	2p	3s	3p	3d
Quantum number, m	0	0	$-1, 0, 1$	0	$-1, 0, 1$	$-2, 1, 0, 1, 2$

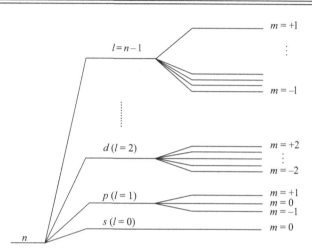

Fig. 10.12 A term diagram for a hydrogenic atom illustrating n^2 degenerate states corresponding to the principal quantum number n. Each of the $n - 1$ quantum numbers l are represented by a horizontal line. Each of the $2l + 1$ quantum numbers m are also represented with a horizontal line.

$$\int_0^\infty e^{-\rho} \rho^{2l} \left| L_{n+l}^{2l+1}(\rho) \right|^2 \rho^2 d\rho = \frac{2n((n+l)!)^3}{(n-l-1)!}, \text{ so that}$$

$$R_{nl}(r) = -\left(\frac{2}{na_B} \right)^{3/2} \left(\frac{(n-l-1)!}{2n((n+l)!)^3} \right)^{1/2} e^{-r/na_B} \left(\frac{2r}{na_B} \right)^l L_{n+l}^{2l+1}\left(\frac{2r}{na_B} \right), \qquad (10.151)$$

where $\rho = 2a_B r$ (Eqs. (10.127) and (10.147)).

Hydrogenic atoms have one free electron, $m_r \approx m_0$, and a nucleus of charge Ze (examples are He$^+$ with $Z = 2$, and Li^{++} with $Z = 3$). They have a hydrogen-like spectrum but with an energy scale changed by the factor Z^2 so that the energy eigenvalues are

$$\boxed{E_n = -\mathrm{Ry} \left(\frac{Z}{n} \right)^2.} \qquad (10.152)$$

Hydrogenic states have the same radial probability density as the hydrogen atom but the length scale is modified by a factor $1/Z$. If the hydrogenic state is an impurity in a semiconductor then the energy and length scale can be modified by the low frequency

453

relative permittivity of the medium, ε_{r0}, and the effective electron mass, m_e^*. This gives rise to an *effective Rydberg constant*,

$$\text{Ry}^* = \text{Ry}\left(\frac{m_e^*/m_0}{\varepsilon_{r0}^2}\right), \tag{10.153}$$

so that the energy eigenvalues of the state are

$$E_n = -\text{Ry}^*\left(\frac{Z}{n}\right)^2. \tag{10.154}$$

There is also an *effective Bohr radius*, a_B^*, that sets the length scale of the hydrogenic state in the semiconductor,

$$a_B^* = \frac{4\pi\varepsilon_0\varepsilon_{r0}\hbar^2}{Zm_e^*e^2}. \tag{10.155}$$

An example of a hydrogenic electron state in a semiconductor such as GaAs is a Si substitutional donor impurity that occupies a Ga lattice site. The extra nonbonding electron forms a hydrogenic state with $Z = 1$. For GaAs with $m_e^* = 0.07 \times m_0$ and $\varepsilon_{r0} = 13.2$, this gives $a_B^* = 10\,\text{nm}$ so that a hydrogenic state in GaAs can be spatially much larger than in an isolated atom.

10.3.2 Hydrogenic Atom Wave Functions

Assuming $\varepsilon_{r0} = 1$ and modifying the Bohr radius so that $a_B \rightarrow a_B/Z$, the solutions for the normalized hydrogenic atom radial wave functions $R_{nl}(r)$ (Eq. (10.151)) become

$$R_{nl}(r) = -\left(\frac{2Z}{na_B}\right)^{3/2}\left(\frac{(n-l-1)!}{2n((n+l)!)^3}\right)^{1/2}e^{-\frac{Zr}{na_B}}\left(\frac{2Zr}{na_B}\right)^l L_{n+1}^{2l+1}\left(\frac{2Zr}{na_B}\right). \tag{10.156}$$

The first few such radial functions are:

$$R_{10}(r) = 2\left(\frac{Z}{a_B}\right)^{3/2}e^{-Zr/a_B}, \tag{10.157}$$

$$R_{20}(r) = 2\left(\frac{Z}{2a_B}\right)^{3/2}\left(1 - \frac{Zr}{2a_B}\right)e^{-Zr/2a_B}, \tag{10.158}$$

$$R_{21}(r) = \frac{2}{\sqrt{3}}\left(\frac{Z}{2a_B}\right)^{3/2}\left(\frac{Zr}{2a_B}\right)e^{-Zr/2a_B}, \tag{10.159}$$

$$R_{30}(r) = 2\left(\frac{Z}{3a_B}\right)^{3/2}\left(1 - \frac{2Zr}{3a_B} + \frac{2}{3}\left(\frac{Zr}{3a_B}\right)^2\right)e^{-Zr/3a_B}, \tag{10.160}$$

$$R_{31}(r) = \frac{2\sqrt{2}}{3}\left(\frac{Z}{3a_B}\right)^{3/2}\left(\frac{Zr}{3a_B}\right)\left(2 - \frac{Zr}{3a_B}\right)e^{-Zr/3a_B}, \tag{10.161}$$

$$R_{32}(r) = \frac{2\sqrt{2}}{3\sqrt{5}} \left(\frac{Z}{3a_B}\right)^{3/2} \left(\frac{Zr}{3a_B}\right)^2 e^{-Zr/3a_B}.$$ (10.162)

Equation (10.114) indicates that the wave functions can be separated into a product of a radial function and spherical harmonics such that $\psi_{nlm}(r, \theta, \phi) = R_{nl}(r)Y_l^m(\theta, \phi)$. Because the solutions for both $R_{nl}(r)$ and $Y_l^m(\theta, \phi)$ are known, it is possible to write down the wave functions for the hydrogenic atom:

$$\psi_{nlm}(r, \theta, \phi) = -\left(\frac{2Z}{na_B}\right)^{3/2} \left(\frac{(n-l-1)!}{2n((n+l)!)^3}\right)^{1/2} e^{-rZ/na_B} \left(\frac{2rZ}{na_B}\right)^l L_{n+l}^{2l+1}\left(\frac{2rZ}{na_B}\right) Y_l^m(\theta, \phi).$$ (10.163)

A few representative hydrogenic wave functions for quantum numbers $n, l,$ and m are:

$$\psi_{100} = 2\left(\frac{Z}{a_B}\right)^{3/2} e^{-Zr/a_B} \left(\frac{1}{4\pi}\right)^{1/2}$$ (10.164)

$$= R_{10}Y_0^0,$$

$$\psi_{200} = 2\left(\frac{Z}{2a_B}\right)^{3/2} \left(1 - \frac{Zr}{2a_B}\right) e^{-Zr/2a_B} \left(\frac{1}{4\pi}\right)^{1/2}$$ (10.165)

$$= R_{20}Y_0^0,$$

$$\psi_{210} = \frac{2}{\sqrt{3}} \left(\frac{Z}{2a_B}\right)^{3/2} \left(\frac{Zr}{2a_B}\right) e^{-Zr/2a_B} \frac{1}{2}\left(\frac{3}{\pi}\right)^{1/2} \cos(\theta)$$ (10.166)

$$= R_{21}Y_1^0,$$

$$\psi_{21\pm1} = \frac{\mp2}{\sqrt{3}} \left(\frac{Z}{2a_B}\right)^{3/2} \left(\frac{Zr}{2a_B}\right) e^{-Zr/2a_B} \frac{1}{2}\left(\frac{3}{2\pi}\right)^{1/2} \sin(\theta)\, e^{\pm i\phi}$$ (10.167)

$$= R_{21}Y_1^{\pm1},$$

$$\psi_{300} = 2\left(\frac{Z}{3a_B}\right)^{3/2} \left(1 - \frac{2Zr}{3a_B} + \frac{2}{3}\left(\frac{Zr}{3a_B}\right)^2\right) e^{-Zr/3a_B} \left(\frac{1}{4\pi}\right)^{1/2}$$ (10.168)

$$= R_{30}Y_0^0,$$

$$\psi_{310} = \frac{2\sqrt{2}}{3} \left(\frac{Z}{3a_B}\right)^{3/2} \left(\frac{Zr}{3a_B}\right) \left(2 - \frac{Zr}{3a_B}\right) e^{-Zr/3a_B} \frac{1}{2}\left(\frac{3}{\pi}\right)^{1/2} \cos(\theta)$$ (10.169)

$$= R_{31}Y_1^0,$$

$$\psi_{31\pm1} = \frac{\mp2\sqrt{2}}{3} \left(\frac{Z}{3a_B}\right)^{3/2} \left(\frac{Zr}{3a_B}\right) \left(2 - \frac{Zr}{3a_B}\right) e^{-Zr/3a_B} \frac{1}{2}\left(\frac{3}{2\pi}\right)^{1/2} \sin(\theta)\, e^{\pm i\phi}$$ (10.170)

$$= R_{30}Y_1^{\pm1},$$

$$\psi_{320} = \frac{2\sqrt{2}}{3\sqrt{5}} \left(\frac{Z}{3a_B}\right)^{3/2} \left(\frac{Zr}{3a_B}\right)^2 e^{-Zr/3a_B} \frac{1}{4}\left(\frac{5}{\pi}\right)^{1/2} \left(3\cos^2(\theta) - 1\right)$$ (10.171)

$$= R_{32}Y_2^0,$$

$$\psi_{32\pm1} = \frac{\mp 2\sqrt{2}}{3\sqrt{5}} \left(\frac{Z}{3a_{\mathrm{B}}}\right)^{3/2} \left(\frac{Zr}{3a_{\mathrm{B}}}\right)^2 \mathrm{e}^{-Zr/3a_{\mathrm{B}}} \frac{1}{2}\left(\frac{15}{2\pi}\right)^{1/2} \sin(\theta)\cos(\theta)\,\mathrm{e}^{\pm i\phi} \tag{10.172}$$

$$= R_{32} Y_2^{\pm1},$$

$$\psi_{32\pm2} = \frac{2\sqrt{2}}{3\sqrt{5}} \left(\frac{Z}{3a_{\mathrm{B}}}\right)^{3/2} \left(\frac{Zr}{3a_{\mathrm{B}}}\right)^2 \mathrm{e}^{-Zr/3a_{\mathrm{B}}} \frac{1}{4}\left(\frac{15}{2\pi}\right)^{1/2} \sin^2(\theta)\,\mathrm{e}^{\pm 2i\phi} \tag{10.173}$$

$$= R_{32} Y_2^{\pm2}.$$

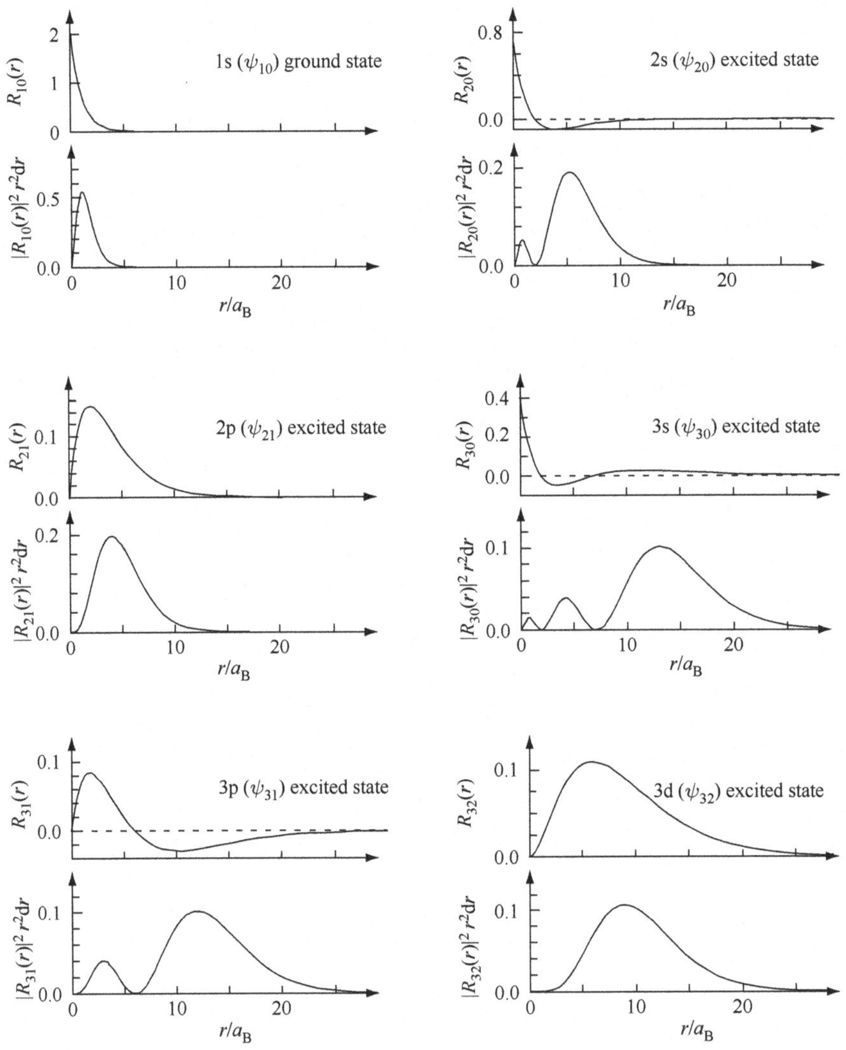

Fig. 10.13 Radial wave function $R_{nl}(r)$ and normalized radial density $|R_{nl}(r)|^2 r^2 \mathrm{d}r$ for a few different states of the hydrogen atom. Radial distance r is normalized to the Bohr radius a_{B}.

The Bohr radius, $a_B = 0.529\,177 \times 10^{-10}$ m (Eq. (10.148)), sets the natural length scale for plotting the radial probability density. In Fig. 10.13, the radial wave function, $R_{nl}(r)$, and normalized radial density probability is plotted for a few different states of the hydrogen atom as a function of radial position r/a_B.

10.3.3 Electromagnetic Radiation

For an atom to emit light via an oscillating dipole mechanism, a dipole $d \sim -e\tilde{z}\cos(\omega t)$ is needed. The electron charge, $-e$, is taken to oscillate in the z direction with frequency ω.

10.3.3.1 No Eigenstate Radiation

For hydrogen eigenstates, $\psi_{nlm}(\mathbf{r}, t) = \psi_{nlm}(\mathbf{r})e^{-i\omega_n t}$ so that the dipole matrix element is of the form

$$\langle \mathbf{r}(t) \rangle = \langle \psi(\mathbf{r}, t)|\hat{\mathbf{r}}|\psi(\mathbf{r}, t) \rangle, \tag{10.174}$$

which has components in the x, y, and z directions with magnitudes $\langle x(t) \rangle$, $\langle y(t) \rangle$, and $\langle z(t) \rangle$, respectively:

$$\langle x(t) \rangle = \int\int\int r \cos(\phi) \sin(\theta)|Y_l^m|^2 R_{nl}^2 r^2 dr \; d(\cos(\theta))d\phi = 0, \tag{10.175}$$

$$\langle y(t) \rangle = \int\int\int r \sin(\phi) \sin(\theta)|Y_l^m|^2 R_{nl}^2 r^2 dr \; d(\cos(\theta))d\phi = 0, \tag{10.176}$$

$$\langle z(t) \rangle = \int\int\int r \cos(\theta)|Y_l^m|^2 R_{nl}^2 r^2 dr \; d(\cos(\theta))d\phi = 0, \tag{10.177}$$

because $\int_0^{2\pi} \cos(\phi)d\phi = 0$ and $|Y_l^m|^2$ is independent of ϕ.

For eigenstates, ψ_{nlm}, the average *vector* $\mathbf{r}$ is *independent of time* and hence zero since the Hamiltonian is isotropic (it has no preferred direction so $\langle \mathbf{r} \rangle$ cannot be a finite constant). Hydrogen does not radiate light if the electron is in an eigenstate.

10.3.3.2 Superposition of Eigenstates

Suppose the state wave function, ψ, is a superposition of two eigenstates such that $\psi = a\psi_n + b\psi_{n'}$ where the principal quantum number $n \neq n'$ so that the energy eigenvalues, E_n and $E_{n'}$, of the two states are different.

Calculating the expectation value of position of the superposition state,

$$\langle \mathbf{r} \rangle = \langle a\psi_n + b\psi_{n'}|\hat{\mathbf{r}}|a\psi_n + b\psi_{n'} \rangle$$
$$= |a|^2 \langle \psi_n|\hat{\mathbf{r}}|\psi_n \rangle + |b|^2 \langle \psi_{n'}|\hat{\mathbf{r}}|\psi_{n'} \rangle + a^* b\langle \psi_n|\hat{\mathbf{r}}|\psi_{n'} \rangle + b^* a\langle \psi_{n'}|\hat{\mathbf{r}}|\psi_n \rangle. \tag{10.178}$$

Since $\psi_n = \varphi_n e^{-iE_n t/\hbar}$, it follows that $\langle \psi_n|\psi_n \rangle = \langle \varphi_n|\varphi_n \rangle = |\varphi_n|^2$ is time-independent. Hence, the first two terms are independent of time and so cannot contribute to the radiation. The last two terms give

$$\langle \mathbf{r}(t) \rangle = a^* b e^{\mathrm{i}(E_n - E_{n'})t/\hbar} \langle \varphi_n | \mathbf{r} | \varphi_{n'} \rangle + b^* a e^{\mathrm{i}(E_{n'} - E_n)t/\hbar} \langle \varphi_{n'} | \mathbf{r} | \varphi_n \rangle$$

$$= 2\mathrm{Re} \left(a^* b \langle \varphi_n | \mathbf{r} | \varphi_{n'} \rangle e^{\mathrm{i}(E_n - E_{n'})t/\hbar} \right) \tag{10.179}$$

$$= 2 |a^* b \langle \varphi_n | \mathbf{r} | \varphi_{n'} \rangle| \cos(\omega_{nn'} t + \delta)$$

$$= 2 |\mathbf{r}_{nn'}| \cos(\omega_{nn'} t + \delta),$$

where it is assumed that $a^* b$ is slowly varying and of order unity, and δ is a phase factor.

Hence, when the atom is undergoing a transition between the states ψ_n and ψ_n the average position of the atom's electron density oscillates with frequency corresponding to the difference in energy between the states. The resulting electromagnetic radiation consists of a photon of energy $\hbar \omega_{nn'}$ and angular frequency $\omega_{nn'}$.

10.3.3.3 Selection Rules for Dipole Radiation

The average radiation from a classical dipole is $P_{\mathrm{rad}} = \frac{Z_0}{12\pi} \frac{\omega^4}{c^2} |\mathbf{d}|^2$ where the dipole moment is $\mathbf{d} = e\mathbf{r}_0$, the expectation value $\langle \mathbf{r}(t) \rangle = \mathbf{r}_0 \cos(\omega t)$, and $Z_0 = \sqrt{\mu_0/\varepsilon_0} = 376.7\,\Omega$ is the impedance of free space.

For a hydrogen atom making a transition from state ψ_{nlm} to state $\psi_{n'l'm'}$, the dipole moment is $\mathbf{d} = 2e\mathbf{r}_{nn'}$ where the factor 2 comes from the expression for $\langle \mathbf{r}(t) \rangle$ and e is the electron charge. Hence, $P_{\mathrm{rad}} = \frac{e^2 Z_0}{3\pi} \frac{\omega^4}{c^2} |\mathbf{r}_{nn'}|^2$.

If the dipole is oscillating in the z direction then the matrix element $\langle \psi_{n'l'm'} | z | \psi_{nlm} \rangle$ is of interest. Because z has odd parity, the states ψ_{nlm} and $\psi_{n'l'm'}$ must have opposite parity or the matrix element will be zero. This implies that the angular momentum quantum numbers l' and l must differ by an odd integer. Separately, for the dipole oscillating in the z direction, the quantum number $m' = m$ or the matrix element will be zero. This follows from evaluating the matrix element by integrating in spherical polar coordinates.

The dipole matrix element selection rule can be obtained by using commutation relations (Problem 10.1). Starting with

$$\left[\hat{L}_z, \hat{z} \right] = 0, \tag{10.180}$$

then

$$0 = \langle \psi_{n'l'm'} | \left[\hat{L}_z, \hat{z} \right] | \psi_{nlm} \rangle = \langle \psi_{n'l'm'} | \hat{L}_z \hat{z} - \hat{z} \hat{L}_z | \psi_{nlm} \rangle, \tag{10.181}$$

$$0 = \langle \psi_{n'l'm'} | m' \hbar \hat{z} - \hat{z} m \hbar | \psi_{nlm} \rangle = \hbar (m' - m) \langle \psi_{n'l'm'} | \hat{z} | \psi_{nlm} \rangle. \tag{10.182}$$

Hence, either $\Delta m = (m' - m) = 0$ or $\langle \psi_{n'l'm'} | \hat{z} | \psi_{nlm} \rangle = 0$. Now, using the commutation relations

$$\left[\hat{L}_z, \hat{x} \right] = \mathrm{i}\hbar \hat{y} \tag{10.183}$$

and

$$\left[\hat{L}_z, \hat{y} \right] = -\mathrm{i}\hbar \hat{x}, \tag{10.184}$$

it follows that

$$\langle \psi_{n'l'm'}| \left[\hat{L}_z, \hat{x} \right] |\psi_{nlm}\rangle = \langle \psi_{n'l'm'}|\hat{L}_z\hat{x} - \hat{x}\hat{L}_z|\psi_{nlm}\rangle, \tag{10.185}$$

$$\langle \psi_{n'l'm'}|m'\hbar\hat{x} - \hat{x}m\hbar|\psi_{nlm}\rangle = i\hbar\langle \psi_{n'l'm'}|\hat{y}|\psi_{nlm}\rangle, \tag{10.186}$$

$$(m' - m)\langle \psi_{n'l'm'}|\hat{x}|\psi_{nlm}\rangle = i\langle \psi_{n'l'm'}|\hat{y}|\psi_{nlm}\rangle, \tag{10.187}$$

and

$$\langle \psi_{n'l'm'}| \left[\hat{L}_z, \hat{y} \right] |\psi_{nlm}\rangle = \langle \psi_{n'l'm'}|\hat{L}_z\hat{y} - \hat{y}\hat{L}_z|\psi_{nlm}\rangle, \tag{10.188}$$

$$\langle \psi_{n'l'm'}|m'\hbar\hat{y} - \hat{y}m\hbar|\psi_{nlm}\rangle = -i\hbar\langle \psi_{n'l'm'}|\hat{x}|\psi_{nlm}\rangle, \tag{10.189}$$

$$(m' - m)\langle \psi_{n'l'm'}|\hat{y}|\psi_{nlm}\rangle = -i\langle \psi_{n'l'm'}|\hat{x}|\psi_{nlm}\rangle. \tag{10.190}$$

Substituting the matrix element for y in Eq. (10.187) into Eq. (10.190) results in

$$(m' - m)^2 \langle \psi_{n'l'm'}|\hat{x}|\psi_{nlm}\rangle = i(m' - m)\langle \psi_{n'l'm'}|\hat{y}|\psi_{nlm}\rangle = \langle \psi_{n'l'm'}|\hat{x}|\psi_{nlm}\rangle. \tag{10.191}$$

From this it is clear that either $\Delta m^2 = (m' - m)^2 = 1$ or $\langle \psi_{n'l'm'}|\hat{x}|\psi_{nlm}\rangle = \langle \psi_{n'l'm'}|\hat{y}|\psi_{nlm}\rangle = 0$. Hence, the dipole selection rule for quantum number m is

$$\boxed{\Delta m = \pm 1, 0.} \tag{10.192}$$

To find the dipole selection rule for quantum number l, use is made of the commutation relation (Problem 10.1),

$$\left[\hat{L}^2, \left[\hat{L}^2, \hat{\mathbf{r}} \right] \right] = 2\hbar^2 \left(\hat{\mathbf{r}}\hat{L}^2 + \hat{L}^2\hat{\mathbf{r}} \right), \tag{10.193}$$

so that

$$2\hbar^2 \langle \psi_{n',l',m'}|(\hat{\mathbf{r}}\hat{L}^2 + \hat{L}^2\hat{\mathbf{r}})|\psi_{nlm}\rangle = 2\hbar^4(l'(l' + 1) + l(l + 1))\langle \psi_{n'l'm'}|\mathbf{r}|\psi_{nlm}\rangle, \tag{10.194}$$

$$\langle \psi_{n',l',m'}|\hat{L}^2 \left[\hat{L}^2, \hat{\mathbf{r}} \right] - \left[\hat{L}^2, \hat{\mathbf{r}} \right] \hat{L}^2|\psi_{nlm}\rangle = \hbar^2(l'(l' + 1) - l(l + 1))\langle \psi_{n'l'm'}| \left[\hat{L}^2, \hat{\mathbf{r}} \right] |\psi_{nlm}\rangle, \tag{10.195}$$

$$\hbar^2(l'(l' + 1) - l(l + 1))\langle \psi_{n',l',m'}|\hat{L}^2\hat{\mathbf{r}} - \hat{\mathbf{r}}\hat{L}^2|\psi_{nlm}\rangle$$
$$= \hbar^4(l'(l' + 1) - l(l + 1))^2 \langle \psi_{n',l',m'}|\hat{\mathbf{r}}|\psi_{nlm}\rangle. \tag{10.196}$$

Hence, either $\langle \psi_{n',l',m'}|\hat{\mathbf{r}}|\psi_{nlm}\rangle = 0$ or

$$2(l'(l' + 1) + l(l + 1)) = (l'(l' + 1) - l(l + 1))^2. \tag{10.197}$$

The latter may be written as

$$(l' + l + 1)^2(l' - l)^2 = (l' + l + 1)^2 + (l' - l)^2 - 1 \tag{10.198}$$

or

$$\left((l' + l + 1)^2 - 1 \right) \left((l' - l)^2 - 1 \right) = 0. \tag{10.199}$$

459

From this it may be deduced that either $l' = l = 0$, in which case symmetry requires $\langle \psi_{n'0m'} | \hat{\mathbf{r}} | \psi_{n0m} \rangle = 0$, or, more interestingly, $\Delta l = l' - l = \pm 1$. Hence, the dipole selection rule for quantum number l is

$$\boxed{\Delta l = \pm 1.} \tag{10.200}$$

The only conditions under which $|r_{nn'}|^2$ is not zero are $\Delta l = l' - l = \pm 1$ and $\Delta m_l = m' - m = 0, \pm 1$. Because a photon carries unit integer of quantized angular momentum $\hbar$, conservation of angular momentum requires orbital quantum number l to change by $\Delta l = \pm l$. Absorbed or emitted radiation is linearly polarized for the matrix element describing a dipole oscillating in the z direction. If the dipole oscillates in the $x-y$ plane there will be interest in evaluating the matrix elements $\langle \psi_{n'l'm'} | \hat{x} \pm i\hat{y} | \psi_{nlm} \rangle = \langle \psi_{n'l'm'} | \hat{L}_\pm | \psi_{nlm} \rangle$, which correspond to circularly polarized radiation with the convention that $\Delta m = +1(-1)$ for absorption of left (right)-handed circularly polarized radiation. The sign of Δm is reversed for emission.

10.3.3.4 Hydrogen 2p to 1s Transition

Consider an electron making a 2p to 1s transition with $\Delta m = 0$. The *initial state* is characterized by principal quantum number $n = 2$, orbital quantum numbers $l = 1, m = 0$, and eigenenergy $E_2 = -3.40\,\text{eV}$. The wave function is

$$\psi_{210} = \frac{1}{\sqrt{3}(2a_B)^{3/2}} \frac{r}{a_B} e^{-r/2a_B} \frac{1}{2} \left(\frac{3}{\pi} \right)^{1/2} \cos(\theta). \tag{10.201}$$

The *final state* has $n = 1, l = 0, m = 0, E_2 = -13.61\,\text{eV}$, and wave function

$$\psi_{100} = \frac{2}{a_B^{3/2}} e^{-r/a_B} \left(\frac{1}{4\pi} \right)^{1/2}. \tag{10.202}$$

Figure 10.14 shows the radial distribution probability for the ψ_{100} and ψ_{210} states. Superposition of ψ_{210} and ψ_{100} causes the hydrogen electron probability density to oscillate at difference energy $\hbar\omega = E_2 - E_1 = 10.20\,\text{eV}$. The oscillation in expectation value of electron position creates a dipole moment and electromagnetic radiation is emitted at wavelength $\lambda = 2\pi c/\omega = 0.122\,\mu\text{m}$, carrying away angular momentum $\hbar$.

The photon field produced by the transition takes energy from the excited state and converts it to photon energy. Typically, the electromagnetic field intensity decays as $e^{-t/\tau}$ where $\tau \sim$ ns so that the *length* of the photon emitted into free space is about 0.3 m. Using the wave functions given by Eqs. (10.164), (10.166), and (10.167) to calculate the matrix element (see Chapter 7), a spontaneous emission time for the 2p to 1s transition in atomic hydrogen of $\tau_{\text{sp}} = 1.6\,\text{ns}$ (with corresponding transition rate $1/\tau_{\text{sp}} = 6.3 \times 10^8\,\text{s}^{-1}$) is obtained. See Problem 10.3.

For the transition from state ψ_{210} to ψ_{100}, the value $\Delta m = 0$ and the photon emitted by the atom is linearly polarized. If $\Delta m = +1$ then the emitted radiation is right-hand circularly polarized (it rotates clockwise in time) and if $\Delta m = -1$, then the emitted radiation is left-hand circularly polarized (it rotates counterclockwise in time).

For the absorption of circularly polarized radiation, the opposite is true. In this case if $\Delta m = +1$ then the absorbed radiation is left-hand circularly polarized (it rotates

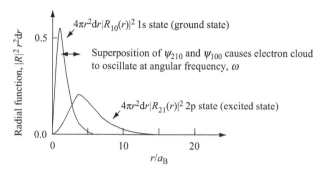

Fig. 10.14 Radial distribution probability for the ψ_{100} and ψ_{210} states. A superposition of these states causes the expectation value of electron position to oscillate at a frequency given by the difference in the value of the state's respective energy eigenvalues.

counterclockwise in time) and if $\Delta m = -1$ then the absorbed radiation is right-hand circularly polarized (it rotates clockwise in time).

10.3.4 Fine Structure of the Hydrogen Atom and Electron Spin

So far, three quantum numbers for the hydrogen atom have been considered. The principal quantum number, n, the orbital angular momentum, l, and the quantum number, m. There is a fourth quantum number, which is the intrinsic angular momentum or spin of the electron, $\sigma_s = \pm\hbar/2$. Total angular momentum is the sum of the orbital angular momentum, **L**, and the spin angular momentum, σ, so, as illustrated in Fig. 10.15, **J** = **L** + σ. For transitions between eigenstates, a new set of rules that include total angular momentum quantum number, j, must be created.

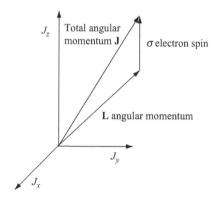

Fig. 10.15 Illustration of contributions to total angular momentum **J** from angular momentum **L** and electron spin σ.

The energy eigenvalues of a hydrogenic atom are $E_n = -m_r(Ze^2)^2/2(4\pi\varepsilon_0)^2\hbar^2 n^2$ with a correction factor for the fact that the spin of the electron interacts with the magnetic field due to the angular momentum, **L**, of the electron wave function. This gives rise to

461

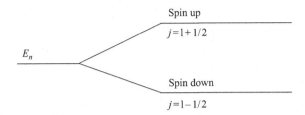

Fig. 10.16 Fine structure in a hydrogenic atom due to electron spin, σ_{rms}.

a small correction to the energy levels of the atom depending upon whether the spin is aligned with **L** or not. This energy correction is illustrated in Fig. 10.16.

For the spin up ($+$) and spin down ($-$) state,

$$E_{nlj}^{+} = -|E_n|\left(1 - \frac{1}{(2l+1)(l+1)}\frac{(Z\alpha)^2}{n}\right), \tag{10.203}$$

$$E_{nlj}^{-} = -|E_n|\left(1 + \frac{1}{(2l+1)(l+1)}\frac{(Z\alpha)^2}{n}\right), \tag{10.204}$$

where α is the *fine structure constant*, $\alpha = e^2/4\pi\varepsilon_0\hbar c = 1/137.037$, and $\alpha^2 = 5.33 \times 10^{-5}$ so this is a small correction. The selection rules for dipole radiation become

$$
\boxed{
\begin{aligned}
\Delta l &= \pm 1, \\
\Delta j &= \pm 1, 0, \\
\Delta m_j &= \pm 1, 0,
\end{aligned}
}
\quad \text{(but } j = 0 \to 0 \text{ is forbidden)}, \tag{10.205}
$$

and, as usual, Δn can be anything.

10.4 Hybridization

The orthonormal basis of stationary electron states of the hydrogen atom are (Eq. (10.114))

$$\psi_{nlm}(r, \theta, \phi) = R_{nl}(r)Y_l^m(\theta, \phi). \tag{10.206}$$

These atomic orbitals have quantum numbers n, l, and m that determine the electron's energy E_n, the square of the angular momentum $\hbar^2 l(l+1)$, and the z component of angular momentum, $\hbar m$.

It is an intrinsic feature of quantum mechanics that new stationary states can be formed from a linear superposition of degenerate $\psi_{nlm}(r, \theta, \phi)$ states. Such a linear superposition of orbitals with the *same* value of n but *different* l and m is called a hybrid orbital. The creation of hybrid orbitals allows electron density to be enhanced in a particular spatial direction compared to the pure atomic orbitals from which it is formed. It is in this way that directional bonds and the resulting geometry of many molecules and crystals can be explained.

By way of example, consider the $n = 2$, $l = 1$ (2p) states of the hydrogen atom. They have orthogonal atomic orbitals,

$$\phi_{2,1,1} = -\frac{1}{2}\sqrt{\frac{3}{2\pi}} R_{2,1}(r) \sin(\theta)\, e^{i\phi}, \tag{10.207}$$

$$\phi_{2,1,0} = \frac{1}{2}\sqrt{\frac{3}{\pi}} R_{2,1}(r) \cos(\theta), \tag{10.208}$$

$$\phi_{2,1,-1} = \frac{1}{2}\sqrt{\frac{3}{2\pi}} R_{2,1}(r) \sin(\theta)\, e^{-i\phi}. \tag{10.209}$$

These can be rewritten as linear combinations of orbitals to form p_x, p_y, and p_z orbitals that are *real* functions of r, θ, ϕ. They are orthogonal and span the same part of Hilbert space:

$$\phi_{2p_x} = \frac{-1}{\sqrt{2}}(\phi_{2,1,1} - \phi_{2,1,-1}) = \sqrt{\frac{3}{4\pi}} R_{2,1}(r)\frac{x}{r}, \tag{10.210}$$

$$\phi_{2p_y} = \frac{1}{\sqrt{2}}(\phi_{2,1,1} + \phi_{2,1,-1}) = -i\sqrt{\frac{3}{4\pi}} R_{2,1}(r)\frac{y}{r}, \tag{10.211}$$

$$\phi_{2p_z} = \frac{1}{\sqrt{2}}\phi_{2,1,0} = \sqrt{\frac{3}{4\pi}} R_{2,1}(r)\frac{z}{r}. \tag{10.212}$$

These functions may be used to illustrate the formation of hybrid orbitals. Consider the carbon atom that has four outer shell electrons in the $2s^2 2p^2$ configuration. Hybrid orbitals for these four electrons can be formed using the p_x, p_y, and p_z orbitals and the 2s hydrogenic orbital. The hybrid orbitals are

$$\psi_1 = \frac{1}{2}(\phi_{2s} + \phi_{2p_x} + \phi_{2p_y} + \phi_{2p_z}), \tag{10.213}$$

$$\psi_2 = \frac{1}{2}(\phi_{2s} + \phi_{2p_x} - \phi_{2p_y} + \phi_{2p_z}), \tag{10.214}$$

$$\psi_3 = \frac{1}{2}(\phi_{2s} + \phi_{2p_x} - \phi_{2p_y} - \phi_{2p_z}), \tag{10.215}$$

$$\psi_4 = \frac{1}{2}(\phi_{2s} - \phi_{2p_x} + \phi_{2p_y} + \phi_{2p_z}). \tag{10.216}$$

The electron probability density have maxima along the four tetrahedral directions, which in Cartesian coordinates are $(1, 1, 1)$, $(-1, -1, 1)$, $(1, -1, -1)$, and $(-1, 1, -1)$. The methane molecule, CH_4, has a hydrogen atom covalently bonded by the electron probability maxima in each of these directions. The angle between adjacent tetrahedrally coordinated bonds is $2 \times \tan^{-1}(\sqrt{2}) = 109.47°$.

10.4.1 sp³ Hybridization to Enhance Electron Density Directivity

Hybrid orbitals can extend electron density in a given spatial direction further than pure orbitals. To see this, consider a linear combination of orbitals

$$\psi = A\,s + B\,p_x + C\,p_y + D\,p_z, \tag{10.217}$$

463

where the orthonormal atomic orbitals are normalized such that $A^2 + B^2 + C^2 + D^2 = 1$ and

$$s = \frac{1}{\sqrt{4\pi}}, \tag{10.218}$$

$$p_x = \sqrt{\frac{3}{4\pi}} \sin(\theta) \cos(\phi) = \sqrt{\frac{3}{4\pi}} \frac{x}{r}, \tag{10.219}$$

$$p_y = \sqrt{\frac{3}{4\pi}} \sin(\theta) \sin(\phi) = \sqrt{\frac{3}{4\pi}} \frac{y}{r}, \tag{10.220}$$

$$p_z = \sqrt{\frac{3}{4\pi}} \cos(\theta) = \sqrt{\frac{3}{4\pi}} \frac{z}{r}, \tag{10.221}$$

where, because only the angular dependence is of interest, the radial part in Eqs. (10.210), (10.211), and (10.212) is ignored. For a pure p_z orbital, the maximum occurs when $A = B = C = 0$, $D = 1$, and $\theta = 0$, so that

$$\psi_{p_z,\max} = \sqrt{\frac{3}{4\pi}}. \tag{10.222}$$

However, for a linear superposition in the z direction it is possible to set $B = C = 0$. Normalization requires $A^2 + D^2 = 1$, so that $D^2 = 1 - A^2$ and

$$\psi_z = As + \sqrt{1 - A^2}\, p_z. \tag{10.223}$$

The maximum is found by taking the derivative,

$$\frac{d\psi_z(\theta = 0)}{dA} = \frac{1}{\sqrt{4\pi}} + \frac{1}{2}\frac{(-2A)}{\sqrt{1 - A^2}}\sqrt{\frac{3}{4\pi}} = 0, \tag{10.224}$$

so that $\sqrt{1 - A^2} = \sqrt{3}\, A$, or $A = \frac{1}{2}$ and

$$\psi_{z,\max} = \frac{1}{2}s + \frac{\sqrt{3}}{2} p_z(\theta = 0). \tag{10.225}$$

Hence, the maximum value in the z direction with $\theta = 0$ is

$$\psi_{z,\max} = \psi_z(\theta = 0) = \frac{1}{2}\frac{1}{\sqrt{4\pi}} + \frac{\sqrt{3}}{2}\sqrt{\frac{3}{4\pi}}, \tag{10.226}$$

and so

$$\psi_{z,\max} = \frac{2}{\sqrt{4\pi}}. \tag{10.227}$$

This is a factor $\psi_{z,\max}/\psi_{p_z,\max} = 2/\sqrt{3} = 1.155$ larger than a purely p_z orbital. Hence, the hybrid orbital can direct about 33% more electron *density* in a given direction and so is capable of forming a stronger bond than a p_z orbital on its own. This is one reason why hybrid orbitals are important.

10.5 Example Exercises

Exercise 10.1
(a) Use the transformation between Cartesian and spherical coordinates,

$$x = r \sin(\theta) \cos(\phi),$$
$$y = r \sin(\theta) \sin(\phi),$$
$$z = r \cos(\theta),$$

to obtain the expression $\hat{L}_z = -i\hbar \frac{\partial}{\partial \phi}$.

(b) Calculate $\left[\phi, \hat{L}_z \right]$.

(c) Show that $\hat{L}^2 = -\hbar^2 \left(\frac{1}{\sin(\theta)} \frac{\partial}{\partial \theta} \left(\sin(\theta) \frac{\partial}{\partial \theta} \right) + \frac{1}{\sin^2(\theta)} \frac{\partial}{\partial \phi^2} \right)$.

(d) Derive $\left[\hat{L}_y, \hat{L}_z \right] = i\hbar \hat{L}_x$ using the fact that $[\hat{x}, \hat{p}_x] = i\hbar$.

(e) If $\hat{L}_{\pm} = \hat{L}_x \pm i\hat{L}_y$ show that $\left[\hat{L}^2, \hat{L}_{\pm} \right] = 0$, $\left[\hat{L}_+, \hat{L}_- \right] = 2\hbar \hat{L}_z$, and $\left[\hat{L}_z, \hat{L}_{\pm} \right] = \pm \hbar \hat{L}_{\pm}$.

(f) Show that $\hat{L}^2 = \hat{L}_{\pm} \hat{L}_{\mp} + \hat{L}_z^2 \mp \hbar \hat{L}_z$ and $\hat{L}^2 = \frac{1}{2} \left(\hat{L}_+ \hat{L}_- + \hat{L}_- \hat{L}_+ \right) + \hat{L}_z^2$.

Exercise 10.2
A hydrogen molecule (H_2) consists of two *identical* hydrogen atoms separated by a center-to-center distance of approximately $2a = 0.074$ nm and having a total moment of inertia I_L.

(a) Show that the solution to the Schrödinger equation gives wave function $\Phi_m = A e^{im\phi}$, where ϕ is the angle of rotation and A is a constant. What values of m are allowed?

(b) At low temperatures the heat capacity of H_2 gas is $c_v = 3k_B/2$, where k_B is the Boltzmann constant and where each spatial degree of freedom in the x, y, and z directions contributes $k_B/2$. At some temperature T, the thermal energy $k_B T$ is high enough to excite rotational modes of the molecule, which has the effect of increasing c_v. If the rotational energy for total angular momentum state quantum number l is $E = \hbar^2 l(l+1)/2I_L$, estimate the temperature at which this occurs. By how much is c_v increased?

Exercise 10.3
Find pairs of two hydrogen atom principal quantum numbers n for which the difference in energy eigenvalue is the same and hence there exist coincident spectral lines.

Exercise 10.4
Use first-order perturbation theory to estimate the energy shifts of the hydrogen 2s and 2p states due to the fact that the proton is not a point charge, treating it (for simplicity) as a uniformly charged hollow spherical shell of radius $b = 5 \times 10^{-14}$ cm. Comment on these results. Compare the results with the expected order of magnitude of the correction due to the finite proton mass. Explain why measuring such energy shifts is not a good way of studying the proton charge distribution and suggest, with reasons, a more effective way.

Exercise 10.5
(a) What is the effect of applying a uniform electric field on the energy spectrum of an atom?

(b) If spin effects are neglected, the four states of the hydrogen atom with quantum number $n = 2$ have the same energy, E^0. Show that when a uniform z-directed electric field $\mathbf{E}$ is applied to hydrogen atoms in these states, the resulting first-order energies are $E^0 \pm 3a_B e|\mathbf{E}|$, E^0, E^0.

Treat the z-directed electric field as a perturbation on the separable, orthonormal, unperturbed electron wave functions $\psi_{nlm}(r, \theta, \phi) = R_n(r)\Theta_l^m(\theta)\Phi_m(\phi)$, where r, θ, and ϕ are the standard spherical coordinates. Use may be made of the unperturbed wave functions

$$\psi_{200} = \frac{2}{(2a_B)^{3/2}} \left(1 - \frac{r}{2a_B}\right) e^{-r/2a_B} \left(\frac{1}{4\pi}\right)^{1/2},$$

$$\psi_{210} = \frac{1}{\sqrt{3}(2a_B)^{3/2}} \frac{r}{a_B} e^{-r/2a_B} \frac{1}{2} \left(\frac{3}{\pi}\right)^{1/2} \cos(\theta).$$

Solutions

Solutions 10.1
(a) For a particle at position $\mathbf{r}$ with momentum $\mathbf{p}$, the classical angular momentum is $\mathbf{L} = \mathbf{r} \times \mathbf{p}$ and so the z-component of angular momentum is $L_z = xp_y - yp_x$. The corresponding quantum mechanical operator in Cartesian coordinates is

$$\hat{L}_z = \hat{x}\hat{p}_y - \hat{y}\hat{p}_x = \hat{x}\left(-i\hbar\frac{\partial}{\partial y}\right) - \hat{y}\left(-i\hbar\frac{\partial}{\partial x}\right) = -i\hbar\left(\hat{x}\frac{\partial}{\partial y} - \hat{y}\frac{\partial}{\partial x}\right).$$

Using the transformation to spherical coordinates,

$$x = r\sin(\theta)\cos(\phi),$$
$$y = r\sin(\theta)\sin(\phi),$$
$$z = r\cos(\theta),$$

the task is to find the expressions for the partial derivatives appearing in the operator $\hat{L}_z$; $\hat{L}_z = -i\hbar\frac{\partial}{\partial\phi}$ is found by taking the derivatives

$$\frac{\partial x}{\partial\phi} = -r\sin(\theta)\sin(\phi) = -y,$$

$$\frac{\partial y}{\partial\phi} = r\sin(\theta)\cos(\phi) = x,$$

$$\frac{\partial z}{\partial\phi} = 0,$$

so that

$$\hat{L}_z = -i\hbar\frac{\partial}{\partial\phi} = -i\hbar\left(\frac{\partial x}{\partial\phi}\frac{\partial}{\partial x} + \frac{\partial y}{\partial\phi}\frac{\partial}{\partial y} + \frac{\partial z}{\partial\phi}\frac{\partial}{\partial z}\right)$$

$$= -i\hbar\left(-\hat{y}\frac{\partial}{\partial x} + \hat{x}\frac{\partial}{\partial y} + 0\right) = -i\hbar\left(\hat{x}\frac{\partial}{\partial y} - \hat{y}\frac{\partial}{\partial x}\right).$$

However, in the absence of the given solution, more calculation is required to find the derivatives $\frac{\partial}{\partial x}$ and $\frac{\partial}{\partial y}$ in spherical coordinates.

Starting with the radial position, $r = \left(x^2 + y^2 + z^2\right)^{1/2}$, the partial derivative

$$\frac{\partial r}{\partial x} = \frac{2x}{2\sqrt{(x^2 + y^2 + z^2)}} = \frac{x}{r},$$

and similarly,

$$\frac{\partial r}{\partial y} = \frac{y}{r}$$

and

$$\frac{\partial r}{\partial z} = \frac{z}{r}.$$

Noting that $\cos(\theta) = \frac{z}{r}$ and taking the partial derivative with respect to x results in

$$-\sin(\theta)\frac{\partial\theta}{\partial x} = z\left(\frac{-1}{r^2}\frac{\partial r}{\partial x}\right),$$

so that

$$\frac{\partial\theta}{\partial x} = \frac{z}{r^2\sin(\theta)}\frac{x}{r} = \frac{\cos(\theta)\cos(\phi)}{r}.$$

Likewise, the partial derivative with respect to y yields

$$-\sin(\theta)\frac{\partial\theta}{\partial y} = z\left(\frac{-1}{r^2}\frac{\partial r}{\partial y}\right)$$

and

$$\frac{\partial\theta}{\partial y} = \frac{z}{r^2\sin(\theta)}\frac{y}{r} = \frac{\cos(\theta)\sin(\phi)}{r}.$$

And for the partial derivative with respect to z,

$$-\sin(\theta)\frac{\partial\theta}{\partial z} = \frac{1}{r} - \frac{z}{r^2}\frac{\partial r}{\partial z} = \frac{1}{r} - \frac{z}{r^2}\left(\frac{z}{r}\right) = \frac{1 - \cos^2(\theta)}{r} = \frac{\sin^2(\theta)}{r}$$

and

$$\frac{\partial\theta}{\partial z} = \frac{-\sin(\theta)}{r}.$$

Since $\tan(\theta) = \frac{y}{x}$, taking the partial derivative with respect to x results in

$$\sec^2(\phi)\frac{\partial\phi}{\partial x} = y\left(\frac{-1}{x^2}\right)$$

and

$$\frac{\partial \phi}{\partial x} = \frac{-y}{x^2} \cos^2(\phi).$$

Likewise, the partial derivative with respect to y yields

$$\sec^2(\phi)\frac{\partial \phi}{\partial y} = \frac{1}{x}$$

and

$$\frac{\partial \phi}{\partial x} = \frac{1}{x} \cos^2(\phi).$$

And the partial derivative with respect to z gives

$$\sec^2(\phi)\frac{\partial \phi}{\partial z} = 0$$

and

$$\frac{\partial \phi}{\partial z} = 0.$$

Therefore,

$$\hat{L}_z = -i\hbar \left(\hat{x}\frac{\partial}{\partial y} - \hat{y}\frac{\partial}{\partial x} \right) = -i\hbar \left(x \left(\frac{\partial r}{\partial y}\frac{\partial}{\partial r} + \frac{\partial \theta}{\partial y}\frac{\partial}{\partial \theta} + \frac{\partial \phi}{\partial y}\frac{\partial}{\partial \phi} \right) - y \left(\frac{\partial r}{\partial x}\frac{\partial}{\partial r} + \frac{\partial \theta}{\partial x}\frac{\partial}{\partial \theta} + \frac{\partial \phi}{\partial x}\frac{\partial}{\partial \phi} \right) \right)$$
$$= -i\hbar \left(\left(\frac{xy}{r} - \frac{yx}{r} \right)\frac{\partial}{\partial r} + \left(x\frac{\cos(\theta)\sin(\phi)}{r} - y\frac{\cos(\theta)\sin(\phi)}{r} \right)\frac{\partial}{\partial \theta} \right.$$
$$\left. + \left(\cos^2(\phi) + \frac{y^2}{x^2}\cos^2(\phi) \right)\frac{\partial}{\partial \phi} \right).$$

Since

$$\left(\frac{xy}{r} - \frac{yx}{r} \right)\frac{\partial}{\partial r} = 0$$

and

$$\left(x\frac{\cos(\theta)\sin(\phi)}{r} - y\frac{\cos(\theta)\sin(\phi)}{r} \right)$$
$$= (\sin(\theta)\cos(\theta)\sin(\phi)\cos(\theta) - \sin(\theta)\cos(\theta)\sin(\phi)\cos(\theta)) = 0,$$

this results in

$$\hat{L}_z = -i\hbar \left(\left(\cos^2(\phi) + \frac{y^2}{x^2}\cos^2(\phi) \right)\frac{\partial}{\partial \phi} \right) = -i\hbar \left(\cos^2(\phi)(1 + \tan^2(\phi))\frac{\partial}{\partial \phi} \right)$$
$$= -i\hbar \left(\cos^2(\phi)\sec^2(\phi)\frac{\partial}{\partial \phi} \right) = -i\hbar\frac{\partial}{\partial \phi}.$$

(b) The calculation of $\left[\hat{\phi}, \hat{L}_z\right]$ proceeds as

$$\left[\hat{\phi}, \hat{L}_z\right]\psi = \left(\hat{\phi}\hat{L}_z - \hat{L}_z\hat{\phi}\right)\psi = \hat{\phi}\left(-i\hbar\frac{\partial}{\partial\phi}\right)\psi - \left(-i\hbar\frac{\partial}{\partial\phi}\right)\hat{\phi}\psi = i\hbar\left(\frac{\partial}{\partial\phi}(\hat{\phi}\psi) - \hat{\phi}\frac{\partial}{\partial\phi}\psi\right)$$

$$= i\hbar\left(\psi\left(\frac{\partial}{\partial\phi}\hat{\phi}\right) + \hat{\phi}\left(\frac{\partial}{\partial\phi}\psi\right) - \hat{\phi}\left(\frac{\partial}{\partial\phi}\psi\right)\right) = i\hbar\psi\frac{\partial}{\partial\phi}\hat{\phi} = i\hbar\psi,$$

so that

$$\left[\hat{\phi}, \hat{L}_z\right] = i\hbar.$$

(c) Because $\hat{L}^2 = \hat{L}_x^2 + \hat{L}_y^2 + \hat{L}_z^2$, it is helpful to find expressions for the x, y, and z components of $\hat{L}$. Starting with $\hat{L}_x$ in Cartesian coordinates,

$$\hat{L}_x = \hat{y}\hat{p}_z - \hat{z}\hat{p}_y = -i\hbar\left(\hat{y}\frac{\partial}{\partial z} - \hat{z}\frac{\partial}{\partial y}\right)$$

$$= -i\hbar\left(\hat{y}\left(\frac{\partial r}{\partial z}\frac{\partial}{\partial r} + \frac{\partial\theta}{\partial z}\frac{\partial}{\partial\theta} + \frac{\partial\phi}{\partial z}\frac{\partial}{\partial\phi}\right) - \hat{z}\left(\frac{\partial r}{\partial y}\frac{\partial}{\partial r} + \frac{\partial\theta}{\partial y}\frac{\partial}{\partial\theta} + \frac{\partial\phi}{\partial y}\frac{\partial}{\partial\phi}\right)\right)$$

$$= -i\hbar\left(\left(\frac{yz}{r} - \frac{zy}{r}\right)\frac{\partial}{\partial r} + \left(\frac{-y\sin(\theta)}{r} - \frac{z\cos(\theta)\sin(\phi)}{r}\right)\frac{\partial}{\partial\theta} + \left(0 - \frac{z}{x}\cos^2(\phi)\right)\frac{\partial}{\partial\phi}\right)$$

$$= -i\hbar\left(-(\sin^2(\theta)\sin(\phi) + \cos^2(\theta)\sin(\phi))\frac{\partial}{\partial\theta} + (-\cot(\theta)\cos(\phi))\frac{\partial}{\partial\phi}\right)$$

$$= i\hbar\left(\sin(\phi)\frac{\partial}{\partial\theta} + \cot(\theta)\cos(\phi)\frac{\partial}{\partial\phi}\right).$$

Similarly, for $\hat{L}_y$ the following is found:

$$\hat{L}_y = \hat{z}\hat{p}_x - \hat{x}\hat{p}_z = -i\hbar\left(\hat{z}\frac{\partial}{\partial x} - \hat{x}\frac{\partial}{\partial z}\right)$$

$$= -i\hbar\left(\hat{z}\left(\frac{\partial r}{\partial x}\frac{\partial}{\partial r} + \frac{\partial\theta}{\partial x}\frac{\partial}{\partial\theta} + \frac{\partial\phi}{\partial x}\frac{\partial}{\partial\phi}\right) - \hat{x}\left(\frac{\partial r}{\partial z}\frac{\partial}{\partial r} + \frac{\partial\theta}{\partial z}\frac{\partial}{\partial\theta} + \frac{\partial\phi}{\partial z}\frac{\partial}{\partial\phi}\right)\right)$$

$$= -i\hbar\left(\left(\frac{zx}{r} - \frac{xz}{r}\right)\frac{\partial}{\partial r} + \left(\frac{z\cos(\theta)\sin(\phi)}{r} + \frac{x\sin(\theta)}{r}\right)\frac{\partial}{\partial\theta} + \left(\frac{-zy}{x^2}\cos^2(\phi) - 0\right)\frac{\partial}{\partial\phi}\right)$$

$$= -i\hbar\left(-(\cos^2(\theta)\cos(\phi) + \sin^2(\theta)\cos(\phi))\frac{\partial}{\partial\theta} + \left(\frac{-r^2\sin(\theta)\cos(\theta)\sin(\phi)}{r^2\sin^2(\theta)\cos^2(\phi)}\right)\frac{\partial}{\partial\phi}\right)$$

$$= i\hbar\left(\cos(\phi)\frac{\partial}{\partial\theta} - \cot(\theta)\sin(\phi)\frac{\partial}{\partial\phi}\right).$$

The expression for $\hat{L}_z$ is given in part (a). $\hat{L}^2$ is found using $\hat{L}^2 = \hat{L}_x^2 + \hat{L}_y^2 + \hat{L}_z^2$.

The expression for $\hat{L}_x^2$ is

$$\hat{L}_x^2 = -\hbar^2 \left(\sin^2(\phi) \frac{\partial^2}{\partial \theta^2} + \cot^2(\theta)\cos^2(\phi) \frac{\partial^2}{\partial \phi^2} \right.$$

$$\left. + \sin(\phi) \frac{\partial}{\partial \theta} \cot(\theta)\cos(\phi) \frac{\partial}{\partial \phi} + \cot(\theta)\cos(\phi) \frac{\partial}{\partial \phi} \sin(\phi) \frac{\partial}{\partial \theta} \right)$$

$$= -\hbar^2 \left(\sin^2(\phi) \frac{\partial^2}{\partial \theta^2} + \cot^2(\theta)\cos^2(\phi) \frac{\partial^2}{\partial \phi^2} + \sin(\phi) \frac{\partial}{\partial \theta} \cot(\theta)\cos(\phi) \frac{\partial}{\partial \phi} \right.$$

$$\left. + \cot(\theta)\cos(\phi)\cos(\phi) \frac{\partial}{\partial \theta} + \cot(\theta)\cos(\phi)\sin(\phi) \frac{\partial \partial}{\partial \phi \partial \theta} \right),$$

and for $\hat{L}_y^2$ it is

$$\hat{L}_y^2 = -\hbar^2 \left(\cos^2(\phi) \frac{\partial^2}{\partial \theta^2} + \cot^2(\theta)\sin^2(\phi) \frac{\partial^2}{\partial \phi^2} \right.$$

$$\left. - \cos(\phi) \frac{\partial}{\partial \theta} \cot(\theta)\sin(\phi) \frac{\partial}{\partial \phi} - \cot(\theta)\sin(\phi) \frac{\partial}{\partial \phi} \cos(\phi) \frac{\partial}{\partial \theta} \right)$$

$$= -\hbar^2 \left(\cos^2(\phi) \frac{\partial^2}{\partial \theta^2} + \cot^2(\theta)\sin^2(\phi) \frac{\partial^2}{\partial \phi^2} + \sin(\phi) \frac{\partial}{\partial \theta} \cot(\theta)\cos(\phi) \frac{\partial}{\partial \phi} \right.$$

$$\left. + \cot(\theta)\sin(\phi)(-\sin(\phi)) \frac{\partial}{\partial \theta} + \cot(\theta)\sin(\phi)\cos(\phi) \frac{\partial \partial}{\partial \phi \partial \theta} \right).$$

Hence,

$$\hat{L}^2 = -\hbar^2 \left(\frac{\partial^2}{\partial \theta^2} + \cot^2(\theta) \frac{\partial^2}{\partial \phi^2} + \cot(\theta)\left(\cos^2(\phi) + \sin^2(\phi)\right) \frac{\partial}{\partial \theta} + \frac{\partial^2}{\partial \phi^2} \right)$$

$$= -\hbar^2 \left(\frac{\partial^2}{\partial \theta^2} + \left(1 + \cot^2(\theta)\right) \frac{\partial^2}{\partial \phi^2} + \cot(\theta) \frac{\partial}{\partial \theta} \right).$$

Noting that $1 + \cot^2(\theta) = \csc^2(\theta) = 1/\sin^2(\theta)$, the expression may be simplified to

$$\hat{L}^2 = -\hbar^2 \left(\frac{\partial^2}{\partial \theta^2} + \cot(\theta) \frac{\partial}{\partial \theta} + \frac{1}{\sin^2(\theta)} \frac{\partial^2}{\partial \phi^2} \right),$$

or, equivalently,

$$\hat{L}^2 = -\hbar^2 \left(\frac{1}{\sin(\theta)} \frac{\partial}{\partial \theta} \left(\sin(\theta) \frac{\partial}{\partial \theta}\right) + \frac{1}{\sin^2(\theta)} \frac{\partial^2}{\partial \phi^2} \right).$$

(d) $[\hat{x}, \hat{p}_x] = i\hbar$ can be used to find

$$\left[\hat{L}_y, \hat{L}_z\right] = [\hat{z}\hat{p}_x - \hat{x}\hat{p}_z, \hat{x}\hat{p}_y - \hat{y}\hat{p}_x] = [\hat{z}\hat{p}_x, \hat{x}\hat{p}_y] - [\hat{z}\hat{p}_x, \hat{y}\hat{p}_x] - [\hat{x}\hat{p}_z, \hat{x}\hat{p}_y] + [\hat{x}\hat{p}_z, \hat{y}\hat{p}_x].$$

The terms $[\hat{z}\hat{p}_x, \hat{y}\hat{p}_x] = [\hat{x}\hat{p}_z, \hat{x}\hat{p}_y] = 0$ so that

$$\left[\hat{L}_y, \hat{L}_z\right] = \hat{z}[\hat{p}_x, \hat{x}]\hat{p}_y + \hat{y}[\hat{x}, \hat{p}_x]\hat{p}_z = i\hbar\hat{y}\hat{p}_z - i\hbar\hat{z}\hat{p}_y = i\hbar(\hat{y}\hat{p}_z - \hat{z}\hat{p}_y) = i\hbar\hat{L}_x.$$

(e) Given $\hat{L}_\pm = \hat{L}_x \pm i\hat{L}_y$, and because $\hat{L}^2$ commutes with $\hat{L}_x$ and $\hat{L}_y$, it follows that

$$\left[\hat{L}^2, \hat{L}_\pm\right] = \left[\hat{L}^2, (\hat{L}_x \pm i\hat{L}_y)\right] = \left[\hat{L}^2, \hat{L}_x\right] + (\pm i)\left[\hat{L}^2, \hat{L}_y\right] = 0.$$

To show that $\left[\hat{L}_+, \hat{L}_-\right] = 2\hbar\hat{L}_z$ the commutator may be expanded and use made of the commutation relation $\left[\hat{L}_x, \hat{L}_y\right] = i\hbar\hat{L}_z$ so that

$$\left[\hat{L}_+, \hat{L}_-\right] = \left[\hat{L}_x + i\hat{L}_y, \hat{L}_x - i\hat{L}_y\right] = \left[\hat{L}_x, \hat{L}_x\right] + \left[\hat{L}_y, \hat{L}_y\right] + \left[\hat{L}_x, -i\hat{L}_y\right] + \left[i\hat{L}_y, \hat{L}_x\right].$$

Since $\left[\hat{L}_x, \hat{L}_x\right] = \left[\hat{L}_y, \hat{L}_y\right] = 0$ and the terms $\left[\hat{L}_x, -i\hat{L}_y\right] = \left[i\hat{L}_y, \hat{L}_x\right] = \hbar\hat{L}_z$ then

$$\left[\hat{L}_+, \hat{L}_-\right] = 2\hbar\hat{L}_z.$$

It also follows that

$$\left[\hat{L}_z, \hat{L}_\pm\right] = \left[\hat{L}_z, \hat{L}_x\right] \pm i\left[\hat{L}_z, \hat{L}_y\right] = i\hbar\hat{L}_y \pm i\left(-i\hbar\hat{L}_x\right) = \pm\hbar\left(\hat{L}_x \pm i\hat{L}_y\right) = \pm\hbar\hat{L}_\pm.$$

(f) Starting from

$$\hat{L}_\pm\hat{L}_\mp = \left(\hat{L}_x \pm i\hat{L}_y\right)\left(\hat{L}_x \mp i\hat{L}_y\right) = \hat{L}_x^2 + \hat{L}_y^2 \pm i\left(\hat{L}_x\hat{L}_y - \hat{L}_y\hat{L}_x\right) = \hat{L}^2 - \hat{L}_z^2 \pm i\left(i\left(\hbar\hat{L}_z\right)\right),$$

which can be rearranged to give

$$\hat{L}^2 = \hat{L}_\pm\hat{L}_\mp + \hat{L}_z^2 \mp \hbar\hat{L}_z.$$

It follows that

$$2\hat{L}^2 = \hat{L}_+\hat{L}_- + \hat{L}_z^2 - \hbar\hat{L}_z + \hat{L}_-\hat{L}_+ + \hat{L}_z^2 + \hbar\hat{L}_z = \hat{L}_+\hat{L}_- + \hat{L}_-\hat{L}_+ + 2\hat{L}_z^2,$$

so that

$$\hat{L}^2 = \frac{1}{2}\left(\hat{L}_+\hat{L}_- + \hat{L}_-\hat{L}_+\right) + \hat{L}_z^2.$$

Solutions 10.2
(a) The Schrödinger equation is

$$\hat{H}\psi_{lm} = \frac{\hat{L}^2}{2I_L}\psi_{lm} = E_l\psi_{lm} = \frac{\hbar^2 l(l+1)}{2I_L}\psi_{lm}, \text{ where } \psi_{lm} = Y_l^m(\theta, \phi) = \Phi_m(\phi)\Theta_m^l(\theta)$$

and $\dfrac{\partial}{\partial\phi}\Phi_m(\phi) = im\Phi_m(\phi)$, so that $\Phi_m = Ae^{im\phi}$.

471

Since single-valueness of the function $\Phi_m(\phi)$ is required, the allowed values of m are $m = 0, \mp 2, \pm 4, \ldots$ because the hydrogen atoms are *identical* and *indistinguishable* particles.

(b) For the ground-state angular momentum, the quantum number is $l = 0$. In this problem, the characteristic absolute temperature $T = E_l/k_B$ when it becomes probable that the angular quantum number $l = 2$ is excited is sought (from part (a), note that $m = \mp 2$ is the *first* excited state). The moment of inertia of the H_2 molecule is $I_L = 2ma^2$, where the mass, $m_p = 1.67 \times 10^{-27}$ kg, can be used for the mass of a hydrogen atom. Hence,

$$T = \frac{\hbar^2 l(l+1)}{2k_B I_L} = \frac{1.1 \times 10^{-68} \times 2(2+1)}{2 \times 1.38 \times 10^{-23} \times 2 \times 1.67 \times 10^{-27} \times (0.037 \times 10^{-10})^2} = 523\,\text{K},$$

and the value of c_V is increased by k_B (the two degrees of freedom are the two angles of rotation) to $5k_B/2$.

Solutions 10.3
Pairs of principal quantum numbers (n_1, n_2) and (n_3, n_4) such that $\frac{1}{n_1^2} - \frac{1}{n_2^2} = \frac{1}{n_3^2} - \frac{1}{n_4^2}$ are sought. Within the first 100 principal quantum numbers, the numbers are:

n_1	n_2	n_3	n_4
5	6	9	90
5	7	7	35
5	9	6	90
6	8	9	72
6	9	8	72
7	8	14	56
7	14	8	56
10	11	22	55
10	14	14	70
10	22	11	55
14	18	21	63
14	21	18	63
30	34	51	85
30	51	34	85
36	45	48	80
36	48	45	80

The way to find these numbers is to write a computer program.

Solutions 10.4
The wave functions for the hydrogen atom are of the form

$$\psi_{nlm} = R_{nl}(r)\Theta_l^m(\theta)\Phi_m(\phi).$$

The 1s state for a hydrogenic state of charge eZ has wave function

$$\psi_{100} = \left(\frac{1}{4\pi}\right)^{1/2}\left(\frac{Z}{a_B}\right)^{3/2} 2\exp\left(-\frac{Zr}{a_B}\right).$$

The 2s state for a hydrogenic state of charge eZ has a wave function given by

$$\psi_{200} = \left(\frac{1}{4\pi}\right)^{1/2} \left(\frac{Z}{a_B}\right)^{3/2} 2\left(1 - \frac{Zr}{2a_B}\right) \exp\left(-\frac{Zr}{2a_B}\right),$$

and the 2p wave function is

$$\psi_{210} = \frac{1}{2}\left(\frac{3}{\pi}\right)^{1/2} \cos(\theta) \left(\frac{Z}{2a_B}\right)^{3/2} \frac{1}{\sqrt{3}}\frac{Zr}{a_B} \exp\left(-\frac{Zr}{2a_B}\right).$$

The Bohr radius is $a_B = 0.0529$ nm.

If the proton is not a point charge but a finite size of radius b, then the wave function has nonzero probability of being within b. There is a change in potential seen by the electron:

$$\Phi = \frac{e^2}{4\pi\varepsilon_0 r}$$

for $r > b$ and

$$\Phi = \frac{e^2}{4\pi\varepsilon_0 b}$$

for $r \leq b$.

Assuming $\hat{H} = \hat{H}_0 + \hat{W}$, then, to first-order, the change in energy is given by

$$E^{(1)} = \langle\psi|\hat{W}|\psi\rangle,$$

where

$$\hat{H}_0 = -\frac{\hbar^2}{2m}\nabla^2 + \frac{e^2}{4\pi\varepsilon_0 r}$$

for all r, and

$$\hat{W} = -\frac{e^2}{4\pi\varepsilon_0 r} + \frac{e^2}{4\pi\varepsilon_0 b}$$

with

$$E^{(1)} = \frac{e^2}{4\pi\varepsilon_0}\langle\psi|\frac{1}{b} - \frac{1}{r}|\psi\rangle$$

for $r < b$ and

$$\hat{W} = 0$$

for $r > b$.

Hence, for the 2s state,

$$E^{(1)} = \frac{e^2}{4\pi\varepsilon_0}\int_0^b \left(\frac{Z}{2a_B}\right)^3 4\left(1 - \frac{Zr}{2a_B}\right)^2 \left(\frac{1}{b} - \frac{1}{r}\right)\exp\left(-\frac{Zr}{a_B}\right)\frac{4\pi}{4\pi}r^2 dr.$$

Since $b \ll a_B$, then $\exp(-Zr/2a_B) \cong 1$ and

$$E_{2s}^{(1)} = \frac{e^2}{\pi\varepsilon_0} \left(\frac{Z}{2a_B}\right)^3 \int_0^b r^2 \left(\frac{1}{b} - \frac{1}{r}\right) dr = \frac{e^2}{\pi\varepsilon_0} \left(\frac{Z}{2a_B}\right)^3 \int_0^b \left(\frac{r^2}{b} - r\right) dr$$

$$= \frac{e^2}{\pi\varepsilon_0} \left(\frac{Z}{2a_B}\right)^3 \left[\frac{1}{3}\frac{r^3}{b} - \frac{1}{2}r^2\right]_0^b = -\frac{e^2}{8\pi\varepsilon_0} \left(\frac{Z}{a_B}\right)^3 \frac{1}{6} b^2 = -\frac{e^2}{48\pi\varepsilon_0} \left(\frac{Z}{a_B}\right)^3 b^2.$$

For hydrogen, $Z = 1$ and, putting in the numbers,

$$E_{2s}^{(1)} = -\frac{e^2}{48\pi\varepsilon_0} \left(\frac{1}{a_B}\right)^3 b^2 = \frac{-(1.602 \times 10^{-19})^2}{48\pi \times 8.854 \times 10^{-12}} \left(\frac{1}{0.5292 \times 10^{-10}}\right)^3 \left(5 \times 10^{-16}\right)^2,$$

giving

$$E_{2s}^{(1)} = 2.0 \times 10^{-10} \, \text{eV}.$$

For the 2p state the result is

$$E_{2p}^{(1)} = 1.5 \times 10^{-21} \, \text{eV}.$$

$E_{2p}^{(1)}$ is expected to be smaller than $E_{2s}^{(1)}$ as the electron spends less time within the region $r < b$.

The eigenenergies of the hydrogenic atom are

$$E_n = -\frac{m_r e^4}{2(4\pi\varepsilon_0)^2\hbar^2} \left(\frac{Z}{n}\right)^2,$$

where the reduced mass is

$$m_r = \frac{m_0 m_p}{m_0 + m_p}.$$

The change in energy eigenvalue introduced by assuming an infinite proton mass is given by the normalized energy,

$$\frac{E^{(1)}}{E_n} = 1 - \frac{m_0}{m_r} = 1 - \frac{m_0(m_0 + m_p)}{m_0 m_p} = 1 - \frac{m_0}{m_p} - 1 = \frac{-m_0}{m_p}.$$

For the 2s state with eigenenergy $E_{2s} \cong \frac{-\text{Ry}}{2^2} = \frac{-13.6}{4} = -3.4 \, \text{eV}$, the energy shift due to the finite mass of the proton is

$$E^{(1)} = \frac{m_0}{m_p} E_{2s}^{(1)} \sim \frac{9 \times 10^{-31}}{1.6 \times 10^{-27}} E_{2s} \sim 5 \times 10^{-4} \times 3.4 \, \text{eV} = 1.5 \times 10^{-3} \, \text{eV}.$$

Therefore, the correction for finite mass of the proton is a much more important energy correction ($\sim 10^{-3} \, \text{eV}$) than the first-order energy shift due to the finite size of the nucleus ($\sim 10^{-10} \, \text{eV}$).

Solutions 10.5

(a) The task is to predict what the effect of applying a uniform electric field on the energy spectrum of an atom is. It is reasonable to expect that the atom is in its lowest-energy ground state. The perturbation due to an electric field applied in the z direction is $\hat{W} = e|\mathbf{E}|\hat{z}$, and the matrix element between the ground state and the first excited state is $W_{01} = e|\mathbf{E}|\langle 0|\hat{z}|1\rangle = e|\mathbf{E}|z_{01}$. The matrix element is the expectation value of the position operator, and so it is at most of the order of the size of an atom $\sim 10^{-8}$ cm. Maximum electric fields in a laboratory are $\sim 10^6$ V cm^{-1} on a macroscopic scale. Transitions can only take place between energy levels that differ at most by the potential difference induced by the field – that is, $\sim 10^{-2}$ eV. This will not be enough to induce transitions between principal quantum numbers n, but may break the degeneracy between states with different l belonging to the same n. The degeneracy of an electronic state ψ_{nlm} in the hydrogen atom is determined by the principal quantum number n to be n^2 since

$$\sum_{l=0}^{l=n-1} (2l+1) = n^2.$$

In the assessment of the influence that an electric field has on the energy levels of an atom, it was assumed that the electric field at the atom is the same as the macroscopic field. However, on a microscopic scale, electric fields can be dramatically enhanced locally depending on the surrounding medium and its geometry.

(b) If the effects of electron spin are ignored, the time-independent Schrödinger equation for an electron in the hydrogen atom is

$$\hat{H}^{(0)}\psi = E^{(0)}\psi,$$

which has the solutions

$$\psi_{nlm}(r, \theta, \phi) = R_{nl}(r)\Theta_l^m(\theta)\Phi_m(\phi) = |nlm\rangle.$$

The principal quantum number n specifies the energy of a state; the orbital quantum number is $l = 0, 1, 2, \ldots, (n-1)$; and the quantum number $m = \pm l, \ldots, \pm 2, \pm 1, 0$.

The electron states specified by n, l, and m are n^2 degenerate, and each state has definite parity. For $n = 2$, there are four states with the same energy. They are $|200\rangle$, $|21-1\rangle$, $|210\rangle$, and $|211\rangle$.

If a uniform electric field $\mathbf{E}$ is applied in the z direction then solutions to the time-independent Schrödinger equation,

$$\left(\hat{H}^{(0)} + \hat{W}\right)\psi = E\psi,$$

are sought in which

$$\hat{W} = e|\mathbf{E}|\hat{z}.$$

Because the degeneracy of the $n = 2$ state is 4, the solutions to the 4×4 matrix,

$$\sum_{j=1, k=1}^{N=4} [H_{jk} - E\delta_{jk}]a_k = 0,$$

are given by the secular equation

$$\begin{vmatrix} \langle 200|\hat{W}|200\rangle - E & \langle 200|\hat{W}|21-1\rangle & \langle 200|\hat{W}|210\rangle & \langle 200|\hat{W}|211\rangle \\ \langle 21-1|\hat{W}|200\rangle & \langle 21-1|\hat{W}|21-1\rangle - E & \langle 21-1|\hat{W}|210\rangle & \langle 21-1|\hat{W}|211\rangle \\ \langle 210|\hat{W}|200\rangle & \langle 210|\hat{W}|21-1\rangle & \langle 210|\hat{W}|210\rangle - E & \langle 210|\hat{W}|211\rangle \\ \langle 211|\hat{W}|200\rangle & \langle 211|\hat{W}|21-1\rangle & \langle 211|\hat{W}|210\rangle & \langle 211|\hat{W}|211\rangle - E \end{vmatrix} = 0.$$

The diagonal matrix elements are zero because the odd parity of z forces the integrand to odd parity:

$$e|\mathbf{E}|\langle nlm|\hat{z}|nlm\rangle = e|\mathbf{E}|\int \psi_{nlm}^* \hat{z}\psi_{nlm}\mathrm{d}^3r = 0.$$

The perturbation is in the z direction, which in spherical coordinates only involves r and θ via the relation $z = r\cos(\theta)$. Hence, because the eigenfunctions are separable into orthonormal functions of r, θ, and ϕ in such a way that $\psi_{nlm}(r,\theta,\phi) = R_{nl}(r)\Theta_l^m(\theta)\Phi_m(\phi)$, it follows that the matrix elements between states with different m for a z-directed perturbation are zero. This is because states described by $\Phi_m(\phi)$ are orthogonal, and the perturbation in z has no ϕ dependence. Hence, the only possible nonzero off-diagonal matrix elements are those that involve states with the *same* value of m:

$$\langle 200|\hat{W}|210\rangle = \langle 210|\hat{W}|200\rangle = e|\mathbf{E}|\int \psi_{210}^* \hat{z}\psi_{200}\mathrm{d}^3r.$$

The wave functions are

$$\psi_{200} = \frac{2}{(2a_{\mathrm{B}})^{3/2}}\left(1 - \frac{r}{2a_{\mathrm{B}}}\right)\mathrm{e}^{-r/2a_{\mathrm{B}}}\left(\frac{1}{4\pi}\right)^{1/2},$$

which has even parity, and

$$\psi_{210} = \frac{1}{\sqrt{3}(2a_{\mathrm{B}})^{3/2}}\frac{r}{a_{\mathrm{B}}}\mathrm{e}^{-r/2a_{\mathrm{B}}}\frac{1}{2}\left(\frac{3}{\pi}\right)^{1/2}\cos(\theta),$$

which has angular dependence and odd parity. Using $z = r\cos(\theta)$, the matrix element is

$$e|\mathbf{E}|\int \psi_{210}^* \hat{z}\psi_{200}\mathrm{d}^3r = \frac{e|\mathbf{E}|}{32\pi}\int_0^\infty \frac{r^4}{a_{\mathrm{B}}^4}\left(2 - \frac{r}{a_{\mathrm{B}}}\right)\mathrm{e}^{-r/a_{\mathrm{B}}}\mathrm{d}r\int_0^\pi \cos^2(\theta)\sin(\theta)\mathrm{d}\theta\int_0^{2\pi}\mathrm{d}\phi$$

$$= -3e|\mathbf{E}|a_{\mathrm{B}}.$$

To solve the integral, use is made of $\int_0^\infty r^n \mathrm{e}^{-r/a_{\mathrm{B}}}\mathrm{d}r = n!a_{\mathrm{B}}^{n+1}$. The secular equation may now be written as

$$\begin{vmatrix} -E & 0 & -3e|\mathbf{E}|a_{\mathrm{B}} & 0 \\ 0 & -E & 0 & 0 \\ -3e|\mathbf{E}|a_{\mathrm{B}} & 0 & -E & 0 \\ 0 & 0 & 0 & -E \end{vmatrix} = 0.$$

This has four solutions: $\pm 3a_B e|\mathbf{E}|$, 0, 0, so that the new first-order corrected energy levels are $E = E^{(0)} \pm 3a_B e|\mathbf{E}|$, $E^{(0)}$, $E^{(0)}$. This partial lifting of degeneracy in hydrogen due to the application of an electric field is called the Stark effect.

The effect of applying an electric field is to break the symmetry of the potential and partially lift the degeneracy of the state. For the states for which degeneracy is lifted, it is as if the atom has an electric dipole moment of magnitude $3e|\mathbf{E}|a_B$.

To find the wave functions for the perturbed states, it is only necessary to consider the 2×2 matrix that relates $|200\rangle$ and $|210\rangle$:

$$\begin{bmatrix} 0 & -3e|\mathbf{E}|a_B \\ -3e|\mathbf{E}|a_B & 0 \end{bmatrix}.$$

This has eigenfunctions $\psi_+ = \frac{1}{\sqrt{2}}(\psi_{210} - \psi_{200})$ with eigenvalue $E_+ = E^{(0)} + 3e|\mathbf{E}|a_B$ and $\psi_- = \frac{1}{\sqrt{2}}(\psi_{210} + \psi_{200})$ with eigenvalue $E_- = E^{(0)} - 3e|\mathbf{E}|a_B$. Because the eigenfunctions $|200\rangle$ and $|210\rangle$ are of different parity, the eigenfunctions ψ_+ and ψ_- are of mixed parity.

10.6 Problems

Problem 10.1
Derive the following commutation relations:

(a) $\left[\hat{L}_z, \hat{x}\right] = i\hbar\hat{y}$

(b) $\left[\hat{L}_z, \hat{y}\right] = -i\hbar\hat{x}$

(c) $\left[\hat{L}_z, \hat{z}\right] = 0$

(d) $\left[\hat{L}^2, \hat{x}\right] = -2\hbar^2\hat{x} + 2i\hbar\left(\hat{L}_z\hat{y} - \hat{L}_y\hat{z}\right)$

(e) $\left[\hat{L}^2, [\hat{L}^2, \hat{\mathbf{r}}]\right] = 2\hbar^2\left(\hat{\mathbf{r}}\hat{L}^2 + \hat{L}^2\hat{\mathbf{r}}\right).$

Problem 10.2
The ground-state wave function of a hydrogenic atom with nuclear charge Ze is $\psi_1(r) = Ae^{-r/r_1}$, where r is the distance between the electron and the nucleus. The electron is subject to a radially symmetric coulomb potential given by

$$V(r) = \frac{-e^2}{4\pi\varepsilon_0\varepsilon_r 0 r}.$$

(a) Find the normalization constant A.
(b) Find the minimized energy expectation value $\langle E_1 \rangle$, and show that

$$\langle E_{\text{kinetic}} \rangle = -\frac{\langle E_{\text{potential}} \rangle}{2},$$

which is a result predicted by the virial theorem.

(c) Show that $r_1 = a_B/Z$ where

$$a_B = \frac{4\pi\varepsilon_0\hbar^2}{m_r e^2},$$

and show that r_1/Z corresponds to the peak in radial probability.
(d) Show that the expectation value $\langle r \rangle = 3a_B/2Z$.
(e) Show that the expectation value of momentum $\langle p \rangle = 0$.

Problem 10.3

To calculate the spontaneous emission rate $A = 1/\tau_{sp}$ for the 2p to 1s ($|n = 2, l = 1, m\rangle \to |n = 1, l = 0, m = 0\rangle$) transition in hydrogen, an average over the three possible values of the quantum number m is taken so that

$$A = \frac{e^2\omega^3}{3\hbar c^3\pi\varepsilon_0} \frac{1}{3} \sum_{m=-1}^{m=1} |\langle 2,1,m|\hat{\mathbf{r}}|1,0,0\rangle|^2,$$

where $\hbar\omega$ is the energy of the emitted photon. Since $r^2 = x^2 + y^2 + z^2$, this equation can be written as

$$A = \frac{e^2\omega^3}{3\hbar c^3\pi\varepsilon_0} \frac{1}{3} \sum_{m=-1}^{m=1} \left(|\langle 2,1,m|\hat{x}|1,0,0\rangle|^2 + |\langle 2,1,m|\hat{y}|1,0,0\rangle|^2 + |\langle 2,1,m|\hat{z}|1,0,0\rangle|^2 \right).$$

(a) Show that

$$x = r\sin(\theta)\cos(\phi) = -\sqrt{\frac{2\pi}{3}} r \left(Y_1^1 - Y_1^{-1} \right),$$

$$y = r\sin(\theta)\sin(\phi) = i\sqrt{\frac{2\pi}{3}} r \left(Y_1^1 + Y_1^{-1} \right),$$

$$z = r\cos(\theta) = \sqrt{\frac{4\pi}{3}} r Y_1^0,$$

and rewrite each matrix element appearing in the expression for spontaneous emission in terms of a radial integral and an angular integral.
(b) Use the standard integral $\int_0^\infty x^n e^{-\mu x} dx = n! \mu^{-n-1}$ to show that the radial integral

$$\int\limits_0^\infty r^3 R_{21}^*(r) R_{10}(r) dr = \frac{2^8}{3^4\sqrt{6}} a_B,$$

where $R_{10}(r) = 2\left(\frac{1}{a_B}\right)^{3/2} e^{-r/a_B}$ and $R_{21}(r) = \frac{2}{\sqrt{3}}\left(\frac{1}{2a_B}\right)^{3/2} \frac{r}{2a_B} e^{-r/2a_B}$,
(c) Show that the angular integrals in (a) are

$$-\sqrt{\frac{2\pi}{3}} \int\limits_{\theta=0}^{\theta=\pi} \int\limits_{\phi=0}^{\phi=2\pi} (Y_1^m)^* \left(Y_1^1 - Y_1^{-1} \right) Y_0^0 \sin(\theta) d\theta d\phi = \frac{-1}{\sqrt{4\pi}}\sqrt{\frac{2\pi}{3}} (\delta_{m,1} - \delta_{m,-1}),$$

$$i\sqrt{\frac{2\pi}{3}} \int\limits_{\theta=0}^{\theta=\pi} \int\limits_{\phi=0}^{\phi=2\pi} (Y_1^m)^* \left(Y_1^1 + Y_1^{-1}\right) Y_0^0 \sin(\theta)d\theta d\phi = \frac{i}{\sqrt{4\pi}}\sqrt{\frac{2\pi}{3}}(\delta_{m,1} + \delta_{m,-1}),$$

$$\sqrt{\frac{4\pi}{3}} \int\limits_{\theta=0}^{\theta=\pi} \int\limits_{\phi=0}^{\phi=2\pi} (Y_1^m)^* Y_1^0 Y_0^0 \sin(\theta)d\theta d\phi = \frac{1}{\sqrt{4\pi}}\sqrt{\frac{4\pi}{3}}(\delta_{m,0}).$$

(d) Combine the results of (b) and (c) to show that

$$\sum_{m=-1}^{m=1} |\langle 21m|\hat{\mathbf{r}}|100\rangle|^2 = 96 \left(\frac{2}{3}\right)^{10} a_B^2 = \left(\frac{32}{27}\right)^3 a_B^2$$

and that the spontaneous emission time for the 2p to 1s transition in hydrogen is $\tau_{sp} = 1.6$ ns.

Problem 10.4
(a) Show, using first-order perturbation theory, that the correction to the 1s ground-state energy of a hydrogen atom subject to a uniform electric field $\mathbf{E}$ in the z direction is zero.

(b) Show that the first-order correction to the ground-state wave function is

$$\left|\psi_0^{(1)}\right\rangle = |1,0,0\rangle - e|\mathbf{E}| \sum_{n\neq 1,l,m} \frac{|n,l,m\rangle\langle n,l,m|\hat{z}|1,0,0\rangle}{E_1 - E_n}.$$

(c) Show to first order in $\mathbf{E}$ that the susceptibility for the 1s state is

$$\chi_{1s} = \frac{\left\langle \psi_0^{(1)} |e\hat{z}| \psi_0^{(1)}\right\rangle}{|\mathbf{E}|} = -2e^2 \sum_{n\neq 1,l,m} \frac{|\langle n,l,m|\hat{z}|1,0,0\rangle|^2}{E_1 - E_n}.$$

Problem 10.5
(a) An electron with zero angular momentum $(l = 0)$ moves in a radial potential $V(r) = 0$ for $r < a$ and $V(r) = \infty$ for $r \geq a$, where a is the radius of a spherical quantum dot. Use the radial Schrödinger equation to find the eigenenergies and normalized eigenstates of the electron.

(b) Find the eigenenergies and normalized eigenstates of an electron with zero orbital angular momentum $(l = 0)$ moving in a radial shell potential with $V(r) = 0$ for $a < r < b$ and $V(r) = \infty$ elsewhere.

Problem 10.6
Consider a hydrogen atom with Bohr radius a_B.

(a) Show that the expectation value $\langle r_{nl}\rangle$ in state ψ_{nlm} is

$$\langle r_{nl}\rangle = \frac{a_B}{2}\left(3n^2 - l(l+1)\right).$$

(b) Show that

$$\langle r_{nl}^2 \rangle = \frac{n^2 a_B^2}{2} \left(5n^2 + 1 - 3l(l+1) \right).$$

(c) Find an analytic expression for the spread in radial probability $\Delta r_{nl} = \sqrt{\langle r_{nl}^2 \rangle - \langle r_{nl} \rangle^2}$.

Problem 10.7

Consider the state ψ_{nlm} of the hydrogen atom with quantum numbers $n = 2$ and $l = 1$.

(a) Calculate the numerical value of radial expectation value, $\langle r_{21} \rangle$.

(b) Calculate the numerical value at which the radial probability density reaches a maximum, r_{21}^{max}.

(c) Calculate the numerical value of spread in radial probability Δr_{21}, and explain why the values $\langle r_{21} \rangle$ and r_{21}^{max} are different.

Problem 10.8

A hydrogen atom, which in free space has eigenstates $\psi_{nlm}(r, \theta, \phi) = R_{nl}(r) Y_l^m(\theta, \phi)$ where $Y_l^m(\theta, \phi)$ are the normalized spherical harmonics and n, l, and m are quantum numbers, is placed at distance $z_0 \geq 0$ in a half-space with potential $V(z \geq 0) = 0$ and $V(z < 0) = \infty$.

(a) Find the analytic expression for the ground-state wave function as $z_0 \to 0$.

(b) Find all the other eigenstates and their degeneracy as $z_0 \to 0$.

11 Toward Quantum Engineering

11.1 Introduction

It is possible to *engineer* properties of materials, devices, and systems by changing experimentally available control parameters to optimally approach a specific objective. The following sections demonstrate some potential applications of quantum engineering and show how this may be achieved by the development of efficient physical models combined with optimization algorithms. The quality of the quantum physical model plays an essential role in determining the range of useful materials, devices, and systems that can be designed. Other, more speculative, potential engineering applications are quantum information processing and quantum sensing, in which the challenge is to exploit quantum phenomena to create systems that are more efficient than their classical counterpart. As with optimal quantum device design, the range of quantum resources available includes identical indistinguishable particles, linear superposition of particle states, entanglement of particles, interference of particles, and measurement whose outcome is noncausal. The engineering challenges associated with these future applications are both significant and interesting.

11.2 Optimal Design of a Heterostructure Tunnel Diode

Atomic layer control of single-crystal heterostructures can be used to create a tunnel diode that, for some range of applied bias voltage, exhibits a desired current–voltage characteristic. The elements required to achieve such a design are a realistic forward model and an optimization algorithm.

11.2.1 Tunnel Diode Model

A single-crystal semiconductor heterostructure tunnel diode can be created in the $GaAs/Al_\xi Ga_{1-\xi}As$ material system. An undoped intrinsic $Al_\xi Ga_{1-\xi}As$ potential energy barrier (the "tunnel barrier") in the semiconductor conduction band separates two bulk GaAs electrodes that each have a doped n-type carrier concentration of n_0. Current can flow when a voltage bias, V_{bias}, is applied across the diode. For a given voltage bias the conduction-band minimum potential energy profile may be calculated in the *depletion approximation*. This is found by relating the electric field, $\mathbf{E}$, to the charge density, ρ, via Gauss' law,

$$\nabla \cdot \mathbf{E} = \frac{\rho}{\varepsilon_0 \varepsilon_{r0}}, \tag{11.1}$$

where ε_{r0} is the low-frequency relative permittivity of the semiconductor structure and ε_0 is the permittivity of free space. See Exercise 11.1. It is assumed that there is no scattering

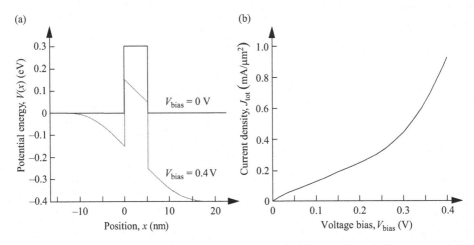

Fig. 11.1 (a) Conduction-band minimum potential energy profile of a GaAs/AlGaAs tunnel diode in the depletion approximation with $V_{bias} = 0\,V$ and $V_{bias} = 0.4\,V$. Carrier concentration $n_0 = 1.5 \times 10^{18}\,cm^{-3}$ in the GaAs contacts, intrinsic $Al_{\xi}Ga_{1-\xi}$ potential energy barrier $V_0 = 0.3\,eV$, barrier thickness $L_b = 5\,nm$ (9 lattice spaces in the [100] direction), $\xi = 0.36$, $m_e^* = 0.07 \times m_0$, and relative permittivity $\varepsilon_{r0} = 13.2$. (b) Calculated current density, J_{tot}, for (a) as a function of V_{bias} when temperature $T = 300\,K$.

into, or contribution to current from, the electrode-barrier charge carrier accumulation region.

The conduction-band minimum potential energy profile of a $GaAs/Al_{\xi}Ga_{1-\xi}As$ tunnel diode in the depletion approximation with $V_{bias} = 0\,V$ and $V_{bias} = 0.4\,V$ is shown in Fig. 11.1(a). The band bending to the left of the $Al_{\xi}Ga_{1-\xi}As$ tunnel barrier creates a two-dimensional potential well into which electrons could scatter. This electrode-barrier region, in which extra electron charge density can accumulate, is ignored in the depletion approximation. A more complete model that includes the additional charge in the electrode-barrier accumulation region should describe electron scattering processes into and tunnelling out of the potential well. Such a model should also simultaneously satisfy the Schrödinger and Poisson equations.

A Fermi–Dirac distribution determines the occupation of electron states that are assumed to behave as if in thermal equilibrium at absolute temperature, T, deep in the electrodes. There is no inelastic electron scattering in the active device region where the band bending and the tunnel barrier exist.

If we adopt a Landauer description of electron transport,[1] each plane wave solution to the Schrödinger equation is a conduction channel that can contribute to current flow. An electron of effective mass m_e^* incident on the tunnel barrier from the contact electrode on the left has a wave vector component in the x direction that is perpendicular to the plane of the $GaAs/Al_{\xi}Ga_{1-\xi}As$ interface, $k_{\perp}$. The energy associated with this is $E_{\perp} = \hbar^2 k_{\perp}^2/2m_e^*$. For a given voltage bias, V_{bias}, the electron transmission, T_e, may be calculated as a function of $E_{\perp}$ using the propagation matrix method. A conventional approach may be used to calculate the current density, $J_{tot}(V_{bias})$, at

---

[1] R. Landauer, *IBM J. Res. Dev.* **1**, 223 (1957).

temperature T for electron concentration n_0 in the contacts.[2] The total current density (see Problem 6.5) is

$$J_{\text{tot}} = \frac{em_e^* k_B T}{2\pi^2 \hbar^3} \int_0^\infty T_e(E_\perp) \ln \left(\frac{1 + e^{(\mu - E_\perp)/k_B T}}{1 + e^{(\mu - E_\perp - eV_{\text{bias}})/k_B T}} \right) dE_\perp. \tag{11.2}$$

As shown in Fig. 11.1(b), the calculated current per square micrometer of junction area as a function of V_{bias} is not exponential. For $V_{\text{bias}} \ll E_F$, current increases approximately linearly. At larger values of V_{bias} there are other features that relate to opening up of Landauer conduction channels and changes in transmission, $T_e(V_{\text{bias}})$.

11.2.2 Optimal Design of a Linear Current–Voltage Characteristic

Atomic layer control of single-crystal $Al_\xi Ga_{1-\xi} As$ alloy composition fraction ξ and substitutional doping concentration, along with a quantitative model of electron transport, are essential elements that, when combined with an optimal design methodology can be used to discover and implement new semiconductor device functions.[3]

The conduction-band offset between AlAs and GaAs is 1.04 eV and the lattice constants differ by 0.14%, allowing single-crystal heterostructures to be grown by molecular beam epitaxy (MBE). In the device to be considered, conduction-band electrons are injected from bulk n-type GaAs contacts. The experimentally accessible design parameters for a device with an intrinsic active region consisting of a number of equal-thickness planar heterostructure layers is the Al alloy fraction, ξ. This determines the conduction-band offset between each heterostructure layer relative to the GaAs contact, and for the jth layer it has a value $U_j = 0.8355 \times \xi$ for $0 \leq \xi \leq 0.42$, corresponding to $0\,\text{eV} \leq U_j \leq 0.35\,\text{eV}$.

The aim is to find the values of the design parameters that give a desired current–voltage characteristic. Formally, the objective function is $s_{\text{obj}}(V_{\text{bias}})$ and the result of the forward physical model simulation is $s_{\text{sim}}(V_{\text{bias}}, \boldsymbol{p})$ for a given set of design parameter values placed in the vector $\boldsymbol{p}$. The number of sample points used to evaluate s_{obj} and s_{sim} is N. The jth layer of an $Al_\xi Ga_{1-\xi}$ heterostructure device has thickness L_j and conduction-band offset U_j. Numerical values of the design parameters L_j and U_j for each layer of the heterostructure are elements of the vector $\boldsymbol{p} = (p_1, p_2, \ldots, p_{\mathcal{P}})$, where $\mathcal{P}$ is the number of elements. A measure of optimality may be established by defining the cost function as

$$C_{\text{cost}}(\boldsymbol{p}) = \sum_{n=1}^{N} w_n |s_{\text{obj}}(V_{\text{bias},n}) - s_{\text{sim}}(V_{\text{bias},n}, \boldsymbol{p})|^2, \tag{11.3}$$

where, for generality, a multiplicative weighting factor, w_n, is included.

The cost function defined by Eq. (11.3) uses a $\mathcal{L}^2$ distance measure between the simulated current density, s_{sim}, and the objective function, s_{obj}. The use of this measure guarantees a continuous derivative for a local minimum in C_{cost}. This is a useful feature

---

[2] R. Tsu and L. Esaki, *Appl. Phys. Lett.* **22**, 562 (1973); K. M. S. V. Bandara and D. D. Coon, *J. Appl. Phys.* **66**, 693 (1989).

[3] A. F. J. Levi and S. Haas, eds. *Optimal Device Design*, Cambridge, Cambridge University Press, 2010 (ISBN 978 0521116602).

for gradient-based optimization. However, other distance measures such as $\mathcal{L}^1$, for which a continuous derivative is *not* guaranteed, change the cost–function landscape and can be helpful in some circumstances.

An optimal design determined by the cost function C_{cost} (Eq. (11.3)) is a minimization problem:

$$\min_{\boldsymbol{p}} C_{\text{cost}}. \tag{11.4}$$

The forward physical model in the optimization problem evaluates current density as a function of voltage bias,

$$s_{\text{sim}}(V_{\text{bias},n}) = \frac{e m_e^* k_B T}{2\pi^2 \hbar^3} \sum_{\alpha} T_e(V_{\text{bias},n}, E_\alpha) \ln \left(\frac{1 + e^{(\mu - E_\alpha)/k_B T}}{1 + e^{(\mu - E_\alpha - eV_{\text{bias}})/k_B T}} \right) \Delta E, \tag{11.5}$$

where the discrete sum used for numerical evaluation is shown explicitly. In this expression, n indexes voltage bias V_{bias}, α indexes energy, and ΔE is a fixed separation between energy samples.

If the design parameters are restricted to those for the material system $Al_\xi Ga_{1-\xi}$ then they are typically subject to the constraints $0 \leq U_j \leq 0.35\,\text{eV}$ and $L_j = n_{\text{int}} \times \delta$ for positive integer n_{int} with atomic layer thickness δ. However, for many cases of interest it is helpful to not only allow positive values of U_j, but also negative values. For heterostructure devices grown on a GaAs substrate, an InGaAs alloy can provide negative values of U_j.

Assuming that the gradient $\nabla_{\boldsymbol{p}} C_{\text{cost}}$ is available, then a solution to Eq. (11.4) may be found using Newtons method. The MATLAB optimization toolbox function fmincon can be used to implement the Newton method.

Although $\nabla_{\boldsymbol{p}} C_{\text{cost}}$ may be evaluated using a finite difference approximation, this method is computationally expensive, requiring $\mathcal{P}$ additional forward solves. Since each optimization often requires several hundred forward solves, efficient calculation of the gradient is desired so the local optimum may be found in a reasonable amount of time. The adjoint method provides a particularly efficient way of evaluating $\nabla_{\boldsymbol{p}} C_{\text{cost}}$.[4]

Optimization has been used to synthesize $GaAs/Al_\xi Ga_{1-\xi}$ heterostructure designs for diodes with linear[5] and nonlinear[6] current-voltage characteristics. Figure 11.2 shows an example of a nonintuitive single-crystal heterostructure diode design with $-0.15\,\text{eV} \leq U_j \leq 0.3\,\text{eV}$ whose objective is a linear current-voltage characteristic in the bias range $0 \leq V_{\text{bias}} \leq 0.4\,\text{V}$. Unlike a conventional resistor that also has a linear current–voltage characteristic, energy dissipation (heating) when current flows does not take place in the active region of the device, rather it occurs deep in the electrodes. The physics determining the feasibility of the design is the existence of abrupt potential steps in the heterostructure profile that create spectrally broad electron transmission resonances. The spectral positioning and effective superposition of these resonances helps to linearize the current-voltage characteristic. The same resonances render the solution stable against perturbations and, hence, are *robust* against small manufacturing imperfections.

[4] A. F. J. Levi and G. Rosen, *SIAM J. Control Optim.* **48**, 3191 (2010).
[5] K. C. Magruder and A. F. J. Levi, *Physica E* **44**, 322 (2011).
[6] K. C. Magruder and A. F. J. Levi, *Physica E* **44**, 1503 (2012).

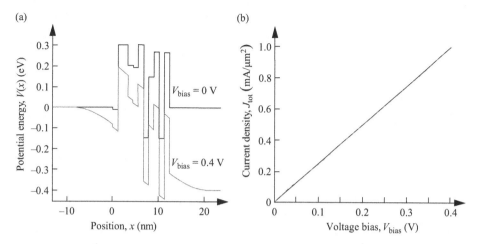

Fig. 11.2 An optimal heterostructure diode design with linear current-voltage characteristic. (a) Non-intuitive potential energy profile $U(x)$ for a linear current-voltage characteristic using the depletion approximation. (b) Simulated current-voltage characteristic (dashed curve) is close to the objective linear behavior (solid line). In the design problem implemented in the AlInGaAs material system: applied bias $0 \leq V_{\text{bias}} \leq 0.4$ V, temperature $T = 300$ K, carrier number concentration in each electrode $n_0 = 1.5 \times 10^{18}$ cm^{-3}, electron effective mass $m_e^* = 0.07 \times m_0$, low-frequency relative permittivity $\varepsilon_{r0} = 13.2$, lattice space $L = 0.565\,32$ nm, and the potential energy of the jth layer -0.15 eV $\leq U_j \leq 0.3$ eV. There are eleven 2-lattice-space-thick layers, each with a potential energy value U_j, that result in an 11-dimensional search space. The total thickness of these layers is 12.4 nm (22 lattice spaces).

11.2.3 The Non-convex Cost Function Landscape

A cost function, such as Eq. (11.3), can be non-convex. It is this fact along with a typically high-dimensional parameter space that can make the minimization problem described by Eq. (11.4) an interesting challenge. It is possible that both multiple local minima exist and that many solutions exist that have a cost value close to optimal.

Some important aspects of the non-convex optimization landscape space can be visualized by considering the specific example of a resonant tunnel diode with just two design variables. The objective function is taken to be the current–voltage characteristic of a device with two Al$_{0.36}$Ga$_{0.64}$As tunnel barriers each of energy 0.3 eV and thickness 2.83 nm with a GaAs potential well of zero energy and thickness 4.52 nm. Bulk GaAs electrodes doped n-type with impurity concentration $n_0 = 1 \times 10^{18}$ cm^{-3} grown either side of the intrinsic active region complete the single-crystal heterostructure device. The voltage bias range under consideration is $0 \leq V_{\text{bias}} \leq 0.4$ V. The two design variables are created by dividing the potential well into two regions of equal thickness. The potential energy on the left side of the well region is V_1 and that on the right is V_2. The allowed values of V_1 and V_2 are in the range -0.15 eV to 0.3 eV.

Figure 11.3(a) plots the objective current–voltage characteristic in which $V_1 = V_2 = 0$ eV. The inset in Fig. 11.3(a) illustrates the conduction-band edge potential energy profile of the device under voltage bias $V_{\text{bias}} = 0.4$ V. Figure 11.3(b) shows an example of the conduction-band edge potential energy profile when the values of V_1 and V_2 are not zero.

Figure 11.4 shows the cost function for the two-dimensional parameter space in which potential barrier design parameters V_1 and V_2 can be independently varied to have an

485

(a)

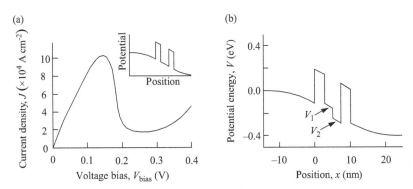

(b)

Fig. 11.3 (a) The objective current-voltage characteristic in which $V_1 = V_2 = 0\,\text{eV}$. The inset illustrates the conduction-band edge potential energy profile of the AlGaAs resonant tunnel diode under voltage bias $V_{\text{bias}} = 0.4\,\text{V}$. Each barrier has 0.3 eV energy and thickness $L_{\text{b}} = 2.83\,\text{nm}$. Well thickness is $L_{\text{w}} = 4.52\,\text{nm}$, effective electron mass is $m_{\text{e}}^* = 0.07 \times m_0$, low-frequency relative permittivity is $\varepsilon_{\text{r}0} = 13.2$, electrodes are doped n-type with impurity concentration $1 \times 10^{18}\,\text{cm}^{-3}$, and temperature is $T = 300\,\text{K}$. (b) An example of the conduction-band edge potential energy profile under voltage bias $V_{\text{bias}} = 0.4\,\text{V}$ when $-0.15\,\text{eV} \leq V_{1,2} \leq 0.3\,\text{eV}$ implemented in the AlInGaAs material system.

energy in the range $-0.15\,\text{eV} \leq V_{1,2} \leq 0.3\,\text{eV}$. In this plot, the optimal design is the point $V_1 = V_2 = 0\,\text{eV}$. It is clear that the cost function landscape has multiple peaks and, while not shown, if the domain is extended beyond $-0.15\,\text{eV} \leq V_{1,2} \leq 0.3\,\text{eV}$, there are also multiple valleys and saddle points corresponding to a two-dimensional design space that has a non-convex cost function. As may be seen in the figure, there is an arc of near optimal designs in which a positive value of V_1 (V_2) is compensated by a negative value of V_2 (V_1). The reason this occurs is that the phase accumulated by an electron propagating in the well region with a positive potential energy can be approximately compensated by the phase accumulated in the other portion of the well with negative potential energy.

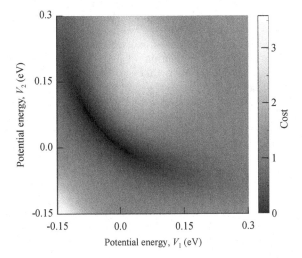

Fig. 11.4 Cost function landscape for the resonant tunnel diode of Fig. 11.3 in the experimentally accessible domain $-0.15\,\text{eV} \leq V_{1,2} \leq 0.3\,\text{eV}$.

11.3 Optimal Design of Density of States

The density of electronic states is a key aspect of any description of a material or device. It has a direct impact on electron transport and electron scattering rates calculated via the golden rule. A peak in the density of electronic states due to a periodic potential formed by atoms in a crystal can enhance electron scattering. And, the density of states of a given material can be influenced by the geometry of a nanoscale device.

A key idea is to seek and control arrangements of atoms that are not constrained by crystal symmetry. Breaking the spatial symmetry of atom positions can create a large number of possibilities, making it feasible to find configurations of atoms that result in a desired density of states. The ability to control the response of a material in a user-defined way is a powerful concept with application to design of materials, devices, and systems.

11.3.1 Tight-Binding Model

A physical model is useful when designing a material to have a particular density of states. Given a forward physical model, an optimization algorithm can seek spatial configurations of atoms characterized by a user-specified or objective density of electronic states, N_{obj}, as a function of energy, E. Key elements of how this is done can be shown using a long-range version of the atomic tight-binding model with Hamiltonian

$$\hat{H}_{\hat{c}^{\dagger}\hat{c}} = -\sum_{i<j} t_{i,j} \left(\hat{c}_i^{\dagger}\hat{c}_j + \hat{c}_i\hat{c}_j^{\dagger} \right), \tag{11.6}$$

where $\hat{c}_i^{\dagger}$ and $\hat{c}_i$ are creation and annihilation operators, respectively, at the atom site $\mathbf{r}_i$. If the directionality of atomic electron wave functions is not included then only s-orbitals need be considered. In this case, the hopping integrals $t_{i,j}$ between an atom at position $\mathbf{r}_i$ and an atom at position $\mathbf{r}_j$ can be parameterized by a power law $t_{i,j} = t_{\text{hop}}/|\mathbf{r}_i - \mathbf{r}_j|^{\alpha}$ in which t_{hop} sets the energy scale. The value of the exponent α determines the spatial range of the interaction. The Hamiltonian matrix in the basis of single-atom states is non-sparse because interaction with all atoms is included. The energy eigenvalues, $E_{\mathbf{k}}$, are solutions of the Schrödinger equation with Hamiltonian given by Eq. (11.6). The density of electron states obtained from the eigenvalues, $E_{\mathbf{k}}$, is

$$N_{\text{sim}}(E) = 2 \sum_{\mathbf{k}} \frac{\gamma/2\pi}{(E - E_{\mathbf{k}})^2 + (\frac{\gamma}{2})^2}, \tag{11.7}$$

where it is assumed that there is a single dominant lifetime, $\hbar/\gamma$, associated with each eigenstate.

11.3.2 Guided Random Walk

The $\mathcal{L}^2$ measure of distance between N sampled values of the forward physical model simulation, $N_{\text{sim}}(E)$, and the objective, $N_{\text{obj}}(E)$, can be used to specify a cost function,

$$C_{\text{cost}} = \sum_{j}^{N} |N_{\text{sim},j} - N_{\text{obj},j}|^2. \tag{11.8}$$

Assuming that the experimentally accessible parameters are atom positions, then variations in atom positions can change C_{cost}. If, after a small random change in atom positions the value of C_{cost} decreases, then the new positions are saved and in this way the system is guided in a series of random steps (a *guided random walk*) to a locally optimal configuration that minimizes the cost function.

As an example, consider nine atoms in a 4×4 two-dimensional domain with periodic boundary conditions and interaction strength exponent $\alpha = 2$. The dashed curve in Fig. 11.5(a) shows a nonsymmetric objective density of states, $N_{\text{obj}}(E)$. The guided random walk optimization algorithm finds a spatial configuration of the atoms that has a density of states, $N_{\text{sim}}(E)$, close to optimal. This is shown as the solid curve in Fig. 11.5(a). The reduction of C_{cost} as a function of the number of guided random walk steps for a typical execution of a guided random walk optimization algorithm is shown in Fig. 11.5(b) using a $\log_{10}$-$\log_{10}$ plot. A straight line in this convergence plot corresponds to power-law behavior. A contour plot of interaction strength for atoms in the optimized positions

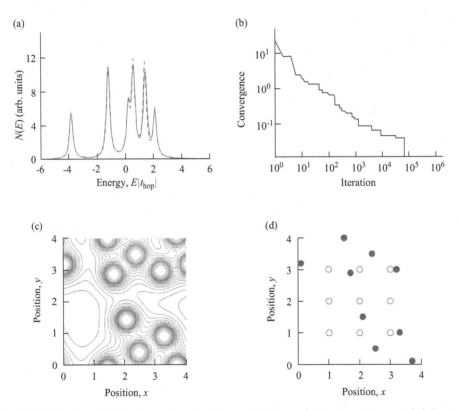

Fig. 11.5 (a) Objective (dash curve) and achieved (solid curve) density of states $N(E)$ for nine atoms in a square domain of dimension 4×4. (b) $\log_{10}$-$\log_{10}$ plot of convergence as a function of number of guided random walk steps. (c) Contour plot of interaction strength for atoms in optimized positions. (d) Positions of atoms giving objective density of states, $N_{\text{obj}}(E)$ (open circles), and positions of atoms with density of states, $N_{\text{sim}}(E)$, close to optimal (solid dots). Energy scale is in units of $|t_{\text{hop}}|$. Periodic boundary conditions are used, $\alpha = 2$, and $\gamma = 0.2828 \times |t_{\text{hop}}|$.

of Fig. 11.5(b) is given in Fig. 11.5(c). The difference in symmetric atom positions of the exact objective density of states $N_{\text{obj}}(E)$ (open circles) and positions of atoms with density of states $N_{\text{sim}}(E)$ close to optimal (solid dots) shown in Fig. 11.5(d) is due to the large number of solutions close to the objective. Apart from translational and rotational degeneracy, there are many near-optimal broken symmetry solutions.

If a material with a specific density of states is required, then optimal design can be used to discover configurations of atoms whose density of states approach the desired values. In general, the objective functionality is obtained by broken symmetry so, in this sense, *broken symmetry is function.*

Often further insight into configurations that result in the objective spectrum $N_{\text{obj}}(E)$ can be obtained. For example, a hierarchy of primitive configurations exists that form the building blocks for an objective density of states.[7] Dimers can be used for symmetric $N(E)$, trimers and larger molecular configurations can provide asymmetry to $N(E)$. These heuristics only apply to the dilute limit in which adjacent atoms uniformly distributed in the domain have weak interaction strength.

Aspects of the tight-binding physical model used have been confirmed experimentally using scanning tunneling microscopy (STM) to precisely position gold atoms on the surface of a nickel–aluminum crystal. These STM measurements[8] show that the splitting in the value of eigenenergies E_i for Au dimers on NiAl depends inversely on Au atom separation corresponding to $\alpha = 1$ in the expression $t_{ij} = t_{\text{hop}}/|\mathbf{r}_i - \mathbf{r}_j|^\alpha$.

11.4 Photon Detection after a Beam Splitter

Controlling the interaction of photons with matter is of both fundamental and practical interest. A basic component in photonics that requires a model of photon–matter interaction is the beam splitter. This is considered next.

As illustrated schematically in Fig. 11.6, a beam splitter has two input ports and two output ports. The reflection and transmission amplitudes experienced by a linearly polarized photon at port 1 are $r_{\text{ph},1}$ and $t_{\text{ph},1}$, respectively. Similarly, at port 2 they are $r_{\text{ph},2}$ and $t_{\text{ph},2}$. If there are integer n_1 photons incident at port 1 and integer n_2 photons incident at port 2, then the input Fock state is $|n_1, n_2\rangle_{\text{in}}$.[9] An input port with zero photons is in the vacuum state. The output of the beam splitter has photons with quantum field amplitude a_3 at port 3 and amplitude a_4 at port 4.

Exercise 11.3 shows how an ideal, lossless, symmetric, 50:50 dielectric beam splitter has $r_{\text{ph},1} = r_{\text{ph},2} = r_{\text{ph}}$ with

$$r_{\text{ph}} = \frac{-1}{\sqrt{2}}, \tag{11.9}$$

and $t_{\text{ph},1} = t_{\text{ph},2} = t_{\text{ph}}$ with

$$t_{\text{ph}} = \frac{i}{\sqrt{2}}. \tag{11.10}$$

---

[7] J. Thalken et al., *Phys. Rev. B* **69**, 195410 (2004).
[8] N. Nilius et al., *Phys. Rev. Lett.* **90**, 196103 (2003).
[9] Ignore, for the moment, that the input state could be a superposition involving quantum field amplitude at both port 1 and 2.

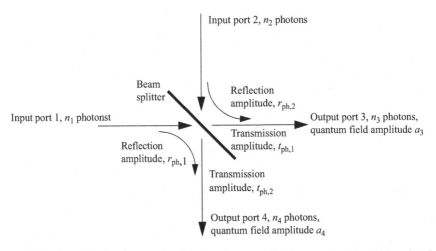

Fig. 11.6 Sketch of a beam splitter showing input ports 1 and 2 and output ports 3 and 4. In the case considered, there are integer n_1 photons incident at port 1 and integer n_2 photons incident at port 2. The single-photon quantum field transmission amplitude is $t_{\mathrm{ph},1,2}$ and the reflection amplitude is $r_{\mathrm{ph},1,2}$.

If the total number of Fock-state photons incident on the 50:50 beam splitter is $n_{\mathrm{tot}} = n_1 + n_2 = 1$, then there is either one photon present at port 1 or one photon present at port 2. This means input state $|n_1 = 1, n_2 = 0\rangle_{\mathrm{in}}$ has only one path to output state $|n_3 = 1, n_4 = 0\rangle_{\mathrm{out}}$ and the quantum field amplitude at port 3 is t_{ph}:

$$|n_1 = 1, n_2 = 0\rangle_{\mathrm{in}} \rightarrow |n_3 = 1, n_4 = 0\rangle_{\mathrm{out}} : t_{\mathrm{ph}} = \frac{i}{\sqrt{2}}. \tag{11.11}$$

Similarly, the input state $|1,0\rangle_{\mathrm{in}}$ has only one path to output state $|0,1\rangle_{\mathrm{out}}$ and the quantum field amplitude at port 4 is r_{ph}:

$$|n_1 = 1, n_2 = 0\rangle_{\mathrm{in}} \rightarrow |n_3 = 0, n_4 = 1\rangle_{\mathrm{out}} : r_{\mathrm{ph}} = \frac{-1}{\sqrt{2}}. \tag{11.12}$$

The photon number detection probability is just the magnitude squared of the quantum field amplitude so that at the output port 3 detector

$$P_{\mathrm{out}}(n_1 = 1, n_2 = 0, n_3 = 1, n_4 = 0) = |t_{\mathrm{ph}}|^2 = \left|\frac{i}{\sqrt{2}}\right|^2 = \frac{1}{2}, \tag{11.13}$$

and at the output port 4 detector

$$P_{\mathrm{out}}(n_1 = 1, n_2 = 0, n_3 = 0, n_4 = 1) = |r_{\mathrm{ph}}|^2 = \left|\frac{-1}{\sqrt{2}}\right|^2 = \frac{1}{2}. \tag{11.14}$$

Placing a single photon at input port 2, so that the input state is $|n_1 = 0, n_2 = 1\rangle_{\mathrm{in}}$, produces similar results.

As indicated in Table 11.1, the probabilities are the same as the flux ratios predicted for a classical electromagnetic wave interacting with the same beam splitter. However,

Table 11.1 Single photon detection probability after a 50:50 beam splitter.

	$|1, 0\rangle_{\text{out}}$	$|0, 1\rangle_{\text{out}}$
$|1, 0\rangle_{\text{in}}$	$\frac{1}{2}$	$\frac{1}{2}$
$|0, 1\rangle_{\text{in}}$	$\frac{1}{2}$	$\frac{1}{2}$

the situation changes dramatically if there are two or more identical indistinguishable photons interacting with the beam splitter and subsequently detected. This is considered next.

11.4.1 Detection of Two Indistinguishable Photons after a Beam Splitter

A bi-photon source is designed to create pairs of indistinguishable photons. The task is to predict what happens if a positive integer number of photons are introduced to the two input ports of the perfect, lossless, symmetric 50:50 dielectric beam splitter illustrated in Fig. 11.6.

If one photon is incident at input port 1 and the other at input port 2, then the input state is $|n_1 = 1, n_2 = 1\rangle_{\text{in}}$. There are different paths from the input state to the output states of the beam splitter so that $|n_1 = 1, n_2 = 1\rangle_{\text{in}} \rightarrow |n_3, n_4\rangle_{\text{out}}$. Noninteracting indistinguishable bosons, such as photons, are exchange symmetric and for the two-photon input state $|1, 1\rangle_{\text{in}}$ there are three exchange-symmetric output product-states. Setting $r_{\text{ph}} = -1$ and $t_{\text{ph}} = i$, the non-normalized quantum field amplitudes for the different output states are:

$$|n_1 = 1, n_2 = 1\rangle_{\text{in}} \rightarrow |n_3 = 2, n_4 = 0\rangle_{\text{out}} : t_{\text{ph}} r_{\text{ph}} = -i, \tag{11.15}$$

$$|n_1 = 1, n_2 = 1\rangle_{\text{in}} \rightarrow |n_3 = 1, n_4 = 1\rangle_{\text{out}} : \frac{r_{\text{ph}} r_{\text{ph}} + t_{\text{ph}} t_{\text{ph}}}{\sqrt{2}} = 0, \tag{11.16}$$

$$|n_1 = 1, n_2 = 1\rangle_{\text{in}} \rightarrow |n_3 = 0, n_4 = 2\rangle_{\text{out}} : r_{\text{ph}} t_{\text{ph}} = -i. \tag{11.17}$$

Note that for the case $|n_1 = 1, n_2 = 1\rangle_{\text{in}} \rightarrow |n_3 = 1, n_4 = 1\rangle_{\text{out}}$ there are two indistinguishable paths the two photons can take from input to output. They are either both reflected or both transmitted through the beam splitter. Each path is equally likely in the 50:50 beam splitter and the indistinguishable photons simultaneously experience both processes. The superposition of product quantum field amplitudes $(r_{\text{ph}} r_{\text{ph}} + t_{\text{ph}} t_{\text{ph}})/\sqrt{2}$ describes this. The quantum field amplitude interference that results in $(r_{\text{ph}} r_{\text{ph}} + t_{\text{ph}} t_{\text{ph}})/\sqrt{2} = 0$ occurs because the *detectors* are unable to distinguish between the two two-photon paths.

The photon-number detection probabilities at the output ports are *proportional* to the magnitude squared of the non-normalized quantum field amplitudes,

$$P_{\text{out}}^{\text{non}}(n_1 = 1, n_2 = 1, n_3 = 2, n_4 = 0) = |t_{\text{ph}} r_{\text{ph}}|^2 = 1, \tag{11.18}$$

$$P_{\text{out}}^{\text{non}}(n_1 = 1, n_2 = 1, n_3 = 1, n_4 = 1) = \left| \frac{r_{\text{ph}} r_{\text{ph}} + t_{\text{ph}} t_{\text{ph}}}{\sqrt{2}} \right|^2 = \frac{|1 - 1|^2}{2} = 0, \tag{11.19}$$

$$P_{\text{out}}^{\text{non}}(n_1 = 1, n_2 = 1, n_3 = 0, n_4 = 2) = |r_{\text{ph}} t_{\text{ph}}|^2 = 1. \tag{11.20}$$

Normalization of $P_{\text{out}}^{\text{non}}$ requires division by the sum

$$P_{\text{sum}} = \sum_j P_{j,\text{out}}^{\text{non}} \tag{11.21}$$

and allows interpretation as probability P_{out}. In the case considered $P_{\text{sum}} = 2$, and so the normalized probability values are

$$P_{\text{out}}(n_1 = 1, n_2 = 1, n_3 = 2, n_4 = 0) = \frac{|t_{\text{ph}} r_{\text{ph}}|^2}{P_{\text{sum}}} = \frac{1}{2}, \tag{11.22}$$

$$P_{\text{out}}(n_1 = 1, n_2 = 1, n_3 = 1, n_4 = 1) = 0, \tag{11.23}$$

$$P_{\text{out}}(n_1 = 1, n_2 = 1, n_3 = 0, n_4 = 2) = \frac{|r_{\text{ph}} t_{\text{ph}}|^2}{P_{\text{sum}}} = \frac{1}{2}. \tag{11.24}$$

Notice that the total probability adds up to unity. Also, importantly, if there is one indistinguishable photon at each input port, then the *quantum amplitudes interfere and exactly cancel* so that there is precisely zero probability of detecting one photon at each output port. This strong quantum interference effect was first measured in experiments by Hong, Ou, and Mandel.[10] In addition, one indistinguishable photon at each input port can *only* result in two photons detected at an output port. This effect is an example of *photon bunching*.

If two photons are introduced at input port 1 and zero at input port 2, then the input state is $|n_1 = 2, n_2 = 0\rangle_{\text{in}}$ and the possible output state quantum field amplitudes are proportional to

$$|n_1 = 2, n_0 = 1\rangle_{\text{in}} \rightarrow |n_3 = 2, n_4 = 0\rangle_{\text{out}} : t_{\text{ph}} t_{\text{ph}} = -1, \tag{11.25}$$

$$|n_1 = 2, n_2 = 0\rangle_{\text{in}} \rightarrow |n_3 = 1, n_4 = 1\rangle_{\text{out}} : \frac{t_{\text{ph}} r_{\text{ph}} + r_{\text{ph}} t_{\text{ph}}}{\sqrt{2}} = \frac{-2i}{\sqrt{2}}, \tag{11.26}$$

$$|n_1 = 2, n_2 = 0\rangle_{\text{in}} \rightarrow |n_3 = 0, n_4 = 2\rangle_{\text{out}} : r_{\text{ph}} r_{\text{ph}} = 1. \tag{11.27}$$

In this case, $P_{\text{sum}} = 4$ and the corresponding photon-number detection probabilities at the output ports are

$$P_{\text{out}}(n_1 = 2, n_2 = 0, n_3 = 2, n_4 = 0) = \frac{|t_{\text{ph}} t_{\text{ph}}|^2}{P_{\text{sum}}} = \frac{1}{4}, \tag{11.28}$$

$$P_{\text{out}}(n_1 = 2, n_2 = 0, n_3 = 1, n_4 = 1) = \frac{|t_{\text{ph}} r_{\text{ph}} + r_{\text{ph}} t_{\text{ph}}|^2}{2P_{\text{sum}}} = \frac{1}{2}, \tag{11.29}$$

$$P_{\text{out}}(n_1 = 2, n_2 = 0, n_3 = 0, n_4 = 2) = \frac{|r_{\text{ph}} r_{\text{ph}}|^2}{P_{\text{sum}}} = \frac{1}{4}. \tag{11.30}$$

Table 11.2 shows the output-state detection probabilities of two indistinguishable photons after a 50:50 beam splitter.

[10]C. K. Hong, Z. Y. Ou, and L. Mandel, *Phys. Rev. Lett.* **59**, 2044 (1987).

Table 11.2 Two photon detection probability after a 50:50 beam splitter

	$|2,0\rangle_{\text{out}}$	$|1,1\rangle_{\text{out}}$	$|0,2\rangle_{\text{out}}$
$|2,0\rangle_{\text{in}}$	$\frac{1}{4}$	$\frac{1}{2}$	$\frac{1}{4}$
$|1,1\rangle_{\text{in}}$	$\frac{1}{2}$	0	$\frac{1}{2}$
$|0,2\rangle_{\text{in}}$	$\frac{1}{4}$	$\frac{1}{2}$	$\frac{1}{4}$

11.4.2 Detection of Multiple Indistinguishable Photons after a Beam Splitter

In general, for integer n_1 and n_2 photons at the input ports 1 and 2, respectively, the quantum field amplitude of integer n_3 and n_4 indistinguishable photons appearing at the output ports of the beam splitter is

$$(-1)^{n_1}\left(\frac{1}{2}\right)^{\frac{n_1+n_2}{2}}\sum_k(-1)^k\sqrt{\binom{n_1}{k}\binom{n_2}{n_3-k}\binom{n_3}{k}\binom{n_4}{n_1-k}}, \tag{11.31}$$

where, because the total number of particles is conserved, $n_4 = n_1 + n_2 - n_3$. Notice that, in this expression, the number of ways of choosing k photons from a set of n photons is given by the binomial coefficient

$$\binom{n}{k} = \frac{n!}{k!(n-k)!}. \tag{11.32}$$

If k is negative or greater than n, the binomial coefficient is set to zero. The terms $(-1)^{n_1}$ and $(-1)^k$ are due to the relative phase difference between a transmitted or reflected photon and it is this that causes the strong quantum interference effects. The probability of detecting photons at the output ports of the beam splitter is the magnitude squared of Eq. (11.31).

Illustrating the richness of the possibilities, Fig. 11.7 plots the probability of photon number detection as a function of input and output states of a perfect lossless symmetric 50:50 beam splitter when the total number of photons is eight, $n_{\text{tot}} = 8$. The system behavior is closest to classical when photons are only present at one input port ($n_1 = n_{\text{tot}}$). The system behavior is most nonclassical when there are equal numbers of photons at the two input ports of the beam splitter ($n_1 = n_2 = n_{\text{tot}}/2$ for n_{tot} an even nonzero positive integer).

For a total number of identical indistinguishable photons $n_{\text{tot}} = 8$, Table 11.3 gives the probability of n_3 detected photons at output port 3 for the cases when $n_1 = 8$ and $n_1 = 4$ at input port 1. Note the zero values for odd-integer n_3 and the rational numbers with denominator $2^{n_{\text{tot}}}$ for the nonzero probability values.

In the preceding, photon number is preserved and the interaction of the optical field with the beam splitter is ideal. The symmetry associated with indistinguishable particles results in quantum amplitude interference between different paths through the system and the probability of detecting a fixed number of photons at an output port can be dramatically different from classical expectations.

Table 11.3 Photon detection probability after a 50:50 beam splitter with $n_{\text{tot}} = 8$ when $n_1 = 8$ and $n_1 = 4$

	$n_1 = 8$	$n_1 = 4$
$n_3 = 0$	$\frac{1}{256}$	$\frac{70}{256}$
$n_3 = 1$	$\frac{8}{256}$	0
$n_3 = 2$	$\frac{28}{256}$	$\frac{40}{256}$
$n_3 = 3$	$\frac{56}{256}$	0
$n_3 = 4$	$\frac{70}{256}$	$\frac{36}{256}$
$n_3 = 5$	$\frac{56}{256}$	0
$n_3 = 6$	$\frac{28}{256}$	$\frac{40}{256}$
$n_3 = 7$	$\frac{8}{256}$	0
$n_3 = 8$	$\frac{1}{256}$	$\frac{70}{256}$

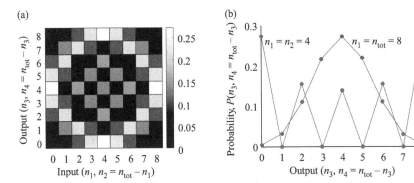

Fig. 11.7 (a) Probability of photon output from a perfect, lossless, symmetric, 50:50 beam splitter showing input of n_1 photons at input port 1 and n_2 photons at input port 2 when there is a total of $n_{\text{tot}} = 8$ indistinguishable photons in the system. (b) Dots show probability of detection at output port 3 for the case when the input port 1 contains $n_1 = n_{\text{tot}} = 8$ and when $n_1 = n_2 = n_{\text{tot}}/2 = 4$. Lines connecting the dots are to guide the eye.

Two-photon quantum interference (the $n_{\text{tot}} = 2$ case) is *not* interference of two separate photons at the beam splitter, but rather it is interference of the two two-photon amplitudes at the *detectors*. Photons do not have to arrive simultaneously at the beam splitter to have their quantum field amplitudes interfere at the detectors; rather it is only that the two two-photon (Feynman) paths must be indistinguishable.[11] It is the existence of indistinguishable two-photon paths that enables quantum-field amplitude interference.

---

[11]T. B. Pittman et al., *Phys. Rev. Lett.* **77**, 1917 (1996).

11.5 Coherent Quantum Control

The equations describing the behavior of a single photon are the same as those for a classical electromagnetic field. The difference is in interpretation. A single photon wave function[12] may be used to describe photon energy density, $U_{ph}(x,t) = |\psi(x,t)|^2$. Because the real and imaginary parts of the single-photon wave function, $\psi(x,t)$, obey Maxwell's equations, all methods used for solving classical electromagnetic field propagation problems can be applied to the case of a single photon.

11.5.1 Control Field

Quantum dynamics can be controlled by generating and applying a control field to the system. Sophisticated control methods exist that require measurement. Examples include measurement-based feedback control and coherent real-time feedback control. Here, a wave-like particle can be controlled using a field generator that controls system dynamics using the simpler method of open-loop control.

Open-loop control has a defined physical objective, model-based creation of control parameters, and a control-field generator that interacts with the quantum system to control quantum dynamics. The basic functional blocks for open-loop control are illustrated in Fig. 11.8 and can be used to control single-photon dynamics in a Fabry–Perot resonator.

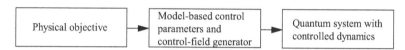

Fig. 11.8 Functional blocks for open-loop control. There is a physical objective, model-based creation of control parameters, and a control-field generator that interacts with the quantum system to control quantum dynamics.

11.5.2 Control of Single-Photon Dynamics in a Fabry–Perot Resonator

The transient dynamics of a photon field interacting with a resonator is of both fundamental and practical interest. A basic property of a resonator is its ability to store photon energy density and release it at a later time. There is therefore motivation to demonstrate control of photon field transient dynamics and hence control of transient response.

The phenomena of interest may be illustrated by considering a linearly polarized photon pulse incident on a resonator with a refractive index profile as illustrated in Fig. 11.9. The symmetric one-dimensional Fabry–Perot resonator consists of two quarter-wavelength lossless dielectric mirrors separated by a cavity of length L_C. Each mirror may be thought of as a flat-plate dielectric beam splitter whose surface is normal to the direction of travel of the photon pulse. Each dielectric mirror has partial transmission, refractive index n_r, and quarter-wave thickness $L_m = \lambda_0/4n_r$, where λ_0 is the resonant photon wavelength in vacuum. At the resonant photon wavelength, the mirror reflection amplitude is $|r_{ph}|e^{i\pi} = -|r_{ph}|$, and the transmission amplitude is $|t_{ph}|e^{i\pi/2} = i|t_{ph}|$.

---

[12]I. Bialynicki-Birula, *Acta Phys. Pol.* **86**, 97 (1994); B. J. Smith and M. G. Raymer, *New J. Phys.* **9**, 414 (2007).

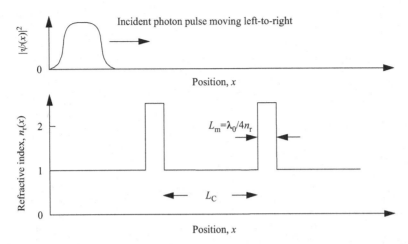

Fig. 11.9 Schematic showing a linearly polarized photon pulse of energy density $|\psi(x)|^2$ moving left-to-right that is incident on a symmetric Fabry–Perot resonator consisting of two quarter-wavelength lossless dielectric mirrors of refractive index n_r separated by cavity of length L_C. The resonant wavelength is λ_0.

The phase of the transmitted field leads the phase of the reflected field by $\pi/2$.[13] Flux conservation in the lossless system requires $|r_\mathrm{ph}|^2 + |t_\mathrm{ph}|^2 = 1$. Transmission through each mirror depends weakly on wavelength such that

$$|t_\mathrm{ph}|^2 = \frac{1}{1 + \left(\frac{k_1^2 - k_2^2}{2 k_1 k_2}\right)^2 \sin^2\left(k_2 L_m\right)}, \tag{11.33}$$

where the propagation constant in vacuum is $k_1 = 2\pi/\lambda$, and in the dielectric mirror it is $k_1 = 2\pi n_\mathrm{r}/\lambda$.

If $L_m = \lambda_0/4 n_\mathrm{r}$ and $n_\mathrm{r} = 1 + \sqrt{2}$, then $|t_\mathrm{ph}^2| = 1/2$ when wavelength $\lambda = \lambda_0$.

A single photon pulse may be described by photon field $\psi(x, t)$ with the interpretation that $|\psi(x, t)|^2$ is proportional to the photon energy density. The unitary dynamics of the photon pulse propagating in the x direction in a lossless dielectric medium may be modeled as a phase-coherent integral of linearly polarized basis states $\phi_\omega(x)$ with amplitudes a_ω,

$$\psi(x, t) = \int \frac{\mathrm{d}\omega}{2\pi} a_\omega \phi_\omega(x) \mathrm{e}^{-i\omega t}, \tag{11.34}$$

where $\phi_\omega(x)$ is a normalized solution of the one-dimensional Helmholtz equation,

$$\frac{\mathrm{d}}{\mathrm{d}x}\left(\frac{1}{\mu_\mathrm{r}(x)} \frac{\mathrm{d}}{\mathrm{d}x} \phi_\omega(x)\right) + \omega^2 \epsilon_\mathrm{r}(x) \epsilon_0 \mu_0 \phi_\omega(x) = 0. \tag{11.35}$$

The permeability of vacuum, μ_0, and permittivity of vacuum, ϵ_0, are related to the speed of light in vacuum via $c = 1/\sqrt{\epsilon_0 \mu_0}$. Assuming a lossless dielectric material, the spatial

---

[13]A. Agnesi and V. Degiorio, *Opt. Laser Tech.* **95**, 72 (2017); V. Degiorio, *Am. J. Phys.* **48**, 81 (1980).

profile may be characterized by piecewise-constant values of relative permeability, μ_r, and relative permittivity, ϵ_r, in each region of the domain; the conditions imposed on $\psi(x)$ at the boundary between regions 1 and 2 at position x_0 are

$$\phi_{\omega,1}(x_0) = \phi_{\omega,2}(x_0) \tag{11.36}$$

and

$$\frac{1}{\mu_{r_{ph,1}}} \frac{d}{dx} \phi_{\omega,1}(x_0) = \frac{1}{\mu_{r_{ph,2}}} \frac{d}{dx} \phi_{\omega,2}(x_0). \tag{11.37}$$

The refractive index is $n_r = \sqrt{\mu_r}\sqrt{\epsilon_r}$. If the coherence time of the photon field is longer than any other characteristic time scale in the system, then solving the Helmholtz equation, Eq. (11.35), completely describes the evolution of the photon field. An efficient and accurate way to solve Eq. (11.35) for the Fabry–Perot resonator subsystem coupled to continuous input and output states is to use the propagation matrix method.

11.5.3 Transient Response

The transient response of a photon pulse with a rectangular envelope function traveling left-to-right and incident on the Fabry–Perot resonator is considered. See Fig. 11.9. The shape of the rectangular pulse with center frequency ω_0 is smoothed by modulating the sinc function of the rectangular pulse by a cosine in order to reduce high-frequency contributions.[14] In this way, a rectangular pulse of duration $2T_0$ (length $2T_0c$) with rise and fall time $\tau_r = 2\pi/\Delta\omega_r$ may be written as

$$\psi(x,t) = \int\limits_{|\omega-\omega_0|\leq\Delta\omega_r} \frac{d\omega}{2\pi} \left(1 + \cos\left(\frac{\pi(\omega-\omega_0)}{\Delta\omega_r}\right)\right) \times \frac{\sin((\omega-\omega_0)T_0)}{(\omega-\omega_0)T_0}\phi_\omega(x)e^{-i\omega t}. \tag{11.38}$$

As a specific example, a cavity with resonant wavelength $\lambda_0 = 1\,500$ nm and resonant frequency $\omega_0 = 2\pi/\tau_0$, where $\tau_0 = 5$ fs, is considered. The refractive index of the quarter-wave mirrors is chosen to be $n_r = 2.5$ and the resonator cavity length is $L_C = 15 \times \lambda_0$. The cavity round-trip time of the photon field in the resonator is τ_{RT}, and the resonator quality factor is Q, where $\tau_Q = Q/\omega_0$ for large values of Q. Typically, if the transient response is dominated by the ring-down time constant $\tau_Q = 1/\gamma$, the photon energy density of a loaded resonator decays as e^{-t/τ_Q} and τ_Q is connected via a Fourier transform to a steady-state Lorentzian energy density spectrum,

$$U_{ph}(\omega) = \frac{U_0}{(\omega-\omega_0)^2 + (\gamma/2)^2}. \tag{11.39}$$

However, the actual transient dynamics of the system to be controlled is more complex than this description would suggest.

Figure 11.10(a) shows the calculated position–time photon energy density plot of a rectangular pulse initially moving left-to-right and incident on the Fabry–Perot resonator. The presence of the resonator imparts temporal structure onto reflected and transmitted photon energy densities. There is a large characteristic reflection at the leading and trailing edge of the photon pulse. The transient burst of reflected energy at the leading edge and

---

[14]The high frequency terms arise from the Gibbs phenomenon.

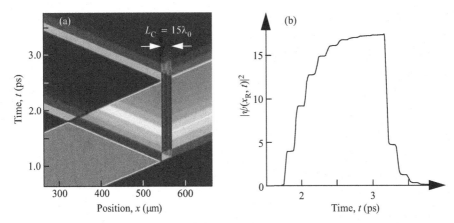

Fig. 11.10 (a) Position-time photon energy density plot of a rectangular pulse incident on a Fabry–Perot resonator. The resonator cavity length is L_C (indicated by two arrows). (b) $|\psi(x_R, t)|^2$ (arbitrary scale) as a function of time detected at position x_R far to the right of the resonator. The photon field decay constant, $2\tau_Q = 229$ fs, is modulated by a stepwise response at the resonant cavity round-trip time $\tau_{RT} = 30\tau_0 = 1\,50$ fs. Parameters are $\lambda_0 = 1\,500$ nm, $L_C = 15 \times \lambda_0$, $n_r = 2.5$, $L_m = \lambda_0/4n_r$, $\tau_0 = 5$ fs, $\omega_0 = 2\pi/\tau_0$ rad s^{-1}, $2\Delta\omega_r = \omega_0/4$ rad s^{-1}, and $T_0\omega_0 = 900$.

trailing edge of the incident pulse is due to nonresonant frequency components associated with the pulse transient rise and fall times and the changing energy density in the resonator. Subsequent reflections decay temporally in a stepwise fashion in time steps of duration τ_{RT}. Figure 11.10(b) shows $|\psi(x_R, t)|^2$ calculated as a function of time detected at position x_R far to the right of the resonator. Photon energy density both in the resonator and transmitted to position x_R does not increase (or decay) as a simple exponential; rather, there is a step-wise buildup (or decay) at each resonant cavity photon round-trip time, τ_{RT}. With increasing rectangular pulse duration, energy density asymptotically approaches the steady-state value, which, on resonance at frequency ω_0, results in unity transmission, zero reflection, and maximum photon energy density in the resonator. However, the steady state is not of interest; rather, it is coherent control of the transient photon–resonator interaction using interference effects and its use to control the transient response of the photon in the system. The shortest timescale on which control is sought is the cavity transit time, $\tau_{RT}/2$.

Physical intuition and development of basic control concepts may be illustrated using a photon pulse whose duration is short compared to the cavity round-trip time, i.e., $2T_0 < \tau_{RT}$. For notational efficiency, in the following $-|r_{ph}| \to -r_{ph}$ and $i|t_{ph}| \to it_{ph}$.

Figure 11.11(a) shows the position-time photon energy density plot of a short rectangular pulse moving left-to-right and incident on the Fabry–Perot resonator. Initially, the photon pulse energy density entering into the cavity shows no indication of wave character. It is only after reflection from the right mirror that self-interference effects are observed and photon resonance inside the cavity begins to build up. The energy stored in the resonator leaks out as forward and backscattered pulses. The shortest time between forward and backscattered pulses is the cavity transit time, $\tau_{RT}/2$. Figure 11.11(b) illustrates the origin of the ring-down using position-time resonant photon ray-tracing of reflected and transmitted amplitudes. The scattered amplitudes at each mirror form a geometric series.

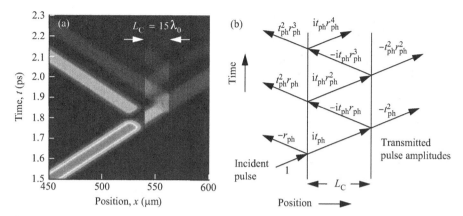

Fig. 11.11 (a) Position-time photon energy density plot of a short rectangular pulse incident on a Fabry–Perot resonator showing ring-down. (b) Position-time resonant photon ray-trace illustrating ring-down in the form of multiple transmitted and reflected amplitudes. For notational efficiency, $-|r_{ph}| \rightarrow -r_{ph}$ and $i|t_{ph}| \rightarrow it_{ph}$. Parameters are $\lambda_0 = 1\,500\,\text{nm}$, $L_C = 15 \times \lambda_0$, $n_r = 2.5$, $L_m = \lambda_0/4n_r$, $\tau_0 = 5\,\text{fs}$, $\omega_0 = 2\pi/\tau_0\,\text{rad s}^{-1}$, $2\Delta\omega_r = \omega_0/4\,\text{rad s}^{-1}$, and $T_0\omega_0 = 60$.

11.5.4 Coherent Control of Transient Response

Coherent control of transient photon response as illustrated in Fig. 11.11 may be achieved using photon control pulses. Similar to Eq. (11.34), the control pulses consist of a coherent integral of basis functions whose amplitudes, a_ω^{cont}, and time delay, t_ω^{cont}, are control parameters that may be optimized. In the following, the use of formal optimization methods is avoided because the geometric series illustrated in Fig. 11.11(b) suggests that a simpler intuitive approach exists.[15]

First, a single control pulse that is just an attenuated, delayed, and phase-shifted version of the lead pulse is considered. Figure 11.12(a) is a position-time photon energy density plot showing a lead pulse and control pulse initially moving left-to-right and incident on the Fabry–Perot resonator. In this example, the control pulse is configured to eliminate ring-down after exactly one photon round-trip time in the cavity. This may be achieved with a control pulse of the same shape that is coherent with the lead pulse, with resonant amplitude $-r_{ph}^2$ relative to the lead pulse, and delayed by a time τ_{RT}. Figure 11.12(b) is a position-time resonant photon ray-trace showing lead and control amplitudes configured to eliminate ring-down. To highlight the difference in the time domain between uncontrolled photon ring-down in the resonator and precise control, Fig. 11.13 shows the transmitted pulse train for the two situations illustrated in Figs. 11.11 and 11.12. Transmitted photon energy density as a function of time for the uncontrolled case (Fig. 11.11(a)) consists of a series of pulses whose peaks occur at equally spaced time intervals τ_{RT} and whose peak value decreases exponentially as e^{-t/τ_Q}. For the controlled case (Fig. 11.13(b)), a coherent control pulse is used to ensure that there is just one transmitted photon energy density pulse.

Figure 11.14 shows the coherent control of a single-photon resonator output pulse using one backward-travelling control pulse. In this case, the control pulse precisely eliminates

---

[15]A. F. J. Levi et al., *Phys. Rev. A* **90**, 022119 (2014).

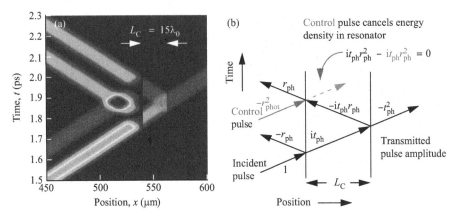

Fig. 11.12 (a) Position-time photon energy density plot showing lead and control pulse. The control pulse eliminates ring-down by removing all photon energy density in the cavity after exactly one round-trip time, τ_{RT}. There is just one transmitted photon pulse. (b) Position-time resonant photon ray-trace showing lead and control amplitudes configured to eliminate ring-down. Parameters are $\lambda_0 = 1\,500\,\text{nm}$, $L_C = 15 \times \lambda_0$, $n_r = 2.5$, $L_m = \lambda_0/4n_r$, $\tau_0 = 5\,\text{fs}$, $\omega_0 = 2\pi/\tau_0\,\text{rad s}^{-1}$, $2\Delta\omega_r = \omega_0/4\,\text{rad s}^{-1}$, and $T_0\omega_0 = 60$.

ring-down by removing all photon energy density in the cavity after exactly one transit time, $\tau_{RT}/2$. There is just one transmitted pulse and one reflected pulse. Viewed far from the resonator, the incident control pulse, which has *exactly* the same energy density as the reflected pulse, appears delayed in time by $\tau_{RT}/2$ relative to the reflected pulse. Absent additional information, the reflected pulse gives the illusion of having advanced in time by $\tau_{RT}/2$ relative to the incident control pulse.

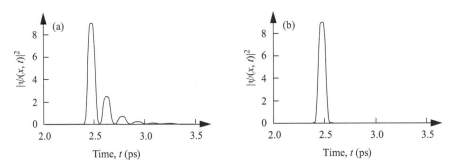

Fig. 11.13 (a) Transmitted photon energy density as a function of time with no control (as in Fig. 11.11). (b) Same as (a) but with a control pulse to eliminate ring-down by removing all photon energy density in the cavity after exactly one round-trip time, τ_{RT}. There is just one transmitted photon pulse. Parameters are $\lambda_0 = 1\,500\,\text{nm}$, $L_C = 15 \times \lambda_0$, $n_r = 2.5$, $L_m = \lambda_0/4n_r$, $\tau_0 = 5\,\text{fs}$, $\omega_0 = 2\pi/\tau_0\,\text{rads}^{-1}$, $2\Delta\omega_r = \omega_0/4\,\text{rad s}^{-1}$, and $T_0\omega_0 = 60$. The photon energy density scale is arbitrary.

A coherent control pulse with amplitude $-r_{\text{ph}}^{2N}$ injected at the Nth photon round trip may be used together with an integrating detector to evaluate a finite geometric sum. Figure 11.15(a) illustrates this for the case $N = 2$. An integrating photon energy detector

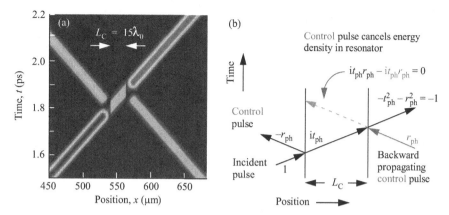

Fig. 11.14 (a) Position-time photon energy density plot showing lead and single backward propagating control pulse. The control pulse eliminates ring-down by removing all photon energy density in the cavity after one half of a round-trip time, τ_{RT}. There is just one transmitted pulse and one reflected pulse. Viewed far from the resonator, the incident control pulse, which has exactly the same energy density as the reflected pulse, appears delayed in time by $\tau_{RT}/2$ relative to the reflected pulse. (b) Position-time resonant optical ray trace showing lead and control amplitudes configured to precisely eliminate ring-down. Parameters are $\lambda_0 = 1\,500$ nm, $L_C = 15 \times \lambda_0$, $n_r = 2.5$, $L_m = \lambda_0/4n_r$, $\tau_0 = 5$ fs, $\omega_0 = 2\pi/\tau_0$ rad s^{-1}, $2\Delta\omega_r = \omega_0/4$ rad s^{-1}, and $T_0\omega_0 = 60$.

at the output measures this geometric sum as

$$\left| \sum_{n=0}^{N-1} a\xi^n \right|^2 = \left| a\frac{1 - \xi^N}{1 - \xi} \right|^2, \tag{11.40}$$

where, on resonance, $\xi = r_{ph}^2$ and $a = t_{ph}^2$. The sum in Eq. (11.40) is guaranteed to converge in the limit $N \to \infty$ because $|r_{ph}| < 1$.

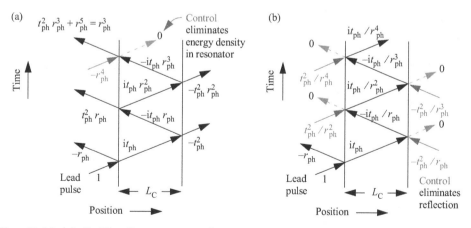

Fig. 11.15 (a) Position-time resonant photon ray-trace showing incident lead and control amplitudes configured to perform a finite geometric sum by integrating detection of right-hand output. (b) Position-time resonant photon ray-trace showing incident lead photon and control amplitudes configured to create a finite divergent geometric series.

501

Figure 11.15(b) illustrates that the finite geometric series in Eq. (11.40) with $|\xi| > 1$ may also be created by using multiple forward- and reverse-propagating control pulses. In this particular example coherent photon control pulses are used to *confine* photon energy density in the resonator. Photon energy density necessarily increases according to Eq. (11.40) because $|r_{\mathrm{ph}}| < 1$. In general, transient photon dynamics in resonators with input and output ports may be used to evaluate arbitrary finite sums of the form

$$\left| \sum_{n=0}^{N-1} a_n \xi^n \right|^2, \tag{11.41}$$

where a_n and ξ are determined by control pulses.

11.5.5 Boolean Logic

Boolean logic functions such as NAND, NOT, OR, XOR, and XNOR can be performed using photon signals and photon control pulses interacting with a beam splitter or a Fabry–Perot resonator. Differential inversion and multiplexing are also possible.

However, to create photonic logic circuits that can be scaled up both in number of devices and logic complexity, additional functionality is necessary. This includes the ability to reshape, retime, and re-amplify optical signals. Reshaping and retiming can be used to remove amplitude and phase noise. Re-amplification is needed to drive multiple inputs (fan-out) and to compensate for optical loss. These "3Rs" are difficult to implement in an energy-*efficient* way. While reshaping and retiming classical light can be achieved using a saturable absorber and re-amplification is possible by employing an active optical gain medium, the single-photon version of 3Rs is more challenging.

11.6 Quantum Information Processing

Just as quantum information is different from classical information, quantum information processing is different from classical computing. Classical computers perform irreversible operations, whereas the operations of an ideal quantum computer are reversible. The unique resources that quantum mechanics provides and for which there is no classical counterpart are wave–particle duality, identical indistinguishable particles, linear superposition of particle states, interference of quantum particle states, entanglement of quantum particle states, and measurement whose outcome is noncausal.

Expanding an arbitrary non-eigenstate of the system, $|\psi\rangle$, in terms of the complete orthonormal basis states of the system gives

$$|\psi\rangle = \sum_0^\infty a_n |n\rangle, \tag{11.42}$$

where, in general, a_n are complex expansion coefficients. The probability of finding a system in the eigenstate $|n\rangle$ after measurement is $|a_n|^2$. Absent subsequent interaction, after measurement the system remains in eigenstate $|n\rangle$. *Measurement is* fundamentally *noncausal* since, in general, for a system initially not in an eigenstate, it is impossible to know the result of measurement beforehand. Between measurements, the time evolution of the complete system described by state vector $|\psi(t)\rangle$ is deterministic and given by Schrödinger's equation,

$$i\hbar\frac{\partial}{\partial t}|\psi(t)\rangle = \hat{H}|\psi(t)\rangle,$$

(11.43)

where $\hat{H}$ is the Hamiltonian operator for the total system.

11.6.1 The Single-Qubit State

The fundamental unit of information processed in a classical computer is a bit. A bit can take on one of two binary values, either a zero or a one, and can be precisely copied and stored in memory.

Information in a quantum system has different properties. The quantum-mechanical analog of a single classical binary bit is a quantum bit (or *qubit*) that is usually described by the bound states of a *two-level system*. The orthonormal basis states $|0\rangle$ and $|1\rangle$ exist in a Hilbert space. Because the single-qubit state $|\psi\rangle$ can be in a superposition of the two basis states $|0\rangle$ and $|1\rangle$ then

$$|\psi\rangle = a_0|0\rangle + a_1|1\rangle = \begin{bmatrix} a_0 \\ a_1 \end{bmatrix},$$

(11.44)

where the magnitude squared of the complex amplitudes a_0 and a_1 is the probability of measuring the system in basis states $|0\rangle$ and $|1\rangle$, respectively. The superposition state and all the information it contains cannot be measured directly. After measurement, the system is either in state $|0\rangle$ or $|1\rangle$ and the probability of finding the system in state $|0\rangle$ is $P_0 = |a_0|^2$ and in state $|1\rangle$ is $P_1 = |a_1|^2$, with the constraint that $|a_0|^2 + |a_1|^2 = 1$. Unlike the classical binary bit, the results of measuring a superposition state in quantum mechanics are probabilistic so that P_0 and P_1 can take on an infinite number of possible values. Probability is established by performing multiple experiments on identically prepared systems. In addition, and in contrast to the classical bit, when the qubit is in a superposition state the no cloning theorem precludes the possibility of making perfect copies of the state.

11.6.2 Representation of a Single Qubit on the Bloch Sphere and Unitary Operations

The superposition state $|\psi\rangle = a_0|0\rangle + a_1|1\rangle$ of basis vectors $|0\rangle$ and $|1\rangle$ is described using two complex amplitudes, a_0 and a_1, that can be written as four real numbers. Measurement collapses the quantum superposition state $|\psi\rangle$ and what remains is eigenstate $|n\rangle$, which is either $|0\rangle$ or $|1\rangle$. The measurement process forces loss of information about the superposition state. Since the probabilities of obtaining the measurement result $|0\rangle$ or $|1\rangle$ are the same for $|\psi\rangle$ or $e^{i\varphi}|\psi\rangle$, it follows that the presence of a global phase angle φ has no physical consequence. Hence, one of the four real numbers associated with a_0 and a_1 is not needed because only the relative phase between basis vectors has physical meaning. It follows that the remaining three real numbers can be used to describe the single-qubit state $|\psi\rangle$ as a point on a sphere of radius $|a_0|^2 + |a_1|^2 = 1$.

As illustrated in Fig. 11.16, the single qubit-state $|\psi\rangle = a_0|0\rangle + a_1|1\rangle$ can be represented as a point on a unit-radius *Bloch sphere*. Restricting a_0 to be real and positive, the state $|\psi\rangle$ on the Bloch sphere can be uniquely expressed as

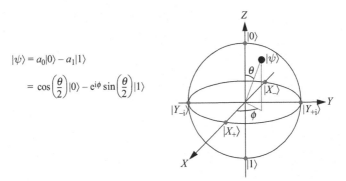

$$|\psi\rangle = a_0|0\rangle - a_1|1\rangle$$

$$= \cos\left(\frac{\theta}{2}\right)|0\rangle - e^{i\phi}\sin\left(\frac{\theta}{2}\right)|1\rangle$$

Fig. 11.16 Geometric representation of a single qubit in state $|\psi\rangle$ that is a point on the Bloch sphere of unit radius. In general, the qubit state is a linear superposition of basis states $|0\rangle$ and $|1\rangle$ in which $|\psi\rangle = a_0|0\rangle + a_1|1\rangle$, a_0 is real and positive, a_1 is complex, and $|a_0|^2 + |a_1|^2 = 1$.

$$|\psi\rangle = \cos\left(\frac{\theta}{2}\right)|0\rangle + e^{i\phi}\sin\left(\frac{\theta}{2}\right)|1\rangle, \tag{11.45}$$

where parameters $0 \le \theta \le \pi$ and $0 \le \phi \le 2\pi$ are the polar and azimuthal angles, respectively. Note that while the value of ϕ is not unique when $|\psi\rangle$ is one of either basis vector $|0\rangle$ or $|1\rangle$, the point on the sphere represented by parameters θ and ϕ *is* unique. The use of such a Bloch sphere to geometrically represent quantum superposition states is a way to visualize unitary operations on a single qubit.

The symmetric and antisymmetric linear superposition of basis states $|0\rangle$ and $|1\rangle$ that correspond to the two points at which the X-axis intersects the surface of the sphere are

$$|X_\pm\rangle = \frac{1}{\sqrt{2}}(|0\rangle \pm |1\rangle). \tag{11.46}$$

On the Y-axis the linear superposition of states $|0\rangle$ and $|1\rangle$ that correspond to the two points at which the Y-axis intersects the surface of the sphere are

$$|Y_{\pm i}\rangle = \frac{1}{\sqrt{2}}(|0\rangle \pm i|1\rangle). \tag{11.47}$$

The Z-axis intersects the surface of the sphere at the basis states

$$|Z_+\rangle = |0\rangle \tag{11.48}$$

and

$$|Z_-\rangle = |1\rangle. \tag{11.49}$$

Written in vector notation, the single-qubit basis states are

$$|0\rangle = \begin{bmatrix} 1 \\ 0 \end{bmatrix} \tag{11.50}$$

and

$$|1\rangle = \left[\begin{array}{c} 0 \\ 1 \end{array} \right].$$
(11.51)

These basis states are eigenvectors of the traceless, unitary, Pauli matrices

$$\hat{U}_X = \left[\begin{array}{cc} 0 & 1 \\ 1 & 0 \end{array} \right],$$
(11.52)

which is a NOT operator that swaps $|0\rangle$ and $|1\rangle$,

$$\hat{U}_Y = \left[\begin{array}{cc} 0 & -i \\ i & 0 \end{array} \right],$$
(11.53)

and the operator

$$\hat{U}_Z = \left[\begin{array}{cc} 1 & 0 \\ 0 & -1 \end{array} \right],$$
(11.54)

which performs a π phase flip on the state $|1\rangle$. Unitary operations on a single qubit that change the state are, most generally, rotations about the X, Y, and Z axes.

In the circuit model of quantum computing, each operation is called a *quantum gate*. The operators are Hermitian and unitary, and hence their own inverse so that $\hat{U}^{\dagger} = \hat{U}^{-1}$. All single-quibit states can be represented on the Bloch sphere. A universal set of single-qubit gates is able to perform any rotation and, in this way, access any state on the Bloch sphere. Because of this, access to any state on the Bloch sphere via a unitary transformation can be reversed so that all quantum gates are *reversible*.

The unitary operation $\hat{U}$ that takes a single-qubit input state $|\psi\rangle = a_0|0\rangle + a_1|1\rangle$ and changes it into an output state $|\psi'\rangle = a_0'|0\rangle + a_1'|1\rangle$ is

$$|\psi'\rangle = \hat{U}|\psi\rangle = \left[\begin{array}{cc} u_{11} & u_{12} \\ u_{21} & u_{22} \end{array} \right] \left[\begin{array}{c} a_0 \\ a_1 \end{array} \right] = \left[\begin{array}{c} a_0' \\ a_1' \end{array} \right].$$
(11.55)

In addition to the unitary single-qubit Pauli gate operators $\hat{U}_X$, $\hat{U}_Y$, $\hat{U}_Z$, there are other commonly used gates. For example, the Hadamard gate,

$$\hat{U}_H = \frac{1}{\sqrt{2}} \left[\begin{array}{cc} 1 & 1 \\ 1 & -1 \end{array} \right],$$
(11.56)

turns the basis states into superposition states.

This may be illustrated by considering the Hadamard gate operating on the state $|0\rangle = \left[\begin{array}{c} 1 \\ 0 \end{array} \right]$, which creates a linear superposition state,

$$\hat{U}_H|\psi\rangle = \frac{1}{\sqrt{2}} \left[\begin{array}{cc} 1 & 1 \\ 1 & -1 \end{array} \right] \left[\begin{array}{c} 1 \\ 0 \end{array} \right] = \frac{1}{\sqrt{2}} \left[\begin{array}{c} 1 \\ 1 \end{array} \right] = \frac{1}{\sqrt{2}} \left(\left[\begin{array}{c} 0 \\ 1 \end{array} \right] + \left[\begin{array}{c} 1 \\ 0 \end{array} \right] \right) = \frac{1}{\sqrt{2}} (|1\rangle + |0\rangle).$$
(11.57)

11.6.3 Multi-Qubit States

Quantum computing makes use of superposition states, entanglement of multi-qubit states, interference, and measurement. A system containing N qubits has 2^N basis states so that a complete simulation of the system requires the solution of a matrix with $2^N \times 2^N$ entries. If N has a large value, then simulation can become difficult using classical computers. For example, if $N = 1\,000$ then the number of matrix elements is greater than 10^{602}. Even if N is as small as 10, the number of matrix elements is $2^{10} \times 2^{10} = 1\,048\,576$.

11.6.4 Two-Qubit States

If $N = 2$ then the system consists of two qubits. The two-qubit states can be written in terms of four basis states that are the product Hilbert space of two single-qubits.

$$|00\rangle = |0\rangle \otimes |0\rangle = \begin{bmatrix} 1 \\ 0 \\ 0 \\ 0 \end{bmatrix}, \tag{11.58}$$

$$|01\rangle = |0\rangle \otimes |1\rangle = \begin{bmatrix} 0 \\ 1 \\ 0 \\ 0 \end{bmatrix}, \tag{11.59}$$

$$|10\rangle = |1\rangle \otimes |0\rangle = \begin{bmatrix} 0 \\ 0 \\ 1 \\ 0 \end{bmatrix}, \tag{11.60}$$

$$|11\rangle = |1\rangle \otimes |1\rangle = \begin{bmatrix} 0 \\ 0 \\ 0 \\ 1 \end{bmatrix}. \tag{11.61}$$

11.6.5 Two-Qubit Superposition States

A general two-qubit state can be written as a superposition of the basis states

$$|\psi\rangle = a_{00}|00\rangle + a_{01}|01\rangle + a_{10}|10\rangle + a_{11}|11\rangle, \tag{11.62}$$

where normalization requires $|a_{00}|^2 + |a_{01}|^2 + |a_{10}|^2 + |a_{11}|^2 = 1$.

As an example, if two single-qubits, one in the *superposition state* $|\psi_\alpha\rangle = \frac{1}{\sqrt{2}}(|0\rangle + |1\rangle)$ and the other in the same superposition state $|\psi_\beta\rangle = \frac{1}{\sqrt{2}}(|0\rangle + |1\rangle)$, do not interact then a product state can be written as

$$|\psi_{\alpha\beta}\rangle = |\psi_\alpha\rangle \otimes |\psi_\beta\rangle = \frac{1}{2}(a_{00}|00\rangle + a_{01}|01\rangle + a_{10}|10\rangle + a_{11}|11\rangle) = \frac{1}{2}\begin{bmatrix} a_{00} \\ a_{01} \\ a_{10} \\ a_{11} \end{bmatrix} = \frac{1}{2}\begin{bmatrix} 1 \\ 1 \\ 1 \\ 1 \end{bmatrix}. \tag{11.63}$$

11.6.6 Two-Qubit Entangled Bell States

If two-qubit states *cannot* be written as a product of the two single-qubit states, then they are *entangled*. For example, $|\psi\rangle = \frac{1}{\sqrt{2}}(|01\rangle - |10\rangle) \neq (a_0|0\rangle + a_1|1\rangle) \otimes (a_0'|0\rangle + a_1'|1\rangle)$. Notice that measurement can destroy entanglement. For example, after measuring the maximally entangled state $|\psi\rangle = \frac{1}{\sqrt{2}}(|01\rangle - |10\rangle)$ and obtaining, say, $|01\rangle = |0\rangle \otimes |1\rangle$, the result is a pure product state.

The particular two-qubit superposition state being considered (Eq. (11.62)) is $|\psi_{\alpha\beta}\rangle = \frac{1}{2}(|00\rangle + |01\rangle + |10\rangle + |11\rangle)$ and this becomes entangled when operated on by the controlled $\hat{Z}$ (CZ) gate, which inverts the sign of a_{11} and leaves the other amplitudes unchanged. The two-qubit $\hat{U}_{CZ}$ unitary matrix operator is

$$\hat{U}_{CZ} = \begin{bmatrix} 1 & 0 & 0 & 0 \\ 0 & 1 & 0 & 0 \\ 0 & 0 & 1 & 0 \\ 0 & 0 & 0 & -1 \end{bmatrix}, \tag{11.64}$$

so that

$$\hat{U}_{CZ}|\psi_{\alpha\beta}\rangle = \frac{1}{2}\begin{bmatrix} 1 & 0 & 0 & 0 \\ 0 & 1 & 0 & 0 \\ 0 & 0 & 1 & 0 \\ 0 & 0 & 0 & -1 \end{bmatrix}\begin{bmatrix} 1 \\ 1 \\ 1 \\ 1 \end{bmatrix} = \frac{1}{2}\begin{bmatrix} 1 \\ 1 \\ 1 \\ -1 \end{bmatrix}. \tag{11.65}$$

Interference using the Hadamard operator,

$$\hat{U}_{1 \otimes H} = \hat{U}_1 \otimes \hat{U}_H = \frac{1}{\sqrt{2}}\begin{bmatrix} 1 & 0 \\ 0 & 1 \end{bmatrix} \otimes \begin{bmatrix} 1 & 1 \\ 1 & -1 \end{bmatrix} = \frac{1}{\sqrt{2}}\begin{bmatrix} 1 & 1 & 0 & 0 \\ 1 & -1 & 0 & 0 \\ 0 & 0 & 1 & 1 \\ 0 & 0 & 1 & -1 \end{bmatrix}, \tag{11.66}$$

has the effect of setting $a_{01} = a_{10} = 0$ and creating a special entangled state,

$$|\psi_{00+11}\rangle = \hat{U}_{1 \otimes H}\left(\hat{U}_{CZ}|\psi_{\alpha\beta}\rangle\right) = \frac{1}{2\sqrt{2}}\begin{bmatrix} 1 & 1 & 0 & 0 \\ 1 & -1 & 0 & 0 \\ 0 & 0 & 1 & 1 \\ 0 & 0 & 1 & -1 \end{bmatrix}\begin{bmatrix} 1 \\ 1 \\ 1 \\ -1 \end{bmatrix} = \frac{1}{\sqrt{2}}\begin{bmatrix} 1 \\ 0 \\ 0 \\ 1 \end{bmatrix}, \tag{11.67}$$

which is the Bell state $|\psi_{00+11}\rangle = \frac{1}{\sqrt{2}}(|00\rangle + |11\rangle)$.

The Bell states are maximally entangled and form an orthonormal basis consisting of the four states

$$|\psi_{00\pm11}\rangle = \frac{1}{\sqrt{2}}(|00\rangle \pm |11\rangle), \tag{11.68}$$

$$|\psi_{01\pm10}\rangle = \frac{1}{\sqrt{2}}(|01\rangle \pm |10\rangle). \tag{11.69}$$

11.6.7 Two-Qubit Controlled Gates

In general, the output of multi-qubit gates are manipulated using operations that have *control* qubits and *target* qubits. The simplest operation acts on a two-qubit state in which the first qubit input is the control and the second qubit input is the target. Only if the first qubit (the control qubit) is 1 does the gate act on the second qubit (the target qubit). A controlled two-qubit unitary gate is a 4×4 matrix,

$$
\hat{U}_{\text{Cgate}} = \begin{bmatrix} 1 & 0 & 0 & 0 \\ 0 & 1 & 0 & 0 \\ 0 & 0 & u_{11} & u_{12} \\ 0 & 0 & u_{21} & u_{22} \end{bmatrix},
\tag{11.70}
$$

where the single-qubit gate, such as $\hat{U}_X$, $\hat{U}_Y$, $\hat{U}_Z$, or $\hat{U}_H$, is the 2×2 unitary matrix

$$
\hat{U}_{\text{gate}} = \begin{bmatrix} u_{11} & u_{12} \\ u_{21} & u_{22} \end{bmatrix}.
\tag{11.71}
$$

For example, setting $\hat{U}_{\text{gate}} = \hat{U}_X$, a two qubit $\hat{U}_{\text{CX}}$ (or controlled NOT, $\hat{U}_{\text{CNOT}}$) gate is

$$
\hat{U}_{\text{CX}} = \begin{bmatrix} 1 & 0 & 0 & 0 \\ 0 & 1 & 0 & 0 \\ 0 & 0 & 0 & 1 \\ 0 & 0 & 1 & 0 \end{bmatrix}.
\tag{11.72}
$$

The symbol and corresponding truth table for the $\hat{U}_{\text{CX}}$ gate with control qubit and target qubit inputs is shown on the left-hand side of Fig. 11.17.

For comparison, the right-hand side of Fig. 11.17 shows the classical exclusive OR (XOR) gate symbol and truth table.

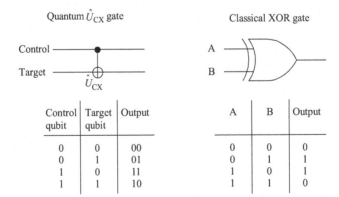

Control qubit	Target qubit	Output		A	B	Output
0	0	00		0	0	0
0	1	01		0	1	1
1	0	11		1	0	1
1	1	10		1	1	0

Fig. 11.17 Quantum $\hat{U}_{\text{CX}}$ gate with control qubit and target qubit inputs and corresponding truth table. The $\hat{U}_{\text{CX}}$ gate has a control and target input. The classical XOR gate is also shown for comparison.

11.6.8 Bell-State Generation

As described in the previous section, starting from a defined initial state it is possible to apply a sequence of operations to create entangled two-qubit Bell states. An example of a quantum circuit that does this for the Bell state $|\psi_{00-11}\rangle = \frac{1}{\sqrt{2}}(|00\rangle - |11\rangle)$ is shown in Fig. 11.18.

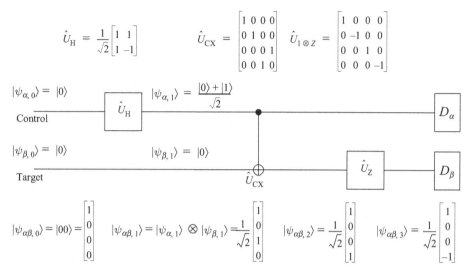

$$\hat{U}_H = \frac{1}{\sqrt{2}}\begin{bmatrix} 1 & 1 \\ 1 & -1 \end{bmatrix} \qquad \hat{U}_{CX} = \begin{bmatrix} 1 & 0 & 0 & 0 \\ 0 & 1 & 0 & 0 \\ 0 & 0 & 0 & 1 \\ 0 & 0 & 1 & 0 \end{bmatrix} \quad \hat{U}_{1 \otimes Z} = \begin{bmatrix} 1 & 0 & 0 & 0 \\ 0 & -1 & 0 & 0 \\ 0 & 0 & 1 & 0 \\ 0 & 0 & 0 & -1 \end{bmatrix}$$

$|\psi_{\alpha,0}\rangle = |0\rangle$ $\hat{U}_H$ $|\psi_{\alpha,1}\rangle = \frac{|0\rangle + |1\rangle}{\sqrt{2}}$ Control D_α

$|\psi_{\beta,0}\rangle = |0\rangle$ $|\psi_{\beta,1}\rangle = |0\rangle$ Target $\hat{U}_{CX}$ $\hat{U}_Z$ D_β

$$|\psi_{\alpha\beta,0}\rangle = |00\rangle = \begin{bmatrix} 1 \\ 0 \\ 0 \\ 0 \end{bmatrix} \quad |\psi_{\alpha\beta,1}\rangle = |\psi_{\alpha,1}\rangle \otimes |\psi_{\beta,1}\rangle = \frac{1}{\sqrt{2}}\begin{bmatrix} 1 \\ 0 \\ 1 \\ 0 \end{bmatrix} \quad |\psi_{\alpha\beta,2}\rangle = \frac{1}{\sqrt{2}}\begin{bmatrix} 1 \\ 0 \\ 0 \\ 1 \end{bmatrix} \quad |\psi_{\alpha\beta,3}\rangle = \frac{1}{\sqrt{2}}\begin{bmatrix} 1 \\ 0 \\ 0 \\ -1 \end{bmatrix}$$

Fig. 11.18 Illustration of Bell state generation starting from control qubit state $|\psi_{\alpha,0}\rangle = |0\rangle$ and target qubit state $|\psi_{\beta,0}\rangle = |0\rangle$. After a sequence of operations that transform the system state, $|\psi_{\alpha\beta,0}\rangle \rightarrow |\psi_{\alpha\beta,1}\rangle \rightarrow |\psi_{\alpha\beta,2}\rangle \rightarrow |\psi_{\alpha\beta,3}\rangle$, detectors D_α and D_β have a probability of measuring each of the four possible outputs of the Bell state $|\psi_{00-11}\rangle = \frac{1}{\sqrt{2}}(|00\rangle - |11\rangle)$ that is $P_{00} = 0.5$, $P_{10} = 0$, $P_{01} = 0$, and $P_{11} = 0.5$.

The system consists of two single qubits that are each initialized in the state $|0\rangle$ so that $|\psi_{\alpha,0}\rangle = |0\rangle$ and $|\psi_{\beta,0}\rangle = |0\rangle$. The state $|0\rangle$ might, for example, be the ground state of a two-level system. The product state of the two noninteracting qubits can be written as $|\psi_{\alpha\beta,0}\rangle = |\psi_{\alpha,0}\rangle \otimes |\psi_{\beta,0}\rangle = |00\rangle$.

For Bell state generation, a Hadamard gate operation, $\hat{U}_H$, is applied to the single qubit $|\psi_{\alpha,0}\rangle = |0\rangle$. This has the effect of creating a single-qubit superposition state $|\psi_{\alpha,1}\rangle = \frac{1}{\sqrt{2}}(|0\rangle + |1\rangle)$. It follows that the input to the $\hat{U}_{CX}$ gate operation is the product state

$$|\psi_{\alpha\beta,1}\rangle = |\psi_{\alpha,1}\rangle \otimes |\psi_{\beta,1}\rangle = \frac{1}{\sqrt{2}}(|0\rangle + |1\rangle) \otimes |0\rangle = \frac{(|00\rangle + |10\rangle)}{\sqrt{2}} = \frac{1}{\sqrt{2}}\begin{bmatrix} 1 \\ 0 \\ 1 \\ 0 \end{bmatrix}. \tag{11.73}$$

The $\hat{U}_{CX}$ gate operation then acts to entangle the two qubits and create an output that is the Bell state $|\psi_{00+11}\rangle = \frac{1}{\sqrt{2}}(|00\rangle + |11\rangle)$. The operation on the $|\psi_{\alpha\beta,1}\rangle = |\psi_{\alpha,1}\rangle \otimes |\psi_{\beta,1}\rangle$ state is

$$\hat{U}_{\text{CX}}\left(|\psi_{\alpha,1}\rangle \otimes |\psi_{\beta,1}\rangle\right) = \frac{1}{\sqrt{2}} \begin{bmatrix} 1 & 0 & 0 & 0 \\ 0 & 1 & 0 & 0 \\ 0 & 0 & 0 & 1 \\ 0 & 0 & 1 & 0 \end{bmatrix} \begin{bmatrix} 1 \\ 0 \\ 1 \\ 0 \end{bmatrix} = \frac{1}{\sqrt{2}} \begin{bmatrix} 1 \\ 0 \\ 0 \\ 1 \end{bmatrix} = \frac{|00\rangle + |11\rangle}{\sqrt{2}} = |\psi_{\alpha\beta,2}\rangle,$$

$$\tag{11.74}$$

so that $|\psi_{\alpha\beta,1}\rangle \to |\psi_{\alpha\beta,2}\rangle$.

The target output of the $\hat{U}_{\text{CX}}$ gate operation is input to a $\hat{U}_Z$ gate whose function is to negate $|11\rangle$. The operation on $|\psi_{\alpha\beta,2}\rangle$ is

$$\hat{U}_{1\otimes Z}|\psi_{\alpha\beta,2}\rangle = \frac{1}{\sqrt{2}} \begin{bmatrix} 1 & 0 & 0 & 0 \\ 0 & -1 & 0 & 0 \\ 0 & 0 & 1 & 0 \\ 0 & 0 & 0 & -1 \end{bmatrix} \begin{bmatrix} 1 \\ 0 \\ 0 \\ 1 \end{bmatrix} = \frac{1}{\sqrt{2}} \begin{bmatrix} 1 \\ 0 \\ 0 \\ -1 \end{bmatrix} = \frac{|00\rangle - |11\rangle}{\sqrt{2}} = |\psi_{\alpha\beta,3}\rangle,$$

$$\tag{11.75}$$

so that $|\psi_{\alpha\beta,2}\rangle \to |\psi_{\alpha\beta,3}\rangle$ and the input to the detectors is the Bell state $|\psi_{\alpha\beta,3}\rangle = \frac{1}{\sqrt{2}}(|00\rangle - |11\rangle) = |\psi_{00-11}\rangle$.

As illustrated in Fig. 11.18, the Bell state can be measured using detectors D_α and D_β. The result of a measurement performed by a perfect detector is a single bit and so either a 1 or a 0. The correlation of the entangled Bell state ensures that if detector D_α measures a 1, then D_β will also register a 1. Likewise, if detector D_α measures a 0, then D_β will also register a 0. The probability of measuring each of the four possible output states of the Bell state $|\psi_{00-11}\rangle = \frac{1}{\sqrt{2}}(|00\rangle - |11\rangle)$ is $P_{00} = 0.5$, $P_{10} = 0$, $P_{01} = 0$, and $P_{11} = 0.5$.

Other Bell states can be generated by inserting or removing gates. For example, removing the $\hat{U}_Z$ gate shown in Fig. 11.18 results in an output that is the Bell state $|\psi_{00+11}\rangle = \frac{1}{\sqrt{2}}(|00\rangle + |11\rangle)$.

11.6.9 Bell's Inequality

Probabilities that are used to describe the outcome of measurements in quantum mechanics cannot be due to local preexisting hidden variables. Here, local means the outcome of an experiment on a system is independent of actions performed on a different system with no causal connection to the first. It is assumed that no causal connection exists between two systems outside of the light cones defined by Einstein's theory of relativity. Preexisting variables (or realism) is a property assigned to the system independently of whether or not the measurement of the property is carried out.

The remarkable fact that local preexisting hidden variables cannot exist in quantum systems may be shown by considering a classical system in which they *do* exist.

Suppose two identical objects have three preexisting measurable properties labeled A, B, and C. The value each property can have is 0 or 1 so that the outcome of a measurement is binary. For example, the two identical objects can have property value $A = 0$, $B = 0$, $C = 1$, and so this outcome could be recorded as $(0, 0, 1)$.

The probability that a pair of properties, say A for the first object and B for the second object, are the same (they both have value 0 or 1) is $P_s(A, B)$ and, of course, $P_s(A, B) = 1 - P_d(A, B)$, where $P_d(A, B)$ is the probability that A and B are different. For the two identical objects, it follows that $P_s(A, A) = P_s(B, B) = P_s(C, C) = 1$.

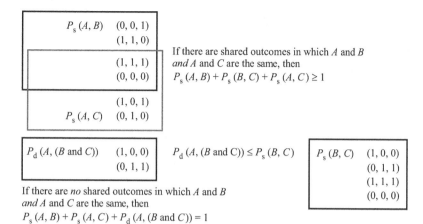

Fig. 11.19 Illustration of the Bell inequality $P_s(A, B) + P_s(B, C) + P_s(A, C) \geq 1$. Two identical objects have three measurable properties labeled A, B, and C. Each measurable property can have a value 0 or 1. If there are no shared outcomes in which both A and B are the same *and* A and C are the same, then the sum $P_s(A, B) + P_s(A, C) + P_d(A, (B \text{ and } C)) = 1$ and, because $P_d(A, (B \text{ and } C)) \leq P_s(B, C)$, the smallest value for $P_s(A, B) + P_s(B, C) + P_s(A, C) = 1$.

For this system, a Bell's inequality is

$$P_s(A, B) + P_s(B, C) + P_s(A, C) \geq 1. \tag{11.76}$$

The number of binary outcomes among three properties is $2^3 = 8$. The eight possible outcomes are $(0, 0, 0)$, $(0, 0, 1)$, $(0, 1, 0)$, $(0, 1, 1)$, $(1, 0, 0)$, $(1, 0, 1)$, $(1, 1, 0)$, and $(1, 1, 1)$.

The outcomes in which A is different from both B and C are $(1, 0, 0)$ and $(0, 1, 1)$. The outcomes in which A and B are the same are $(0, 0, 1)$, $(1, 1, 0)$, $(1, 1, 1)$, and $(0, 0, 0)$. The outcomes in which A and C are the same are $(1, 0, 1)$, $(0, 1, 0)$, $(1, 1, 1)$, and $(0, 0, 0)$. The shared outcomes in which both A and B are the same *and* A and C are the same are $(1, 1, 1)$ and $(0, 0, 0)$. If the probability of such a shared outcome is nonzero, then the sum of probabilities $P_s(A, B) + P_s(B, C) + P_s(A, C) > 1$.

Figure 11.19 is a graphical representation of Bell's inequality.[16] If there are *no* shared outcomes in which both A and B are the same *and* A and C are the same, then the sum $P_s(A, B) + P_s(A, C) + P_d(A, (B \text{ and } C)) = 1$ and, because $P_d(A, (B \text{ and } C)) \leq P_s(B, C)$, then the smallest value for $P_s(A, B) + P_s(B, C) + P_s(A, C) = 1$. If there *are* shared outcomes in which both A and B are the same *and* A and C are the same, then the sum $P_s(A, B) + P_s(B, C) + P_s(A, C) \geq 1$.

Only one example is required to prove that quantum systems violate Bell's inequality. Suppose two qubits with three properties are in the Bell state

$$|\psi_{00+11}\rangle = \frac{|00\rangle + |11\rangle}{\sqrt{2}}, \tag{11.77}$$

and that the three properties associated with each qubit are defined by orthogonal eigenstates such that

---

[16]Closely following the elegant description of L. Maccone, *Am. J. Phys.* **81**, 854 (2013).

$$A : |a_0\rangle = |0\rangle, \tag{11.78}$$
$$|a_1\rangle = |1\rangle,$$

$$B : |b_0\rangle = \frac{|0\rangle + \sqrt{3}|1\rangle}{2}, \tag{11.79}$$

$$|b_1\rangle = \frac{\sqrt{3}|0\rangle - |1\rangle}{2},$$

$$C : |c_0\rangle = \frac{|0\rangle - \sqrt{3}|1\rangle}{2}, \tag{11.80}$$

$$|c_1\rangle = \frac{\sqrt{3}|0\rangle + |1\rangle}{2}.$$

The probability $P_{\mathrm{s}}(A, B)$ is found by expressing the state

$$|\psi_{00+11}\rangle = \frac{|00\rangle + |11\rangle}{\sqrt{2}}$$

in terms of the two qubit states. For example, since for measurable property A and measurable property B, the state $|00\rangle$ may be written as

$$|00\rangle = \frac{|a_0\rangle(|b_0\rangle + \sqrt{3}|b_1\rangle)}{4}$$

and the state $|11\rangle$ as

$$|11\rangle = \frac{|a_1\rangle(\sqrt{3}|b_0\rangle - |b_1\rangle)}{4},$$

then the $|\psi_{00+11}\rangle$ state is

$$|\psi_{00+11}\rangle = \frac{|00\rangle + |11\rangle}{\sqrt{2}} = \frac{|a_0\rangle \left(|b_0\rangle + \sqrt{3}|b_1\rangle\right) + |a_1\rangle \left(\sqrt{3}|b_0\rangle - |b_1\rangle\right)}{2\sqrt{2}}. \tag{11.81}$$

The magnitude squared of the coefficient for $|a_0\rangle|b_0\rangle$ is the probability of measuring both objects with property value 0. In this case the probability is $|\frac{1}{2\sqrt{2}}|^2 = \frac{1}{8}$. The magnitude squared of the coefficient for $|a_1\rangle|b_1\rangle$ is the probability of measuring both objects with property value 1. The result is $|\frac{1}{2\sqrt{2}}|^2 = \frac{1}{8}$. It follows that the probability $P_{\mathrm{s}}(A, B) = \frac{1}{8} + \frac{1}{8} = \frac{1}{4}$.

In the same way, the results are $P_{\mathrm{s}}(A, C) = \frac{1}{4}$ and $P_{\mathrm{s}}(B, C) = \frac{1}{4}$ so that

$$P_{\mathrm{s}}(A, B) + P_{\mathrm{s}}(B, C) + P_{\mathrm{s}}(C, A) = \frac{3}{4} < 1, \tag{11.82}$$

which *violates* Bell's inequality.

If a quantum system violates Bell's inequality, then the outcome of measurements in quantum mechanics cannot be due to local preexisting hidden variables. At a minimum, quantum mechanics excludes either local behavior or hidden variables. The standard *Copenhagen interpretation* of quantum mechanics maintains the locality of Einstein's relativistic causality and excludes preexisting hidden variables. In this view, the properties of a system are only obtained via interaction between the quantum system and the measurement instrument. In stark contrast to our everyday (classical) experience, any measurement of the system does not reveal values of preexisting properties.

The fact that quantum systems violate Bell's inequality has been proven by experiment.[17] The strong, nonclassical correlation that occurs in the measurement of Bell states is a resource that can be exploited to create unique quantum system functionality.

11.6.10 Teleportation

The existence of entangled quantum states, such as the Bell states, enables nonclassical capabilities. An example is teleportation in which a single-qubit state, $|\psi\rangle$, can be destroyed in one location and created in another.

Without using quantum resources it is, in general, not possible for Alice to send a single-qubit $|\psi\rangle = a_0|0\rangle + a_1|1\rangle$ to Bob via a classical communication channel because the qubit amplitudes a_0 and a_1 can, in principle, require infinite precision. However, it is possible to transmit one qubit between Alice and Bob by sending only two classical bits of information *if* Alice and Bob initially share one entangled state.

A protocol to achieve this is illustrated in Fig. 11.20. Alice has a single qubit she wants to transmit to Bob. To do so, she first prepares and then shares the Bell state $|\psi_{00+11}\rangle = \frac{1}{\sqrt{2}}(|00\rangle + |11\rangle)$ with Bob. Alice has the single qubit in state $|\psi\rangle$ and the first qubit of the Bell state and Bob has the second qubit of the Bell state. Initially, their joint state is

$$|\psi_0\rangle = |\psi\rangle|\psi_{00+11}\rangle = \frac{a_0|0\rangle(|00\rangle + |11\rangle) + a_1|1\rangle(|00\rangle + |11\rangle)}{\sqrt{2}}. \tag{11.83}$$

In this equation, and as illustrated in Fig. 11.20, the first two qubits belong to Alice and the third qubit belongs to Bob. Alice performs a $\hat{U}_{\text{CX}}$ on her two qubits and then a Hadamard transformation. The $\hat{U}_{\text{CX}}$ gate performs a NOT on the target Bell state when the control is 1. After the $\hat{U}_{\text{CX}}$ gate, the state of the system is

$$|\psi_1\rangle = \frac{a_0|0\rangle(|00\rangle + |11\rangle) + a_1|1\rangle(|10\rangle + |01\rangle)}{\sqrt{2}}. \tag{11.84}$$

After the Hadamard gate creates a superposition state, the state of the system becomes

$$|\psi_2\rangle = \frac{a_0(|0\rangle + |1\rangle)(|00\rangle + |11\rangle) + a_1(|0\rangle - |1\rangle)(|10\rangle + |01\rangle)}{2}, \tag{11.85}$$

which may be rewritten as[18]

$$|\psi_2\rangle = \frac{1}{2}|00\rangle(a_0|0\rangle + a_1|1\rangle)$$

$$+ \frac{1}{2}|01\rangle(a_0|1\rangle + a_1|0\rangle)$$

$$+ \frac{1}{2}|10\rangle(a_0|0\rangle - a_1|1\rangle)$$

$$+ \frac{1}{2}|11\rangle(a_0|1\rangle - a_1|0\rangle). \tag{11.86}$$

[17]For a review, see J-A. Larsson, *J. Phys. A.* **47**, 424003 (2014).
[18]M. A. Nielsen and I. L. Chuang, *Quantum Computation and Quantum Information*, Cambridge, Cambridge University Press, 2010 (ISBN 9781 107 00217 3), pp. 26–27.

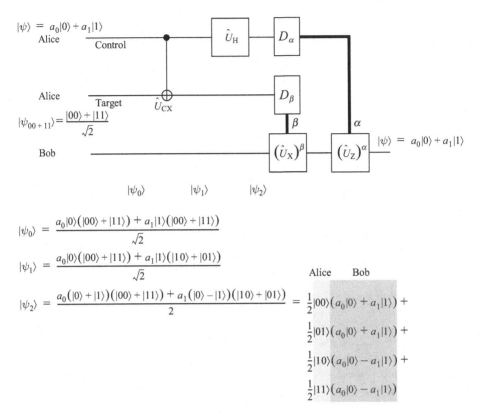

$$|\psi_0\rangle = \frac{a_0|0\rangle(|00\rangle + |11\rangle) + a_1|1\rangle(|00\rangle + |11\rangle)}{\sqrt{2}}$$

$$|\psi_1\rangle = \frac{a_0|0\rangle(|00\rangle + |11\rangle) + a_1|1\rangle(|10\rangle + |01\rangle)}{\sqrt{2}}$$

$$|\psi_2\rangle = \frac{a_0(|0\rangle + |1\rangle)(|00\rangle + |11\rangle) + a_1(|0\rangle - |1\rangle)(|10\rangle + |01\rangle)}{2}$$

Alice   Bob

$$= \frac{1}{2}|00\rangle(a_0|0\rangle + a_1|1\rangle) +$$

$$\frac{1}{2}|01\rangle(a_0|0\rangle + a_1|1\rangle) +$$

$$\frac{1}{2}|10\rangle(a_0|0\rangle - a_1|1\rangle) +$$

$$\frac{1}{2}|11\rangle(a_0|0\rangle - a_1|1\rangle)$$

Fig. 11.20 Illustration of teleportation. Alice has a single qubit, $|\psi\rangle = a_0|0\rangle + a_1|1\rangle$, that she wants to transmit to Bob. To do so, she first prepares and then shares the Bell state $|\psi_{00+11}\rangle = \frac{1}{\sqrt{2}}(|00\rangle + |11\rangle)$ with Bob. Alice has the single qubit in state $|\psi\rangle$ and the first qubit of the Bell state and Bob has the second qubit of the Bell state. The initial state of the system is $|\psi_0\rangle$. Alice performs a $\hat{U}_{CX}$ gate on her two qubits and then a Hadamard gate $\hat{U}_H$. After the $\hat{U}_{CX}$ gate the system is in state $|\psi_1\rangle$ and after the Hadamard operation the system is in state $|\psi_2\rangle$. Alice then measures her two qubits (left-hand column with heading "Alice") and obtains one-bit binary values α and β from detectors D_α and D_β, respectively. These two bits of information are communicated to Bob over a classical channel (thick lines), which he then uses to create the qubit state $|\psi\rangle = a_0|0\rangle + a_1|1\rangle$ at his location.

Alice then performs a measurement on her two qubits using the two detectors that have a one-bit binary output value α for detector D_α and a one-bit binary output value β for detector D_β. Upon completion of the measurement, the two qubits in Alice's possession are in the state $|\alpha\beta\rangle$. The results of the measurement are one of four possibilities and can be transmitted via a classical channel as two classical bits, (00), (01), (10), or (11) to Bob. After receiving the two classical bits α and β, Bob applies gates $\left(\hat{U}_X\right)^\beta$ and then $\left(\hat{U}_Z\right)^\alpha$ to his qubit and in doing so creates the state $|\psi\rangle = a_0|0\rangle + a_1|1\rangle$ at his location. See Eq. (11.86). It is in this way that the state $|\psi\rangle$ is *teleported* from Alice to Bob.

Because Alice is no longer in possession of $|\psi\rangle$, the state has *not* been duplicated and so the no cloning theorem is not violated.

Because Alice transmits the results of her measurement to Bob over a classical channel as two bits of information α and β, teleportation cannot be used to transmit information faster than the speed of light.

11.7 Example Exercises

Exercise 11.1
Find the conduction-band minimum potential energy profile, $V(x)$, of a planar GaAs/Al$_\xi$Ga$_{1-\xi}$As heterostructure tunnel diode with a rectangular potential barrier of energy V_0 and barrier thickness L_b as a function of voltage bias $V_{bias} = 0.4$ V in the depletion approximation.

The carrier concentration in the GaAs electrodes is $n_0 = 1 \times 10^{18}$ cm^{-3} and the relative low-frequency permittivity is $\varepsilon_{r0} = 13.2$. The intrinsic Al$_\xi$Ga$_{1-\xi}$As rectangular potential barrier energy with Al alloy fraction $\xi = 0.36$ is $\Delta E_{CB} = V_0 = \xi \times 0.84$ eV $= 0.3$ eV in the conduction band of the single-crystal heterostructure and the barrier thickness is $L_b = 5$ nm (18 atomic monolayers).

Exercise 11.2
The guided random walk gradient descent is an algorithm that can be used to minimize a cost function. For a given step size, a random direction in parameter space is taken and accepted if it results in a reduced cost. This iterative process is repeated until a minimum in cost is found. For cases where the cost landscape is smooth and slowly varying across successive iterations, it is often a good policy to implement a line search. In this case, the direction of the guided random walk is held fixed if an improvement in the cost is found in a given iteration. The trajectory is then continued until an increase in cost is found, at which point a random direction and increment step are selected that yields a lower cost.

Starting from a random initial configuration, implement a guided random walk with line search to find the minimum in the resonant tunnel diode cost function illustrated in Fig. 11.4. Comment on how the efficiency of the algorithm is expected to change with increasing dimensionality of the search space.

Exercise 11.3
Show that photon reflection and transmission amplitude at an ideal lossless 50:50 dielectric beam splitter can be written as

$$r_{ph} = \frac{-1}{\sqrt{2}}$$

and

$$t_{ph} = \frac{i}{\sqrt{2}},$$

respectively.

Solutions

Solutions 11.1
The rectangular potential barrier of intrinsic Al$_\xi$Ga$_{1-\xi}$As with Al alloy fraction $\xi = 0.36$ has energy $\Delta E_{CB} = V_0 = \xi \times 0.84$ eV $= 0.3$ eV and thickness $L_b = 5$ nm. Carrier

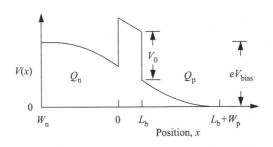

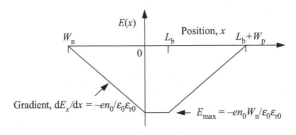

Fig. 11.E21

concentration in the GaAs electrodes, $n_0 = 1 \times 10^{18}\,\mathrm{cm}^{-3}$. When a voltage bias of $V_{\mathrm{bias}} = 0.4\,\mathrm{V}$ is applied, the conduction-band minimum potential profile changes and an accumulation and depletion region form. As shown in Fig. 11.E21, the charge per unit area in the accumulation region is $Q_{\mathrm{n}} = -en_0 W_{\mathrm{n}}$ and in the depletion region it is $Q_{\mathrm{p}} = en_0 W_{\mathrm{p}}$, where the depletion region thickness is $W_{\mathrm{n}} = W_{\mathrm{p}} = W_{\mathrm{np}}$. Charge neutrality requires $Q_{\mathrm{n}} + Q_{\mathrm{p}} = 0$. The depletion thickness, W_{np}, is the same because of the symmetry of the structure and the use of the same doping concentration profile in the left and right contacts.

The gradient of the electric field in the uniformly doped depletion region of the planar structure is

$$\frac{\mathrm{d}E_x}{\mathrm{d}x} = \frac{\rho}{\varepsilon_0 \varepsilon_{\mathrm{r}0}} = \frac{-en_0}{\varepsilon_0 \varepsilon_{\mathrm{r}0}}.$$

The maximum electric field at $x = 0$ is the gradient multiplied by the depletion thickness,

$$E_{\mathrm{max}} = \frac{-en_0 W_{\mathrm{np}}}{\varepsilon_0 \varepsilon_{\mathrm{r}0}}.$$

The electric field in the planar structure is related to the potential via

$$E_x = -\frac{\mathrm{d}}{\mathrm{d}x} V(x),$$

which upon integration from $x = -W_{\mathrm{np}}$ to $x = 0$ gives

$$\int_{-W_{\mathrm{np}}}^{0} E_x \mathrm{d}x = \int_{0}^{E_{\mathrm{max}}} E_x \mathrm{d}x = -V(x=0) = V_{\mathrm{bias}} - \frac{en_0 W_{\mathrm{np}}^2}{2\varepsilon_0 \varepsilon_{\mathrm{r}0}},$$

where, using simple geometry, the integral is

$$\int_{L_b}^{L_b+W_{np}} E_x dx = \frac{E_{max}W_{np}}{2}.$$

Integration from $x = L_b + W_{np}$ to $x = 0$ gives

$$\int_{L_b+W_{np}}^{L_b} E_{max} dx + \int_{L_b}^{0} E_x dx = \frac{en_0 W_{np}^2}{2\varepsilon_0\varepsilon_{r0}} + \frac{en_0 W_{np}L_b}{\varepsilon_0\varepsilon_{r0}} = V_{bias} - \frac{en_0 W_{np}^2}{2\varepsilon_0\varepsilon_{r0}},$$

which is a quadratic equation

$$W_{np}^2 + W_{np}L_b - \frac{\varepsilon_0\varepsilon_{r0}V_{bias}}{en_0} = 0$$

with solution

$$W_{np} = \frac{-L_b}{2} \pm \sqrt{\frac{L_b^2}{4} - \frac{\varepsilon_0\varepsilon_{r0}V_{bias}}{en_0}}.$$

The potential is $\int E_x dx' = -V(x)$, so that

$$V(-W_{np} < x < 0) = V_{bias} - \frac{en_0}{2\varepsilon_0\varepsilon_{r0}}(x + W_{np})^2,$$

$$V(0 \leq x \leq L_b) = \left(V_{bias} - \frac{en_0}{2\varepsilon_0\varepsilon_{r0}}W_{np}^2\right) - \frac{en_0}{\varepsilon_0\varepsilon_{r0}}W_{np}x + V_0,$$

$$V(L_b < x < L_b + W_{np}) = \frac{en_0}{2\varepsilon_0\varepsilon_{r0}}(W_{np} - x + L_b)^2.$$

The resulting potential profile of a GaAs/Al$_\xi$Ga$_{1-\xi}$As tunnel diode in the depletion approximation and for $V_{bias} = 0.4$ V is shown in Fig. 11.E21. For the case considered, relative low-frequency permittivity $\varepsilon_{r0} = 13.2$ and carrier concentration in the GaAs electrodes $n_0 = 1 \times 10^{18}$ cm^{-3}.

Solutions 11.2

Figure 11.E22 shows an example of a trajectory (white path) in the two-dimensional search space of the resonant tunnel diode device optimization problem illustrated in Fig. 11.4.

After 100 iterations, values of the potential energy V_1 and V_2 are very close to the optimal solution that, in this case, is known to be $V_1 = V_2 = 0$ eV.

Figure 11.E23 shows cost on a $\log_{10}$ scale as a function of iteration. Horizontal portions of the curve are iterations that do not result in a reduction of cost. The calculations performed that do not yield a reduction in cost contribute directly to the inefficiency of the search algorithm.

In this example there are just two parameters, V_1 and V_2, and so the optimization problem is two-dimensional. If instead of two potential values there are N, then the problem becomes N-dimensional. The search space becomes larger and the efficiency of the optimization algorithm, as measured by the time it takes to find an optimal solution, will be impacted.

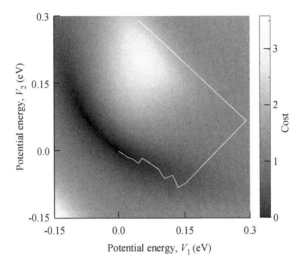

Fig. 11.E22

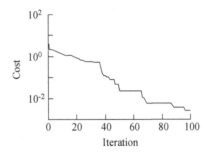

Fig. 11.E23

Solutions 11.3

As indicated in Fig. 11.E24, a lossless beam splitter has input ports 1 and 2 and output ports 3 and 4. Quantum field reflection and transmission amplitudes for a single photon entering port 1 or 2 are $r_{\mathrm{ph},1}$, $r_{\mathrm{ph},2}$, $t_{\mathrm{ph},1}$, and $t_{\mathrm{ph},2}$, respectively.

For a single-photon input state $|n_1 = 1, n_2 = 0\rangle_{\mathrm{in}}$, the quantum field input at port 1 can be set to $a_1 = 1$ and at port 2 it is $a_2 = 0$. The quantum field amplitude output is $a_3 = t_{\mathrm{ph},1}$ at port 3 and $a_4 = r_{\mathrm{ph},1}$ at port 4. The single-photon input state is a product state so that $|1,0\rangle_{\mathrm{in}} = |1\rangle_1 \otimes |0\rangle_2$, where $|0\rangle$ is the vacuum state. For input state $|n_1 = 0, n_2 = 1\rangle_{\mathrm{in}}$, the quantum field output is $a_3 = r_{\mathrm{ph},2}$ at port 3 and $a_4 = t_{\mathrm{ph},2}$ at port 4. The single-photon input and output *quantum field amplitudes* are related via

$$\left[\begin{array}{c} a_3 \\ a_4 \end{array}\right]_{\mathrm{out}} = \left[\begin{array}{cc} t_{\mathrm{ph},1} & r_{\mathrm{ph},2} \\ r_{\mathrm{ph},1} & t_{\mathrm{ph},2} \end{array}\right] \left[\begin{array}{c} a_1 \\ a_2 \end{array}\right]_{\mathrm{in}} = \hat{U}_{\mathrm{B}} \left[\begin{array}{c} a_1 \\ a_2 \end{array}\right]_{\mathrm{in}},$$

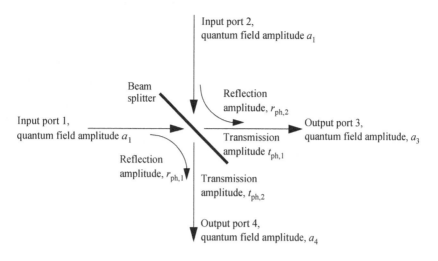

Fig. 11.E24

where $\hat{U}_\mathrm{B}$ is the 2×2 matrix describing the beam splitter.[19]

Photon probability is conserved because a lossless beam splitter is being considered. This means that the matrix $\hat{U}_\mathrm{B}$ is unitary and, consequently, its Hermitian adjoint is its inverse, $\hat{U}_\mathrm{B}^\dagger = \hat{U}_\mathrm{B}^{-1}$. Hence,

$$\hat{U}_\mathrm{B}^\dagger = \begin{bmatrix} t_{\mathrm{ph},1}^* & r_{\mathrm{ph},1}^* \\ r_{\mathrm{ph},2}^* & t_{\mathrm{ph},2}^* \end{bmatrix} = \frac{1}{t_{\mathrm{ph},1}t_{\mathrm{ph},2} - r_{\mathrm{ph},1}r_{\mathrm{ph},2}} \begin{bmatrix} t_{\mathrm{ph},2} & -r_{\mathrm{ph},2} \\ -r_{\mathrm{ph},1} & t_{\mathrm{ph},1} \end{bmatrix} = \hat{U}_\mathrm{B}^{-1}.$$

Because the determinant of a unitary matrix has unit modulus, in general,

$$t_{\mathrm{ph},1}t_{\mathrm{ph},2} - r_{\mathrm{ph},1}r_{\mathrm{ph},2} = e^{i\varphi},$$

where φ is a global phase factor that has no impact on relative phase between matrix elements and may be set to $\varphi = \pi$, so that

$$t_{\mathrm{ph},1}t_{\mathrm{ph},2} - r_{\mathrm{ph},1}r_{\mathrm{ph},2} = -1$$

since $-1 = e^{i\pi}$. Inserting this into the expression for $\hat{U}_\mathrm{B}^\dagger$ gives

$$r_{\mathrm{ph},1} = r_{\mathrm{ph},2}^*$$

and

$$t_{\mathrm{ph},1} = -t_{\mathrm{ph},2}^*,$$

from which it may be concluded that $|r_{\mathrm{ph},1}| = |r_{\mathrm{ph},2}|$ and $|t_{\mathrm{ph},1}| = |t_{\mathrm{ph},2}|$. Re-expressing the complex terms for $r_{\mathrm{ph},1}$ and $t_{\mathrm{ph},1}$ gives

---

[19]For more on the matrix representation of a beam splitter, see, for example, F. Bucholtz and J. M. Singley, *Opt. Eng.* **59**, 120801 (2020); R. A. Campos, B. E. Saleh, and M. C. Teich, *Phys Rev. A* **40**, 1371 (1989); B. Yurke, S. L. McCall, and J. R. Klauder, *Phys Rev. A* **33**, 4033 (1986).

$$|r_{\text{ph},1}|e^{i\theta_{r_{\text{ph},1}}} = |r_{\text{ph},2}|e^{-i\theta_{r_{\text{ph},2}}}$$

and

$$|t_{\text{ph},1}|e^{i\theta_{t_{\text{ph},1}}} = -|t_{\text{ph},2}|e^{-i\theta_{t_{\text{ph},2}}}.$$

Dividing these equations gives

$$\frac{|t_{\text{ph},1}|e^{i\theta_{t_{\text{ph},1}}}}{|r_{\text{ph},1}|e^{i\theta_{r_{\text{ph},1}}}} = \frac{-|t_{\text{ph},2}|e^{-i\theta_{t_{\text{ph},2}}}}{|r_{\text{ph},2}|e^{-i\theta_{r_{\text{ph},2}}}} \rightarrow e^{i\theta_{t_{\text{ph},1}} - i\theta_{r_{\text{ph},1}}} = -e^{-i\theta_{t_{\text{ph},2}} + i\theta_{r_{\text{ph},2}}} = e^{-i\theta_{t_{\text{ph},2}} + i\theta_{r_{\text{ph},2}} + i\pi},$$

where use is made of $-1 = e^{i\pi}$. Hence,

$$(\theta_{t_{\text{ph},1}} - \theta_{r_{\text{ph},1}}) + (\theta_{t_{\text{ph},2}} - \theta_{r_{\text{ph},2}}) = \pi.$$

For the perfect, lossless, symmetric, 50:50 beam splitter $r_{\text{ph},1} = r_{\text{ph},2}$ and $t_{\text{ph},1} = t_{\text{ph},2}$. Therefore, the phase difference between transmission and reflection at each port is the same,

$$(\theta_{t_{\text{ph},1}} - \theta_{r_{\text{ph},1}}) = (\theta_{t_{\text{ph},2}} - \theta_{r_{\text{ph},2}}) = \frac{\pi}{2},$$

and it is clear that the phase of the transmitted field leads the phase of the reflected field by $\pi/2$. For the perfect, lossless, symmetric, 50:50 dielectric beam splitter $|r_{\text{ph},1}| = |r_{\text{ph},2}| = |t_{\text{ph},1}| = |t_{\text{ph},2}|$, which can only be satisfied if $r_{\text{ph},1} = r_{\text{ph},2} = r_{\text{ph}}$ is real and $t_{\text{ph},1} = t_{\text{ph},2} = t_{\text{ph}}$ is pure imaginary. Finally, given the fact that the determinant requires $r_{\text{ph},1}r_{\text{ph},2} - t_{\text{ph},1}t_{\text{ph},2} = 1$, then

$$r_{\text{ph}} = \frac{-1}{\sqrt{2}}$$

and

$$t_{\text{ph}} = \frac{i}{\sqrt{2}},$$

so that for the single photon,

$$\hat{U}_{\text{B}} = \frac{1}{\sqrt{2}}\begin{bmatrix} i & -1 \\ -1 & i \end{bmatrix} = \begin{bmatrix} t_{\text{ph}} & r_{\text{ph}} \\ r_{\text{ph}} & t_{\text{ph}} \end{bmatrix},$$

which satisfies the unitary requirement $\hat{U}_{\text{B}}^{\dagger} = \hat{U}_{\text{B}}^{-1}$.

11.8 Problems

Problem 11.1
Adopting the material parameters and constraints used to generate Fig. 11.2, find a heterostructure diode design that minimizes current density for $0 \leq V_{\text{bias}} \leq 0.2\,\text{V}$ and has current density $J(0.2\,\text{V} < V_{\text{bias}} \leq 0.4\,\text{V}) = (V_{\text{bias}} - 0.2) \times 5 \times 10^{-3}\,\text{A}\,\mu\text{m}^{-2}$. Comment on

the feasibility of the design objective and, in particular, what physical phenomena limits device performance.

Problem 11.2
Given integer n_1 photons at port 1 and integer n_2 photons at port 2 of a perfect, lossless, 50:50 dielectric beam splitter, derive the probability of finding n_3 photons at output port 3 and n_4 photons at output port 4.

Problem 11.3
The quantum field amplitudes at input port 1 and input port 2 of a perfect, lossless, 50:50 dielectric beam splitter are a_1 and a_2, respectively. The quantum field amplitudes at output ports 3 and 4 are a_3 and a_4, respectively. The single-photon input amplitudes are related to the output amplitudes via

$$\begin{bmatrix} a_3 \\ a_4 \end{bmatrix}_{out} = \frac{1}{\sqrt{2}} \begin{bmatrix} i & -1 \\ -1 & i \end{bmatrix} \begin{bmatrix} a_1 \\ a_2 \end{bmatrix}_{in} = \hat{U}_B \begin{bmatrix} a_1 \\ a_2 \end{bmatrix}_{in},$$

where $\hat{U}_B$ is a 2×2 unitary matrix describing the beam splitter.

Find the probability of detecting the photon at output port 3 and the probability of detecting the photon at output port 4 if the single-photon input state is the linear superposition:

(a) $\left| a_1 = \sqrt{\frac{1}{2}},\ a_2 = \sqrt{\frac{1}{2}} \right\rangle$,

(b) $\left| a_1 = i\sqrt{\frac{1}{2}}, a_2 = \sqrt{\frac{1}{2}} \right\rangle$,

(c) $\left| a_1 = \sqrt{\frac{1}{4}},\ a_2 = \sqrt{\frac{3}{4}} \right\rangle$,

(d) $\left| a_1 = i\sqrt{\frac{1}{4}}, a_2 = \sqrt{\frac{3}{4}} \right\rangle$.

Problem 11.4
Show that Eq. (11.85) may be written as Eq. (11.86).

Appendix A   Physical Values

SI-MKS[1, 2]

Speed of light in free space $\qquad c = 2.997\ 924\ 58 \times 10^8\ \mathrm{m\,s}^{-1}$

Planck's constant $\qquad \hbar = 6.582\ 119\ 569 \ldots \times 10^{-16}\ \mathrm{eV\,s}$

$\qquad = 1.054\ 571\ 817 \ldots \times 10^{-34}\ \mathrm{J\,s}$

Electron charge $\qquad e = 1.602\ 176\ 634 \times 10^{-19}\ \mathrm{C}$

Electron mass $\qquad m_0 = 9.109\ 383\ 7015(28) \times 10^{-31}\ \mathrm{kg}$

Neutron mass $\qquad m_\mathrm{n} = 1.674\ 927\ 498\ 04(95) \times 10^{-27}\ \mathrm{kg}$

Proton mass $\qquad m_\mathrm{p} = 1.672\ 621\ 923\ 69(51) \times 10^{-27}\ \mathrm{kg}$

Boltzmann constant $\qquad k_\mathrm{B} = 1.380\ 649 \times 10^{-23}\ \mathrm{J\,K}^{-1}$

Permittivity of free space $\qquad \varepsilon_0 = 8.854\ 187\ 8128(13) \times 10^{-12}\ \mathrm{F\,m}^{-1}$

Permeability of free space $\qquad \mu_0 = 4\pi \times 10^{-7}\ \mathrm{H\,m}^{-1}$

Speed of light in free space $\qquad c = 1/\sqrt{\varepsilon_0 \mu_0}$

Avogadro's number $\qquad N_\mathrm{A} = 6.022\ 140\ 76 \times 10^{23}\ \mathrm{mol}^{-1}$

Bohr radius $\qquad a_\mathrm{B} = 0.529\ 177\ 210\ 903(80) \times 10^{-10}\ \mathrm{m}$

$$= \frac{4\pi\varepsilon_0\hbar^2}{m_0 e^2}$$

Inverse fine-structure constant $\quad \alpha^{-1} = 137.035\ 999\ 084(21)$

$$= \frac{4\pi\varepsilon_0\hbar c}{e^2}$$

A.1   Constants in Quantum Mechanics

The above lists physical values.[1,2] In this book, it has been assumed that Planck's constant, the speed of light in free space, the value of the electron charge, etc., may be treated as constants. Notice, however, that this appendix is titled "Physical Values" and not "Physical Constants." The reason for this is that it is not possible to prove theoretically or experimentally that, for example, Planck's constant does in fact have the same value

---

[1] See http://physics.nist.gov/constants.
[2] The numbers in parentheses are one standard deviation uncertainty in the last two digits.

in all parts of space or that it has been the same over all time. On the contrary, there seems to be some tentative evidence that the values of these constants may have changed with time.[3] However, because of the way SI defines some units of measure and the use of quantum mechanics to find relationships between quantities, some physical values are absolute (there is no uncertainty in their value).

A.2 The MKS and SI Units of Measurement

To ensure uniform standards in commercial transactions involving weights and measures and to facilitate information exchange within the science and technology community, it is useful to agree on a system of units of measurement. The foundation of the MKS and SI measurement system may be traced to the creation of the decimal metric method during the time of the French Revolution (1787–99). An important step in the development of this standardized system of measurement occurred in 1799 when physical objects representing the meter and the kilogram were placed in the Archives de la République in Paris, France. In 1874, the British Association for the Advancement of Science (BAAS) introduced a system of measurement called CGS in which the centimeter is used for measurement of distance, the gram for mass, and the second for time. This was rapidly adopted by the mainstream experimental physical science community. Then, in 1901, Giorgi suggested that if the metric system of meters for distance, kilograms for mass, and seconds for time were used instead of centimeter, gram, and second, a consistent system of electromagnetic units could be developed. This meters, kilograms, seconds approach is called the MKS system, and it was adopted by the International Electrotechnical Commission in 1935.

Of course, there are other units that should also be defined, such as force, energy, and power. Recognizing this need, in 1948 the General Conference on Weights and Measures introduced a number of units, including the newton for force, the joule for energy, and the watt for power. The newton, joule, and watt are named after scientists. The first letter of each name serves as the abbreviation for the unit and is written in uppercase.

Internationally agreed upon units of measure continue to evolve over time. In 1960, the General Conference on Weights and Measures extended the MKS scheme using seven basic units of measure from which all other units may be derived. Called Système Internationale d'Unités, it is commonly known as SI.

The seven basic units of measure are:

m	The meter for length
kg	The kilogram for mass
s	The second for time
A	The ampere for current
cd	The candela for light intensity
mol	The mol for the amount of a substance
K	The kelvin for thermodynamic temperature

The definition of these and derived units often seems a little arbitrary and, in some cases, has changed with time.

The *meter* (symbol m) is the basic unit of length in SI. One meter is equal to approximately 39.37 inches. The meter was first defined by the French Academy of Sciences

---

[3] J. K. Webb, M. T. Murphy, V. V. Flambaum, V. A. Dzuba, J. D. Barrow, C. W. Churchill, J. X. Prochaska, and A. M. Wolfe, *Phys. Rev. Lett.* **87**, 091301/1–4 (2001).

in 1791 as $1/10^7$ of the quadrant of the Earth's circumference running from the North Pole through Paris to the equator. In 1889, the International Bureau of Weights and Measures defined the meter as the distance between two lines on a particular standard bar of 90 percent platinum and 10 percent iridium. In 1960, the definition of the meter changed again and it was defined as being equal to 1 650 763.73 wavelengths of the orange–red line in the spectrum of the krypton-86 atom in a vacuum. The definition of the meter changed yet again in 1983, when the General Conference on Weights and Measures defined the meter as the distance traveled by light in a vacuum in 1/299 792 458 of one second. This definition of the meter in terms of the speed of light has the consequence that the speed of light is an *exact* quantity in SI.

The *kilogram* (symbol kg) is the basic unit of mass in SI. It is equal to the mass of a particular platinum–iridium cylinder kept at the International Bureau of Weights and Measures laboratory (Bureau International des Poids et Mesures, or BIPM) at Sèvres, near Paris. Nowadays, the BIPM kilogram is the only SI unit that is based on a physical object (sometimes called an artifact). The cylinder was supposed to have a mass equal to a cubic decimeter of water at its maximum density. The cylinder was later discovered to be 28 parts per million too large. Unfortunately, since the same mass of water at maximum density defines the liter, one liter is $1000.028\,\mathrm{cm}^3$. Compounding this lack of elegance, in 1964 the General Conference on Weights and Measures redefined the liter to be a cubic decimeter while, at the same time recommending that the unit not be used in work requiring great precision. To make connection to the imperial unit of weight, the pound is now defined as being exactly 0.453 592 37 kg.

The *second* (symbol s) is the basic unit of time in SI. The second was defined as 1/86 400 of the mean solar day, which is the average period of rotation of the Earth on its axis relative to the Sun. In 1956, the International Committee on Weights and Measures defined the ephemeris second as 1/31 556 925.9747 of the duration of the tropical year for 1900. This definition was ratified by the General Conference on Weights and Measures in 1960. In 1964, the International Committee on Weights and Measures suggested a second be defined as 9 192 631 770 periods of radiation from the transition between the two hyperfine levels of the ground state of the cesium-133 atom when unperturbed by external fields. In 1967, this became the official definition of the second in SI.

The *ampere* (symbol A) is the basic unit of electric current in SI. The ampere corresponds to the flow of one coulomb of electric charge per second. A flow of one ampere is produced in a resistance of one ohm by a potential difference of one volt. Since 1948, the ampere has been defined as the constant current that, if maintained in two straight parallel conductors of infinite length of negligible circular cross-section and placed one meter apart in a vacuum, would produce between these conductors a force equal to 2×10^{-7} newtons per meter of length.

The *candela* (symbol cd) is the unit of luminous intensity in SI. The candela replaces the international candle. One candela is 0.982 international candles. One candela is defined as the luminous intensity in a given direction of a source that emits monochromatic radiation of frequency 540×10^{12} Hz and has a radiant intensity in that same direction of 1/683 watt per steradian.

The *mol* (symbol mol) is defined as the amount of substance containing the same number of chemical units (atoms, molecules, ions, electrons, or other specified entities or groups of entities) as exactly 12 grams of carbon-12. The number of units in a mol, also known as Avogadro's number, is $6.022\,141\,76 \times 10^{23}$.

The *kelvin* (symbol K) is the unit for thermodynamic temperature in SI. Absolute zero temperature (0 K) is the temperature at which a thermodynamic system has the lowest energy. It corresponds to -273.16 on the Celsius scale and to -459.67 on the Fahrenheit scale. The kelvin is defined as $1/273.16$ of the triple point of pure water. The triple point of pure water is the temperature at which the liquid, solid, and gaseous forms can be maintained simultaneously.

Another absolute temperature scale used by engineers in the United States of America is the Rankine scale. Although the zero point of the Rankine scale is also absolute zero, each rankine is $5/9$ of the kelvin.

The *newton* (symbol N) is the unit of force in SI. It is defined as that force necessary to provide a mass of one kilogram with an acceleration of one meter per second per second. One newton is equal to a force of 100 000 dynes in the CGS system, or a force of about 0.2248 pound in the foot-pound-second system.

The *joule* (symbol J) is the unit of energy in SI. It is equal to the work done by a force of one newton acting through one meter. It is also equal to one watt-second, which is the energy dissipated in one second by a current of one ampere through a resistance of one ohm. One joule is equal to 10^7 ergs in CGS.

The *watt* (symbol W) is the unit of power in SI. One watt is equal to one joule of work per second, $1/746$ horsepower, or the power dissipated in an electrical conductor carrying one amp over a potential drop of one volt.

A.3 Other Units of Measurement

While SI has been in use for many years, there are still a number of other schemes that may be found in the scientific literature. One of the more common is CGS (centimeter, gram, second). For reference, the list of physical values is given below.

CGS

Speed of light	$c = 2.997\ 924\ 58 \times 10^{10}\ \mathrm{cm\,s^{-1}}$
Planck's constant	$\hbar = 6.582\ 1169\ 569 \ldots \times 10^{-16}\ \mathrm{eV\,s}$
	$= 1.054\ 571\ 817 \ldots \times 10^{-27}\ \mathrm{erg\,s}$
Electron charge	$e = 1.602\ 176\ 634 \times 10^{-12}\ \mathrm{erg}$
Electron mass	$m_0 = 9.109\ 383\ 7015(28) \times 10^{-28}\ \mathrm{g}$
Neutron mass	$m_\mathrm{n} = 1.674\ 927\ 489\ 04(95) \times 10^{-24}\ \mathrm{g}$
Proton mass	$m_\mathrm{p} = 1.672\ 621\ 923\ 69(51) \times 10^{-24}\ \mathrm{g}$
Boltzmann constant	$k_\mathrm{B} = 1.380\ 649 \times 10^{-16}\ \mathrm{erg\,K^{-1}}$
	$= 8.617\ 342 \times 10^{-5}\ \mathrm{V\,K^{-1}}$
Bohr radius	$a_\mathrm{B} = 0.529\ 177\ 210\ 903(80) \times 10^{-8}\ \mathrm{cm}$
	$= \dfrac{\hbar \alpha^{-1}}{m_0 c}$
Inverse fine-structure constant	$\alpha^{-1} = 137.035\ 999\ 084(21)$
	$= \dfrac{\hbar c}{e^2}$

In addition to internationally agreed upon units of measurement, theorists sometimes adopt a shorthand notation in which Planck's constant is unity. On occasions the speed of light is also set to unity.

A.4   Use of Quantum-Mechanical Effects to Define Units of Measure

Over the years there has been a trend to try and use the properties of atoms and quantum-mechanical phenomena to define units of measure. In the 1960s, the second was defined using radiation from the transition between the two hyperfine levels of the ground state of the cesium-133 atom. Such radiation is, of course, quantum mechanical in origin. More recently, the macroscopic quantum phenomena of Josephson tunneling and the quantum Hall effect have been used to precisely define voltage and resistance values, respectively. In 1988, the International Committee on Weights and Measures adopted the Josephson constant, K_J, and the von Klitzing constant, R_K, for electrical measurements. The Josephson constant $(K_J = e/\pi\hbar = 483\,597.848\,4\ldots\,\mathrm{GHz\,V^{-1}})$ relates frequency and voltage by way of the ac Josephson effect. The von Klitzing constant $(R_K = 2\pi\hbar/e^2 = 25\,812.807\,45\ldots\,\Omega)$ is the unit of resistance in the integer quantum Hall effect. This leads directly to an expression for Planck's constant, $\hbar = 2/\pi K_J^2 R_K$.

In the future, it may be possible to use the Einstein expressions for energy $E = mc^2$ and $E = \hbar\omega$ to eliminate the need for a physical BIPM artifact in France to define the kilogram mass in SI. If this approach were used, then Planck's constant would become an exact quantity. The definition of *one kilogram* would define Planck's constant as $\hbar = c^2/\omega$ in SI.

Appendix B   Geometry

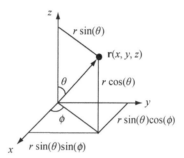

Fig. B.1

The position of a point particle in three-dimensional Euclidean space is given by the vector $\mathbf{r}(x, y, z)$ in Cartesian coordinates, where x, y, and z are measured in the $\mathbf{x}^\sim$, $\mathbf{y}^\sim$, and $\mathbf{z}^\sim$ unit-vector directions, respectively (see Fig. B.1). The Cartesian differential volume element is just $dx\,dy\,dz$. In spherical coordinates, the position vector is $\mathbf{r}(r, \theta, \phi)$, where r, θ, and ϕ are the radial and the two angular coordinates, respectively. The coordinate θ is the polar angle from the z-axis with $0 \le \theta \le \pi$, and ϕ is the azimuthal angle in the $x-y$ plane with $0 \le \phi \le 2\pi$. The differential volume element is $d^3 r = r^2 \sin(\theta)\,d\phi\,d\theta\,dr$. The relationship between the two coordinate systems is given by

$$x = r\sin(\theta)\cos(\phi),$$
$$y = r\sin(\theta)\sin(\phi),$$
$$z = r\cos(\theta).$$

The value of r is related to Cartesian coordinates through

$$r^2 = x^2 + y^2 + z^2.$$

B.2 Useful Trigonometric Relations

Functions of real variable x

$$\sin(x) = \tfrac{1}{2i}(e^{ix} - e^{-ix})$$

$$\cos(x) = \tfrac{1}{2}(e^{ix} + e^{-ix})$$

$$e^{ix} = \cos(x) + i\,\sin(x)$$

$$e^{-ix} = \cos(x) - i\,\sin(x)$$

$$\cos^2(x) + \sin^2(x) = 1$$

$$|e^{ix} - 1| = 4\sin^2(\tfrac{x}{2})$$

$$2\sin(x)\cos(y) = \sin(x+y) + \sin(x-y)$$

$$2\cos(x)\cos(y) = \cos(x+y) + \cos(x-y)$$

$$2\sin(x)\sin(y) = \cos(x-y) - \cos(x+y)$$

$$\sin(x \pm y) = \sin(x)\cos(y) \pm \cos(x)\sin(y)$$

$$\cos(x \pm y) = \cos(x)\cos(y) \mp \sin(x)\sin(y)$$

$$\tan(x \pm y) = \frac{\tan(x) \pm \tan(y)}{1 \mp \tan(x)\tan(y)}$$

$$1 + \tan^2(x) = \sec^2(x) = \frac{1}{\cos^2(x)}$$

$$1 + \cot^2(x) = \csc^2(x) = \frac{1}{\sin^2(x)}$$

$$\operatorname{sech}(z) = \frac{1}{\cosh(z)}$$

Hyperbolic functions of complex variable z

$$\sinh(z) = \tfrac{1}{2}(e^z - e^{-z})$$

$$\cosh(z) = \tfrac{1}{2}(e^z + e^{-z})$$

$$\cosh^2(z) - \sinh^2(z) = 1$$

$$\operatorname{cosech}(z) = \frac{1}{\sinh(z)}$$

For the triangle illustrated in Fig. B.2,

$$\frac{a}{\sin(A)} = \frac{b}{\sin(B)} = \frac{c}{\sin(C)},$$

$$a^2 = b^2 + c^2 - 2bc\,\cos(A),$$

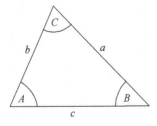

Fig. B.2

where A is the angle opposite side a, etc.

For similar triangles of the type illustrated in Fig. B.3,

$$\frac{a}{b} = \frac{a'}{b'}.$$

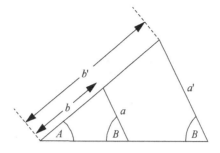

Fig. B.3

And for a circle bisected as shown in Fig. B.4,

$$a^2 = bc.$$

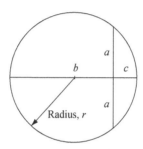

Fig. B.4

B.3   Mensuration

Area of triangle side a, b, c is $\frac{1}{2}bc\sin(A) = \sqrt{s(s-a)(s-b)(s-c)}$ where $2s = a+b+c$

Area of circle radius r is πr^2

Circumference (perimeter) of circle radius r is $2\pi r$

Area of ellipse axes $2a$ and $2b$ is πab

Perimeter of ellipse axes $2a$ and $2b$ is $\sim 2\pi\sqrt{\frac{a^2+b^2}{2}}$

Area of cylinder radius r and height h_0 is $2\pi r(h_0 + r)$

Volume of cylinder radius r and height h_0 is $\pi r^2 h_0$

Area of sphere radius r is $4\pi r^2$

Volume of sphere radius r is $\frac{4}{3}\pi r^3$

Volume of cone or pyramid base area A and height h_0 is $\frac{1}{3}Ah_0$

Appendix C Useful Mathematical Relations

C.1 Factorials and Series Expansions

The **factorial** function is

$$n! = \prod_{k=1}^{k=n} k, \qquad \text{for all integers } n \geq 1,$$

so, for example, $5! = 5 \times 4 \times 3 \times 2 \times 1 = 120$. The factorial $0! = 1$. The number of ways of choosing k distinct objects from a given set of n objects is given by the binomial coefficient

$$\binom{n}{k} = \frac{n!}{k!(n-k)!}.$$

Multifactorials are the product of integers in steps of two $(n!!)$, three $(n!!!)$, and so on. The double factorial $n!!$ is defined recursively for integer n as

$$n!! = 1, \qquad\qquad \text{if } n = 0 \text{ or } n = -1,$$
$$n!! = n(n-2)\dots, \qquad \text{if } n > 0,$$

so, for example, $5!! = 5 \times 3 \times 1 = 15$ and $6!! = 6 \times 4 \times 2 = 48$.

The **Taylor series expansion** for a smooth function $f(x)$ about x_0 is

$$f(x) = \sum_{n=0}^{n=\infty} \frac{1}{n!} f^{(n)}(x_0)(x - x_0)^n,$$

where $f^{(n)}$ is the nth derivative of the function $f(x)$ at the position x_0. A Maclaurin expansion is a special case of the Taylor expansion in which $x_0 = 0$.

Well-known expansions are

$$e^x = 1 + x + \frac{x^2}{2!} + \frac{x^3}{3!} + \cdots = \sum_{n=0}^{n=\infty} \frac{x^n}{n!},$$

$$\frac{1}{1-a^x} = \sum_{n=0}^{n=\infty} a^{nx}, \qquad \text{for } a > 1 \text{ and } x < 0 \text{ or } 0 < a < 1 \text{ and } x > 0,$$

$$\sin(x) = x - \frac{x^3}{3!} + \frac{x^5}{5!} - \cdots = \sum_{n=0}^{n=\infty} \frac{(-1)^n x^{2n+1}}{(2n+1)!},$$

530

$$\cos(x) = 1 - \frac{x^2}{2!} + \frac{x^4}{4!} - \cdots = \sum_{n=0}^{n=\infty} \frac{(-1)^n x^{2n}}{(2n)!},$$

$$\frac{1}{1-x} = 1 + x + x^2 + x^3 + \cdots = \sum_{n=0}^{n=\infty} x^n, \qquad \text{for } |x| < 1,$$

$$\frac{1}{1+x} = 1 - x + x^2 - x^3 + \cdots = \sum_{n=0}^{n=\infty} (-1)^n x^n, \qquad \text{for } |x| < 1,$$

$$\log(1+x) = x - \frac{x^2}{2} + \frac{x^3}{3} - \frac{x^4}{4} + \cdots = \sum_{n=0}^{n=\infty} \frac{(-1)^n x^{n+1}}{n+1}, \qquad \text{for } |x| < 1.$$

The **Binomial series expansion** requires $|x| < 1$ and is

$$(1+x)^n = 1 + nx + \frac{n(n-1)}{2!}x^2 + \frac{n(n-1)(n-2)}{3!}x^3 + \cdots + \frac{n(n-1)(n-2)\cdots 2 \times 1}{n!}x^n,$$

or, in slightly more compact notation,

$$(1+x)^n = 1 + \binom{n}{1}x + \binom{n}{2}x^2 + \binom{n}{3}x^3 + \cdots + \binom{n}{n-1}x^n + \binom{n}{n}x^n.$$

Stirling's formula is $\log(n!) \cong n\log(n) - n$ for $n \gg 1$.

Sum of n natural numbers

$$\sum_{k=1}^{n} k = \frac{n(n+1)}{2}$$

Sum of the first N terms of a geometric series

$$\sum_{n=0}^{n=N-1} ax^n = a\frac{1-x^N}{1-x}$$

C.2 Differentiation

The derivative of the product of two differentiable functions $f(x)$ and $g(x)$ may be expressed as:

$$\frac{d}{dx}(f(x)g(x)) = \left(\frac{d}{dx}(f(x))\right)g(x) + f(x)\left(\frac{d}{dx}g(x)\right),$$

or using the notation where $'$ indicates a derivative, then

$$(fg)' = (f')g + f(g').$$

If the ratio of two differentiable functions $f(x)$ and $g(x)$ takes on the indeterminate form $0/0$ at position $x = x_0$, then it can be shown, using a Taylor expansion, that

$$\left.\frac{f(x)}{g(x)}\right|_{x \to x_0} = \left.\frac{\frac{d}{dx}f(x)}{\frac{d}{dx}g(x)}\right|_{x \to x_0},$$

which is L'Hospital's rule.

C.3 Integration

If an integrand may be written as the product of two functions UV', where $'$ indicates a derivative, it is often useful to consider the method of *integration by parts*. This follows from the product rule for which $(UV)' = U'V + UV'$ so that $UV' = (UV)' - U'V$. Integrating both sides yields

$$\int UV' \, \mathrm{d}x = UV - \int U'V \, \mathrm{d}x.$$

Useful *standard indefinite integrals* include:

$$\int \cos(x) \, \mathrm{d}x = \sin(x),$$

$$\int \sin(x) \, \mathrm{d}x = -\cos(x),$$

$$\int \left(\frac{1}{x}\right) \, \mathrm{d}x = \ln(x),$$

$$\int a^x \, \mathrm{d}x = \frac{a^x}{\ln(a)},$$

$$\int \sinh(x) \, \mathrm{d}x = \cosh(x),$$

$$\int \cosh(x) \, \mathrm{d}x = \sinh(x),$$

$$\int e^{ax} \cos(bx) \, \mathrm{d}x = e^{ax} \frac{(a \, \cos(bx) + b \, \sin(bx))}{a^2 + b^2},$$

$$\int e^{ax} \sin(bx) \, \mathrm{d}x = e^{ax} \frac{(a \, \sin(bx) - b \, \cos(bx))}{a^2 + b^2}.$$

Useful *standard definite integrals* include:

$$\int_0^\infty e^{-ax} \, \mathrm{d}x = \frac{1}{a}, \qquad \text{for } \mathrm{Re}(a > 1),$$

$$\int_0^\infty x^n e^{-\mu x} \, \mathrm{d}x = n! \mu^{-n-1}, \qquad \text{for } \mathrm{Re}(\mu > 1),$$

$$\int_0^\infty x^{2n} e^{-ax^2} \, \mathrm{d}x = \frac{1}{2} \int_{-\infty}^\infty x^{2n} e^{-ax^2} \, \mathrm{d}x = \frac{(2n-1)!!}{2(2a)^n} \sqrt{\frac{\pi}{a}}, \qquad \text{for } a > 0 \text{ and } n = 0, 1, 2, \ldots,$$

$$\int_0^\infty x^{2n+1} e^{-ax^2} \, \mathrm{d}x = \frac{n!}{2a^{n+1}}, \qquad \text{for } a > 0,$$

$$\int_{-\infty}^\infty x^{2n+1} e^{-ax^2} \, \mathrm{d}x = 0,$$

$$\int_0^\infty e^{-ax^2}\,dx = \frac{1}{2}\sqrt{\frac{\pi}{a}},$$

$$\int_0^\infty x e^{-ax^2}\,dx = \frac{1}{2a},$$

$$\int_0^\infty x^2 e^{-ax^2}\,dx = \frac{1}{4}\sqrt{\frac{\pi}{a^3}},$$

$$\int_0^\infty x^3 e^{-ax^2}\,dx = \frac{1}{2a^2},$$

$$\int_0^\infty x^{\nu-1} e^{-\mu x^p}\,dx = \frac{1}{p}\mu^{\frac{-\nu}{p}}\Gamma\left(\frac{\nu}{p}\right), \qquad \text{for } \operatorname{Re}\mu > 0, \operatorname{Re}\nu > 0, p > 0,$$

$$\int_0^\infty \frac{x^3}{e^x - 1}\,dx = \frac{\pi^4}{15},$$

$$\int_0^\infty \frac{x^{p-1}}{e^{rx} - q}\,dx = \frac{1}{qr^p}\Gamma(p)\sum_{k=1}^\infty \frac{q^k}{k^p}, \qquad \text{for } p > 0, r > 0, -1 < q < 1,$$

$$\int_0^\infty \frac{x^{2n-1}}{e^{px} - 1}\,dx = (-1)^{n-1}\left(\frac{2\pi}{p}\right)^{2n}\frac{B_{2n}}{4n}, \qquad \text{for } n = 1, 2, \ldots,$$

where the first few Bernoulli numbers are $B_0 = 1$, $B_1 = -1/2$, $B_2 = 1/6$, $B_4 = -1/30$, $B_6 = 1/42$, $B_8 = -1/30$, such that[1]

$$\zeta(2n) = 2^{2n-1}\frac{\pi^{2n}}{(2n)!}|B_{2n}|,$$

and Riemann's zeta function is defined as

$$\zeta(n) = \sum_{k=1}^{k=\infty} \frac{1}{k^n}.$$

$\operatorname{Re} n > 1$ is required for the series to converge.

C.4 The Dirac Delta Function

In one dimension: $\qquad \int_{-\infty}^\infty \delta(x - x')\,dx = 1 \quad$ and $\quad \delta(x - x') = \int_{-\infty}^\infty \frac{dk}{2\pi} e^{ik(x-x')}$.

In three dimensions: $\qquad \int_{-\infty}^\infty dr^3 \delta(\mathbf{r} - \mathbf{r}') = 1 \quad$ and $\quad \delta(\mathbf{r} - \mathbf{r}') = \int_{-\infty}^\infty \frac{d^3k}{(2\pi)^3} e^{i\mathbf{k}\cdot(\mathbf{r}-\mathbf{r}')}$.

The average value of the integral is assumed in the limit $\pm\infty$.

---

[1] I. S. Gradshteyn and I. M. Ryzhik, *Table of Integrals, Series, and Products*, San Diego, Academic Press, 1980, p. 1074 (ISBN 0 12 294760 6).

Other expressions of the delta function in one dimension are:

$$\delta(x - x') = \int\limits_{0}^{\infty} \frac{1}{\pi} \cos(k(x - x'))\, dk,$$

$$\delta(x - x') = \frac{1}{\pi} \lim_{\eta \to \infty} \frac{\sin(\eta(x - x'))}{x - x'},$$

$$\delta(x - x') = \frac{1}{\pi} \lim_{\eta \to \infty} \frac{\sin^2(\eta(x - x'))}{\eta(x - x')^2},$$

$$\delta(x - x') = \frac{1}{\pi} \lim_{\varepsilon \to 0} \frac{\varepsilon}{(x - x')^2 + \varepsilon^2}.$$

The delta function has the property

$$\delta(ax) = \frac{1}{|a|} \delta x \quad \text{for } a \neq 0 \text{ and} \quad \delta(x^2 - a^2) = \frac{1}{2|a|}(\delta(x + a) + \delta(x - a)) \text{ for } a \neq 0.$$

C.5 Root of a Quadratic Equation

The roots of $ax^2 + bx + c = 0$ are $x = \frac{-b \pm \sqrt{b^2 - 4ac}}{2a}$.

C.6 Fourier Integral

$$F(k) = \frac{1}{\sqrt{2\pi}} \int\limits_{x=-\infty}^{x=\infty} f(x) e^{-ikx}\, dx,$$

Table C.1 Fourier integral pairs

$F(k)$	$f(x)$
$\dfrac{1}{\sqrt{2\pi}}$	$\delta(x)$
$\dfrac{1}{|k|}$	$\dfrac{1}{|x|}$
$\sqrt{2\pi}\,\delta(k + a)$	e^{iax}, Re a
$\dfrac{a\sqrt{2/\pi}}{a^2 + k^2}$	$e^{-a|x|}$, $a > 0$
$\dfrac{\sqrt{\pi/2}\,e^{-a|k|}}{a}$	$\dfrac{1}{a^2 + x^2}$, Re $a > 0$
$\dfrac{e^{-k^2/4a^2}}{a\sqrt{2}}$	$e^{-a^2 x^2}$, $a > 0$
$\sqrt{\dfrac{\pi}{2}}$, for $k < a$, 0 for $|k| > a$	$\dfrac{\sin(ax)}{x}$

$$f(x) = \frac{1}{\sqrt{2\pi}} \int\limits_{k=-\infty}^{k=\infty} F(k)e^{ikx} \, dk,$$

where $f(x)$ satisfies the condition that $\int_{-\infty}^{\infty} |f(x)|^2 \, dx$ is finite.

Examples of the Fourier integral pairs are given in Table C.1. Notice that the Fourier transform of a Gaussian is another Gaussian function.

C.7 Discrete Fourier Transform

Consider a complex function $f(x)$ uniformly sampled at N locations such that $f(x_n)$ is the value at position x_n. The series outside the sampling range is periodic such that $x_n = x_{n+N}$ for all n. The discrete Fourier transform is defined as

$$F(k_j) = \sum_{n=0}^{N-1} f(x_n)e^{-ik_j x_n},$$

for $j = 0, 1, \ldots, N-1$. The inverse transform is

$$f(x_n) = \frac{1}{N} \sum_{j=0}^{N-1} F(k_j)e^{ik_j x_n},$$

for $n = 0, 1, \ldots, N-1$.

C.8 Correlation Functions

$$f(t) = \langle E_1^*(t)E_2(t+\tau) \rangle = \int\limits_{\tau=-\infty}^{\tau=\infty} E_1^*(t)E_2(t+\tau)d\tau \quad \text{and} \quad F(\omega) = E_1^*(\omega)E_2(\omega),$$

$$g^{(1)}(\tau) = \frac{\langle E^*(t)E(t+\tau) \rangle}{\langle E^*(t)E(t) \rangle},$$

$$g^{(2)}(\tau) = \frac{\langle E^*(t)E^*(t+\tau)E(t+\tau)E(t) \rangle}{\langle E^*(t)E(t) \rangle^2}.$$

Appendix D   Matrices

D.1   Matrices and Determinants

A rectangular array of real or complex numbers of the form

$$
\mathbf{A} = \begin{bmatrix} a_{11} & a_{12} & a_{13} & \cdots & a_{1N} \\ a_{21} & a_{22} & a_{23} & & a_{2N} \\ \vdots & & & & \vdots \\ a_{M1} & \cdot & \cdot & \cdot & a_{MN} \end{bmatrix} = [a_{ij}]
$$

is a matrix. Matrix $\mathbf{A}$ is square if $M = N$. A horizontal line of numbers is called a row or row vector and a vertical line of numbers is called a column or column vector. The $M \times N$ matrix has elements a_{ij} in which the first subscript denotes the row and the second subscript denotes the column.

The transpose $\mathbf{A}^{\mathrm{T}}$ of matrix $\mathbf{A}$ is obtained by interchanging the rows and columns:

$$
\mathbf{A}^{\mathrm{T}} = \begin{bmatrix} a_{11} & a_{21} & a_{31} & \cdots & a_{M1} \\ a_{12} & a_{22} & a_{32} & & \\ \vdots & & & & \vdots \\ a_{1N} & \cdot & \cdot & \cdot & a_{MN} \end{bmatrix} = [a_{ji}].
$$

A real square matrix is symmetric if it is equal to its transpose, so that $\mathbf{A} = \mathbf{A}^{\mathrm{T}}$. A real square matrix is skew symmetric if $\mathbf{A} = -\mathbf{A}^{\mathrm{T}}$, in which case the elements $a_{ij} = -a_{ji}$ and $a_{ii} = 0$.

Multiplication of an $M \times N$ matrix $\mathbf{A}$ with an $R \times P$ matrix $\mathbf{B}$ is only defined when $R = N$. The resulting $M \times P$ matrix $\mathbf{C}$ consists of elements $c_{ik} = \sum_{i=1}^{N} a_{ji}b_{ik}$ that is the (inner) product of the jth row vector of the matrix $\mathbf{A}$ and the kth column vector of the matrix $\mathbf{B}$. Matrix multiplication is associative and distributive, so that

$\mathbf{A}(\mathbf{B}\mathbf{C}) = (\mathbf{A}\mathbf{B})\mathbf{C},$
$(\mathbf{A} + \mathbf{B})\mathbf{C} = \mathbf{A}\mathbf{C} + \mathbf{B}\mathbf{C},$

but is not, in general, commutative. Hence, in general,

$\mathbf{A}\mathbf{B} \neq \mathbf{B}\mathbf{A}.$

536

Further, $\mathbf{AB} = 0$ does not require $\mathbf{A} = 0$ or $\mathbf{B} = 0$. So, for example $\mathbf{AB} = 0$ does not imply $\mathbf{BA} = 0$.

The determinant of a 2×2 matrix is $|\mathbf{A}| = \det(\mathbf{A}) = a_{11}a_{22} - a_{12}a_{21}$.

The inverse of matrix $\mathbf{A} = \begin{bmatrix} a_{11} & a_{12} \\ a_{21} & a_{22} \end{bmatrix}$ is $\mathbf{A}^{-1} = \frac{1}{|\mathbf{A}|} \begin{bmatrix} a_{22} & -a_{12} \\ -a_{21} & a_{11} \end{bmatrix}$, so that $\mathbf{AA}^{-1} = 1$, where $\mathbf{1}$ is the unit or identity matrix.

For a 3×3 matrix, $\mathbf{A} = \begin{bmatrix} a_{11} & a_{12} & a_{13} \\ a_{21} & a_{22} & a_{23} \\ a_{31} & a_{32} & a_{33} \end{bmatrix}$. Expanding along the first column gives

$$|\mathbf{A}| = a_{11} \begin{vmatrix} a_{22} & a_{23} \\ a_{32} & a_{33} \end{vmatrix} - a_{21} \begin{vmatrix} a_{12} & a_{13} \\ a_{32} & a_{33} \end{vmatrix} + a_{31} \begin{vmatrix} a_{12} & a_{13} \\ a_{22} & a_{23} \end{vmatrix} = a_{11}M_{11} - a_{21}M_{21} + a_{31}M_{31},$$

where M_{ik} is the minor of the element a_{ik}.

In general, the determinant of an $N \times N$ matrix $\mathbf{A} = \sum_{i,k}^{N} a_{ik}$ is $|\mathbf{A}| = \sum_{k=1}^{N} (-1)^{i+k} a_{ik} M_{ik}$ where $i = 1, 2, \ldots, N$. The inverse is $\mathbf{A}^{-1} = \frac{1}{|\mathbf{A}|} \left[\sum_{i,k}^{N} (-1)^{i+k} M_{ik} \right]$, where $(-1)^{i+k} M_{ki}$ is the cofactor.

Appendix E   Vector Calculus and Maxwell's Equations

E.1   Vector Calculus

In **Cartesian coordinates** (x, y, z):

$$\nabla V = \mathbf{\tilde{x}}\frac{\partial V}{\partial x} + \mathbf{\tilde{y}}\frac{\partial V}{\partial y} + \mathbf{\tilde{z}}\frac{\partial V}{\partial z},$$

$$\nabla \cdot \mathbf{A} = \frac{\partial A_x}{\partial x} + \frac{\partial A_y}{\partial y} + \frac{\partial A_z}{\partial z},$$

$$\nabla \times \mathbf{A} = \begin{bmatrix} \mathbf{\tilde{x}} & \mathbf{\tilde{y}} & \mathbf{\tilde{z}} \\ \frac{\partial}{\partial x} & \frac{\partial}{\partial y} & \frac{\partial}{\partial z} \\ A_x & A_y & A_z \end{bmatrix} = \mathbf{\tilde{x}}\left(\frac{\partial A_z}{\partial y} - \frac{\partial A_y}{\partial z}\right) + \mathbf{\tilde{y}}\left(\frac{\partial A_x}{\partial z} - \frac{\partial A_z}{\partial x}\right) + \mathbf{\tilde{z}}\left(\frac{\partial A_y}{\partial x} - \frac{\partial A_x}{\partial y}\right),$$

$$\nabla^2 V = \frac{\mathrm{d}^2 V}{\mathrm{d}x^2} + \frac{\mathrm{d}^2 V}{\mathrm{d}y^2} + \frac{\mathrm{d}^2 V}{\mathrm{d}z^2},$$

where $\mathbf{\tilde{x}}, \mathbf{\tilde{y}}$, and $\mathbf{\tilde{z}}$ are unit vectors in the x, y, and z directions, respectively.

In **Spherical coordinates** $(\mathrm{r}, \theta, \phi)$:

$$\nabla V = \mathbf{\tilde{r}}\frac{\partial V}{\partial x} + \boldsymbol{\tilde{\theta}}\frac{1}{r}\frac{\partial}{\partial \theta}(V) + \boldsymbol{\tilde{\phi}}\frac{1}{r\sin(\theta)}\frac{\partial}{\partial \phi}(V),$$

$$\nabla \cdot \mathbf{A} = \frac{1}{r^2}\frac{\partial}{\partial r}\left(r^2 A_r\right) + \frac{1}{r\sin(\theta)}\frac{\partial}{\partial \theta}(A_\theta \sin(\theta)) + \frac{1}{r\sin(\theta)}\frac{\partial A_\phi}{\partial \phi},$$

$$\nabla \times \mathbf{A} = \frac{1}{r^2 \sin(\theta)} \begin{vmatrix} \mathbf{\tilde{r}} & r\boldsymbol{\tilde{\theta}} & r\sin(\theta)\boldsymbol{\tilde{\phi}} \\ \frac{\partial}{\partial r} & \frac{\partial}{\partial \theta} & \frac{\partial}{\partial \phi} \\ A_r & rA_\theta & r\sin(\theta)A_\phi \end{vmatrix},$$

$$\nabla \times \mathbf{A} = \frac{\mathbf{\tilde{r}}}{r\sin(\theta)}\left(\frac{\partial}{\partial \theta}A_\phi \sin(\theta) - \frac{\partial A_\theta}{\partial \phi}\right) + \frac{\boldsymbol{\tilde{\theta}}}{r}\left(\frac{1}{\sin(\theta)}\frac{\partial A_r}{\partial \phi} - \frac{\partial}{\partial r}(rA_\phi)\right) + \frac{\boldsymbol{\tilde{\phi}}}{r}\left(\frac{\partial}{\partial r}(rA_\theta) - \frac{\partial A_r}{\partial \theta}\right),$$

$$\nabla^2 V = \frac{1}{r^2}\frac{\partial}{\partial r}\left(r^2\frac{\partial V}{\partial r}\right) + \frac{1}{r^2 \sin(\theta)}\frac{\partial}{\partial \theta}\left(\sin(\theta)\frac{\partial V}{\partial \theta}\right) + \frac{1}{r^2 \sin^2(\theta)}\frac{\partial^2 V}{\partial \phi^2},$$

where the first term on the right-hand side,

$$\frac{1}{r^2}\frac{\partial}{\partial r}\left(r^2\frac{\partial V}{\partial r}\right) \equiv \frac{1}{r}\frac{\mathrm{d}^2}{\mathrm{d}r^2}(rV).$$

Useful vector relationships for the vector fields $\mathbf{a}, \mathbf{b}$, and $\mathbf{c}$ are

$$\nabla \cdot (\nabla \times \mathbf{a}) = 0,$$
$$\nabla \times \nabla \times \mathbf{a} = \nabla(\nabla \cdot \mathbf{a}) - \nabla^2 \mathbf{a},$$
$$\nabla \cdot (\mathbf{a} \times \mathbf{b}) = \mathbf{b} \cdot (\nabla \times \mathbf{a}) - \mathbf{a} \cdot (\nabla \times \mathbf{b}),$$
$$\mathbf{a} \times (\mathbf{b} \times \mathbf{c}) = (\mathbf{a} \cdot \mathbf{c})\mathbf{b} - (\mathbf{a} \cdot \mathbf{b})\mathbf{c},$$
$$\mathbf{a} \cdot \mathbf{b} \times \mathbf{c} = \mathbf{b} \cdot \mathbf{c} \times \mathbf{a} = \mathbf{c} \cdot \mathbf{a} \times \mathbf{b}.$$

Other useful relations in vector calculus are the **divergence theorem**, which relates volume and surface integrals,

$$\int_{V_{\text{vol}}} \nabla \cdot \mathbf{a}\, d^3r = \int_{S_{\text{surf}}} \mathbf{a} \cdot \mathbf{n}^{\sim} dS_{\text{surf}},$$

where $\mathbf{n}^{\sim}$ is the unit normal vector to the surface S_{surf}. **Stokes's theorem** relates surface and line integrals:

$$\int_{V_{\text{vol}}} (\nabla \times \mathbf{a}) \cdot \mathbf{n}^{\sim} dS_{\text{surf}} = \oint_{C_{\text{loop}}} \mathbf{a} \cdot d\mathbf{l},$$

where $d\mathbf{l}$ is the vector line element on the closed loop C_{loop}.

E.2 Maxwell's Equations
In SI-MKS units:

$$\nabla \cdot \mathbf{D} = \rho \qquad \text{Coulomb's law}$$

$$\nabla \cdot \mathbf{B} = 0 \qquad \text{No magnetic monopoles}$$

$$\nabla \times \mathbf{E} = -\frac{\partial \mathbf{B}}{\partial t} \qquad \text{Faraday's law}$$

$$\nabla \times \mathbf{H} = \mathbf{J} + \frac{\partial \mathbf{D}}{\partial t} \qquad \text{Modified Ampere's law}$$

Current continuity requires that $\nabla \cdot \mathbf{J} + \frac{\partial \rho}{\partial t} = 0$, and in these equations for linear media, $\mathbf{B} = \mu \mathbf{H}$ and $\mathbf{D} = \varepsilon \mathbf{E} = \varepsilon_0(1 + \chi_e)\mathbf{E} = \varepsilon_0 \mathbf{E} + \mathbf{P}$.

In SI units, the permittivity of free space is $\varepsilon_0 = 8.854\,187\,8 \times 10^{-12}\,\text{F m}^{-1}$ exactly, and the permeability of free space is $\mu_0 = 4\pi \times 10^{-7}\,\text{H m}^{-1}$.

In Gaussian or CGS units, Maxwell's equations take on a different form. In this case,

$$\nabla \cdot \mathbf{D} = 4\pi\rho, \text{ where } \mathbf{D} = \varepsilon \mathbf{E} = (1 + 4\pi\chi_e)\mathbf{E} = \mathbf{E} + 4\pi\mathbf{P},$$

$$\nabla \cdot \mathbf{B} = 0,$$

$$\nabla \times \mathbf{E} = -\frac{1}{c}\frac{\partial \mathbf{B}}{\partial t},$$

$$\nabla \times \mathbf{H} = \frac{4\pi}{c}\mathbf{J} + \frac{1}{c}\frac{\partial \mathbf{D}}{\partial t}.$$

The way to convert Maxwell's equations from CGS to SI-MKS units is to use the conversion factors in Table E.1.

Table E.1 Conversion factors

Quantity	CGS	SI
Electric field	$\mathbf{E}$	$(4\pi\varepsilon_0)^{1/2}\,\mathbf{E}$
Displacement vector field	$\mathbf{D}$	$\left(\dfrac{4\pi}{\varepsilon_0}\right)^{1/2}\mathbf{D}$
Magnetic flux density	$\mathbf{B}$	$\left(\dfrac{4\pi}{\mu_0}\right)^{1/2}\mathbf{B}$
Magnetic field vector	$\mathbf{H}$	$(4\pi\mu_0)^{1/2}\,\mathbf{H}$
Charge density	ρ_e	$\left(\dfrac{1}{4\pi\varepsilon_0}\right)^{1/2}\rho_e$
Current density	$\mathbf{J}_e$	$\left(\dfrac{1}{4\pi\varepsilon_0}\right)^{1/2}\mathbf{J}_e$
Electrical conductivity	σ_e	$\left(\dfrac{1}{4\pi\varepsilon_0}\right)^{1/2}\sigma_e$
Speed of light	c	$\left(\dfrac{1}{\varepsilon_0\mu_0}\right)^{1/2}$

Appendix F   The Greek Alphabet

F.1   The Greek Alphabet

A	α	alpha	= a
B	β	beta	= b
Γ	γ	gamma	= g
Δ	δ	delta	= d
E	ε	epsilon	= e
Z	ζ	zeta	= z
H	η	eta	= e
Θ	θ	theta	= th(th)
I	ι	iota	= i
K	κ	kappa	= k
Λ	λ	lambda	= l
M	μ	mu	= m
N	ν	nu	= n
Ξ	ξ	xi	= x(ks)
O	o	omicron	= o
Π	π	pi	= p
P	ρ	rho	= r
Σ	σ	sigma	= s
T	τ	tau	= t
Y	υ	upsilon	= u
Φ	ϕ	phi	= ph(f)
X	χ	chi	= kh(hh)
Ψ	ψ	psi	= ps
Ω	ω	omega	= ō

Appendix G   Crystal Structure

G.1   Solids Classified by Atomic Arrangement

Solids are made of atoms. As illustrated in Fig. G.1, there are different ways of arranging atoms spatially to form a solid. Atoms in a crystalline solid, for example, are arranged in a spatially periodic fashion. Because this periodicity extends over many periods, crystalline solids are said to be characterized by long-range spatial order. The position of atoms that form amorphous solids do not have long-range spatial order. Polycrystalline solids are made up of small regions of crystalline material with boundaries that break the spatial periodicity of atom positions.

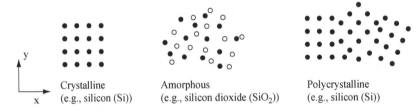

Crystalline
(e.g., silicon (Si))

Amorphous
(e.g., silicon dioxide (SiO_2))

Polycrystalline
(e.g., silicon (Si))

Fig. G.1 Different types of solids according to atomic arrangement. In the figure, a dot represents the position of an atom.

Atoms in a crystalline solid are located in space on a *lattice*. The *unit cell* is a lattice volume representative of the entire lattice; it is repeated throughout the crystal and entirely fills the crystal volume. The smallest unit cell that can be used to form the lattice is called a *primitive cell*.

G.1.1   Two-Dimensional Square Lattice

Figure G.2 shows a square lattice of atoms and a square unit cell translated by vector **R**. In general, the unit vectors $\mathbf{a}_j$ do not have to be in the **x** and **y** directions.

G.1.2   Three-Dimensional Crystals

Three-dimensional crystals are made up of a periodic array of atoms. For a given lattice there exists a basic unit cell that can be defined by the three vectors $\mathbf{a}_1$, $\mathbf{a}_2$, and $\mathbf{a}_3$. Crystal structure is defined as a real-space translation of basic points throughout space via

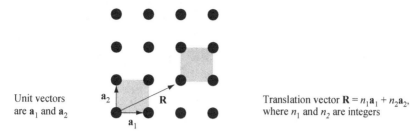

Fig. G.2 A two-dimensional square lattice can be created by translating the unit vectors $\mathbf{a}_1$ and $\mathbf{a}_2$ through space according to $\mathbf{R} = n_1\mathbf{a}_1 + n_2\mathbf{a}_2$, where n_1 and n_2 are integers.

$$\boxed{\mathbf{R} = n_1\mathbf{a}_1 + n_2\mathbf{a}_2 + n_3\mathbf{a}_3,} \tag{G.1}$$

where n_1, n_2, and n_3 are integers. This complete real-space lattice is called the Bravais lattice. The volume of the basic unit cell (the primitive cell) is

$$\Omega_{\text{cell}} = \mathbf{a}_1 \cdot (\mathbf{a}_2 \times \mathbf{a}_3). \tag{G.2}$$

A good choice for the vectors $\mathbf{a}_1, \mathbf{a}_2$, and $\mathbf{a}_3$ that defines the primitive unit cell is due to Wigner and Seitz. The Wigner–Seitz cell about a lattice reference point is specified in such a way that any point of the cell is closer to that lattice point than any other. The Wigner–Seitz cell may be found by bisecting with perpendicular planes all vectors connecting a reference atom position to all atom positions in the crystal. The smallest volume enclosed is the Wigner–Seitz cell.

G.1.3 Cubic Lattices in Three Dimensions

Possibly the simplest three-dimensional lattice to visualize is one in which the unit cell is cubic. In Fig. G.3, L is the *lattice constant*.

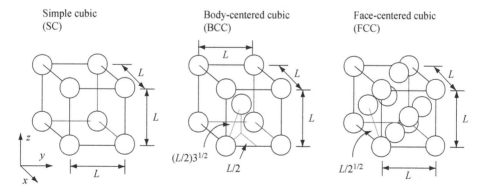

Fig. G.3 Illustration of the indicated three-dimensional cubic unit cells, each of lattice constant L. In the figure, each sphere represents an atom.

543

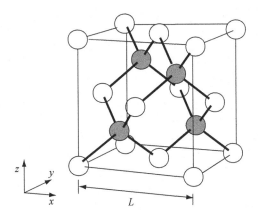

Fig. G.4 The diamond lattice cubic unit cell with lattice constant L. The tetrahedrally coordinated nearest-neighbor bonds are shown as thick lines. GaAs is an example of a III–V compound semiconductor with the zinc-blende crystal structure. This is the same as diamond, except that instead of one atom type populating the lattice, the atom type alternates between Ga (dark spheres) and As (white spheres).

The simple cubic (SC) unit cell has an atom located on each corner of a cube of side L. The body-centered cubic (BCC) unit cell is the same as the SC but with an additional atom in the center of the cube. Elements with the BCC crystal structure include Fe ($L = 0.287\,\mathrm{nm}$), Cr ($L = 0.288\,\mathrm{nm}$), and W ($L = 0.316\,\mathrm{nm}$). The face-centered cubic (FCC) unit cell is the same as the SC but with an additional atom on each face of the cube. Elements with the FCC crystal structure include Al ($L = 0.405\,\mathrm{nm}$), Ni ($L = 0.352\,\mathrm{nm}$), Au ($L = 0.408\,\mathrm{nm}$), and Cu ($L = 0.361\,\mathrm{nm}$). The diamond crystal structure shown in Fig. G.4 consists of two interpenetrating FCC lattices off-set from each other by $L(\mathbf{x}^{\sim} + \mathbf{y}^{\sim} + \mathbf{z}^{\sim})/4$. Elements with the *diamond crystal structure* include Si ($L = 0.543\,\mathrm{nm}$), Ge ($L = 0.566\,\mathrm{nm}$), and C ($L = 0.357\,\mathrm{nm}$). The compound GaAs has the *zinc-blende crystal structure*, which is the same as the diamond lattice except that the atom type alternates between Ga and As. Group III–V compound semiconductors with the zinc-blende crystal structure include GaAs ($L = 0.565\,\mathrm{nm}$), AlAs ($L = 0.566\,\mathrm{nm}$), AlGaAs, InP ($L = 0.587\,\mathrm{nm}$), InAs ($L = 0.606\,\mathrm{nm}$), InGaAs, and InGaAsP.

G.1.4 The Reciprocal Lattice

Because the properties of crystals are often studied using wave-scattering experiments, it is important to consider the reciprocal lattice that exists in reciprocal space (also known as wave vector space or k-space). Given the basic unit cell defined by the vectors $\mathbf{a}_1, \mathbf{a}_2$, and $\mathbf{a}_3$ in real space, it is possible to construct three fundamental reciprocal vectors, $\mathbf{g}_1, \mathbf{g}_2$, and $\mathbf{g}_3$, in reciprocal space defined by $\mathbf{a}_i \cdot \mathbf{g}_j = 2\pi \delta_{ij}$, so that $\mathbf{g}_1 = 2\pi(\mathbf{a}_2 \times \mathbf{a}_3)/\Omega_{\mathrm{cell}}, \mathbf{g}_2 = 2\pi(\mathbf{a}_3 \times \mathbf{a}_1)/\Omega_{\mathrm{cell}}$, and $\mathbf{g}_3 = 2\pi(\mathbf{a}_1 \times \mathbf{a}_2)/\Omega_{\mathrm{cell}}$.

Crystal structure may be defined as a reciprocal-space translation of basic points throughout the space, in which

$$\boxed{\mathbf{G} = n_1 \mathbf{g}_1 + n_2 \mathbf{g}_2 + n_3 \mathbf{g}_3,} \tag{G.3}$$

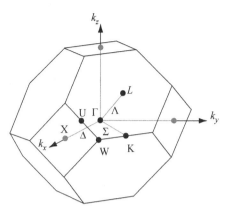

Fig. G.5 The Brillouin zone for the face-centered cubic lattice with lattice constant L. Some high-symmetry points are $\Gamma = (0, 0, 0)$, $X = (2\pi/L)(1, 0, 0)$, $L = (2\pi/L)(0.5, 0.5, 0.5)$, $W = (2\pi/L)(1, 0.5, 0)$. The high-symmetry line between the points Γ and X is labeled Δ, the line between the points Γ and L is Λ, and the line between Γ and K is Σ.

where n_1, n_2, and n_3 are integers. The complete space spanned by **G** is called the *reciprocal lattice*. The volume of the three-dimensional reciprocal-space unit cell is

$$\Omega_k = \mathbf{g}_1 \cdot (\mathbf{g}_2 \times \mathbf{g}_3) = \frac{(2\pi)^3}{\Omega_{\text{cell}}}. \tag{G.4}$$

The *Brillouin zone* of the reciprocal lattice has the same definition as the Wigner–Seitz cell in real space. The first Brillouin zone may be found by bisecting with perpendicular planes all reciprocal-lattice vectors. The smallest volume enclosed is the first Brillouin zone.

To find the basic reciprocal lattice vectors for a face-centered cubic lattice, note that the basic unit cell vectors in real space are $\mathbf{a}_1 = (0, 1, 1)(L/2)$, $\mathbf{a}_2 = (1, 0, 1)(L/2)$, and $\mathbf{a}_3 = (1, 1, 0)(L/2)$, so that $\mathbf{g}_1 = 2\pi(-1, 1, 1)/L$, $\mathbf{g}_2 = 2\pi(1, -1, 1)/L$, and $\mathbf{g}_3 = 2\pi(1, 1, -1)/L$. Hence, the reciprocal lattice of a face-centered cubic lattice in real space is a body-centered cubic lattice.

Figure G.5 is an illustration of the first Brillouin zone for the face-centered cubic lattice with lattice constant L. As may be seen, in this case the Brillouin zone is a truncated octahedron.

Appendix H   Classical Mechanics and Classical Electromagnetism

H.1   Motivation

Before learning about quantum mechanics, it is worth spending a little time and effort reviewing some key elements of classical mechanics and classical electrodynamics.

H.2   Classical Mechanics

H.2.1   Introduction

Classical mechanics is often used to predict the motion of large (macroscopic) objects. This could be a very difficult task since such large objects tend to have many degrees of freedom[1] and so, in principle, could be described by a corresponding number of parameters. The remarkable success of classical mechanics is due to the fact that powerful concepts can be exploited to simplify the way systems are modeled. Various constraints may be used to reduce the description of motion to a simple set of differential equations. Assuming an object is rigid is a constraint and constants of the motion often include conservation of energy and momentum.[2]

Consider a rock dropped from a significant height. Classical mechanics initially ignores the internal degrees of freedom of the rock (it is assumed to be rigid), but instead defines a center of mass so that the rock can be described as a point particle of mass, m. Angular momentum is decoupled from the center of mass motion. Why is this all possible? The answer is neither simple nor obvious.

The assumed rigidity is possible because variables describing small oscillatory relative motion of tightly bound atoms that form the rock usually time-average to zero in a time that is short compared to the time-scale describing center-of-mass motion. This *separation of scales* enables *adiabatic elimination* of a very large number of degrees of freedom.

It is known from experiments that atomic-scale particle motion can be very different from the predictions of classical mechanics. Because large objects are made up of many atoms, one approach is to suggest that quantum effects are somehow averaged out in large objects. The underlying notion of finding a means to link quantum mechanics to classical mechanics is so important it is called the *correspondence principle*. Formally, it is required that the results of classical mechanics be obtained in the limit $\hbar \to 0$. While a simple and

---

[1] For example, an object may be able to vibrate in many different ways.

[2] In 1915, Emmy Noether showed that the existence of a symmetry due to a local interaction gives rise to a conserved quantity. For example, conservation of energy is due to time translation symmetry, conservation of linear momentum is due to space translational symmetry, and angular momentum conservation is due to rotational symmetry.

546

convenient test, this approach misses the point. The results of classical mechanics are obtained because the quantum mechanical wave nature of objects is usually averaged out by a mechanism called *decoherence*. However, sometimes even large (macroscopic) objects can show quantum effects. For example, they can tunnel through a thin potential barrier if the constituents are properly prepared and the wave nature of the entire object is maintained.[3]

In practice, the motion of macroscopic material bodies is *usually* described by classical mechanics. In this case, the linear momentum of a rigid object of mass m is $\mathbf{p} = m\,d\mathbf{x}/dt$, where $\mathbf{v} = d\mathbf{x}/dt$ is the velocity of the object moving in the direction of the unit vector $\mathbf{x}^{\sim} = \mathbf{x}/|\mathbf{x}|$. Time is measured in units of seconds (s), and distance is measured in units of meters (m). The magnitude of momentum is measured in units of kilogram meters per second (kg m s^{-1}), and the magnitude of velocity (speed) is measured in units of meters per second (m s^{-1}). Classical mechanics assumes that there exists an inertial frame of reference for which the motion of the object is described by the differential equation

$$\mathbf{F} = \frac{d\mathbf{p}}{dt} = m\,\frac{d^2\mathbf{x}}{dt^2} = m\,\frac{d\mathbf{v}}{dt}, \tag{H.1}$$

where the vector $\mathbf{F}$ is the force whose magnitude is measured in units of newtons (N). Force is a real vector field such that a particle can be subject to a force the magnitude and direction of which are different in different parts of space.

The definition of *work* is the integral of the force applied to a particle multiplied by the infinitesimal distance moved in the direction of the force for the complete path from position $\mathbf{r}_1$ to $\mathbf{r}_2$, where $\mathbf{r}$ is spatial vector coordinate. See Fig. H.1.

Fig. H.1 Illustration of a classical particle trajectory from position $\mathbf{r}_1$ to $\mathbf{r}_2$, where $\mathbf{r}$ is a spatial vector coordinate.

For a *conservative* force field, the work W_{12} is the same for any path between points 1 and 2. Hence,

$$W_{12} = \int_{\mathbf{r}=\mathbf{r}_1}^{\mathbf{r}=\mathbf{r}_2} \mathbf{F} \cdot d\mathbf{r} = m \int \frac{d\mathbf{v}}{dt} \cdot \mathbf{v}\,dt = \frac{m}{2} \int \frac{d}{dt}\left(v^2\right) dt, \tag{H.2}$$

so that $W_{12} = m\left(v_2^2 - v_1^2\right)/2 = T_2 - T_1$, where $v^2 = \mathbf{v} \cdot \mathbf{v}$ and the scalar $T = mv^2/2$ is called the kinetic energy of the particle. The work done around any *closed path* in a conservative force field, such as the one illustrated in Fig. H.2, is zero, or

$$\oint \mathbf{F} \cdot d\mathbf{r} = 0. \tag{H.3}$$

---

[3] In other words, the wave function is coherent throughout the domain. For an introduction to this see A. J. Leggett, *Physics World* **12**, 73 (1999).

Fig. H.2 Illustration of a closed-path classical particle trajectory.

This is *always* true if force is the gradient of a *single-valued* spatial scalar field where

$$\mathbf{F} = -\nabla V(\mathbf{r}) \tag{H.4}$$

since $\oint \mathbf{F} \cdot d\mathbf{r} = - \oint \nabla V \cdot d\mathbf{r} = - \oint dV = 0$, where the scalar field $V(\mathbf{r})$ is the potential. Potential energy is measured in joules (J) or electron volts (eV). If the forces acting on the object are conservative, then total energy, which is the sum of kinetic and potential energy, is a constant of the motion. In other words, total energy $T + V$ is conserved.

Since kinetic and potential energy can be expressed as functions of position and time, it is possible to define a *Hamiltonian* function for the system, which is

$$\boxed{H = T + V.} \tag{H.5}$$

This function may then be used to describe the dynamics of particles in the system.

For a nonconservative force, such as a particle subject to frictional forces, the work done around any closed path is not zero, and $\oint \mathbf{F} \cdot d\mathbf{r} \neq 0$.

The concept of force helps ensure that the motion of objects can be described as a simple process of *cause and effect*. A force field in three-dimensional space is represented mathematically as a continuous, integrable vector field, $\mathbf{F}(\mathbf{r})$. Assuming that time is also continuous and integrable, a conservative force field energy may be conveniently partitioned between a kinetic and potential term and total energy is conserved. By representing the total energy as a function or Hamiltonian, $H = T + V$, a differential equation may be found that describes the dynamics of the object. Integration of the differential equation of motion gives the trajectory of the object as it moves through space.

In practice, these ideas are very powerful and may be applied to many problems involving the motion of macroscopic objects. As an example, consider the problem of finding the motion of a particle mass, m, attached to a spring. Experience suggests that the solution will be oscillatory and so characterized by a frequency and amplitude of oscillation. However, the power of the theory is that relationships among all the parameters that govern the behavior of the system may be found.

H.2.2 The Simple Harmonic Oscillator

Figure H.3(a) illustrates a classical particle mass, m, constrained to motion in one dimension and attached to a lightweight spring. The displacement, x, from the position of lowest potential energy at $x = 0$, is proportional to the force on the particle[4] such that $F = -\kappa_0 x$. The spring constant κ_0 sets the scale relating force to displacement. The finite mass of the spring is ignored.

---

[4] This is Hooke's law, which was published in 1678.

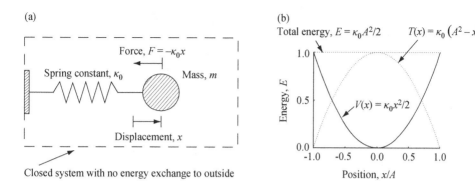

(a)

Force, $F = -\kappa_0 x$

Spring constant, κ_0

Mass, m

Displacement, x

Closed system with no energy exchange to outside

(b)

Total energy, $E = \kappa_0 A^2/2$ $T(x) = \kappa_0 \left(A^2 - x^2\right)/2$

$V(x) = \kappa_0 x^2/2$

Position, x/A

Energy, E

Fig. H.3 (a) Illustration of the closed system one-dimensional harmonic oscillator showing particle mass m attached to a lightweight spring with spring constant κ_0 and displacement x of the particle from its position of lowest potential energy at $x = 0$. (b) Kinetic energy T, potential energy V, and total energy E as functions of position, x.

Potential energy, obtained by integrating Eq. (H.4), is

$$V = \frac{1}{2}\kappa_0 x^2 = \int_0^x \kappa_0 x' \mathrm{d}x', \tag{H.6}$$

and kinetic energy is

$$T = \frac{1}{2}m \left(\frac{\mathrm{d}x}{\mathrm{d}t}\right)^2, \tag{H.7}$$

so that the total energy function is

$$H = T + V = \frac{1}{2}m \left(\frac{\mathrm{d}x}{\mathrm{d}t}\right)^2 + \frac{1}{2}\kappa_0 x^2. \tag{H.8}$$

The system is *closed* with no exchange of energy outside the system. There is no dissipation, the total energy in the system is a constant, and[5]

$$\frac{\mathrm{d}H}{\mathrm{d}t} = 0 = m\frac{\mathrm{d}x}{\mathrm{d}t}\frac{\mathrm{d}^2 x}{\mathrm{d}t^2} + \kappa_0 x\frac{\mathrm{d}x}{\mathrm{d}t}, \tag{H.9}$$

so that the *equation of motion* can be written as

$$\boxed{\kappa_0 x + m\frac{\mathrm{d}^2 x}{\mathrm{d}t^2} = 0.} \tag{H.10}$$

The solutions for this second-order linear differential equation are

$$x(t) = A \cos(\omega_0 t + \phi), \tag{H.11}$$

$$\frac{\mathrm{d}x}{\mathrm{d}t} = -\omega_0 A \sin(\omega_0 t + \phi), \tag{H.12}$$

---

[5] Note: $\frac{\mathrm{d}}{\mathrm{d}t}\left(\frac{\mathrm{d}x}{\mathrm{d}t}\right)^2 = \frac{\mathrm{d}}{\mathrm{d}t}\left(\frac{\mathrm{d}x}{\mathrm{d}t}\frac{\mathrm{d}x}{\mathrm{d}t}\right) = \frac{\mathrm{d}^2 x}{\mathrm{d}t^2}\frac{\mathrm{d}x}{\mathrm{d}t} + \frac{\mathrm{d}x}{\mathrm{d}t}\frac{\mathrm{d}^2 x}{\mathrm{d}t^2} = 2\frac{\mathrm{d}x}{\mathrm{d}t}\frac{\mathrm{d}^2 x}{\mathrm{d}t^2}.$

$$\frac{d^2x}{dt^2} = -\omega_0^2 A \cos(\omega_0 t + \phi), \tag{H.13}$$

where A is the amplitude of oscillation, ω_0 is the angular frequency of oscillation measured in radians per second (rad s^{-1}), and ϕ is a fixed phase. The velocity leads the displacement in phase by $\pi/2$ and the acceleration is in antiphase with the displacement.

The potential energy and kinetic energy may be written as

$$V = \frac{1}{2}\kappa_0 A^2 \cos^2(\omega_0 t + \phi) \tag{H.14}$$

and

$$T = \frac{1}{2}m\omega_0^2 A^2 \sin^2(\omega_0 t + \phi), \tag{H.15}$$

respectively. The total energy is

$$E = T + V = \frac{m\omega_0^2 A^2}{2} = \frac{\kappa_0 A^2}{2}, \tag{H.16}$$

since $\sin^2(\theta) + \cos^2(\theta) = 1$ and $\kappa_0 = m\omega_0^2$. An increase in total energy increases the amplitude to $A = \sqrt{2E/\kappa_0} = \sqrt{2E/m\omega_0^2}$, and an increase in κ_0, corresponding to an increase in the stiffness of the spring, decreases A. The theory determines the relationships among all the parameters of the classical harmonic oscillator: κ_0, m, A, and total energy.

The classical simple harmonic oscillator vibrates in a single *mode* with frequency ω_0. The vibrational energy in the mode can be changed continuously by varying the amplitude of vibration, A.

Velocity is zero when $x = \pm A$, and the particle changes its direction of motion and starts moving back towards the position of lowest potential energy at $x = 0$. The position $x = \pm A$, where velocity is zero, is the *classical turning point* of the motion. Peak velocity, $v_{max} = \pm A\omega_0$, occurs as the particle crosses position $x = 0$. Maximum acceleration, $a_{max} = \pm A\omega_0^2$, occurs when $x = \pm A$.

H.2.3   Generalization and the Lagrangian

Generalizations and unifying concepts have been helpful in developing efficient ways to solve problems in classical mechanics. The kinetic energy of a particle with Cartesian coordinates $x_j (j = 1, 2, 3)$ only depends on the time derivative $\dot{x}_j \equiv dx_j/dt$. Rather than consider the Hamiltonian that is the sum of kinetic and potential energy, a Lagrangian, defined as the difference between kinetic energy, $T = T(\dot{x}_j)$, and potential energy, $V = V(x_j)$, can be written as

$$\mathcal{L}(x_j, \dot{x}_j) = T - V. \tag{H.17}$$

In classical mechanics, the path followed by a dynamical system moving from one point to another in configuration space within a given time interval, t_1 to t_2, minimizes the time integral of the Lagrangian. In the formalism of calculus of variations, an extrema exists if the action integral

$$\delta \int_{t_1}^{t_2} \mathcal{L}(x_j, \dot{x}_j) \mathrm{d}t = 0. \tag{H.18}$$

Since action is a quantity with units of length multiplied by momentum or energy multiplied by time, Eq. (H.18) finds stationary action. This also leads directly to the Lagrange equations of motion,

$$\frac{\partial \mathcal{L}}{\partial x_j} - \frac{\mathrm{d}}{\mathrm{d}t} \frac{\partial \mathcal{L}}{\partial \dot{x}_j} = 0. \tag{H.19}$$

Cartesian coordinates may be replaced by *generalized* coordinates, q_j, and *generalized* velocities, $\dot{q}_j$, in the Lagrange equations. For a system of N particles, each with three degrees of freedom, there are $x_j (j = 1, 2, ..., 3N)$ position coordinates that are functions of generalized coordinates so that $x_j = x_j(q_1, q_2, ..., q_{3N})$. Any set of coordinates that completely specifies the state of the system are called generalized coordinates. A proper set of independent generalized coordinates is limited in number to the degrees of freedom in the system. The ability to choose a convenient coordinate system and then solve a calculus of variations problem, entirely avoiding Newtonian forces, is a generalization of great practical importance because it can simplify many problems.

Classical mechanics requires, but does not explain why, the trajectory taken between a fixed initial and final state in $\{q, t\}$ configuration space minimizes the action integral, Eq. (H.18). Quantum mechanics addresses this by allowing the linear system to explore *all* possible paths. Often this results in the classical path emerging as a result of the properties of stationary phase in which the *other* paths with quantum amplitude and phase interfere with each other and cancel.

H.2.4 Increasing Complexity to Discover New Phenomena

The simple harmonic oscillator model is popular in part because of the ease with which it may be described to others. Introducing complexity to the model is a strategy to discover new phenomena. Complexity might be increased by considering the oscillatory behavior of several particles coupled by springs. Another possibility is to drive the oscillator in the presence of frictional forces and study transient behavior in different, increasingly complex, potentials.

As a step towards more complex systems, the equations of motion of an isolated linear chain of particles, each with mass m, connected by identical springs is considered next. This particular problem is a common starting point for the study of lattice vibrations in crystals.

H.2.5 The Monatomic Linear Chain

Part of an isolated linear chain of particles, each of mass m, connected by springs is shown in Fig. H.4. Each particle, labeled with an integer, j, occupies a lattice site with position $x_j = jL$, where L is the lattice constant. Displacement from the lattice site of the jth particle is u_j, and there are a large number of particles in the chain.

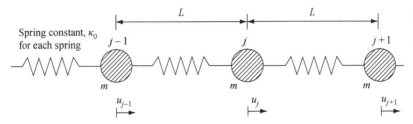

Fig. H.4 Illustration of part of an isolated linear chain of particles, each of mass m, connected by identical springs with spring constant κ_0. Each particle is labeled with an integer, j, and occupies a lattice site with position jL, where L is the lattice constant. Displacement from the lattice site of the jth particle is u_j. It is assumed that there are a large number of particles in the linear chain.

Assuming small deviations u_j, the Hamiltonian of the linear chain is

$$H = \sum_j \frac{m}{2}\left(\frac{du_j}{dt}\right)^2 + V_0(0) + \frac{1}{2!}\sum_{jk}\frac{\partial^2 V_0(0)}{\partial u_j \partial u_k}u_j u_k + \frac{1}{3!}\sum_{jkl}\frac{\partial^3 V_0(0)}{\partial u_j \partial u_k \partial u_l}u_j u_k u_l + \cdots .$$

$$\text{(H.20)}$$

The first term on the right-hand side is a sum over kinetic energy of each particle, and $V_0(0)$ is the potential energy when all particles are stationary in their lattice-site positions. The remaining terms come from a Taylor expansion of the potential about the lattice site positions. In general, each particle oscillates about its lattice-site position and is coupled to other oscillators via the potential.

If the force constant $\kappa_{jk} = (\partial^2 V_0/\partial u_j \partial u_k)|_0$ is real and symmetric then $\kappa_{jk} = \kappa_{kj}$, and if all springs are identical then $\kappa_0 = \kappa_{jk}$. Restricting the sum in Eq. (H.20) to relative motion of nearest neighbors, keeping only the κ_0 term, and setting $V_0(0) = 0$, the Hamiltonian may be written

$$H = \sum_j \frac{m}{2}\left(\frac{du_j}{dt}\right)^2 + \frac{\kappa_0}{2}\sum_j\left(2u_j^2 - u_j u_{j+1} - u_j u_{j-1}\right).$$

$$\text{(H.21)}$$

Displacement from the jth lattice site is related to that of its nearest neighbor by

$$u_{j\pm1} = u_j e^{\pm iqL},$$

$$\text{(H.22)}$$

where $q = 2\pi/\lambda$ is the magnitude of the wave vector of a vibration of wavelength λ. If there is no dissipation in the system, $dH/dt = 0$, and the equation of motion is

$$m\frac{d^2 u_j}{dt^2} = \kappa_0(u_{j+1} + u_{j-1} - 2u_j).$$

$$\text{(H.23)}$$

Substituting time dependence of the form $e^{-i\omega t}$ gives

$$-m\omega^2 u_j = \kappa_0\left(e^{iqL} + e^{-iqL} - 2\right)u_j = -4\kappa_0 \sin^2\left(\frac{qL}{2}\right)u_j,$$

$$\text{(H.24)}$$

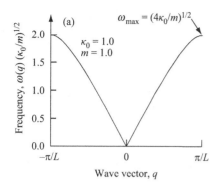
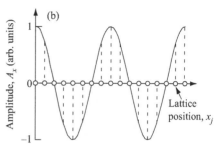

Fig. H.5 (a) Dispersion relation for lattice vibrations of a one-dimensional monatomic linear chain. The dispersion relation is linear at low values of q. The maximum frequency of oscillation is $\omega_{max} = (4\kappa_0/m)^{1/2}$. Particles are of mass $m = 1.0$, and the spring constant is $\kappa_0 = 1.0$. (b) Amplitude of vibrational motion in the x direction on a portion of the linear chain for a particular mode of frequency ω. Lattice-site position x_j is indicated as an open circle.

from which it follows that

$$\omega(q) = \sqrt{\frac{4\kappa_0}{m}} \sin\left(\frac{qL}{2}\right). \tag{H.25}$$

The *dispersion relation* $\omega = \omega(q)$ for the monatomic linear chain plotted in Fig. H.5(a) consists of a single *acoustic branch*, $\omega = \omega_{acoustic}(q)$, with maximum frequency $\omega_{max} = \sqrt{4\kappa_0/m}$. The vibration frequency approaches $\omega \to 0$ *linearly* as $q \to 0$. In the long wavelength limit ($q \to 0$), the acoustic branch dispersion relation describing lattice dynamics of a monatomic linear chain predicts that vibrational waves propagate at constant group velocity $v_g = \partial\omega/\partial q$. This is the velocity of sound waves in the system.

Each normal mode of frequency ω and wave vector q can involve harmonic motion of all the particles in the chain. As illustrated in Fig. H.5(b), not all particles have the same amplitude. The total energy in a mode is proportional to the sum of the amplitudes squared of all particles in the chain.

Simplifications imposed by *symmetry* constrain the type of oscillatory motion that is possible. The motion of a given atom is determined by forces due to the relative position of its nearest neighbors. Forces from displacements of more distant neighbors are not included. The facts that there is only one spring constant and that there is only one type of atom are additional constraints on the system. Such constraints eliminate all but a few of the possible solutions to the equations of motion.

The motion of coupled oscillators can be described using the idea that a given radial frequency of oscillation, ω, corresponds to definite wave vector $\mathbf{q}$. Hence, the dispersion relationship $\omega = \omega(\mathbf{q})$ determines how vibration waves and pulses propagate through the system. For example, in one dimension the phase velocity of a wave is

$$v_p = \omega/q, \tag{H.26}$$

and a pulse made up of wave components near a value q_0 propagates at the group velocity,

$$v_g = \frac{\partial \omega}{\partial q}\Big|_{q=q_0}. \tag{H.27}$$

If the dispersion relation is known then quantities of practical importance such as v_p and v_g may be determined.[6]

H.2.6 The Diatomic Linear Chain

Part of a diatomic linear chain with lattice constant L is shown in Fig. H.6. There are *two atoms per unit cell* with mass m_1 and mass m_2, respectively, and spaced by $L/2$. The displacement from the jth lattice site is u_j.

The motion of one atom is related to that of its nearest equal-mass neighbor by

$$u_{j\pm 2} = u_j e^{\pm iqL}, \tag{H.28}$$

where $q = 2\pi/\lambda$ is the magnitude of the wave vector of a vibration of wavelength, λ. Assuming forces are only due to the relative position of nearest neighbors, the equations of motion for the two types of atoms are

$$m_1 \frac{d^2 u_j}{dt^2} = \kappa_0(u_{j+1} + u_{j-1} - 2u_j), \tag{H.29}$$

$$m_2 \frac{d^2 u_{j-1}}{dt^2} = \kappa_0(u_j + u_{j-2} - 2u_{j-1}), \tag{H.30}$$

or

$$m_1 \frac{d^2 u_j}{dt^2} = \kappa_0 \left(1 + e^{iqL}\right) u_{j-1} - 2\kappa_0 u_j, \tag{H.31}$$

$$m_2 \frac{d^2 u_{j-1}}{dt^2} = \kappa_0 \left(1 + e^{-iqL}\right) u_j - 2\kappa_0 u_{j-1}. \tag{H.32}$$

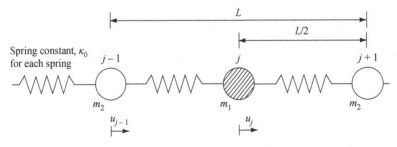

Fig. H.6 Part of an isolated linear chain of particles of alternating mass m_1 and m_2 connected by identical springs with spring constant κ_0. The lattice constant is L and there are two particles per *unit cell* spaced by $L/2$. The displacement from the jth lattice site is u_j.

[6] A nice discussion of these and other velocities may be found in L. Brillouin, *Wave Propagation and Group Velocity*, New York, Academic Press, 1960.

Solutions for u_j and u_{j-1} have time dependence of the form $e^{-i\omega t}$, giving

$$\left(2\kappa_0 - m_1\omega^2\right)u_j - \kappa_0\left(1 + e^{iqL}\right)u_{j-1} = 0, \tag{H.33}$$

$$-\kappa_0\left(1 + e^{-iqL}\right)u_j + \left(2\kappa_0 - m_2\omega^2\right)u_{j-1} = 0. \tag{H.34}$$

This linear set of equations can be expressed as the product of a 2×2 matrix $\mathbf{A}$ and a column vector $\mathbf{u}$ where $\mathbf{Au} = \mathbf{0}$. If $\mathbf{A}^{-1}$ exists then $\mathbf{A}^{-1}\mathbf{Au} = \mathbf{u} = \mathbf{0}$, which is of no interest. If $\mathbf{A}^{-1}$ does *not* exist then $\mathbf{u} \neq \mathbf{0}$, there is vibrational motion of atoms, and $|\mathbf{A}| = 0$ with eigensolutions given by the characteristic equation

$$\begin{vmatrix} 2\kappa_0 - m_1\omega^2 & -\kappa_0\left(1 + e^{iqL}\right) \\ -\kappa_0\left(1 + e^{-iqL}\right) & 2\kappa_0 - m_2\omega^2 \end{vmatrix} = 0, \tag{H.35}$$

so that the characteristic polynomial is

$$\omega^4 - 2\kappa_0\left(\frac{m_1 + m_2}{m_1 m_2}\right)\omega^2 + \frac{2\kappa_0^2}{m_1 m_2}(1 - \cos(qL)) = 0, \tag{H.36}$$

with roots, or *eigenvalues*, ω, that depend on q.

In the *long wavelength limit* $q \to 0$ ($\lambda \to \infty$),

$$\omega^2\left(\omega^2 - 2\kappa_0\left(\frac{m_1 + m_2}{m_1 m_2}\right)\right) = 0. \tag{H.37}$$

The solutions are $\omega = 0$ and $\omega = \sqrt{2\kappa_0\left(\frac{m_1+m_2}{m_1 m_2}\right)}$, with the latter corresponding to both atom types beating against each other.

In the *short wavelength limit* $q \to \pi/L$,

$$\omega^4 - 2\kappa_0\left(\frac{m_1 + m_2}{m_1 m_2}\right)\omega^2 + \frac{4\kappa_0^2}{m_1 m_2} = 0 \tag{H.38}$$

and solutions are $\omega_1 = \sqrt{2\kappa_0/m_1}$, with only atoms of mass m_1 vibrating, and $\omega_2 = \sqrt{2\kappa_0/m_2}$, with only atoms of mass m_2 vibrating.

As shown in Fig. H.7, there is an *acoustic branch*, $\omega = \omega_{\text{acoustic}}(q)$, for which vibration frequency linearly approaches $\omega \to 0$ as $q \to 0$, and there is an *optic branch*, $\omega = \omega_{\text{optic}}(q)$, for which $\omega \neq 0$ as $q \to 0$. The acoustic branch can propagate low-frequency sound waves with constant group velocity, $v_g = \partial\omega/\partial q$.[7]

In three dimensions there are extra degrees of freedom, resulting in a total of three acoustic and three optic branches. For a wave propagating in a given direction there is one *longitudinal acoustic* and one *longitudinal optic* branch with atom motion parallel to the wave propagation direction. There are also two *transverse acoustic* and two *transverse optic* branches with atom motion normal to the direction of wave propagation.

---

[7] Typical values for the velocity of sound waves in a semiconductor at room temperature are $v_g = 8.4 \times 10^3$ m s^{-1} in (100)-oriented Si and $v_g = 4.7 \times 10^3$ m s^{-1} in (100)-oriented GaAs. For comparison, the speed of sound in air at temperature $0\,^\circ$C at sea level is 331.3 m s^{-1} or 741 mph.

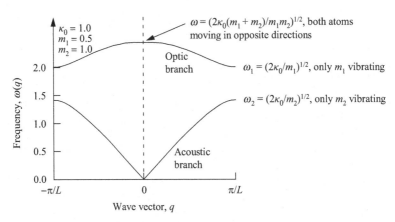

Fig. H.7 Dispersion relation for lattice vibrations of a one-dimensional diatomic linear chain. Particles are mass $m_1 = 0.5$ and $m_2 = 1.0$. The spring constant is $\kappa_0 = 1.0$.

Figure H.8 shows the dispersion relation along crystal symmetry directions of bulk GaAs.[8] As predicted, the highest vibrational frequency in GaAs is a $q = 0$ longitudinal optic (LO) mode of frequency $\nu_{LO} = 8.78 \times 10^{12}$ Hz = 8.78 THz.

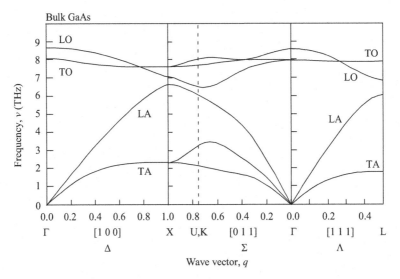

Fig. H.8 Lattice vibration dispersion relation along crystal symmetry directions of bulk GaAs, which has the zinc-blende crystal structure, lattice constant $L = 0.565$ nm, and two atoms per unit cell. The longitudinal acoustic (LA), transverse acoustic (TA), longitudinal optic (LO), and transverse optic (TO) branches are indicated.[8]

---

[8] Lattice vibration dispersion relations for additional semiconductor crystals may be found in H. Bilz and W. Kress, *Phonon Dispersion Relations in Insulators*, Springer Series in Solid-State Sciences **10**, Berlin, Springer-Verlag, 1979 (ISBN 3 540 09399 0).

H.2.7 The Damped Oscillator Subject to External Harmonic Force

A particle mass m is constrained to motion in the x direction and is attached to a lightweight spring obeying Hooke's law such that the force on the particle due to the spring is the product of the displacement from lattice site position $x = 0$ and the spring constant κ_0. The particle is also subject to a small x-directed external harmonic force $F(t)$ of amplitude F_0 oscillating at angular frequency ω and a small dissipative force term due to friction $\gamma m\, dx/dt$, where the constant γ is a characteristic damping rate. Particle displacement, applied force, frequency, and dissipation due to friction are all treated as continuous variables.

The system is nonconservative (open) because the friction and the driving force cause energy exchange with the environment.

The equation of motion for the system that includes dissipation and a forcing term may be written as

$$\frac{d^2x}{dt^2} + \gamma\frac{dx}{dt} + \omega_0^2 x = \frac{F(t)}{m}, \tag{H.39}$$

where $\omega_0^2 = \kappa_0/m$. The solution is a superposition of the sum of the homogeneous term $x_h(t)$ and the particular term $x_p(t)$. For small damping, $\gamma \ll \omega_0$ so that $x_h(t) = x_{h0}e^{-\gamma t/2}\cos(\omega_0 t + \phi_h)$ where x_{h0} is an oscillation amplitude and ϕ_h a phase.

The transient displacement, $x_h(t)$, involving the natural angular frequency of the oscillator $\omega_0 = \sqrt{\kappa_0/m}$ is, after a sufficiently long time, damped out by the term $e^{-\gamma t/2}$ and the remaining steady-state response is a harmonic oscillation that has the same frequency as the forcing term $F(t) = F_0 e^{-i\omega t}$. Substituting a particular solution of the form

$$x_p(t) = x_{p0}e^{-i\omega t} \tag{H.40}$$

into the equation of motion results in

$$-\omega^2 x_{p0}e^{-i\omega t} - i\omega\gamma x_{p0}e^{-i\omega t} + \omega_0^2 x_{p0}e^{-i\omega t} = \frac{F_0}{m}e^{-i\omega t}, \tag{H.41}$$

so that oscillation amplitude as a function of frequency is

$$x_{p0}(\omega) = \frac{F_0}{m}\frac{1}{(\omega_0^2 - \omega^2) - i\omega\gamma}, \tag{H.42}$$

and

$$|x_{p0}(\omega)| = \frac{F_0}{m}\frac{1}{\sqrt{(\omega_0^2 - \omega^2)^2 + \omega^2\gamma^2}}, \tag{H.43}$$

$$\mathrm{Re}(x_{p0}(\omega)) = \frac{F_0}{m}\frac{(\omega_0^2 - \omega^2)}{(\omega_0^2 - \omega^2)^2 + \omega^2\gamma^2}, \tag{H.44}$$

$$\mathrm{Im}(x_{p0}(\omega)) = \frac{F_0}{m}\frac{\omega\gamma}{(\omega_0^2 - \omega^2)^2 + \omega^2\gamma^2}, \tag{H.45}$$

$$\frac{\mathrm{Im}(x_{p0}(\omega))}{\mathrm{Re}(x_{p0}(\omega))} = \tan(\phi) = \frac{\omega\gamma}{\omega_0^2 - \omega^2}. \tag{H.46}$$

General features include the facts that the real part of $x_{p0}(\omega)$ is zero at angular frequency ω_0, the imaginary part of $x_{p0}(\omega)$ peaks near ω_0 when γ is small, and the phase ϕ is independent of the driving force. The displacement amplitude of oscillation is $|x_{p0}(\omega)|$ and the velocity, dx/dt, varies sinusoidally with magnitude $\omega|x_{p0}(\omega)|$. Amplitude, $|x_{p0}(\omega)|$, has a peak if $\gamma < \sqrt{2}\omega_0$ and is critically damped if $\gamma = 2\omega_0$. For small γ, a maximum in amplitude occurs when $\left(\omega_0^2 - \omega^2\right)^2 + \omega^2\gamma^2$ is a minimum and this results in a renormalized resonance angular frequency,

$$\omega_0' = \sqrt{\omega_0^2 - \frac{\gamma^2}{2}}, \tag{H.47}$$

whose value is less than ω_0 due to the presence of damping. The phase of the displacement, $x(t)$, *always* lags the driving force, $F(t)$. The phase ϕ in Eq. (H.46) is found using the four-quadrant inverse tangent. At low angular frequency, $\omega \ll \omega_0$, the displacement is nearly in phase with the driving force and motion is controlled by the stiffness of the spring, independent of the particle mass. On resonance, $\omega = \omega_0$ and particle *velocity* is precisely in phase with the driving force, particle displacement lags the driving force by $\pi/2$ phase, and motion is limited by the frictional force. When $\omega \gg \omega_0$, the displacement lags the driving force by almost π phase and motion is controlled by the mass of the particle.

The steady-state behavior of amplitude, $|x_{p0}|$, and phase, ϕ, as a function of radial frequency, ω, for a damped oscillator subject to an external harmonic force is shown in Fig. H.9 for the indicated values of γ normalized to natural resonant frequency ω_0. Peak amplitude occurs at resonant frequency $\omega_0' < \omega_0$. The real and imaginary values of amplitude, x_{p0}, in Eq. (H.42) are shown in Fig. H.10 for the indicated values of γ. In steady state, the average power absorbed by the damped system is the instantaneous power integrated over many oscillations. The instantaneous frictional force is $\gamma m\, dx/dt$. When the mass moves from x to $x + dx$ in time dt, the work done by the frictional force is the force times the distance moved,

$$dW_\gamma = \gamma m \frac{dx}{dt} dx. \tag{H.48}$$

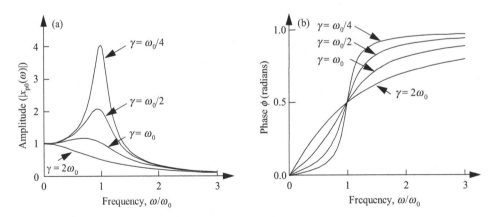

Fig. H.9 (a) Steady-state amplitude, $|x_{p0}|$, and (b) phase, ϕ, of damped oscillator driven by external harmonic force. The values of γ are indicated. The particle has unit mass and the external force has unit amplitude.

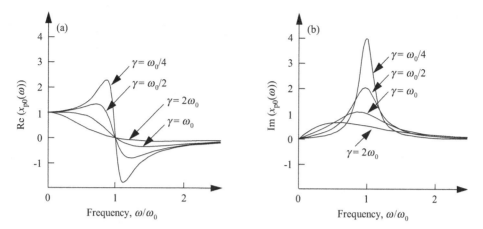

Fig. H.10 (a) Real and (b) imaginary amplitude of damped oscillator driven by external harmonic force. The values of γ are indicated. The particle has unit mass and the external force has unit amplitude.

The rate at which energy is dissipated is

$$\frac{\mathrm{d}W_\gamma}{\mathrm{d}t} = P_\gamma = \gamma m \frac{\mathrm{d}x}{\mathrm{d}t} \frac{\mathrm{d}x}{\mathrm{d}t}, \tag{H.49}$$

which is the work done divided by the time $\mathrm{d}t$. Because $\mathrm{d}x/\mathrm{d}t$ varies sinusoidally and has magnitude $\omega|x_{\mathrm{p}0}|$, the average value of $(\mathrm{d}x/\mathrm{d}t)^2$ per oscillation period is

$$\left\langle \left(\frac{\mathrm{d}x}{\mathrm{d}t} \right)^2 \right\rangle = \frac{1}{2} (\omega|x_{\mathrm{p}0}|)^2. \tag{H.50}$$

Hence, the average steady-state power that is supplied by the external harmonic force and absorbed by the damped system is

$$\langle P(\omega) \rangle = \frac{\gamma m}{2} (\omega|x_{\mathrm{p}0}|)^2 = \frac{1}{2} \frac{F_0^2}{m\gamma} \frac{\gamma^2 \omega^2}{(\omega_0^2 - \omega^2)^2 + \omega^2 \gamma^2}, \tag{H.51}$$

and so is inversely proportional to γ.

As expected, the average power absorbed by the damped system is zero at $\omega = 0$, peaks at the natural frequency, ω_0, and approaches zero as $\omega \to \infty$. The absorbed power will heat a mechanical system and since, according to Eq. (H.51), this absorption is inversely proportional to γ, reducing damping will *increase* heating when the system is driven at a frequency near ω_0. If the driving force is turned off at time $t = 0$, the system, starting with amplitude $|x_{\mathrm{h}0}(t = 0)|$, will evolve in time according to the homogeneous solution and will dissipate total energy due to frictional forces. The solution to Eq. (H.39) in the absence of the force term $F(t)$ is of the form $x_{\mathrm{h}}(t) = x_{\mathrm{h}0}e^{-i\omega t - i\phi}$ with

$$\omega_0^2 - i\omega\gamma - \omega^2 = 0. \tag{H.52}$$

If $\omega_0^2 > \gamma^2/4$ then the real and imaginary parts of the complex frequency are

$$\omega_{\mathrm{Re}} = \sqrt{\omega_0^2 - \gamma^2/4} \tag{H.53}$$

and

$$\omega_{\mathrm{Im}} = -\gamma^2/2. \tag{H.54}$$

Particle position,

$$x_{\mathrm{h}}(t) = x_{\mathrm{h}0}e^{-\gamma t/2}e^{-i\omega_{\mathrm{Re}}t - i\phi_h}, \tag{H.55}$$

is oscillatory at frequency ω_{Re} and decreases in *amplitude* to e^{-1} of its initial value in a characteristic time $\tau_{\mathrm{amp}} = 2/\gamma$. It follows that the time dependence of total oscillator *energy*, $m\omega_{\mathrm{Re}}^2|x_{\mathrm{h}0}|^2/2$, is proportional to $e^{-\gamma t}$. The quality factor of the oscillator is defined as

$$\boxed{Q \equiv \frac{1}{2}\left|\frac{\omega_{\mathrm{Re}}}{\omega_{\mathrm{Im}}}\right|.} \tag{H.56}$$

For the case being considered, $Q = (\sqrt{\omega_0^2 - \gamma^2/4})/\gamma$. When damping is small $\gamma \ll \omega_0$ so that $Q \to \omega_0/\gamma$ and, after the driving force is turned off at time $t = 0$, the energy stored in the system decreases exponentially in time according to $e^{-\gamma t} = e^{-\omega_0 t/Q} = e^{-t/\tau}$, where $\tau = 1/\gamma$ is a characteristic lifetime of *energy* stored in the oscillator.

A pendulum oscillating at its resonant frequency is an example of a simple harmonic oscillator that can be used to keep track of time and act as a clock. Analogous electrical circuits exist. For example, a capacitor (spring), an inductor (mass), and a resistor (damping term) can be configured as a resonator. If a circuit has a capacitor, an inductor, a voltage supply, and a component with negative *differential* resistance, it is possible to create bias conditions under which the circuit will oscillate without an external time-dependent input. The voltage range for which the differential resistance is negative determines the amplitude of the oscillator output.

H.2.8 Coupled Oscillator Normal Modes and Beats

A monatomic linear chain of N coupled harmonic oscillators constrained to motion in one dimension is described by N linear differential equations. There is one differential equation for each variable, so if the motion is in three dimensions then there are $3N$ differential equations. The solution to the linear differential equations is a set of normal modes in which all oscillators move at the same frequency and have fixed amplitude ratios. If the initial state of a conservative system is a normal mode then the system remains in that eigenmode. However, in such a system any linear combination of normal modes can also be an initial state that subsequently evolves according to the coupled linear differential equations. This coupled non-normal mode behavior can give rise to beats. To see how this works, consider two identical oscillators each of mass m and spring constant κ_0, coupled by spring constant κ, and constrained to longitudinal motion in one dimension. Displacement from the position of lowest potential energy for each particle is u_1 and u_2, respectively, where the subscript labels the oscillator. The system is illustrated in Fig. H.11.

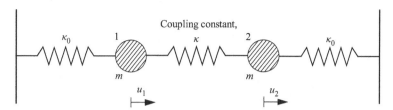

Fig. H.11 Two identical oscillators each of mass m and spring constant κ_0, coupled by spring constant κ. The displacement from the position of lowest potential energy for each particle is indicated as u_1 and u_2, respectively.

The equations of motion are

$$-\frac{d^2 u_1}{dt^2} + \left(-\frac{\kappa_0}{m} - \frac{\kappa}{m}\right) u_1 + \frac{\kappa}{m} u_2 = 0, \tag{H.57}$$

$$\frac{\kappa}{m} u_1 - \frac{d^2 u_2}{dt^2} + \left(-\frac{\kappa_0}{m} - \frac{\kappa}{m}\right) u_2 = 0, \tag{H.58}$$

and these equations may be written in matrix form as $\mathbf{Au} = \mathbf{0}$,

$$\begin{bmatrix} -\frac{d^2}{dt^2} - \left(\frac{\kappa_0}{m} + \frac{\kappa}{m}\right) & \frac{\kappa}{m} \\ \frac{\kappa}{m} & -\frac{d^2}{dt^2} - \left(\frac{\kappa_0}{m} + \frac{\kappa}{m}\right) \end{bmatrix} \begin{bmatrix} u_1 \\ u_2 \end{bmatrix} = \mathbf{Au} = \mathbf{0}. \tag{H.59}$$

If the inverse $\mathbf{A}^{-1}$ exists then $\mathbf{A}^{-1}\mathbf{Au} = \mathbf{1u} = \mathbf{u} = \mathbf{0}$, where $\mathbf{1}$ is the identity matrix. The solution $\mathbf{u} = \mathbf{0}$ is the trivial result involving no motion, and so particle displacements $u_1 = u_2 = 0$. However, the interesting solutions where there *is* motion of the particles requires that the inverse $\mathbf{A}^{-1}$ does *not* exist which happens if the characteristic determinant is zero. The solutions to $|\mathbf{A}| = 0$ are normal modes in which all particles oscillate with the same eigenfrequency, ω.

Substituting frequency dependence of the form $e^{-i\omega t}$ into Eq. (H.59) gives

$$\begin{bmatrix} \omega^2 - \left(\frac{\kappa_0}{m} + \frac{\kappa}{m}\right) & \frac{\kappa}{m} \\ \frac{\kappa}{m} & \omega^2 - \left(\frac{\kappa_0}{m} + \frac{\kappa}{m}\right) \end{bmatrix} \begin{bmatrix} u_1 \\ u_2 \end{bmatrix} = \begin{bmatrix} 0 \\ 0 \end{bmatrix} = \mathbf{Au} = \mathbf{0} \tag{H.60}$$

and, by Cramers theorem, this homogeneous system of linear equations has a nontrivial solution if the characteristic determinant is zero, $|\mathbf{A}| = 0$. Hence,

$$\begin{vmatrix} \omega^2 - \left(\frac{\kappa_0}{m} + \frac{\kappa}{m}\right) & \frac{\kappa}{m} \\ \frac{\kappa}{m} & \omega^2 - \left(\frac{\kappa_0}{m} + \frac{\kappa}{m}\right) \end{vmatrix} = 0, \tag{H.61}$$

giving the characteristic polynomial

$$\left(\omega^2 - \left(\frac{\kappa_0}{m} + \frac{\kappa}{m}\right)\right)^2 - \left(\frac{\kappa}{m}\right)^2 = 0 \tag{H.62}$$

or

$$\omega^2 - \left(\frac{\kappa_0}{m} + \frac{\kappa}{m}\right) = \pm\frac{\kappa}{m}, \tag{H.63}$$

with normal-mode frequencies $\omega_1 = \pm\sqrt{\kappa_0/m}$ and $\omega_2 = \pm\sqrt{(\kappa_0 + 2\kappa)/m}$. Substituting ω_1 into Eq. (H.60) gives

$$\frac{\kappa}{m}\begin{bmatrix} -1 & 1 \\ 1 & -1 \end{bmatrix}\begin{bmatrix} u_1 \\ u_2 \end{bmatrix} = \begin{bmatrix} 0 \\ 0 \end{bmatrix}, \tag{H.64}$$

with solution $u_1 = u_2$ corresponding to a symmetric longitudinal mode with the particles moving in the same direction. Similarly, substitution of ω_2 has solution $u_1 = -u_2$ corresponding to an antisymmetric longitudinal mode with the particles moving in opposite directions.

In general, each normal mode k of the N-particle system has an amplitude A_k. Hence, the solution for displacement of each particle in a one-dimensional system with $N = 2$ is a linear superposition of normal mode amplitudes A_1 and A_2. This gives

$$u_1 = A_1 \cos(\omega_1 t + \phi_1) + A_2 \cos(\omega_2 t + \phi_2) \tag{H.65}$$

and

$$u_2 = A_1 \cos(\omega_1 t + \phi_1) - A_2 \cos(\omega_2 t + \phi_2), \tag{H.66}$$

where $\phi_{1,2}$ is a phase that depends on initial conditions.

H.2.8.1 Coupled Oscillator Beats

Beats arise when the system is excited in a non-normal mode. As an example, consider the initial condition at time $t = 0$ with both particles at rest, $u_1(t = 0) = a_1$ and $u_2(t = 0) = 0$. The particle displacements at $t = 0$ are a superposition of normal modes and, because the particles are initially at rest, the initial conditions can only be met if both normal modes are excited such that $A_1 = A_2 = a_1/2$ and $\phi_1 = \phi_2 = 0$. Therefore, in this particular case,

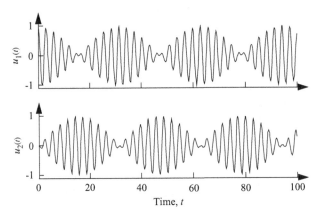

Fig. H.12 Beats in a two-mode system with $\omega_2/\omega_1 = 1.1$ and initial conditions $u_1(t = 0) = a_1$, $u_2(t = 0) = 0$, and $du_1(t = 0)/dt = du_2(t = 0)/dt = 0$.

$$u_1(t) = \frac{a_1}{2}(\cos(\omega_1 t) + \cos(\omega_2 t)) = a_1 \cos\left(\frac{\omega_1 + \omega_2}{2}t\right)\cos\left(\frac{\omega_1 - \omega_2}{2}t\right), \tag{H.67}$$

$$u_2(t) = \frac{a_1}{2}(\cos(\omega_1 t) - \cos(\omega_2 t)) = a_1 \sin\left(\frac{\omega_1 + \omega_2}{2}t\right)\sin\left(\frac{\omega_1 - \omega_2}{2}t\right). \tag{H.68}$$

The amplitude of the term oscillating at half the sum frequency $\omega_1 + \omega_2$ is modulated by half the difference frequency $\omega_1 - \omega_2$. The beat frequency is $(\omega_1 - \omega_2)/2$. See Fig. H.12 for an example with $\omega_2/\omega_1 = 1.1$. Energy flows between the oscillating particles at the difference frequency $\omega_1 - \omega_2$. The total energy in the system is a constant and, in agreement with Parsevals theorem, is given by the sum of the energies of the normal modes.

H.2.9 Transient Dynamics of Driven Oscillation

The transient dynamics of driven oscillators may be studied numerically. If the numerical simulations are accurate and stable, then a wide range of phenomena may be explored. This includes the study of systems not easily accessible by analytic methods such as nonlinear behavior.

Ordinary differential equations may always be expressed in terms of first-order differential equations. For example,

$$\frac{d^2 y}{dt^2} + a(t)\frac{dy}{dt} = b(t) \tag{H.69}$$

may be written as two first-order coupled differential equations,

$$\frac{dy}{dt} = f(t), \tag{H.70}$$

$$\frac{df}{dt} = b(t) - a(t)f(t), \tag{H.71}$$

N coupled first-order differential equations for the functions y_j have the form

$$\frac{d}{dt}y_j(t) = f_j(t, \, y_1, \ldots, y_N), \tag{H.72}$$

where the right-hand side is a known function and $j = 1, 2, \ldots, N$.

After boundary conditions and an initial value are applied, a finite step h_0 is used to numerically solve the equations. For example, Euler's method gives

$$y_{n+1} = y_n + h_0 f(t_n, y_n) + O\left(h_0^2\right) \tag{H.73}$$

to advance the solution from t_n to $t_{n+1} \equiv t_n + h_0$. It advances the solution through an interval $h_0 = t_{n+1} - t_n$, using only derivative information contained in $f(t, y)$ at the beginning of the interval. Unfortunately, the error in each step is $O(h_0^2)$, which is only one power of h_0 smaller than the estimate function $h_0 f(t_n, y_n)$.

By first taking a trial step to the midpoint of the interval and then using the value of both t and y at the midpoint, a more accurate step may be computed across the complete interval. Evaluating $f(t, y)$ in such a way that first-order and some higher-order terms cancel, a very accurate numerical integrator can be implemented. The fourth-order Runge–Kutta method does just this and can be used to integrate nonhomogeneous differential

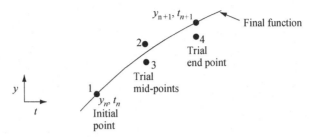

Fig. H.13 Fourth-order Runge–Kutta method used to numerically estimate the integration of a function.

equations. As illustrated in Fig. H.13, each step along $f(t, y)$ is evaluated four times, once at the initial point, twice at the midpoints, and once at the trial end point.

The fourth-order Runge–Kutta method is summarized by the equation

$$y_{n+1} = y_n + \frac{k_1}{6} + \frac{k_2}{3} + \frac{k_3}{3} + \frac{k_4}{6} + O\left(h_0^5\right),$$ (H.74)

where

$$k_1 = h_0 f(t_n, y_n)$$ (H.75)

is used to evaluate at the initial point,

$$k_2 = h_0 f\left(t_n + \frac{h_0}{2}, y_n + \frac{k_1}{2}\right)$$ (H.76)

is used to estimate the midpoint using $k_1/2$,

$$k_3 = h_0 f\left(t_n + \frac{h_0}{2}, y_n + \frac{k_2}{2}\right)$$ (H.77)

is used to estimate the midpoint using $k_2/2$, and

$$k_4 = h_0 f(t_n + h_0, y_n + k_3)$$ (H.78)

is used to evaluate the end point using k_3.

When implementing the Runge–Kutta method, care has to be taken to make sure that the estimates for trial points k_1, k_2, k_3, and k_4 are updated using the most current values.

H.2.10 Phasor Diagram of a Harmonically-Driven Damped Oscillator

A phasor diagram plots particle velocity as a function of position. In the case of a simple harmonic oscillator with unit amplitude and unit frequency this is a circle since particle position is of the form $x = \cos(\omega_0 t)$ and particle velocity is $dx/dt = \omega_0 \sin(\omega_0 t)$. For a lightly damped ($\gamma \ll \omega_0$), sinusoidally forced harmonic oscillator the homogeneous term $x_h(t) = x_{h0} e^{-\gamma t/2} \cos(\omega_0 t + \phi_h)$ decays away in a characteristic time $\tau_{amp} = 2/\gamma$ leaving the particular term $x_p(t) = x_{p0} \cos(\omega t + \phi_p)$. The transient behavior of an oscillator system initially at rest and driven harmonically at the natural frequency ω_0 has an amplitude

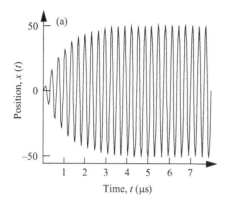
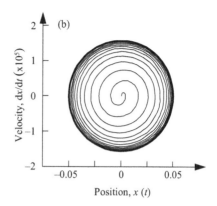

Fig. H.14 (a) Transient response of an oscillator system initially at rest and driven harmonically at the natural frequency $f_0 = \omega_0/2\pi = 3.14$ MHz. The damping rate is $\gamma = \omega_0/10$. The amplitude of oscillations builds up monotonically and the system reaches steady state after a characteristic time given by the inverse of the friction damping term. (b) Phasor diagram of the transient response in (a).

that builds up monotonically. This is illustrated in Fig. H.14. The greater the Q of the resonator, the larger the steady-state amplitude and the longer it takes to reach steady-state.

H.2.11 Control of a Harmonically-Driven Damped Oscillator

The mechanically driven damped harmonic oscillator may be controlled to within limits determined by the available range of experimental parameter values. A basic level of control may be demonstrated by using a force to stop harmonic oscillation. A harmonically driven damped oscillator can be subjected to a precisely timed impulsive force, after which no more forces are applied and the oscillator is left motionless. In effect, the mechanical oscillator is hit in a precise way to stop oscillation.[9] The lowest-energy motionless state of the oscillator is *guaranteed stable* and hence a controlled point from which to subsequently time-evolve the driven system. The same approach may be used to control nonlinear dynamical systems in which motion can be chaotic.

As shown in Fig. H.15, application of a large force applied for a fixed amount of time can be used to stop harmonic motion in a time much less than one period of oscillation. Since the time at which the force starts to be applied is critical, it is necessary to know and predict the exact position and velocity of the mechanically oscillating particle.

Driven coupled oscillators can be controlled in more subtle ways. For example, a very small transfer of mechanical energy per period of two driven oscillators can result in synchronized motion. Famously, Huygens,[10] described such synchronization between two pendulum clocks in 1665, which was 22 years *before* Newton published the *Principia*.

---

[9] There is an analogous control mechanism in quantum mechanics. See A. F. J. Levi, L. Campos Venuti, T. Albash, and S. Haas, *Phys. Rev. A* **90**, 022119 (2014).

[10] C. Huygens, Letters to de Sluse (letters; no. 1333 of 24 February 1665, no. 1335 of 26 February 1665, no. 1345 of 6 March 1665) (*Société hollandaise des sciences*, Martinus Nijhoff, La Haye, 1895).

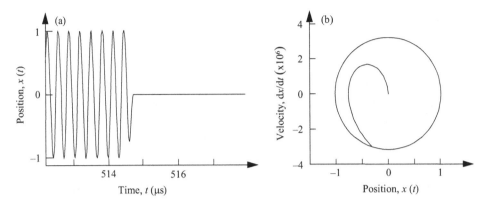

Fig. H.15 (a) A damped harmonic oscillator driven at natural resonant frequency $f_0 = \omega_0/2\pi = 3.14$ MHz, reaches steady state, and is then subject to a precisely timed impulsive force, after which no more forces are applied and the oscillator is stationary. The damping rate is $\gamma = \omega_0/200$. (b) Phasor diagram of transient response in (a).

H.2.12 Transient Dynamics of a Harmonically-Driven Damped Nonharmonic Oscillator

Solutions to a forced damped nonharmonic oscillator may be considered by adding a nonlinear term such that the restoring force on a particle of mass m due to the spring is of the form $F = -\kappa_1 x - \kappa_3 x^3$. The equation of motion may be written as

$$\frac{d^2 x}{dt^2} + \gamma \frac{dx}{dt} + \frac{\kappa_1}{m} x + \frac{\kappa_3}{m} x^3 = \frac{F(t)}{m}, \tag{H.79}$$

where γ is the friction damping rate, the spring constant $\kappa_1 = m\omega_0^2$, and $d^2 x/dt^2$ is acceleration. If the driving force, $F(t)$, is harmonic, and so of the form $\cos(\omega t)$, integration of Eq. (H.79) can give oscillatory solutions, bifurcation to a frequency halved behavior, and chaotic trajectories. The nonharmonic one-dimensional potential is

$$V(x) = \frac{\kappa_1}{2} x^2 + \frac{\kappa_3}{4} x^4. \tag{H.80}$$

Depending on the sign of the coefficients κ_1 and κ_3 the potential can be that due to a spring that gets stiffer with displacement from minimum potential energy or a double well (Duffing) potential.[11] Position $x(t)$ is obtained by numerically integrating

$$\frac{d^2 x}{dt^2} = -\frac{\kappa_1}{m} x - \frac{\kappa_3}{m} x^3 - \gamma \frac{dx}{dt} + \frac{F_0}{m} \cos(\omega t). \tag{H.81}$$

The results may be visualized using a position-time plot or a phasor diagram in which velocity is plotted as a function of position.

---

[11]The nonlinear equation describing an oscillator with a cubic nonlinearity in the force is called the Duffing equation and is named after Georg Duffing, a German engineer. He wrote a book, whose title when translated into English is: Forced oscillators with variable natural frequency and technical significance, *Series: Sammlung Vieweg* **41/42**, Vieweg and Sohn, Braunschweig (1918).

There exist many driven nonlinear dynamical systems, both in nature and man-made, that can exhibit a transition from periodic to chaotic behavior as a function of system control parameters. A common transition from periodic motion to chaos is via an infinite series of period-doubling bifurcations. Remarkably, a universal ratio characterizing the transition to chaos via such period-doubling bifurcations exists.[12]

H.3 Classical Electromagnetism

H.3.1 Electrostatics

The study of static distributions of charges and fields requires slow introduction of charge into the physical system. This type of adiabatic integration is a standard approach used to find solutions to many practical problems.

The electrostatic force due to a point charge Q_c in vacuum (free space) separated by a distance r from charge $-Q_c$ is

$$\mathbf{F}(\mathbf{r}) = \frac{-Q_c^2}{4\pi\varepsilon_0 r^2}\mathbf{r}^\sim, \tag{H.82}$$

where $\varepsilon_0 = 8.854\,1878 \times 10^{-12}$ F m^{-1} is the permittivity of free space[13] measured in units of farads per meter. Force is a vector field the direction of which, in this case, is given by the unit vector $\mathbf{r}^\sim$. It is a central force because it has no angular dependence. Electrostatic force is measured in units of newtons (N), and charge is measured in coulombs (C). An electron has charge $Q_c = -e = -1.602\,1765 \times 10^{-19}$ C.

The force experienced by a charge e in an electric field is $\mathbf{F} = e\mathbf{E}$, where the vector field $\mathbf{E}$ is the electric field whose magnitude is measured in units of volts per meter (V m^{-1}). The (negative) potential energy is the product of force and distance moved,

$$V = \int e\mathbf{E} \cdot \mathrm{d}\mathbf{x}. \tag{H.83}$$

Electrostatic force can be related to potential via $\mathbf{F} = -\nabla V$ (Eq. (H.4)) and hence the coulomb potential energy due to a point charge e in vacuum (free space) separated by a distance r from charge $-e$ is

$$V(r) = \frac{-e^2}{4\pi\varepsilon_0 r}. \tag{H.84}$$

The coulomb potential is a scalar field that has no angular dependence and is classified as a central-force potential. Coulomb potential energy is measured in joules (J) or electron volts (eV).

When there are no currents or time-varying magnetic fields in a nonmagnetic medium, the Maxwell equations for electric field $\mathbf{E}$ and magnetic flux density $\mathbf{B}$ are

$$\nabla \cdot \mathbf{E} = \frac{\rho}{\varepsilon_0 \varepsilon_r} \tag{H.85}$$

---

[12]M. J. Feigenbaum, *J. Stat. Phys.* **21**, 669 (1979).
[13]There is speculation that virtual quantum fluctuations of charged particle-antiparticle pairs may give rise to the polarizability of the vacuum. See, for example, G. Leuchs, M Hawton, and L. L. Sanchez-Soto, *Physics* **2**, 14 (2020).

and

$$\nabla \cdot \mathbf{B} = 0. \tag{H.86}$$

In Eq. (H.85), ε_r is the relative permittivity of the medium and ρ is charge density. The electric field is defined as $\mathbf{E} = -\nabla V$ and leads directly to Poisson's equation,

$$\boxed{\nabla^2 V = -\frac{\rho}{\varepsilon_0 \varepsilon_r},} \tag{H.87}$$

which relates potential to the local charge density.

Because electric field is the negative gradient of the potential, only *differences* in the potential are important. The direction of the electric field is positive from positive electric charge to negative electric charge. Sometimes it is useful to visualize electric field as field lines originating on a positive charge and terminating on a negative charge. The divergence of the electric field is the local charge density. It follows that the flux of electric field lines flowing out of a closed surface is equal to the charge enclosed. This is Gauss' law, which may be expressed as

$$\int_{V_{\text{vol}}} \nabla \cdot \mathbf{E} \, dV_{\text{vol}} = \oint_{S_{\text{surf}}} \mathbf{E} \cdot \mathbf{n}^{\sim} dS_{\text{surf}} = \int_{V_{\text{vol}}} \frac{\rho}{\varepsilon_0 \varepsilon_r} dV_{\text{vol}}, \tag{H.88}$$

where the first two integrals are expressions for the net electric flux out of the region of volume V_{vol} enclosed by surface S_{surf} with unit-normal vector $\mathbf{n}^{\sim}$ (Stokes' theorem) and the right-hand side is charge enclosed in volume V_{vol}.

Maxwell's expression for the divergence of the magnetic flux density given in Eq. (H.86) is interpreted physically as there being no magnetic monopoles (leaving the possibility of dipole and higher-order magnetic fields). Magnetic flux density $\mathbf{B}$ is a vector field, and its magnitude is measured in units of tesla (T).

Sometimes it is useful to define the displacement vector field, $\mathbf{D} = \varepsilon_0 \varepsilon_r \mathbf{E}$. In this expression, ε_r is the average value of the relative permittivity of the material in which the electric field exists. It is also useful to define the quantity $\mathbf{H} = \mathbf{B}/\mu_0 \mu_r$, which is the magnetic field vector where μ_0 is called the permeability of free space and μ_r is called the relative permeability.

H.3.1.1 *The Parallel Plate Capacitor*

Electric charge and energy can be stored by doing work to spatially separate charge Q_c and $-Q_c$ in a capacitor. Capacitance is the proportionality constant relating the voltage bias V_{bias} applied to the amount of charge stored. Capacitance is defined as

$$\boxed{C = \frac{Q_c}{V_{\text{bias}}}} \tag{H.89}$$

and is measured in units of farads (F).

A capacitor is a very useful device for storing electric charge. It is an essential part of the field-effect transistor used in silicon integrated circuits and thus is of great interest to electrical engineers.

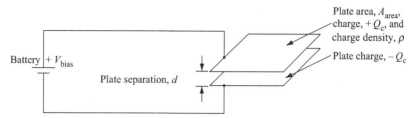

Fig. H.16 A parallel-plate capacitor attached to a battery supplying voltage V_{bias}. The capacitor consists of two thin, square, metal plates each of area A_{area} facing each other a distance d apart.

Maxwell's equations may be used to calculate how much charge can be stored for every volt of potential applied to a capacitor. Figure H.16 is an illustration of a parallel plate capacitor. Two thin, square, metal plates each of area A_{area} are placed facing each other a small distance d apart. One plate is attached to the positive terminal of a battery, and the other plate is attached to the negative terminal of the same battery, which supplies a voltage V_{bias}. The capacitance of the device may be calculated by noting that the charge per unit area on a plate is ρ and the voltage is the integral of the uniform electric field between the plates, so that $V_{bias} = |\mathbf{E}| \times d = \rho d / \varepsilon_0 \varepsilon_{r0}$. Hence,

$$C = \frac{Q_c}{V_{bias}} = \frac{\rho A_{area}}{\rho d / \varepsilon_0 \varepsilon_{r0}} = \frac{\varepsilon_0 \varepsilon_{r0} A_{area}}{d}, \tag{H.90}$$

where ε_{r0} is the low-frequency relative permittivity or dielectric constant of the material between the plates. This is an accurate measure of the capacitance, and errors due to fringing fields at the edges of the plates are necessarily small since $d \ll \sqrt{A_{area}}$.

A typical parallel-plate capacitor has $d = 100\,\text{nm}$ and $\varepsilon_{r0} = 10$, so the amount of extra charge per unit area per volt of potential difference applied is $Q_c = C V_{bias} = \varepsilon_0 \varepsilon_{r0} V_{bias} / d = 8.8 \times 10^{-4}\,\text{C m}^{-2}\,\text{V}^{-1}$ or, in terms of number of electrons per square centimeter per volt, $Q_c = 5.5 \times 10^{11}$ electrons $\text{cm}^{-2}\,\text{V}^{-1}$. This corresponds to one electron per $(13.5\,\text{nm})^2\,\text{V}^{-1}$. In a metal, this charge might sit in the first $0.5\,\text{nm}$ from the surface, giving a density of $\sim 10^{19}\,\text{cm}^{-3}$ or 10^{-4} of the typical bulk charge density in a metal of $10^{23}\,\text{cm}^{-3}$. A device of area $1\,\text{mm}^2$ with $d = 100\,\text{nm}$ and $\varepsilon_{r0} = 10$ has capacitance $C = 88\,\text{nF}$.

The extra charge sitting on the metal plates creates an electric field between the plates and stores energy in the capacitor. The energy is found by calculating the current that flows when a battery is attached that maintains voltage V_{bias}. The current flow, I, measured in amperes, is $dQ_c/dt = C\,dV_{bias}/dt$, so the instantaneous power supplied at time t to the capacitor is $I V_{bias}$, which is just $dQ_c/dt = C\,dV_{bias}/dt$ times the voltage. Hence, the instantaneous power is $C V_{bias}\,(dV_{bias}/dt)$. The energy stored in the capacitor is the integral of the instantaneous power from a time when there is no extra charge on the plates, say time $t' = -\infty$, to time $t' = t$. At $t = -\infty$ the voltage is zero and so the stored energy is

$$\Delta E_c = \int_{t'=-\infty}^{t'=t} C V_{bias} \frac{dV_{bias}}{dt'} dt' = \int_{V'_{bias}=0}^{V'_{bias}=V_{bias}} C V'_{bias} dV'_{bias} = \frac{1}{2} C V_{bias}^2 = \frac{Q_c^2}{2C}, \tag{H.91}$$

569

$$\boxed{\Delta E_{\mathrm{c}} = \frac{1}{2}CV_{\mathrm{bias}}^2.} \tag{H.92}$$

Since the capacitance of the parallel plate capacitor is $C = Q_{\mathrm{c}}/V_{\mathrm{bias}} = \varepsilon_0\varepsilon_{\mathrm{r}0}A_{\mathrm{area}}/d$ and the magnitude of the electric field is $|\mathbf{E}| = V_{\mathrm{bias}}/d$, the stored energy *per unit volume* may be written in terms of the electric field to give a stored *energy density* $\Delta U_{\mathrm{E}} = \varepsilon_0\varepsilon_{\mathrm{r}0}|\mathbf{E}|^2/2$. Finally, substituting in the expression $\mathbf{D} = \varepsilon_0\varepsilon_{\mathrm{r}0}\mathbf{E}$ gives

$$\boxed{\Delta U_{\mathrm{E}} = \frac{1}{2}\mathbf{E}\cdot\mathbf{D}} \tag{H.93}$$

for the energy stored per unit volume in the electric field. A similar result,

$$\boxed{\Delta U_{\mathrm{H}} = \frac{1}{2}\mathbf{B}\cdot\mathbf{H},} \tag{H.94}$$

holds for the energy stored per unit volume in a magnetic field.

Magnetic flux density can be stored in an inductor. Inductance is measured in units of henrys (H) and is defined in terms of magnetic flux linkage such that

$$L_{\mathrm{ind}} = \frac{1}{I}\int_{S_{\mathrm{surf}}}\mathbf{B}\cdot\mathbf{n}^{\sim}\mathrm{d}S_{\mathrm{surf}}, \tag{H.95}$$

where I is the current and S_{surf} is a specified surface through which magnetic flux passes. The unit normal vector to the surface S_{surf} is $\mathbf{n}^{\sim}$.

H.3.1.2 The Coulomb Blockade

The Coulomb blockade effect may be important in determining the operation of very small electronic devices. Figure H.17(a) shows an initially distant electron being moved through space and placed onto a small metal sphere. Figure H.17(b) is the corresponding energy-position diagram. The Coulomb blockade is the value of *charging energy* ΔE_{c} needed to place an extra electron onto a capacitor. This charging energy is discrete because the electron has a single value of charge.

The capacitance of a small sphere can be found by considering two spherical conducting metal shells of radii r_1 and r_2, where $r_1 < r_2$. Assuming that there is a charge $+Q_{\mathrm{c}}$ on the inner surface of the shell radius r_2 and a charge $-Q_{\mathrm{c}}$ on the outer surface of the shell radius r_1 and applying Gauss's law (Eq. (H.88)) it follows that at radius $r_1 < r < r_2$ the electric flux flowing out of a closed surface is equal to the charge enclosed, so that electric field in the radial direction is

$$E_r = \frac{Q_{\mathrm{c}}}{4\pi\varepsilon_0\varepsilon_{\mathrm{r}0}r^2}. \tag{H.96}$$

Since $\mathbf{E} = -\nabla V$, the potential is found by integration with the result

$$C = \frac{4\pi\varepsilon_0\varepsilon_{\mathrm{r}0}}{\left(\frac{1}{r_1} - \frac{1}{r_2}\right)}. \tag{H.97}$$

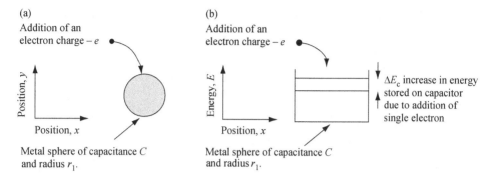

Fig. H.17 (a) An electron of charge $-e$ being placed onto a small metal sphere of radius r_1 and capacitance C. (b) Energy-position diagram for the system in (a). The vertical axis is energy stored by the capacitor and the horizontal axis is distance moved by the electron and size of the metal sphere. The increase in energy stored in the capacitor due to the addition of single electron is $\Delta E_{\rm c} = e^2/2C$.

For an isolated conducting sphere $r_2 \to \infty$, so that the capacitance is

$$C = 4\pi\varepsilon_0\varepsilon_{\rm r0}r_1. \tag{H.98}$$

The energy stored in the capacitor is the integral of the instantaneous power from time $t' = -\infty$ to time $t' = t$. At $t = -\infty$ the voltage is zero and the charging energy is the expression given by Eq. (H.91), $\Delta E_{\rm c} = Q_{\rm c}^2/2C$. If $Q_{\rm c} = -e$ then

$$\Delta E_{\rm c} = \frac{e^2}{2C} = \frac{e^2}{8\pi\varepsilon_0\varepsilon_{\rm r0}r_1} \tag{H.99}$$

is the charging energy for one electron placed onto a metal sphere of radius r_1 embedded in a dielectric with low-frequency relative permittivity $\varepsilon_{\rm r0}$. Because electron charge has a single value, an electron added to the metal sphere must have enough energy to overcome the coulomb charging energy, $\Delta E_{\rm c}$. The effect is called the *Coulomb blockade*. Very small single-electron devices such as single-electron transistors[14] and single-electron memory cells[15] exploit this effect. If a metal sphere embedded in a dielectric with $\varepsilon_{\rm r0} = 10$ has a radius $r_1 = 1\,{\rm nm}$ ($2\,{\rm nm}$ diameter $\sim 7\,{\rm atoms}$ diameter $\sim 180\,{\rm atoms}$ total), then $\Delta E_{\rm c} = 72\,{\rm meV}$. If $r_1 = 1\,{\rm nm}$ and $\varepsilon_{\rm r} = 1$, then $\Delta E_{\rm c} = 720\,{\rm meV}$.

Notice that explicit use is made of the fact that electron charge has a single value while a continuously variable charge was used to derive the charging energy, $\Delta E_{\rm c}$. A more rigorous theory should avoid this inconsistency.

H.3.1.3 *Charged Particle in a Harmonic Oscillator Potential Subject to Constant Electric Field* **E**

A particle of mass m and charge e with motion constrained to one dimension is attached to a spring with force constant $\kappa_0 = m\omega_0^2$ and is subject to a constant applied electric field

---

[14]K. Uchida et al. *IEEE Trans. Electron Devices* **50**, 1623 (2003); A. Fujiwara et al. *Appl. Phys. Lett.* **88**, 053121 (2006).

[15]K. Yano et al. *Proc. IEEE* **87**, 633 (1999); S. Huang et al. *IEEE Trans. Nanotech.* **3**, 210 (2004).

$\mathbf{E}$ in the positive x direction. The Hamiltonian of the system, which contains contributions from both the oscillator and the static electric field, is

$$\hat{H} = \frac{p^2}{2m} + \frac{m\omega_0^2}{2}x^2 + e|\mathbf{E}|x, \tag{H.100}$$

where $\mathbf{p}$ is particle momentum.

The force on the charged particle due to the electric field is $e\mathbf{E}$ and the new position of minimum potential energy of the particle is displaced from $x = 0$ to $x = x_\mathrm{d}$ with $e\mathbf{E} = \kappa_0 x_\mathrm{d}$. The total potential energy is

$$\begin{aligned} V &= \frac{1}{2}\kappa_0(x - x_\mathrm{d})^2 + e|\mathbf{E}|(x - x_\mathrm{d}) \\ &= \frac{1}{2}\kappa_0 x^2 - \kappa_0 x x_\mathrm{d} + \frac{1}{2}\kappa_0 x_\mathrm{d}^2 + e|\mathbf{E}|x - e|\mathbf{E}|x_\mathrm{d}. \end{aligned} \tag{H.101}$$

Now, since $e\mathbf{E} = \kappa_0 x_\mathrm{d}$, this may be written as

$$V = \frac{1}{2}\kappa_0 x^2 + \frac{1}{2}\kappa_0 x_\mathrm{d}^2 - e|\mathbf{E}|x_\mathrm{d} = \frac{m\omega_0^2}{2}x^2 - \frac{e^2|\mathbf{E}|^2}{2m\omega_0^2}. \tag{H.102}$$

The potential energy changes an amount $-e^2|\mathbf{E}|^2/2m\omega_0^2 = -\kappa_0 x_\mathrm{d}^2/2$ by storing energy in the spring. The potential energy always decreases because it is proportional to x_d^2 and so independent of the displacement direction.

H.3.2 Electrodynamics

Classical electrodynamics describes the spatial and temporal behavior of electric and magnetic fields. Although Maxwell published his paper on electrodynamics in 1864, it took quite some time before the predictions of the theory were confirmed. Maxwell's achievement was to show that time-varying electric fields are intimately coupled with time-varying magnetic fields. In fact, it is not possible to separate magnetic and electric fields – there are only *electromagnetic* fields.

The idea that $\mathbf{E}$ and $\mathbf{B}$ are *ordinary vector fields* is an essential element of the classical theory of electromagnetism. Just as with the description of large macroscopic bodies in classical mechanics, quantum effects are assumed to average out. The classical electromagnetic field may be viewed as a *classical macroscopic limit* of a quantum description in terms of photons. In many situations, there are large numbers of uncorrelated photons that contribute to the electromagnetic field over the time scales of interest, and so it is not necessary to consider the discrete (quantum) nature of photons.

Often, analysis of a complicated field in a linear system is simplified by decomposing the field into plane wave spatial components of angular frequency ω such that

$$Ae^{i(\mathbf{k}\cdot\mathbf{r} - \omega t)}, \tag{H.103}$$

where $\mathbf{r}$ is position, A is the amplitude of the wave, which may be a complex number, and $\mathbf{k}$ is the wave vector of magnitude $|\mathbf{k}| = k$ propagating in the $\mathbf{k}^\sim = \mathbf{k}/k$ direction. The fact that Eq. (H.103) is a complex quantity is a mathematical convenience. Only real values are measured. Because $\mathbf{E}$ and $\mathbf{B}$ are represented by ordinary vectors, the physical field is found by taking the real part of Eq. (H.103).

Table H.1 Maxwell equations

$\nabla \cdot \mathbf{D} = \rho$	Coulomb's law
$\nabla \cdot \mathbf{B} = 0$	No magnetic monopoles
$\nabla \times \mathbf{E} = -\dfrac{\partial \mathbf{B}}{\partial t}$	Faraday's law
$\nabla \times \mathbf{H} = \mathbf{J} + \dfrac{\partial \mathbf{D}}{\partial t}$	Modified Ampère's law

Maxwell's equations completely describe classical electric and magnetic fields. The differential form of the equations in SI-MKS units are given in Table H.1.

In these expressions, $\mathbf{D}$ is the displacement vector field and in linear media it is related to the electric field $\mathbf{E}$ by $\mathbf{D} = \varepsilon \mathbf{E} = \varepsilon_0 \varepsilon_r \mathbf{E} = \varepsilon_0 (1 + \chi_e) \mathbf{E} = \varepsilon_0 \mathbf{E} + \mathbf{P}$. The displacement vector field $\mathbf{D}$ may also be thought of as the electric flux density, which is measured in C m^{-2}, χ_e is the electric susceptibility, $\mathbf{P}$ is the electric polarization field. The magnetic field vector is $\mathbf{H}$ and $\mathbf{B}$ is the magnetic flux density. The convention is that $\mathbf{B} = \mu \mathbf{H} = \mu_0 \mu_r \mathbf{H} = \mu_0 (1 + \chi_m) \mathbf{H} = \mu_0 (\mathbf{H} + \mathbf{M})$, where μ is the permeability, μ_r is the relative permeability, χ_m is the magnetic susceptibility, and $\mathbf{M}$ is the magnetization. The current density measured in A m^{-2} is $\mathbf{J}$, and $\partial \mathbf{D}/\partial t$ is the displacement current measured in A m^{-2}. The permeability of free space, measured in henrys per meter, is $\mu_0 = 4\pi \times 10^{-7}$ H m^{-1}. The speed of light in free space is $c = 1/\sqrt{\varepsilon_0 \mu_0} = 2.99792458 \times 10^8$ m s^{-1}, and the impedance in free space is $Z_0 = \sqrt{\mu_0/\varepsilon_0}$ or approximately $376.73\,\Omega$.

The use of vector calculus to describe Maxwell's equations enables a very compact and efficient description and derivation of relationships between fields.

In vector calculus, the divergence of a vector field is a source, so $\nabla \cdot \mathbf{B} = 0$ in Maxwell's equations is interpreted as the absence of sources of magnetic flux density, or, equivalently, the absence of magnetic monopoles. By the same interpretation, the electric charge density is the source of the displacement vector field, $\mathbf{D}$.

The divergence of the modified Ampère's law given in Table H.1 is

$$\nabla \cdot (\nabla \times \mathbf{H}) = \nabla \cdot \mathbf{J} + \nabla \cdot \frac{\partial \mathbf{D}}{\partial t} = 0, \tag{H.104}$$

since from vector calculus $\nabla \cdot (\nabla \times \mathbf{a}) = 0$. Using Coulomb's law, $\nabla \cdot \mathbf{D} = \rho$, results in

$$\boxed{0 = \nabla \cdot \mathbf{J} + \frac{\partial \rho}{\partial t},} \tag{H.105}$$

which is an expression of *current continuity*. Physically, an increase in charge density in some volume of space is caused by current flowing through the surface enclosing the volume. The current continuity equation expresses the idea that electric charge is a conserved quantity. According to Eq. (H.105), charge does not spontaneously appear or disappear, but rather it is transported from one region of space to another by current.

To further illustrate the usefulness of vector calculus, consider Ampère's circuital law, which states that a line integral of static magnetic field taken about any given closed path must equal the current, I, enclosed by that path. The sign convention is that current is positive if advancing in a right-hand screw sense, where the screw rotation is in the direction of circulation for line integration. Vector calculus allows Ampère's law to be derived almost trivially using Maxwell's equations and Stokes' theorem, as follows:

$$\oint \mathbf{H} \cdot d\mathbf{l} = \int_{S_{\text{surf}}} (\nabla \times \mathbf{H}) \cdot \mathbf{n}^{\sim} dS_{\text{surf}} = \int_{S_{\text{surf}}} \mathbf{J} \cdot \mathbf{n}^{\sim} dS_{\text{surf}} = I. \tag{H.106}$$

H.3.2.1 Light Propagation in a Dielectric Medium

In a dielectric, current density $\mathbf{J} = 0$ because the dielectric has no mobile charge, and if $\mu_r = 1$ at optical frequencies then $\mathbf{H} = \mathbf{B}/\mu_0$. Hence, using the equations in Table H.1, taking the curl of the expression of Faraday's law and using the modified Ampère's law,

$$\nabla \times (\nabla \times \mathbf{E}) = -\frac{\partial}{\partial t}(\nabla \times \mathbf{B}) = -\mu_0 \frac{\partial}{\partial t}(\nabla \times \mathbf{H}) = -\mu_0 \frac{\partial^2}{\partial t^2}\mathbf{D}. \tag{H.107}$$

The left-hand term may be rewritten by making use of the relationship $\nabla \times \nabla \times \mathbf{a} = \nabla(\nabla \cdot \mathbf{a}) - \nabla^2 \mathbf{a}$, so that

$$\nabla(\nabla \cdot \mathbf{E}) - \nabla^2 \mathbf{E} = -\mu_0 \frac{\partial^2}{\partial t^2}\mathbf{D}. \tag{H.108}$$

For a source-free dielectric $\nabla \cdot \mathbf{D} = 0$ (and assuming ε in the medium is not a function of space – it is isotropic – so that $\nabla \cdot \mathbf{E} = 0$), this becomes

$$\nabla^2 \mathbf{E}(\mathbf{r}, t) = \frac{\partial^2}{\partial t^2}\mu_0 \varepsilon \mathbf{E}(\mathbf{r}, t), \tag{H.109}$$

where $\varepsilon = \varepsilon_0 \varepsilon_r$ is the complex permittivity function that results in $\mathbf{D}(\mathbf{r}, t) = \varepsilon_0 \varepsilon_r \mathbf{E}(\mathbf{r}, t)$. Taking the Fourier transform with respect to time gives a *wave equation* for electric field:

$$\nabla^2 \mathbf{E}(\mathbf{r}, \omega) = -\omega^2 \mu_0 \varepsilon_0 \varepsilon_r(\omega) \mathbf{E}(\mathbf{r}, \omega). \tag{H.110}$$

Since the speed of light is $c = 1/\sqrt{\varepsilon_0 \mu_0}$, the wave equation can be written as

$$\boxed{\nabla^2 \mathbf{E}(\mathbf{r}, \omega) = \frac{-\omega^2}{c^2}\varepsilon_r(\omega)\mathbf{E}(\mathbf{r}, \omega).} \tag{H.111}$$

If $\varepsilon_r(\omega)$ is real and positive, the solutions to this wave equation for an electric field propagating in a homogeneous isotropic medium are just plane waves. The speed of wave propagation is $c/n_r(\omega)$, where $n_r(\omega) = \sqrt{\varepsilon_r(\omega)}$ is the *refractive index* of the material. In the more general case, when relative permeability $\mu_r \neq 1$, the refractive index is

$$n_r(\omega) = \sqrt{\varepsilon_r(\omega)}\sqrt{\mu_r(\omega)} = \frac{\sqrt{\varepsilon(\omega)}\sqrt{\mu(\omega)}}{\sqrt{\varepsilon_0 \mu_0}}. \tag{H.112}$$

If ε and μ are both real and positive, the refractive index is real-positive and electromagnetic waves propagate in a bulk medium. In nature, the refractive index in a transparent material usually takes a positive value. If one of either ε or μ is negative, the refractive index is imaginary and electromagnetic waves cannot propagate in a bulk medium.

It is common for metals to have negative values of ε. Free electrons of mass m in a bulk metal can collectively oscillate at a long-wavelength natural frequency called the plasma frequency, ω_p. In a three-dimensional gas of electrons of density n the plasma frequency is $\omega_p = \sqrt{ne^2/\varepsilon_0 m}$, which for $m = m_0$ and $10^{21}\,\text{cm}^{-3} < n < 10^{23}\,\text{cm}^{-3}$ gives $1\,\text{eV} < \hbar\omega_p < 12\,\text{eV}$. At long wavelengths a good approximation for relative permittivity of a bulk metal is $\varepsilon_r(\omega) = 1 - \omega_p^2/\omega^2$. At frequencies above the plasma frequency ε is positive and electromagnetic waves can propagate through the metal. For frequencies below ω_p permittivity is negative, the refractive index is imaginary, and electromagnetic waves cannot propagate in a bulk metal and are reflected. This is the reason why bulk metals are usually not transparent to electromagnetic radiation of frequency less than ω_p.

Another possibility is that both ε and μ are real and negative. Such *metamaterial*, artificial structures with behavior not normally occurring in nature, may have negative relative permittivity and negative relative permeability simultaneously over some frequency range.

In the simple situation, relative permeability $\mu_r = 1$ and relative permittivity function $\varepsilon_r(\omega) = 1$, an electric-field plane-wave propagating in the $\mathbf{k}^\sim$-direction with real wavevector $\mathbf{k}$ and constant complex vector $\mathbf{E}_0$ has a spatial dependence of the form $\mathbf{E}(\mathbf{r}) = \mathbf{E}_0 e^{i\mathbf{k}\cdot\mathbf{r}}$. The magnitude of the amplitude of the wave is $|\mathbf{E}_0|$. The electromagnetic field wave has a linear *dispersion relation* obtained from the wave equation which, in free space, is $\omega = ck$. When dispersion $\omega = \omega(\mathbf{k})$ is nonlinear, *phase velocity ω/k, group velocity $\partial\omega/\partial k$*, and the energy velocity of waves can all be different.

If the electromagnetic wave propagates in a homogeneous isotropic dielectric medium characterized by $\mu_r = 1$ and a complex relative permittivity function, then $\varepsilon(\omega) = \varepsilon_0\varepsilon_r(\omega) = \varepsilon_0(\varepsilon_r'(\omega) + i\varepsilon_r''(\omega))$, where $\varepsilon_r'(\omega)$ and $\varepsilon_r''(\omega)$ are the real and imaginary parts, respectively, of the frequency-dependent relative permittivity function. In this situation, the wave vector $\mathbf{k}(\omega)$ may become complex, giving an electric field

$$\mathbf{E}(\mathbf{r}, \omega) = \mathbf{E}_0(\omega)e^{i\mathbf{k}(\omega)\cdot\mathbf{r}} = \mathbf{E}_0(\omega)e^{i(k'(\omega)+ik''(\omega))\mathbf{k}^\sim\cdot\mathbf{r}}, \tag{H.113}$$

where $k'(\omega)$ and $k''(\omega)$ are the real and imaginary parts, respectively, of the frequency-dependent wave number. The ratio of $k'(\omega)$ in the medium and $k = \omega/c$ in free space is the refractive index. Because $\mu_r = 1$ in the dielectric being considered, the refractive index is

$$n_r(\omega) = \sqrt{\frac{1}{2}\left(\varepsilon_r'(\omega) + \sqrt{\varepsilon_r'^2(\omega) + \varepsilon_r''^2(\omega)}\right)}. \tag{H.114}$$

Equation (H.114) is obtained by substituting Eq. (H.113) into Eq. (H.111) and separating the real and imaginary parts of the resulting expression.

If the imaginary part of $\mathbf{k}(\omega)$ is positive, then this physically corresponds to an exponential spatial decay in field amplitude $e^{-k''r}$ due to absorption processes.

If $k''(\omega) = 0$ and $\mu_r = 1$, the refractive index is $n_r(\omega) = \sqrt{\varepsilon_r'(\omega)}$ and there is no spatial decay in the electric field, so that

$$\mathbf{E}(\mathbf{r}, \omega) = \mathbf{E}_0 e^{-i\omega t}e^{i\mathbf{k}(\omega)\cdot\mathbf{r}} \tag{H.115}$$

and

$$\mathbf{H}(\mathbf{r}, \omega) = \mathbf{H}_0 e^{-i\omega t}e^{i\mathbf{k}(\omega)\cdot\mathbf{r}} \tag{H.116}$$

for the magnetic field vector. In this case, Maxwell's equations for electromagnetic waves in free space are:

$$\nabla \cdot \mathbf{D} = 0, \tag{H.117}$$

$$\nabla \cdot \mathbf{B} = 0, \tag{H.118}$$

$$\nabla \times \mathbf{E} = -\frac{\partial \mathbf{B}}{\partial t}, \tag{H.119}$$

$$\nabla \times \mathbf{H} = \frac{\partial \mathbf{D}}{\partial t}. \tag{H.120}$$

The first two equations require that $\mathbf{k} \cdot \mathbf{E} = 0$ and $\mathbf{k} \cdot \mathbf{B} = 0$. This means that $\mathbf{E}$ and $\mathbf{B}$ are perpendicular (transverse) to the direction of propagation $\mathbf{k}^{\sim}$. To find the relationship between transverse electric and magnetic field vectors in free space, consider the plane-wave expressions for $\mathbf{E}(\mathbf{r}, \omega)$ and $\mathbf{H}(\mathbf{r}, \omega)$ given by Eqs. (H.115) and (H.116). Substituting them into the first curl equation (Eq. (H.119)) and recalling that in free space $\mathbf{H} = \mathbf{B}/\mu_0$ gives

$$\nabla \times \mathbf{E}_0 e^{-i\omega t} e^{i\mathbf{k}(\omega)\cdot\mathbf{r}} = -\mu_0 \frac{\partial}{\partial t} \mathbf{H}_0 e^{-i\omega t} e^{i\mathbf{k}(\omega)\cdot\mathbf{r}}, \tag{H.121}$$

$$i\mathbf{k} \times \mathbf{E}_0 e^{-i\omega t} e^{i\mathbf{k}(\omega)\cdot\mathbf{r}} = i\omega \, \mu_0 \mathbf{H}_0 e^{-i\omega t} e^{i\mathbf{k}(\omega)\cdot\mathbf{r}}, \tag{H.122}$$

$$i\mathbf{k} \times \mathbf{E} = i\omega \, \mu_0 \mathbf{H}. \tag{H.123}$$

Using the fact that the dispersion relation for plane waves in free space is $\omega = ck$ and the speed of light is $c = 1/\sqrt{\varepsilon_0 \mu_0}$, leads directly to

$$\boxed{\mathbf{H} = \sqrt{\frac{\varepsilon_0}{\mu_0}} \mathbf{k}^{\sim} \times \mathbf{E},} \tag{H.124}$$

or $\mathbf{B}c = \mathbf{k}^{\sim} \times \mathbf{E}$, where $\mathbf{k}^{\sim} = \mathbf{k}/|\mathbf{k}|$ is the unit vector for $\mathbf{k}$.

It is not possible to separate out an oscillating electric or magnetic field; they are related to each other in such a way that there are only *electromagnetic* fields. Figure H.18 illustrates the magnetic field and the electric field for a linearly polarized plane-wave propagating in free space in the x direction. The shading is to help guide the eye. In general, oscillating transverse electromagnetic waves can change both in time and in space.

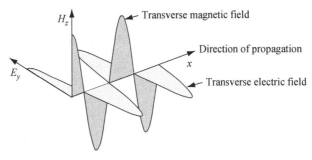

Fig. H.18 Illustration of transverse magnetic field H_z and electric field E_y of a plane wave propagating in free space in the x direction with wave vector $\mathbf{k}$. The shading is to guide the eye. The electric field is linearly polarized in the y direction.

Electrodynamics is related to electrostatics by the fact that a static electric field can only be formed by movement of charge, and hence by a current. When a transient current flows, a magnetic field is produced. It is not possible to form a static electric field without a transient current and an associated magnetic field.

H.3.2.2 Power and Momentum in an Electromagnetic Wave

The power in an electromagnetic wave can be obtained by considering the response of a test charge e moving at velocity $\mathbf{v}$ in an external electric field $\mathbf{E}$. The rate of work or power is $e\mathbf{v} \cdot \mathbf{E}$, where $e\mathbf{v}$ is a current. This may be generalized to a continuous distribution of current density $\mathbf{J}$ so that the total power in a given volume is

$$\int_{V_{\text{vol}}} d^3r \, \mathbf{J} \cdot \mathbf{E} = \int_{V_{\text{vol}}} \left(\mathbf{E} \cdot (\nabla \times \mathbf{H}) - \mathbf{E} \cdot \frac{\partial \mathbf{D}}{\partial t} \right) d^3r, \tag{H.125}$$

where Maxwell's equation, $\nabla \times \mathbf{H} = \mathbf{J} + \partial \mathbf{D}/\partial t$, has been used. Because this is the power that is extracted from the electromagnetic field, energy conservation requires that there must be a corresponding reduction in electromagnetic field energy in the same volume. Making use of the result from vector calculus, $\mathbf{E} \cdot (\nabla \times \mathbf{H}) = \mathbf{H} \cdot (\nabla \times \mathbf{E}) - \nabla \cdot (\mathbf{E} \times \mathbf{H})$, and Maxwell's equation, $\nabla \times \mathbf{E} = -\partial \mathbf{B}/\partial t$,

$$\int_{V_{\text{vol}}} d^3r \, \mathbf{J} \cdot \mathbf{E} = -\int_{V_{\text{vol}}} \left(\nabla \cdot (\mathbf{E} \times \mathbf{H}) + \mathbf{E} \cdot \frac{\partial \mathbf{D}}{\partial t} + \mathbf{H} \cdot \frac{\partial \mathbf{B}}{\partial t} \right) d^3r, \tag{H.126}$$

or, in differential form,

$$\mathbf{E} \cdot \frac{\partial \mathbf{D}}{\partial t} + \mathbf{H} \cdot \frac{\partial \mathbf{B}}{\partial t} = -\mathbf{J} \cdot \mathbf{E} - \nabla \cdot (\mathbf{E} \times \mathbf{H}). \tag{H.127}$$

Generalizing the result for energy density stored in static electric and magnetic fields (Eqs. (H.93) and (H.94), respectively) to electromagnetic waves in a medium with linear response and no dispersion, the total energy density is

$$U_{\text{S}} = \frac{1}{2}(\mathbf{E} \cdot \mathbf{D} + \mathbf{B} \cdot \mathbf{H}). \tag{H.128}$$

(The time averaged energy density is half this value.) After substitution into the differential expression (Eq. (H.127)), this gives

$$\boxed{\frac{\partial U_{\text{S}}}{\partial t} = -\mathbf{J} \cdot \mathbf{E} - \nabla \cdot \mathbf{S},} \tag{H.129}$$

where

$$\boxed{\mathbf{S} = \mathbf{E} \times \mathbf{H}} \tag{H.130}$$

is the *Poynting vector*, which is the energy flux density in the electromagnetic field. The magnitude of $\mathbf{S}$ is measured in units of $\text{J m}^{-2} \text{ s}^{-1}$, and the negative divergence of $\mathbf{S}$ is the flow of electromagnetic energy out of the system. The rate of change of electromagnetic energy density is given by the rate of loss due to work done by the electromagnetic field density on sources given by $-\mathbf{J} \cdot \mathbf{E}$ and the rate of loss due to electromagnetic energy flow given by $-\nabla \cdot \mathbf{S}$.

In the absence of sources of current, the energy per unit volume is the energy flux density divided by the speed of light. Thus,

$$U_S = \frac{|\mathbf{S}|}{c}. \tag{H.131}$$

Energy density, U_S, is measured in units of J m^{-3} or, equivalently, in units of kg m^{-1} s^{-2}.

Substituting $\mathbf{H} = \sqrt{\varepsilon_0/\mu_0}\, \mathbf{k}^\sim \times \mathbf{E}$ (Eq. (H.124)) into the expression for the Poynting vector gives

$$\mathbf{S} = \mathbf{E} \times \mathbf{H} = \sqrt{\frac{\varepsilon_0}{\mu_0}}\mathbf{E} \times \mathbf{k}^\sim \times \mathbf{E} = \sqrt{\frac{\varepsilon_0}{\mu_0}}((\mathbf{E} \cdot \mathbf{E})\mathbf{k}^\sim - (\mathbf{E} \cdot \mathbf{k}^\sim)\mathbf{E}) \tag{H.132}$$

since, from vector calculus, $\mathbf{a} \times (\mathbf{b} \times \mathbf{c}) = (\mathbf{a} \cdot \mathbf{c})\mathbf{b} - (\mathbf{a} \cdot \mathbf{b})\mathbf{c}$. For transverse electromagnetic waves, the second term on the right-hand side $(\mathbf{E} \cdot \mathbf{k}^\sim)$ is zero, and so the energy flux in the electromagnetic field becomes

$$\mathbf{S} = \sqrt{\frac{\varepsilon_0}{\mu_0}}(\mathbf{E} \cdot \mathbf{E})\mathbf{k}^\sim = \frac{(\mathbf{E} \cdot \mathbf{E})}{Z_0}\mathbf{k}^\sim, \tag{H.133}$$

where $Z_0 \equiv \sqrt{\mu_0/\varepsilon_0}$ is defined as the *impedance of free space*.[16]

Electromagnetic waves also carry momentum and so can exert a force on a charged particle. The classical Lorentz force on a test charge e moving at velocity $\mathbf{v}$ is

$$\mathbf{F} = e(\mathbf{E} + \mathbf{v} \times \mathbf{B}). \tag{H.134}$$

Using Newton's second law for mechanical motion, which relates the rate of change of momentum to force (Eq. (H.1)), it is possible to show that the momentum density in an electromagnetic wave is the energy flux density multiplied by $1/c^2$, so that

$$\mathbf{p} = \frac{\mathbf{E} \times \mathbf{H}}{c^2} = \frac{\mathbf{S}}{c^2} \tag{H.135}$$

and, for a plane wave,

$$\boxed{\mathbf{p} = \frac{U_S}{c}\mathbf{k}^\sim.} \tag{H.136}$$

In this equation, $\mathbf{k}^\sim$ is the unit vector in the direction of propagation of the wave, U_S is the energy density, and c the speed of light. The magnitude of the momentum is

$$\boxed{|\mathbf{p}| = \frac{1}{c}\frac{|\mathbf{S}|}{c} = \frac{U_S}{c} = p.} \tag{H.137}$$

---

[16] $Z_0 \simeq 376.73\,\Omega$. It is often convenient to use the approximation $Z_0 = 120 \times \pi\,\Omega$.

H.3.2.3 Choosing a Potential

In general, Maxwell's equations allow electric and magnetic fields to be described in terms of a scalar potential, $V(\mathbf{r}, t)$, and a vector potential, $\mathbf{A}(\mathbf{r}, t)$.

Because Maxwell's equations state that $\nabla \cdot \mathbf{B} = 0$, it must be possible to choose a vector field $\mathbf{A}$ for which $\mathbf{B} = \nabla \times \mathbf{A}$ since any vector field $\mathbf{a}$ satisfies $\nabla \cdot (\nabla \times \mathbf{a}) = 0$. Using this expression for $\mathbf{B}$, Faraday's law can now be rewritten as

$$\nabla \times \mathbf{E} = -\frac{\partial \mathbf{B}}{\partial t} = -\frac{\partial}{\partial t} \nabla \times \mathbf{A} \tag{H.138}$$

or

$$\nabla \times \left(\mathbf{E} + \frac{\partial \mathbf{A}}{\partial t} \right) = 0. \tag{H.139}$$

The curl of the gradient of any scalar field is zero, so the last equation may be equated with the gradient of a scalar field, V, where

$$\mathbf{E} + \frac{\partial \mathbf{A}}{\partial t} = -\nabla V. \tag{H.140}$$

In general, the scalar and vector fields are functions of space and time, and there is freedom choosing functions $V(\mathbf{r}, t)$ and $\mathbf{A}(\mathbf{r}, t)$, giving

$$\mathbf{E}(\mathbf{r}, t) = -\nabla V(\mathbf{r}, t) - \frac{\partial}{\partial t} \mathbf{A}(\mathbf{r}, t), \tag{H.141}$$

$$\mathbf{B}(\mathbf{r}, t) = \nabla \times \mathbf{A}(\mathbf{r}, t). \tag{H.142}$$

The form, or *gauge*, used for $V(\mathbf{r}, t)$ and $\mathbf{A}(\mathbf{r}, t)$ is chosen to simplify a specific calculation.

Consider a *static* electric field, $\mathbf{E}(\mathbf{r})$. Previously, the gauge where $\mathbf{E}(\mathbf{r}) = -\nabla V(\mathbf{r})$ was chosen. An interesting consequence of this choice is that, because the static electric field is expressed as a gradient of a potential, the absolute value of the potential need not be known to within a constant. Only differences in potential have physical consequences and therefore meaning. Another choice of gauge is where $\mathbf{A} = -\mathbf{E}t$. In this case, there is a remarkable degree of latitude in the choice of $\mathbf{A}$, since any time-independent function can be added to $\mathbf{A}$ without changing the electric field. Other possibilities for the gauge involve combinations of the scalar and vector potential.

H.3.2.4 Dipole Radiation

When a current flows through a conductor, a magnetic field exists in space around the conductor. This fact is described by the modified Ampère's law in Maxwell's equations in Table H.1. If the magnitude of the current varies over time, then so does the magnetic field. Faraday's law indicates that a changing magnetic field coexists with a changing electric field. Hence, a conductor carrying an oscillating current is always surrounded by interdependent oscillating $\mathbf{E}$ and $\mathbf{H}$ fields. As the oscillating current changes its direction in the conductor, the magnetic field also tries to change direction. To do so, the magnetic field that exists in space must first try to disappear by collapsing into the conductor before growing again in the opposite direction. Typically, not all of the magnetic field disappears before current flows in the opposite direction. The portions of the magnetic field and its related electric field that are unable to return to the conductor before the current starts

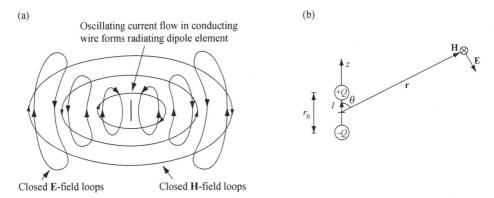

Fig. H.19 (a) Illustration of a short conductor carrying an oscillating harmonic time-dependent current surrounded by interdependent oscillating **E** and **H** fields. The oscillating current flowing up and down the conducting wire produces electromagnetic waves that propagate in free space. (b) A small length of conducting wire connects two conducting spheres oriented in the z direction that have center-to-center spacing of r_0. Oscillatory current, I, flows in the wire, charging and discharging the spheres. The magnetic and electric field at position r is indicated.

to increase in the opposite direction are propagated away as electromagnetic radiation. In this way, an oscillating current (associated with acceleration and deceleration of charge in a conductor) produces electromagnetic waves. This is illustrated in Fig. H.19(a) for a short conducting wire carrying a harmonic time-dependent current.

Radiation due to changes in current in a conductor may be described by considering an element of length r_0 carrying oscillating current $I(t)$. An appropriate arrangement is shown schematically in Fig. H.19(b). A small length of conducting wire connects two conducting spheres separated by distance r_0 and oriented in the $\mathbf{z}^{\sim}$ direction. Oscillatory current flows in the wire so that $I(t) = I_0 e^{iwt}$, where measurable current is the real part of this function. The harmonic time-dependent current is related to the charge on the spheres by $I(t) = \pm dQ(t)/dt$, where $Q(t) = Q_0 e^{iwt}$. The plus sign is for the upper sphere, and the minus sign is for the lower sphere. It follows that $I(t) = \pm d\left(Q_0 e^{iwt}\right)/dt = \pm i\omega Q(t)$, so that $Q(t) = \pm I(t)/i\omega$. For equal and opposite charges separated by a small distance, the dipole moment for the harmonic time-dependent source may be defined as

$$\mathbf{d} = Q\mathbf{z}^{\sim} r_0 = \frac{Ir_0\mathbf{z}^{\sim}}{i\omega}. \tag{H.143}$$

If either the current, I, or the current density, $\mathbf{J}$, is known, then the other quantities of interest, such as the total radiated electromagnetic power, P_r, may be found.

The total time-averaged far-field radiated electromagnetic power P_r in free space from a sinusoidally oscillating dipole source is

$$\boxed{P_r = \frac{Z_0}{12\pi} \frac{\omega^4 |\mathbf{d}|^2}{c^2}.} \tag{H.144}$$

Approximating $Z_0 = 120 \times \pi\,\Omega$, the time-averaged radiated power is $P_r = 10\,\omega^4 |\mathbf{d}|^2/c^2$.

H.4   Example Exercises

Exercise H.1
Two intelligent players seated at a round table alternately place round beer mats on the table. The beer mats are not allowed to overlap, and the last player to place a mat on the table·wins. Who wins and what is the strategy?

Exercise H.2
Visitors from another planet wish to measure the circumference of the Earth. To do this, they run a tape measure around the equator. How many extra meters are needed if the tape measure is raised one meter above the ground?[17]

Exercise H.3
A plug fits exactly *into* a square hole of side 2 cm, a circular hole of radius $r = 1$ cm, and an isosceles triangular hole with a base 2 cm wide and a height of $h = 2$ cm. What is the smallest and largest *convex* solid volume of the plug?

Exercise H.4
A molecule consists of two atoms of mass m_1 and mass m_2 with position $\mathbf{r}_1$ and $\mathbf{r}_2$, respectively. Show that the Hamiltonian can be separated into center of mass motion and relative motion of the two atoms. If the potential for relative atom motion is quadratic in space, show that the frequency of oscillation about the position of minimum potential energy is $\omega = \sqrt{\kappa_0/m_\mathrm{r}}$, where κ_0 is the spring constant and m_r is the *reduced mass* such that $1/m_\mathrm{r} = 1/m_1 + 1/m_2$.

Exercise H.5
A micro-electromechanical systems (MEMS) structure is a cantilever beam shown in Fig. H.E20. The lowest-frequency vibrational mode of a long, thin cantilever beam attached at one end is[18]

$$\nu = \frac{3.52}{2\pi} \frac{d}{l^2} \sqrt{\frac{E_\mathrm{Young}}{12\rho}},$$

where l is the length, d is the thickness, ρ is the density of the beam, and E_Young, defined as uniaxial tensile stress divided by strain in bulk material, is Young's modulus.

(a) A cantilever made of silicon using a MEMS process has dimension $l = 100\,\mu$m, $d = 0.1\,\mu$m, $\rho = 2.328 \times 10^3\,\mathrm{kg\,m^{-3}}$, and Young's modulus $E_\mathrm{Young} = 1.96 \times 10^{11}\,\mathrm{N\,m^{-2}}$. Calculate the natural frequency of vibration for the cantilever.

(b) The vibrational energy of a cantilever with width w and free-end displacement amplitude A is

[17]The visitors were not aware that the meter was first defined by the French Academy of Sciences in 1791 as $1/10^7$ of the quadrant of the Earth's circumference running from the North Pole through Paris to the equator.

[18]L. D. Landau and E. M. Lifshitz, *Theory of Elasticity*, Oxford, Butterworth–Heinemann, 1986 (ISBN 0 7506 2633 X).

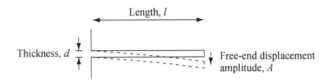

Fig. H.E20

$$\frac{wd^3 A^2 E_{\text{Young}}}{6l^3}.$$

How much vibrational energy is there in the cantilever of (a) when width $w = 5\,\mu\text{m}$ and amplitude $A = 1\,\mu\text{m}$?

Exercise H.6
A spherical dendrimer structure has a core that is a redox active molecule. The charge state of an iron atom in the core can be neutral or $+e$. Highly fluorescent rhodamine dye molecules incorporated into the dendrimer are sensitive to the local electric field and can sense the charged state of the redox core.

The dendrimer, 4 nm in diameter with $\varepsilon_{\text{r0}} = 2$, is placed on a flat, perfectly conducting metal sheet maintained at zero electrical potential. Calculate the maximum force on the ion core when the fluorescence emission spectrum indicates the iron atom is in the charged state.

Exercise H.7
The circuit shown in Fig. H.E21 is energized by an input current, I, from a current source at resonant frequency $\omega = 1/\sqrt{LC}$. The values of the series inductor and capacitor are L and C, respectively. On resonance, the Q of the circuit is large and is given by $Q = \omega L/r$, where r is the value of a series resistor. Show that $|V_2| \sim Q|V_1|$ when resistance R is nonzero.

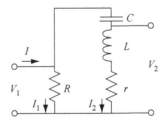

Fig. H.E21

Exercise H.8
The amplitude of magnetic flux density, B, in a monochromatic plane-polarized electromagnetic wave traveling in a vacuum is 10^{-6} T. Calculate the value of the total energy density. How is the total energy density divided between the electric and magnetic components?

582

Solutions

Solutions H.1
The first player to place a beer mat on the table is guaranteed to win the game if he or she uses the *symmetry* of the problem to their advantage. Only the first player need be intelligent to guarantee winning. The second player must merely abide by the rules of the game. The first player places the first beer mat in the center of the table and then always places a beer mat symmetrically opposite the position chosen by the other player.

Solutions H.2
Since the circumference of a sphere of radius R is just $2\pi R$, the extra length needed is only $2\pi = 6.28$ m. Because the circumference of a sphere is linear in the radius, the same amount of extra tape is needed *independent* of the Earth's radius.

Solutions H.3
The plug has a circular base of radius $r = 1$ cm, a height of $h = 2$ cm, and a straight top edge of length 2 cm (see Fig. H.E22). The *smallest* convex solid is made of isosceles triangle cross-sections perpendicular to the circular base and straight edge. Since all the triangular cross-sections combine to make up the volume of the plug, the volume is half that of a cylinder height h and radius r. The cylinder's volume is $\pi r^2 h = 2\pi$ cm^3, so the answer is π cm^3.

The *largest* convex volume is $(2\pi - 8/3)$ cm^3 and found by slicing the cylinder with two plane cuts to obtain the needed isosceles triangle cross-section.

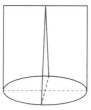

Fig. H.E22

Solutions H.4
A diatomic molecule consists of two atoms with mass m_1 and m_2 and position $\mathbf{r}_1$ and $\mathbf{r}_2$, respectively. The center of mass coordinate is $\mathbf{R}$ and relative position vector is $\mathbf{r}$ (see Fig. H.E23). If the forces, and hence the potential, governing relative motion depend only on the magnitude of the difference vector then $|\mathbf{r}| = |\mathbf{r}_2 - \mathbf{r}_1|$. Choosing the origin as the center of mass gives $m_1\mathbf{r}_1 + m_2\mathbf{r}_2 = 0$, so that

$$\mathbf{r}_1 = \frac{-m_2}{(m_1 + m_2)}\mathbf{r} \quad \text{and} \quad \mathbf{r}_2 = \frac{m_1}{(m_1 + m_2)}\mathbf{r}.$$

That this is so is easy to see since, for example,

$$\mathbf{r}_1 = \frac{-m_2}{m_1}\mathbf{r}_2 = \frac{-m_2}{m_1}(\mathbf{r}_1 + \mathbf{r}),$$

583

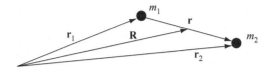

$$\text{Fig. H.E23}$$

$$\mathbf{r}_1\left(1 + \frac{m_2}{m_1}\right) = \frac{-m_2}{m_1}\mathbf{r},$$

$$\mathbf{r}_1(m_1 + m_2) = -m_2\mathbf{r},$$

$$\mathbf{r}_1 = \frac{-m_2}{(m_1 + m_2)}\mathbf{r}.$$

Combining the center of mass motion and relative motion, the Hamiltonian is the sum of kinetic and potential energy terms T and V, respectively, and

$$H = T + V$$

$$= \frac{1}{2}m_1\left(\frac{d\mathbf{R}}{dt} - \frac{m_2}{(m_1 + m_2)}\frac{d\mathbf{r}}{dt}\right)^2 + \frac{1}{2}m_2\left(\frac{d\mathbf{R}}{dt} + \frac{m_1}{(m_1 + m_2)}\frac{d\mathbf{r}}{dt}\right)^2 + V(|\mathbf{r}|),$$

where the total kinetic energy is

$$T = \frac{1}{2}(m_1 + m_2)\left(\frac{d\mathbf{R}}{dt}\right)^2 + \frac{1}{2}\frac{m_1 m_2}{(m_1 + m_2)}\left(\frac{d\mathbf{r}}{dt}\right)^2 = \frac{1}{2}M\left(\frac{d\mathbf{R}}{dt}\right)^2 + \frac{1}{2}m_\mathrm{r}\left(\frac{d\mathbf{r}}{dt}\right)^2.$$

The total mass is

$$M = m_1 + m_2$$

and the *reduced mass* is

$$m_\mathrm{r} = \frac{m_1 m_2}{(m_1 + m_2)}.$$

Because the potential energy of the two interacting particles depends only on the separation between them, $V = V(|\mathbf{r}_2 - \mathbf{r}_1|) = V(|\mathbf{r}|)$. If the potential is quadratic in space then the molecule oscillation frequency is $\omega = \sqrt{\kappa_0/m_\mathrm{r}}$, where κ_0 is the spring constant and m_r is the reduced mass.

The same conclusion may be reached assuming motion is restricted to one dimension. The force governing relative motion of particle mass m_1 and mass m_2 is given by relative displacement u_1 and u_2, respectively, multiplied by spring constant κ_0 (see Fig. H.E24). The equations of motion are

$$m_1\frac{d^2 u_1}{dt^2} = \kappa_0(u_2 - u_1),$$

$$m_2\frac{d^2 u_2}{dt^2} = \kappa_0(u_1 - u_2),$$

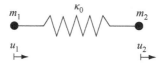

Fig. H.E24

which has solution of the form $e^{-i\omega t}$ giving

$$(\kappa_0 - m_1\omega^2)u_1 - \kappa_0 u_2 = 0,$$
$$-\kappa_0 u_1 + (\kappa_0 - m_2\omega^2)u_2 = 0,$$

or

$$\begin{bmatrix} a_{11} & a_{12} \\ a_{21} & a_{22} \end{bmatrix} \begin{bmatrix} u_1 \\ u_2 \end{bmatrix} = \mathbf{A}\mathbf{u} = \mathbf{0}.$$

If $\mathbf{A}^{-1}$ does *not* exist then $\mathbf{u} \neq \mathbf{0}$ and $|\mathbf{A}| = 0$. The solutions are given by the characteristic equation,

$$\begin{vmatrix} \kappa_0 - m_1\omega^2 & -\kappa_0 \\ -\kappa_0 & \kappa_0 - m_2\omega^2 \end{vmatrix} = \left(\kappa_0 - m_1\omega^2\right)\left(\kappa_0 - m_2\omega^2\right) - \kappa_0^2 = 0.$$

Hence,

$$\kappa_0 = \left(\frac{m_1 m_2}{m_1 + m_2}\right)\omega^2 = m_r\omega^2$$

and, as before, the frequency of oscillation is just

$$\omega = \sqrt{\kappa_0\left(\frac{m_1 + m_2}{m_1 m_2}\right)} = \sqrt{\kappa_0/m_r}.$$

Solutions H.5
(a) The frequency of vibration of the cantilever beam with dimensions $l = 100\,\mu\text{m}$, $d = 0.1\,\mu\text{m}$, density $\rho = 2.328 \times 10^3\,\text{kg m}^{-3}$, and Young's modulus $E_{\text{Young}} = 1.96 \times 10^{11}\,\text{N m}^{-2}$, is found using the equation

$$\nu = \frac{3.52}{2\pi}\frac{d}{l^2}\sqrt{\frac{E_{\text{Young}}}{12\rho}} = \frac{3.52 \times 0.1 \times 10^{-6}}{(100 \times 10^{-6})^2}\sqrt{\frac{1.96 \times 10^{11}}{12 \times 2.328 \times 10^3}} = \frac{9.3 \times 10^4\,\text{rad s}^{-1}}{2\pi} = 15\,\text{kHz}.$$

(b) The vibrational energy of the cantilever in (a), with width $w = 5\,\mu\text{m}$ and free-end displacement amplitude $A = 1\,\mu\text{m}$ is very small:

$$\frac{wd^3 A^2 E_{\text{Young}}}{6l^3} = \frac{5 \times 10^{-6} \times (0.1 \times 10^{-6})3 \times (1 \times 10^{-6})^2 \times 1.96 \times 10^{11}}{6 \times (100 \times 10^{-6})^3} = 1.63 \times 10^{-16}\,\text{J}.$$

Solutions H.6

The solution to this exercise is found by considering a point charge $+e$ placed at a distance d from a large perfectly conducting metal sheet (see Fig. H.E25). No current can flow in the conductor, so electric field lines from the charge intersect normally to the surface of the conductor. This boundary condition and the symmetry of the problem suggest using the method of images, in which an image charge is placed at a distance d from the position of the sheet and opposite the original charge. The force is then calculated for two oppositely charged point particles separated by distance $2d$ and embedded in a medium with relative dielectric permittivity ε_r. The attractive force when $\varepsilon_{r0} = 2$ and $d = 2$ nm is

$$F = \frac{-e^2}{4\pi\varepsilon_0\varepsilon_{r0}4d^2} = \frac{-(1.602 \times 10^{-19})^2}{4\pi \times 8.854 \times 10^{-12} \times 2 \times 4 \times (2 \times 10^{-9})^2} = -7.2 \times 10^{-12}\,\text{N},$$

where the negative sign indicates an attractive force. A force of a few pN is very small and will have little effect on the relative position of the Fe core atom with respect to the metal sheet. The maximum static force is, absent a dielectric response from the dedrimer, when $\varepsilon_{r0} = 1$.

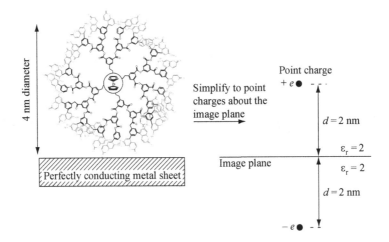

Fig. H.E25

Solutions H.7

The circuit is driven by an input current, I, from a current source at resonant frequency $\omega = 1/\sqrt{LC}$. It is assumed that on resonance the Q of the circuit is large and is given by $Q = \omega L/r$, where r is the value of a series resistor.

The solution for current is of the form $I = I_1 e^{i\omega t}$, where $|I_1|$ is the amplitude of the current and $\omega = 1/\sqrt{LC}$ when on resonance. The impedance of the LCr part of the circuit is

$$\frac{-i}{\omega C} + i\omega L + r,$$

which on resonance has the value

$$i\omega L - \frac{i}{\omega C} + r = \frac{i\left(\omega^2 LC - 1\right) + \omega Cr}{\omega C} = r,$$

so the equivalent circuit consists of two resistors in parallel with an impedance seen by the input of $rR/(R+r)$. Hence, the modulus of voltage V_1 is

$$|V_1| = |I_1|\left(\frac{rR}{R+r}\right).$$

The impedance seen at V_2 is $Z_2 = -i\omega L + r$ and the voltage V_2 is given by the current I_2 multiplied by Z_2. The current I_2 flowing through the LCr part of the circuit is given by

$$I_2 = I\left(\frac{R}{R+r}\right),$$

so the voltage V_2 is

$$V_2 = I\left(\frac{R}{R+r}\right)(i\omega L + r).$$

The modulus of this voltage is

$$|V_2| = |I_1|\left(\frac{R}{R+r}\right)\sqrt{\omega^2 L^2 + r^2} = |V_1|\sqrt{\frac{\omega^2 L^2}{r^2} + 1} = |V_1|\sqrt{Q^2 + 1} \sim |V_1|Q.$$

This shows that the Q of the circuit can be used to amplify an oscillating voltage.

Solutions H.8

Given that the amplitude of magnetic flux density, B, in a monochromatic plane-polarized electromagnetic wave travelling in a vacuum is 10^{-6} T, the task is to calculate the value of the total energy density. The energy density is given by

$$U = \frac{1}{2}\varepsilon_0 E^2 + \frac{1}{2\mu_0}B^2,$$

and $E/B = c$ for plane waves in a vacuum. Thus, $U = \frac{1}{2}\varepsilon_0 c^2 B^2 + \frac{1}{2\mu_0}B^2$, and because $c^2 = 1/\varepsilon_0\mu_0$,

$$U = \frac{1}{2\mu_0}B^2 + \frac{1}{2\mu_0}B^2 = \frac{1}{\mu_0}B^2.$$

For the magnetic field given $U = 10^{-12}/\left(4\pi \times 10^{-7}\right) = 8 \times 10^{-7}$ J m^{-3}.

The total energy density is divided equally between the electric and magnetic components. The time-averaged energy density is half this value.

H.5   Problems

Problem H.1

Find the largest and smallest volume of a convex plug manufactured from a sphere of radius $r = 1$ cm that fits exactly into a circular hole of radius $r = 1$ cm, an isosceles triangle with base 2 cm and a height $h = 1\,\text{cm}$, and a half circle radius $r = 1\,\text{cm}$ and base 2 cm.

Problem H.2
An initially stationary particle mass m_1 is on a frictionless table surface and another particle mass m_2 is positioned vertically below the edge of the table. The distance from the particle mass m_1 to the edge of the table is l. The two particles are connected by a taut, light, inextensible string of length $L > l$.
(a) How much time elapses before particle mass m_1 is launched off the edge of the table?
(b) What is the subsequent motion of the particles?
(c) How is the answer for (a) and (b) modified if the string has spring constant κ_0?

Problem H.3
The velocity of waves in shallow water may be approximated as $v = \sqrt{gh}$ where g is the acceleration due to gravity and h is the depth of the water. Sketch the lowest frequency standing water wave in a 5 m long garden pond that is 0.9 m deep and estimate its frequency.

Problem H.4
What is the dispersion relation of a wave whose group velocity is (a) half the phase velocity, (b) twice the phase velocity, (c) four times the phase velocity, and (d) the negative of phase velocity?

Problem H.5
(a) If *complex* field $\mathbf{G} = (\mathbf{D}/\sqrt{\epsilon_0} + i\mathbf{B}/\sqrt{\mu_0})/\sqrt{2}$, show that Maxwell's equations in free space and with no free charges is

$$\nabla \cdot \mathbf{G} = 0$$

with

$$i\frac{\partial \mathbf{G}}{\partial t} = \frac{1}{\sqrt{\epsilon_0 \mu_0}} \nabla \times \mathbf{G}.$$

(b) Show that the energy flux density in the electromagnetic field is

$$\mathbf{S} = \mathbf{E} \times \mathbf{H} = \frac{-i}{\sqrt{\epsilon_0 \mu_0}} (\mathbf{G}^* \times \mathbf{G}).$$

(c) If the field $\mathbf{G}$ is purely real, what is the value of $\mathbf{S}$?
(d) Show that the electromagnetic energy density is $U_S = |\mathbf{G}|^2$.
(e) How would Maxwell's equations be modified if magnetic charge g (magnetic monopoles) exists? Derive an expression for conservation of magnetic current and write down a generalized Lorentz force law that includes magnetic charge. Write Maxwell's equations with magnetic charge in terms of the field $\mathbf{G}$.

Problem H.6
(a) A thin dielectric film with relative permittivity $\varepsilon_{r0} = 10$ uniformly coats a small metal sphere and doubles the capacitance to 2.2×10^{-18} F. What is the thickness of the dielectric film and the single electron charging energy of the dielectric coated metal sphere?

(b) The dielectric coated metal sphere of part (a) is now coated with metal. What is the new value of the single-electron charging energy for the central metal sphere?

(c) Compare the result in (b) to the charging energy of a metal sphere radius $0.5\,\mathrm{nm} < r_0 < 10\,\mathrm{nm}$ embedded in a dielectric with $\varepsilon_{r0} = 10$ and surrounded by metal shell of internal radius $r_1 = 2r_0$. Plot single-electron charging energy ΔE_c as a function of r_0.

Problem H.7
(a) A diatomic molecule has atoms with mass m_1 and m_2. An isotopic form of the molecule has atoms with mass m_1' and m_2'. Find the ratio of vibration oscillation frequency ω/ω' of the two molecules.

(b) What is the ratio of vibrational frequencies for carbon monoxide isotope 12 ($^{12}\mathrm{C}^{16}\mathrm{O}$) and carbon monoxide isotope 13 ($^{13}\mathrm{C}^{16}\mathrm{O}$)?

Problem H.8
A centimeter-long linear chain of spherical atoms has nearest neighbor spacing of 0.25 nm.

(a) What is the minimum diameter of a one atom thick disk made of these atoms?

(b) What is the minimum diameter of a sphere made of these atoms?

Problem H.9
An electromagnetic wave has electric field $\mathbf{E}(\mathbf{r}, \omega) = \mathbf{E}_0(\omega)e^{i(k'(\omega)+ik''(\omega))\mathbf{k}^\sim \cdot \mathbf{r}}$ where $k'(\omega)$ and $k''(\omega)$ are the real and imaginary parts, respectively, of the frequency-dependent wave number. The wave propagates in a homogeneous isotropic dielectric characterized by relative permeability $\mu_r = 1$ and complex permittivity function $\varepsilon(\omega) = \varepsilon_0 \varepsilon_r(\omega) = \varepsilon_0(\varepsilon_r'(\omega) + i\varepsilon_r''(\omega))$, where $\varepsilon_r'(\omega)$ and $\varepsilon_r''(\omega)$ are the real and imaginary parts, respectively, of the frequency-dependent relative permittivity function.

(a) Derive the expression for refractive index, $n_r(\omega) = \sqrt{\frac{1}{2}\left(\varepsilon_r'(\omega) + \sqrt{\varepsilon_r'^2(\omega) + \varepsilon_r''^2(\omega)}\right)}$.

(b) Introduce absorption coefficient $\alpha(\omega) = 2k''(\omega)$ and show that $\alpha(\omega) = \frac{\omega \varepsilon_r''(\omega)}{cn_r(\omega)}$.

Problem H.10
A particle moves between two points A and B in a vertical plane as illustrated in Fig. H.P26. If acceleration due to gravity is g and the velocity is initially zero, find the

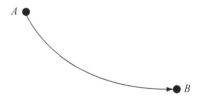

A

B

Fig. H.P26

shape of the frictionless surface on which the particle must move to give a trajectory that takes the shortest time.

Problem H.11
Materials with negative relative permeability and negative relative permittivity can have negative refractive index. In this situation group velocity is the negative of phase velocity. Suppose a point source of electromagnetic radiation in air is placed at a distance z_1 normal to the surface of a slab of negative refractive index material of thickness $z_2 > z_1$. The value of the negative refractive index material is $n_r = -1$. Use ray tracing to find the positions at which electromagnetic radiation from the point source is focused to a point. Comment on the statement that this slab of negative index material makes a "perfect lens."

Problem H.12
An electromagnetic field of wavelength $\lambda_0 = 1\,500$ nm in free space propagates around the inside circumference of a silica disk resonator of density $\rho = 2.2$ g cm^{-3} and refractive index $n_r = 1.5$. The disk has radius $R = \frac{0.2}{2\pi}$ mm and thickness $d = 1$ µm. Electromagnetic field loss in the disk is dominated by surface roughness with average value $\alpha = 0.016$ cm^{-1}.

(a) Calculate the lowest natural radial mechanical oscillation frequency of the disk in the absence of light.

(b) Calculate maximum resonant enhancement in electric field and electric field intensity and the full-with half-maximum (FWHM) in the field intensity frequency spectrum.

(c) Repeat the calculation in (b) but for $R = \frac{0.02}{2\pi}$ mm.

(d) Compare and explain the results obtained in (b) and (c).

(e) 10 µW of optical power at $\lambda_0 = 1\,500$ nm wavelength is coupled into the dielectric disk in (b). Estimate the force exerted on the disk due to radiation pressure and estimate the change in disk radius if Young's modulus for the dielectric material is 73 GPa. Compare the resulting shift in resonant frequency to the optical FWHM. What optical modulation depth might be achievable in the system?

Problem H.13
The initial length $l = 2$ µm of a thin horizontal microbeam increases by Δ causing the beam to describe the arc of a circle between fixed endpoints separated by distance $L = l$ (see Fig. H.P27).

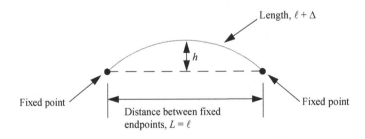

Fig. H.P27

(a) Calculate and plot the midpoint height h as a function of Δ for 0 nm $\leq \Delta \leq$ 10 nm.

(b) Calculate and plot the mechanical gain $g \equiv h/\Delta$ for 0 nm $< \Delta \le 10$ nm.

(c) Setting $\Delta = 0$, find and plot h as a function of fixed endpoint separation L in the continuous range 0 to l. Figure H.P28 shows examples for $L = 0$, $L = 2l/\pi$, and $L = l$.

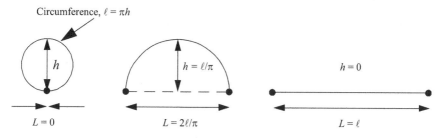

Circumference, $\ell = \pi h$

h

$L = 0$

$h = \ell/\pi$

$L = 2\ell/\pi$

$h = 0$

$L = \ell$

Fig. H.P28

Problem H.14

Johnson (*Phys. Rev.* **32**, 97 (1928)) and Nyquist (*Phys. Rev.* **32**, 110 (1928)) showed that thermal fluctuations create RMS voltage noise $V_{\text{RMS}} = \sqrt{4Rk_{\text{B}}T\Delta f}$ in a macroscopic resistor of value R (ohms) at absolute temperature T (kelvin) measured over a frequency bandwidth Δf, so long as the frequencies considered $f \ll k_{\text{B}}T/(2\pi\hbar)$. This noise can limit the sensitivity of an RF receiver. Bruccoleri et al.[19] showed how the circuit in Fig. H.P29, in which the current-source transconductance amplifier (g_{m} cell) is an inverter, could be used to cancel thermal noise generated by the input load resistor R.

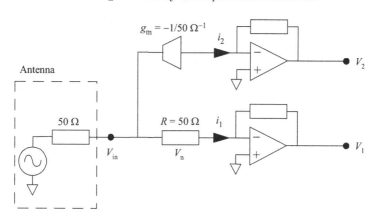

$g_{\text{m}} = -1/50\ \Omega^{-1}$

i_2

V_2

Antenna

$50\ \Omega$

V_{in}

$R = 50\ \Omega$

i_1

V_{n}

V_1

Fig. H.P29

Explain how this noise cancellation works by evaluating the current i_1 and i_2 for a voltage signal V_{in} at the input and voltage noise V_{n} generated in the resistor R. What physical principals and conservation laws are exploited to analyze the circuit? What limits the performance of the noise cancellation circuit?

[19]F. Bruccoleri, E. A. M. Klumperink, and B. Nauta, *IEEE J. Solid-State Circuits* **39**, 275 (2004).

Problem H.15

Suppose that the dipole radiation energy-loss rate dU/dt from an electron of charge e, mass m_0, and acceleration a, obeys the Larmor formula,

$$\frac{dU}{dt} = \frac{-2e^2 a^2 \epsilon_0 \mu_0}{4\pi \epsilon_0 3c},$$

where c is the speed of light and ϵ_0 and μ_0 are the permittivity and permeability of free-space, respectively.

(a) If it is possible to describe electron motion around a proton classically and if the electron is initially in a circular orbit of radius r around the proton, what is the acceleration and velocity of the electron and how long does it take the electron to complete one round trip assuming $r = a_B = 0.0529\,\text{nm}$?

(b) What is the total (nonrelativistic) energy, U, of the electron in (a)?

(c) If energy loss due to electromagnetic radiation occurs slowly compared to the round-trip time, τ_r, an adiabatic approximation assumes the orbit radius, r, remains almost circular at all times. Use the time derivative of the total electron energy at radius r calculated in (b) and the Larmor formula to find the time it takes the electron to radiate away all its kinetic energy and arrive at the origin, $r = 0$ (corresponding to a classical collapse of the hydrogen atom).

(d) Within the approximation of (c), at what radius is the electron velocity 10% of the speed of light? How do relativistic effects influence the calculation in (c)?

Problem H.16

Two identical antennas, labeled 1 and 2, are separated in free space by distance L (see Fig. H.P30). If the antennas receive an electromagnetic signal from a stationary source S_n that has angular frequency ω_n then, as a function of time t, a unit-amplitude source signal is $S_n(t) = e^{i\omega_n t}$. If the nth signal, $S_n(t)$, is a plane wave and has an angle of arrival θ_n measured anticlockwise from normal incidence then there is a relative phase difference of ϕ_n between the contribution of $S_n(t)$ arriving at antenna 1 and 2. In general, the relationship between the angle of arrival θ_n of the nth signal and the phase difference ϕ_n is

$$\phi_n = \frac{2\pi L}{\lambda_n} \sin(\theta_n)$$

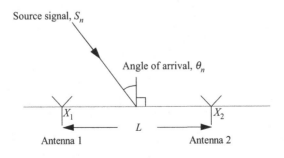

Source signal, S_n

Angle of arrival, θ_n

X_1 X_2

L

Antenna 1   Antenna 2

Fig. H.P30

for a signal of wavelength $\lambda_n = 2\pi c/\omega_n$, where c is the speed of light. If there are only two plane-wave sources, $S_1(t)$ and $S_2(t)$, then each antenna receives the sum of the two signals. At any given time, this sum may be written in matrix form as $\mathbf{X} = \mathbf{AS}$ where vector

$$\mathbf{X} = \begin{bmatrix} X_1 \\ X_2 \end{bmatrix}$$

describes signal $X_1(t)$ received at antenna 1 and signal $X_2(t)$ received at antenna 2,

$$\mathbf{S} = \begin{bmatrix} S_1 \\ S_2 \end{bmatrix}$$

describes sources $S_1(t)$ and $S_2(t)$, and the time-independent complex mixing matrix is

$$\mathbf{A} = \begin{bmatrix} a_{11} & a_{12} \\ a_{21} & a_{22} \end{bmatrix}.$$

(a) Find the matrix elements of the mixing matrix, $\mathbf{A}$.

(b) Find the inverse mixing matrix $\mathbf{A}^{-1}$ and find the conditions when it not possible to separate the source signals using $\mathbf{S} = \mathbf{A}^{-1}\mathbf{X}$.

(c) In a typical wireless receiver system implementation, the complex signals are separated into their real (in-phase, I) and imaginary (quadrature, Q) components at each antenna relative to a reference local oscillator. This doubles the size of the mixing matrix. Find the matrix elements for the mixing matrix in this case.

(d) Discuss how the ability to separate source signals is changed if the position of the antennas and sources vary in time so that $L = L(t)$ and $\theta_n = \theta_n(t)$?

(e) Can an RF receiver be used to measure the electromagnetic field?

Problem H.17

The Drude model of electrical conduction predicts a zero-frequency (DC) normal metal conductivity, $\sigma_0 = e^2 n\tau/m$, where e is the electron charge, n the electron density, m the electron mass, and the characteristic electron collision time is τ. The frequency dependent (AC) conductivity is

$$\sigma(\omega) = \frac{\sigma_0}{1 - i\omega\tau} = \frac{\omega_p^2 \epsilon_0 \tau}{1 - i\omega\tau},$$

where $\omega_p = 2\pi f_p$ is the electron plasma frequency and $1/\tau = \omega_\tau = 2\pi f_\tau$ is the electron collision rate. A linearly polarized electromagnetic plane-wave propagating in the x direction and normally incident on a planar metal interface at $x = x_0$ has electric field $E_x(x) = E_{x0}e^{ikx}$ in the metal, where

$$k = \frac{\omega}{c}\sqrt{\epsilon_r(\omega)} = \frac{\omega}{c}\sqrt{1 - \frac{\sigma_0 \tau}{\epsilon_0(1 + \omega^2\tau^2)} + i\frac{\sigma_0}{\epsilon_0\omega(1 + \omega^2\tau^2)}}$$

and ϵ_r is the permittivity of the metal. For copper, $\sigma_0(\text{Cu}) = 5.9 \times 10^7\,\text{S}\,\text{m}^1$, $n(\text{Cu}) = 8.46 \times 10^{28}\,\text{m}^3$, $\tau(\text{Cu}) = 25\,\text{fs}$, $f_p(\text{Cu}) = 2\,600\,\text{THz}$, $f_\tau(\text{Cu}) = 6.4\,\text{THz}$.

(a) Find an expression for the frequency dependent skin depth $\delta_x(\omega) = 1/(\text{Im}(\text{k}(\omega))$ of copper in the low-frequency limit, $\omega\tau \ll 1$. What is the value of δ_x for an electromagnetic field oscillating at frequency $f = 100\,\text{GHz}$? How does the value of surface resistance $R_s = 1/(\sigma_0\delta_x)$ vary in the frequency range $1\,\text{GHz} < f < 1\,\text{THz}$?

(b) Find an expression for the skin depth of copper in the frequency range $f_\tau < f < f_\text{p}$. What is the value of δ_x at frequency $f < 100\,\text{THz}$ and how does it vary with frequency in the range $10\,\text{THz} < f < 1\,000\,\text{THz}$?

(c) What happens to the propagation of electromagnetic field when $f > f_\text{p}$?

Problem H.18

(a) A uniform disk of mass m, initially at rest and with point-of-contact on a frictionless surface at position A, moves down the indicated surface in the presence of acceleration due to gravity (Fig. H.P31). How long does it take the point of contact to reach point B? What value of h/L minimizes the time to travel from A to B?

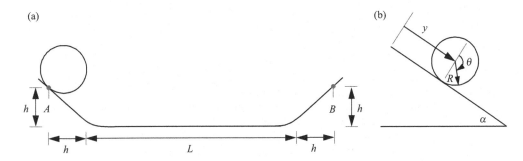

(a)

(b)

Fig. H.P31

(b) Suppose there is just enough friction to ensure that a uniform disk radius R and mass m rolls down an inclined plane set at angle α without slipping. Show that its acceleration is $2/3$ of the value it would have if there were no friction.

(c) How far does the disk travel on the surface specified in (a) with friction described in (b)?

Problem H.19

Show, using calculus of variations, that the Lagrange equations of motion (Eq. (H.19)) may be obtained from the extremum of the action integral (Eq. (H.18)) describing a dynamical system moving between two points in configuration space in given time interval, t_1 to t_2.

Problem H.20

A conservative system of N particles and $n \leq 3N$ degrees of freedom has particle positions $x_j = x_j(q_1, q_2, ..., q_n)$ and $j = 1, 2, ..., n$, where the q_j form a proper set of independent generalized coordinates. Show that

$$\frac{\partial \mathcal{L}}{\partial q_j} - \frac{\text{d}}{\text{dt}}\frac{\partial \mathcal{L}}{\partial \dot{q}_j} = 0.$$

Problem H.21

Two identical oscillators with motion in one dimension, each of mass m and spring constant κ_0, are coupled by a damping piston with velocity-dependent friction force, $-\gamma m(du_1/dt - du_2/dt)$, where γ is a damping rate and $u_{1,2}$ is particle displacement from the position of minimum potential energy, as shown in Fig. H.P32.

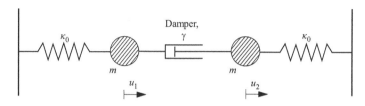

Fig. H.P32

(a) Find equations of motion and eigenfrequencies for displacements u_1 and u_2.

(b) Solve for transient evolution of particle position when $\kappa_0/m = 1$, $\gamma = 0.1$, and the initial condition is $u_1 = 2$ and $u_2 = 1$ at time $t = 0$.

(c) Explain the motion remaining in the system in the limit $t \to \infty$. What percentage of initial energy is dissipated in the limit $t \to \infty$?

Problem H.22

A standard forward finite-difference first derivative of a smooth real-valued function of real variable x is

$$\frac{df(x)}{dx} = \frac{f(x+h_0) - f(x)}{h_0}.$$

This loses accuracy due to machine rounding error as h_0 becomes small. If u is the unit round-off (e.g., $u \sim 10^{-16}$ for double precision) then $h_0 \sim \sqrt{u}$ minimizes both truncation error and rounding error. This difficulty in using the finite-difference approximation may be avoided by considering the analytic function of complex variable z that is real on the real axis with $f(z) = f(x + ih_0)$.

(a) For small h_0, expand $f(x + ih_0)$ in a Taylor series about x to show that

$$\frac{df(x)}{dx} = \text{Im}\left(\frac{f(x+ih_0)}{h_0}\right) + err_1$$

and find the truncation error, err_1.

(b) Similarly, find the second derivative,

$$\frac{d^2 f(x)}{dx^2}, \quad \text{and truncation error, } err_2.$$

(c) For function $f(x) = e^x / \sqrt{\cos^3(x) + \sin^3(x)}$, compute

$$\frac{df(x = \pi/4)}{dx}$$

using the complex method and the finite-difference method. Use MATLAB's Symbolic toolbox to find the actual derivative. Compare and plot the errors in both methods for the range $10^{-1} < h_0 < 10^{-16}$.

Index

Printed in the United States
by Baker & Taylor Publisher Services